FUNDAMENTALS OF INTERNAL COMBUSTION ENGINES

THIRD EDITION

H.N. GUPTA
Former Professor of Mechanical Engineering
Indian Institute of Technology
(Banaras Hindu University)
Varanasi

PHI Learning Private Limited
Delhi-110092
2025

*In fond memory of **Shri Asoke K. Ghosh** (October 1942–February 2024), Founder Chairman and Managing Director of PHI Learning, whose vision endlessly inspires.*

The Legacy Continues... .

Published by Asoke K. Ghosh, PHI Learning Private Limited, Rimjhim House, 111, Patparganj Industrial Estate, Delhi-110092 and Printed by Syndicate Binders, A-20, Hosiery Complex, Noida, Phase-II Extension, Noida-201305 (N.C.R. Delhi).

₹1075.00

FUNDAMENTALS OF INTERNAL COMBUSTION ENGINES, Third Edition

H.N. Gupta

ISBN-978-93-5443-974-2 (Print Book)
ISBN-978-93-5443-854-7 (e-Book)

The export rights of the book are vested solely with the publisher.

Contents

Preface

Although the second edition of this book was very well received by individuals, academic institutions and professionals, the author felt the need, and the publisher also agreed, to enrich the book and help the readers by revising some of the materials and also adding some new contents to keep them abreast of the new technologies and advancement in the recent past. With this in mind, the author pursued to revise the book.

Retaining objectives of the previous editions, the third edition presents basic concepts of Internal Combustion Engines (ICEs). The book is simple, easy to understand and comprehensive in its coverage. Two new chapters have been added and some less useful and irrelevant topics, such as Equilibrium Charts, detailed Carburettor parts, Superchargers and Turbocharger Controls are deleted. Chapter 19 is a new Chapter, entitled "Advanced Combustion Engines", which includes advanced low temperature combustion model, such as Homogeneous Charge Compression Ignition (HCCI) Engine, Premixed Charge Compression Ignition (PCCI) Engine, Reactivity Controlled Compression Ignition (RCCI) Engine, Flexible Fuel Vehicle and Gasoline Direct-injection Compression Ignition (GDCI) Engine. Finally, a new Chapter entitled "Introduction to Electric Vehicles" (Chapter 20) has been added, which takes a perspective on Electric Vehicle (EV), Hybrid Electric Vehicle (HEV) and Fuel Cell Electric Vehicle (FCEV). HEVs involve the use of Internal Combustion Engines (ICEs) in addition to electric machines, power electronics, and control systems.

Various advanced technologies are discussed to improve the efficiency and environmental performance of ICEs, including design and operating characteristics of main subsystems, such as fuel injection, conventional and advanced low-temperature combustion, turbo-charging, exhaust gas recirculation (EGR), variable valve actuation (VVA), cooling, after-treatment, waste heat recovery (VHR), and electronic controls.

I am grateful to PHI Learning, who invited me to produce this third edition. Also, I sincerely appreciate the contribution of my students, and convey my sincere thanks to Dr. J.V. Tirkey, Professor, Indian Institute of Technology (BHU) for having useful discussions and suggestions on the subject.

I hope the new edition will continue to serve well at the undergraduate level and that the advanced topics will prove useful to the students at postgraduate level. Constructive criticism and suggestions for improvement of the book will be highly appreciated.

H.N. Gupta
hngupta@gmail.com

CHAPTER 01

Introduction to Internal Combustion Engines

1.1 AN OVERVIEW

An engine is a device which transforms the chemical energy of a fuel into thermal energy and uses this energy to produce mechanical work. Engines normally convert thermal energy into mechanical work and, therefore, they are called heat engines. When fuel burns in the presence of atmospheric air, a tremendous amount of heat energy is released. The products of combustion attain a very high temperature. A heat engine converts the released heat energy into useful work with the help of a working fluid.

Heat engines are broadly classified into:

(a) External Combustion Engines (EC Engines)
(b) Internal Combustion Engines (IC Engines)

In external combustion engines, the combustion of the fuel in presence of air takes place outside the engine cylinder. The heat energy released from the fuel is utilized to raise the high-pressure steam in a boiler from water. Steam is a working fluid, which enters into the cylinder of a steam engine to perform mechanical work. Here, the prof combustion of fuel do not enter into the engine cylinder and hence they do not form the working fluid.

The steam turbine in a steam power plant is another example of an external combustion engine. The steam engine may be called an intermittent external combustion engine and the steam turbine a continuous external combustion engine. A closed cycle gas turbine plant is also an example of an external combustion engine. Here, normally the air is a working substance which completes a thermodynamic cycle. It receives heat from products of combustion of fuel in a heat exchanger and rejects heat from another heat exchanger to the surroundings. Here also the products of combustion do not enter into the turbine. Stirling engine is also an external combustion engine.

In internal combustion engines, either the combustion of the fuel takes place inside the engine cylinder or the products of combustion enter into the cylinder as a working fluid. In reciprocating engines having cylinder and piston, the combustion of the fuel takes place inside the cylinder and such engines may be called intermittent internal combustion engines. In an open cycle gas turbine plant, the products of combustion of fuel enter into the gas turbine and work is obtained in the form of rotation of the turbine shaft. Such a turbine is an example of a continuous internal combustion engine.

The intermittent internal combustion engines are most popular because of their use as the primemover in motor vehicles, and usually these engines are reciprocating engines. The reciprocating engine mechanism consists of a piston which moves in a cylinder and forms a movable gas-tight seal. By means of a connecting rod and a crankshaft arrangement, the reciprocating motion of the piston is converted to rotary motion at the crankshaft.

The steam turbine plant is the most popular continuous external combustion engine used for large electric power generation. The essential components are boiler, steam turbine, condenser, and feed pump.

The main advantages of an internal combustion engine over an external combustion engine are:

(a) Greater mechanical simplicity
(b) Higher power output per unit weight because of the absence of auxiliary units like boiler, condenser, and feed pump
(c) Lower initial cost
(d) Higher brake thermal efficiency as only a small fraction of heat energy of the fuel is dissipated to the cooling system.

The advantages of internal combustion engine accrue from the fact that they work at an average temperature which is much below the maximum temperature of the working fluid in the cycle. The disadvantages of the internal combustion engine over the external combustion engine are:

(a) The IC engines cannot use solid fuels which are cheaper. Only liquid or gaseous fuels of given specifications can be efficiently used. These fuels are relatively more expensive.
(b) The IC engines are not self-starting whereas the EC engines have a high-starting torque.
(c) The intermittent IC engines have reciprocating parts and hence they are susceptible to the problems of vibration.

Wankel engine is a rotary intermittent internal combustion engine. Jet engines and rockets are also internal combustion engines. They fall under the category of rotary continuous internal combustion engine. This book deals with the intermittent internal combustion engines, mainly of reciprocating type and thus excludes the gas turbine. Reciprocating IC engines are used in automobiles, motorcycles and scooters, power boats, ships, locomotives, and small aircraft. Due to scarcity of electric power, the use of IC engines as a portable power output unit has gained tremendous momentum. These engines are also used in farm tractors, lawnmowers and in many other devices.

1.2 HISTORICAL DEVELOPMENT

The first IC engine for commercial use was developed by a Frenchman, J.J.E. Lenoir (1822–1900) in the year 1860. Coal gas and air mixture were drawn into the engine cylinder during the first-half of the piston stroke. At this point the charge was ignited by a spark. This caused rise in pressure and the burned gases, the so-called products of combustion, delivered power to the piston for the second-half of the stroke. On the return stroke, the gases were discharged from the cylinder. The return stroke was possible by using a large flywheel which stored energy during the power stroke and dissipated energy during the return stroke, exactly in the same manner as in the case of a steam engine. By the year 1865 about 5000 engines were built in sizes up to 6 hp providing efficiency, however, not exceeding 5 per cent, but it was better than the efficiency of a small steam engine of those times.

Nicolaus A. Otto (1832–1891) and Eugen Langen (1833–1895) developed a free piston engine in 1867 in Germany. Air-fuel mixture was taken in a cylinder and ignited by a gas flame during the early part of the outward stroke to accelerate a free piston, and a vacuum was thus generated in the cylinder. The piston was brought inward by atmospheric pressure acting on the piston from the other side. During the inward stroke the burned gas was exhausted through a slide valve. The piston rod was connected by a ratchet and a rack and pinion device to the flywheel mounted on the output shaft. The inertia of the flywheel moved the piston outwards and induced a fresh charge through a slide valve to repeat the cycle. The thermal efficiency of this engine was found to be 11 per cent.

In 1862, Alphonse Beau de Rochas (1815–1893), a Frenchman, described the principles of four-stroke cycle and the conditions under which maximum efficiency could be obtained in IC engines. It gave an idea of igniting fuel at higher pressures, nearly at the end of compression instead of burning the fuel at atmospheric pressure. Increased cylinder volume with minimum surface-to-volume ratio, maximum possible speed and higher expansion ratio were also suggested for higher thermal efficiency.

Beau de Roches could not, however, build any engine himself based on his principles. In 1876, Nicolaus August Otto built an engine based on these principles. This engine worked on the four-stroke principle—intake, compression, expansion or power and exhaust strokes. Ignition was nearly at the end of compression. Otto engine resulted in reduced weight and volume and gave higher thermal efficiency. This was the breakthrough that effectively founded the IC engine industry. By 1890, almost 50,000 of these engines had been sold in Europe and the USA.

By the 1880s, Dugald Clerk and James Robson of the UK and Karl Benz of Germany developed the two-stroke internal combustion engine. In this engine, compression of the charge takes place during the inward or upward stroke and expansion or power is obtained during the outward or downward stroke. Exhaust and intake processes occur during the end of the power stroke and at the beginning of the compression stroke. In 1885, James Atkinson of England developed an engine with an expansion stroke larger than the compression stroke. The larger expansion stroke was used to increase the efficiency of the engine. Efficiency could also have been increased by increasing the compression ratio, but in order to avoid knocking problems the compression ratio at that time with the quality of fuel available had to be kept below four.

In 1892, Rudolf Diesel (1858–1913) developed a different type of engine in which a high compression ratio was used to ignite the fuel. Fuel was injected nearly at the end of compression which was then ignited by hot compressed air. The efficiency of the engine was increased due to higher compression and expansion ratios. The knocking problem was also overcome. The present diesel engine is designed on the same working principle. Both four-stroke and two-stroke diesel engines have been developed.

1.3 MODERN DEVELOPMENTS

In 1957, Wankel's rotary IC engine was tested successfully after being under research and development for many years. It was based on the design of Felix Wankel of Germany. It used an equilaterally shaped triangular piston that moved in a chamber. Other modern developments are the revival of stirling engine, free piston engine, stratified charge engine, variable compression ratio engine, variable valve timing engine, etc.

Fuels and air pollution problems have also had a major impact on engine development. Gasoline and lighter fractions of crude oil were used in IC engines. During the First World War, serious crude oil shortage was experienced due to high demand. William Burton and his associates of Standard Oil of Indiana developed a thermal cracking process where heavier oils were heated under pressure and decomposed into more volatile compounds. These thermally cracked gasolines satisfied the demand of that time.

During the recent years there has been a growing demand for petroleum fuels and as time goes, this demand will grow further causing greater scarcity of conventional fuels. Apart from this problem, gasoline powered engines discharge significant amount of pollutants like carbon monoxide, unburned hydrocarbons, oxides of nitrogen and lead compounds if the petrol is not unleaded. Diesel engines are a significant source of small soot or smoke particles as well as hydrocarbons and oxides of nitrogen. Much work is being done on the use of alternative fuels. The most prominent among these are natural gas, liquefied petroleum gas, methanol, ethanol and hydrogen. Recently, the biodiesel fuel has been tested in the engine. It is a fuel made from renewable fats and oils, such as vegetable oils, through a simple refining process. One of the principal commodities used as a source for biodiesel is soyabeans. From an environmental point of view these gases are far less polluting than the conventional fuels. New materials now becoming available offer the possibilities of reduced engine weight, less cost, reduced heat losses, and increased efficiency.

1.4 ENGINE CLASSIFICATIONS

There are different types of internal combustion engines that can be classified on the following basis:

Thermodynamic cycle

- Constant volume heat supplied or Otto cycle
- Constant pressure heat supplied or Diesel cycle
- Partly constant volume and partly constant pressure heat supplied or dual cycle
- Joule or Brayton cycle.

Working cycle

- Four-stroke cycle—naturally aspirated, supercharged, and turbocharged
- Two-stroke cycle—crankcase scavenged, supercharged, and turbocharged.

Fuel

- Light oil engines using kerosene or petrol
- Heavy oil engines using diesel or mineral oils
- Gas engines using gaseous fuels. The gas used may be natural gas, liquefied petroleum gas (LPG), hydrogen, etc.
- Bi-fuel engines. In these engines the gas is used as the basic fuel and the liquid fuel is used for starting the engine.

Method of fuel supply

- Fuel supply through carburettor. In the petrol engine the fuel is mixed with air in the carburettor and the charge enters into the cylinder during the suction stroke.
- Multi-point port injection (MPI), used in modern spark-ignition engines.
- Single point throttle body injection—this method is also applied to spark-ignition engines.
- Fuel injection at high pressure into the engine cylinder—it is used in diesel engines or compression-ignition engines.

Method of ignition

- Spark ignition (SI) used in conventional petrol engines, known as SI engines.
- Compression ignition (CI) used in conventional diesel engines, known as CI engines
- Pilot injection of fuel oil in gas engines.

Method of cooling

- Water-cooled engine—cylinder walls are cooled by circulating water in the jacket surrounding the cylinder.
- Air-cooled engines—atmospheric air blows over the hot surfaces (motor cycles, scooters, etc.).

Speed

- Low speed engine
- Medium speed engine
- High speed engine.

Field of application

- Stationary engines for power generation
- Marine engines for propulsion of ships
- Automotive engines for land transport
- Aero-engines for aircraft
- Locomotive engines for Railways.

Lubrication system

- Wet sump lubrication
- Dry sump lubrication
- Pressure lubrication.

Method of control under variable load

- Quantity control engines—the air-fuel ratio remains almost constant. The quantity of the charge increases as the load is increased. An example is the engine that uses a carburettor.
- Quality control engines—as the load increases, the quantity of fuel is increased, thus changing the quality of air/fuel ratio. An example is the engine using injectors.
- Combined control engines—in these engines, both the quantity and quality of the charge are changed depending on the variation in load.

Basic engine design

- Reciprocating engines—subdivided by arrangement of cylinders, for example, in-line engines, V-engines, opposed cylinder engine, opposed piston engine, radial engine.
- Rotary engines—Wankel engines.

Number of cylinders

- Single-cylinder engines
- Multi-cylinder engines.

Valve or port design

- Valves—overhead valves (I-head), underhead valves (L-head), rotary valves
- Ports—cross-scavenged, loop-scavenged, through- or uniflow-scavenged.

Combustion chamber design

- Open chamber, for example, bath-tub, wedge, bowl-in-position, hemispherical
- Divided chamber—small and large auxiliary chambers, for example, swirl chambers, prechambers.

1.5 CLASSIFICATION OF RECIPROCATING ENGINES BY APPLICATION

Table 1.1 shows the classification of reciprocating engines by application. For different applications, the approximate range of engine power, predominant types such as compression-ignition or spark-ignition, number of working strokes, and methods of cooling are listed.

1.6 CLASSIFICATIONS OF ENGINES BY CYLINDER ARRANGEMENT

Multi-cylinder reciprocating engines are commonly classified by cylinder arrangement. Some of the basic types of these engines are shown in Figure 1.1.

In-line engines

All cylinders are arranged linearly in one cylinder bank and transmit power to a single crankshaft. Four-cylinder, in-line engines are common for passenger cars. 800 cc Maruti cars have three cylinders in-line. Six-cylinder, in-line engines are also common. More than six cylinders tend to be very long and usually have problems involved with crankshaft torsional vibration. Up to 12 cylinders in-line engines are used for large ships.

V-engines

The 'V' type is essentially two 'in-line' engines set at an angle, and utilizes a common crankshaft. V-engines of eight cylinders or more have excellent balance and also relatively free from vibrational problems. V-8 and V-12 arrangements are commonly used to produce more power from a compact engine. They run smoothly without much vibration.

Table 1.1 Classification of reciprocating engines by application

Class	*Service*	*Approximate range of engine power (kW)*	*Predominant type*		
			CI or SI	*No. of strokes*	*Method of cooling*
Road vehicles	Motor cycles, scooters	0.75–37.5	SI	2,4	A
	Small passenger cars	15–75	SI	4	A, W
	Heavy passenger cars	75–375	SI	4	W
	Light commercial	37.5–150	SI, CI	4	W
	Heavy (long distance) commercial	112.5–375	CI	4	W
Off-road vehicles	Light vehicles (Factory, airport, etc.)	1.5–15	SI	2, 4	A, W
	Agriculture	3–150	SI, CI	2, 4	A, W
	Earth moving	37.5–750	CI	2, 4	W
	Military	37.5–1875	CI	2, 4	A, W
Railroad	Rail cars	150–375	CI	2, 4	W
	Locomotives	375–3000	CI	2, 4	W
Marine	Outboard	0.4–75	SI	2	W
	Inboard motorboats	4–750	SI, CI	4	W
	Light naval craft	30–2250	CI	2, 4	W
	Ships	3750–22500	CI	2, 4	W
	Ships' auxiliaries	75–750	CI	4	W
Airborne vehicles	Airplanes	50–2600	SI	4	A
	Helicopters	50–1500	SI	4	A
Home use	Lawnmowers	0.75–3	SI	2, 4	A
	Snow blowers	2–4.5	SI	2, 4	A
	Light tractors	2–7.5	SI	4	A
Stationary	Building service	7.5–375	CI	2, 4	W
	Electric power	40–22,500	CI	2, 4	W
	Gas pipe line	750–3750	SI	2, 4	W
Special for racing	Vehicles and boats	75–1500	SI	4	W
Toys	Model airplanes, autos, etc.	0.01–0.4	HW	2	A

SI—Spark ignition, CI—Compression ignition, HW—hot-wire ignition carburetted mixture.

Source: Taylor, C.F.: *The Internal Combustion Engine in Theory and Practice*, Vol. 2, Table 10-1, MIT Press, Cambridge, Massachusetts, 1968.

Opposed cylinder engines

The opposed cylinder engines consist of two or more cylinders on opposite sides of a common crankshaft. An opposed cylinder engine can be visualized as two in-line engines—180 degrees apart. The opposed cylinder engines are used where light weight and short length are important, as in light aircraft.

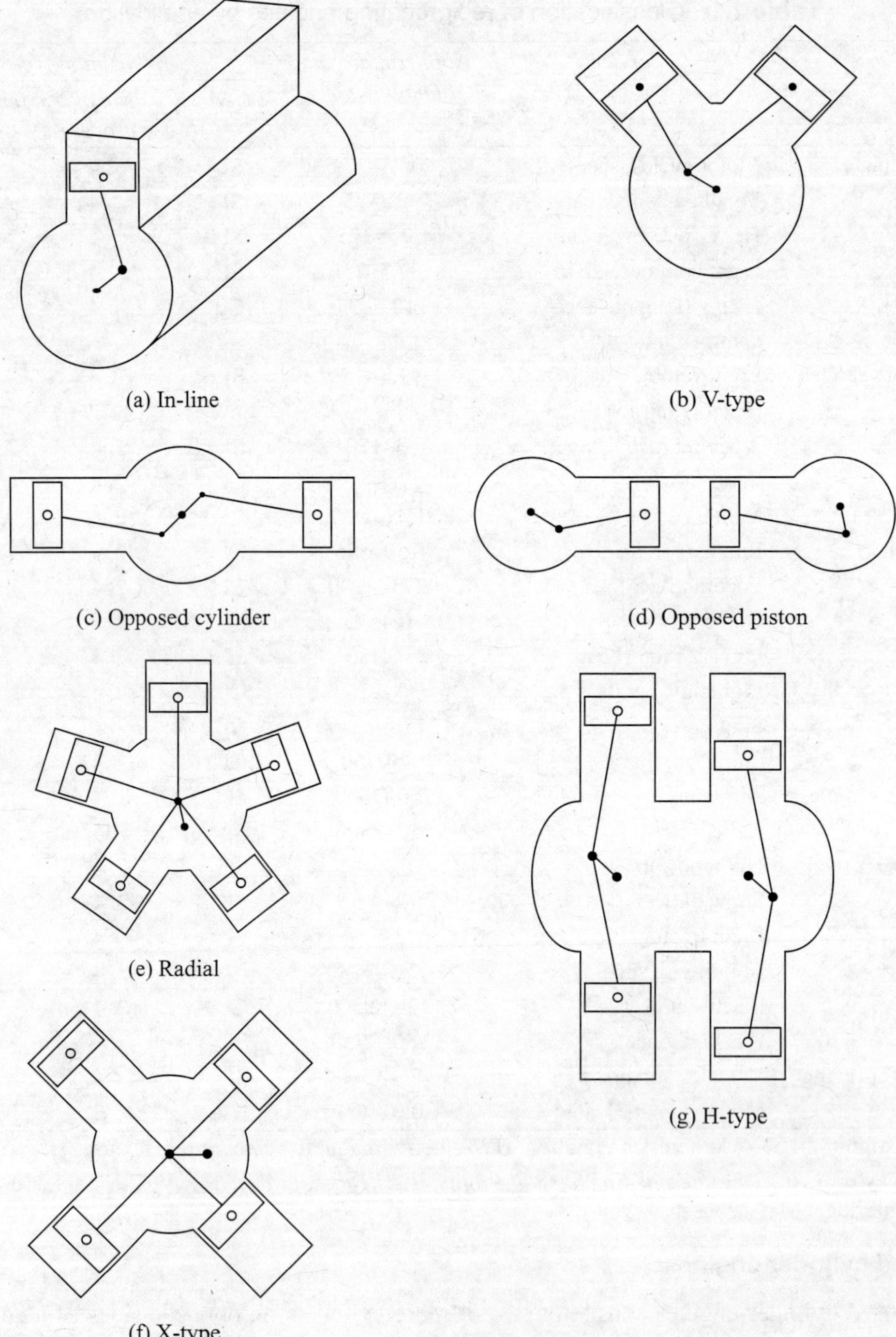

Figure 1.1 Classification of engines by cylinder arrangement.

Opposed piston engines

When a single cylinder houses two pistons, each of which drives a separate crankshaft, it is called an opposed piston engine. The advantages of this engine type include the absence of cylinder heads leading to reduction in heat loss, and piston controlled ports for through-scavenging. The disadvantages include the requirement of two crankshafts geared together, and a shape that may be awkward in many installations. These engines are used in large diesel installations.

Radial engines

The radial type of engine consists of cylinders placed radially, and equally spaced, around a common crankshaft. These engines are used mostly for aircraft. Basically, the radial arrangement gives the lowest weight per unit displacement because the material used in crankshaft and crankcase is minimum for a given number of cylinders.

X-type engine

The X-type engine is a more complicated version of the V-type, essentially four in-line arrangements utilizing a single crankshaft. The X-types do not presently find wide use, except in some diesel installations.

H-type engine

The H-type engine essentially comprises two opposed cylinders, utilizing two separate but interconnected crankshafts. The H-types do not presently find wide use.

1.7 ENGINE COMPONENTS

Figure 1.2 shows the cutaway view of Chrysler 2.2 litre, four-cylinder, four-stroke, spark-ignition engine. The arrangement of different engine parts is shown clearly.

The engine cylinders are contained in a cylinder block. The cylinder block and the upper-half of the crankcase are generally cast together. The bottom-half of the crankcase consists of a combined transmission casing and an oil sump. The block is normally made of gray cast iron because of its good wear resistance and low cost implementation. It also has a water jacket surrounding the engine cylinder for cooling purposes. The cylinder liner is used to increase the life of the cylinder. When the cylinder surfaces wear out, the liners can be renewed easily. The wear resistance of the liner material is more than that of the cylinder block material. Metals suitable for liners are nickel alloy steels, heat treated chromium, and other alloy cast irons.

The piston reciprocates inside the engine cylinder and transmits the gas force to the connecting rod and then to the crank, which in turn rotates the crankshaft from where the power output is obtained. The pistons are made from anodized aluminium alloy. Three compression rings and a slotted oil control ring are generally provided over the piston. The piston rings prevent the escape of the expanding gases from the combustion chamber to the crankcase. An oil control ring scrapes the excess oil from the cylinder wall. The connecting rod and the crank-arm of the crankshaft translate the linear motion of the piston into rotational motion of the crankshaft. The small-end of the connecting rod reciprocates and the large-end follows the rotational motion. A connecting rod is normally a drop forging of steel with a brass small-end bush and a detachable white metal big-end

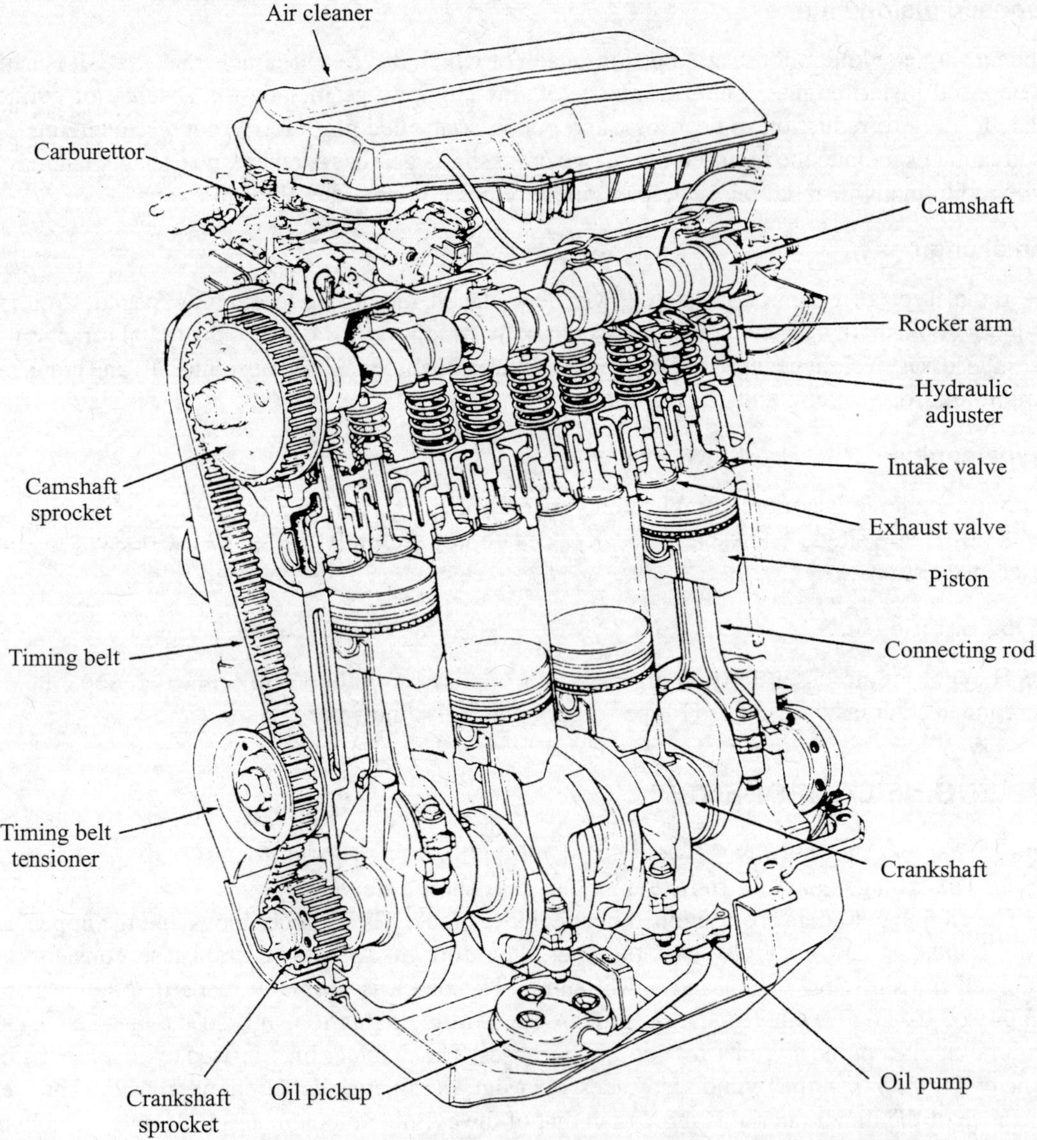

Source: Heywood, J.B.: *Internal Combustion Engine Fundamentals,* McGraw-Hill, New York, 1988.

Figure 1.2 Cutaway view of Chrysler 2.2 litre four-cylinder spark-ignition engine.

shell bearing. Figure 1.3 shows the details of piston, piston rings, connecting rod and bearings. The crank and crankshafts are steel forged and machined to a smooth finish. The crankshaft is supported in main bearings and is statically and dynamically balanced. The crankshaft carries a flywheel to even out the fluctuations of torque.

1. Piston
2. Piston ring—scraper
3. Piston rings—taper
4. Piston ring—parallel
5. Small-end bush
6. Gudgeon pin
7. Circlip
8. Gudgeon pin lubricating hole
9. Connecting rod
10. Cylinder wall lubricating jet
11. Connecting rod cap
12. Lock washer
13. Bolts
14. Connecting rod bearings
15. Connecting rod and cap marking

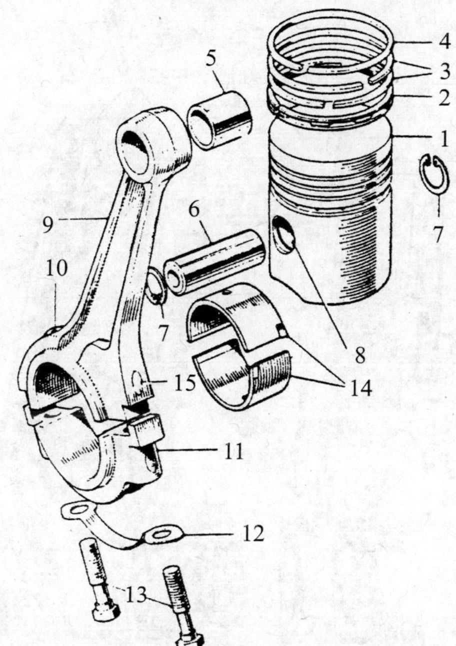

Source: Haynes, J.H.: *Owner's Workshop Manual*, J.H. Haynes & Co., England, 1971.

Figure 1.3 Piston, piston rings, and connecting rod.

An exploded view of the cylinder head is shown in Figure 1.4. The cylinder head seals off the cylinders and is made of cast iron or aluminium. The cylinder head contains spark plugs for an SI engine or a fuel-injector for a CI engine. In overhead valve engines, the cylinder head also contains parts of the valve mechanism. Normally there are two valves per cylinder—an inlet valve for admitting the charge into the cylinder and an exhaust valve for exhausting the burnt gas out from the cylinder. The inlet valve is of larger diameter than that of the exhaust valve. Valves are made from forged alloy steel. The exhaust valve operates at about 700°C, as the hot burned gas passes through it. The cooling of the exhaust valve is therefore required. It can be done by filling partially a hollow stem of the valve with sodium. Evaporation and condensation of sodium provide heat flow from the hot valve head to the cooler stem. The overhead valves are operated by means of rocker arms mounted on the rocker shaft running along the top of the cylinder head. The rocker arms are activated by pushrods and tappets which in turn rise and fall in accordance with the cams on the camshaft. The valves are held closed by small powerful springs. A camshaft made of cast iron or forged steel with one cam per valve is used to open and close the valve. In four-stroke engines, two rotations of the crankshaft provide one rotation of the camshaft. The two shafts are connected together by gears, belt, or chain. The exploded view of crankshaft and camshaft is shown in Figure 1.5.

The crankcase is sealed at the bottom with a pressed steel or cast aluminium oil pan which acts as an oil reservoir for the lubricating system. The oil pump is driven from the gear of the camshaft. The centrifugal water pump and radiator cooling fan are driven together along with the dynamo from the crankshaft pulley wheel by a belt.

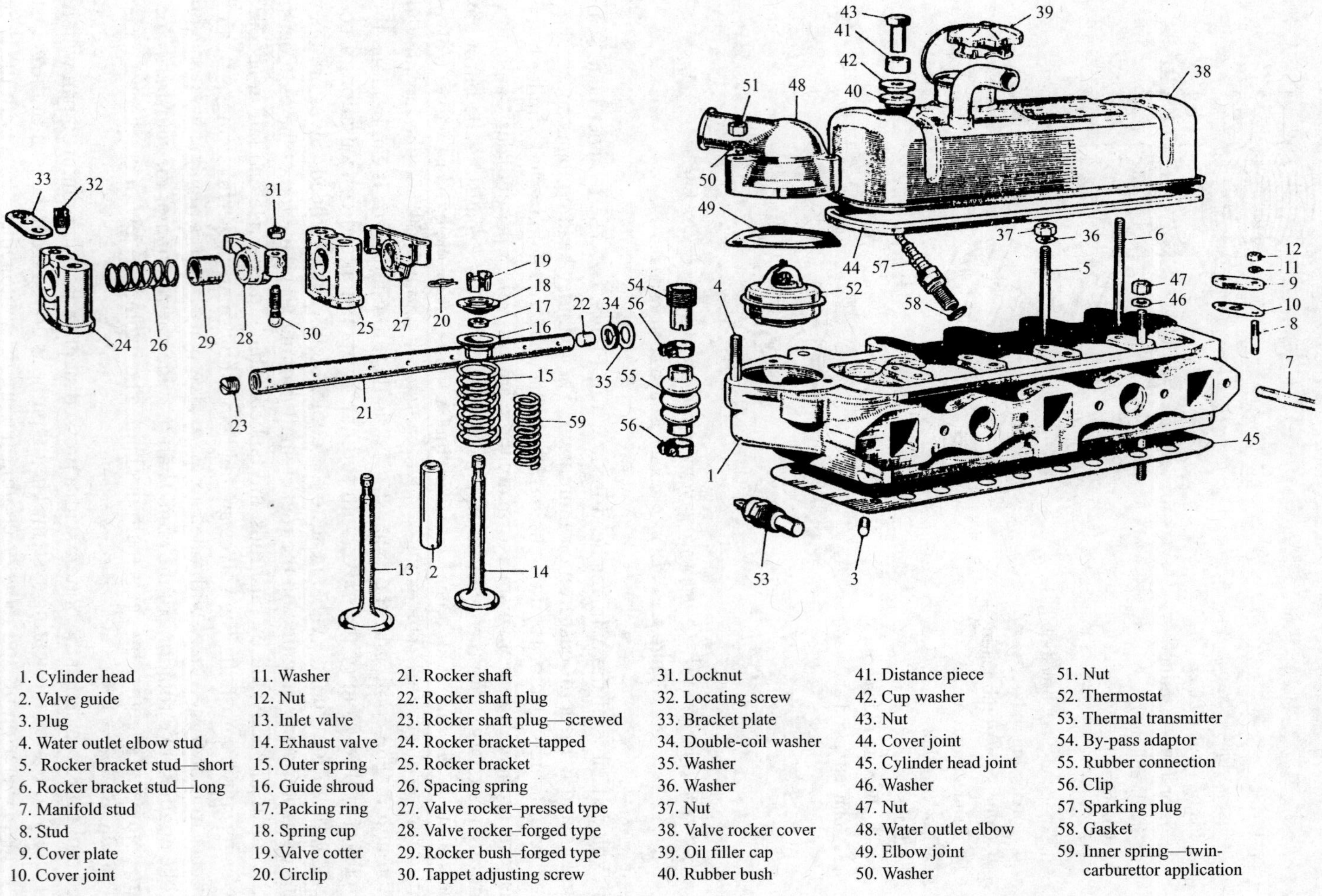

Source: Haynes, J.H.: *Owner's Workshop Manual*, J.H. Haynes & Co., England, 1971.

Figure 1.4 Exploded view of the cylinder head.

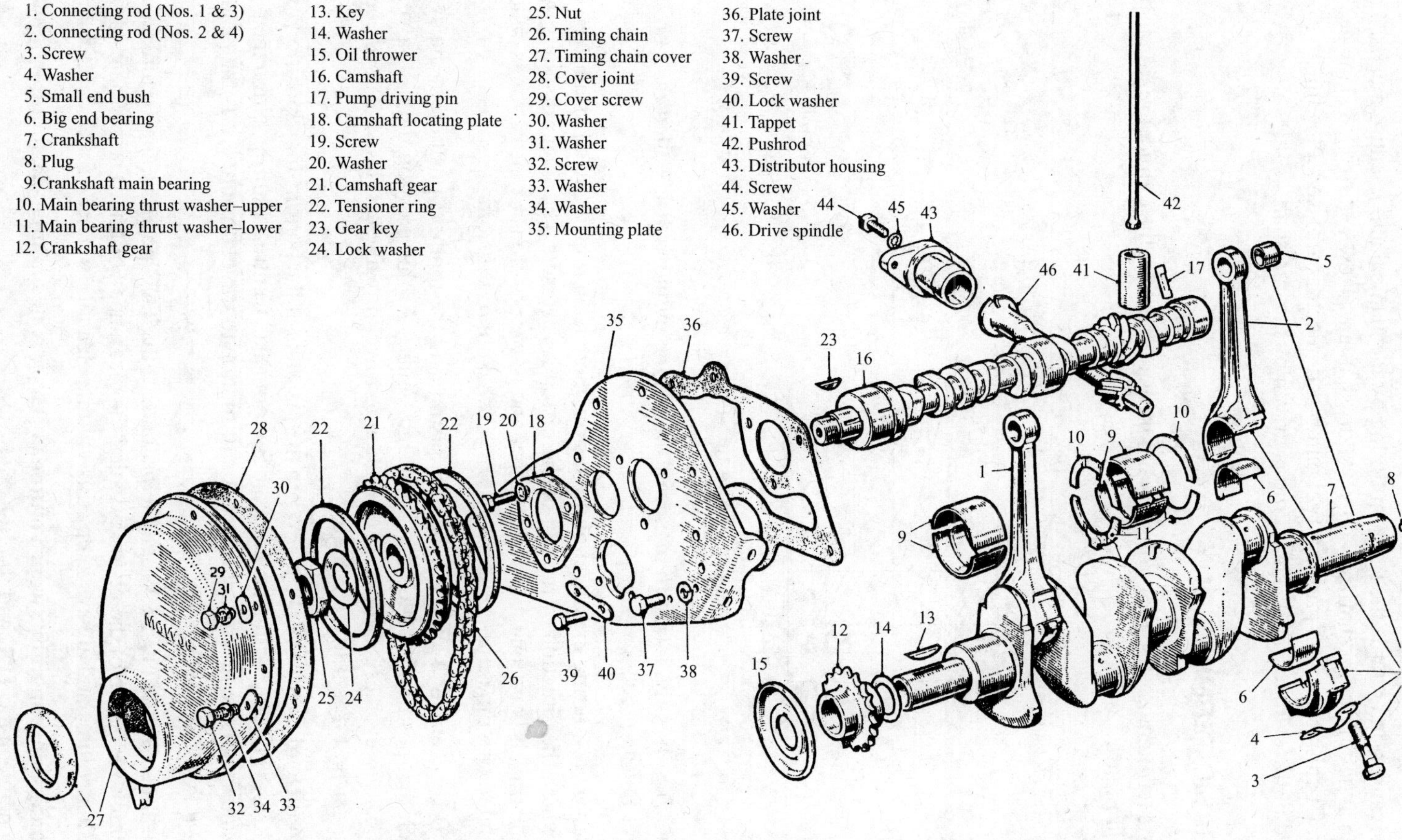

Source: Haynes, J.H.: *Owner's Workshop Manual*, J.H. Haynes & Co., England, 1971.

Figure 1.5 Exploded view of the engine internal components.

In multi-cylinder SI engines, a distributor is mounted towards the rear of the cylinder block and it advances and retards the ignition timing by mechanical and vacuum means. The distributor is driven at half the crankshaft speed by a short shaft and skew gear from a skew gear on the camshaft. In CI engines, a high pressure fuel pump and injectors are used. One injector per cylinder is mounted in the cylinder head. The pump is driven by the camshaft.

1.8 BASIC TERMINOLOGY

The basic terminology used for volumes and measurements in the cylinder region is presented and shown in Figure 1.6.

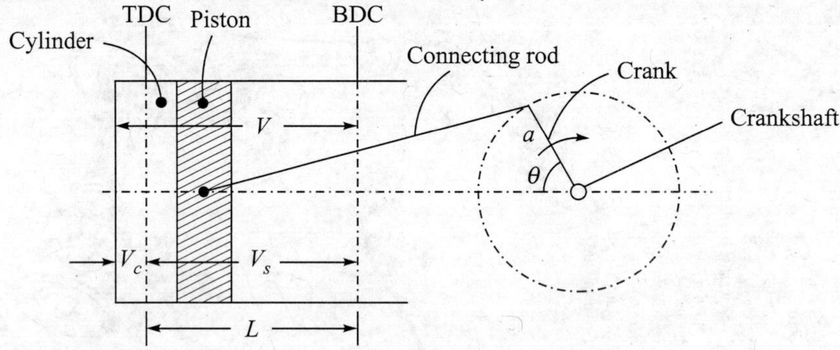

Figure 1.6 Basic terminology.

1. Bore (*d*): It is the inside diameter of the engine cylinder. It is called the bore as it is made through a boring process.

2. Stroke (*L*): During the travel of the piston, there is an upper as well as a lower limiting position at which the direction of motion of the piston is reversed. The linear distance through which the piston travels between the extreme upper and lower positions of the piston is called the stroke. It is equal to two times the crank radius, $L = 2a$, where a is the crank radius.

3. Top Dead Centre (TDC): When the piston is at the topmost position of the cylinder during its travel, that position is called the Top Dead Centre. At this position the piston velocity is zero and the piston reverses its direction of motion to travel downwards. It is the dead centre when the piston is farthest from the crankshaft.

4. Bottom Dead Centre (BDC): When the piston is at the bottom-most position of the cylinder during its travel, that position is called the Bottom Dead Centre. At this position the piston velocity is zero and the piston reverses its direction of motion to travel upwards. It is the dead centre when the piston is nearest to the crankshaft.

5. Clearance volume (V_c): When the piston is at the TDC position, the volume contained in the cylinder above the top of the piston is called the clearance volume. The piston cannot occupy any part of this volume and always keeps this volume clear.

6. Piston displacement or swept volume (V_s): It is the volume swept through by the piston in moving between the TDC and the BDC, i.e.

$$V_s = \frac{\pi}{4} d^2 L \tag{1.1}$$

7. Cylinder volume (V): The cylinder volume includes both the clearance volume and the swept volume, i.e.

$$V = V_c + V_s \tag{1.2}$$

8. Compression ratio (r): It is the ratio of the volume when the piston is at BDC to the volume when the piston is at TDC. Hence, it is the ratio of total cylinder volume to clearance volume.

$$r = \frac{V_c + V_s}{V_c} \tag{1.3}$$

9. Mean piston speed: As the piston moves inside the engine cylinder its speed changes continuously. It is zero at TDC and BDC and maximum nearly at the mid-position of TDC and BDC. The crank angle θ is zero at TDC, it is almost 90° when the piston speed is maximum and 180° at BDC. In the limit of infinitely long connecting rod, the motion is simple harmonic and maximum speed will be at 90° crank angle. Thus in a half rotation of the crank, the piston moves a distance equal to the length of the stroke, L. In full rotation, the distance travelled by piston will be $2L$. If N is the engine speed in revolutions per minute (rpm) and L is in metres, the mean piston speed will be $2LN/60$ m/s.

$$\text{Mean or Average piston speed, } \bar{u}_p = \frac{2LN}{60} \tag{1.4}$$

Maximum mean piston speed is kept within the range of 8–15 m/s. This keeps the resistance to the gas flow into the engine intake system and stresses due to the inertia of the moving parts within safe limit. Automobile engines operate at the higher end and large marine diesel engines operate at lower end of this range.

1.9 FOUR-STROKE SPARK-IGNITION ENGINE

In a four-stroke cycle engine, the cycle of operation is completed in four-strokes of the piston or two revolutions of the crankshaft. Thus 720°CA (crank angle) is required to complete a cycle. Figure 1.7 shows the movement of the piston and the position of the valves during each stroke. The individual strokes are:

- Induction or suction stroke
- Compression stroke
- Expansion or power stroke
- Exhaust stroke.

Figure 1.8 shows the p–V diagram indicating the pressure variation during a four-stroke cycle operation with respect to the volume of the cylinder.

Induction or suction stroke

The inlet valve opens at point 0 just before the TDC. As the piston moves from TDC to BDC, a mixture of air and fuel (charge) is introduced into the cylinder through the inlet valve. Due to the movement of the piston the pressure in the cylinder is reduced to a value below the atmospheric pressure and the charge flows through the induction system because of this pressure difference.

Ideally, the inlet valve should close at point 1, but in fact it does not happen so until the piston has moved part of the way along the return stroke (point 1′).

Compression stroke

With both the valves closed, the charge which is taken into the cylinder during the suction stroke is compressed by the return stroke of the piston. At the TDC position the charge occupies the volume space above the piston, which is called the clearance volume. Just before the end of the compression stroke, the spark is timed to occur at a point *S*. There is a time delay between *S* and the actual commencement of combustion. The combustion process occurs mainly at almost constant volume, and there is a large increase in pressure and temperature of the charge during this process (2–3).

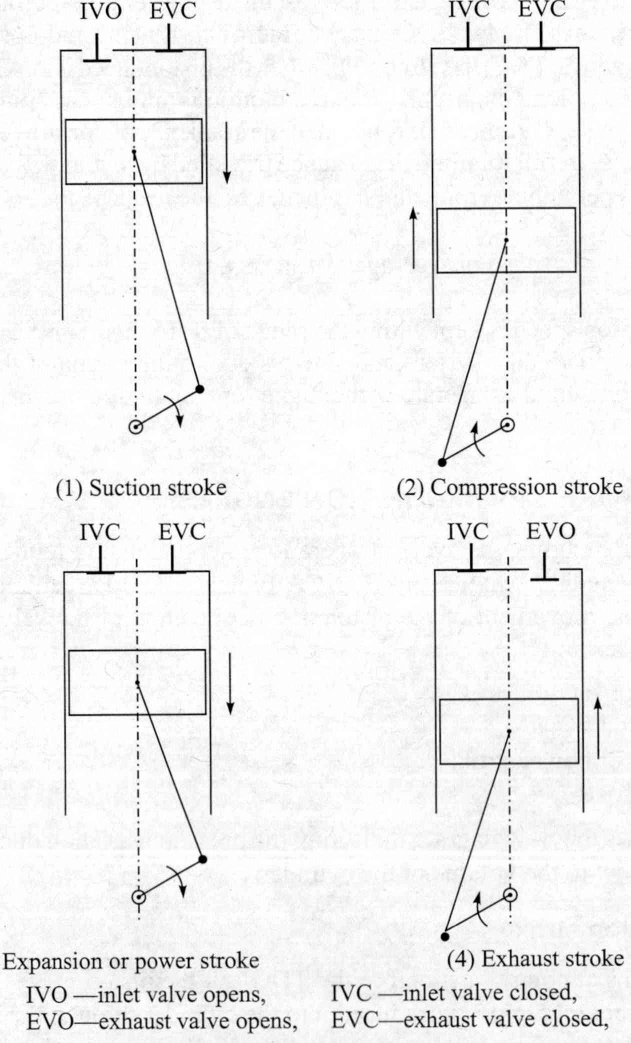

Figure 1.7 Four-strokes of IC engines.

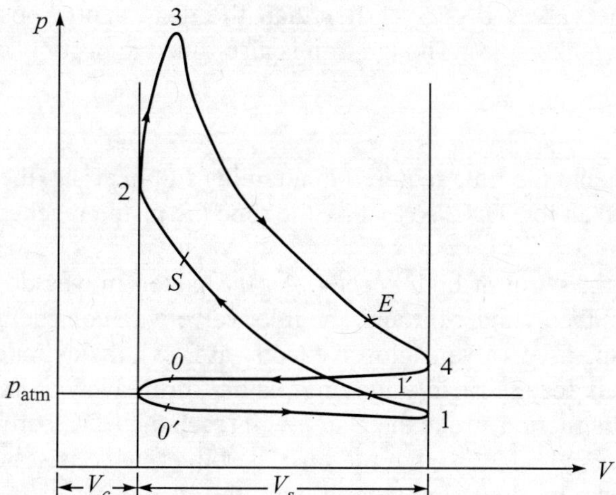

Figure 1.8 p–V diagram of a four-stroke SI engine.

Expansion or power stroke

The hot high pressure burnt gases force the piston towards the BDC. Both the inlet and exhaust valves remain closed. It appears that this expansion proceeds to completion at point 4, but in order to assist in exhausting the gaseous products the exhaust valve opens at some point *E*, which is before the BDC. Power is obtained during this stroke. Both pressure and temperature decrease during expansion.

Exhaust stroke

As the piston further moves from BDC to TDC it sweeps out the burnt gases through the exhaust valve. The inlet valve remains closed. The pressure during this stroke is slightly higher than the atmospheric pressure. The exhaust valve closes nearly at the end of the exhaust stroke (point 0′). The clearance volume cannot be exhausted and at the commencement of the next cycle this volume is full of exhaust gas from the previous cycle at about the atmospheric pressure. The exhaust gases remaining in the clearance space are called the *residual gases*. The mixture which is further compressed consists of both the fresh charge and the residual gases.

Each cylinder of a four-stroke engine completes all the operations in two engine revolutions. Thus for one complete cycle, there is only one power stroke while the crankshaft makes two revolutions.

1.10 VALVE-TIMING OF FOUR-STROKE SI ENGINE

For the successful running of the engine, the precise timing of the opening and closing of inlet and exhaust valves and the exact point at which ignition takes place are very important. Theoretically, valves should open and close and the spark should occur at the dead centre positions of the piston strokes. Actually, these events are displaced somewhat from the dead centre positions. The main factors that affect the timing of valves are the high velocity of the charge at the entry to the cylinder and the high velocity of the exhaust gases at the exit from the cylinder.

The opening of the valves occurs earlier than the dead centre position and the closing continues even at later crank angles. The ignition is also timed to occur earlier.

1.10.1 Inlet Valve

Due to the inertia effect and the time required in attaining full opening, the inlet valve is made to open somewhat earlier than the TDC, so that by the time the piston reaches the TDC, the valve is fully open.

Majority of IC engines run at high speeds. As the piston moves down during the suction stroke, the charge enters the cylinder through the inlet valve with considerable momentum due to pressure difference. If the inlet valve is closed exactly at BDC, the cylinder would not be full of charge as the tendency is for the rapidly moving piston to run away from the incoming charge. Consequently, during the suction stroke the piston will reach the BDC before the charge could get enough time to enter the cylinder through the restricted inlet valve passages. Moreover, there is considerable resistance to the flow of charge through the air-cleaner, carburettor, inlet manifold, and ports. This means that if the inlet valve is closed at BDC the cylinder for each cycle would receive charge less than its capacity and the pressure inside the cylinder would remain somewhat less than the atmosphere.

Consequently, in actual operation, the inlet valve is kept open till the cylinder pressure equals the atmospheric pressure. As the piston reverses and commences the compression stroke, it appears that some of the charge may be sent back to the induction pipe. So, it looks to be impossible to leave the inlet valve open after BDC. However, there will be a period around both the dead centre positions when the crank could swing through a wide angle while the piston remains relatively stationary. Therefore, this extra period during which the piston does not move very far up on the stroke, makes it possible to keep the inlet valve open after BDC. The momentum of the incoming charge helps it to continue to fill and run itself into the cylinder, thereby improving the volumetric efficiency of the engine.

1.10.2 Exhaust Valve

In order to exhaust as much of the products of combustion as possible, the exhaust valve opens somewhat before the BDC position. Thus, some of the exhaust gas leaves by virtue of its excess pressure above atmospheric and hence the exhaust gas is already flowing freely from the cylinder by the time the piston commences the exhaust stroke. During the exhaust stroke the piston has to do less work in order to push the exhaust gas out.

By closing the exhaust valve late, the kinetic energy of the exhaust gas can be utilized to assist in maximum exhausting of the cylinder before the exhaust valve closes and thus reducing the amount of residual gases.

Figure 1.9 shows the valve timing diagram for a four-stroke cycle engine. The diagram must actually consist of two circles, one superimposed over the other, since the four-stroke cycle is completed in two revolutions of the crankshaft. But to present a clear concept, it is usual to draw the diagram as a spiral.

It should be noted that the values of the angular positions shown in the diagram are average ones, and considerable differences occur between different engines. Spark is provided before the end of the compression stroke by the ignition system.

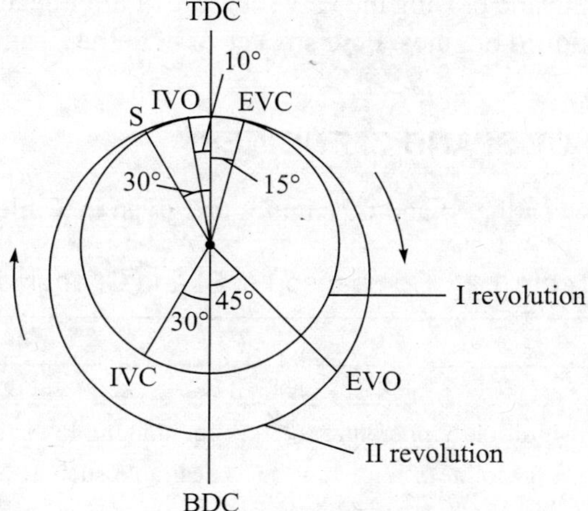

Figure 1.9 Valve-timing diagram of a four-stroke SI engine.

Note that there may be a period when both the valves are open at the same time. This happens near the TDC. The inlet valve has opened 10° before the TDC and the exhaust valve has not yet closed, it may close 15° after the TDC. This period is called the *valve overlap period.* This period should not be excessive, otherwise the burnt gases will be sucked into the cylinder, or it may allow the fresh charge to escape through the exhaust valve.

1.11 FOUR-STROKE COMPRESSION IGNITION (CI) ENGINES

The four-stroke CI engine is similar to the four-stroke SI engine except that a high compression ratio is used in the CI engines. During the suction stroke, the inlet valve opens and only the air enters into the cylinder as the piston moves from TDC to BDC.

With the inlet and exhaust valves closed, the piston compresses the air. Both the air pressure and temperature rise. When the piston almost reaches the TDC, fuel is injected in a finely divided form into the hot swirling air in the combustion space. Ignition occurs after a short delay, the gas pressure rises rapidly and a pressure wave is set up.

Work is done by the gas pressure on the piston as the piston sweeps the maximum cylinder volume. During this expansion or power stroke, the temperature and pressure of the burnt gas fall.

As the piston approaches the BDC, the exhaust valve opens and the products of combustion are rejected from the cylinder during the exhaust stroke. Near the TDC, the inlet valve opens again and the cycle is repeated.

The typical valve timings for a four-stroke CI engine are as follows:

- Inlet valve opens about 30° before TDC (bTDC)
- Inlet valve closes about 50° after BDC (aBDC)
- Exhaust valve opens about 45° before BDC (bBDC)
- Exhaust valve closes about 30° after TDC (aTDC)
- Injection of fuel is about 15° before TDC.

The reasons for opening and closing the valves before or after dead centres for CI engines are exactly the same as those for SI engines. Four strokes of both the SI and CI engines are the same.

1.12 COMPARISON OF SI AND CI ENGINES

The basic differences between the SI and CI engines are given in Table 1.2.

Table 1.2 Comparison of SI and CI engines

SI engine	*CI engine*
• It works on Otto cycle.	It works on Diesel or Dual combustion cycle.
• A fuel having higher self-ignition temperature is desirable, such as petrol (gasoline).	A fuel having lower self-ignition temperature is desirable such as diesel oil.
• Air and fuel mixture in gaseous form is inducted through the carburettor in the cylinder during the suction stroke.	Only air is introduced into the cylinder during the suction stroke and therefore the carburettor is not required. Fuel is injected at high pressure through fuel injectors direct into the combustion chamber.
• The throttle valve of the carburettor controls the quantity of the charge. The quality of the charge remains almost fixed during normal running conditions at variable speed and load. So it is a quantity governed engine.	The amount of air inducted is fixed but the amount of fuel injected is varied by regulating the quantity of fuel in the pump. The air-fuel ratio is varied at varying load. So, it is a quality governed engine.
• Spark is required to burn the fuel. For this, an ignition system with spark plugs is required. Because of this it is called a spark-ignition (SI) engine.	Combustion of fuel takes place on its own without any external ignition system. Fuel burns in the presence of highly compressed air inside the engine cylinder.
• A compression ratio of 6 to 10.5 is employed. The upper limit is fixed by the anti-knock quality of fuel. The engine tends to knock at higher compression ratios.	A compression ratio of 14 to 22 is employed. The upper limit of compression ratio is limited by the rapidly increasing weight of the engine. Engine tends to knock at lower compression ratios.
• Part load efficiency is poor, since even at part load the air/fuel ratio is not much varied. In order to improve the part load efficiency of the SI engine, the MPFI technique of fuel supply is used in modern engines.	Part load efficiency is good. As the load decreases, the fuel supply to the engine can also be reduced and lean mixture to the engine is then supplied.
• The cost of the petrol is higher than that of the diesel oil.	The cost of diesel oil is less than that of petrol. Moreover, as fuel is sold on volume basis and diesel oil has higher specific gravity, more weight is obtained in one litre.
• Noise and vibration are less because of less engine weight.	Noise and vibrations are more because of heavier engine components due to higher compression ratio.
• The main pollutants are carbon monoxide (CO), oxides of nitrogen (NO_x) and hydrocarbons (HC).	Apart from CO, NO_x, and HC, soot or smoke particles are also emitted to the atmosphere.

1.13 TWO-STROKE ENGINES

The two strokes of the cycle are completed once during each revolution of the crankshaft. Figure 1.10 represents a two-stroke SI engine with crankcase compression. As the piston moves upwards in the cylinder and covers both the transfer port (T) and exhaust port (E), there is compression of the charge into the cylinder. At the same time in the crankcase the charge expands and the pressure is reduced below the atmospheric pressure. So, fresh charge enters into the crankcase through the spring-loaded automatic valve (S). Ignition occurs before the TDC. The gas pressure rises rapidly and pushes the piston downwards to produce power.

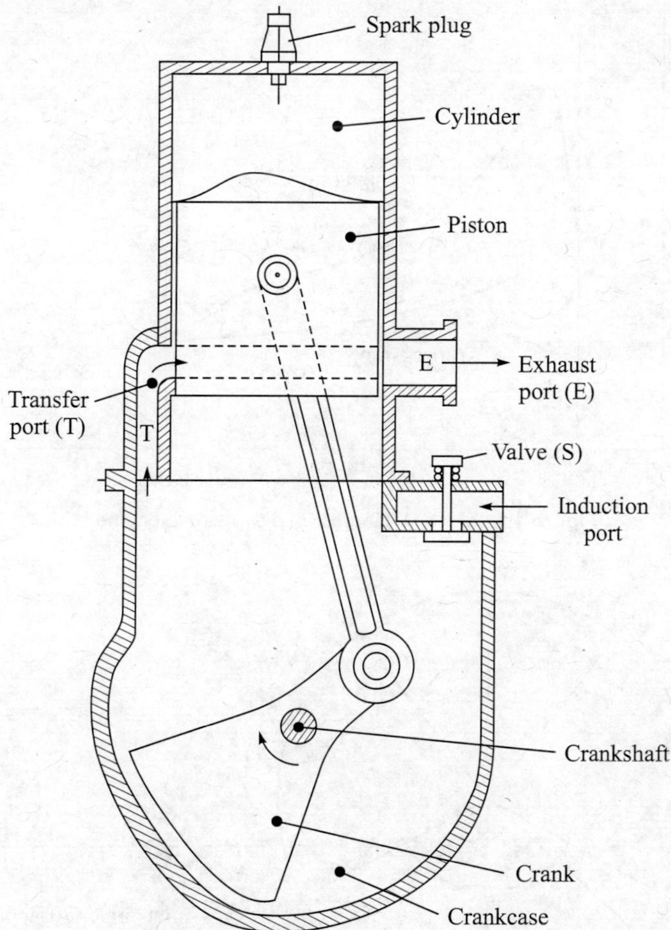

Figure 1.10 Two-stroke SI engine.

As the piston descends through about 80 per cent of the working stroke, the exhaust port (E) is uncovered by the piston and exhaust begins. The transfer port (T) is uncovered a little later in the stroke due to its position in relation to the exhaust port (E). The charge in the crankcase, which has been compressed by the descending piston, enters the cylinder through the transfer port (T).

The piston is shaped to deflect the fresh charge upwards in the cylinder and prevent its escape through the exhaust port. As the piston rises, the transfer port is closed slightly before the exhaust port and after the exhaust port is closed, compression of the charge in the cylinder begins. The cycle is then repeated.

The pressure diagram and the timing diagram for a two-stroke SI engine are shown in Figures 1.11(a) and 1.11(b) respectively.

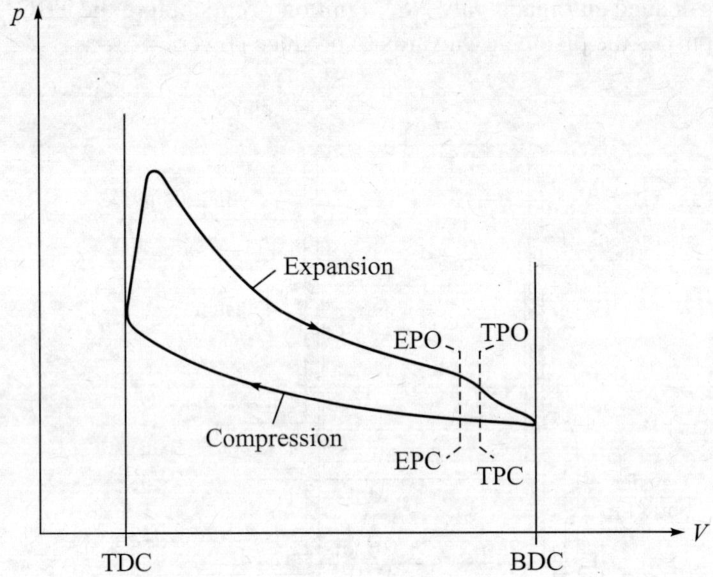

Figure 1.11(a) *p–V* diagram of a two-stroke engine.

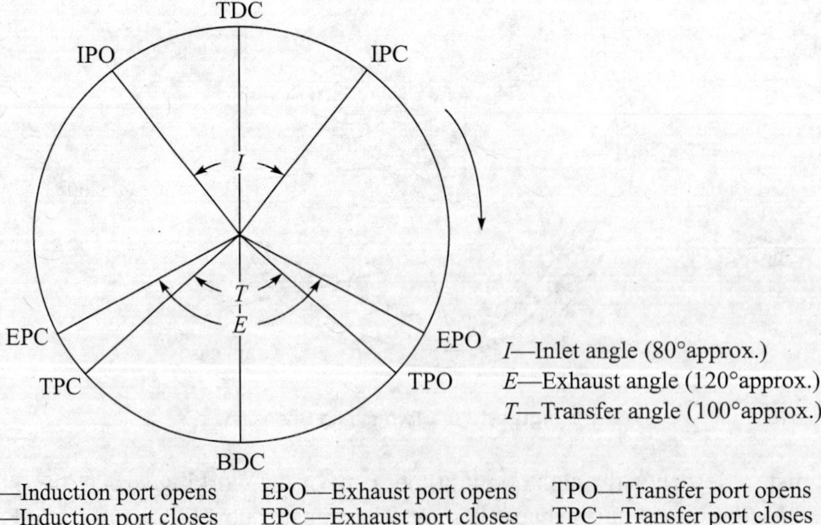

Figure 1.11(b) Timing diagram of a two-stroke engine.

The description of the two-stroke cycle as given in this section also applies to the CI engines with the exception that only air is compressed and the spark plug is replaced by the fuel injector. A higher compression ratio is also used in the case of CI engines.

1.14 COMPARISON OF FOUR-STROKE AND TWO-STROKE ENGINES

A comparison of four-stroke and two-stroke engines indicating their relative merits and demerits is presented in Table 1.3.

Table 1.3 Comparison of four-stroke and two-stroke engines

Four-stroke engine	*Two-stroke engine*
• One power stroke is obtained in every two revolutions of the crankshaft as the cycle is completed in four-strokes of the piston or in two revolutions of the crankshaft.	One power stroke is obtained in each revolution of the crankshaft as the cycle is completed in two strokes or in one revolution of the crankshaft.
• One power stroke in two revolutions of the crankshaft makes the turning movement of the shaft non-uniform and hence a heavier flywheel is needed to rotate the shaft uniformly.	The turning movement of the shaft is more uniform and hence a lighter flywheel is needed to rotate the shaft uniformly.
• Power produced for the same size of the engine is less and for the same power output, the engine is larger in size, because only one power stroke is obtained in two revolutions.	Power produced for the same size of the engine is more and for the same power output, the engine is smaller in size, because one power stroke is obtained in every revolution.
• It contains valves and valve mechanism.	It has ports. Some engines are fitted with exhaust valve or reed valve.
• Because of heavy weight and complicated valve mechanism, the initial cost is high.	It is light in weight and has no valve mechanism. Its initial cost is therefore low.
• Due to positive scavenging and greater time of induction, its thermal efficiency and volumetric efficiency are higher.	It has lower thermal efficiency and volumetric efficiency. Some of the fresh charge escapes unburnt during scavenging in petrol engines.
• Used where high efficiency is important as in automobiles, power generation and aeroplanes.	Used where low cost, low weight and compactness are important as in scooters, mopeds, lawnmowers, motor cycles, etc. Two-stroke diesel engines are used in very large sizes for ship propulsion because of low weight and compactness.
• Normally water-cooled, the wear and tear are therefore less. It consumes less amount of lubricant. The lubricant is placed in the crankcase. It is not mixed with the fuel.	Normally air-cooled, the wear and tear are therefore more. It requires more amount of lubricant. Usually, mobil oil is mixed with fuel.

It may be noted that the two-stroke cycle has twice as many power strokes per crank revolution as the four-stroke cycle. However in two-stroke cycle engines, the power output per unit displaced volume is less than twice the power output of an equivalent four-stroke cycle engine at the same engine speed. This is because of (a) the reduction in the effective expansion stroke and (b) some fresh charge escaping unburnt during the scavenging process.

1.15 GEOMETRY OF RECIPROCATING ENGINE

The geometry of the piston, the connecting rod and the crank is shown in Figure 1.12. From the figure,

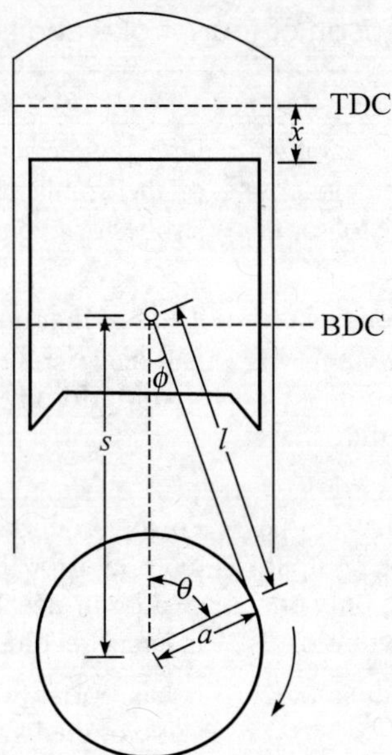

Figure 1.12 Geometry of reciprocating engine.

$$l \sin\phi = a \sin\theta$$

or

$$\sin\phi = \frac{a}{l}\sin\theta$$

where

ϕ = angle which the connecting rod makes with the centre line of the cylinder
θ = crank angle from the TDC position
a = crank radius
l = connecting rod length

$$\cos\phi = \sqrt{1-\sin^2\phi}$$

$$= \sqrt{1-\frac{a^2}{l^2}\sin^2\theta} \tag{1.5}$$

The distance s between the crank axis and the wrist pin axis is given by

$$s = a\cos\theta + l\cos\phi$$

$$= a\cos\theta + \sqrt{l^2 - a^2\sin^2\theta} \tag{1.6}$$

The piston displacement from the TDC position is given by

$$x = l + a - s$$

$$= l + a - a\cos\theta - \sqrt{l^2 - a^2\sin^2\theta} \tag{1.7}$$

Instantaneous cylinder volume,

$$V = V_c + V_x$$

$$= V_c + \frac{\pi}{4}d^2 x$$

$$\therefore \quad V = \frac{V_s}{(r-1)} + \frac{\pi}{4}d^2[l + a - a\cos\theta - (l^2 - a^2\sin^2\theta)^{1/2}]$$

$$= \frac{V_s}{(r-1)} + \frac{\pi}{4}d^2 a\left[\frac{l}{a} + 1 - \cos\theta - \left(\frac{l^2}{a^2} - \sin^2\theta\right)^{1/2}\right]$$

where r is the compression ratio.
Since crank radius a is half of the stroke length L,

$$a = \frac{L}{2}$$

$$V = \frac{V_s}{(r-1)} + \frac{V_s}{2}[R + 1 - \cos\theta - (R^2 - \sin^2\theta)^{1/2}] \tag{1.8}$$

where R is the ratio of connecting rod length to crank radius $\left(\frac{l}{a}\right)$

$$\frac{dV}{d\theta} = \frac{V_s}{2}\sin\theta[1 + \cos\theta(R^2 - \sin^2\theta)^{-1/2}] \tag{1.9}$$

Instantaneous piston speed

$$u_p = \frac{dx}{dt} = \frac{dx}{d\theta}\cdot\frac{d\theta}{dt} = \omega\frac{dx}{d\theta} = 2\pi N\frac{dx}{d\theta}$$

$$\therefore \quad u_p = 2\pi N\left[a\sin\theta - \frac{1}{2}(l^2 - a^2\sin^2\theta)^{-1/2}(-2a^2\sin\theta\cos\theta)\right]$$

$$= 2\pi N\left[a\sin\theta + \frac{a^2\sin\theta\cos\theta}{\sqrt{l^2 - a^2\sin^2\theta}}\right] \tag{1.10}$$

Mean piston speed

$$\bar{u}_p = 2LN$$

$$\frac{u_p}{\bar{u}_p} = \frac{\pi}{L} a \sin\theta \left[1 + \frac{a \cos\theta}{\sqrt{l^2 - a^2 \sin^2\theta}} \right]$$

$$\therefore \quad = \frac{\pi}{2} \sin\theta \left[1 + \frac{\cos\theta}{\sqrt{R^2 - \sin^2\theta}} \right] \tag{1.11}$$

Approximate values of piston displacement, piston speed and acceleration can be obtained as follows:

From Eq. (1.6)

$$s = a \cos\theta + \sqrt{l^2 - a^2 \sin^2\theta}$$

$$= a \cos\theta + l \left(1 - \frac{a^2}{l^2} \sin^2\theta \right)^{1/2}$$

$$s \simeq a \cos\theta + l \left(1 - \frac{a^2}{2l^2} \sin^2\theta \right)$$

Instantaneous piston displacement,

$$x = a + l - s$$

$$= a + l - a \cos\theta - l \left(1 - \frac{a^2}{2l^2} \sin^2\theta \right)$$

$$= a \left[(1 - \cos\theta) + \frac{a}{2l} \sin^2\theta \right] \tag{1.12}$$

Instantaneous piston speed,

$$u_p = \frac{dx}{dt} = \left(a \sin\theta + \frac{a^2}{2l} \cdot 2 \sin\theta \cos\theta \right) \frac{d\theta}{dt}$$

$$= \omega a \sin\theta \left(1 + \frac{a}{l} \cos\theta \right)$$

$$= \omega a \sin\omega t \left(1 + \frac{a}{l} \cos\omega t \right)$$

$$= \omega a \left(\sin\omega t + \frac{a}{2l} \sin 2\omega t \right) \tag{1.13}$$

Instantaneous acceleration is

$$a_p = \frac{du_p}{dt} = \omega a\left(\omega \cos \omega t + \frac{a}{2l} \cdot 2\omega \cos 2\omega t\right)$$

$$= \omega^2 a\left(\cos \omega t + \frac{a}{l} \cos 2\omega t\right) \tag{1.14}$$

1.16 ENGINE PERFORMANCE PARAMETERS

In order that different types of engines or different engines of the same type may be compared, certain performance criteria must be defined.

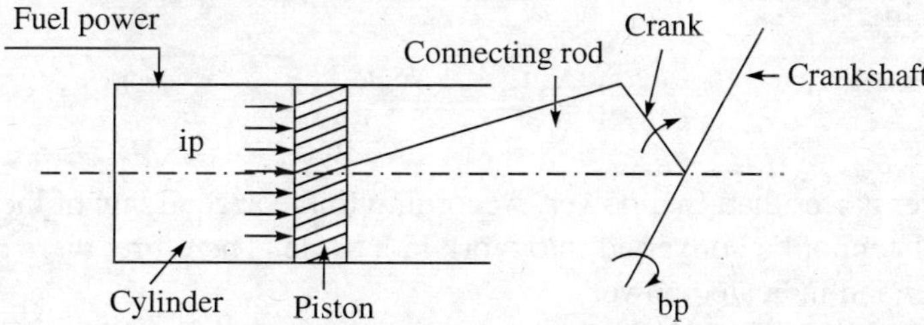

Figure 1.13 Power terminology.

Figure 1.13 shows the fuel power, indicated power and brake power developed at different locations in the engine.

1.16.1 Fuel Power

When a certain mass of fuel is introduced into the engine cylinder, the combustion of fuel takes place in the presence of air. The rate of heat energy which is contained in the fuel and may liberate during the combustion process may be termed fuel power.

$$\text{Fuel power} = \dot{m}_f \times \text{CV (kW)} \tag{1.15}$$

where

$\dot{m}_f$ = mass flow rate of fuel (kg/s)
CV = calorific value or heating value of fuel (kJ/kg)

1.16.2 Indicated Power (ip)

It is the power developed inside the engine cylinder. With this power, the rate of work is done by the gas on the piston.

$$\text{Work done per cycle} = p_{mi} \times A \times L \quad \text{(J)}$$

where

p_{mi} = indicated mean effective pressure (N/m^2)
A = area of piston (m^2)
L = Stroke length (m)

If n is the number of cylinder

$$\text{ip} = \frac{p_{m_i} L\, A(\text{number of cycles/minute}) \cdot n}{60} \text{ (W)}$$

If N = engine speed in revolution per minute (RPM)

For four-stroke engine, number of cycles/min = $\frac{N}{2}$
For two-stroke engine, number of cycles/min = N
For four-stroke engine,

$$\text{ip} = \frac{p_{m_i} L A N \cdot n}{120} \quad (1.16)$$

For two-stroke engine,

$$\text{ip} = \frac{p_{m_i} L A N \cdot n}{60} \quad (1.17)$$

Indicated power is less than fuel power. According to the second law of thermodynamics, all the heat of the fuel cannot be converted into work in a cycle. Therefore, there are thermal losses between fuel power and indicated power.

1.16.3 Brake Power (bp)

This is the actual power available at the crankshaft and may be called power output of the engine, which is utilized for useful purpose. Brake power is less than indicated power owing to mechanical friction and auxiliary component losses. Auxiliary components are accessories which the engine drives.

The brake power is given by

$$\text{bp} = \omega T = \frac{2\pi N T}{60} \text{ (watt)} \quad (1.18)$$

where

ω = angular velocity (rad/s)
N = engine speed (RPM)
T = torque (N·m)

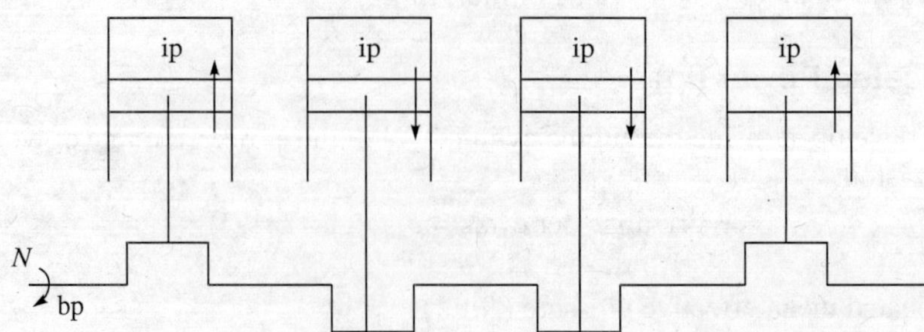

Figure 1.14 Brake power developed in a multicylinder engine.

If there are n cylinders, the bp in Eq. (1.18) is not multiplied by the number of cylinders as it is done in the case of ip. Figure 1.14 shows bp developed in a multicylinder engine. It is produced at the crankshaft. The crankshaft is a single shaft which gets power from all the cylinders. The torque T and engine speed N are measured at the crankshaft.

No load on the engine implies that there is no useful power output from the crankshaft. The bp is zero and all the ip is equal to the friction power. At half load, bp is reduced to half the value of bp produced at full load.

1.16.4 Friction Power (fp)

The difference between the ip and the bp is the friction power (fp), and it is the power required to overcome the frictional resistance of the engine parts, including power absorbed by auxiliary components.

$$\text{fp} = \text{ip} - \text{bp} \tag{1.19}$$

At zero bp, at the same speed the engine develops just enough power to overcome frictional resistance and drives engine accessories.

1.16.5 Indicated Thermal Efficiency (η_i)

It is the ratio of ip to fuel power. It may also be defined as the ratio of work of one cycle to the heat supplied by the fuel in that cycle.

$$\eta_i = \frac{\text{ip}}{\dot{m}_f \times \text{CV}} = \frac{W}{Q_s} \tag{1.20}$$

where

W = work of one cycle

Q_s = heat supplied by fuel in the cycle

The typical value of η_i is in the range of 35–45%.

1.16.6 Brake Thermal Efficiency (η_b)

It is the ratio of bp to fuel power. It may also be called overall efficiency of the engine.

$$\eta_b = \frac{\text{bp}}{\dot{m}_f \times \text{CV}} \tag{1.21}$$

Usually engines have brake thermal efficiency about 30%. Some large, slow CI engines can have higher brake thermal efficiency.

1.16.7 Mechanical Efficiency (η_m)

It is the ratio of bp to ip.

$$\eta_m = \frac{\text{bp}}{\text{ip}} = \frac{\eta_b}{\eta_i} \tag{1.22}$$

Mechanical efficiency usually lies between 80 and 90%.

1.16.8 Combustion Efficiency (η_c)

The time available for the combustion process of an engine cycle is very short. All the fuel molecules may not find oxygen molecules, or the local temperature may not favour the reaction. Consequently, a small fraction of fuel does not react and goes out with the exhaust flow. A combustion efficiency accounts for the fraction of fuel which burns. The typical value of η_c is in the range of 0.95 – 0.98.

$$\eta_c = \frac{\dot{Q}_s}{\dot{m}_f \text{ CV}} \tag{1.23}$$

where

$\dot{Q}_s$ = heat supplied to the gas per unit time

$$\dot{Q}_s = \dot{m}_f \cdot CV \cdot \eta_c \tag{1.24}$$

Using η_c,

$$\eta_i = \frac{\text{ip}}{\dot{m}_f \cdot \text{CV} \cdot \eta_c} \quad \text{and} \quad \eta_b = \frac{\text{bp}}{\dot{m}_f \cdot \text{CV} \cdot \eta_c} \tag{1.25}$$

1.16.9 Relative Efficiency or Efficiency Ratio (η_{rel})

It is the ratio of thermal efficiency of actual cycle to the air-standard efficiency of ideal cycle. It indicates the degree of development of the engine.

$$\eta_{\text{rel}} = \frac{\text{thermal efficiency of the actual cycle}}{\text{air-standard efficiency of ideal cycle}} \tag{1.26}$$

1.16.10 Volumetric Efficiency (η_v)

The power output of an IC engine depends upon the amount of air which can be induced into the cylinder. More air means more fuel can be burnt and more energy can be converted into output power. The intake system (the air filter, carburettor and throttle plate (SI engine), intake manifold, intake valve) restricts the amount of air which an engine of given displacement can ideally induct. Volumetric efficiency is the parameter used to measure the effectiveness of a four-stroke engine's induction process. It is the ratio of volume of air-induced per cycle measured at the free air delivered (FAD) conditions (at atmospheric pressure and temperature) to the swept volume of the cylinder.

$$\eta_v = \frac{(V_a)_{\text{FAD}}}{V_s} \tag{1.27}$$

It may also be defined as the mass of air actually induced inside the cylinder per cycle to the mass of air which could be present ideally in the swept volume at atmospheric conditions.

$$\eta_v = \frac{(m_a)_{\text{actual}}}{(m_a)_{\text{ideal}}} = \frac{(\dot{m}_a)_{\text{actual}}}{(\dot{m}_a)_{\text{ideal}}} = \frac{(\dot{m}_a)_{\text{actual}}}{\rho_a V_s (N/2)} \tag{1.28}$$

where

$(m_a)_{\text{actual}}$ = mass of air that actually entered into the cylinder per cycle
$(m_a)_{\text{ideal}}$ = $\rho_a V_s$ = ideal mass of air in the swept volume at atmospheric conditions
$(\dot{m}_a)_{\text{actual}}$ = actual mass flow rate of air into the cylinder
$(\dot{m}_a)_{\text{ideal}}$ = ideal mass flow rate of air = $\rho_a \cdot V_s\,(N/2)$
ρ_a = density of air at atmospheric conditions
N = engine speed

Typical maximum values of volumetric efficiency for naturally aspirated engines at wide open throttle (WOT) are in the range of 80–90%. It reduces to much lower values as the throttle is closed. The volumetric efficiency for CI engines is little higher than that for SI engines.

1.16.11 Indicated Mean Effective Pressure (imep or p_{mi})

Pressure in the cylinder of an engine continuously changes during the cycle. An average or mean effective pressure is the height of a rectangle having the same length and area as the cycle plotted on a p–V diagram. This is shown for an Otto cycle in Figure 1.15.

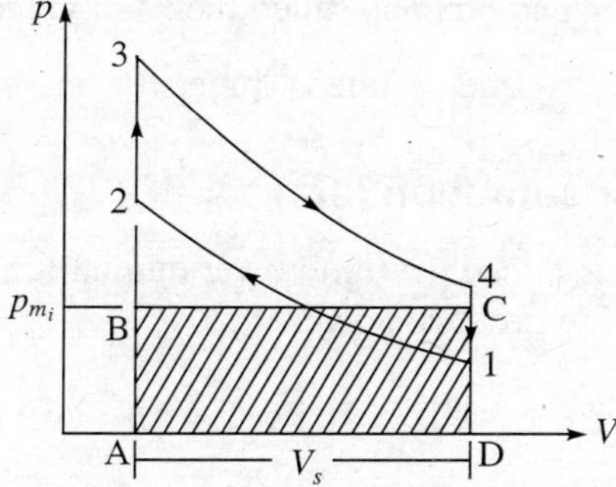

Figure 1.15 Indicated mean effective pressure on p–V diagram.

The rectangle ABCDA has the same length as the cycle 12341, and area ABCDA is equal to area 12341. Then the indicated mean effective pressure, p_{mi}, is the height AB of the rectangle. Area of the cycle 12341 or area ABCDA represents the indicated work done in a cycle, W_i.

$$W_i = p_{m_i} \cdot V_s$$

$$\therefore \qquad p_{m_i} = \frac{W_i}{V_s} \tag{1.29}$$

1.16.12 Brake Mean Effective Pressure (bmep or p_{m_b})

The bmep is defined as that part of the mean effective pressure acting on the pistons which would give the measured bp. There is no contribution of this mean effective pressure towards friction power. It is computed by measuring the brake power.

$$\text{bp} = \frac{p_{m_b} L A N n}{60} \quad \text{for two-stroke engine} \tag{1.30}$$

and $$\text{bp} = \frac{p_{m_b} L A N n}{120} \quad \text{for four-stroke engine} \tag{1.31}$$

$$\therefore \quad \frac{p_{m_b} L A N n}{60 \text{ or } 120} = \frac{2\pi NT}{60}$$

For a given engine, stroke length L, area of cross section A and the number of cylinders n remain fixed.

$$\therefore \quad p_{m_b} \propto T \tag{1.32}$$

The brake mean effective pressure is proportional to the torque and independent of the engine speed. The bmep is a useful criterion for comparing engine performance.

Typical maximum values of bmep for naturally aspirated SI engines are in the range 8.5–10.5 bar.

For CI engines, typical maximum values are 7.0–9.0 bar for naturally aspirated engines and 10.0–12.0 bar for turbocharged engines. The maximum bmep of good engine designs is well established. It is a measure of effectiveness with which the engine designer has used the swept volume.

The difference between imep and bmep is called the friction mean effective pressure (fmep).

$$\text{fmep} = \text{imep} - \text{bmep} \tag{1.33}$$

1.16.13 Specific Fuel Consumption (sfc)

It is the mass flow rate of fuel consumed per unit power output. It measures how economically an engine is using the fuel supplied to produce power.

$$\text{sfc} = \frac{\dot{m}_f}{\text{power}}$$

where

$\dot{m}_f$ = mass flow rate of fuel (kg/h)

Power is in kW and sfc is in kg/kWh.

Brake power gives brake specific fuel consumption (bsfc) and indicated power gives indicated specific fuel consumption (isfc).

$$\therefore \quad \text{bsfc} = \frac{\dot{m}_f}{\text{bp}} \quad \text{and} \quad \text{isfc} = \frac{\dot{m}_f}{\text{ip}} \tag{1.34}$$

The ratio of isfc to bsfc gives mechanical efficiency, η_m.

$$\eta_m = \frac{\text{bp}}{\text{ip}} = \frac{\text{isfc}}{\text{bsfc}} \tag{1.35}$$

For given fuel consumption, bsfc is greater than isfc. Low values of sfc are desirable. For SI engines, typical lowest value of bsfc is about 0.27 kg/kWh. For CI engines, the best value is lower.

1.16.14 Specific Power, Output per Displacement, Engine Specific Volume and Engine Specific Weight

These parameters are important in transportation, where keeping weight to a minimum is necessary.

Specific power is the power per unit piston area. It measures the effectiveness with which the piston area is used, regardless of the cylinder size.

$$\text{Specific power, SP} = \frac{\text{bp}}{A_p} \tag{1.36}$$

Output per displacement indicates the effectiveness of power output from the given size of the engine.

$$\text{Output per displacement, OPD} = \frac{\text{bp}}{V_{s\,\text{Total}}} \tag{1.37}$$

Engine specific volume indicates the effectiveness with which engine space is utilized. It is the reciprocal of output per displacement.

$$\text{Engine specific volume, ESV} = \frac{V_{s\,\text{Total}}}{\text{bp}} \tag{1.38}$$

Engine specific weight indicates the relative economy with which materials are used.

$$\text{Engine specific weight, ESW} = \frac{\text{Engine weight}}{\text{bp}} \tag{1.39}$$

where

A_p = piston face area of all pistons

$V_{s\,\text{Total}}$ = total displaced or swept volume in all cylinders or engine capacity.

1.16.15 Air/Fuel and Fuel/Air Ratios

Air/fuel ratio (A/F) is the ratio of air mass flow rate $\dot{m}_a$ to the fuel mass flow rate $\dot{m}_f$. Fuel/air (F/A) ratio is the reciprocal of air/fuel ratio.

$$\text{Air/fuel ratio (A/F)} = \frac{\dot{m}_a}{\dot{m}_f} \tag{1.40}$$

$$\text{Fuel/air ratio (F/A)} = \frac{\dot{m}_f}{\dot{m}_a} \tag{1.41}$$

The normal operating range for a conventional SI engine using gasoline as a fuel is $12 \leq \text{A/F} \leq 18$ ($0.056 \leq \text{F/A} \leq 0.083$). For CI engine with diesel fuel, it is $18 \leq \text{A/F} \leq 70$ ($0.014 \leq \text{F/A} \leq 0.056$).

EXAMPLE 1.1 A 3*l* spark-ignition V6 engine operates on a four-stroke cycle at 3000 rpm. The compression ratio is 9.0, the length of the connecting rod is 17.2 cm and engine is square. At this speed, combustion ends at 20° aTDC.

Determine:

(a) The cylinder bore and the stroke length
(b) The average piston speed
(c) The clearance volume of one cylinder
(d) The distance the piston travels from TDC at the end of combustion
(e) The volume in the combustion chamber at the end of combustion
(f) The piston speed and acceleration at the end of combustion

Solution:

(a) Since the engine is square, L = d.
Swept volume per cylinder

$$V_s = \frac{3}{6} = 0.5\ l$$

$$= 0.0005\ \text{m}^3$$

$$V_s = \frac{\pi}{4} d^2 \cdot L = \frac{\pi}{4} d^3 = 0.0005$$

$$\therefore \quad d = 0.086\ \text{m} = \boxed{8.6\ \text{cm}} \quad \textbf{Ans.}$$

$$L = d = \boxed{8.6\ \text{cm}} \quad \textbf{Ans.}$$

(b) Average piston speed,

$$\bar{u}_p = 2LN$$

$$= 2 \times 0.086 \times \frac{3000}{60} = \boxed{8.6\ \text{m/s}} \quad \textbf{Ans.}$$

(c) Compression ratio,

$$r = \frac{V_s + V_c}{V_c}$$

$$\therefore \quad V_c = \frac{V_s}{r-1} = \frac{0.0005}{(9-1)} = 62.5 \times 10^{-6}\ \text{m}^3 = \boxed{62.5\ \text{cm}^3} \quad \textbf{Ans.}$$

(d) Refer Figure 1.12.

$$\text{Crank radius, } a = \frac{L}{2} = \frac{8.6}{2} = 4.3\ \text{cm}$$

Distance between the crank axis and the wrist pin axis is given by

$$s = a\cos\theta + \sqrt{l^2 - a^2\sin^2\theta}$$

$$= 4.3\cos 20° + \sqrt{(17.2)^2 - (4.3)^2 \sin^2 20}$$

$$= 21.18\ \text{cm}$$

$$x = l + a - s$$

$$= 17.2 + 4.3 - 21.18 = \boxed{0.32 \text{ cm}} \quad \textbf{Ans.}$$

(e) Instantaneous volume,

$$V = V_c + \frac{\pi}{4} d^2.x$$

$$= 62.5 + \frac{\pi}{4}(8.6)^2 \times 0.32 = \boxed{81.1 \text{ cm}^3} \quad \textbf{Ans.}$$

(f)
$$\frac{u_p}{\bar{u}_p} = \frac{\pi}{2} \sin\theta \left(1 + \frac{\cos\theta}{\sqrt{R^2 - \sin^2\theta}}\right)$$

$$R = \frac{l}{a} = \frac{17.2}{4.3} = 4$$

$$\frac{u_p}{\bar{u}_p} = \frac{\pi}{2} \sin 20° \left(1 + \frac{\cos 20°}{\sqrt{16 - \sin^2 20°}}\right) = 0.6639$$

$$u_p = 0.6639 \times 8.6 = \boxed{5.71 \text{ m/s}} \quad \textbf{Ans.}$$

or
$$u_p = \omega a \left(\sin \omega t + \frac{a}{2l} \sin 2\omega t\right)$$

$$= \frac{2\pi N}{60} \cdot a \left(\sin\theta + \frac{a}{2l} \sin 2\theta\right)$$

$$= 2\pi \times \frac{3000}{60} \times 4.3 \left(\sin 20° + \frac{4.3}{2 \times 17.2} \cdot \sin 40°\right)$$

$$= 570.6 \text{ cm/s} = \boxed{5.71 \text{ m/s}} \quad \textbf{Ans.}$$

Acceleration,
$$a_p = \omega^2 a \left(\cos \omega t + \frac{a}{l} \cos 2\omega t\right)$$

$$= \left(\frac{2\pi N}{60}\right)^2 \cdot a \left(\cos\theta + \frac{a}{l} \cos 2\theta\right)$$

$$= \left(\frac{2\pi \times 3000}{60}\right)^2 \times 4.3 \left(\cos 20° + \frac{4.3}{17.2} \cos 40°\right)$$

$$= 480,075 \text{ cm/s}^2 = \boxed{4801 \text{ m/s}^2} \quad \textbf{Ans.}$$

EXAMPLE 1.2 A six-cylinder, four-stroke CI engine with 10 cm bore and 12 cm stroke runs at 2000 rpm. The compression ratio is 18. The engine is connected to a dynamometer which gives a brake output torque reading of 300 N·m. Air enters the cylinder at 0.9 bar and 39°C and the mechanical efficiency of the engine is 85%. The engine is running with an air-fuel ratio 20. Heating value of the fuel is 42,000 kJ/kg. Combustion efficiency is 98%. Standard density of air is 1.181 kg/m^3. Calculate: (i) brake power, (ii) indicated power, (iii) brake mean effective pressure, (iv) indicated mean effective pressure, (v) friction mean effective pressure, (vi) power lost to friction, (vii) brake work per unit mass of air in the cylinder, (viii) brake specific power, (ix) brake output per displacement, (x) engine specific volume, (xi) rate of fuel flow into the engine, (xii) brake thermal efficiency, (xiii) indicated thermal efficiency, (xiv) volumetric efficiency and (xv) brake specific fuel consumption.

Solution:

(i) Brake power,
$$\text{bp} = \frac{2\pi NT}{60}$$
$$= \frac{2\pi \times 2000 \times 300}{60} = 62.83 \times 10^3 \text{ W}$$
$$= \boxed{62.83 \text{ kW}} \quad \textbf{Ans.}$$

(ii) Indicated power,
$$\text{ip} = \frac{\text{bp}}{\eta_m} = \frac{62.83}{0.85} = \boxed{73.92 \text{ kW}} \quad \textbf{Ans.}$$

(iii)
$$\text{bp} = \frac{p_{m_b} LANn}{120}$$

$$\text{bmep; } p_{m_b} = \frac{120 \times \text{bp}}{LANn}$$

$$= \frac{120 \times 62.83 \times 10^3}{0.12 \times \frac{\pi}{4}(0.10)^2 \times 2000 \times 6} = 6.67 \times 10^5 \text{ N/m}^2 = \boxed{6.67 \text{ bar}} \quad \textbf{Ans.}$$

(iv)
$$\text{imep, } p_{m_i} = \frac{p_{m_b}}{\eta_m} = \frac{6.67}{0.85} = \boxed{7.85 \text{ bar}} \quad \textbf{Ans.}$$

(v)
$$\text{fmep} = p_{m_i} - p_{m_b}$$
$$= 7.85 - 6.67 = \boxed{1.18 \text{ bar}} \quad \textbf{Ans.}$$

(vi) Friction power,
$$\text{fp} = \text{ip} - \text{bp}$$
$$= 73.92 - 62.83 = \boxed{11.09 \text{ kW}} \quad \textbf{Ans.}$$

(vii) Brake work from one cylinder in one cycle,

$$W_b = p_{m_b} \cdot V_s = p_{m_b} \frac{\pi}{4} d^2 L$$

$$= 6.67 \times 10^5 \times \frac{\pi}{4}(0.10)^2 \times 0.12 = 0.629 \times 10^3 \text{ J}$$

$$= 0.629 \text{ kJ}$$

Swept volume, $$V_s = \frac{\pi}{4} d^2 L$$

$$= \frac{\pi}{4}(0.10)^2(0.12) = 0.9425 \times 10^{-3} \text{ m}^3$$

Clearance volume, $$V_c = \frac{V_s}{(r-1)} = \frac{0.9425 \times 10^{-3}}{(18-1)} = 0.0554 \times 10^{-3} \text{ m}^3$$

V_{BDC} = total cylinder volume = $V_s + V_c$

$$= (0.9425 + 0.0554) \times 10^{-3}$$

$$= 0.9979 \times 10^{-3} \text{ m}^3$$

Mass of air in one cylinder per cycle,

$$m_a = \frac{p V_{\text{BDC}}}{RT}$$

$$= \frac{0.9 \times 10^5 \times 0.9979 \times 10^{-3}}{287 \times (39 + 273)} = 1.002 \times 10^{-3} \text{ kg}$$

Brake specific work per unit mass of air,

$$w_b = \frac{W_b}{m_a} = \frac{0.629}{1.002 \times 10^{-3}} = \boxed{628 \text{ kJ/kg}}$$ **Ans.**

(viii) Brake specific power,

$$\text{SP} = \frac{\text{bp}}{A_p}$$

Piston face area, $$A_p = \frac{\pi}{4} d^2 \times n$$

$$= \frac{\pi}{4}(0.10)^2 \times 6 = 0.015\pi \text{ m}^2$$

$$\text{SP} = \frac{62.83}{0.015\pi} = \boxed{1333 \text{ kW/m}^2}$$ **Ans.**

(ix) Brake output per displacement,

$$\text{OPD} = \frac{\text{bp}}{V_{s_{\text{Total}}}} = \frac{\text{bp}}{V_s \times n}$$

$$= \frac{62.83}{0.9425 \times 10^{-3} \times 6} = 11.11 \times 10^3 \text{ kW/m}^3$$

$$= \boxed{11.11 \text{ kW}/l} \quad \textbf{Ans.}$$

(x) Engine specific volume,

$$\text{ESV} = \frac{V_{s_{\text{Total}}}}{\text{bp}} = \frac{1}{\text{OPD}} = \frac{1}{11.11} = \boxed{0.09\ l/\text{kW}} \quad \textbf{Ans.}$$

(xi) $\dfrac{m_a}{m_f} = 20$

$\therefore$ mass of fuel in one cylinder for one cycle, $m_f = \dfrac{m_a}{20}$

$$m_f = \frac{1.002 \times 10^{-3}}{20} = 0.0501 \times 10^{-3} \text{ kg}$$

The rate of fuel flow into the engine is given by

$$\dot{m}_f = m_f \times \frac{N}{120} \times n$$

$$= 0.0501 \times 10^{-3} \times \frac{2000}{120} \times 6 = \boxed{0.00501 \text{ kg/s}} \quad \textbf{Ans.}$$

(xii) Brake thermal efficiency,

$$\eta_b = \frac{\text{bp}}{\dot{m}_f \times \text{CV} \times \eta_c}$$

$$= \frac{62.83}{0.00501 \times 42{,}000 \times 0.98} = 0.305 = \boxed{30.5\%} \quad \textbf{Ans.}$$

(xiii) Indicated thermal efficiency,

$$\eta_i = \frac{\eta_b}{\eta_m} = \frac{0.305}{0.85} = 0.359 = \boxed{35.9\%} \quad \textbf{Ans.}$$

(xiv) Volumetric efficiency,

$$\eta_v = \frac{m_a}{\rho_a \cdot V_s}$$

ρ_a is the standard air density which is equal to 1.181 kg/m^3.

$$\eta_v = \frac{1.002 \times 10^{-3}}{1.181 \times 0.9425 \times 10^{-3}} = 0.9 = \boxed{90\%} \quad \textbf{Ans.}$$

(xv) Brake specific fuel consumption

$$\text{bsfc} = \frac{\dot{m}_f}{\text{bp}}$$

$$= \frac{0.00501 \times 3600}{62.83} = \boxed{0.287 \text{ kg/kWh}} \quad \textbf{Ans.}$$

EXAMPLE 1.3 A two-cylinder, four-stroke gas engine has a bore of 350 mm and a stroke of 575 mm. At 250 rpm, the torque developed is 5.0 kNm. The air/fuel ratio is 7:1 by volume. The estimated volumetric efficiency is 85% and the calorific value of the coal gas is 16,800 kJ/m^3. Calculate (a) the bp, (b) the mean piston speed, (c) the bmep and (d) the brake thermal efficiency of the engine.

Solution:

(a) $$\text{bp} = \frac{2\pi NT}{60} = \frac{2\pi \times 250 \times 5.0}{60} = \boxed{130.9 \text{ kW}} \quad \textbf{Ans.}$$

(b) $$\bar{u}_p = \frac{2LN}{60} = \frac{2 \times 0.575 \times 250}{60} = \boxed{4.79 \text{ m/s}} \quad \textbf{Ans.}$$

(c) $$\text{bp} = \frac{p_{m_b} L A N n}{120}$$

$$\therefore \quad p_{m_b} = \frac{\text{bp} \times 120}{L A N n}$$

$$= \frac{130.9 \times 120}{0.575 \times \frac{\pi}{4}(0.35)^2 \times 250 \times 2} = 568 \text{ kN/m}^2 = \boxed{5.68 \text{ bar}} \quad \textbf{Ans.}$$

(d) Swept volume per cylinder,

$$V_s = \frac{\pi}{4} d^2 L$$

$$= \frac{\pi}{4}(0.35)^2 \times 0.575 = 0.05532 \text{ m}^3$$

Volumetric efficiency, $$\eta_v = \frac{V_a}{V_s}$$

Volume of air induced, $$V_a = \eta_v \times V_S = 0.85 \times 0.05532$$

$$= 0.04702 \text{ m}^3/\text{cycle}$$

Air/fuel ratio by volume = 7 : 1

$\therefore$ Volume of fuel induced, $V_f = \dfrac{V_a}{7} = \dfrac{0.04702}{7} = 0.00672 \text{ m}^3/\text{cycle}$

Heat supplied by the fuel, $\dot{Q}_f = V_f \times \dfrac{N}{120} \times n \times \text{CV}$

$$= \frac{0.00672 \times 250 \times 2 \times 16,800}{120}$$

$$= 470.4 \text{ kJ/s}$$

Brake thermal efficiency,

$$\eta_b = \frac{\text{bp}}{\dot{Q}_f} = \frac{130.9}{470.4} = 0.278 = \boxed{27.8\%} \quad \textbf{Ans.}$$

EXAMPLE 1.4 A 2 l SI engine operating on a four-stroke cycle develops a brake power of 60 kW. At this condition, air–fuel ratio is 15:1, volumetric efficiency is 80%, indicated thermal efficiency is 35%, mechanical efficiency is 85%, calorific value of fuel is 44 MJ/kg and density of air is 1.181 kg/m^3. Determine the engine speed.

Solution: Swept volume, $V_s = 2000$ cc

Volume of air taken in per cycle, $V_a = \eta_v \cdot V_s$

$\therefore$ $$V_a = 0.8 \times 2000 = 1600 \text{ cc}$$

Mass of air, $$m_a = \rho_a \cdot V_a = 1.181 \times 1600 \times 10^{-6} \text{ k}$$

$$= 18.896 \times 10^{-4} \text{ kg}$$

$$\frac{m_a}{m_f} = 15$$

$\therefore$ Mass of fuel, $m_f = \dfrac{m_a}{15} = \dfrac{18.896 \times 10^{-4}}{15} = 1.26 \times 10^{-4} \text{ kg}$

$$\dot{m}_f = m_f \times \frac{N}{120} = \frac{1.26 \times 10^{-4} \times N}{120} = 1.05 \times 10^{-6} \times N \text{ kg/s}$$

Indicated power, $$\text{ip} = \frac{bp}{\eta_m} = \frac{60}{0.85} = 70.59 \text{ kW}$$

Indicated thermal efficiency, $$\eta_i = \frac{\text{ip}}{\dot{m}_f \times \text{CV}}$$

$$\therefore \qquad \dot{m}_f = \frac{\text{ip}}{\eta_i \times \text{CV}}$$

$$1.05 \times 10^{-6} \times N = \frac{70.59}{0.35 \times 44{,}000}$$

$$N = \frac{70.59}{1.05 \times 10^{-6} \times 0.35 \times 44{,}000} = \boxed{4365 \text{ RPM}} \quad \textbf{Ans.}$$

EXAMPLE 1.5 A 3*l*, four-stroke, four-cylinder CI engine running at 3000 rpm produces 48 kW of brake power. Brake thermal efficiency is 30% and the calorific value of fuel is 42,000 kJ/kg. Assume density of air at ambient condition as 1.181 kg/m^3, air–fuel ratio 21:1 and mechanical efficiency 0.8.

Calculate:

(i) Rate of fuel flow into the engine (kg/s)
(ii) Rate of air flow into the engine (kg/s)
(iii) Mass rate of exhaust flow (kg/s)
(iv) Brake output per displacement (kW/*l*)
(v) Brake specific fuel consumption (kg/kWh)
(vi) Indicated thermal efficiency
(vii) Volumetric efficiency
(viii) Brake mean effective pressure (bar)

Solution:

(i) $$\eta_b = \frac{\text{bp}}{\dot{m}_f \times \text{CV}}$$

$$\therefore \qquad \dot{m}_f = \frac{\text{bp}}{\eta_b \times \text{CV}} = \frac{48}{0.3 \times 42{,}000} = \boxed{3.81 \times 10^{-3} \text{ kg/s}} \quad \textbf{Ans.}$$

(ii) $$\frac{\dot{m}_a}{\dot{m}_f} = 21$$

$$\therefore \qquad \dot{m}_a = 21 \times \dot{m}_f = 21 \times 3.81 \times 10^{-3} = \boxed{80.01 \times 10^{-3} \text{ kg/s}} \quad \textbf{Ans.}$$

(iii) $$\dot{m}_{\text{exh}} = \dot{m}_a + \dot{m}_f = (80.01 + 3.81) \times 10^{-3} = \boxed{83.82 \times 10^{-3} \text{ kg/s}} \quad \textbf{Ans.}$$

(iv) $$\text{Brake output per displacement} = \frac{\text{bp}}{V_{s\,\text{Total}}}$$

$$= \frac{48}{3} = \boxed{16 \text{ kW}/l} \quad \textbf{Ans.}$$

(v) Brake specific fuel consumption,

$$\text{bsfc} = \frac{\dot{m}_f}{\text{bp}}$$

$$= \frac{3.81 \times 10^{-3} \times 3600}{48} = \boxed{0.286 \text{ kg/kWh}} \quad \textbf{Ans.}$$

(vi) $\text{ip} = \dfrac{\text{bp}}{\eta_m} = \dfrac{48}{0.8} = 60 \text{ kW}$

$$\eta_i = \frac{\text{ip}}{\dot{m}_f \times \text{CV}} = \frac{60}{3.81 \times 10^{-3} \times 42{,}000} = 0.375 = \boxed{37.5\%} \quad \textbf{Ans.}$$

or $$\eta_i = \frac{\eta_b}{\eta_m} = \frac{0.3}{0.8} = 0.375 = \boxed{37.5\%} \quad \textbf{Ans.}$$

(vii) Volumetric efficiency, $\eta_v = \dfrac{(\dot{m}_a)_{\text{actual}}}{(\dot{m}_a)_{\text{ideal}}}$

Total swept volume = 3×10^{-3} m^3
Mass of air in this volume,

$$(m_a)_{\text{ideal}} = 3 \times 10^{-3} \times 1.181 = 3.543 \times 10^{-3} \text{ kg}$$

Rate of mass of air, $(\dot{m}_a)_{\text{ideal}} = (m_a)_{\text{ideal}} \times \dfrac{N}{120}$

$$(\dot{m}_a)_{\text{ideal}} = 3.543 \times 10^{-3} \times \frac{3000}{120} = 0.0886 \text{ kg/s} = 88.6 \times 10^{-3} \text{ kg/s}$$

$$(\dot{m}_a)_{\text{actual}} = 80.01 \times 10^{-3} \text{ kg/s}$$

$$\eta_v = \frac{80.01 \times 10^{-3}}{88.6 \times 10^{-3}} = 0.903 = \boxed{90.3\%} \quad \textbf{Ans.}$$

(viii) $\text{bp} = \dfrac{p_{m_b} L A N n}{120}$

$\therefore$ $$p_{m_b} = \frac{120 \times \text{bp}}{L A N n} = \frac{120 \times \text{bp}}{V_{s_{\text{Total}}} \cdot N}$$

$$= \frac{120 \times 48}{3 \times 10^{-3} \times 3000} = 640 \text{ k Pa} = \boxed{6.4 \text{ bar}} \quad \textbf{Ans.}$$

EXAMPLE 1.6 A four-cylinder, four-stroke SI engine having compression ratio 9 delivers a brake power of 40 kW at 4000 rpm. The air-fuel ratio is 15:1 and the calorific value of fuel is 44 MJ/kg. Volumetric efficiency is 90% and brake thermal efficiency is 32%. The engine is under square and the ratio is 1.1 ($L/d = 1.1$). Density of air is 1.181 kg/m^3. Determine:

(a) Engine capacity (cm^3)
(b) Cylinder bore and stroke length (cm)
(c) Average piston speed (m/s)
(d) Clearance volume of each cylinder (cm^3)

Solution:

(a)
$$\eta_b = \frac{\text{bp}}{\dot{m}_f \times \text{CV}}$$

$$\dot{m}_f = \frac{\text{bp}}{\eta_b \times \text{CV}} = \frac{40}{0.32 \times 44{,}000} = 2.84 \times 10^{-3} \text{ kg/s}$$

$$\dot{m}_a = \frac{A}{F} \times \dot{m}_f = 15 \times 2.84 \times 10^{-3} = 42.6 \times 10^{-3} \text{ kg/s}$$

$$\dot{m}_a \text{ per cylinder} = \frac{42.6 \times 10^{-3}}{4} = 10.65 \times 10^{-3} \text{ kg/s}$$

$\because$
$$\dot{m}_a \text{ per cylinder} = m_a \times \frac{N}{120}$$

$\therefore$ Mass of air per cylinder per cycle, $m_a = \dfrac{120 \times \dot{m}_a \text{ per cylinder}}{N}$

$\therefore$
$$m_a = \frac{120 \times 10.65 \times 10^{-3}}{4000} = 0.3195 \times 10^{-3} \text{ kg}$$

Ideal mass of air per cylinder per cycle,

$$(m_a)_{\text{ideal}} = \frac{m_a}{\eta_v} = \frac{0.3195 \times 10^{-3}}{0.9} = 0.355 \times 10^{-3} \text{ kg}$$

Swept volume,
$$V_s = \frac{(m_a)_{\text{ideal}}}{\rho_{\text{air}}} = \frac{0.355 \times 10^{-3}}{1.181}$$

$$= 0.3 \times 10^{-3} \text{ m}^3$$

$$= 300 \text{ cm}^3$$

Engine capacity $= n \times V_s = 4 \times 300 = \boxed{1200 \text{ cm}^3}$ **Ans.**

(b) $V_s = \frac{\pi}{4} d^2 \cdot L = \frac{\pi}{4} d^2 \times (1.1d) = 300$

$$\therefore \quad d^3 = \frac{300 \times 4}{\pi \times 1.1} = 347$$

$$d = \boxed{7.03 \text{ cm}} \quad \textbf{Ans.}$$

$$L = 1.1d = 1.1 \times 7.03 = \boxed{7.73 \text{ cm}} \quad \textbf{Ans.}$$

(c) Average piston speed, $\overline{u}_p = \dfrac{2LN}{60}$

$$\therefore \quad \overline{u}_p = \frac{2 \times 0.0773 \times 4000}{60} = \boxed{10.31 \text{ m/s}} \quad \textbf{Ans.}$$

(d) Compression ratio, $r = \dfrac{V_s + V_c}{V_c}$

$$\therefore \quad V_c = \frac{V_s}{(r-1)} = \frac{300}{(9-1)} = \boxed{37.5 \text{ cm}^3} \quad \textbf{Ans.}$$

EXAMPLE 1.7 Consider two engines with the following details:

Engine 1: Six cylinder, four-stroke SI engine, indicated power is 60 kW, average piston speed is 12 m/s

Engine 2: Two-cylinder, two-stroke SI engine, indicated power is 12 kW

Assume the indicated mean effective pressure of both the engines to be the same. Ratio of bore of engines 1 and 2 = 2:1. Determine the average piston speed of engine 2.

Solution:

For four-stroke engine, $\text{ip} = \dfrac{p_{m_i} L A N n}{120}$

For two-stroke engine, $\text{ip} = \dfrac{p_{m_i} L A N n}{60}$

Average piston speed, $\overline{u}_p = 2LN$

$$\therefore \quad LN = \frac{\overline{u}_p}{2}$$

$$\therefore \quad \text{ip}_1 = p_{m_i} A_1 \frac{\overline{u}_{p1}}{2} \times \frac{6}{120}$$

$$\text{ip}_2 = p_{m_i} A_2 \frac{\overline{u}_{p2}}{2} \times \frac{2}{60}$$

$$\frac{\text{ip}_1}{\text{ip}_2} = \frac{A_1}{A_2}\frac{\bar{u}_{p1}}{\bar{u}_{p2}} \times \frac{6}{120} \times \frac{60}{2}$$

$$\bar{u}_{p2} = \frac{3}{2}\left(\frac{\text{ip}_2}{\text{ip}_1}\right)\cdot\left(\frac{A_1}{A_2}\right)\cdot\bar{u}_{p1}$$

$$= \frac{3}{2} \times \frac{12}{60} \times \left(\frac{2}{1}\right)^2 \times 12 = \boxed{14.4 \text{ m/s}} \quad \textbf{Ans.}$$

EXAMPLE 1.8 A 12 cm^3 single cylinder two-stroke SI engine operates at 6000 rpm with a volumetric efficiency of 0.85. The engine is square (bore = stroke). The fuel-air ratio is 0.067 and density of air is 1.181 kg/m^3.
Determine:

(a) Average piston speed (m/s)
(b) Flow rate of air into the engine (kg/s)
(c) Flow rate of fuel into the engine (kg/s)
(d) Fuel input for one cycle (kg/cycle)

Solution:

(a) $V_s = \frac{\pi}{4}d^2 L = \frac{\pi}{4}d^3 = 12 \text{ cm}^3$

$\therefore$ $$d = 2.48 \text{ cm}$$
$$L = d = 0.0248 \text{ m}$$

Average piston speed, $\bar{u}_p = \frac{2LN}{60}$

$$= \frac{2 \times 0.0248 \times 6000}{60} = \boxed{4.96 \text{ m/s}} \quad \textbf{Ans.}$$

(b) Actual volume of air per cycle, $V_a = V_s \times \eta_v$

$$= 12 \times 0.85 = 10.2 \text{ cm}^3$$

$$m_a = \rho_a \times V_a = 1.181 \times 10.2 \times 10^{-6} = 12.046 \times 10^{-6} \text{ kg/cycle}$$

For two-stroke, $\dot{m}_a = m_a \times \frac{N}{60} = 12.046 \times 10^{-6} \times \frac{6000}{60}$

$$= \boxed{12.046 \times 10^{-4} \text{ kg/s}} \quad \textbf{Ans.}$$

(c) $\dot{m}_f = \frac{F}{A} \times \dot{m}_a$

$$= 0.067 \times 12.046 \times 10^{-4} = \boxed{0.8071 \times 10^{-4} \text{ kg/s}} \quad \textbf{Ans.}$$

(d) $\because \quad \dot{m}_f = m_f \times \dfrac{N}{60}$

$$\therefore \quad m_f = \frac{60 \times \dot{m}_f}{N} = \frac{60 \times 0.8071 \times 10^{-4}}{6000} = \boxed{0.8071 \times 10^{-6} \text{ kg/cycle}} \quad \textbf{Ans.}$$

EXAMPLE 1.9 An SI engine develops indicated power of 50 kW at full load. Its brake thermal efficiency is 30% and brake specific fuel consumption is 0.286 kg/kWh. At 75% of load, it has a mechanical efficiency of 70%. Assuming constant frictional losses, calculate brake power, friction power, mechanical efficiency, indicated thermal efficiency and indicated specific fuel consumption. Also, calculate mechanical efficiency at half load.

Solution: Let bp at full load = x kW

$\therefore$ bp at 75% of load = $0.75x$ kW

ip at 75% of load = $(0.75x + \text{fp})$ kW

At 75% load,
$$\eta_m = \frac{0.75x}{0.75x + \text{fp}} = 0.7$$

$$\therefore \quad 0.75x = 0.525x + 0.7\text{fp}$$

$$\therefore \quad \text{fp} = \frac{0.225x}{0.7} = 0.3214x$$

It remains constant at all loads.

At full load,
$$\text{ip} = \text{bp} + \text{fp} = 50$$

$$x + 0.3214x = 50$$

$$x = \frac{50}{1.3214} = 37.84 \text{ kW}$$

$$\therefore \quad \boxed{\text{bp} = 37.84 \text{ kW}} \quad \textbf{Ans.}$$

$$\text{fp} = 0.3214 \times 37.84 = \boxed{12.16 \text{ kW}} \quad \textbf{Ans.}$$

(or fp = ip – bp = 50 – 37.84 = 12.16 kW)

$$\eta_m = \frac{\text{bp}}{\text{ip}} = \frac{37.84}{50} = 0.7568 = \boxed{75.68\%} \quad \textbf{Ans.}$$

$$\eta_i = \frac{\eta_b}{\eta_m} = \frac{0.30}{0.7568} = 0.396 = \boxed{39.6\%} \quad \textbf{Ans.}$$

Indicated specific fuel consumption,

$$\text{isfc} = \text{bsfc} \times \eta_m = 0.286 \times 0.7568 = \boxed{0.216 \text{ kg/kWh}} \quad \textbf{Ans.}$$

At half load,
$$\text{bp} = \frac{37.84}{2} = 18.92 \text{ kW}$$

$$\text{fp} = 12.16 \text{ kW}$$

$$\therefore \quad \eta_m = \frac{bp}{bp + fp} = \frac{18.92}{18.92 + 12.16} = 0.609 = \boxed{60.9\%} \quad \textbf{Ans.}$$

REVIEW QUESTIONS

1. Define a heat engine. How are heat engines classified?
2. What do you understand by an external combustion engine? Give some examples of this type of the engine.
3. What do you understand by an internal combustion engine? Give some examples of this type of engine.
4. Distinguish between internal combustion and external combustion engines. What are the relative merits and demerits of internal combustion engines over the external combustion engines?
5. Distinguish between intermittent and continuous IC engines. Give some examples of these types of engines.
6. Give an account of historical development of IC engines.
7. Give an account of the modern development of IC engines.
8. How are the reciprocating IC engines classified? Briefly describe the each type.
9. How are the reciprocating IC engines classified according to their applications? Mention the predominant type of engines used in each case.
10. How are the reciprocating IC engines classified according to cylinder arrangement? Briefly describe the each type with the help of suitable diagrams.
11. Describe the functions of important engine components in a four-stroke IC engine. Also mention the materials used for these engine components.
12. Define swept volume, clearance volume, compression ratio and mean piston speed.
13. Describe with the help of diagrams, the working principle of the four-stroke SI engine.
14. Describe the valve timing of a four-stroke SI engine. Draw the *p–V* diagram and valve timing diagram for an SI engine.
15. Describe the working principle of the four-stroke CI engine. Mention the typical values of valve timings for a four-stroke CI engine.
16. Distinguish between spark-ignition and compression-ignition engines.
17. Describe a two-stroke SI engine with the help of a diagram. What modifications are required for the two-stroke CI engine?
18. Distinguish between four-stroke and two-stroke IC engines. Mention their relative merits and demerits.
19. Draw the *p–V* diagram and the typical valve-timing diagram for a two-stroke IC engine.
20. What are the major pollutants from the exhaust of SI and CI engines?
21. Derive expressions to evaluate instantaneous piston displacement, cylinder volume, piston speed and acceleration with respect to crank angle.
22. Define the following power terms used in IC Engines: fuel power, indicated power, brake power and friction power.

23. Define the following efficiency terms used in IC Engines: indicated thermal efficiency, brake thermal efficiency, mechanical efficiency, combustion efficiency, relative efficiency and volumetric efficiency.
24. Explain the following terms used in IC Engines: indicated mean effective pressure, brake mean effective pressure, brake specific fuel consumption and indicated specific fuel consumption.
25. Define the following terms and give their significances: specific power, output per displacement, Engine specific volume and Engine specific weight.

PROBLEMS

1.1 A 3*l* V6 engine operating on a four-stroke cycle worn out after 160,000 km. The average speed over its lifetime was 40 km/h at an engine speed of 2000 rpm.
Calculate:
(a) How many revolutions has the engine experienced?
(b) How many intake strokes have occurred in one cycle?
(c) How many spark plug firings have occurred in the entire engine?

[Ans. (a) 48×10^7 (b) 24×10^7 (c) 144×10^7]

1.2 A four-cylinder, 2*l* engine operates on a four-stroke cycle at 3000 rpm. The compression ratio is 9:1. The stroke to bore ratio is 1.1:1 and connecting rod length is 18 cm.
Calculate:
(a) Clearance volume of one cylinder
(b) Bore and stroke
(c) Average piston speed
(d) Piston speed when crank angle $\theta = 60°$ aTDC

[Ans. (a) 62.5 cm^3 (b) 8.33 cm, 9.17 cm (c) 9.17 m/s (d) 14.1 m/s]

1.3 An SI V6 engine with 8 cm bore and 9 cm stroke operates on a four-stroke cycle at 3500 rpm. The compression ratio is 9.2 and the length of the connecting rods is 17 cm. At this speed, the combustion completes at 22° aTDC. The engine is connected to a dynamometer which gives a brake output torque reading of 210 Nm. Air enters the cylinders at 90 kPa and 57°C, and the mechanical efficiency of the engine is 85%. Calculate
(a) The distance that the piston travels from TDC at the end of combustion
(b) The piston speed at the end of combustion
(c) The volume in the combustion chamber at the end of combustion
(d) The brake power
(e) The indicated power
(f) The brake mean effective pressure
(g) The indicated mean effective pressure
(h) The friction mean effective pressure
(i) The power lost to friction

[Ans. (a) 0.41 cm, (b) 7.7 m/s, (c) 75.8 cm^3, (d) 77 kW, (e) 90.6 kW, (f) 9.72 bar, (g) 11.44 bar, (h) 1.72 bar, (i) 13.6 kW]

1.4 The bore and stroke of a four-cylinder four-stroke engine are 90 mm and 100 mm, respectively, and the torque measured is 120 N·m. Calculate the brake mean effective pressure. [**Ans.** 5.93 bar]

1.5 A four-stroke single cylinder square CI engine has indicated thermal efficiency of 35% and its mechanical efficiency is 80%. The rate of fuel consumption is 20 kg/h. The engine is running at a speed of 1800 rpm. Mean piston speed is 12 m/s. Calorific value of fuel is 42,000 kJ/kg. Determine the brake specific fuel consumption and the brake mean effective pressure developed by the engine. [**Ans.** 0.306 kg/kWh, 6.93 bar]

1.6 A V6 four-stroke SI engine is required to give 190 kW at 4500 rpm. The brake thermal efficiency is 32%. The air/fuel ratio is 12.5:1 and the volumetric efficiency at this speed is 68%. If the stroke/bore ratio is 0.8, determine the engine displacement required and the dimensions of the bore and stroke. The calorific value of the fuel is 44,200 kJ/kg and the free air conditions are 1.013 bar and 15°C. [**Ans.** 5.37 *l*, 112.5 mm, 90 mm]

1.7 A six-cylinder, 4.2 *l* square SI engine operates on a four-stroke cycle at 2400 rpm. At this condition, 900 J of indicated work per cycle per cylinder is produced. Mechanical efficiency of the engine is 65%. Calculate

(i) indicated mean effective pressure
(ii) brake mean effective pressure
(iii) friction mean effective pressure
(iv) brake power
(v) torque
(vi) specific power
(vii) output per displacement
(viii) engine specific volume and
(ix) power lost to friction.

[**Ans.** (i) 12.86 bar, (ii) 8.36 bar, (iii) 4.5 bar, (iv) 70.22 kW, (v) 279.4 Nm, (vi) 0.161 kW/cm^2, (vii) 0.01672 kW/cm^3, (viii) 59.8 cm^3/kW, (ix) 37.8 kW]

1.8 A 2 *l* four-stroke four-cylinder CI engine operating at 2800 rpm produces 52 kW of brake power. Air-fuel ratio is 20:1 and volumetric efficiency of the engine is 90%. Take density of air as 1.181 kg/m^3. Calculate

(a) Mass rate of air flow into the engine (kg/s)
(b) Brake specific fuel consumption (kg/kWh)
(c) Mass rate of exhaust flow (kg/h)
(d) Brake output per displacement (kW/*l*)

[**Ans.** (a) 0.0496 kg/s, (b) 0.1717 kg/kWh, (c) 187.5 kg/h, (d) 26 kW/*l*]

1.9 A 10 cm^3 single cylinder two-stroke over-square SI engine operates at 6500 rpm with a volumetric efficiency of 0.82. The over-square ratio is 1.1 (d/L). The air-fuel ratio is 14.7:1. The ambient conditions are 1.013 bar and 25°C.

Calculate:

(a) Average piston speed (m/s)
(b) Flow rate of air into the engine (kg/s)

(c) Flow rate of fuel into the engine (kg/s)
(d) Fuel input for one cycle (mg/cycle)

[**Ans.** (a) 4.75 m/s, (b) 1.052×10^{-3} kg/s, (c) 0.07156×10^{-3} kg/s, (d) 0.6606 mg/cycle]

1.10 A three-cylinder, four-stroke cycle SI engine having bore 68.5 mm and stroke 72.0 mm runs at 3000 rpm with air-fuel ratio 15:1 and develops indicated mean effective pressure 10 bar. Calorific value of fuel is 44,000 kJ/kg. Mechanical efficiency is 0.85, combustion efficiency is 0.97 and volumetric efficiency is 0.9. Density of air is 1.18 kg/m^3. Determine

(a) Engine capacity (cm^3)
(b) Indicated power (kW)
(c) Rate of air flow into the engine (kg/s)
(d) Heat supplied by fuel (kJ/s)
(e) Brake thermal efficiency
(f) Brake specific fuel consumption (kg/kWh)

[**Ans.** (a) 797 cm^3 (b) 19.9 kW (c) 21.16×10^{-3} kg/s (d) 60.21 kJ/s (e) 28.07% (f) 0.3005 kg/kWh]

1.11 An engine develops an indicated power of 50 kW at full load at a certain speed. It has mechanical efficiency of 80%. Assuming that the frictional losses remain constant for the given speed at all loads, calculate mechanical efficiency at three-fourths load, half load and quarter load. [**Ans.** 75%, 66.7%, 50%]

2.1 INTRODUCTION

In internal combustion engines, the conversion of heat energy into mechanical work is a complicated process. As the working fluid passes through the engine and combustion of fuel takes place, complicated chemical, thermal, and physical changes occur. Friction and heat transfer between the gases and cylinder walls in actual engines, make the analysis more complicated. To examine all these changes quantitatively and to account for all the variables, creates a very complex problem. The usual method of approach is through the use of certain theoretical approximations. The two commonly employed approximations of an actual engine in order of their increasing accuracy are (a) the air-standard cycle and (b) the fuel-air cycle. They give an insight into some of the important parameters that influence engine performance.

In the air-standard cycle the working fluid is assumed to be air. The values of the specific heat of air are assumed to be constant at all temperatures. This ideal cycle represents the upper limit of the performance, which an engine may theoretically attain. One step closer to the conditions existing in the actual engine is to consider the fuel-air cycle. This cycle considers the effect of variation of specific heat with temperature and the dissociation of some of the lighter molecules that occur at high temperatures. The analysis of the fuel-air cycle is presented in Chapter 4.

2.2 AIR-STANDARD CYCLE

The analysis of the air-standard cycle is based upon the following assumptions:

1. The working fluid in the engine is always an ideal gas, namely pure air with constant specific heats.
2. A fixed mass of air is taken as the working fluid throughout the entire cycle. The cycle is considered closed with the same air remaining in the cylinder to repeat the cycle. The intake and exhaust processes are not considered.
3. The combustion process is replaced by a heat transfer process from an external source.
4. The cycle is completed by heat rejection to the surrounding until the air temperature and pressure correspond to initial conditions. This is in contrast to the exhaust and intake processes in an actual engine.

5. All the processes that constitute the cycle are reversible.
6. The compression and expansion processes are reversible adiabatic.
7. The working medium does not undergo any chemical change throughout the cycle.
8. The operation of the engine is frictionless.

Because of the above simplified assumptions, the peak temperature, the pressure, the work output, and the thermal efficiency calculated by the analysis of an air-standard cycle are higher than those found in an actual engine. However, the analysis shows the relative effects of the principal variables, such as compression ratio, inlet pressure, inlet temperature, etc. on the engine performance.

In the present chapter, the following air-standard cycles are described and their work output, thermal efficiency, and mean effective pressure are evaluated:

(a) Otto cycle
(b) Diesel cycle
(c) Dual combustion cycle
(d) Atkinson cycle.

The first three cycles are particularly relevant to the reciprocating internal combustion engine and the fourth one is relevant to a combination of engine and turbine. Some shortcomings of these ideal cycles are obvious, but these cycles give a valuable insight into real effects and possibilities.

2.3 CARNOT CYCLE

This cycle was first proposed by a French engineer Sadi Carnot in 1824. The Carnot cycle is composed of four reversible processes, two isothermal and two adiabatic. It is the most efficient cycle operating between two specified temperature limits. An engine operating on this cycle is called the Carnot heat engine. It is a theoretical heat engine.

The working of this engine can be understood by considering a piston–cylinder device as shown in Figure 2.1. The cylinder wall is insulated. The cylinder head is such that its insulation may be removed to allow the heat exchange with reservoirs during isothermal processes and insulation may be brought back during reversible adiabatic processes.

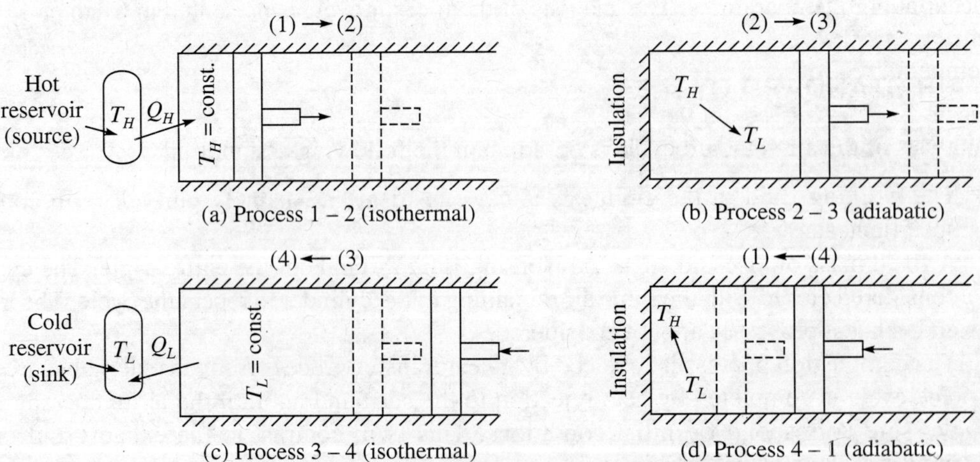

Figure 2.1 Working principle of Carnot engine.

The p–V and T–s diagrams of this cycle are shown in Figure 2.2 (a) and Figure 2.2(b), respectively.

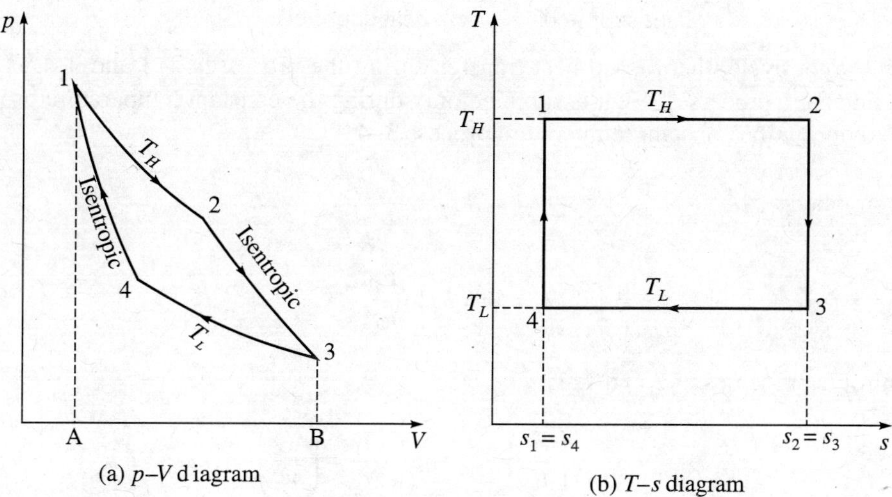

Figure 2.2 Carnot cycle.

The Carnot engine operates on the following principles:

Process 1–2 (reversible isothermal expansion):

The gas is initially contained in a cylinder at a temperature T_H and the cylinder head is in close contact with a hot reservoir (source), which is also at the temperature T_H. The gas pressure at state 1 is higher than that of surrounding, the gas expands slowly and work is done by the gas on the surroundings. As soon as the temperature of the gas drops by an infinitesimal amount dT, some heat is transferred from the reservoir into the gas, raising gas temperature to T_H. Thus, the gas temperature is kept constant.

Process 2–3 (reversible adiabatic expansion):

At state 2, hot reservoir is removed and insulation is placed on the cylinder head so that the system becomes adiabatic. The gas continues to expand. The work is further done on the surroundings. The temperature drops from T_H to T_L. No heat transfer occurs during this process.

Process 3–4 (reversible isothermal compression):

At state 3, the insulation at the cylinder head is removed. The cylinder head is brought in contact with a sink at temperature T_L. The piston is pushed inward by an external force. Now the work is done on the gas by surroundings. As the gas is compressed, its temperature rises. As soon as the temperature rises by an infinitesimal amount dT, heat is transferred from the gas to the sink and the gas temperature drops back to T_L. Thus, the temperature during this process is kept constant.

Process 4–1 (reversible adiabatic compression):

At state 4, sink is removed and insulation is placed back on the cylinder head. The compression follows the reversible adiabatic process. The temperature of the gas rises from T_L to T_H. No heat transfer occurs during this process. Thus the cycle is completed.

Thermal efficiency of a cycle,

$$\eta = \frac{\text{work done}}{\text{heat supplied}} = \frac{\text{heat supplied–heat rejected}}{\text{heat supplied}} \tag{2.1}$$

In the Carnot cycle, there is no heat transfer during the processes 2–3 and 4–1 as they are reversible adiabatic processes. Heat is supplied only during the constant temperature process 1–2 and rejected during the constant temperature process 3–4.

∴ Heat supplied,
$$Q_s = m R T_H \ln \frac{V_2}{V_1} \tag{2.2}$$

and heat rejected,
$$Q_R = m R T_L \ln \frac{V_3}{V_4} \tag{2.3}$$

From isentropic processes 2–3 and 4–1,

$$\frac{T_2}{T_3} = \left(\frac{V_3}{V_2}\right)^{\gamma-1} \quad \text{and} \quad \frac{T_1}{T_4} = \left(\frac{V_4}{V_1}\right)^{\gamma-1}$$

or
$$\frac{T_H}{T_L} = \left(\frac{V_3}{V_2}\right)^{\gamma-1} \quad \text{and} \quad \frac{T_H}{T_L} = \left(\frac{V_4}{V_1}\right)^{\gamma-1}$$

∴
$$\frac{V_3}{V_2} = \frac{V_4}{V_1}$$

or
$$\frac{V_3}{V_4} = \frac{V_2}{V_1} \tag{2.4}$$

∴
$$\eta = \frac{Q_s - Q_R}{Q_s} = \frac{m R (T_H - T_L) \ln (V_2 / V_1)}{m R T_H \ln (V_2 / V_1)} = \frac{T_H - T_L}{T_H} = 1 - \frac{T_L}{T_H} \tag{2.5}$$

Thermal efficiency of a Carnot cycle can also be evaluated from T–s diagram.

$$Q_s = T_H (s_2 - s_1) \tag{2.6}$$

$$Q_R = T_L (s_3 - s_4) = T_L (s_2 - s_1) \tag{2.7}$$

∴
$$\eta = \frac{Q_s - Q_R}{Q_s} = \frac{(T_H - T_L)(s_2 - s_1)}{T_H (s_2 - s_1)} = \frac{T_H - T_L}{T_H} = 1 - \frac{T_L}{T_H} \tag{2.8}$$

The net work output of the cycle is given by the area 12341. The gross work output during expansion is the area 123BA1. The work of compression is the area 143BA1. The ratio of the net work output to the gross work output is called the work ratio. The Carnot cycle has a high thermal ratio but a low work ratio.

$$\text{Work ratio} = \frac{\text{network output}}{\text{gross work output}} \tag{2.9}$$

Reversible isothermal heat transfer is very difficult to achieve in reality, because it would require very large heat exhangers and take a long time. In contrast, a power cycle in an engine is completed in a fraction of a second. Therefore, it is not practical to build an engine that would operate on a cycle that closely approximates the Carnot cycle.

2.4 STIRLING CYCLE

Recently, a number of experimental engines that operate on Stirling cycle have been successfully developed. It is not an internal combustion engine, but it is discussed here because it is a heat engine used to propel vehicles as one of the applications. The Stirling cycle consists of four reversible processes—two isothermal and two constant volume. The two constant volume processes are regeneration processes. Regeneration is a process during which heat is transferred to a thermal energy storage device (called a regenerator) during one part of the cycle and is transferred back to the working fluid during another part of the cycle.

Figure 2.3 shows the *p–V* and *T–s* diagrams of the Stirling cycle, which consists of the following four reversible processes:

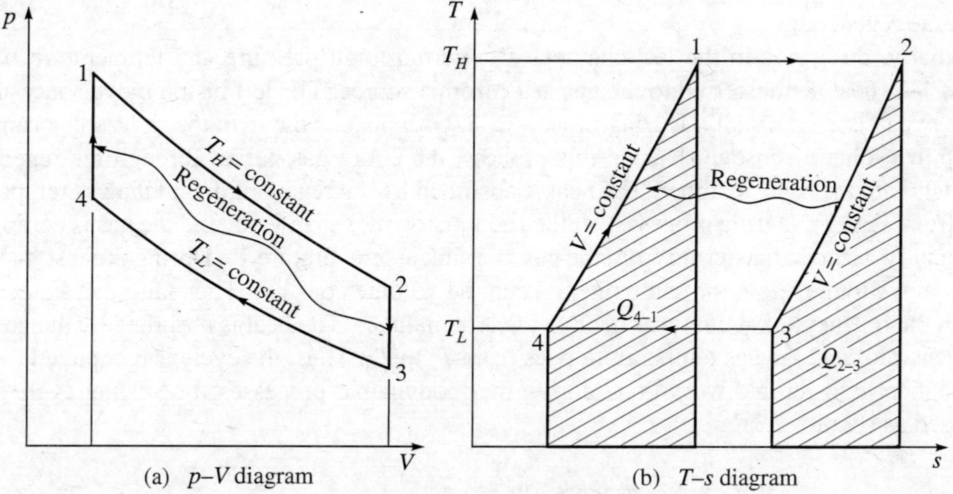

Figure 2.3 Stirling cycle.

1. Process 1–2 is a reversible isothermal expansion process. During this process, the temperature remains constant at T_H. Heat is added from the external source.
2. Process 2–3 is a reversible constant volume regeneration process. Heat is transferred internally from the working fluid to the regenerator. The temperature falls from T_H to T_L.
3. Process 3–4 is a reversible isothermal compression process. The temperature remains constant at T_L. Heat is rejected to the external sink.
4. Process 4–1 is a constant volume regeneration process. Heat is transferred internally from the regenerator back to the working fluid. The temperature rises from T_L to T_H. This completes the cycle.

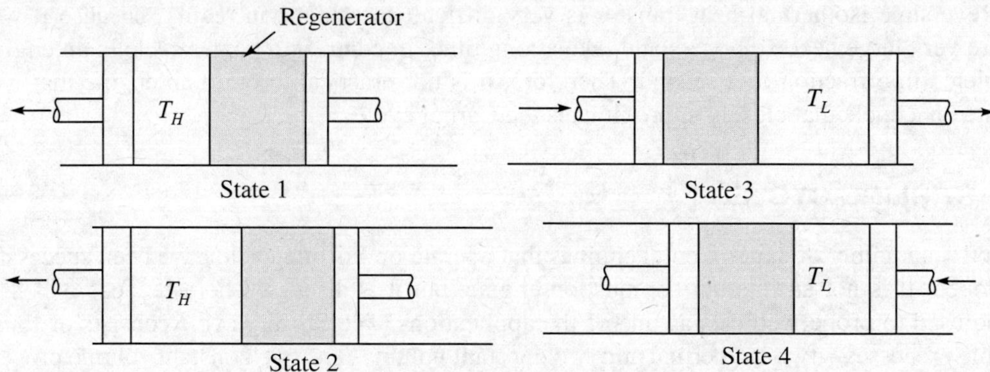

Figure 2.4 Working principle of Stirling engine.

A hypothetical engine working on Stirling cycle is shown in Figure 2.4. It may be considered as a cylinder with two pistons on each side and a regenerator in the middle. The regenerator can be a ceramic mesh or porous plug which is used for temporary storage of thermal energy. Thus, there are two chambers, one on the left-hand side of the regenerator and the other on the right-hand side of the regenerator.

Initially, the gas is in the left chamber at a certain high pressure and temperature. During process 1–2, heat is transferred to the gas at T_H from a source. The left piston moves outward and the gas expands isothermally. During process 2–3, both pistons move to the right at the same rate to keep the volume constant. During this process, the entire gas passes through the regenerator and enters into the right chamber. The heat is absorbed by the regenerator and the gas temperature drops from T_H to T_L. During process 3–4, the right piston moves inward, and the gas is compressed isothermally. Heat is transferred from the gas to a sink at temperature T_L. During process 4–1, both pistons move to the left at the same rate to keep the volume constant. The entire gas again passes through the regenerator and returns back to the left chamber. The heat is absorbed by the gas from the regenerator and the gas temperature rises from T_L to T_H. Thus, the cycle is completed.

The heat transfer and work done during thermodynamic processes for Stirling cycle can be expressed as (Figure 2.3)

$$Q_{1-2} = W_{1-2} = m\,R\,T_H \ln\frac{V_2}{V_1} \tag{2.10}$$

$$Q_{2-3} = m\,c_v(T_L - T_H) = -m\,c_v(T_H - T_L),\; W_{2-3} = 0 \tag{2.11}$$

$$Q_{3-4} = W_{3-4} = m\,R\,T_L \ln\frac{V_4}{V_3} = -m\,R\,T_L \ln\frac{V_2}{V_1} \tag{2.12}$$

$$Q_{4-1} = m\,c_v\,(T_H - T_L),\; W_{4-1} = 0 \tag{2.13}$$

Negative sign indicates that heat is rejected.

Using 100% efficiency of regenerator,

$$|Q_{2-3}| = |Q_{4-1}|$$

Heat supplied, $$Q_s = m R T_H \ln \frac{V_2}{V_1} \tag{2.14}$$

Heat rejected, $$Q_R = m R T_L \ln \frac{V_2}{V_1} \tag{2.15}$$

Thermal efficiency, $$\eta = \frac{Q_s - Q_R}{Q_s} = \frac{m R (T_H - T_L) \ln \frac{V_2}{V_1}}{m R T_H \ln \frac{V_2}{V_1}}$$

$$= \frac{T_H - T_L}{T_H} = 1 - \frac{T_L}{T_H} \tag{2.16}$$

It is the same as Carnot efficiency.

In practice, regenerator efficiency cannot be 100%. Hence, the Stirling cycle efficiency will be less than the Carnot efficiency. It can be expressed as

$$\eta = \frac{R(T_H - T_L)\ln(V_2 / V_1)}{R T_H \ln(V_2 / V_1) + (1 - \varepsilon) c_v (T_H - T_L)} \tag{2.17}$$

where ε is the regenerator effectiveness. For 100% efficiency of regenerator, $\varepsilon = 1$. In this case, the above expression will reduce to Carnot efficiency.

2.5 ERICSSON CYCLE

The p–V and T–s diagrams of the Ericsson cycle are shown in Figure 2.5. It consists of two reversible isothermal processes and two reversible constant pressure processes.

Process 1–2 is a reversible isothermal expansion process.

$$Q_{1-2} = W_{1-2} = m R T_H \ln \frac{p_1}{p_2} \tag{2.18}$$

Process 2–3 is a reversible constant pressure process.

$$Q_{2-3} = m c_p (T_L - T_H) = -m c_p (T_H - T_L) \tag{2.19}$$

$$W_{2-3} = p_2 (V_3 - V_2) = m R (T_L - T_H) = -m R (T_H - T_L) \tag{2.20}$$

Negative sing indicates that heat is rejected and work is done on the system.

Process 3–4 is a reversible isothermal compression process.

$$Q_{3-4} = W_{3-4} = m R T_L \ln \frac{p_3}{p_4} = -m R T_L \ln \frac{p_1}{p_2} \tag{2.21}$$

Process 4–1 is a reversible constant pressure process.

$$Q_{4-1} = m c_p (T_H - T_L) \tag{2.22}$$

$$W_{4-1} = p_1 (V_1 - V_4) = m R (T_H - T_L) \tag{2.23}$$

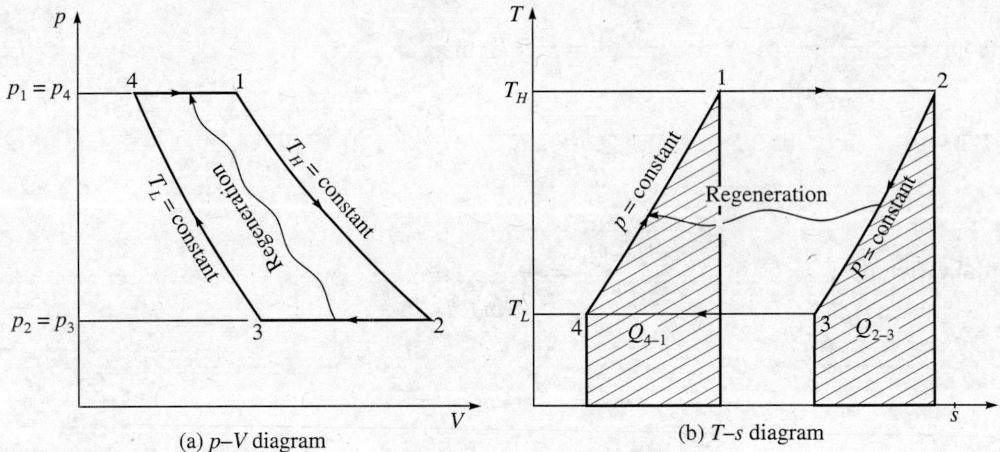

Figure 2.5 Ericsson cycle.

Using 100% efficiency of regenerator, heat rejected during process 2–3 is transferred through regenerator to heat supplied during process 4–1.

$$|Q_{2-3}| = |Q_{4-1}|$$

Heat supplied from external source, $Q_s = m\,RT_H \ln \dfrac{p_1}{p_2}$ (2.24)

Heat rejected to external sink, $Q_R = m\,RT_L \ln \dfrac{p_1}{p_2}$ (2.25)

Thermal efficiency, $\eta = \dfrac{Q_s - Q_R}{Q_s} = \dfrac{m\,R(T_H - T_L)\ln(p_1/p_2)}{m\,RT_H \ln(p_1/p_2)}$

$$= \frac{T_H - T_L}{T_H} = 1 - \frac{T_L}{T_H} \quad (2.26)$$

For perfect effectiveness of regenerator, thermal efficiency of Ericsson cycle is the same as Carnot efficiency for given source and sink temperatures.

A steady-flow Ericsson engine system operating on Ericsson cycle is shown in Figure 2.6.

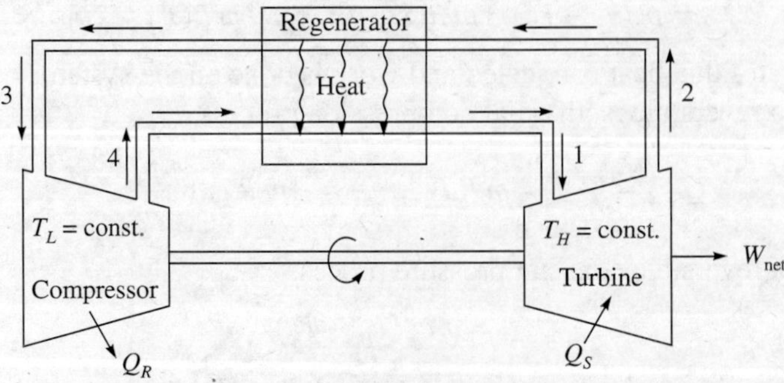

Figure 2.6 Working principle of Ericsson engine.

The isothermal expansion process 1–2 and isothermal compression process 3–4 are executed in a turbine and a compressor, respectively. A counter-flow heat exchanger serves as a regenerator. Both Stirling and Ericsson engines are external combustion engines. They offer the following advantages:

1. A variety of fuels can be used as a source of thermal energy.
2. The combustion process is almost complete as there is enough time for combustion. Thus, air pollution is reduced and more energy can be extracted from the fuel.
3. These engines operate on closed cycles. Working fluid should be stable, chemically inert and possess high thermal conductivity. Hydrogen and helium can be used in these engines.

2.6 LENOIR CYCLE

The p–V and T–s diagrams of air-standard Lenoir cycle are shown in Figure 2.7. On the basis of this cycle, Lenoir developed the first commercial internal combustion engine of this type (1860). Gas and air are drawn into the cylinder during the first half of the piston stroke at the atmospheric pressure (process 0–1). At about halfway through the first stroke, the intake valve is closed and the charge is then ignited with a spark. Combustion raises the temperature and pressure in the cylinder almost at constant volume (process 1–2). The burned gases then delivers power to the piston for the second half of the stroke. The gas expands adiabatically (process 2–3). The pressure drops to atmospheric pressure. Finally, the piston is brought back by using heavy flywheel from BDC to TDC during exhaust stroke (process 3–0). The pressure remains atmospheric.

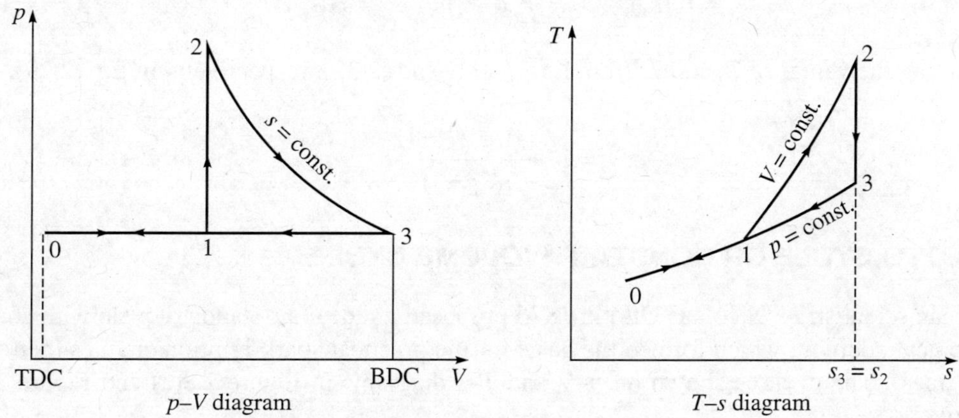

Figure 2.7 Lenoir cycle.

The intake process 0–1 and the latter half of the exhaust stroke process 1–0 cancel each other. The cycle becomes 1–2–3–1.

Lenoir cycle consists of the following three reversible processes:

Process 1–2 : Heat supplied at constant volume

Process 2–3 : Isentropic expansion. There is no heat transfer during this process

Process 3–1 : Heat rejected at constant pressure

Heat supplied, $$Q_s = m c_v (T_2 - T_1) \tag{2.27}$$

Heat rejected, $$Q_R = m c_p (T_3 - T_1) \tag{2.28}$$

Thermal efficiency, $$\eta = \frac{Q_s - Q_R}{Q_s} = 1 - \frac{Q_R}{Q_S}$$

$$\therefore \quad \eta = 1 - \gamma \left(\frac{T_3 - T_1}{T_2 - T_1} \right) \tag{2.29}$$

Let pressure ratio, $$\frac{p_2}{p_1} = r_p \tag{2.30}$$

For constant volume process 1–2,

$$\frac{T_2}{T_1} = \frac{p_2}{p_1} = r_p$$

$$T_2 = r_p \cdot T_1 \tag{2.31}$$

For isentropic process 2–3,

$$\frac{T_3}{T_2} = \left(\frac{p_3}{p_2} \right)^{(\gamma-1)/\gamma} = \left(\frac{p_1}{p_2} \right)^{(\gamma-1)/\gamma} = \left(\frac{1}{r_p} \right)^{(\gamma-1)/\gamma}$$

$$\therefore \quad T_3 = T_2 (r_p)^{(1-\gamma)/\gamma} = r_p T_1 (r_p)^{(1-\gamma)/\gamma} = T_1 (r_p)^{1/\gamma} \tag{2.32}$$

Substituting the values of T_2 and T_3 from Eqs. (2.31) and (2.32), respectively, in Eq. (2.29),

$$\eta = 1 - \frac{\gamma (r_p^{(1/\gamma)} - 1)}{(r_p - 1)} \tag{2.33}$$

2.7 OTTO CYCLE OR CONSTANT VOLUME CYCLE

A German scientist, A. Nicolaus Otto in 1876 proposed an ideal air-standard cycle with constant volume heat addition, which formed the basis for the practical spark-ignition engines (petrol and gas engines). The cycle is shown on *p–V* and *T–s* diagrams in Figure 2.8(a) and Figure 2.8(b) respectively.

At point 1, the piston is at the bottom dead centre (BDC) position and air is trapped inside the engine cylinder. As the piston moves upwards with valves closed, air is compressed isentropically, represented by process 1–2. At point 2, the piston reaches the top dead centre (TDC) position. Heat is supplied to the air from an outer source during the constant volume process 2–3. In an actual engine, it is equivalent to burning of fuel instantaneously by an electric spark. At point 3, air is at its highest pressure and temperature. It is now able to push the piston from TDC to BDC and hence produces the work output. This process of expansion is an isentropic process represented

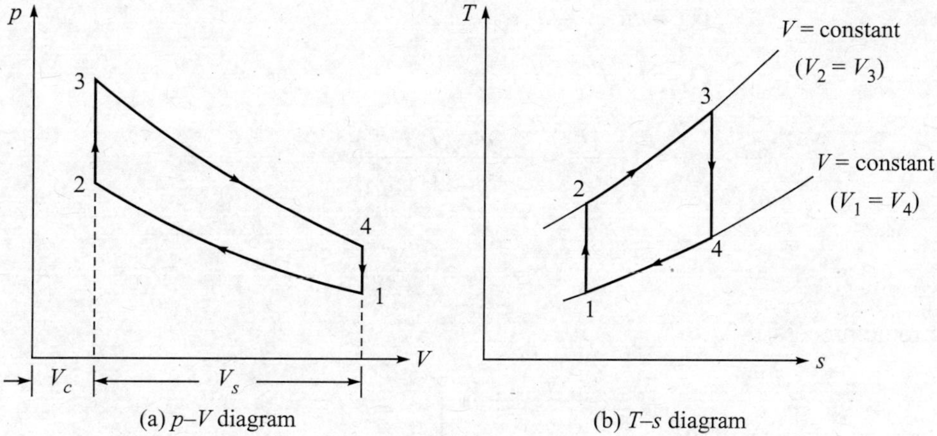

Figure 2.8 Otto cycle.

by process 3–4. At the end of this expansion process, the heat is rejected at constant volume represented by process 4–1. The cycle is thus completed.

Let us summarize:

Process 1–2 is reversible adiabatic or isentropic compression. There is no heat transfer.
Process 2–3 is reversible constant volume heating.
Process 3–4 is reversible adiabatic or isentropic expansion. There is no heat transfer.
Process 4–1 is reversible constant volume heat rejection.

V_s = swept volume
V_c = clearance volume
V_1 = total cylinder volume = $V_s + V_c$

Compression ratio,

$$r = \frac{\text{total cylinder volume}}{\text{clearance volume}}$$

$$= \frac{V_s + V_c}{V_c} = \frac{V_1}{V_2} \tag{2.34}$$

Thermal efficiency of any cycle,

$$\eta = \frac{\text{work done}}{\text{heat supplied}} = \frac{\text{heat supplied} - \text{heat rejected}}{\text{heat supplied}}$$

$$= 1 - \frac{\text{heat rejected}}{\text{heat supplied}} = 1 - \frac{Q_2}{Q_1} \tag{2.35}$$

where

Q_1 = heat supplied during the cycle
Q_2 = heat rejected.

In the Otto cycle, there is no heat transfer during the processes 1–2 and 3–4 as they are reversible adiabatic processes. Heat is supplied only during the constant volume process 2–3 and heat is rejected only during the constant volume process 4–1.

∴ $$Q_1 = mc_v(T_3 - T_2)$$

and $$Q_2 = mc_v(T_4 - T_1)$$

∴ $$\eta = 1 - \frac{Q_2}{Q_1} = 1 - \frac{mc_v(T_4 - T_1)}{mc_v(T_3 - T_2)}$$

$$= 1 - \frac{T_4 - T_1}{T_3 - T_2} \tag{2.36}$$

For isentropic process 1–2,

$$\frac{T_2}{T_1} = \left(\frac{V_1}{V_2}\right)^{\gamma-1} = r^{\gamma-1}$$

∴ $$T_2 = T_1 r^{\gamma-1} \tag{2.37}$$

Also for isentropic process 3–4,

$$\frac{T_3}{T_4} = \left(\frac{V_4}{V_3}\right)^{\gamma-1} = \left(\frac{V_1}{V_2}\right)^{\gamma-1} = (r)^{\gamma-1}$$

∴ $$T_3 = T_4 r^{\gamma-1} \tag{2.38}$$

Substituting the values of T_2 and T_3 from Eqs. (2.37) and (2.38) in Eq. (2.36), the thermal efficiency or the air-standard efficiency of Otto cycle is

$$\eta = 1 - \frac{T_4 - T_1}{T_4 \cdot r^{\gamma-1} - T_1 \cdot r^{\gamma-1}} = 1 - \frac{T_4 - T_1}{r^{\gamma-1}(T_4 - T_1)}$$

$$= 1 - \frac{1}{r^{\gamma-1}} \tag{2.39}$$

It is clear from Eq. (2.39) that the thermal efficiency of the Otto cycle depends only on the compression ratio, r and it increases with an increase in the compression ratio, where γ for air is taken as constant and equal to 1.4. For the combustion products of the fuel-air mixture, γ is approximately taken as 1.3.

The effect of compression ratio on the air-standard efficiency for two values of adiabatic exponent, $\gamma = 1.4$ and $\gamma = 1.3$, is shown in Figure 2.9.

The variation of compression ratio is taken from 4 to 12. These are the possible values in spark-ignition engines. Figure 2.9 shows that the thermal efficiency of the cycle increases with the increase in compression ratio. At a higher value of adiabatic exponent γ, the efficiency also increases. For a given value of compression ratio and adiabatic exponent, the thermal efficiency is a constant and does not depend upon the amount of heat supplied. Heat supplied is proportional to the engine load and hence the thermal efficiency of the Otto cycle does not depend upon the engine load.

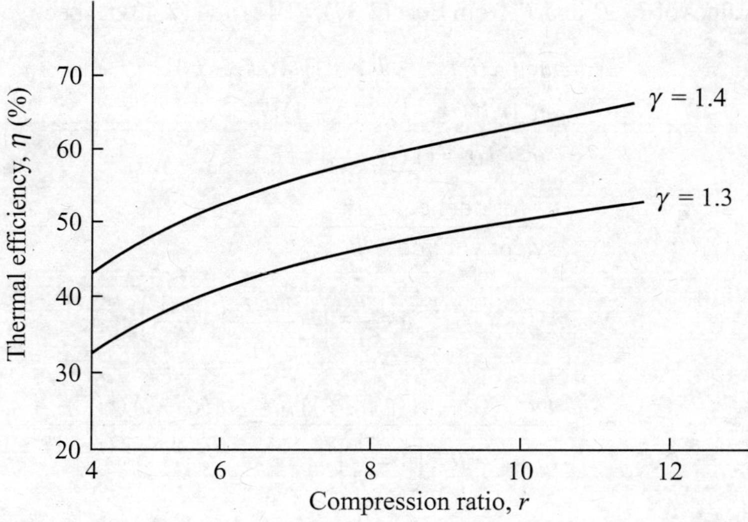

Figure 2.9 Thermal efficiency vs. compression ratio for different values of the adiabatic exponent γ.

From the first law of thermodynamics, for a closed system the work done during a cycle,

$$W = \text{heat supplied} - \text{heat rejected}$$

$$\therefore \quad W = Q_1 - Q_2 = mc_v(T_3 - T_2) - mc_v(T_4 - T_1)$$

$$= mc_v[(T_3 - T_2) - (T_4 - T_1)] \tag{2.40}$$

For constant volume process 2–3,

$$\frac{p_2}{T_2} = \frac{p_3}{T_3}$$

$$\therefore \quad T_3 = \frac{p_3}{p_2} T_2 = \alpha T_2 \tag{2.41}$$

where $\alpha = \dfrac{p_3}{p_2}$; it is the pressure ratio with heat added in a process at constant volume.

Substituting the value of T_2 in Eq. (2.41) from Eq. (2.37),

$$T_3 = \alpha T_1 r^{\gamma-1} \tag{2.42}$$

From Eq. (2.38),

$$T_4 = \frac{T_3}{r^{\gamma-1}}$$

Substituting in the above equation the value of T_3 from Eq. (2.42),

$$T_4 = \frac{\alpha T_1 r^{\gamma-1}}{r^{\gamma-1}} = \alpha T_1 \tag{2.43}$$

Substituting the values of T_2, T_3 and T_4 from Eqs. (2.37), (2.42) and (2.43) respectively in Eq. (2.40)

$$\begin{aligned} W &= mc_v[(\alpha T_1 r^{\gamma-1} - T_1 r^{\gamma-1}) - (\alpha T_1 - T_1)] \\ &= mc_v[r^{\gamma-1}(\alpha - 1) - (\alpha - 1)]T_1 \\ &= mc_v T_1(\alpha - 1)(r^{\gamma-1} - 1) \end{aligned} \tag{2.44}$$

Mean effective pressure, $$p_m = \frac{\text{work done}}{\text{swept volume}} = \frac{W}{V_s} \tag{2.45}$$

Swept volume, $$V_s = V_1 - V_2 = V_1\left(1 - \frac{V_2}{V_1}\right) = \frac{mRT_1}{p_1}\left(1 - \frac{1}{r}\right) \tag{2.46}$$

$\therefore$ $$p_m = \frac{mc_v T_1(\alpha - 1)(r^{\gamma-1} - 1)\,p_1}{mRT_1\left(1 - \frac{1}{r}\right)} = \frac{p_1(\alpha - 1)(r^{\gamma-1} - 1)\,r}{(\gamma - 1)(r - 1)} \tag{2.47}$$

Using Eq. (2.39),

$$p_m = \frac{p_1(\alpha - 1)\,\eta r^{\gamma}}{(\gamma - 1)(r - 1)} \tag{2.48}$$

Equation (2.44) shows that the work output of the cycle increases with the increase in pressure ratio α, and compression ratio r. The increase in pressure ratio is the result of an increase in the amount of heat received. Equation (2.48) shows that the mean effective pressure, p_m, increases in proportion to initial pressure, p_1. An increase in the compression ratio r, also increases the mean effective pressure p_m.

Figure 2.10 shows the variation of mean effective pressure versus pressure ratio for various values of the compression ratio. The initial pressure p_1 is taken as 1 bar and the value of the adiabatic exponent γ, is taken as 1.3. It is observed that the mean effective pressure increases with an increase in pressure ratio and also with an increase in compression ratio.

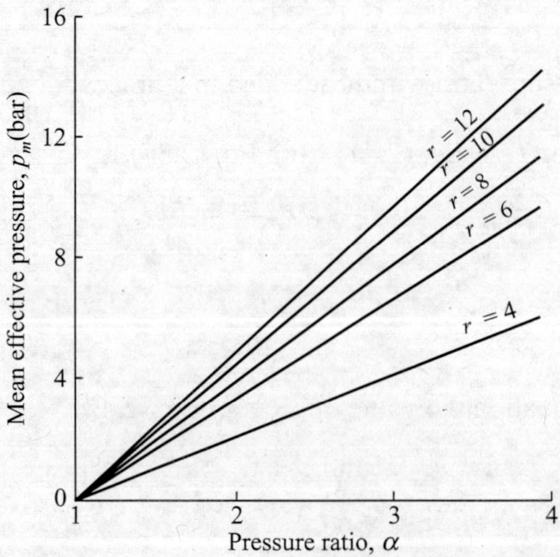

Figure 2.10 Mean effective pressure vs. pressure ratio for different values of compression ratio *r*.

The analysis of the expressions used to determine the thermal efficiency and the mean effective pressure of the cycle shows that the use of higher compression ratios is the most effective way of improving the performance of the engine. However, the maximum value of the compression ratio in an actual engine is restricted, so that normal combustion of the fuel in SI engines is ensured without creating problems of detonation and self-ignition.

2.8 DIESEL CYCLE

Rudolf Diesel in 1892 introduced this cycle. It is a theoretical cycle for slow speed compression-ignition Diesel engine. In this cycle, heat is added at constant pressure and rejected at constant volume. The compression and expansion processes are isentropic. The *p-V* and *T–s* diagrams are shown in Figure 2.11(a) and Figure 2.11(b) respectively.

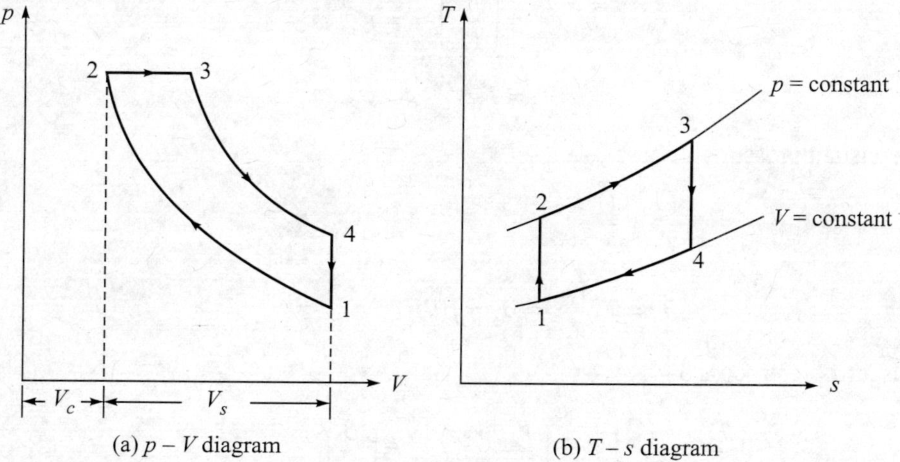

(a) $p - V$ diagram (b) $T - s$ diagram

Figure 2.11 Diesel cycle.

Let us summarize:

Process 1–2 is isentropic compression. There is no heat transfer.

Process 2–3 is reversible constant pressure process. Heat is supplied during this process.

Process 3–4 is isentropic expansion. There is no heat transfer.

Process 4–1 is reversible constant volume process. Heat is rejected during this process.

Heat supplied, $$Q_1 = mc_p(T_3 - T_2) \tag{2.49}$$

Heat rejected, $$Q_2 = mc_v(T_4 - T_1) \tag{2.50}$$

Thermal efficiency,
$$\eta = 1 - \frac{Q_2}{Q_1}$$
$$= 1 - \frac{mc_v(T_4 - T_1)}{mc_p(T_3 - T_2)}$$
$$= 1 - \frac{1}{\gamma}\left(\frac{T_4 - T_1}{T_3 - T_2}\right) \tag{2.51}$$

Two ratios are used to analyse the Diesel cycle:

1. *Compression ratio, r.* It is the ratio of the total cylinder volume to the clearance volume, i.e.

$$r = \frac{V_1}{V_2} \tag{2.52}$$

2. *Cut-off ratio, β.* At point 3, the heat supplied (i.e. the fuel supply in an actual engine) is cut-off. The ratio of the volume at the point of cut-off to the clearance volume or the volume from where the heat supplied begins is called the cut-off ratio, i.e.

Cut-off ratio,
$$\beta = \frac{V_3}{V_2} \tag{2.53}$$

For isentropic process 1–2,

$$\frac{T_2}{T_1} = \left(\frac{V_1}{V_2}\right)^{\gamma-1} = r^{\gamma-1}$$

$$\therefore \quad T_2 = T_1 r^{\gamma-1} \tag{2.54}$$

For constant pressure process 2–3,

$$\frac{V_2}{T_2} = \frac{V_3}{T_3}$$

$$\therefore \quad T_3 = T_2 \frac{V_3}{V_2} = \beta T_2 = \beta T_1\, r^{\gamma-1} \tag{2.55}$$

For isentropic process 3–4,

$$\frac{T_4}{T_3} = \left(\frac{V_3}{V_4}\right)^{\gamma-1} = \left(\frac{V_3}{V_2}\cdot\frac{V_2}{V_1}\right)^{\gamma-1} = \left(\frac{\beta}{r}\right)^{\gamma-1}$$

$$\therefore \quad T_4 = T_3\left(\frac{\beta}{r}\right)^{\gamma-1} = \beta T_1\, r^{\gamma-1}\left(\frac{\beta}{r}\right)^{\gamma-1} = T_1\beta^{\gamma} \tag{2.56}$$

Substituting the values of T_2, T_3 and T_4 from Eqs. (2.54), (2.55) and (2.56) respectively in Eq. (2.51), the thermal efficiency or air-standard efficiency of a Diesel cycle can be expressed as

$$\eta = 1 - \frac{1}{\gamma}\left[\frac{T_1\beta^{\gamma} - T_1}{\beta T_1\, r^{\gamma-1} - T_1\, r^{\gamma-1}}\right]$$

$$= 1 - \frac{1}{\gamma}\,\frac{\beta^{\gamma} - 1}{r^{\gamma-1}(\beta - 1)}$$

$$= 1 - \frac{1}{r^{\gamma-1}}\left[\frac{\beta^{\gamma} - 1}{\gamma(\beta - 1)}\right] \tag{2.57}$$

Equation (2.57) shows that the thermal efficiency of a Diesel cycle depends on the compression ratio r, the adiabatic exponent γ, and the cut-off ratio β. By comparing Eqs. (2.57) and (2.39), it is observed that the efficiency of a Diesel cycle differs from the efficiency of an Otto cycle by the term $(\beta^{\gamma} - 1)/\gamma(\beta - 1)$, which is always greater than unity. Therefore, the air-standard efficiency of the Diesel cycle is less than that of the Otto cycle for the same compression ratio. For the given value of the cut-off ratio β, the efficiency of the Diesel cycle increases with increase in compression ratio r. In actual practice, the normal range of compression ratio for diesel engines is from 16 to 20, whereas for spark-ignition engines, it is 6 to 10. Because of the higher compression ratio used in diesel engines, the efficiency of a diesel engine is more than that of a petrol engine.

Figure 2.12 shows that the thermal efficiency of a Diesel cycle increases with the increase in compression ratio r, and adiabatic exponent γ. However an increase in the cut-off ratio β, reduces this efficiency. An increase in the value of the cut-off ratio means that the heat addition at constant pressure is increased, which corresponds to an increase in load on the engine. Therefore, as the load increases, the thermal efficiency drops. The maximum value of thermal efficiency will be obtained at no load, i.e. at idle running condition of the engine. At this condition, the entire work that the engine performs goes to overcome friction in mating parts and to drive the auxiliary mechanisms. However, this conclusion will be further examined in the light of actual engines with parameters other than those mentioned at this stage to avoid misleading results.

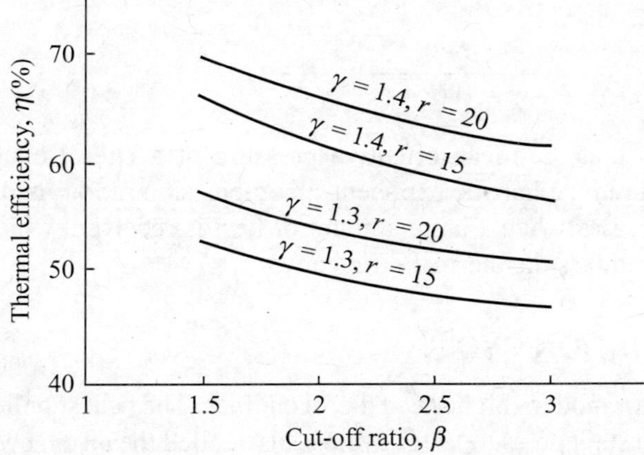

Figure 2.12 Thermal efficiency vs. cut-off ratio at different compression ratios and adiabatic exponents.

Work done during a cycle,

$$W = Q_1 - Q_2$$

$$= mc_p(T_3 - T_2) - mc_v(T_4 - T_1) \tag{2.58}$$

Substituting the values of T_2, T_3 and T_4 from Eqs. (2.54), (2.55) and (2.56), respectively, in Eq. (2.58),

$$W = mc_p(\beta T_1 r^{\gamma-1} - T_1 r^{\gamma-1}) - mc_v(T_1\beta^{\gamma} - T_1)$$

$$= mc_p T_1\left[r^{\gamma-1}(\beta - 1) - \frac{1}{\gamma}(\beta^{\gamma} - 1)\right] \tag{2.59}$$

Swept volume, $$V_s = V_1 - V_2 = V_1\left(1 - \frac{1}{r}\right)$$

$$= \frac{mRT_1}{p_1}\frac{r-1}{r} \tag{2.60}$$

Mean effective pressure, $p_m = \dfrac{W}{V_s}$

$$\therefore \quad p_m = \frac{c_p}{R} p_1 \left[r^{\gamma-1}(\beta - 1) - \frac{1}{\gamma}(\beta^\gamma - 1) \right] \frac{r}{r-1}$$

$$= \frac{\gamma}{\gamma - 1} p_1 \left[r^{\gamma-1}(\beta - 1) - \frac{1}{\gamma}(\beta^\gamma - 1) \right] \frac{r}{r-1}$$

$$= \frac{p_1}{(\gamma - 1)(r - 1)} [\gamma r^\gamma(\beta - 1) - r(\beta^\gamma - 1)] \tag{2.61}$$

$$= \frac{p_1}{(\gamma - 1)(r - 1)} r^\gamma \gamma(\beta - 1) \left[1 - \frac{1}{r^{\gamma-1}} \cdot \frac{\beta^\gamma - 1}{\gamma(\beta - 1)} \right]$$

$$= \frac{p_1 r^\gamma}{(\gamma - 1)(r - 1)} \eta\gamma(\beta - 1) \tag{2.62}$$

Equation (2.62) shows that the mean effective pressure of a Diesel cycle increases with the increase in initial pressure p_1, adiabatic exponent γ, compression ratio r, and thermal efficiency η. The cut-off ratio β increases when a large amount of heat is received, which increases the mean effective pressure and reduces the thermal efficiency.

2.9 DUAL COMBUSTION CYCLE

It is a theoretical cycle for modern high speed diesel engines. The heat supplied is partly at constant volume and partly at constant pressure. This cycle is also called the *mixed cycle* or *limited pressure cycle*. The compression and expansion processes are isentropic and heat is rejected at constant volume. The p–V and T–s diagrams are shown in Figures 2.13 (a) and 2.13 (b) respectively.

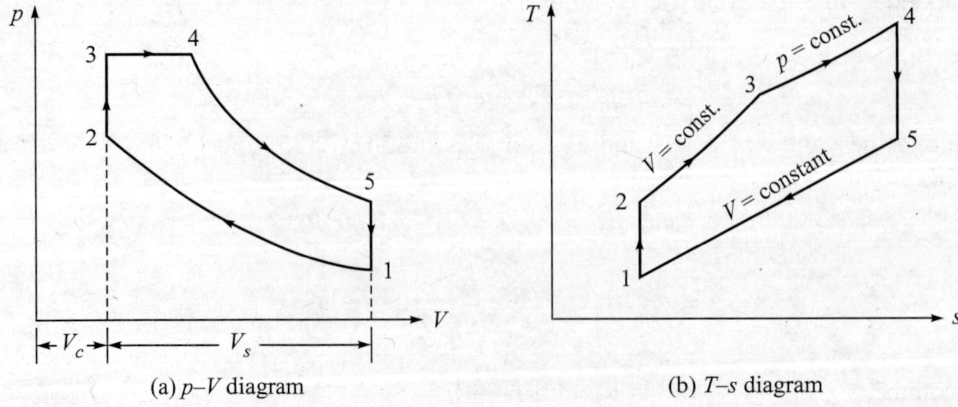

Figure 2.13 Dual combustion cycle.

Here:

Process 1–2 is isentropic compression. There is no heat transfer. Work is done on the system.

Process 2–3 is reversible constant volume process. Part of the heat is supplied during this process.

Process 3–4 is reversible constant pressure process. The remaining part of the heat is supplied during this process.

Process 4–5 is isentropic expansion. There is no heat transfer. Work is done by the system.

Process 5–1 is reversible constant volume process. Heat is rejected during this process.

Heat supplied during the process 2–3 $= mc_v(T_3 - T_2)$

Heat supplied during the process 3–4 $= mc_p(T_4 - T_3)$

Total heat supplied, $Q_1 = mc_v(T_3 - T_2) + mc_p(T_4 - T_3)$ (2.63)

Heat rejected during process 5–1, $Q_2 = mc_v(T_5 - T_1)$ (2.64)

Thermal efficiency,
$$\eta = 1 - \frac{Q_2}{Q_1}$$

$$= 1 - \frac{mc_v(T_5 - T_1)}{mc_v(T_3 - T_2) + mc_p(T_4 - T_3)}$$

$$= 1 - \frac{T_5 - T_1}{(T_3 - T_2) + \gamma(T_4 - T_3)} \quad (2.65)$$

Three ratios are used to analyse the Dual combustion cycle:

(1) Compression ratio, $r = \frac{V_1}{V_2}$ (2.66)

(2) Pressure ratio, $\alpha = \frac{p_3}{p_2}$ (2.67)

(3) Cut-off ratio, $\beta = \frac{V_4}{V_3}$ (2.68)

These ratios are always greater than 1.

For isentropic process 1–2,

$$\frac{T_2}{T_1} = \left(\frac{V_1}{V_2}\right)^{\gamma-1} = r^{\gamma-1}$$

$$\therefore \quad T_2 = T_1 r^{\gamma-1} \quad (2.69)$$

For constant volume process 2–3,

$$\frac{p_2}{T_2} = \frac{p_3}{T_3}$$

$$\therefore \quad T_3 = \frac{p_3}{p_2} T_2 = \alpha T_2 = \alpha T_1 r^{\gamma-1} \quad (2.70)$$

For constant pressure process 3–4,

$$\frac{V_3}{T_3} = \frac{V_4}{T_4}$$

$$\therefore \quad T_4 = \frac{V_4}{V_3} T_3 = \beta T_3 = \beta\alpha T_1 r^{\gamma-1} \tag{2.71}$$

For isentropic process 4–5,

$$\frac{T_5}{T_4} = \left(\frac{V_4}{V_5}\right)^{\gamma-1} = \left(\frac{V_4}{V_3}\cdot\frac{V_3}{V_5}\right)^{\gamma-1} = \left(\frac{V_4}{V_3}\cdot\frac{V_2}{V_1}\right)^{\gamma-1} = \left(\frac{\beta}{r}\right)^{\gamma-1}$$

$$\therefore \quad T_5 = T_4\left(\frac{\beta}{r}\right)^{\gamma-1} = \beta\alpha T_1 r^{\gamma-1}\left(\frac{\beta}{r}\right)^{\gamma-1} = \alpha\beta^\gamma T_1 \tag{2.72}$$

Substituting the values of T_2, T_3, T_4 and T_5 from Eqs. (2.69), (2.70), (2.71) and (2.72) respectively in Eq. (2.65),

$$\eta = 1 - \frac{(\alpha\beta^\gamma T_1 - T_1)}{(\alpha T_1 r^{\gamma-1} - T_1 r^{\gamma-1}) + \gamma(\beta\alpha T_1 r^{\gamma-1} - \alpha T_1 r^{\gamma-1})}$$

$$= 1 - \frac{\alpha\beta^\gamma - 1}{r^{\gamma-1}[(\alpha-1) + \alpha\gamma(\beta-1)]} \tag{2.73}$$

Equation (2.73) shows that the increase in the compression ratio r, and the higher values of the adiabatic exponent cause an increase in the thermal efficiency. With a constant amount of heat added, the values of α and β depend on what part of the heat is added at constant volume and what part at constant pressure. An increase in the value of α and the corresponding reduction in β results in a higher thermal efficiency. $\beta = 1$ leads to Otto cycle and $\alpha = 1$ leads to Diesel cycle,

Work done during a cycle,

$$W = Q_1 - Q_2$$

$$= mc_v(T_3 - T_2) + mc_p(T_4 - T_3) - mc_v(T_5 - T_1)$$

$$= mc_v T_1[(\alpha r^{\gamma-1} - r^{\gamma-1}) + \gamma(\beta\alpha r^{\gamma-1} - \alpha r^{\gamma-1}) - (\alpha\beta^\gamma - 1)]$$

$$= \frac{mRT_1}{\gamma-1}[r^{\gamma-1}(\alpha-1) + \gamma\alpha r^{\gamma-1}(\beta-1) - (\alpha\beta^\gamma - 1)] \tag{2.74}$$

Swept volume, $V_s = V_1 - V_2 = V_1\left(1 - \frac{1}{r}\right) = \frac{mRT_1}{p_1}\left(\frac{r-1}{r}\right)$ (2.75)

Mean effective pressure, $p_m = \frac{W}{V_s}$

$$\therefore \quad p_m = p_1 \frac{[r^\gamma(\alpha-1) + \gamma\alpha r^\gamma(\beta-1) - r(\alpha\beta^\gamma - 1)]}{(\gamma-1)(r-1)}$$

$$= \frac{p_1}{(\gamma-1)(r-1)} r^\gamma\left[(\alpha-1) + \gamma\alpha(\beta-1) - \frac{\alpha\beta^\gamma - 1}{r^{\gamma-1}}\right] \tag{2.76}$$

$$= \frac{p_1}{(\gamma - 1)(r - 1)} r^{\gamma} [(\alpha - 1) + \gamma\alpha(\beta - 1)] \left[1 - \frac{\alpha\beta^{\gamma} - 1}{r^{\gamma - 1}\{(\alpha - 1) + \gamma\alpha(\beta - 1)\}} \right]$$

$$= \frac{p_1 r^{\gamma}}{(\gamma - 1)(r - 1)} \eta[(\alpha - 1) + \gamma\alpha(\beta - 1)] \tag{2.77}$$

2.10 COMPARISON OF OTTO, DIESEL AND DUAL COMBUSTION CYCLES

The significant parameters in cycle analysis are compression ratio, peak pressure, peak temperature, heat addition, heat rejection, and the net work. In order to compare the performance of these cycles, some of the parameters are kept fixed.

For the same compression ratio and heat addition

For the comparison of Otto, Diesel and Dual combustion cycles, these cycles are drawn on a single *p–V* and *T–s* diagram, as shown in Figures 2.14(a) and 2.14(b) respectively. In drawing these diagrams, the compression ratio and the heat input are kept the same for the three cycles.

Here,

1–2–3–4–1 represents the Otto cycle.

1–2–2′–3′–4′–1 represents the Dual combustion cycle.

1–2–3″–4″–1 represents the Diesel cycle.

$\frac{V_1}{V_2}$ is the compression ratio for all the three cycles.

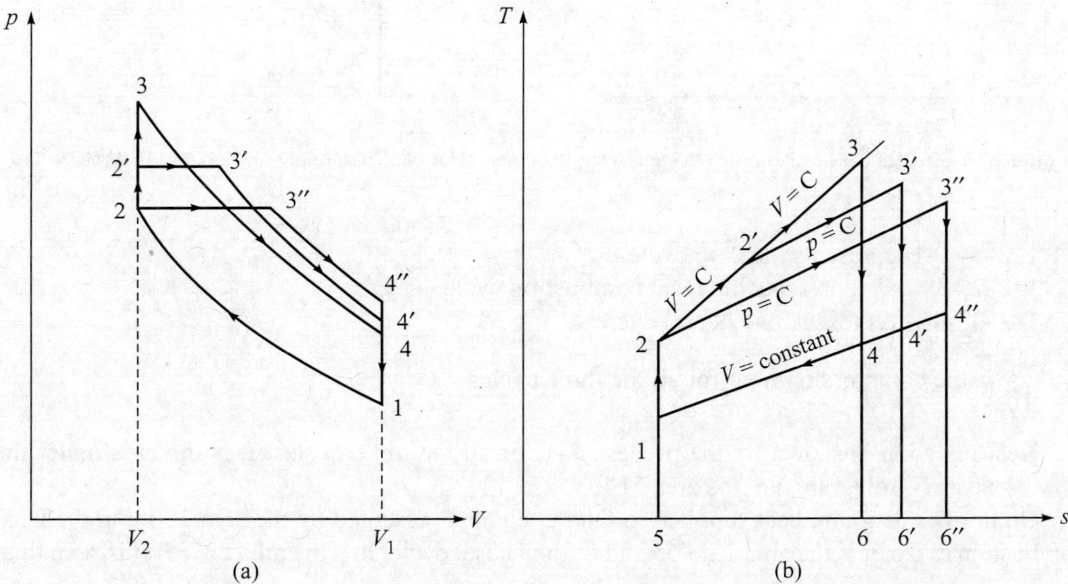

Figure 2.14 *p–V* and *T–s* diagrams having the same compression ratio and heat addition for the three cycles.

The T–s diagram is drawn in such a way, so that the area 5236 = area 522′3′6′ = area 523″6″. These areas represent heat input which is the same for the three cycles. Area 5146, area 514′6′ and area 514″6″ represent heat rejection in Otto cycle, Dual combustion cycle and Diesel cycle respectively. As the area 5146 < area 514′6′ < area 514″6″ and

$$\eta = 1 - \frac{\text{heat rejected}}{\text{heat supplied}}$$

therefore, η of the Otto cycle > η of the Dual combustion cycle > η of the Diesel cycle.

For the same compression ratio and heat rejection

Figure 2.15 shows the p–V and T–s diagrams for the three cycles having the same compression ratio and heat rejection.

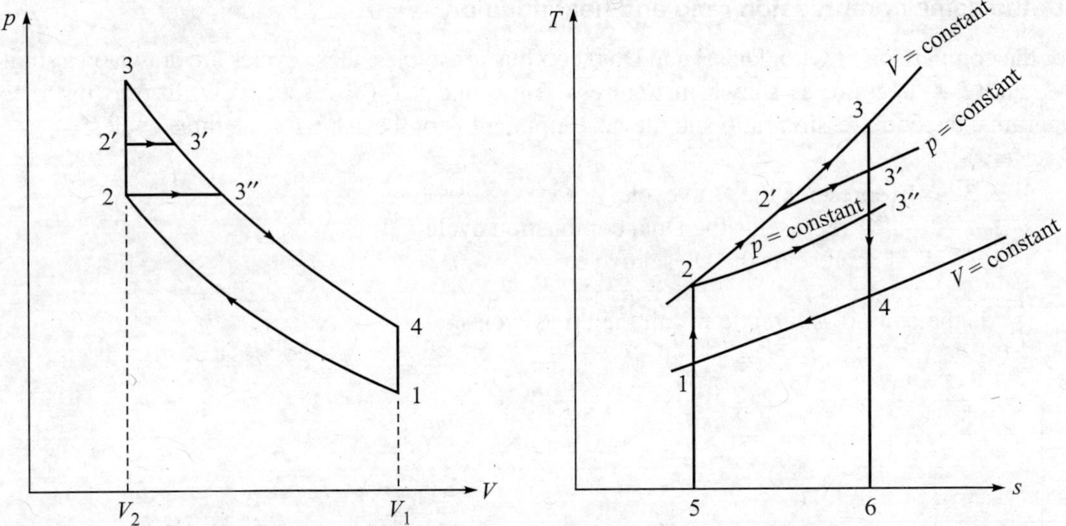

Figure 2.15 p–V and T–s diagrams having the same compression ratio and heat rejection for the three cycles.

Here:

1–2–3–4–1 represents the Otto cycle.

1–2–2′–3′–4–1 represents the Dual combustion cycle.

1–2–3″–4–1 represents the Diesel cycle.

$\frac{V_1}{V_2}$ is the compression ratio for all the three cycles.

Heat rejection is shown by the process 4–1 for all the three cycles. It is the area under the curve 4–1 on T–s diagram, i.e. the area 5146.

On the T–s diagram, heat supplied in Otto cycle is represented by the area 2365; in the Dual combustion cycle, it is the area 22′3′65 and in the Diesel cycle, it is the area 23″65. It is seen that:

$$\text{area } 2365 > \text{area } 22'3'65 > \text{area } 23''65$$

$$\therefore \quad Q_{1\text{Otto cycle}} > Q_{1\text{Dual cycle}} > Q_{1\text{Diesel cycle}}$$

$$\eta = 1 - \frac{Q_2}{Q_1}$$

∴ For the same value of heat rejection Q_2,

$$\eta_{\text{Otto cycle}} > \eta_{\text{Dual cycle}} > \eta_{\text{Diesel cycle}}$$

The above relation is applicable for the same compression ratio and heat rejection.

For the same peak pressure, peak temperature and heat rejection

Figure 2.16 represents the *p–V* and *T–s* diagrams for the three cycles having the same peak pressure, peak temperature, and heat rejection.

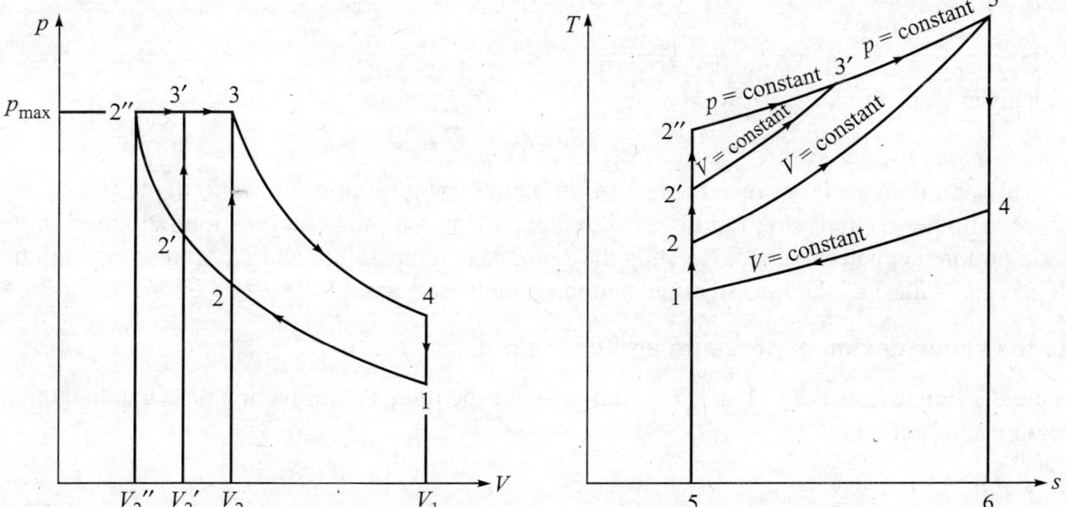

Figure 2.16 *p–V* and *T–s* diagrams having the same peak pressure, peak temperature and heat rejection for the three cycles.

Here:

1–2–3–4–1 represents the Otto cycle.
1–2′–3′–3–4–1 represents the Dual combustion cycle.
1–2″–3–4–1 represents the Diesel cycle.

In all the three cycles, the maximum pressure p_{max} is the same. The peak temperature is the temperature at point 3, i.e. T_3 is also the same for the three cycles. Heat rejection is during the process 4–1—this is also the same for the three cycles.

The compression ratio will now be different for the three cycles, such as:

$$r = \frac{V_1}{V_2} \text{ for Otto cycle}$$

$$r = \frac{V_1}{V_2'} \text{ for Dual combustion cycle}$$

$$r = \frac{V_1}{V_2''} \text{ for Diesel cycle}$$

$\therefore$ $$r_{\text{Diesel cycle}} > r_{\text{Dual cycle}} > r_{\text{Otto cycle}}$$

Heat supplied is as follows:

$$Q_{1\text{Diesel cycle}} = \text{area } 2''365 \text{ on the } T\text{–}s \text{ diagram}$$

$$Q_{1\text{Dual cycle}} = \text{area } 2'3'365 \text{ on the } T\text{–}s \text{ diagram.}$$

$$Q_{1\text{Otto cycle}} = \text{area } 2365 \text{ on the } T\text{–}s \text{ diagram.}$$

Now, $$\text{area } 2''365 > \text{area } 2'3'365 > \text{area } 2365$$

$\therefore$ $$Q_{1\text{Diesel cycle}} > Q_{1\text{Dual cycle}} > Q_{1\text{Otto cycle}}.$$

$$\eta = 1 - \frac{Q_2}{Q_1}$$

For a given value of Q_2,

$$\eta_{\text{Diesel cycle}} > \eta_{\text{Dual cycle}} > \eta_{\text{Otto cycle}}$$

This comparison is more realistic. In an actual compression-ignition engine, i.e. diesel engine, a higher compression ratio is used in comparison with the compression ratio used in the sparkignition, i.e. petrol engine. Keeping the same peak temperature and peak pressure, both the engines can withstand the same thermal and mechanical stresses.

For the same maximum pressure and heat input

Figure 2.17 represents the *p*–*V* and *T*–*s* diagrams for the three cycles having the same maximum pressure and heat input.

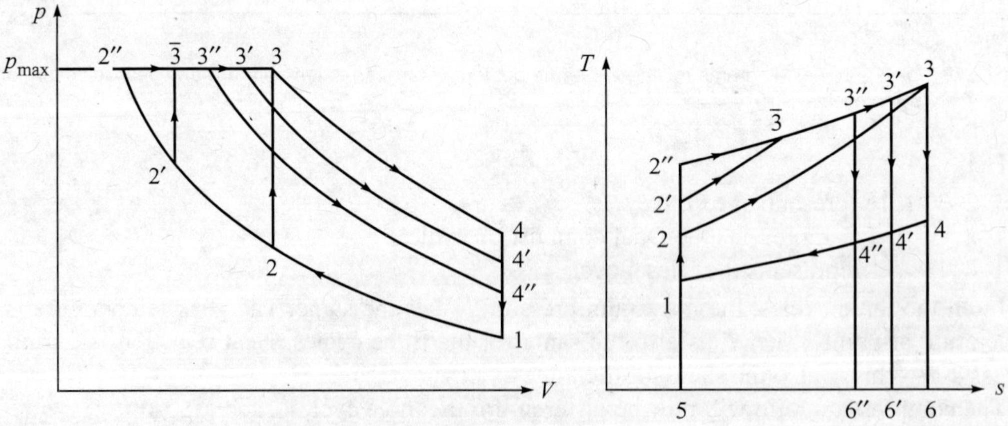

Figure 2.17 *p*–*V* and *T*–*s* diagrams having the same maximum pressure and heat input for the three cycles.

Here:

1–2–3–4–1 represents the Otto cycle.

1–2′– $\bar{3}$ –3′–4′–1 represents the Dual combustion cycle.

1–2″–3″–4″–1 represents the Diesel cycle.

The maximum pressure p_{max} is the same for all the three cycles. As the heat input is also the same for the three cycles, therefore, on the T–s diagram,

$$\text{area } 2''3''6''5(Q_{1\text{Diesel cycle}}) = \text{area } 2'\overline{3}3'6'5(Q_{1\text{Dual cycle}})$$
$$= \text{area } 2365(Q_{1\text{Otto cycle}})$$

Heat rejected for the Otto cycle is the area 1465, heat rejected for the Dual combustion cycle is the area 14′6′5, and the heat rejected for the Diesel cycle is the area 14″6″5.

$$\therefore \quad \text{area } 1465 > \text{area } 14'6'5 > \text{area } 14''6''5.$$

$$\therefore \quad Q_{2\text{Otto cycle}} > Q_{2\text{Dual cycle}} > Q_{2\text{Diesel cycle}}$$

$$\eta = 1 - \frac{Q_2}{Q_1}$$

∴ For the same value of Q_1,

$$\eta_{\text{Diesel cycle}} > \eta_{\text{Dual cycle}} > \eta_{\text{Otto cycle}}$$

For the same maximum pressure and work output

Here,

$$\eta = \frac{\text{work done}}{\text{heat supplied}} = \frac{\text{work done}}{\text{work done + heat rejected}}$$

From the T–s diagram of Figure 2.17, it is clear that if the work done for the three cycles is the same then:

$$\text{area } 12341(\text{work done in Otto cycle}) = \text{area } 12'\overline{3}3'4'1(\text{work done in Dual combustion cycle})$$
$$= \text{area } 12''3''4''1(\text{work done in Diesel cycle})$$

These areas will be equal, only when the heat rejected

$$Q_{2\text{Otto cycle}} > Q_{2\text{Dual cycle}} > Q_{2\text{Diesel cycle}}$$

$$\eta = \frac{W}{W + Q_2}$$

∴ For the same work output W,

$$\eta_{\text{Diesel cycle}} > \eta_{\text{Dual cycle}} > \eta_{\text{Otto cycle}}$$

2.11 ATKINSON CYCLE

Atkinson cycle is shown on p–V and T–s diagrams in Figure 2.18. This is an ideal cycle for Otto engine exhausting to a gas turbine. If the cylinder contents could be expanded to the initial pressure p_1, more work will be obtained and the efficiency will be increased. The incomplete expansion in Otto cycle (process 3–4) is further allowed to proceed to the lowest cycle pressure. A large increase in volume is required to reduce the pressure to p_1.

Here:

Process 1–2 is reversible adiabatic compression. There is no heat transfer.

Process 2–3 is reversible constant volume process. Heat is supplied during this process.

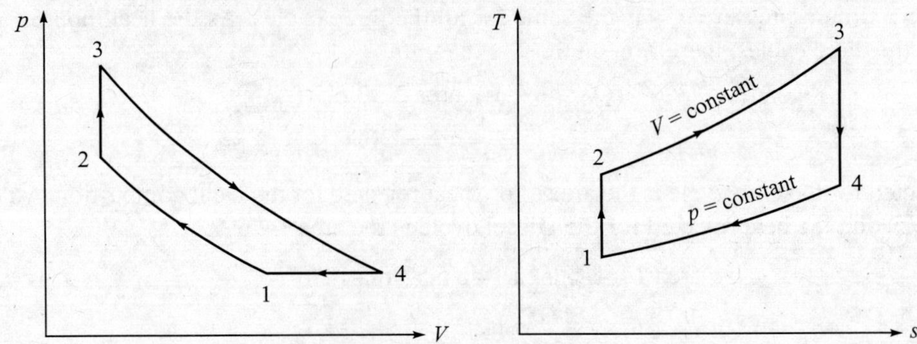

Figure 2.18 Atkinson cycle.

Process 3–4 is reversible adiabatic expansion. There is no heat transfer.

Process 4–1 is reversible constant pressure process. Heat is rejected during this process.

Compression ratio, $$r = \frac{V_1}{V_2} \tag{2.78}$$

Expansion ratio, $$e = \frac{V_4}{V_3} \tag{2.79}$$

Pressure ratio, $$\alpha = \frac{p_3}{p_2} = \frac{p_3}{p_4} \times \frac{p_4}{p_2}$$

$$= \frac{p_3}{p_4} \times \frac{p_1}{p_2} = \left(\frac{V_4}{V_3}\right)^{\gamma} \left(\frac{V_2}{V_1}\right)^{\gamma} = \left(\frac{e}{r}\right)^{\gamma} \tag{2.80}$$

Air standard thermal efficiency,

$$\eta = 1 - \frac{Q_2}{Q_1}$$

Heat supplied, $$Q_1 = mc_v(T_3 - T_2) \tag{2.81}$$

Heat rejected, $$Q_2 = mc_p(T_4 - T_1) \tag{2.82}$$

$\therefore$ $$\eta = 1 - \gamma \frac{T_4 - T_1}{T_3 - T_2} \tag{2.83}$$

For isentropic process 1–2,

$$\frac{T_2}{T_1} = \left(\frac{V_1}{V_2}\right)^{\gamma - 1} = r^{\gamma - 1}$$

$\therefore$ $$T_2 = T_1 r^{\gamma - 1} \tag{2.84}$$

For constant volume process 2–3,

$$\frac{p_2}{T_2} = \frac{p_3}{T_3}$$

$$\therefore \qquad T_3 = \frac{p_3}{p_2} T_2 = \alpha T_2 = \alpha T_1\, r^{\gamma-1} = \left(\frac{e}{r}\right)^{\gamma} r^{\gamma-1} T_1$$

$$\therefore \qquad T_3 = \frac{e^{\gamma}}{r} T_1 \tag{2.85}$$

For isentropic process 3–4,

$$\frac{T_4}{T_3} = \left(\frac{V_3}{V_4}\right)^{\gamma-1} = \left(\frac{1}{e}\right)^{\gamma-1}$$

$$\therefore \qquad T_4 = T_3\left(\frac{1}{e}\right)^{\gamma-1} = \frac{e^{\gamma}}{r} T_1 \left(\frac{1}{e}\right)^{\gamma-1} = \frac{e}{r} T_1 \tag{2.86}$$

Substituting the values of T_2, T_3 and T_4 from Eqs. (2.84), (2.85) and (2.86) respectively in Eq. (2.83),

$$\eta = 1 - \frac{\gamma\left(\frac{e}{r} - 1\right) T_1}{\left(\frac{e^{\gamma}}{r} - r^{\gamma-1}\right) T_1}$$

$$= 1 - \gamma \frac{e - r}{e^{\gamma} - r^{\gamma}} \tag{2.87}$$

Work done,

$$W = Q_1 - Q_2$$

$$= m[c_v(T_3 - T_2) - c_p(T_4 - T_1)]$$

$$= mc_v T_1 \left[r^{\gamma-1}(\alpha - 1) - \gamma\left(\frac{e}{r} - 1\right)\right]$$

$$= \frac{mR}{\gamma - 1} T_1 \left[r^{\gamma-1}\left(\frac{e^{\gamma}}{r^{\gamma}} - 1\right) - \gamma\left(\frac{e}{r} - 1\right)\right]$$

$$= \frac{mR}{\gamma - 1} T_1 \left[\frac{e^{\gamma}}{r} - r^{\gamma-1} - \frac{\gamma}{r}(e - r)\right]$$

$$= \frac{mR}{\gamma - 1} T_1 \frac{1}{r} \,[e^{\gamma} - r^{\gamma} - \gamma(e - r)] \tag{2.88}$$

Swept volume,

$$V_s = V_1 - V_2 = V_1 = V_1\left(1 - \frac{1}{r}\right) = \frac{mRT_1}{p_1} \cdot \frac{r - 1}{r} \tag{2.89}$$

Mean effective pressure,

$$p_m = \frac{W}{V_s}$$

$$\therefore \qquad p_m = \frac{p_1[(e^{\gamma} - r^{\gamma}) - \gamma(e - r)]}{(\gamma - 1)(r - 1)} \tag{2.90}$$

$$= \frac{p_1}{(\gamma-1)(r-1)}(e^\gamma - r^\gamma)\left[1 - \frac{\gamma(e-r)}{e^\gamma - r^\gamma}\right]$$

$$= \frac{p_1\eta(e^\gamma - r^\gamma)}{(\gamma-1)(r-1)} \tag{2.91}$$

2.12 MILLER CYCLE

The Miller cycle (R.H. Miller, 1890–1967) is a modification of the Atkinson cycle. It has also an expansion ratio greater than the compression ratio. However, it is obtained in a much different way. The air intake is unthrottled and the amount of air inducted into the cylinder is controlled by closing the intake valve at the proper time. The inlet valve closing time depends on engine speed and load. This control is now possible by using variable valve timing (VVT) systems. As air flow through the intake system is unthrottled, the pressure drop is not much experienced in the intake manifold. The Miller cycle engine has essentially no negative pump work. It is similar to a CI engine. The shorter compression stroke which absorbs work, combined with the longer expansion stroke which produces work, results in a greater net indicated work per cycle.

There are two variations of this cycle. One uses early inlet valve closing before BDC and the other uses late inlet valve closing after BDC. As the piston moves downward on the intake stroke from TDC to BDC, the cylinder pressure follows the constant pressure line from point 6 to point 1. The cycle for early inlet valve closing is shown in Figure 2.19(a). The inlet valve is closed at point 1 and the cylinder pressure decreases during the expansion along process 1–7. When the piston reaches BDC and starts back towards TDC on the compression stroke, the cylinder pressure retraces the path 7–1 and then follows the path 1–2. The resulting cycle is 6–1–7–1–2–3–4–5–6. The work in the first part of the intake process 6–1 is cancelled by part of the exhaust stroke 1–6. Also, the net work done along the two paths 1–7 and 7–1 cancels. The net indicated work is the area within the loop 1–2–3–4–5–1. There is no pump work.

For late inlet valve closing (Figure 2.19(b)), air is inducted during the entire intake stroke. As the piston returns back from BDC towards TDC, a portion of the intake air is then forced

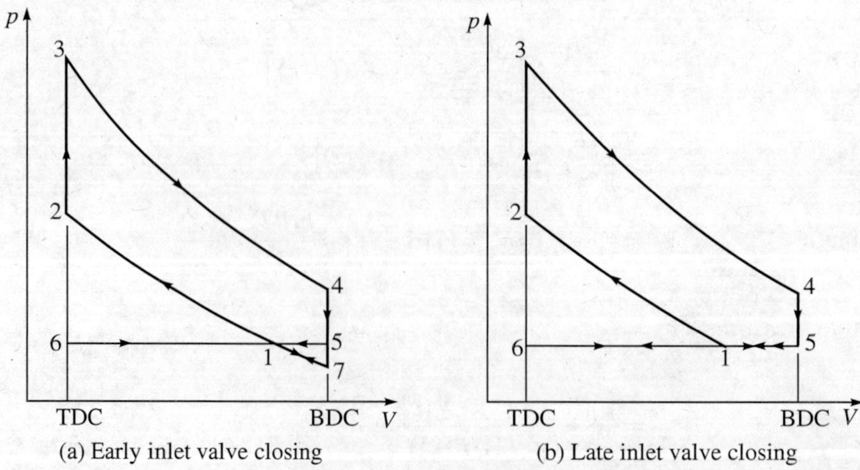

Figure 2.19 Miller cycle.

back into the intake manifold before the intake valve closes at point 1. This results in the cycle 6–1–5–1–2–3–4–5–1–6. The net indicated work is again the area within the loop 1–2–3–4–5–1.

Process 1–2 is reversible adiabatic compression. There is no heat transfer.

Process 2–3 is reversible constant volume process. Heat is supplied during this process.

Process 3–4 is reversible adiabatic expansion. There is no heat transfer.

Process 4–5 is reversible constant volume process. A part of the heat is rejected during this process.

Process 5–1 is reversible constant pressure process. The remaining part of heat is rejected during the process.

Heat supplied during the process 2–3,

$$Q_1 = mc_v\,(T_3 - T_2) \tag{2.92}$$

Heat rejected during processes 4–5 and 5–1,

$$Q_2 = mc_v\,(T_4 - T_5) + mc_p(T_5 - T_1) \tag{2.93}$$

Thermal efficiency,

$$\eta = 1 - \frac{Q_2}{Q_1}$$

$$= 1 - \frac{mc_v(T_4 - T_5) + mc_p(T_5 - T_1)}{mc_v(T_3 - T_2)}$$

$$= 1 - \frac{(T_4 - T_5) + \gamma(T_5 - T_1)}{(T_3 - T_2)} \tag{2.94}$$

Compression ratio, $$r = \frac{V_1}{V_2} \tag{2.95}$$

Expansion ratio, $$e = \frac{V_4}{V_3} = \frac{V_5}{V_2} \tag{2.96}$$

$$\lambda = \frac{e}{r} = \frac{V_5}{V_1} = \frac{T_5}{T_1} \tag{2.97}$$

For isentropic process 1–2, $$T_2 = T_1(r)^{\gamma-1} \tag{2.98}$$

From process 2–3, $$T_3 = T_2\left(\frac{p_3}{p_2}\right) = \alpha T_2 = \alpha T_1(r)^{\gamma-1} \tag{2.99}$$

where α is the pressure ratio = $\frac{p_3}{p_2}$.

From process 3–4, $$\frac{T_4}{T_3} = \left(\frac{V_3}{V_4}\right)^{\gamma-1}$$

$$T_4 = T_3\left(\frac{1}{e}\right)^{\gamma-1} = \alpha T_1\left(\frac{r}{e}\right)^{\gamma-1} = \alpha T_1\left(\frac{1}{\lambda}\right)^{\gamma-1} \tag{2.100}$$

From process 5–1, $\dfrac{T_5}{T_1} = \dfrac{V_5}{V_1} = \lambda$

∴ $$T_5 = \lambda T_1 \tag{2.101}$$

Substituting the values of $T_2 - T_5$ from Eqs. (2.98 – 2.101), respectively, in Eq. (2.94),

$$\eta = 1 - \frac{\{\alpha(r/e)^{\gamma-1} - \lambda\} + \gamma(\lambda - 1)}{r^{\gamma-1}(\alpha - 1)}$$

$$= 1 - \frac{\alpha\lambda^{1-\gamma} - \lambda(1-\gamma) - \gamma}{r^{\gamma-1}(\alpha - 1)}$$

$$= 1 - \frac{(\alpha - 1)\lambda^{1-\gamma} + \lambda^{1-\gamma} - \lambda(1-\gamma) - \gamma}{r^{\gamma-1}(\alpha - 1)}$$

$$= 1 - \frac{\lambda^{1-\gamma}}{r^{\gamma-1}} - \frac{\lambda^{1-\gamma} - \lambda(1-\gamma) - \gamma}{r^{\gamma-1}(\alpha - 1)}$$

$$= 1 - (\lambda r)^{1-\gamma} - \frac{\lambda^{1-\gamma} - \lambda(1-\gamma) - \gamma}{r^{\gamma-1}(\alpha - 1)} \tag{2.102}$$

Now, $$Q_1 = m c_v (T_3 - T_2)$$

$$= m c_v T_1 r^{\gamma-1}(\alpha - 1)$$

∴ $$r^{\gamma-1}(\alpha - 1) = \frac{Q_1}{m c_v T_1} = \frac{Q_1}{(p_1 V_1 / R)\cdot\{R/(\gamma - 1)\}} = \frac{Q_1}{p_1 V_1}(\gamma - 1) \tag{2.103}$$

Substituting the value of $r^{\gamma-1}(\alpha - 1)$ from Eq. (2.103) into Eq. (2.102),

$$\eta = 1 - (\lambda r)^{1-\gamma} - \frac{[\lambda^{1-\gamma} - \lambda(1-\gamma) - \gamma]}{(\gamma - 1)}\cdot\frac{p_1 V_1}{Q_1} \tag{2.104}$$

The Miller cycle is more efficient than the Otto cycle. Equation (2.104) reduces to the Otto cycle thermal efficiency if $\lambda = 1$.

Work done, $$W = \eta Q_1 \tag{2.105}$$

Swept volume, $$V_s = V_5 - V_2$$

$$= V_1\left(\frac{V_5}{V_1} - \frac{V_2}{V_1}\right) = V_1\left(\lambda - \frac{1}{r}\right) = V_1\left(\frac{\lambda r - 1}{r}\right) \tag{2.106}$$

Mean effective pressure,

$$p_m = \frac{W}{V_s} = \frac{\eta Q_1 r}{V_1(\lambda r - 1)} \tag{2.107}$$

As λ increases, the indicated mean effective pressure (imep) decreases significantly, since the fraction of the displacement volume filled with the inlet fuel-air mixture decreases. This decrease in imep is a disadvantage of the Miller cycle. It can be improved by supercharging.

2.13 BRAYTON CYCLE

This is the ideal cycle for the closed cycle gas turbine unit (Figure 2.20). The working substance is air.

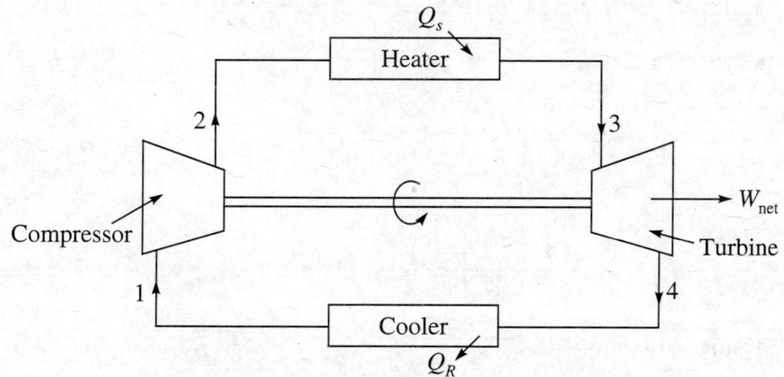

Figure 2.20 A closed cycle gas turbine unit.

It consists of the following four internally reversible processes:

Process 1–2: Isentropic compression in a compressor
Process 2–3: Constant pressure heat addition in a heater
Process 3–4: Isentropic expansion in a turbine
Process 4–1: Constant pressure heat rejection in a cooler

The p–V and T–s diagrams for the Brayton cycle are shown in Figure 2.21. All four processes of the Brayton cycle are executed in steady-flow units.

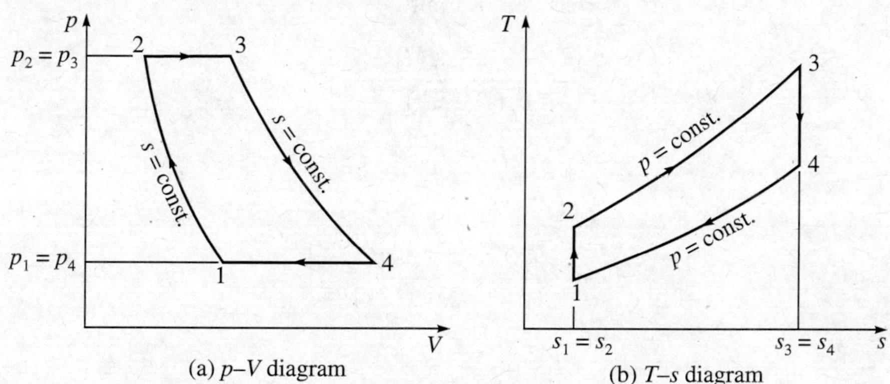

Figure 2.21 Brayton cycle.

Applying the steady-flow energy equation to each part of the cycle and neglecting the change in kinetic energy, we have for unit mass of air,

Work input to compressor $= h_2 - h_1 = c_p(T_2 - T_1)$ (2.108)

Heat supplied in heater, $Q_s = h_3 - h_2 = c_p(T_3 - T_2)$ (2.109)

Work output from turbine $= h_3 - h_4 = c_p(T_3 - T_4)$ (2.110)

Heat rejected from cooler, $Q_R = h_4 - h_1 = c_p(T_4 - T_1)$ (2.111)

Thermal efficiency,
$$\eta = 1 - \frac{Q_R}{Q_s} = 1 - \frac{T_4 - T_1}{T_3 - T_2} \tag{2.112}$$

For isentropic process 1–2,

$$T_2 = T_1\left(\frac{p_2}{p_1}\right)^{(\gamma-1)/\gamma} = T_1(r_p)^{(\gamma-1)/\gamma} \tag{2.113}$$

For isentropic process 3–4,

$$T_3 = T_4\left(\frac{p_3}{p_4}\right)^{(\gamma-1)/\gamma}$$

$$= T_4\left(\frac{p_2}{p_1}\right)^{(\gamma-1)/\gamma} = T_4(r_p)^{(\gamma-1)/\gamma} \tag{2.114}$$

where r_p is the pressure ratio $= \dfrac{p_2}{p_1}$

Substituting the values of T_2 and T_3 from Eqs. (2.113) and (2.114), respectively, in Eq. (2.112).

$$\eta = 1 - \frac{T_4 - T_1}{(T_4 - T_1)r_p^{(\gamma-1)/\gamma}} = \frac{1}{r_p^{(\gamma-1)/\gamma}} \tag{2.115}$$

For the Brayton cycle, the cycle efficiency depends only on the pressure ratio; γ for air is taken as 1.4.

$$\text{Work ratio} = \frac{\text{Network output}}{\text{grosswork output}} = \frac{(h_3 - h_4) - (h_2 - h_1)}{(h_3 - h_4)}$$

$$= \frac{c_p(T_3 - T_4) - c_p(T_2 - T_1)}{c_p(T_3 - T_4)}$$

$$= 1 - \frac{T_2 - T_1}{T_3 - T_4}$$

$$= 1 - \frac{T_1(r_p^{(\gamma-1)/\gamma} - 1)}{T_3[1 - \{1/r_p^{(\gamma-1)/\gamma}\}]}$$

$$= 1 - \frac{T_1}{T_3}\cdot r_p^{(\gamma-1)/\gamma} \tag{2.116}$$

The work ratio depends both on the pressure ratio and on the ratio of the minimum and the maximum temperature. For a given inlet temperature T_1, the maximum temperature T_3 should be as high as possible for a high work ratio.

EXAMPLE 2.1 A hot reservoir at 307 °C and a cold reservoir at 17 °C are available. Calculate the thermal efficiency and the work ratio of a Carnot cycle using air as the working fluid. The maximum and minimum pressures in the cycles are 62.4 bar and 1.04 bar, respectively.

Solution: The cycle is shown in *T–s* diagram in Figure 2.22.

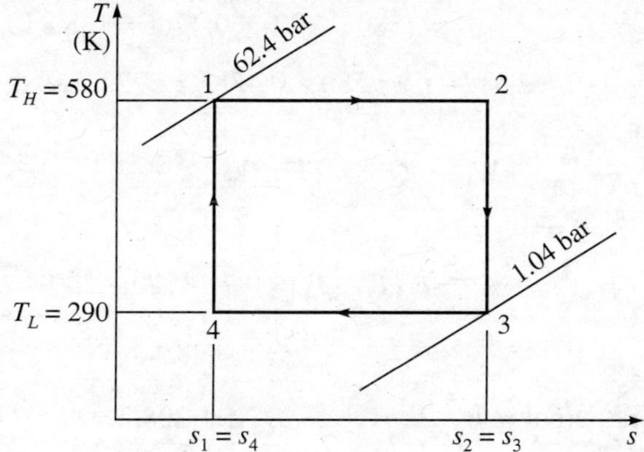

Figure 2.22 Example 2.1

$$T_H = 307 + 273 = 580 \text{ K}$$

$$T_L = 17 + 273 = 290 \text{ K}$$

$$\eta_{\text{Carnot}} = 1 - \frac{T_L}{T_H}$$

$$= 1 - \frac{290}{580} = 0.5$$

$$= \boxed{50\%} \quad \textbf{Ans.}$$

To find the net work and work ratio, the entropy change $s_2 - s_1$ is required. From general property relation,

$$Tds = dh - v\,dp$$

$$ds = \frac{dh}{T} - \frac{v\,dp}{T} = c_p \frac{dT}{T} - \frac{R}{p} dp$$

$$\Delta s = s_2 - s_1 = c_p \ln \frac{T_2}{T_1} - R \ln \frac{p_2}{p_1}$$

Since entropy is a property, it does not depend on path.

$$\therefore \quad s_2 - s_1 = s_3 - s_1 = c_p \ln\frac{T_3}{T_1} - R\ln\frac{p_3}{p_1}$$

$$= 1.005\ln\frac{290}{580} - 0.287\ln\frac{1.04}{62.4}$$

$$= -0.6966 + 1.1751 = 0.4785 \text{ kJ/(kg K)}$$

$$\text{Net work output} = (T_H - T_L)(s_2 - s_1)$$

$$= (580 - 290) \times 0.4785 = 138.8 \text{ kJ/kg}$$

Q_{1-2} = area under line 1–2 = $T_H(s_2 - s_1) = 580 \times 0.4785 = 277.5$ kJ/kg

Process 1–2 is isothermal,

$$W_{1-2} = Q_{1-2} = 277.5 \text{ kJ/kg}$$

For isentropic process 2–3, $Q = 0$

$$W_{2-3} = -\Delta u = c_v(T_2 - T_3) = 0.718(580 - 290)$$

$$= 208.2 \text{ kJ/kg}$$

$$\therefore \quad \text{Gross work output} = W_{1-2} + W_{2-3} = 277.5 + 208.2 = 485.7 \text{ kJ/kg}$$

$$\text{Work ratio} = \frac{\text{network output}}{\text{gross work output}} = \frac{138.8}{485.7} = \boxed{0.286} \quad \textbf{Ans.}$$

EXAMPLE 2.2 An ideal Stirling cycle operates with 1 kg of air between thermal energy reservoirs at 27°C and 527°C. The maximum and minimum pressures of the cycle are 2000 kPa and 100 kPa, respectively. Determine the net work done in a cycle and the thermal efficiency. How much heat is stored and recovered in the regenerator per cycle? When the cycle is repeated 300 times per minute, determine the external rate of heat input and power produced.

Solution: Refer to Figure 2.23.

Given:

$m = 1$ kg air

$T_L = 27 + 273 = 300$ K

$T_H = 527 + 273 = 800$ K

$p_1 = 2000$ kPa

$p_3 = 100$ kPa

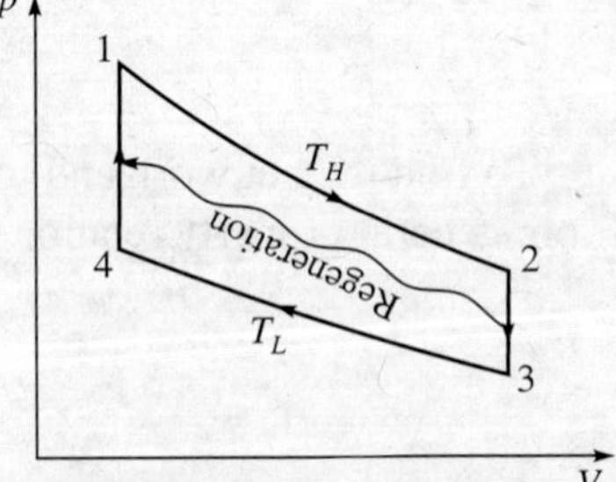

Figure 2.23 Example 2.2.

For constant volume process 2–3,

$$p_2 = p_3 \times \frac{T_2}{T_3} = p_3 \times \frac{T_H}{T_L} = 100 \times \frac{800}{300}$$

$$= 266.7 \text{ kPa}$$

Net work done, $W_{net} = mR(T_H - T_L)\ln\frac{V_2}{V_1} = mR(T_H - T_L)\ln\frac{p_1}{p_2}$

$$= 1 \times 0.287(800 - 300)\ln\frac{2000}{266.7} = \boxed{289.1 \text{ kJ}} \quad \textbf{Ans.}$$

Assuming perfect regeneration,

$$\text{External heat supplied per cycle} = Q_s = Q_{1-2} = mRT_H \ln\frac{p_1}{p_2}$$

$$= 1 \times 0.287 \times 800 \ln\frac{2000}{266.7} = 462.6 \text{ kJ}$$

Thermal efficiency,

$$\eta = \frac{\text{network done}}{\text{heat supplied}} = \frac{289.1}{462.6} = 0.625 = \boxed{62.5\%} \quad \textbf{Ans.}$$

$$\left(\text{or} \quad \eta = 1 - \frac{T_L}{T_H} = 1 - \frac{300}{800} = 0.625 = 62.5\%\right)$$

Heat transfer in the regenerator during constant volume process is

$$mc_v(T_H - T_L) = 1 \times 0.718(800 - 300) = \boxed{359 \text{ kJ/cycle}} \quad \textbf{Ans.}$$

$$\text{External rate of heat input} = 462.6 \times \frac{300}{60} = \boxed{2313 \text{ kW}} \quad \textbf{Ans.}$$

$$\text{Power produced} = 289.1 \times \frac{300}{60} = \boxed{1445.5 \text{ kW}} \quad \textbf{Ans.}$$

EXAMPLE 2.3 An Ericsson cycle operates between thermal energy reservoirs at 627 °C and 7 °C while producing 500 kW of power. Determine net work done, external heat supplied per cycle, heat transfer in the regenerator per kg of air per cycle and thermal efficiency of the cycle. Also, determine the external rate of heat input when the cycle is repeated 2000 times per minute.

Solution: Refer to Figure 2.24

Given:

$T_H = 627 + 273 = 900$ K

$T_L = 7 + 273 = 280$ K

Number of cycles per second $= \frac{2000}{60} = 33.33$

Net work done per cycle,

$$W_{net} = \frac{500}{33.33} = \boxed{15 \text{ kJ}} \quad \textbf{Ans.}$$

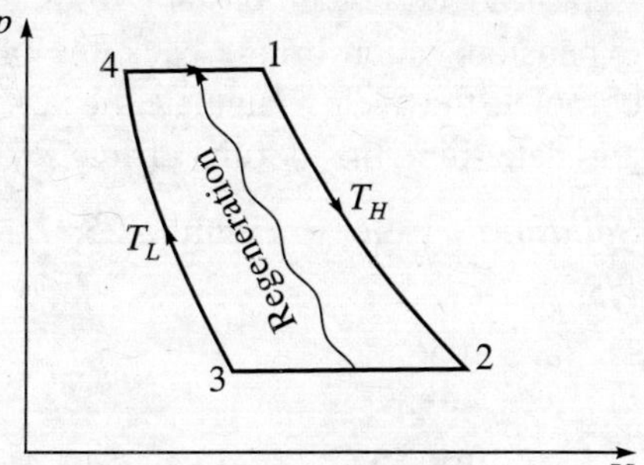

Figure 2.24 Example 2.3.

External heat supplied per cycle,

$$Q_s = mRT_H \ln\frac{p_1}{p_2}$$

$$W_{\text{net}} = mR(T_H - T_L)\ln\frac{p_1}{p_2}$$

$\therefore$
$$mR(900-280)\ln\frac{p_1}{p_2} = 15$$

$$mR\ln\frac{p_1}{p_2} = \frac{15}{620} = 0.0242$$

$\therefore$
$$Q_s = 0.0242 \times 900 = \boxed{21.78 \text{ kJ/cycle}} \quad \textbf{Ans.}$$

Assuming perfect regeneration, heat transfer in the regenerator per kg of air per cycle is

$$c_p(T_H - T_L) = 1.005(900-280) = \boxed{623.1 \text{ kJ/kg}} \quad \textbf{Ans.}$$

Thermal efficiency,

$$\eta = \frac{W_{\text{net}}}{Q_s} = \frac{15}{21.78} = 0.689 = \boxed{68.9\%} \quad \textbf{Ans.}$$

$$\left(\text{or } \eta = 1 - \frac{T_L}{T_H} = 1 - \frac{280}{900} = 0.689 = 68.9\%\right)$$

External rate of heat input,

$$\dot{Q}_s = Q_s \times \text{number of cycles per second}$$

$$\dot{Q}_s = 21.78 \times 33.33 = \boxed{726 \text{ kW}} \quad \textbf{Ans.}$$

EXAMPLE 2.4 Lenoir air-standard cycle is executed in a closed system with 0.003 kg of air. Heat is added at the constant volume from 95 kPa and 17 °C to 380 kPa. It follows an isentropic expansion which brings back the pressure to 95 kPa. Heat is rejected at constant pressure to complete the cycle. Calculate the net work done per cycle and determine the thermal efficiency.

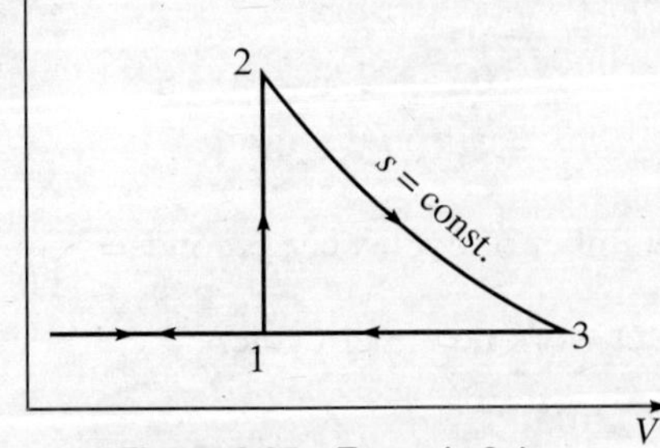

Figure 2.25 Example 2.4.

Solution: Refer to Figure 2.25.

Given:

$m = 0.003$ kg

$p_1 = 95 \text{ kPa} = p_3$

$T_1 = 17 + 273 = 290$ K

$p_2 = 380$ kPa

For constant volume process 1–2,

$$T_2 = T_1 \times \frac{p_2}{p_1} = 290 \times \frac{380}{95} = 1160 \text{ K}$$

For isentropic process 2–3,

$$T_3 = T_2 \left(\frac{p_3}{p_2}\right)^{(\gamma-1)/\gamma} = 1160\left(\frac{95}{380}\right)^{0.286} = 780.3 \text{ K}$$

Heat supplied, $Q_s = mc_v(T_2 - T_1) = 0.003 \times 0.718(1160 - 290)$

$= 1.874$ kJ

Heat rejected, $Q_R = mc_p(T_3 - T_1) = 0.003 \times 1.005(780.3 - 290)$

$= 1.478$ kJ

Net work done in a cycle $= Q_s - Q_R$

$= 1.874 - 1.478 = \boxed{0.396 \text{ kJ}}$ **Ans.**

Thermal efficiency, $\eta = \dfrac{\text{net work done}}{\text{heat supplied}}$

$= \dfrac{0.396}{1.874} = 0.211 = \boxed{21.1\%}$ **Ans.**

EXAMPLE 2.5 A gas engine working on the Otto cycle has cylinder bore of 200 mm and stroke length of 250 mm. The clearance volume is 1570 cm³. The pressure and temperature at the beginning of compression are 1 bar and 27°C respectively. The maximum temperature of the cycle is 1400°C. Determine the pressure and temperature at the salient points, the air-standard efficiency, the work done and the mean effective pressure. (For air, take c_v = 0.718 kJ/(kg K) and R = 0.287 kJ/(kg K). Also calculate the ideal power developed by the engine, if the number of working cycles per minute is 500.

Solution: Refer to Figure 2.26.

Given data:

Cylinder bore, $d = 20$ cm

Stroke, $L = 25$ cm

Clearance volume, $V_c = 1570 \text{ cm}^3$

$p_1 = 1$ bar

$T_1 = 27 + 273 = 300$ K

Maximum temperature, $T_3 = 1400 + 273 = 1673$ K

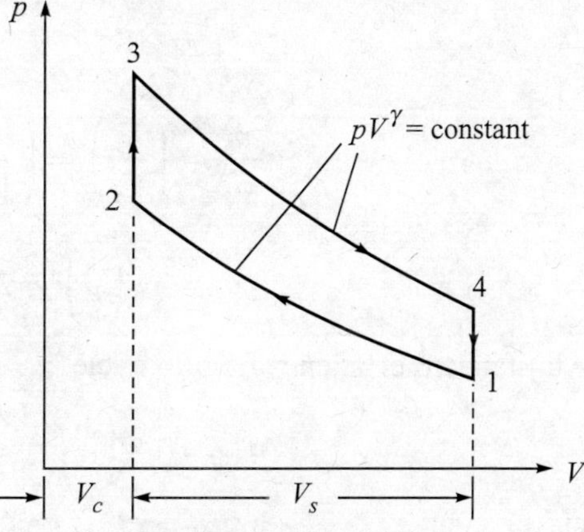

Figure 2.26 Example 2.5.

Swept volume, $V_s = \frac{\pi}{4} d^2 L = \frac{\pi}{4}(20)^2 25$

$$= 7854 \text{ cm}^3$$

$$V_1 = V_s + V_c = 7854 + 1570$$

$$= 9424 \text{ cm}^3$$

$$V_2 = V_c = 1570 \text{ cm}^3$$

Compression ratio, $r = \frac{V_1}{V_2} = \frac{9424}{1570} = 6$

For isentropic process 1–2,

$$T_2 = T_1 r^{\gamma-1} = 300 \times (6)^{0.4} = 614.3 \text{ K}$$

$$= \boxed{341.3°\text{C}} \quad \textbf{Ans.}$$

$$\frac{p_2}{p_1} = \left(\frac{V_1}{V_2}\right)^{\gamma} \quad p_2 = p_1\, r^{\gamma} = 1 \times 6^{1.4} = \boxed{12.286 \text{ bar}} \quad \textbf{Ans.}$$

For constant volume process 2–3,

$$\frac{p_2}{T_2} = \frac{p_3}{T_3}$$

$\therefore$ $$p_3 = p^2 \times \frac{T_3}{T_2} = 12.286 \times \frac{1673}{614.3} = \boxed{33.46 \text{ bar}} \quad \textbf{Ans.}$$

For isentropic process 3–4,

$$\frac{T_4}{T_3} = \left(\frac{V_3}{V_4}\right)^{\gamma-1} = \left(\frac{V_2}{V_1}\right)^{\gamma-1}$$

$\therefore$ $$T_4 = T_3 \left(\frac{1}{r}\right)^{\gamma-1} = 1673\left(\frac{1}{6}\right)^{0.4} = 817 \text{ K} = \boxed{544°\text{C}} \quad \textbf{Ans.}$$

$$\frac{p_4}{p_3} = \left(\frac{V_3}{V_4}\right)^{\gamma} = \left(\frac{V_2}{V_1}\right)^{\gamma} = \left(\frac{1}{r}\right)^{\gamma}$$

$\therefore$ $$p_4 = p_3 \left(\frac{1}{r}\right)^{\gamma} = 33.46 \times \left(\frac{1}{6}\right)^{1.4} = \boxed{2.723 \text{ bar}} \quad \textbf{Ans.}$$

Air-standard efficiency of Otto cycle,

$$\eta_{\text{Otto}} = 1 - \left(\frac{1}{r}\right)^{\gamma-1} = 1 - \left(\frac{1}{6}\right)^{0.4} = 0.5116 = \boxed{51.16\%} \quad \textbf{Ans.}$$

Heat supplied, $Q_1 = c_v(T_3 - T_2)$ for unit mass

$$= 0.718(1673 - 614.3) = 760.15 \text{ kJ/kg}$$

Heat rejected, $Q_2 = c_v(T_4 - T_1)$ for unit mass

$$= 0.718(817 - 300) = 371.2 \text{ kJ/kg}$$

Work done per unit mass,

$$W = Q_1 - Q_2 = 760.15 - 371.2 = 389 \text{ kJ/kg}$$

$$m = \frac{p_1 V_1}{R T_1} = \frac{1 \times 10^5 \times 9424 \times 10^{-6}}{287 \times 300} = 0.01095 \text{ kg}$$

$\therefore$ Work done $= 389 \times 0.01095 = \boxed{4.26 \text{ kJ}}$ **Ans.**

This is the work done for one Otto cycle.

Mean effective pressure, $p_m = \dfrac{\text{work done}}{\text{swept volume}}$

$\therefore$ $p_m = \dfrac{4.26 \times 10^3}{7854 \times 10^{-6}} = 5.424 \times 10^5 \text{ N/m}^2 = \boxed{5.424 \text{ bar}}$ **Ans.**

Power developed = work done per cycle × no. of cycles per second

$$= 4.26 \times \frac{500}{60} = \boxed{35.5 \text{ kW}}$$ **Ans.**

EXAMPLE 2.6 A petrol engine is supplied with fuel having calorific value 42,000 kJ/kg. The pressures in the cylinder at 5 % and 75% of the compression stroke are 1.2 bar and 4.8 bar respectively. Assume that the compression follows the law $pV^{1.3}$ = constant. Find the compression ratio of the engine. If the relative efficiency of the engine compared with the air standard efficiency is 60 %, calculate the specific fuel consumption in kg/kWh.

Solution: In Figure 2.27, the process 1–2 represents compression, following the law $pV^{1.3}$ = constant, for a given petrol engine. The point a and the point b, respectively, represent the 5 % and 75% of the compression stroke. The other processes are completed on theoretical basis.

From Figure 2.27:

$$V_a = V_1 - 0.05V_s$$
$$= V_s + V_c - 0.05V_s$$
$$= V_c + 0.95V_s$$
$$V_b = V_1 - 0.75V_s$$
$$= V_s + V_c - 0.75V_s$$
$$= V_c + 0.25V_s$$
$$p_a V_a^{1.3} = p_b V_b^{1.3}$$

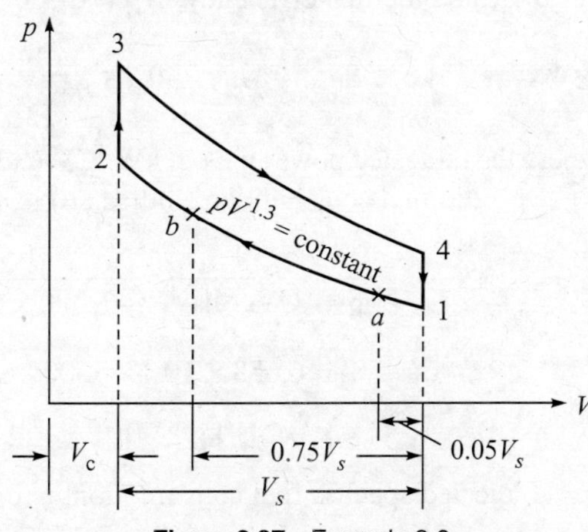

Figure 2.27 Example 2.6.

$$\therefore \quad \frac{V_a}{V_b} = \left(\frac{p_b}{p_a}\right)^{1/1.3} = \left(\frac{4.8}{1.2}\right)^{1/1.3} = 2.905$$

or $$\frac{V_c + 0.95V_s}{V_c + 0.25V_s} = 2.905$$

or $$\frac{1 + 0.95\dfrac{V_s}{V_c}}{1 + 0.25\dfrac{V_s}{V_c}} = 2.905$$

Compression ratio, $$r = \frac{V_1}{V_2} = \frac{V_c + V_s}{V_c} = 1 + \frac{V_s}{V_c}$$

$\therefore$ $$\frac{V_s}{V_c} = r - 1$$

or $$\frac{1 + 0.95(r-1)}{1 + 0.25(r-1)} = 2.905$$

or $$0.95r + 0.05 = 2.905(0.25r + 0.75)$$

$\therefore$ $$r = \boxed{9.5} \quad \textbf{Ans.}$$

Theoretical cycle for the petrol engine is the Otto cycle.

$\therefore$ Air-standard efficiency $$\eta = 1 - \left(\frac{1}{r}\right)^{\gamma-1}$$

$$= 1 - \frac{1}{9.5^{0.4}} = 0.5936 = 59.36\%$$

$$\text{Relative efficiency} = \frac{\text{indicated thermal efficiency}}{\text{air-standard efficiency}}$$

$\therefore$ Indicated thermal efficiency, $\eta_{\text{th}i} = 0.6 \times 0.5936 = 0.356$

Now, $$\eta_{\text{th}i} = \frac{\text{ip}}{\text{CV} \times \dot{m}}$$

where the indicated power ip is in kW, CV (calorific value) of the fuel is in kJ/kg and $\dot{m}$ is in kg/s.

Specific fuel consumption (indicated) is $\dot{m}$/ip.

$\therefore$ $$\frac{\dot{m}}{\text{ip}} = \frac{1}{\text{CV} \times \eta_{\text{th}_i}} = \frac{1}{42{,}000 \times 0.356} = 6.688 \times 10^{-5} \text{ kg/kWs}$$

$$= 6.688 \times 10^{-5} \times 3600 \text{ kg/kWh}$$

$$= 0.241 \text{ kg/kWh}$$

$\therefore$ Indicated specific fuel consumption, isfc = $\boxed{0.241 \text{ kg/kWh}}$ **Ans.**

EXAMPLE 2.7 Show that the compression ratio for the maximum work in an Otto cycle is given by

$$r = \left(\frac{T_3}{T_1}\right)^{1/[2(\gamma-1)]}$$

where T_1 and T_3 are the lower and upper limits of absolute temperature respectively.
Also, prove that the intermediate temperatures for this condition are:

$$T_2 = T_4 = \sqrt{T_1 T_3}$$

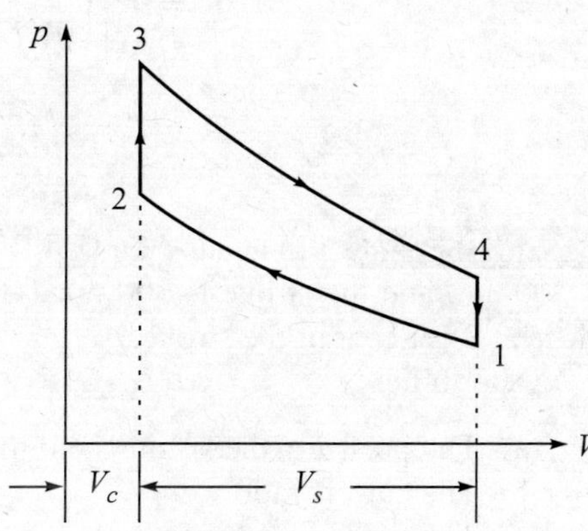

Figure 2.28 Example 2.7.

Solution: Refer to Figure 2.28.

Heat supplied, $Q_1 = mc_v(T_3 - T_2)$

Heat rejected, $Q_2 = mc_v(T_4 - T_1)$

$$\text{Work done} = Q_1 - Q_2$$

$$= mc_v[(T_3 - T_2) - (T_4 - T_1)] \quad \text{(i)}$$

Compression ratio, $r = \dfrac{V_1}{V_2}$

From the adiabatic process 1–2,

$$T_2 = T_1 r^{\gamma-1} \quad \text{(ii)}$$

and from the adiabatic process 3–4

$$T_3 = T_4 r^{\gamma-1}$$

$$\therefore \quad T_4 = T_3 r^{1-\gamma} \quad \text{(iii)}$$

$\therefore$ Work done, $W = mc_v\,[(T_3 - T_1 r^{\gamma-1}) - (T_3 r^{1-\gamma} - T_1)]$

T_1 and T_3 are fixed, therefore, for maximum work,

$$\frac{1}{mc_v}\cdot\frac{dW}{dr} = -T_1(\gamma-1)r^{\gamma-2} - T_3(1-\gamma)r^{-\gamma} = 0$$

or $\quad T_3(r)^{-\gamma} = T_1(r)^{\gamma-2}$

$$\therefore \quad \frac{T_3}{T_1} = r^{2(\gamma-1)}$$

$$\therefore \quad r = \left(\frac{T_3}{T_1}\right)^{1/[2(\gamma-1)]} \quad \text{Proved.}$$

From Eq. (ii),

$$T_2 = T_1\left[\left(\frac{T_3}{T_1}\right)^{1/[2(\gamma-1)]}\right]^{\gamma-1} = T_1\left(\frac{T_3}{T_1}\right)^{1/2} = \sqrt{T_1 T_3}$$

From Eq. (iii),

$$T_4 = T_3\left[\left(\frac{T_3}{T_1}\right)^{1/[2(\gamma-1)]}\right]^{1-\gamma} = T_3\left(\frac{T_1}{T_3}\right)^{1/2} = \sqrt{T_1T_3}$$

$\therefore$ $$T_2 = T_4 = \sqrt{T_1T_3} \quad \text{Proved.}$$

EXAMPLE 2.8 In an air-standard Diesel cycle, the compression ratio is 16, the cylinder bore is 200 mm and the stroke is 300 mm. Compression begins at 1 bar and 27°C. The cut-off takes place at 8 per cent of the stroke.

Determine:

(a) The pressure, the volume and the temperature at all salient points.
(b) The cut-off ratio
(c) The work done per cycle
(d) The air-standard efficiency
(e) The mean effective pressure

Solution: *Given:* $d = 0.2$ m, $L = 0.3$ m, $p_1 = 1$ bar, $T_1 = 27 + 273 = 300$ K, $V_3 - V_2 = 0.08V_s$, $r = 16$

Refer to Figure 2.29:

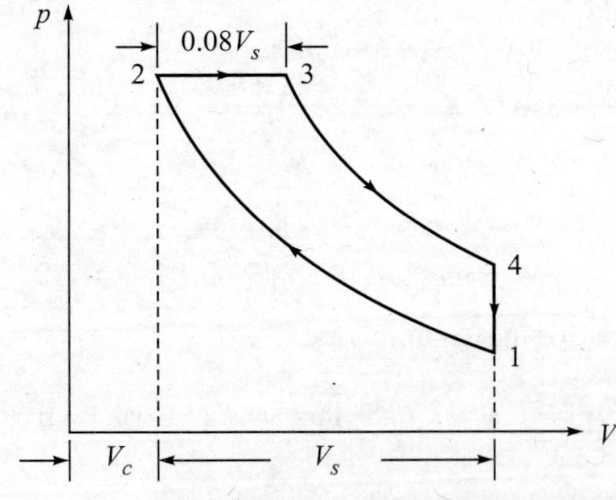

Figure 2.29 Example 2.8.

(a) Pressure, volume and temperature at salient points:

Swept volume, $$V_s = \frac{\pi}{4}d^2L = \frac{\pi}{4}(0.2)^2(0.3) = 0.00942 \text{ m}^3$$

Compression ratio, $$r = \frac{V_1}{V_2} = \frac{V_s + V_c}{V_c} = \frac{V_s}{V_c} + 1$$

$\therefore$ $$\frac{V_s}{V_c} = r - 1$$

or $$V_c = \frac{V_s}{r-1} = \frac{0.00942}{16-1} = 0.000628 \text{ m}^3$$

$$V_2 = V_c = \boxed{0.000628 \text{ m}^3} \quad \textbf{Ans.}$$

$$V_1 = V_s + V_c = 0.00942 + 0.000628 = \boxed{0.010048 \text{ m}^3} \quad \textbf{Ans.}$$

$$m = \frac{p_1 V_1}{RT_1} = \frac{1 \times 10^5 \times 0.010048}{287 \times 300} = 0.01167 \text{ kg}$$

$$p_2 = p_1\left(\frac{V_1}{V_2}\right)^\gamma = p_1 r^\gamma = 1 \times 16^{1.4} = \boxed{48.5 \text{ bar}} \quad \textbf{Ans.}$$

$$p_3 = p_2 = 48.5 \text{ bar} \quad \textbf{Ans.}$$

$$T_2 = T_1 r^{\gamma-1} = (16)^{0.4} \times 300 = \boxed{909.4 \text{ K}} \quad \textbf{Ans.}$$

(b) Cut-off ratio, $$\beta = \frac{V_3}{V_2}$$

Given: $$V_3 - V_2 = 0.08 V_s$$

or $$\frac{V_3}{V_2} - 1 = 0.08\frac{V_s}{V_2} = 0.08\frac{V_s}{Vc} = 0.08(r-1)$$

∴ $$\beta = \frac{V_3}{V_2} = 1 + 0.08(16-1) = \boxed{2.2} \quad \textbf{Ans.} \text{ part (b)}$$

Now, $$V_3 = \beta V_2 = 2.2 \times 0.000628 = 0.001382 \text{ m}^3$$

[or $V_3 = 0.08V_s + V_c = 0.08 \times 0.00942 + 0.000628 = 0.001382 \text{ m}^3$]

$$T_3 = \beta T_2 = 2.2 \times 909.4 = \boxed{2001 \text{ K}} \quad \textbf{Ans.}$$

$$p_4 = p_3\left(\frac{V_3}{V_4}\right)^\gamma = p_3\left(\frac{V_3}{V_2} \times \frac{V_2}{V_1}\right)^\gamma = p_3\left(\frac{\beta}{r}\right)^\gamma$$

∴ $$p_4 = 48.5\left(\frac{2.2}{16}\right)^{1.4} = \boxed{3.016 \text{ bar}} \quad \textbf{Ans.}$$

$$T_4 = T_3\left(\frac{\beta}{r}\right)^{\gamma-1} = 2001\left(\frac{2.2}{16}\right)^{0.4} = \boxed{904.8 \text{ K}} \quad \textbf{Ans.}$$

$$V_4 = V_1 = \boxed{0.010048 \text{ m}^3} \quad \textbf{Ans.}$$

(c) Heat supplied, $$Q_1 = mc_p(T_3 - T_2)$$
$$= 0.01167 \times 1.005(2001 - 909.4) = 12.8 \text{ kJ}$$

Heat rejected, $$Q_2 = mc_v(T_4 - T_1)$$
$$= 0.01167 \times 0.718(904.8 - 300) = 5.068 \text{ kJ}$$

∴ Work done per cycle, $W = Q_1 - Q_2 = 12.8 - 5.068 = \boxed{7.732 \text{ kJ}}$ **Ans.**

(d) Air-standard efficiency $= \dfrac{W}{Q_1} = \dfrac{7.732}{12.8} = 0.604$

$$= \boxed{60.4\%} \quad \textbf{Ans.}$$

(e) Mean effective pressure, $p_m = \dfrac{W}{V_s} = \dfrac{7.732 \times 10^3}{0.00942} \text{ N/m}^2$

$$= 8.21 \times 10^5 \text{ N/m}^2 = \boxed{8.21 \text{ bar}} \quad \textbf{Ans.}$$

EXAMPLE 2.9 The mean effective pressure of a Diesel cycle is 7 bar, the compression ratio is 12 and the initial pressure is 1 bar. Determine the cut-off ratio and the air-standard efficiency.

Solution: The mean effective pressure is given by

$$p_m = \frac{p_1}{(\gamma - 1)(r - 1)} [\gamma r^\gamma(\beta - 1) - r(\beta^\gamma - 1)]$$

Given: $p_m = 7$ bar, $r = 12$, $p_1 = 1$ bar

∴ $$7 = \frac{1}{(1.4 - 1)(12 - 1)} [1.4 \times 12^{1.4}(\beta - 1) - 12(\beta^{1.4} - 1)]$$

or $$30.8 = 45.4\beta - 45.4 - 12\beta^{1.4} + 12$$

or $$45.4\beta - 12\beta^{1.4} - 64.2 = 0$$

Solving by trial-and-error method, the cut-off ratio

$$\beta = \boxed{2.2} \quad \textbf{Ans.}$$

Air-standard efficiency,

$$\eta = 1 - \frac{1}{r^{\gamma-1}} \left[\frac{\beta^\gamma - 1}{\gamma(\beta - 1)}\right]$$

$$= 1 - \frac{1}{(12)^{0.4}} \frac{2.2^{1.4} - 1}{1.4(2.2 - 1)}$$

$$= 0.556 = \boxed{55.6\%} \quad \textbf{Ans.}$$

EXAMPLE 2.10 In an engine working on the Diesel cycle, the air-fuel ratio is 30:1. The temperature of air at the beginning of the compression is 27°C, the compression ratio is 16:1. What is the ideal efficiency of the engine based on the air-standard cycle? The calorific value of the fuel used is 42,000 kJ/kg.

Solution: Refer to Figure 2.30.

Given: $$\frac{\text{mass of air}}{\text{mass of fuel}} = \frac{m_{\text{air}}}{m_{\text{fuel}}} = 30$$

$$T_1 = 27 + 273 = 300 \text{ K}$$
$$r = 16$$
$$\text{CV} = 42{,}000 \text{ kJ/kg}$$

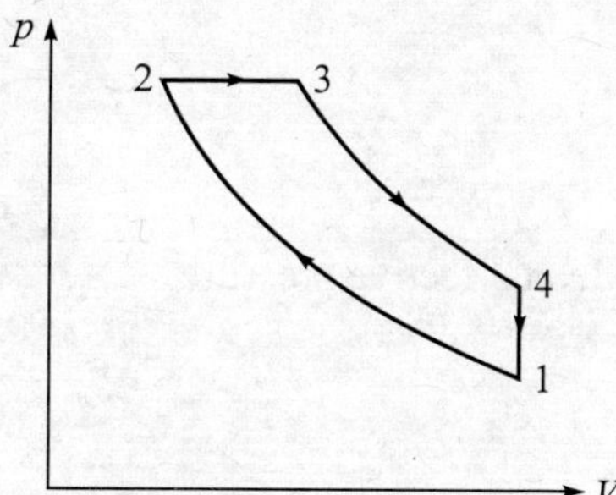

Figure 2.30 Example 2.10.

From the process 1–2,

$$T_2 = T_1 r^{\gamma-1} = 300 \times (16)^{0.4}$$
$$= 909.4 \text{ K}$$

For constant pressure process 2–3,

$$\text{Heat supplied} = m_{\text{air}} c_p (T_3 - T_2) = m_{\text{fuel}} \times \text{CV}$$

$$\therefore \quad T_3 - T_2 = \frac{m_{\text{fuel}}}{m_{\text{air}}} \times \frac{\text{CV}}{c_p} = \frac{1}{30} \times \frac{42{,}000}{1.005} = 1393$$

or $\quad T_3 = 1393 + 909.4 = 2302 \text{ K}$

Cut-off ratio, $\quad \beta = \frac{V_3}{V_2} = \frac{T_3}{T_2} = \frac{2302}{909.4} = 2.53$

$$\eta = 1 - \frac{1}{r^{\gamma-1}} \frac{\beta^{\gamma} - 1}{\gamma(\beta - 1)}$$

$$= 1 - \frac{1}{16^{0.4}} \frac{2.53^{1.4} - 1}{1.4(2.53 - 1)}$$

$$= 0.589 = \boxed{58.9\%} \quad \textbf{Ans.}$$

EXAMPLE 2.11 In an engine working on Diesel cycle, the inlet pressure is 1 bar. The pressure at the end of isentropic compression is 32.425 bar. The ratio of expansion is 6. Calculate the air-standard efficiency and the mean effective pressure of the cycle.

Solution: Refer to Figure 2.30.
Consider the process 1–2,

Compression ratio, $\quad r = \frac{V_1}{V_2} = \left(\frac{p_2}{p_1}\right)^{1/\gamma} = \left(\frac{32.425}{1}\right)^{1/1.4} = 12$

Expansion ratio, $\quad r_e = \frac{V_4}{V_3} = 6 \text{ (given)}$

$\therefore$ Cut-off ratio, $\quad \beta = \frac{V_3}{V_2} = \frac{V_1}{V_2} \times \frac{V_3}{V_4} \quad (\because V_1 = V_4)$

$$= \frac{r}{r_e} = \frac{12}{6} = 2$$

Air-standard efficiency, $\quad \eta = 1 - \frac{1}{r^{\gamma-1}} \frac{\beta^{\gamma} - 1}{\gamma(\beta - 1)}$

$$= 1 - \frac{1}{12^{0.4}} \frac{2^{1.4} - 1}{1.4(2 - 1)}$$

$$= 0.5667 = \boxed{56.67\%} \quad \textbf{Ans.}$$

Mean effective pressure,

$$p_m = \frac{p_1 r^{\gamma}}{(\gamma - 1)(r - 1)} \eta\gamma(\beta - 1)$$

$$= \frac{1 \times 12^{1.4} \times 0.5667 \times 1.4(2 - 1)}{(1.4 - 1)(12 - 1)}$$

$$= \boxed{5.846 \text{ bar}} \quad \textbf{Ans.}$$

EXAMPLE 2.12 An air-standard Dual cycle has a compression ratio of 15, and the compression begins at 1 bar, 27°C. The maximum pressure is limited to 60 bar. The heat transferred to air at constant volume is twice that at constant pressure. Calculate (a) the pressures and temperatures at the cardinal points of the cycle, (b) the cycle efficiency, and (c) the mean effective pressure of the cycle.

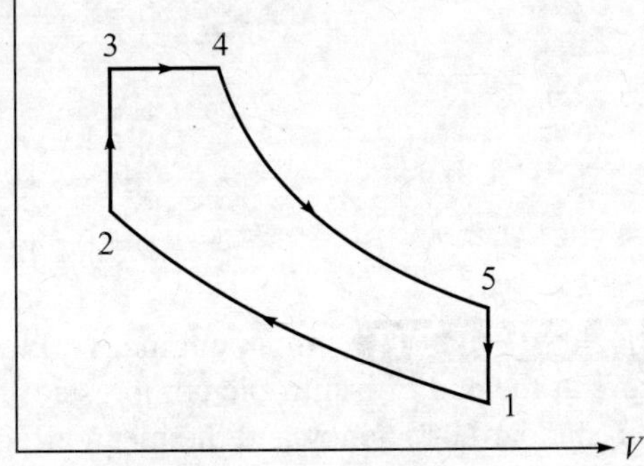

Figure 2.31 Example 2.12.

Solution: Refer to Figure 2.31.

Given: Compression ratio, $r = 15$

$$p_1 = 1 \text{ bar}$$

$$T_1 = 27 + 273 = 300 \text{ K}$$

$$p_3 = p_4 = 60 \text{ bar}$$

$$Q_{2-3} = 2 \times Q_{3-4}$$

(a) Considering the process 1–2,

$$\frac{T_2}{T_1} = \left(\frac{V_1}{V_2}\right)^{\gamma-1} = (r)^{\gamma-1}$$

$\therefore$

$$T_2 = T_1 r^{\gamma-1} = 300 \times (15)^{0.4} = \boxed{886 \text{ K}} \quad \textbf{Ans.}$$

Also,

$$p_1 V_1^{\gamma} = p_2 V_2^{\gamma}$$

$\therefore$

$$p_2 = p_1 \left(\frac{V_1}{V_2}\right)^{\gamma} = p_1 r^{\gamma} = 1 \times (15)^{1.4} = \boxed{44.3 \text{ bar}} \quad \textbf{Ans.}$$

Considering the process 2–3,

$$\frac{p_2 V_2}{T_2} = \frac{p_3 V_3}{T_3} \quad (\because V_2 = V_3)$$

$\therefore$

$$T_3 = T_2 \times \frac{p_3}{p_2} = 886 \times \frac{60}{44.3} = \boxed{1200\text{K}} \quad \textbf{Ans.}$$

$$Q_{2-3} = 2 \times Q_{3-4} \quad \text{(Given)}$$

$\therefore$ $$c_v(T_3 - T_2) = 2c_p(T_4 - T_3)$$

or $$0.718(1200 - 886) = 2 \times 1.005(T_4 - 1200)$$

$\therefore$ $$T_4 = \boxed{1312 \text{ K}} \quad \textbf{Ans.}$$

Considering the process 3–4,

$$\frac{p_3V_3}{T_3} = \frac{p_4V_4}{T_4}$$

or $$T_4 = T_3 \times \frac{V_4}{V_3} \quad (\because p_3 = p_4)$$

or $$T_4 = \beta T_3$$

$\therefore$ Cut-off ratio, $$\beta = \frac{T_4}{T_3} = \frac{1312}{1200} = 1.093$$

Considering the process 4–5,

$$\frac{T_5}{T_4} = \left(\frac{V_4}{V_5}\right)^{\gamma-1} = \left(\frac{V_4}{V_3} \times \frac{V_3}{V_5}\right)^{\gamma-1} = \left(\frac{V_4}{V_3} \times \frac{V_2}{V_1}\right)^{\gamma-1} = \left(\frac{\beta}{r}\right)^{\gamma-1}$$

$\therefore$ $$T_5 = 1312\left(\frac{1.093}{15}\right)^{0.4} = \boxed{460 \text{ K}} \quad \textbf{Ans.}$$

Considering the process 5–1 ($V_5 = V_1$),

$$\frac{p_5}{T_5} = \frac{p_1}{T_1}$$

$\therefore$ $$p_5 = p_1 \times \frac{T_5}{T_1} = 1 \times \frac{460}{300} = \boxed{1.53 \text{ bar}} \quad \textbf{Ans.}$$

(b) Heat supplied per unit mass,

$$Q_1 = c_v(T_3 - T_2) + c_p(T_4 - T_3)$$

$\therefore$ $$Q_1 = 0.718(1200 - 886) + 1.005(1312 - 1200)$$
$$= 338 \text{ kJ/kg}$$

Heat rejected per unit mass,

$$Q_2 = c_v(T_5 - T_1)$$

$\therefore$ $$Q_2 = 0.718(460 - 300) = 115 \text{ kJ/kg}$$

Work done, $$W = Q_1 - Q_2$$
$$= 338 - 115 = 223 \text{ kJ/kg}$$

Air-standard efficiency,

$$\eta = \frac{W}{Q_1} = \frac{223}{338} = 0.66$$

$$= \boxed{66\%} \quad \textbf{Ans.}$$

(c) Swept volume, $V_s = V_1 - V_2 = V_1\left(1 - \frac{1}{r}\right)$

$$= \frac{m\,RT_1}{p_1}\left(1 - \frac{1}{r}\right)$$

For unit mass, $V_s = \frac{1 \times 287 \times 300}{1 \times 10^5}\left(1 - \frac{1}{15}\right)$

$$= 0.804 \text{ m}^3/\text{kg}$$

Mean effective pressure, $p_m = \frac{W}{V_s} = \frac{223 \times 10^3}{0.804} \text{N/m}^2 = 2.77 \times 10^5 \text{ N/m}^2$

$$= \boxed{2.77 \text{ bar}} \quad \textbf{Ans.}$$

EXAMPLE 2.13 A Dual combustion cycle operates with a compression ratio of 12 and with a cut-off ratio of 1.615. The maximum pressure is 52.17 bar. The pressure and temperature before compression are 1 bar and 62°C respectively. Assuming indices of compression and expansion of 1.35, calculate (a) the temperatures at cardinal points, (b) the cycle efficiency, and (c) the mean effective pressure of the cycle.

Solution: Refer to Figure 2.32.

Given: Compression ratio, $r = 12$

Cut-off ratio, $\beta = 1.615$

Maximum pressure, $p_3 = p_4 = 52.17$ bar

$p_1 = 1$ bar

$T_1 = 62 + 273 = 335$ K

Indices of compression and expansion,

$n = 1.35$

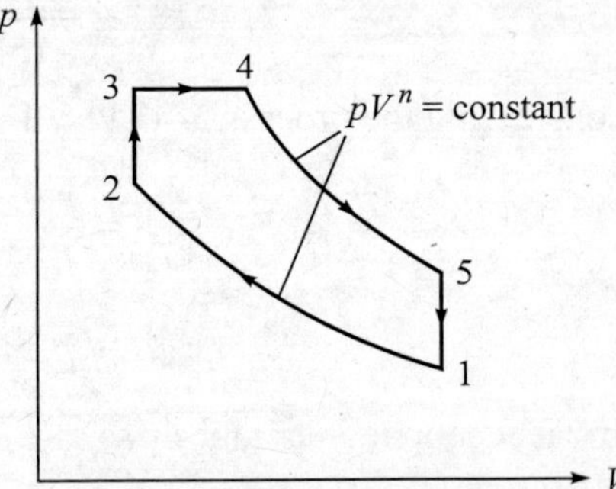

Figure 2.32 Example 2.13.

(a) Considering the process 1–2,

$$T_2 = T_1 r^{n-1} = 335 \times (12)^{0.35} = \boxed{799 \text{ K}} \quad \textbf{Ans.}$$

$$p_2 V_2^n = p_1 V_1^n$$

$$\therefore \quad p_2 = p_1\left(\frac{V_1}{V_2}\right)^n = p_1 r^n = 1 \times (12)^{1.35} = \boxed{28.63 \text{ bar}}$$

Considering the process 2–3,

$$T_3 = T_2 \times \frac{p_3}{p_2} = 799 \times \frac{52.17}{28.63} = \boxed{1456 \text{ K}} \quad \textbf{Ans.}$$

Considering the process 3–4,

$$T_4 = T_3 \times \frac{V_4}{V_3} = T_3 \times \beta = 1456 \times 1.615 = \boxed{2351\text{ K}} \quad \textbf{Ans.}$$

Considering the process 4–5,

$$\frac{T_5}{T_4} = \left(\frac{V_4}{V_5}\right)^{n-1} = \left(\frac{V_4}{V_3} \times \frac{V_3}{V_5}\right)^{n-1} = \left(\frac{V_4}{V_3} \times \frac{V_2}{V_1}\right)^{n-1} = \left(\frac{\beta}{r}\right)^{n-1}$$

$$\therefore \qquad T_5 = 2351 \times \left(\frac{1.615}{12}\right)^{0.35} = \boxed{1165\text{ K}} \quad \textbf{Ans.}$$

(b) Processes 1–2 and 4–5 are polytropic, therefore, there will be heat transfer during these processes as well.

Heat transfer during the process 1–2,

$$Q_{1\text{-}2} = \left(\frac{\gamma - n}{\gamma - 1}\right)\left(\frac{p_1V_1 - p_2V_2}{n-1}\right)$$

$$= \left(\frac{\gamma - n}{\gamma - 1}\right)(mR)\left(\frac{T_1 - T_2}{n-1}\right)$$

For unit mass,

$$Q_{1\text{-}2} = \left(\frac{1.4 - 1.35}{1.4 - 1}\right)(0.287)\left(\frac{335 - 799}{1.35 - 1}\right)$$

$$= -47.56\text{ kJ/kg}$$

Heat transfer during the process 2–3,

$$Q_{2\text{-}3} = c_v(T_3 - T_2) = 0.718(1456 - 799)$$

$$= 471.7\text{ kJ/kg}$$

Heat transfer during the process 3–4,

$$Q_{3\text{-}4} = c_p(T_4 - T_3) = 1.005(2351 - 1456)$$

$$= 899.5\text{ kJ/kg}$$

Heat transfer during the process 4–5,

$$Q_{4\text{-}5} = \left(\frac{\gamma - n}{\gamma - 1}\right)(R)\left(\frac{T_4 - T_5}{n-1}\right)$$

$$= \left(\frac{1.4 - 1.35}{1.4 - 1}\right)(0.287)\left(\frac{2351 - 1165}{1.35 - 1}\right)$$

$$= 121.6\text{ kJ/kg}$$

Heat transfer during the process 5–1,

$$Q_{5\text{-}1} = c_v(T_1 - T_5) = 0.718(335 - 1165)$$

$$= -596\text{ kJ/kg}$$

The positive sign of heat transfer represents the heat supplied and the negative sign represents the heat rejected.

$\therefore$ Heat supplied, $Q_1 = Q_{2-3} + Q_{3-4} + Q_{4-5}$

$= 471.7 + 899.5 + 121.6$

$= 1492.8$ kJ/kg

and heat rejected, $Q_2 = Q_{1-2} + Q_{5-1}$

$= 47.56 + 596$

$= 643.56$ kJ/kg

Work done, $W = Q_1 - Q_2$

$= 1492.8 - 643.56 = 849.2$ kJ/kg

Efficiency, $\eta = \frac{W}{Q_1} = \frac{849.2}{1492.8} = 0.569 = \boxed{56.9\%}$ **Ans.**

(c) Swept volume, $V_s = V_1 - V_2 = V_1\left(1 - \frac{1}{r}\right) = \frac{mRT_1}{p_1}\left(\frac{r-1}{r}\right)$

For unit mass, $V_s = \frac{RT_1}{p_1}\left(\frac{r-1}{r}\right) = \frac{287 \times 335}{1 \times 10^5} = \frac{11}{12}$

$= 0.8813$ m³/kg

Mean effective pressure, $p_m = \frac{W}{V_s}$

$= \frac{849.2 \times 10^3}{0.8813} = 9.635 \times 10^5$ N/m²

$= \boxed{9.635 \text{ bar}}$ **Ans.**

EXAMPLE 2.14 A perfect gas undergoes an Atkinson cycle. The gas is compressed adiabatically from 1 bar, 27°C to 4 bar. The maximum pressure of the cycle is 16 bar. Calculate (a) the work done per kg of gas, (b) the efficiency of the cycle, and (c) the mean effective pressure of the cycle.

Take $c_p = 0.761$ kJ/(kg K) and $c_v = 0.573$ kJ/(kg K)

Solution: Refer to Figure 2.33.

Given:

$p_1 = 1 \text{ bar} = p_4$

$T_1 = 27 + 273 = 300$ K

$p_2 = 4$ bar

$p_3 = 16$ bar

$\gamma = c_p/c_v = 0.761/0.573$

$= 1.328$

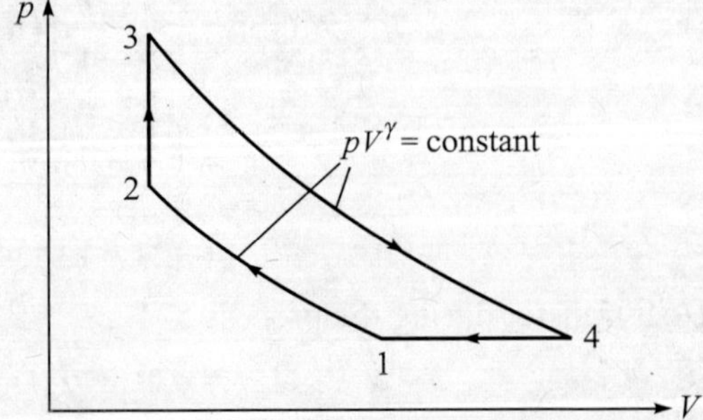

Figure 2.33 Atkinson cycle: Example 2.14.

For adiabatic compression process 1–2,

$$\frac{T_2}{T_1} = \left(\frac{p_2}{p_1}\right)^{(\gamma-1)/\gamma} = (4)^{(1.328-1)/1.328}$$

$$= (4)^{0.247}$$

$$= 1.408$$

$$\therefore \quad T_2 = 300 \times 1.408 = 422.5 \text{ K}$$

For constant volume process 2–3,

$$\frac{p_2}{T_2} = \frac{p_3}{T_3}$$

$$\therefore \quad T_3 = \frac{p_3}{p_2} \times T_2 = \frac{16}{4} \times 422.5 = 1690 \text{ K}$$

For adiabatic expansion process 3–4,

$$\frac{T_4}{T_3} = \left(\frac{p_4}{p_3}\right)^{(\gamma-1)/\gamma}$$

$$\therefore \quad T_4 = T_3\left(\frac{1}{16}\right)^{0.247} = 1690 \times 0.5042 = 852 \text{ K}$$

(a) Work done per kg, W:

Heat supplied per kg, $Q_1 = c_v(T_3 - T_2)$

$$= 0.573(1690 - 422.5)$$

$$= 726.3 \text{ kJ/kg}$$

Heat rejected per kg, $Q_2 = c_p(T_4 - T_1)$

$$= 0.761(852 - 300)$$

$$= 420.1 \text{ kJ/kg}$$

$$W = 726.3 - 420.1 = \boxed{306.2 \text{ kJ/kg}} \quad \textbf{Ans.}$$

(b) Efficiency of the cycle,

$$\eta = \frac{W}{Q_1} = \frac{306.2}{726.3} = 0.422 = \boxed{42.2\%} \quad \textbf{Ans.}$$

(c) Mean effective pressure,

$$p_m = \frac{W}{V_s}$$

Swept volume, $V_s = V_1 - V_2 = V_1\left(1 - \frac{1}{r}\right) = \frac{mRT_1}{p_1}\left(\frac{r-1}{r}\right)$

For adiabatic process 1–2,

$$p_1 V_1^{\gamma} = p_2 V_2^{\gamma}$$

$\therefore$ Compression ratio, $r = \frac{V_1}{V_2} = \left(\frac{p_2}{p_1}\right)^{1/\gamma} = (4)^{1/1.328} = 2.84$

$$R = c_p - c_v = 0.761 - 0.573 = 0.188 \text{ kJ/(kg K)}$$

$\therefore$ Swept volume for unit mass,

$$V_s = \frac{0.188 \times 10^3 \times 300}{1 \times 10^5}\left(\frac{2.84 - 1}{2.84}\right)$$

$$= 0.3654 \text{ m}^3\text{/kg}$$

$$\therefore \quad p_m = \frac{306.2 \times 10^3}{0.3654} = 8.38 \times 10^5 \text{ N/m}^2$$

$$= \boxed{8.38 \text{ bar}}$$ **Ans.**

EXAMPLE 2.15 A single cylinder, 0.3 l SI engine operates on an air-standard Miller cycle with early valve closing. It has a compression ratio of 9 and an expansion ratio of 11. When the intake valve closes, conditions in the cylinder are 1.0 bar and 27 °C. Heat supplied is 0.5187 kJ.

Calculate:

(a) Temperature and pressure at all points in the cycle
(b) Work done per cycle
(c) Work ratio
(d) Thermal efficiency
(e) Mean effective pressure

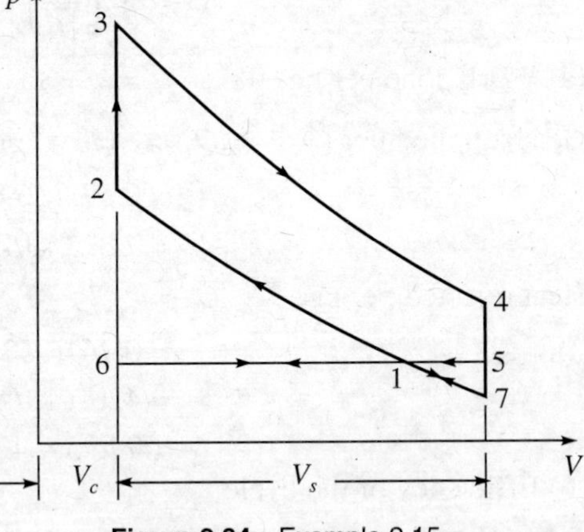

Figure 2.34 Example 2.15.

Solution: Refer to Figure 2.34.

Swept volume, $V_s = 0.3 l$

Expansion ratio, $e = \frac{V_4}{V_3} = \frac{V_s + V_c}{V_c} = 11$

$$\therefore \quad V_c = \frac{V_s}{10} = \frac{0.3}{10} = 0.03 \, l$$

$$V_c = V_2 = V_3 = V_6 = 0.03 l$$

$$V_4 = 11V_3 = 11 \times 0.03 = 0.33 l = V_5 = V_7$$

Compression ratio,

$$r = \frac{V_1}{V_2} = 9$$

$$\therefore \quad V_1 = 9 \times 0.03 = 0.27 \, l$$

$$T_1 = 27 + 273 = 300 \text{ K}$$

$$p_1 = p_5 = 1.0 \text{ bar}$$

(a) Temperature and pressure

$$p_7 = p_1\left(\frac{V_1}{V_7}\right)^{\gamma} = 1.0\times\left(\frac{0.27}{0.33}\right)^{1.4} = \boxed{0.755 \text{ bar}} \quad \textbf{Ans.}$$

$$T_7 = T_1\left(\frac{V_1}{V_7}\right)^{\gamma-1} = 300\left(\frac{0.27}{0.33}\right)^{0.4} = \boxed{276.9 \text{ K}} \quad \textbf{Ans.}$$

$$T_2 = T_1(r)^{\gamma-1} = 300\,(9)^{0.4} = \boxed{722.5 \text{ K}} \quad \textbf{Ans.}$$

$$p_2 = p_1(r)^{\gamma} = 1.0\times(9)^{1.4} = \boxed{21.67 \text{ bar}} \quad \textbf{Ans.}$$

Mass of air, $m = \dfrac{p_1V_1}{RT_1}$

$$= \frac{1\times10^5\times0.27\times10^{-3}}{287\times300} = 0.3136\times10^{-3} \text{ kg}$$

Heat supplied during process 2–3,

$$Q_s = m\,c_v(T_3 - T_2)$$

$$0.3136\times10^{-3}\times0.718(T_3 - 722.5) = 0.5187$$

$$\therefore \qquad T_3 = \boxed{3026 \text{ K}} \quad \textbf{Ans.}$$

$$p_3 = p_2\left(\frac{T_3}{T_2}\right) = 21.67\times\left(\frac{3026}{722.5}\right) = \boxed{90.76 \text{ bar}} \quad \textbf{Ans.}$$

$$T_4 = T_3\left(\frac{V_3}{V_4}\right)^{\gamma-1} = 3026\times\left(\frac{1}{11}\right)^{0.4} = \boxed{1160 \text{ K}} \quad \textbf{Ans.}$$

$$p_4 = p_3\left(\frac{V_3}{V_4}\right)^{\gamma} = 90.76\times\left(\frac{1}{11}\right)^{1.4} = \boxed{3.162 \text{ bar}} \quad \textbf{Ans.}$$

$$p_5 = p_1 = \boxed{1.0 \text{ bar}} \quad \textbf{Ans.}$$

$$T_5 = T_4\left(\frac{p_5}{p_4}\right) = 1160\times\frac{1}{3.162} = \boxed{367 \text{ K}} \quad \textbf{Ans.}$$

(b)
$$W_{1-2} = \frac{m\,R(T_1 - T_2)}{(\gamma - 1)} = \frac{0.3136\times10^{-3}\times287(300 - 722.5)}{0.4} = -95.07 \text{ J}$$

$$W_{2-3} = 0$$

$$W_{3-4} = \frac{mR(T_3 - T_4)}{(\gamma - 1)} = \frac{0.3136 \times 10^{-3} \times 287(3026 - 1160)}{0.4} = 419.9 \text{ J}$$

$$W_{4-5} = 0$$

$$W_{5-1} = p_1(V_1 - V_5) = 1 \times 10^5 (0.27 - 0.33) \times 10^{-3} = -6.0 \text{ J}$$

Net work done per cycle,

$$W_{net} = -95.07 + 419.9 - 6.0 = 318.8 \text{ J} = \boxed{0.3188 \text{ kJ}} \quad \textbf{Ans.}$$

(c) Work ratio $= \dfrac{\text{net work}}{\text{gross work}} = \dfrac{318.8}{419.9} = \boxed{0.759}$ **Ans.**

(d) Thermal efficiency,

$$\eta = \frac{W_{net}}{Q_s} = \frac{0.3188}{0.5187} = 0.615 = \boxed{61.5\%} \quad \textbf{Ans.}$$

(e) Mean effective pressure,

$$p_m = \frac{W_{net}}{V_s} = \frac{318.8}{0.3 \times 10^{-3}} = 10.63 \times 10^5 \text{ N/m}^2$$

$$= \boxed{10.63 \text{ bar}} \quad \textbf{Ans.}$$

EXAMPLE 2.16 In a closed-cycle gas turbine unit (Brayton cycle), air is drawn at 1.02 bar and 20 °C and is compressed to 7.14 bar. The maximum cycle temperature is limited to 760°C. Calculate the ideal cycle thermal efficiency and the work ratio.

Solution: Refer to Figure 2.35.

Given:

$$T_1 = 20 + 273 = 293 \text{ K}$$
$$p_1 = 1.02 \text{ bar} = p_4$$
$$T_3 = 760 + 273 = 1033 \text{ K}$$
$$p_2 = 7.14 \text{ bar} = p_3$$

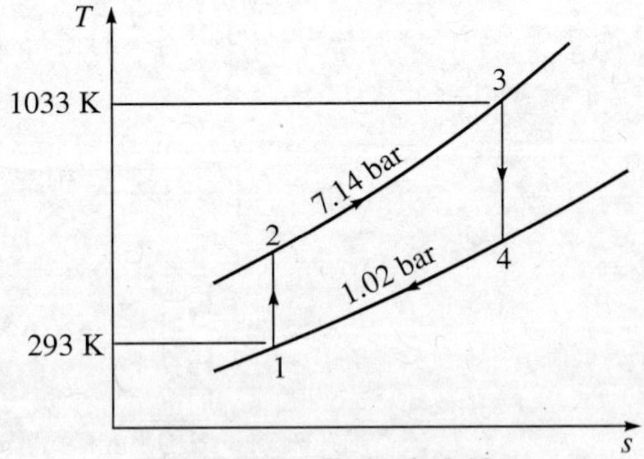

Figure 2.35 Example 2.16.

Pressure ratio,

$$r_p = \frac{p_2}{p_1} = \frac{7.14}{1.02} = 7$$

$$\gamma = 1.4$$

Thermal efficiency,

$$\eta = 1 - \frac{1}{(r_p)^{(\gamma-1)/\gamma}}$$

$$= 1 - \frac{1}{(7)^{0.286}} = 1 - 0.573 = 0.427 = \boxed{42.7\%} \quad \textbf{Ans.}$$

$$\frac{T_2}{T_1} = (r_p)^{(\gamma-1)/\gamma} = \frac{T_3}{T_4} = (7)^{0.286} = 1.745$$

$$T_2 = 1.745 T_1 = 1.745 \times 293 = 511 \text{ K}$$

$$T_4 = \frac{T_3}{1.745} = \frac{1033}{1.745} = 592 \text{ K}$$

$$\text{Net work output} = c_p(T_3 - T_4) - c_p(T_2 - T_1)$$

$$= 1.005(1033 - 592) - 1.005(511 - 293)$$

$$= 224.1 \text{ kJ/kg}$$

$$\text{Gross work output} = c_p(T_3 - T_4)$$

$$= 1.005(1033 - 592) = 443.2 \text{ kJ/kg}$$

$$\text{Work ratio} = \frac{\text{network output}}{\text{gross work output}} = \frac{224.1}{443.2} = \boxed{0.506}$$ **Ans.**

REVIEW QUESTIONS

1. Mention the two commonly employed approximations of an actual engine. How are they different?
2. What are the assumptions made in analyzing the air-standard cycle?
3. Why is the analysis of air-standard cycle important, though the results obtained are much higher than the actual results?
4. Describe thermodynamic processes involved in a Carnot cycle. Represent the cycle on p–V and T–s diagrams. Derive expression to evaluate thermal efficiency of the cycle. What is a work ratio?
5. Describe theoretically the working principle of a Carnot engine. Why is it not possible to build a Carnot engine?
6. Describe Stirling cycle with the help of p–V and T–s diagrams. Derive an expression to evaluate thermal efficiency of this cycle with perfect regeneration.
7. Describe the working principle of Stirling engine with perfect regeneration.
8. Describe Ericsson cycle with the help of p–V and T–s diagrams. Derive an expression to evaluate thermal efficiency of this cycle with perfect regeneration.
9. Describe the working principle of Ericsson engine with perfect regeneration.
10. Describe Lenoir cycle with the help of p–V and T–s diagrams. Derive an expression to evaluate thermal efficiency of this cycle.
11. Describe the Otto cycle with the help of p–V and T–s diagrams. Where does this cycle find its application?

12. Define the terms compression ratio and pressure ratio. Derive the expressions to evaluate efficiency, work done and mean effective pressure of an Otto cycle using the terms of pressure ratio and compression ratio.
13. Show and explain the variation of thermal efficiency of an Otto cycle with compression ratio at different adiabatic exponents. Does the thermal efficiency of the Otto cycle depend upon the engine load?
14. Show and explain the variation of mean effective pressure of an Otto cycle with pressure ratio at different compression ratios.
15. Describe a Diesel cycle with the help of *p–V* and *T–s* diagrams. Where does this cycle find its application?
16. Define the term cut-off ratio. Derive expressions to evaluate thermal efficiency, work done and mean effective pressure using compression ratio and cut-off ratio.
17. Show and explain the variation of thermal efficiency of a Diesel cycle with cut-off ratio at different compression ratios and adiabatic exponents.
18. Describe a Dual combustion cycle with the help of *p–V* and *T–s* diagrams. Where does this cycle find its application?
19. Derive expressions to evaluate thermal efficiency, work done and mean effective pressure of a dual combustion cycle using compression ratio, pressure ratio and cut-off ratio.
20. Compare the thermal efficiencies of Otto, Diesel and Dual combustion cycles under the following conditions:
    (a) For the same compression ratio and heat addition
    (b) For the same compression ratio and heat rejection
    (c) For the same peak pressure, peak temperature and heat rejection
    (d) For the same maximum pressure and heat input
    (e) For the same maximum pressure and work output.
21. Describe the Atkinson cycle with the help of *p–V* and *T–s* diagrams. Where does this cycle find its application?
22. Define compression ratio and expansion ratio. Derive expressions to evaluate thermal efficiency, work done and mean effective pressure of an Atkinson cycle using the terms compression ratio and expansion ratio.
23. Describe Miller cycle. Represent the cycle on *p–V* diagram for early inlet value closing and for late inlet valve closing. Derive expressions to evaluate thermal efficiency and mean effective pressure of the cycle.
24. Describe Brayton cycle. Derive expressions to evaluate thermal efficiency and work ratio of the cycle.

PROBLEMS

2.1 In a Carnot cycle operating between 800°C and 15°C, the maximum and minimum pressure are 210 bar and 1 bar, respectively. Calculate the cycle efficiency and the work ratio. Assume air to be the working fluid. **[Ans.** 73.2%, 0.212]

2.2 An air-standard Stirling cycle with 100% regeneration operates with a maximum pressure of 60 bar and a minimum pressure of 1.0 bar. The maximum volume of the air is 10 times the minimum volume. The temperature during heat rejection is 27°C. Calculate net work produced by the cycle per kg of air and thermal efficiency of the cycle.

[**Ans.** 991.3 kJ/kg, 83.3%]

2.3 An Ericsson cycle comprising flow processes operates between thermal energy reservoirs at 1100 K and 300 K and between pressure limits of 100 kPa and 500 kPa. Assuming complete regeneration, calculate net work output, thermal efficiency and work ratio during the cycle. [**Ans.** 369.4 kJ/kg, 72.7%, 0.727]

2.4 The pressure range in a Lenoir air-standard cycle is from 1.0 bar to 5.0 bar. Initial temperature is 27°C. Calculate the net work done per kg of air and thermal efficiency of the cycle.

[**Ans.** 211.8 kJ/kg, 24.6%]

2.5 An engine working on Otto cycle has a pressure and temperature at the beginning of compression as 1 bar and 20°C respectively. After compression, the temperature becomes 380°C. The maximum temperature of the cycle is limited to 1500°C. Calculate the compression ratio, the temperature and pressure at all cardinal points, the air-standard efficiency, the work done per kg, and the mean effective pressure.

[**Ans.** $r = 7.415$, $p_1 = 1$ bar, $T_1 = 293$ K, $p_2 = 16.53$ bar, $T_2 = 653$ K, $p_3 = 44.87$ bar, $T_3 = 1773$ K, $p_4 = 2.715$ bar, $T_4 = 795.5$ K, $\eta = 55.13\%$, $W = 443.3$ kJ/kg, $p_m = 6.09$ bar]

2.6 An engine working on Otto cycle has cylinder bore of 210 mm and stroke length of 240 mm. The clearance volume is 1550 cc. The pressure and temperature at the beginning of compression are 1 bar and 17°C respectively. The maximum pressure of the cycle is 50 bar. Determine the pressure and temperature at the salient points, the air-standard efficiency, the work done, and the mean effective pressure. What will be the ideal power developed by the engine, if the working cycle per minute is 600?

[**Ans.** $p_1 = 1$ bar, $T_1 = 290$ K, $p_2 = 13.34$ bar, $T_2 = 607.9$ K, $p_3 = 50$ bar, $T_3 = 2278$ K, $p_4 = 3.75$ bar, $T_4 = 1087$ K, $\eta = 52.3\%$, $W = 7.429$ kJ, $p_m = 8.94$ bar, power = 74.29 kW]

2.7 A Diesel cycle operates at a pressure of 1 bar at the beginning of compression and the air is compressed to 1/18th of the initial volume. Heat is supplied until the volume is twice that of the clearance volume. Calculate the air-standard efficiency and the mean effective pressure of the cycle. [**Ans.** $\eta = 63.16\%$, $p_m = 7.438$ bar]

2.8 A four-cylinder engine operating on a Diesel cycle has cylinder bore of 200 mm and stroke length of 250 mm. The engine speed is 3000 rpm. The pressure and temperature of the air at the beginning of compression are 1 bar and 27°C respectively. The clearance volume is 1/15th of the swept volume. The maximum temperature attained by the cycle is 1800°C. Determine the pressure and temperature at the salient points, the compression ratio, the air-standard efficiency, the work done per kg of air, the mean effective pressure, and the power developed by the engine.

[**Ans.** $p_1 = 1$ bar, $T_1 = 300$ K, $p_2 = 48.5$ bar, $T_2 = 909.4$ K, $p_3 = 48.5$ bar, $T_3 = 2073$ K, $p_4 = 3.17$ bar, $T_4 = 950.9$ K, $r = 16$, $\eta = 60\%$, $W = 702.1$ kJ/kg, $p_m = 8.7$ bar, power = 683.2 kW]

2.9 For an engine working on a Diesel cycle, the pressures in the cylinder at 20 per cent and 75 per cent of the compression stroke are 1.5 bar and 4 bar respectively. Assuming that the compression follows the law $pV^{1.3}$ = constant, find the compression ratio. The cut-off takes place at 5% of the stroke. What will be the cut-off ratio? If the relative efficiency of the engine compared with the air-standard efficiency is 60%, evaluate the fuel consumption in kg/kWh, if the calorific value of the fuel is 40,000 kJ/kg.

[**Ans.** $r = 5.2$, $\beta = 1.21$, fuel consumption = 0.3246 kg/kWh]

2.10 An oil engine operating on a Dual cycle has cylinder bore of 250 mm and stroke length of 300 mm. The compression and expansion ratios are 10 and 6 respectively. The initial pressure and temperature of the air are 1 bar and 27°C respectively. Determine the pressure and temperature at all salient points, the air-standard efficiency of the cycle, the mean effective pressure, and the power of the engine, if the working cycles per second are 12. Assume that the heat added at constant pressure is twice the heat added at constant volume.

[**Ans.** $p_1 = 1$ bar, $T_1 = 300$ K, $p_2 = 25.12$ bar, $T_2 = 753.6$ K, $p_3 = p_4 = 47.1$ bar, $T_3 = 1413$ K, $T_4 = 2355$ K, $p_5 = 3.83$ bar, $T_5 = 1150$ K, $\eta = 57\%$, $p_m = 10.45$ bar, power = 184.7 kW]

2.11 An oil engine operating on a Dual cycle has a compression ratio of 15 and compression begins at 1 bar, 30°C. The maximum pressure reached is 65 bar. The heat transferred to air at constant pressure is equal to that at constant volume. The expansion and compression follow the law $pV^{1.25}$ = constant. Determine the pressure and temperature at cardinal points, the cycle efficiency, and the mean effective pressure of the cycle.

[**Ans.** $p_1 = 1$ bar, $T_1 = 303$ K, $p_2 = 29.52$ bar, $T_2 = 596.3$ K, $p_3 = 65$ bar, $T_3 = 1313$ K, $p_4 = 65$ bar, $T_4 = 1825$ K, $p_5 = 3.32$ bar, $T_5 = 1007$ K, $\eta = 54.3\%$, $p_m = 9.235$ bar]

2.12 Determine the percentage increase in efficiency of an Atkinson cycle in comparison with the Otto cycle for a compression ratio of 8.5. The pressure and temperature of air at the beginning of compression are 1 bar and 27°C respectively. The peak pressure is 45 bar for both cycles. [**Ans.** 9.09%]

2.13 An air-standard Dual combustion cycle has a maximum cycle pressure of 65 bar. The minimum pressure and temperature are 1 bar and 22°C respectively. The compression ratio is 17 and the mean effective pressure of the cycle is 11 bar. Determine the maximum cycle temperature when the thermal efficiency of the cycle is 65%. [**Ans.** 2046°C]

2.14 Air undergoes an Atkinson cycle. The air is compressed adiabatically from 1 bar, 17°C to 5 bar. The maximum pressure of the cycle is 20 bar. Calculate the work done per kg of gas, the efficiency of the cycle, and the mean effective pressure.

[**Ans.** $W = 497$ kJ/kg, $\eta = 50.2\%$, $p_m = 8.74$ bar]

2.15 A four-cylinder, 2.0 *l* SI engine operates on an air-standard Miller cycle late valve closing. It has a compression ratio of 8 and an expansion ratio of 10. The maximum temperature of the cycle is 3500 K. When inlet valve closes, conditions in the cylinder are 100 kPa and 60°C. Calculate

(a) Temperature and pressure at all points in the cycle
(b) Power output when the cycle is repeated 600 times per minute
(c) Thermal efficiency
(d) Mean effective pressure

[**Ans.** (a) $T_2 = 765$ K, $p_2 = 1838$ kPa, $p_3 = 8409$ kPa, $T_4 = 1393$ K, $p_4 = 334.8$ kPa, $T_5 = 416$ K (b) 5.49 kW, (c) 60%, (d) 10.98 bar]

2.16 A closed-cycle gas turbine unit working on Brayton cycle operates with maximum and minimum temperatures of 800°C and 15°C and has a pressure ratio of 6. Calculate the ideal cycle efficiency and the work ratio. [**Ans.** 40.1%, 0.553]

3.1 INTRODUCTION

Combustion of the fuel-air mixture inside the engine cylinder is one of the processes that controls engine power, efficiency, and emissions. It is therefore essential to understand the combustion phenomenon. There are two types of chemical reactions. One is exothermic, in which heat energy is liberated and the other is endothermic, in which heat energy is absorbed. The combustion of fuel in internal combustion engines is a fast exothermic reaction in the gaseous phase where oxygen obtained from air is usually one of the reactants.

In the present chapter the thermodynamic analysis of the combustion process of fuel is carried out. The change of chemical energy into thermal energy is important for producing power in IC engines. The law of thermodynamics and the conservation of mass are used for the analysis.

In IC engines, the liquid and gaseous fuels are used. The liquid hydrocarbons such as gasoline, kerosene and diesel fuel are the common fuels. Any fuel, such as gasoline is actually a mixture of many hydrocarbons. Alcohols are sometimes used as fuels in IC engines. Gaseous hydrocarbon fuels are also a mixture of various constituent hydrocarbons.

3.2 PROPERTIES OF AIR

In IC engines, the combustion of fuel takes place in the presence of air and not pure oxygen. Air contains many constituents, particularly oxygen, nitrogen, argon and other vapours and inert gases. Its volumetric composition is approximately 21% O_2, 78% N_2, and 1% argon. Since neither nitrogen nor argon enters into the chemical reaction, it is sufficiently accurate to assume that the volumetric air proportions are 21% oxygen and 79% nitrogen and that for 100 moles of air, there are 21 moles of oxygen and 79 moles of nitrogen. That is,

$$\frac{\text{moles of } N_2}{\text{moles of } O_2} = \frac{79}{21} = 3.76$$

Therefore, for each mole of oxygen in air, there are 3.76 moles of nitrogen, the molecular weight of dry air is taken as 28.967 or 29. To account for argon, which is clubbed with nitrogen, the equivalent molecular weight of nitrogen is taken as 28.16. However, for the analysis, nitrogen in air may be considered as pure and the molecular weight of nitrogen can be taken as 28. In terms of mass, air contains approximately 23% oxygen and 77% nitrogen.

3.3 COMBUSTION WITH AIR

The general formula for the fuel used in IC engines can be taken as $C_mH_nO_p$, where m, n, and p represent the number of moles of carbon, hydrogen and oxygen atoms in a mole of fuel.

The basic stoichiometric (chemically correct) equation for fuel-air reaction is

$$C_mH_nO_p + Y_{cc}O_2 + 3.76Y_{cc}N_2 \rightarrow mCO_2 + \frac{n}{2}H_2O + 3.76Y_{cc}N_2 \quad (3.1)$$

where Y_{cc} is the chemically correct moles of O_2 per mole of fuel. The N_2 does not take part in the reaction. There are 3.76 moles of N_2 per mole of O_2 and since Y_{cc} moles of O_2 are theoretically necessary for the oxidation of the fuel, $3.76Y_{cc}$ moles of N_2 are present.
By balancing the number of moles of O_2 on both sides of Eq. (3.1),

$$\frac{p}{2} + Y_{cc} = m + \frac{n}{4}$$

$$\therefore \quad Y_{cc} = m + \frac{n}{4} - \frac{p}{2} \quad (3.2)$$

The stoichiometric equation now becomes

$$C_mH_nO_p + \left(m + \frac{n}{4} - \frac{p}{2}\right)O_2 + 3.76\left(m + \frac{n}{4} - \frac{p}{2}\right)N_2 \rightarrow mCO_2 + \frac{n}{2}H_2O + 3.76\left(m + \frac{n}{4} - \frac{p}{2}\right)N_2 \quad (3.3)$$

The stoichiometric equation for pure hydrocarbons (C_mH_n), where the oxygen atom is not present ($p = 0$) can be written as

$$C_mH_n + \left(m + \frac{n}{4}\right)O_2 + 3.76\left(m + \frac{n}{4}\right)N_2 \rightarrow mCO_2 + \frac{n}{2}H_2O + 3.76\left(m + \frac{n}{4}\right)N_2 \quad (3.4)$$

The combustion equation for octane (C_8H_{18}) can be written with $m = 8$ and $n = 18$ as

$$C_8H_{18} + 12.5O_2 + 3.76(12.5)N_2 \rightarrow 8CO_2 + 9H_2O + 3.76(12.5)N_2$$

or

$$C_8H_{18} + 12.5O_2 + 47N_2 \rightarrow 8CO_2 + 9H_2O + 47N_2 \quad (3.5)$$

In the above chemical equations, only a chemically correct amount of oxygen is included. The stoichiometric or the chemically correct amount of air required to oxidize the reactants is called the *theoretical air*. When combustion is achieved with theoretical air, no oxygen is obtained in the products. In practice, this is not possible. More oxygen than is theoretically necessary is required to achieve complete combustion of reactants. The excess air is needed because the fuel is of finite size, and each droplet must be surrounded by more than the necessary number of oxygen molecules to assure oxidation of all the fuel molecules. The excess air is usually expressed as a percentage of the theoretical air. Thus, if 20% more air than is theoretically required is used, this is expressed as 120% theoretical air or 20% excess air. There is 1.2 times as much air actually used than is theoretically required. The combustion of octane with 20% excess air can be written as

$$C_8H_{18} + 1.2(12.5)O_2 + 1.2 \times 47N_2 \rightarrow 8CO_2 + 9H_2O + 56.4N_2 + 2.5O_2$$

or

$$C_8H_{18} + 15O_2 + 56.4N_2 \rightarrow 8CO_2 + 9H_2O + 56.4N_2 + 2.5O_2 \quad (3.6)$$

In general, if Y moles of O_2 are supplied for complete combustion of 1 mole of fuel $C_mH_nO_p$ and $Y \geq Y_{cc}$ (excess air), the combustion equation can be written as

$$C_mH_nO_p + YO_2 + 3.76YN_2 \rightarrow mCO_2 + \frac{n}{2}H_2O + 3.76YN_2 + (Y - Y_{cc})O_2 \quad (3.7)$$

If the amount of air is insufficient to provide complete combustion, then all the carbon will not be oxidized to carbon dioxide but some carbon monoxide will also be formed. When incomplete combustion occurs, information about the products is necessary in order to balance the combustion equation. The products of combustion can be obtained by experimental methods. However, with certain assumptions the combustion equation can be written. These assumptions are: (i) all hydrogen appears as H_2O and (ii) all carbon is oxidized to CO. If any oxygen remains, part of the CO is oxidized to CO_2.

Let Y_{min} denotes the moles of the minimum allowable O_2 content in the reactants per mole of fuel, so that all H_2 is converted to H_2O and all C is converted to CO. There is no formation of CO_2 with Y_{min} mole of O_2. The combustion equation becomes

$$C_mH_nO_p + Y_{min}O_2 + 3.76Y_{min}N_2 \rightarrow mCO + \frac{n}{2}H_2O + 3.76Y_{min}N_2 \quad (3.8)$$

By balancing the moles of O_2 on both sides of Eq. (3.8),

$$\frac{p}{2} + Y_{min} = \frac{m}{2} + \frac{n}{4}$$

$$\therefore \quad Y_{min} = \frac{m}{2} + \frac{n}{4} - \frac{p}{2} = Y_{cc} - \frac{m}{2} \quad (3.9)$$

For insufficient amount of air, such that $Y_{min} \leq Y \leq Y_{cc}$, the combustion equation under the above assumptions can be written as

$$C_mH_nO_p + YO_2 + 3.76YN_2 = XCO + (m - X)CO_2 + \frac{n}{2}H_2O + 3.76YN_2 \quad (3.10)$$

By balancing the moles of O_2 on both sides of Eq. (3.10),

$$\frac{p}{2} + Y = \frac{X}{2} + (m - X) + \frac{n}{4}$$

$$\therefore \quad X = 2\left(m + \frac{n}{4} - \frac{p}{2} - Y\right) = 2(Y_{cc} - Y) \quad (3.11)$$

The moles of CO formed in the combustion product is therefore equal to $2(Y_{cc} - Y)$ and the moles of CO_2 formed in the combustion product is $m - 2(Y_{cc} - Y)$, which is equal to $2(Y - Y_{min})$.

Therefore, the combustion equation becomes

$$C_mH_nO_p + YO_2 + 3.76YN_2 \rightarrow 2(Y_{cc} - Y)CO + 2(Y - Y_{min})CO_2 + \frac{n}{2}H_2O + 3.76YN_2 \quad (3.12)$$

For a reactant mixture containing 1 mole of $C_mH_nO_p$, Y moles of O_2 and $3.76Y$ moles of N_2, the number of moles of products of combustion for the two cases are shown in Table 3.1.

Table 3.1 Moles of products of combustion

Name of the product constituent	*Number of moles of the product constituent*	
	Case 1 : Y ≥ Ycc [From Eq. (3.7)]	Case 2: Ymin ≤ Y ≤ Ycc [From Eq. (3.12)]
CO	N(1) = 0	N(1) = 2(Ycc – Y)
CO_2	N(2) = m	N(2) = 2 (Y – Ymin)
H_2O	$N(3) = \frac{n}{2}$	$N(3) = \frac{n}{2}$
N_2	N(4) = 3.76 Y	N(4) = 3.76 Y
O_2	N(5) = Y – Ycc	N(5) = 0

In Table 3.1, the subscripting system is adopted such that the correspondence between mole numbers and gases is

$$1 = CO, \quad 2 = CO_2, \quad 3 = H_2O, \quad 4 = N_2 \quad \text{and} \quad 5 = O_2$$

3.4 EQUIVALENCE RATIO

The stoichiometric equation defines theoretically the correct mixture of fuel and air for complete combustion. To allow for a mixture different from the correct mixture, the equivalence ratio ϕ is introduced. This is defined as the ratio of the actual fuel/air ratio $(F/A)_a$ to the stoichiometric fuel/air ratio $(F/A)_s$. It may also be defined as stoichiometric air/fuel ratio $(A/F)_s$ to actual air/fuel ratio $(A/F)_a$. Thus,

$$\phi = \frac{(F/A)_a}{(F/A)_s} = \frac{(A/F)_s}{(A/F)_a} \quad (3.13)$$

If the equivalence ratio ϕ is greater than unity, the mixture is said to be rich and if ϕ is less than unity the mixture is said to be weak. The spark-ignition engines may normally run with both rich and weak mixtures but the compression-ignition engines normally run with weak mixtures only. The inverse of equivalence ratio ϕ is called the relative air/fuel ratio λ. Therefore,

$$\lambda = \frac{1}{\phi} = \frac{(A/F)_a}{(A/F)_s} \quad (3.14)$$

For lean mixtures, $\phi < 1$, $\lambda > 1$

For stoichiometric mixtures, $\phi = \lambda = 1$

For rich mixtures, $\phi > 1$, $\lambda < 1$

The fuel/air ratio or the air/fuel ratio is normally expressed in the ratios of corresponding masses.

For octane (C_8H_{18}), the stoichiometric air/fuel ratio can be obtained from Eq. (3.5) as

$$(\text{Air/fuel ratio})_{\text{stoichiometric}} = \frac{\text{mass of air}}{\text{mass of fuel}} = \frac{(\text{mass of } O_2 + \text{mass of } N_2) \text{ per mole fuel}}{\text{mass of fuel per mole fuel}}$$

$$= \frac{(12.5 \times 32 + 47 \times 28)}{(12 \times 8 + 1 \times 18)} = \frac{400 + 1316}{114} = \frac{1716}{114} = 15.05$$

If 20% excess air is used in actual practice, then from Eq. (3.6)

$$(\text{Air/fuel ratio})_{\text{actual}} = \frac{15 \times 32 + 56.4 \times 28}{114} = 18.06$$

Equivalence ratio, $$\phi = \frac{15.05}{18.06} = 0.833$$

As the equivalence ratio is less than 1, it indicates a lean mixture, i.e. excess air is used for the combustion of fuel.

EXAMPLE 3.1 Methanol is burned with 20% excess air. Determine the stoichiometric air/fuel ratio and the actual air/fuel ratio. If the air is supplied at 1 bar and 27°C, calculate the volume of air supplied per kmole of fuel. Determine the molecular weight of the reactants and products. Also, determine the dew point of the products.

Solution: The stoichiometric combustion equation is

$$CH_3OH + 1.5O_2 + 1.5(3.76)N_2 \rightarrow CO_2 + 2H_2O + 5.64N_2$$

$$\therefore \quad \text{Stoichiometric air/fuel ratio} = \frac{1.5 \times 3.2 + 5.64 \times 28}{32} = 6.435$$

The combustion equation with 20% excess air is

$$CH_3OH + 1.2 \times 1.5O_2 + 1.2 \times 5.64N_2 \rightarrow CO_2 + 2H_2O + 6.768N_2 + 0.3O_2$$

or $$CH_3OH + 1.8O_2 + 6.768N_2 \rightarrow CO_2 + 2H_2O + 6.768N_2 + 0.3O_2$$

$$\therefore \quad \text{Actual air/fuel ratio} = \frac{1.8 \times 32 + 6.768 \times 28}{32} = 7.722$$

1 kmole of fuel (CH_3OH) reacts with 1.8 + 6.768 = 8.568 kmole of air.

$$\text{Volume of air, } V = \frac{n\bar{R}T}{p} = \frac{8.568 \times 8314 \times 300}{1 \times 10^5} = \boxed{213.7 \text{ m}^3/\text{kmole fuel}} \quad \textbf{Ans.}$$

The total number of moles in the reactants when excess air is supplied

$$= 1 \text{ mole of } CH_3OH + 1.8 \text{ moles of } O_2 + 6.768 \text{ moles of } N_2$$

$$= 9.568 \text{ moles}$$

Mole fraction of the species are:

$$CH_3OH = \frac{1}{9.568} = 0.1045$$

$$O_2 = \frac{1.8}{9.568} = 0.1881$$

$$N_2 = \frac{6.768}{9.568} = 0.7074$$

Molecular weight of reactants $= \Sigma x_i M_i$

$$= 0.1045 \times 32 + 0.1881 \times 32 + 0.7074 \times 28 = \boxed{29.17} \quad \textbf{Ans.}$$

Total number of moles in products

$$= 1 \text{ mole } CO_2 + 2 \text{ moles } H_2O + 6.768 \text{ moles } N_2 + 0.3 \text{ moles } O_2$$
$$= 10.068 \text{ moles}$$

Mole fraction of the products are:

$$CO_2 = \frac{1}{10.068} = 0.0993$$

$$H_2O = \frac{2}{10.068} = 0.1987$$

$$N_2 = \frac{6.768}{10.068} = 0.6722$$

$$O_2 = \frac{0.3}{10.068} = 0.0298$$

Molecular weight of products $= \Sigma x_i M_i$

$$= 0.0993 \times 44 + 0.1987 \times 18 + 0.6722 \times 28 + 0.0298 \times 32$$

$$= \boxed{27.72} \quad \textbf{Ans.}$$

Partial pressure of water vapour $= 0.1987 \times 1$ bar

$= 0.1987$ bar

The dew point is the saturation temperature corresponding to partial pressure of 0.1987 bar

$= \boxed{60°C}$ (obtained from steam tables). **Ans.**

EXAMPLE 3.2 A fuel mixture of 40% C_7H_{16} and 60% C_8H_{18} by volume is used in an engine having a bore of 120 mm and stroke length of 145 mm. The compression ratio is 8.5. The percentage composition of dry products of combustion by volume is CO_2 = 12%, CO = 1.5%, O_2 = 2.5% and the rest is N_2. Calculate the air/fuel ratio and the mass of the residual gases left in the cylinder at the end of the exhaust stroke if the pressure and temperature are 1.1 bar and 720 K.

Solution: Percentage of N_2 in the dry products of combustion

$$= 100 - (12 + 1.5 + 2.5) = 100 - 16 = 84\%$$

The combustion equation can be written as

$$0.4XC_7H_{16} + 0.6XC_8H_{18} + Y(O_2 + 3.76N_2) \rightarrow 12CO_2 + 1.5CO + 2.5O_2 + 84N_2 + ZH_2O$$

N_2 is not reacting,

$$\therefore \quad 3.76Y = 84$$

or

$$Y = 22.34$$

$\therefore$ 22.34 moles of O_2 is supplied.

By balancing C on both sides of the reaction,

$$2.8X + 4.8X = 13.5$$

$$\therefore \quad X = 1.776$$

By O_2 balance:

$$Y = 12 + 0.75 + 2.5 + \frac{Z}{2}$$

or

$$22.34 = 15.25 + \frac{Z}{2}$$

$$\therefore \quad Z = 14.18$$

The combustion equation now becomes

$$0.4(1.776)C_7H_{16} + 0.6(1.776)C_8H_{18} + 22.34(O_2 + 3.76N_2) \rightarrow 12CO_2 + 1.5CO + 2.5O_2 + 84N_2 + 14.18H_2O.$$

The combustion equation per mole of fuel can be written as

$$0.4C_7H_{16} + 0.6C_8H_{18} + 12.58(O_2 + 3.76N_2) \rightarrow 6.757CO_2 + 0.8446CO + 1.408O_2 + 47.3N_2 + 7.98H_2O$$

The above equation must be checked for H_2 balance as well.

No. of moles of H_2 on L.H.S. $= \dfrac{6.4 + 10.8}{2} = 8.6$

No. of moles of H_2 on R.H.S. = 7.98

Difference of H_2 moles = 8.6 – 7.98 = 0.62

This difference must be added to R.H.S. to balance the above combustion equation.

Let us assume that this H_2 is also present in the form of H_2O in products; corresponding to this 0.31 moles of O_2 is assumed to be present in reactants from the other sources. The balanced equation now becomes

$$0.4C_7H_{16} + 0.6C_8H_{18} + 12.58(O_2 + 3.76N_2) + 0.31O_2 \rightarrow 6.757CO_2 + 0.8446CO + 1.408O_2 + 47.3N_2 + 8.6H_2O$$

$$\therefore \quad \text{Air/fuel ratio} = \frac{12.58\,(32 + 3.76 \times 28)}{0.4 \times 100 + 0.6 \times 114} = \boxed{15.93} \quad \textbf{Ans.}$$

Swept volume of cylinder, $V_s = \dfrac{\pi}{4} d^2 L = \dfrac{\pi}{4}(0.12)^2(0.145) = 1.64 \times 10^{-3}\ \text{m}^3$

Clearance volume, $V_c = \frac{V_s}{r-1} = \frac{1.64 \times 10^{-3}}{7.5} = 0.2187 \times 10^{-3}\ \text{m}^3$

Molecular weight of the product = $\Sigma x_i M_i$

$$= \frac{6.757 \times 44 + 0.8446 \times 28 + 1.408 \times 32 + 47.3 \times 28 + 8.6 \times 18}{6.757 + 0.8446 + 1.408 + 47.3 + 8.6} = \frac{1845.2}{64.91}$$

$= 28.43$

Gas constant, $R = \frac{\bar{R}}{\text{mol.wt.}} = \frac{8314}{28.43} = 292.4$ J/(kg K)

Mass of the exhaust gases in the clearance space,

$$m = \frac{pV_c}{RT} = \frac{1.1 \times 10^5 \times 0.2187 \times 10^{-3}}{292.4 \times 720} = \boxed{0.1143 \times 10^{-3}\ \text{kg}}$$ **Ans.**

EXAMPLE 3.3 The analysis of a fuel is found to be carbon 86%, hydrogen 5%, oxygen 2%, sulphur 0.5% by weight and the remainder is nitrogen. Determine the weight of stoichiometric air required per kg of fuel for complete combustion. If the actual supply of air is 25% in excess of this, estimate the percentage of dry products of combustion by weight and by volume.

Solution: The combustion reactions can be written as

$$C + O_2 \rightarrow CO_2$$
$$2H_2 + O_2 \rightarrow 2H_2O$$
$$S + O_2 \rightarrow SO_2$$

1 kg of fuel contains 0.86 kg C, 0.05 kg H_2, 0.02 kg O_2, and 0.005 kg S.

$$N_2 \text{ in 1 kg fuel} = 1 - (0.86 + 0.05 + 0.02 + 0.005) = 1 - 0.935 = 0.065 \text{ kg}$$

12 kg C needs 32 kg O_2 and produces 44 kg CO_2; therefore, 0.86 kg C needs $\frac{32}{12} \times 0.86 = 2.293$ kg O_2 and produces $\frac{44}{12} \times 0.86 = 3.153$ kg CO_2.

4 kg H_2 needs 32 kg O_2 and produces 36 kg H_2O; therefore, 0.05 kg H_2 needs $\frac{32}{4} \times 0.05 = 0.4$ kg O_2 and produces $\frac{36}{4} \times 0.05 = 0.45$ kg H_2O.

32 kg S requires 32 kg O_2 and produces 64 kg SO_2; therefore, 0.005 kg S requires 0.005 kg O_2 and produces 0.01 kg SO_2.

Total oxygen required for the complete combustion of fuel

$$= 2.293 + 0.4 + 0.005 = 2.698 \text{ kg } O_2$$

The amount of oxygen required per kg of fuel for complete combustion (theoretically)

$$= 2.698 - 0.02 = 2.678 \text{ kg}$$

Amount of theoretical air required per kg of fuel

$$= \frac{2.678 \times 100}{23} = 11.64 \text{ kg}$$

$\therefore$ Stoichiometric air/fuel ratio = $\boxed{11.64}$ **Ans.**

Since 25% excess air is supplied, the actual quantity of air supplied per kg of fuel

$$= 11.64 \times 1.25 = 14.55 \text{ kg}$$

The combustion products will be:

$$CO_2 = 3.153 \text{ kg}$$
$$H_2O = 0.450 \text{ kg}$$
$$SO_2 = 0.010 \text{ kg}$$
$$O_2 \text{ in excess air} = 0.23 \times 0.25 \times 11.64 = 0.6693 \text{ kg}$$
$$N_2 \text{ in air} = 0.77 \times 1.25 \times 11.64 = 11.2035 \text{ kg}$$
$$N_2 \text{ in fuel} = 0.065 \text{ kg}$$
$$\text{Total } N_2 \text{ in exhaust} = 11.2035 + 0.065 = 11.2685 \text{ kg}$$

Omitting the wet gases, water vapour and SO_2 which are condensed, the products of combustion of dry gases will contain only CO_2, N_2, and O_2.

$$CO_2 = 3.1530 \text{ kg}$$
$$N_2 = 11.2685 \text{ kg}$$
$$O_2 = 0.6693 \text{ kg}$$
$$\text{Total weight} = 15.0908 \text{ kg}$$

Percentage composition by weight:

$$CO_2 = \frac{3.1530}{15.0908} \times 100 = \boxed{20.89\%}$$
$$N_2 = \frac{11.2685}{15.0908} \times 100 = \boxed{74.67\%}$$
$$O_2 = \frac{0.6693}{15.0908} \times 100 = \boxed{4.44\%}$$

Ans.

Proportionate moles of:

$$CO_2 = \frac{3.1530}{44} = 0.07166$$
$$N_2 = \frac{11.2685}{28} = 0.40245$$
$$O_2 = \frac{0.6693}{32} = 0.02092$$
$$\text{Total} = 0.49503 \text{ moles}$$

Volumetric analysis of dry products:

$$\left.\begin{aligned} CO_2 &= \frac{0.07166}{0.49503} \times 100 = \boxed{14.48\%} \\ N_2 &= \frac{0.40245}{0.49503} \times 100 = \boxed{81.30\%} \\ O_2 &= \frac{0.02092}{0.49503} \times 100 = \boxed{4.22\%} \end{aligned}\right\} \textbf{Ans.}$$

EXAMPLE 3.4 A hydrocarbon fuel has the following composition of dry products of combustion by volume:

$$CO_2 = 12\%, \qquad CO = 0.5\%, \qquad O_2 = 4\%, \qquad \text{and the rest } N_2.$$

Determine the air/fuel ratio, the per cent theoretical air and the percentage composition of fuel on a mass basis.

Solution: The combustion equation of an unknown hydrocarbon fuel can be written as

$$C_mH_n + xO_2 + yN_2 \rightarrow 12CO_2 + 0.5CO + 4O_2 + 83.5N_2 + zH_2O$$

C balance : $m = 12 + 0.5 = 12.5$

N_2 balance : $y = 83.5; \quad y = 3.76x, \quad \therefore x = \dfrac{83.5}{3.76} = 22.21$

O_2 balance : $22.21 = 12 + \dfrac{0.5}{2} + 4 + \dfrac{z}{2}$

$\therefore \quad z = 11.92$

H_2 balance: $\dfrac{n}{2} = z = 11.92$

$\therefore \quad n = 23.84$

The combustion equation becomes

$$C_{12.5}H_{23.84} + 22.21O_2 + 83.5N_2 \rightarrow 12CO_2 + 0.5CO + 4O_2 + 83.5N_2 + 11.92H_2O$$

$$\text{Air/fuel ratio} = \frac{(22.21)(32) + (83.5)(28)}{(12)(12.5) + 1(23.84)} = \frac{3048.72}{173.84} = \boxed{17.54} \quad \textbf{Ans.}$$

The stoichiometric combustion equation is

$$C_{12.5}H_{23.84} + 18.46O_2 + (18.46)(3.76)N_2 \rightarrow 12.5CO_2 + 11.92H_2O + 69.41N_2$$

$$\therefore \quad \text{Air/fuel ratio (stoichiometric)} = \frac{(18.46)(32) + (69.41)(28)}{173.84}$$

$$= 14.58$$

$$\text{Per cent theoretical air} = \frac{17.54}{14.58} \times 100 = \boxed{120.3\%} \quad \textbf{Ans.}$$

Fuel composition:

$$\left.\begin{aligned} C &= \frac{12.5 \times 12}{173.84} = 0.863 = \boxed{86.3\%} \\ H &= \frac{23.84}{173.84} = 0.137 = \boxed{13.7\%} \end{aligned}\right\} \textbf{Ans.}$$

EXAMPLE 3.5 A spark-ignition engine fuel has a composition of 86% carbon and 14% hydrogen by weight. The engine is supplied with a fuel having equivalence ratio of 1.25. Assuming that all the hydrogen is burnt and that the carbon burns to carbon monoxide and carbon dioxide so that there is no free carbon left, calculate the percentage analysis of dry exhaust gases by volume.

Solution: The fuel can be represented by a general formula C_mH_n. The molecular weight of the fuel = $12m + n$

$$\text{Fraction of C} = \frac{12m}{12m+n} = 0.86$$

or $$12m = 10.32m + 0.86n$$

or $$1.68m = 0.86n$$

$\therefore$ $$\frac{m}{n} = \frac{0.86}{1.68}$$

As the percentage composition of exhaust gas is required, therefore, the proportionate formula may be used to obtain the result

$$n = 1.953m$$

The proportionate formula for the fuel will therefore be $CH_{1.953}$.
For complete combustion,

$$CH_{1.953} + 1.488O_2 + 5.595N_2 \rightarrow CO_2 + \frac{1.953}{2}H_2O + 5.595N_2$$

Equivalence ratio, $$\phi = 1.25$$

Relative air/fuel ratio, $$\lambda = \frac{1}{\phi} = \frac{1}{1.25} = 0.8$$

It means 80% air is supplied in comparison to stoichiometric air.
The chemical equation becomes

$$CH_{1.953} + (1.488O_2 + 5.595N_2)(0.8) \rightarrow xCO + (1-x)CO_2 + 0.9765H_2O + 4.476N_2$$

By oxygen balance,

$$(1.488)(0.8) = \frac{x}{2} + (1-x) + \frac{0.9765}{2}$$

$\therefore$ $$x = 0.5957$$

$\therefore$ $$CH_{1.953} + 1.1904O_2 + 4.476N_2 \rightarrow 0.5957CO + 0.4043CO_2 + 0.9765H_2O + 4.476N_2$$

$\therefore$ Total number of moles of dry exhaust gas

$$= 0.5957 + 0.4043 + 4.476 = 5.476$$

Volumetric analysis of dry products of combustion:

$$\left.\begin{aligned} CO &= \frac{0.5957}{5.476} \times 100 = \boxed{10.88\%} \\ CO_2 &= \frac{0.4043}{5.476} \times 100 = \boxed{7.38\%} \\ N_2 &= \frac{4.476}{5.476} \times 100 = \boxed{81.74\%} \end{aligned}\right\} \textbf{Ans.}$$

Alternative Solution: Proportionate formula for the fuel = $C_{0.86}H_{1.68}$

$$m = 0.86, \quad n = 1.68$$

$$C_mH_n + Y_{cc}O_2 + 3.76Y_{cc}N_2 \rightarrow m\,CO_2 + mCO_2 + \frac{n}{2}H_2O + 3.76Y_{cc}N_2$$

Y_{cc} is the chemically correct moles of O_2 for complete combustion of fuel.

By O_2 balance: $\quad Y_{cc} = m + \frac{n}{4} = 0.86 + \frac{1.68}{4} = 0.86 + 0.42 = 1.28$

$$C_mH_n + Y_{min}O_2 + 3.76Y_{min}N_2 \rightarrow mCO + \frac{n}{2}H_2O + 3.76Y_{min}N_2$$

Y_{min} is the minimum number of moles of O_2 to convert all hydrogen to H_2O and all carbon to CO.

$$Y_{min} = \frac{m}{2} + \frac{n}{4} = \frac{0.86}{2} + \frac{1.68}{4} = 0.43 + 0.42 = 0.85$$

Equivalence ratio, $\quad \phi = 1.25$

Relative air/fuel ratio, $\quad \lambda = \frac{1}{\phi} = \frac{1}{1.25} = 0.8$

It means 80% air is supplied in comparison to stoichiometric air.

Therefore, moles of oygen actually supplied, $Y = 0.8Y_{cc}$

$\therefore$ $\quad Y = 0.8 \times 1.28 = 1.024$

Exhaust gas analysis of dry products:

$$\text{Number of moles of CO} = 2(Y_{cc} - Y) = 2(1.28 - 1.024) = 0.512$$

$$\text{Number of moles of } CO_2 = 2(Y - Y_{min}) = 2(1.024 - 0.85) = 0.348$$

$$\text{Number of moles of } N_2 = 3.76Y = 3.76 \times 1.024 = 3.85$$

$$\text{Total number of moles of dry products} = 0.512 + 0.348 + 3.85 = 4.71$$

Percentage analysis by volume,

$$\left.\begin{aligned} CO &= \frac{0.512}{4.71}\times 100 = \boxed{10.87\%} \\ CO_2 &= \frac{0.348}{4.71}\times 100 = \boxed{7.39\%} \\ N_2 &= \frac{3.85}{4.71}\times 100 = \boxed{81.74\%} \end{aligned}\right\} \textbf{Ans.}$$

3.5 ENTHALPY OF FORMATION

In dealing with the thermodynamic properties of a fixed chemical composition, tables were developed describing the properties of the substances. In each of these tables the thermodynamic properties are given relative to some arbitrary datum. In the steam tables, for example, the enthalpy of saturated liquid water at 0°C is assumed to be zero. This procedure is quite adequate when no change in compositions is involved, because we are concerned with the changes in the properties of a given substance. However, this procedure is not adequate when dealing with a chemical reaction, because of composition changes during the process. To overcome this difficulty, the enthalpy of all elements is assumed to be zero at an arbitrary reference state.

Enthalpy of formation $\bar{h}_f^\circ$ is the enthalpy of reaction for the formation of a substance from its elements in their most stable forms at an arbitrary reference state of 25°C and 1 atm. By stable forms of elements, we mean forms such as H_2, O_2 for hydrogen and oxygen instead of H and O. The stable form of carbon is graphite instead of diamond. The enthalpy of formation of any element in its most stable form at an arbitrary reference state of 25°C and 1 atm. is taken as zero. The justification of arbitrarily assigning the value of zero to the enthalpy of elements at 25°C and 1 atm. rests on the fact that in the absence of nuclear reactions the mass of each element is conserved in a chemical reaction. No conflicts arise with this choice of reference state and it proves to be very convenient in studying the chemical reactions from a thermodynamic point of view.

The enthalpy of formation is denoted by $\bar{h}_f^\circ$, where the superscript° refers to the standard pressure. The bar over h represents the molar enthalpy, i.e. the enthalpy per unit mole.

Consider a steady-state combustion process in which 1 mole of CO_2 is formed from its elements of 1 mole of C and 1 mole of O_2 at the reference state of 25°C and 1 atm. Heat is transferred, so the CO_2 finally exists at the reference state. The reaction equation can be written as

$$C + O_2 \rightarrow CO_2 \tag{3.15}$$

Let H_R be the total enthalpy of all the reactants and H_P be the total enthalpy of all the products. Q is the heat transfer required to carry out the reaction. The first law of thermodynamics for this process, having no work transfer and no kinetic energy change is

$$Q + H_R = H_P \tag{3.16}$$

For the reaction Eq. (3.15), the enthalpy of all the reactants is zero, since they are all elements, hence $H_R = 0$.

$$\therefore \qquad Q = H_P = -\,393.52 \text{ MJ/kmol} \tag{3.17}$$

Heat transfer Q has been carefully measured and found to be –393.52 MJ/kmol. Actually, the enthalpy of formation is usually found by the application of statistical thermodynamics, using the observed spectroscopic data. The negative sign indicates that heat is liberated in the formation of CO_2 from its elements. Therefore, the enthalpy of formation of CO_2 at the standard state of 25°C and 1 atm. is – 393.52 MJ/kmol. That is,

$$(\bar{h}_f^\circ)_{CO_2} = -\ 393.52 \text{ MJ/kmol} \tag{3.18}$$

The negative sign for the enthalpy of formation is due to the reaction being exothermic. The enthalpy of reactants is zero at 25°C and 1 atm. Therefore, the enthalpy of CO_2 at 25°C and 1 atm. must be negative.

Table 3.2 lists the enthalpy of formation for several different substances at 25°C and 1 atm. pressure. In front of the formula, g and l within parantheses represent the gaseous and liquid state respectively.

Table 3.2 Enthalpy of formation at 25°C and 1 atm.

Substance	*Formula*	$\bar{h}_f^\circ$ *(MJ/kmol)*	*Substance*	*Formula*	$\bar{h}_f^\circ$ *(MJ/kmol)*
Carbon monoxide	$CO(g)$	–110.52	Cetane	$C_{16}H_{34}(l)$	– 45.45
Carbon dioxide	$CO_2(g)$	–393.52	Acetylene	$C_2H_2(g)$	+ 226.87
Water	$H_2O(l)$	–285.8	Ethene	$C_2H_4(g)$	+ 52.32
Water	$H_2O(g)$	–241.82	Propylene	$C_3H_6(g)$	+ 20.41
Methane	$CH_4(g)$	–74.87	Benzene	$C_6H_6(g)$	+ 82.98
Ethane	$C_2H_6(g)$	– 84.68	Methanol	$CH_3OH(g)$	– 201.17
Propane	$C_3H_8(g)$	–103.85	Methanol	$CH_3OH(l)$	– 238.58
Butane	$C_4H_{10}(g)$	–126.22	Ethanol	$C_2H_5OH(g)$	– 208.45
Isooctane	$C_8H_{18}(g)$	–224.1	Ethanol	$C_2H_5OH(l)$	– 249.35
Isooctane	$C_8H_{18}(l)$	– 259.28	Ammonia	$NH_3(g)$	– 45.93
n-Octane	$C_8H_{18}(g)$	– 208.58	Hydrazine	$N_2H_4(g)$	+ 95.41
n-Octane	$C_8H_{18}(l)$	–250.10	Hydrogen peroxide	$H_2O_2(g)$	– 136.31
Dodecane	$C_{12}H_{26}(g)$	–290.97			

3.6 FIRST LAW ANALYSIS FOR STEADY-STATE REACTING SYSTEMS

The steady-state combustion process will transfer heat and may produce work. An energy balance would give

$$Q + H_R = W + H_P$$

or

$$Q + \sum_R n_i \bar{h}_i = W + \sum_P n_j \bar{h}_j \tag{3.19}$$

The enthalpies of formation may be used in this analysis, since they are all relative to the same reference base.

Table 3.3 Change in enthalpy $(\bar{h}^\circ - \bar{h}^\circ_{298})$ between the reference state and the actual state for different substances (MJ/kmol)

Temp, T (K)	CO	CO_2	H_2O	N_2	O_2
0	– 8.669	– 9.364	– 9.904	– 8.669	– 8.682
100	– 5.770	– 6.456	– 6.615	– 5.770	– 5.778
200	– 2.858	– 3.414	– 3.280	– 2.858	– 2.866
298	0.000	0.000	0.000	0.000	0.000
300	0.054	0.067	0.063	0.054	0.054
400	2.975	4.008	3.452	2.971	3.029
500	5.929	8.314	6.920	5.912	6.088
600	8.941	12.916	10.498	8.891	9.247
700	12.021	17.761	14.184	11.937	12.502
800	15.175	22.815	17.991	15.046	15.841
900	18.397	28.041	21.924	18.221	19.246
1000	21.686	33.405	25.978	21.460	22.707
1100	25.033	38.894	30.167	24.757	26.217
1200	28.426	44.484	34.476	28.108	29.765
1300	31.865	50.158	38.903	31.501	33.351
1400	35.338	55.907	43.447	34.936	36.966
1500	38.848	61.714	48.095	38.405	40.610
1600	42.384	67.580	52.844	41.903	44.279
1700	45.940	73.492	57.685	45.430	47.970
1800	49.522	79.442	62.609	48.982	51.689
1900	53.124	85.429	67.613	52.551	55.434
2000	56.739	91.450	72.689	56.141	59.199
2100	60.375	97.500	77.831	59.748	62.986
2200	64.019	103.575	83.036	63.371	66.802
2300	67.676	109.671	88.295	67.007	70.634
2400	71.340	115.788	93.604	70.651	74.492
2500	75.023	121.926	98.964	74.312	78.375
2600	78.714	128.085	104.370	77.973	82.274
2700	82.408	134.256	109.813	81.659	86.199
2800	86.115	140.444	115.294	85.345	90.144
2900	89.826	146.645	120.813	89.036	94.111
3000	93.542	152.862	126.361	92.738	98.098
3200	100.998	165.331	137.553	100.161	106.127
3400	108.479	177.849	148.854	107.608	114.232
3600	115.976	190.405	160.247	115.031	122.399
3800	123.495	202.999	171.724	122.570	130.629
4000	131.026	215.635	183.280	130.076	138.913
4200	138.578	228.304	194.903	137.603	147.248
4400	146.147	241.003	206.585	145.143	155.628
4600	153.724	253.734	218.325	152.699	164.046
4800	161.322	266.500	230.120	160.272	172.502
5000	168.929	279.295	241.957	167.858	180.987

In most of the cases, the reactants and the products are not at the reference condition of 25°C and 1 atm. In these cases the property change between the reference state and the actual state must be accounted for. Table 3.3 shows the change in enthalpy between the reference state and the actual state, i.e. $(\bar{h}^\circ - \bar{h}^\circ_{298})$. The temperature 298 K is the temperature of the reference state. The superscript ° denotes that the pressure is 1 atm. If these tabulated values are not available, then the ideal gas law, $\bar{h} = \bar{c}_p T$, can be used.

The first law for a steady-state, steady-flow reaction may now be written as

$$Q + \sum_{R} n_i [\bar{h}^\circ_f + (\bar{h}^\circ - \bar{h}^\circ_{298})]_i = W + \sum_{P} n_j [\bar{h}^\circ_f + (\bar{h}^\circ - \bar{h}^\circ_{298})]_j \tag{3.20}$$

The pressure effect is very small and need not be included.

EXAMPLE 3.6 Gaseous propane (C_3H_8) undergoes a steady-state, steady-flow reaction with atmospheric air. Determine the heat transfer per mole of fuel when the reactants and products are both at 25°C and 1 atm. pressure.

Solution: The combustion equation can be written as

$$C_3H_8 + 5O_2 + 5(3.76)N_2 \rightarrow 3CO_2 + 4H_2O(l) + 18.8N_2$$

At 25°C, H_2O will be in the liquid state.

The first law, with no work being done in the control volume is

$$Q + \sum_{R} n_i \bar{h}_i = \sum_{P} n_j \bar{h}_j$$

Since the enthalpy of all the elements at 25°C and 1 atm. is zero, the summation terms contain only the enthalpies of formation for the compounds. Using Table 3.2,

$$\sum_{R} n_i \bar{h}_i = (\bar{h}^\circ_f)_{C_3H_8} = -\,103.85 \text{ MJ/kmol}$$

and

$$\sum_{P} n_j \bar{h}_j = 3(\bar{h}^\circ_f)_{CO_2} + 4(\bar{h}^\circ_f)_{H_2O(l)} = 3(-393.52) + 4(-285.8)$$

$$= -\,2323.7 \text{ MJ/kmol}$$

$$Q = \sum_{P} n_j \bar{h}_j - \sum_{R} n_i \bar{h}_i = -\,2323.7 + 103.85$$

$$= -\,2219.85 \text{ MJ/kmol fuel}$$

$$= \boxed{-2219.85 \text{ kJ/mol fuel}}$$ **Ans.**

EXAMPLE 3.7 The gaseous fuel $C_{12}H_{26}$ and the air enter the diesel engine at 25°C. The products of combustion leave at 600 K, and 200% theoretical air is used. The heat loss from the engine is 93 MJ/kmol fuel. Determine the work for a fuel rate of 1 kmol/h.

Solution: The combustion equation can be written as

$$C_{12}H_{26}(g) + 2(18.5)O_2 + 2(18.5)(3.76)N_2 \rightarrow 12CO_2 + 13H_2O + 18.5O_2 + 139.12N_2$$

H_2O at 600 K and at standard pressure will be in gaseous state.

The first law for open system can be written as

$$Q + \sum_{R} n_i [\bar{h}_f^o + (\bar{h}^o - \bar{h}_{298}^o)]_i = W + \sum_{P} [\bar{h}_f^o + (\bar{h}^o - \bar{h}_{298}^o)]_j$$

$$Q = -93 \text{ MJ/kmol fuel} \qquad \text{(given)}$$

The negative sign indicates heat loss from the system.

Using Tables 3.2 and 3.3,

$$\sum_{R} n_i [\bar{h}_f^o + (\bar{h}^o - \bar{h}_{298}^o)]_i = 1(\bar{h}_f^o)_{C_{12}H_{26}} = -290.97 \text{ MJ/kmol} = H_R$$

and

$$\begin{aligned}\sum_{P} n_j [\bar{h}_f^o + (\bar{h}^o - \bar{h}_{298}^o)]_j &= 12[\bar{h}_f^o + (\bar{h}^o - \bar{h}_{298}^o)]_{CO_2} + 13[\bar{h}_f^o + (\bar{h}^o - \bar{h}_{298}^o)]_{H_2O} \\ &\quad + 18.5[\bar{h}_f^o + (\bar{h}^o - \bar{h}_{298}^o)]_{O_2} + 139.12[\bar{h}_f^o + (\bar{h}^o - \bar{h}_{298}^o)]_{N_2} \\ &= 12(-393.52 + 12.916) + 13(-241.82 + 10.498) \\ &\quad + 18.5(0 + 9.247) + 139.12(0 + 8.891) \\ &= -4567 - 3007 + 171 + 1237 \\ &= -6166 \text{ MJ/kmol} = H_P\end{aligned}$$

$$Q + H_R = W + H_P$$

or

$$-93 - 290.97 = W - 6166$$

$\therefore$

$$W = 5782 \text{ MJ/kmol fuel}$$

$$\dot{W} = \dot{n}_f W = 1(5782) = 5782 \text{ MJ/h}$$

where

$$\dot{n}_f = 1 \text{ kmol/h}$$

$\therefore$

$$\dot{W} = \frac{5782 \times 10^3}{3600} = \boxed{1606 \text{ kW}}$$ **Ans.**

EXAMPLE 3.8 An engine produces 600 kW of power and uses isooctane $C_8H_{18}(l)$ as a fuel at 25°C; 150% theoretical air is used and the air enters at 400 K. The products of combustion leave at 700 K. The heat loss from the engine is 150 kW. Determine the fuel consumption for complete combustion.

Solution: The combustion equation can be written as

$$C_8H_{18} + 1.5(12.5)O_2 + 1.5(12.5)(3.76)N_2 \rightarrow 8CO_2 + 9H_2O + 6.25O_2 + 70.5N_2$$

Using Tables 3.2 and 3.3,

$$\begin{aligned}H_R = \sum_{R} n_i [\bar{h}_f^o + (\bar{h}^o - \bar{h}_{298}^o)]_i &= 1(-259.28) + (1.5)(12.5)(3.029) + 70.5(2.971) \\ &= -259.28 + 56.79 + 209.46 = 6.97 \text{ MJ/kmol fuel}\end{aligned}$$

$$H_P = \sum_P n_j[\bar{h}_f^o + (\bar{h}^o - \bar{h}_{298}^o)]_j = 8(-393.52 + 17.761) + 9(-241.82 + 14.184)$$

$$+ 6.25(12.502) + 70.5(11.937)$$

$$= -3005.9 + (-2048.5) + 78.1 + 841.6$$

$$= -4134.7 \text{ MJ/kmol fuel}$$

$$\therefore \quad H_P - H_R = -4134.7 - 6.97 = -4141.67 \text{ MJ/kmol fuel}$$

The first law for open system can be written as

$$\dot{Q} + \dot{n}_f H_R = \dot{W} + \dot{n}_f H_P$$

or
$$\dot{Q} - \dot{W} = \dot{n}_f (H_P - H_R)$$

or
$$\dot{n}_f = \frac{\dot{Q} - \dot{W}}{H_P - H_R} = \frac{-150 - 600}{-4141.67 \times 10^3} \text{ kmol/s}$$

$$= \frac{750 \times 3600}{4141.67 \times 10^3} = 0.6519 \text{ kmol/h}$$

Molecular weight of fuel $= (12)(8) + 18 = 114$

$$\therefore \quad \dot{m}_f = \dot{n}_f \times M_f = 0.6519 \times 114 = \boxed{74.32 \text{ kg/h}}$$ **Ans.**

EXAMPLE 3.9 A mixture of propane and oxygen, in the proper ratio, for the complete combustion and at 25°C and 1 atm., reacts in a constant volume bomb calorimeter. Heat is transferred until the products of combustion are at 400 K. Determine the heat transfer per mole of propane.

Solution: The combustion equation can be written as

$$C_3H_8 + 5O_2 \rightarrow 3CO_2 + 4H_2O$$

The first law for closed system, V = constant, can be written as

$$Q = U_P - U_R = \sum_P n_j \bar{u}_j - \sum_R n_i \bar{u}_i$$

$$\bar{u} = \bar{h} - \bar{R}T$$

Now, $$U_R = \sum_R n_i \bar{u}_i = \sum_R n_i [\bar{h}_f^o + (\bar{h}^o - \bar{h}_{298}^o) - \bar{R}T]_i$$

$$= 1[-103.85 + 0 - 8.314 \times 10^{-3} \times 298] + 5[0 + 0 - 8.314 \times 10^{-3} \times 298]$$

$$= -106.33 - 12.39 = -118.72 \text{ MJ/kmol fuel}$$

and $$U_P = \sum_P n_j \bar{u}_j = \sum_P n_j [\bar{h}_f^o + (\bar{h}^o - \bar{h}_{298}^o) - \bar{R}T]_j$$

$$= 3(-393.52 + 4.008 - 8.314 \times 10^{-3} \times 400) + 4(-241.82 + 3.452 - 8.314 \times 10^{-3} \times 400)$$

$$= -1178.4 - 966.7 = -2145.1 \text{ MJ/kmol fuel}$$

$$\therefore \quad Q = U_P - U_R = -2145.1 + 118.72 = -2026.4 \text{ MJ/kmol fuel}$$

The negative sign indicates that heat is liberated and the reaction is exothermic.

Heat liberated = 2026.4 MJ/kmol propane

= 2026.4 kJ/mol propane **Ans.**

3.7 ENTHALPY OF COMBUSTION, INTERNAL ENERGY OF COMBUSTION AND HEATING VALUES

The enthalpy of combustion ΔH is the difference between the enthalpy of products and the enthalpy of the reactants when complete combustion of unit quantity of fuel occurs at a given temperature and pressure, i.e.

$$\Delta H = H_P - H_R$$

$$\therefore \quad \Delta H = \sum_{P} n_j [\bar{h}_f^o + (\bar{h}_T^o - \bar{h}_{298}^o)]_j - \sum_{R} n_i [\bar{h}_f^o + (\bar{h}_T^o - \bar{h}_{298}^o)]_i \tag{3.21}$$

The standard values of enthalpy of combustion $\Delta\bar{h}_c^o$ of various fuels at a temperature of 25°C and pressure of 1 atm. are given in Table 3.4.

Table 3.4 Enthalpy of vaporization $\bar{h}_{fg}$ and enthalpy of combustion $\Delta\bar{h}_c^o$ at 25°C and 1 atm.

Fuel	$\bar{h}_{fg}$ (kJ/kmol)	$\Delta\bar{h}_c^o$ (kJ/kmol)	
		$H_2O(l)$	$H_2O(g)$
Hydrogen, $H_2(g)$	–	– 285812	– 241826
Carbon, C(*s*)	–	– 393552	– 393552
Carbon monoxide, CO(*g*)	–	– 283022	– 283022
Methane, $CH_4(g)$	–	– 890303	– 802231
Acetylene, $C_2H_2(g)$	–	– 1299650	– 1255660
Ethylene, $C_2H_4(g)$	–	– 1411200	– 1323220
Ethane, $C_2H_6(g)$	5021	– 1560690	– 1428730
Propane, $C_3H_8(g)$	14820	– 2219230	– 2043290
n-Butane, $C_4H_{10}(g)$	21066	– 2877620	– 2657690
n-Pentane, $C_5H_{12}(g)$	26426	– 3556980	– 3293060
n-Hexane, $C_6H_{14}(g)$	31552	– 4195050	– 3887150
n-Heptane, $C_7H_{16}(g)$	36547	– 4853710	– 4501820
n-Octane, $C_8H_{18}(g)$	41484	– 5511900	–5116030
n-Octane, $C_8H_{18}(l)$	41484	– 5470420	– 5074540
Benzene, $C_6H_6(g)$	33849	– 3301680	– 3169730
Methanol, $CH_3OH(g)$	39790	– 764242	– 676234
Methanol, $CH_3OH(l)$	39790	– 726271	– 638264
Ethanol, $C_2H_5OH(g)$	42560	–1409584	–1277572
Ethanol, $C_2H_5OH(l)$	42560	– 1367024	– 1235012
Hydrazine, $N_2H_4(l)$	44760	– 622250	– 534278

Equation (3.21) can also be written as-

$$\Delta H = \left(\sum_{P} n_j \bar{h}^{\circ}_{f_j} - \sum_{R} n_i \bar{h}^{\circ}_{f_i}\right) + \sum_{P} n_j (\bar{h}^{\circ}_T - \bar{h}^{\circ}_{298})_j - \sum_{R} n_i (\bar{h}^{\circ}_T - \bar{h}^{\circ}_{298})_i$$

or
$$\Delta H = \Delta \bar{h}^{\circ}_c + \sum_{P} n_j (\bar{h}^{\circ}_T - \bar{h}^{\circ}_{298})_j - \sum_{R} n_i (\bar{h}^{\circ}_T - \bar{h}^{\circ}_{298})_i \quad (3.22)$$

where
$$\Delta \bar{h}^{\circ}_c = \sum_{P} n_j \bar{h}^{\circ}_{f_j} - \sum_{R} n_i \bar{h}^{\circ}_{f_i} \quad (3.23)$$

Equation (3.23) suggests that standard enthalpy of combustion can be obtained with the help of enthalpies of formation.

Take an example of methane, CH_4. The combustion equation for methane can be written as

$$CH_4 + 2O_2 \rightarrow CO_2 + 2H_2O(l)$$

$$\begin{aligned} \Delta \bar{h}^{\circ}_c &= \sum_{P} n_j \bar{h}^{\circ}_{f_j} - \sum_{R} n_i \bar{h}^{\circ}_{f_i} \\ &= n_{CO_2} \bar{h}^{\circ}_{f_{CO_2}} + n_{H_2O} \bar{h}^{\circ}_{f_{H_2O(l)}} - n_{CH_4} \bar{h}^{\circ}_{f_{CH_4}} \\ &= (1)(-393.52) + 2(-285.8) - (1)(-74.87) \\ &= -890.25 \text{ MJ/kmol} \end{aligned}$$

This value of $\Delta \bar{h}^{\circ}_c$ for methane, when H_2O in the product is liquid, almost matches with the corresponding value given in Table 3.4.

The internal energy of combustion, ΔU, is the difference between the internal energy of products and the internal energy of the reactants, when complete combustion of unit quantity of fuel occurs at a given temperature and volume. That is,

$$\Delta U = U_P - U_R = [H - pV]_P - [H - pV]_R$$

$$\therefore \quad \Delta U = \sum_{P} n_j [\bar{h}^{\circ}_f + (\bar{h}^{\circ}_T - \bar{h}^{\circ}_{298}) - p\bar{v}]_j - \sum_{R} n_i [\bar{h}^{\circ}_f + (\bar{h}^{\circ}_T - \bar{h}^{\circ}_{298}) - p\bar{v}]_i \quad (3.24)$$

If all the products and reactants are gases,

$$\Delta U = \sum_{P} n_j [\bar{h}^{\circ}_f + (\bar{h}^{\circ}_T - \bar{h}^{\circ}_{298}) - \bar{R}T]_j - \sum_{R} n_i [\bar{h}^{\circ}_f + (\bar{h}^{\circ}_T - \bar{h}^{\circ}_{298}) - \bar{R}T]_i \quad (3.25)$$

Equations (3.21) and (3.25) give the relation between ΔH and ΔU, i.e.

$$\Delta U = \Delta H - \bar{R}T\left(\sum_{P} n_j - \sum_{R} n_i\right)$$

or
$$\Delta H = \Delta U + \Delta n \bar{R}T \quad (3.26)$$

where
$$\Delta n = \sum_{P} n_j - \sum_{R} n_i \quad (3.27)$$

It should be noted that for fuels both ΔH and ΔU are negative. These reactions are exothermic as heat is liberated by combustion of fuels, and by convention the heat released is taken as negative and the heat added as positive. ΔH may also be called as heat of reaction at constant pressure in a closed system or heat of reaction in a flow process and ΔU as heat of reaction at constant volume in a closed system. ΔH and ΔU are the properties of the fuel; they refer specifically to chemically correct oxygen mixtures reacting at 25°C and are measured quantities.

The heating value or calorific value of a fuel is equal to the enthalpy of combustion but is of opposite sign. The heating value is a positive number. There are two types of heating values, depending on the phase of water formed in the products of combustion. The higher heating value (HHV) is measured when all the water vapour is condensed at the reference temperature of 25°C. If the water is in the vapour phase, the lower heating value (LHV) is measured. The relation between the two heating values is given by

$$\text{HHV} - \text{LHV} = m_{H_2O} h_{fg25°C}$$

where m_{H_2O} is the mass of water in kg formed in the combustion product per kg of fuel. The value of h_{fg} at 25°C is obtained as 2442 kJ/kg from steam tables. The heating values are in kJ/kg of fuel.

EXAMPLE 3.10 Determine the enthalpy of combustion (enthalpy of reaction) for

$$CO + \frac{1}{2}O_2 \rightarrow CO_2 \quad \text{at 1500 K}$$

Solution: Figure 3.1 shows the ΔH at 1500 K on the *H*–*T* diagram.

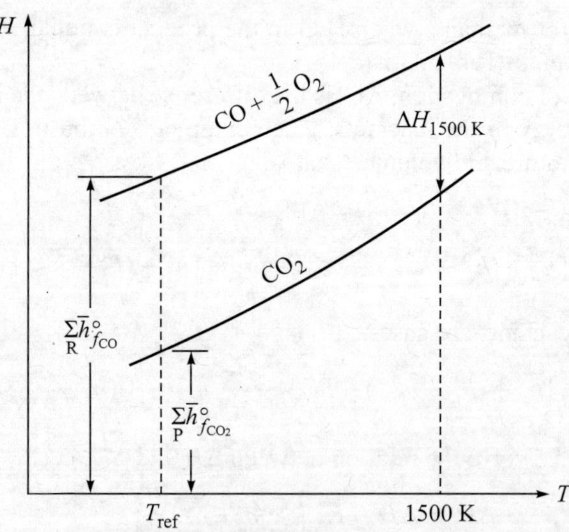

Figure 3.1 Enthalpy–Temperature diagram: Example 3.10.

$$\Delta H = H_P - H_R$$

$$H_P = \sum_P n_j \bar{h}_j = \sum_P n_j [\bar{h}_f^\circ + (\bar{h}_T^\circ - \bar{h}_{298}^\circ)]_j$$

$$= n_{CO_2} [\bar{h}_f^\circ + (\bar{h}_T^\circ - \bar{h}_{298}^\circ)]_{CO_2}$$

$$= -393.52 + 61.714 = -331.806 \text{ MJ/kmol}$$

$$H_R = \sum_R n_i \bar{h}_i = \sum_R n_i [\bar{h}_f^\circ + (\bar{h}_T^\circ - \bar{h}_{298}^\circ)]_i$$

$$= n_{CO}[\bar{h}_f^\circ + (\bar{h}_T^\circ - \bar{h}_{298}^\circ)]_{CO} + n_{O_2}[\bar{h}_f^\circ + (\bar{h}_T^\circ - \bar{h}_{298}^\circ)]_{O_2}$$

$$= (-110.52 + 38.848) + 0.5(0 + 40.610)$$

$$= -51.367 \text{ MJ/kmol}$$

$$\therefore \quad \Delta H = -331.806 + 51.367 = \boxed{-280.439 \text{ MJ/kmol CO}} \quad \textbf{Ans.}$$

EXAMPLE 3.11 Determine the amount of heat transfer per kg of fuel during the complete combustion of propane in an open steady-flow system with 30% excess air. Propane enters at 400 K and air enters at 300 K and the products leave at 900 K. The average molar specific heat of propane at constant pressure may be taken as 83.7 kJ/(kmol K).

Solution: Figure 3.2 shows the flow process in an open, steady-flow system.

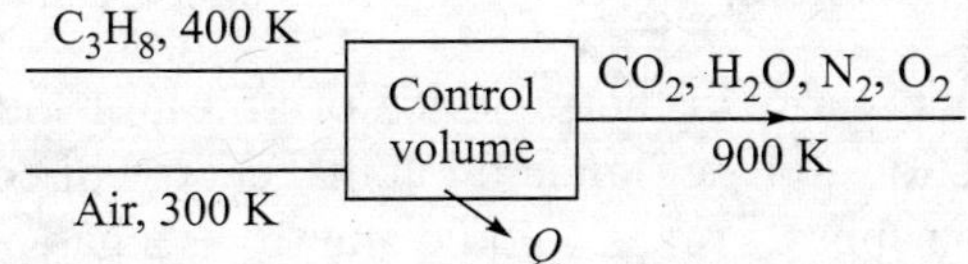

Figure 3.2 Open, steady-flow system: Example 3.11.

The reaction equation is

$$C_3H_8 + (1.3)5O_2 + (1.3)(5)(3.76)N_2 \rightarrow 3CO_2 + 4H_2O(g) + 24.44N_2 + 1.5O_2$$

or

$$C_3H_8 + 6.5O_2 + 24.44N_2 \rightarrow 3CO_2 + 4H_2O(g) + 24.44N_2 + 1.5O_2$$

The value of pressure is not given, therefore, we assume ideal gas behaviour. Since the products leave at 900 K, the water will be considered in the gaseous state.

Energy balance gives, $Q = H_P - H_R$, where

$$H_P = \sum_P n_j [\bar{h}_f^\circ + (\bar{h}_T^\circ - \bar{h}_{298}^\circ)]_j$$

$$= n_{CO_2}[\bar{h}_f^\circ + (\bar{h}_{900}^\circ - \bar{h}_{298}^\circ)]_{CO_2} + n_{H_2O}[\bar{h}_f^\circ + (\bar{h}_{900}^\circ - \bar{h}_{298}^\circ)]_{H_2O}$$

$$+ n_{N_2}[\bar{h}_f^\circ + (\bar{h}_{900}^\circ - \bar{h}_{298}^\circ)]_{N_2} + n_{O_2}[\bar{h}_f^\circ + (\bar{h}_{900}^\circ - \bar{h}_{298}^\circ)]_{O_2}$$

$$= 3(-393.52 + 28.041) + 4(-241.82 + 21.924)$$

$$+ 24.44(0 + 18.221) + 1.5(0 + 19.246)$$

$$= -1501.83 \text{ MJ/kmol}$$

$$H_R = \sum_R n_i [\bar{h}_f^\circ + (\bar{h}_T^\circ - \bar{h}_{298}^\circ)]_i$$

$$= n_{C_3H_8}[\bar{h}_f^o + (\bar{h}_{400}^o - \bar{h}_{298}^o)]_{C_3H_8} + n_{O_2}[\bar{h}_f^o + (\bar{h}_{300}^o - \bar{h}_{298}^o)]_{O_2}$$

$$+ n_{N_2}[\bar{h}_f^o + (\bar{h}_{300}^o - \bar{h}_{298}^o)]_{N_2}$$

$$= 1[-103.85 + 0.0837(400 - 298)] + 6.5(0 + 0.054) + 24.44(0 + 0.054)$$

$$= -93.64 \text{ MJ/kmol}$$

$$\therefore \quad Q = H_P - H_R$$

$$= -1501.83 + 93.64 = -1408.19 \text{ MJ/kmol}$$

$$= \frac{-1408.19}{44} = -32 \text{ MJ/kg}$$

The negative sign shows that heat is liberated.

$\therefore$ Heat liberated = $\boxed{32 \text{ MJ/kg}}$ **Ans.**

Note: This heat liberated is not the heat of reaction or enthalpy of reaction, as the products are not brought back to the initial temperature. Moreover, the reactants are also at different temperatures.

EXAMPLE 3.12 Methane burns in the presence of 150% of theoretical air. Calculate the standard enthalpy of combustion and the standard internal energy of combustion, when the water in the products of combustion appears (a) as a liquid and (b) as a gas.

Solution: The chemical equation is

$$CH_4 + (1.5)(2)O_2 + 3(3.76)N_2 \rightarrow CO_2 + 2H_2O + O_2 + 11.28N_2$$

or

$$CH_4 + 3O_2 + 11.28N_2 \rightarrow CO_2 + 2H_2O + O_2 + 11.28N_2$$

(a) Water appears as a liquid:

$$H_P = \sum_P n_j[\bar{h}_f^o + (\bar{h}_T^o - \bar{h}_{298}^o)]_j$$

At standard condition, $\bar{h}_T^o = \bar{h}_{298}^o$

$$\therefore \quad H_P = \sum_P n_j \bar{h}_{f_j}^o = 1 \cdot \bar{h}_{f_{CO_2}}^o + 2 \cdot \bar{h}_{f_{H_2O(l)}}^o \qquad (\text{For } O_2 \text{ and } N_2,\ \bar{h}_f^o = 0)$$

$$= -393.52 + 2(-285.8) = -965.12 \text{ MJ/kmol}$$

and

$$H_R = \sum_R n_i \bar{h}_{f_i}^o = 1 \cdot \bar{h}_{f_{CH_4}}^o = -74.87 \text{ MJ/kmol}$$

$$\therefore \quad \Delta H = \Delta \bar{h}_c^o = H_P - H_R = -965.12 + 74.87$$

$$= \boxed{-890.25 \text{ MJ/kmol}} \quad \textbf{Ans.}$$

This value is approximately the same as given in Table 3.4.

Now,

$$\Delta U = \Delta H - \Delta n \bar{R} T$$

No. of moles of product = 2 (1 for CO_2 and 1 for O_2, no. of moles of N_2 cancells out from the two sides and the no. of moles of H_2O being liquid is neglected)

No. of moles of reactant = 4 (1 for CH_4 and 3 for O_2)

$$\Delta n = 2 - 4 = -2$$

$$\therefore \quad \Delta U = -890.25 - (-2)(8.314 \times 10^{-3})(298)$$

$$= \boxed{-885.29 \text{ MJ/kmol}} \quad \textbf{Ans.}$$

(b) Water appears as a gas:

$$H_P = 1.(-393.52) + 2 \cdot (-241.82)$$

$$= -877.16 \text{ MJ/kmol}$$

$$\therefore \quad \Delta H = \Delta \bar{h}_c^{\circ} = H_P - H_R = -877.16 + 74.87$$

$$= \boxed{-802.29 \text{ MJ/kmol}} \quad \textbf{Ans.}$$

This value of $\Delta \bar{h}_c^{\circ}$ is approximately the same as given in Table 3.4.

Now, $\Delta U = \Delta H - \Delta n \bar{R} T$

No. of moles of product = 4 (1 for CO_2, 2 for H_2O and 1 for O_2)

No. of moles of reactant = 4

$$\Delta n = 4 - 4 = 0$$

$$\therefore \quad \Delta U = \Delta H = \boxed{-802.29 \text{ MJ/kmol}} \quad \textbf{Ans.}$$

EXAMPLE 3.13 The lower calorific value of a liquid fuel at constant pressure is 44,000 kJ/kg. The analysis of fuel by mass is 84% carbon and 16% hydrogen. Determine the higher calorific value at constant pressure and the lower and higher calorific values at constant volume. At 25°C, h_{fg} for H_2O is 2442 kJ/kg.

Solution: The combustion equations are:

$$C + O_2 \rightarrow CO_2$$

and

$$H_2 + \frac{1}{2} O_2 \rightarrow H_2O$$

0.84 kg of C produces $0.84 \times \frac{44}{12} = 3.08$ kg of CO_2 per kg of fuel,

and 0.16 kg of H_2 produces $0.16 \times \frac{18}{2} = 1.44$ kg of H_2O per kg of fuel

$$(\text{HHV})_p - (\text{LHV})_p = m_{H2O} h_{fg}$$

$$= 1.44 \times 2442 = 3516 \text{ kJ/kg fuel}$$

$$\therefore \quad (\text{HHV})_p = 44{,}000 + 3516$$

$$= \boxed{47{,}516 \text{ kJ/kg fuel}} \quad \textbf{Ans.}$$

The combustion equation becomes,

$$0.84 \text{ kg of C} + 0.16 \text{ kg of H}_2 + \left(0.84 \times \frac{32}{12} + 0.16 \times \frac{16}{2}\right) \text{kg of O}_2 \rightarrow 3.08 \text{ kg CO}_2 + 1.44 \text{ kg H}_2\text{O } (l)$$

where H_2O is taken as liquid at 25°C.

or $$0.84 \text{ kg C} + 0.16 \text{ kg H}_2 + 3.52 \text{ kg O}_2 \rightarrow 3.08 \text{ kg CO}_2 + 1.44 \text{ kg H}_2\text{O}(l)$$

Note that mass is conserved. On both sides of reaction, the total mass is 4.52 kg. The volume occupied by liquid is negligible in comparison to gases, therefore the moles of liquid H_2O and liquid fuels are neglected.

No. of moles of product, $$n_P = \frac{3.08}{44} = 0.07 \text{ kmol/kg fuel}$$

and no. of moles of reactant, $$n_R = \frac{3.52}{32} = 0.11 \text{ kmol/kg fuel}$$

$$\Delta n = n_P - n_R = 0.07 - 0.11 = -0.04 \text{ kmol}$$

$$\Delta n\bar{R}T = -0.04 \times 8.314 \times 298 = -99 \text{ kJ/kg fuel}$$

$$\Delta H = \Delta U + \Delta n\bar{R}T$$

If the calorific vaue is denoted by CV,

$$-(\text{CV})_p = -(\text{CV})_v + \Delta n\bar{R}T$$

or $$(\text{CV})_v = (\text{CV})_p + \Delta n\bar{R}T$$

∴ $$(\text{HHV})_v = (\text{HHV})_p + \Delta n\bar{R}T$$

$$= 47{,}516 - 99 = \boxed{47{,}417 \text{ kJ/kg fuel}}$$ **Ans.**

and $$(\text{LHV})_v = (\text{LHV})_p + \Delta n\bar{R}T$$

$$= 44{,}000 - 99 = \boxed{43{,}901 \text{ kJ/kg fuel}}$$ **Ans.**

3.8 ADIABATIC COMBUSTION TEMPERATURE

The calculation of the temperature of the products of a combustion reaction is very important for the design of internal combustion engines. If no work, no heat transfer, no change in potential energy and kinetic energy occur, then all the thermal energy goes to raise the temperature of the products of combustion. When the combustion is complete, the maximum amount of chemical energy is converted into heat energy and the temperature of the products reaches its maximum. We can use this temperature as the upper limit for the actual combustion temperature and then this temperature is called the maximum adiabatic combustion or flame temperature.

If the combustion is incomplete or excess air is used, the temperature of the products of combustion will be less than the maximum adiabatic combustion temperature. Incomplete combustion restricts the conversion of all the chemical energy into thermal energy and excess air will have cooling effect, hence the temperature will be lower than the maximum temperature in both the cases. Dissociation of the combustion products at high temperature also reduces the flame temperature.

The adiabatic flame temperature can be calculated as follows:

Consider a steady-flow adiabatic combustion reaction in which there is no work transfer. Neglecting the changes in potential energy and kinetic energy and applying the first law of thermodynamics,

$$Q = H_P - H_R$$

As the process is adiabatic, $Q = 0$.

$$\therefore \quad H_P - H_R = 0$$

or

$$H_P = H_R$$

$$\therefore \quad \sum_P n_j[\bar{h}_f^\circ + (\bar{h}_T^\circ - \bar{h}_{298}^\circ)]_j = \sum_R n_i[\bar{h}_f^\circ + (\bar{h}_T^\circ - \bar{h}_{298}^\circ)]_i \tag{3.28}$$

Figure 3.3 shows the process on *H*–*T* diagram for steady-flow adiabatic combustion.

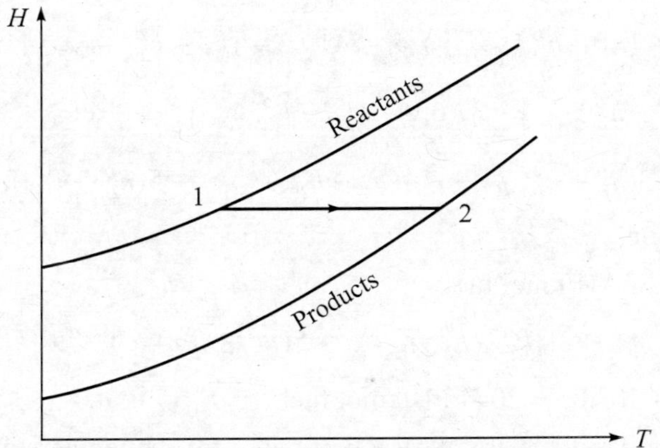

Figure 3.3 Steady-flow adiabatic combustion.

In a closed system for constant pressure process, the adiabatic flame temperature can be obtained from $H_P = H_R$ and for a constant volume process it can be obtained from $U_P = U_R$.

A trial-and-error method can be used to obtain adiabatic flame temperature. Initially, some product temperature is guessed and then the validity of Eq. (3.28) is checked. If this equation is not satisfied, a new value of temperature is guessed. When Eq. (3.28) is satisfied, the guessed temperature is the adiabatic flame temperature.

A computer program using the Newton–Raphson iteration technique can also be used to obtain the adiabatic flame temperature.

EXAMPLE 3.14 Gaseous propane is burned with 100 per cent excess air. Determine the adiabatic flame temperature for steady-flow process.

Solution: The combustion equation with 100% excess air is

$$C_3H_8 + 2(5)O_2 + 2(5)(3.76)N_2 \rightarrow 3CO_2 + 4H_2O + 5O_2 + 37.6N_2$$

or $$C_3H_8 + 10O_2 + 37.6N_2 \rightarrow 3CO_2 + 4H_2O + 5O_2 + 37.6N_2$$

The product temperature is normally very high, so water may be considered in a gaseous state. The temperature of the reactants is not given; we will therefore assume that the reactants are at 25°C ($T = 298$ K).

$$H_R = \sum_R n_i[\bar{h}_f^o + (\bar{h}_T^o - \bar{h}_{298}^o)]_i$$

$$= \sum_R n_i \bar{h}_{f_i}^o = n_{C_3H_8}\, \bar{h}_{f_{C_3H_8}}^o = 1\cdot(-103.85) = -103.85 \text{ MJ/kmol fuel}$$

$$H_P = \sum_P n_j[\bar{h}_f^o + (\bar{h}_T^o - \bar{h}_{298}^o)]_j$$

$$= n_{CO_2}[\bar{h}_f^o + (\bar{h}_T^o - \bar{h}_{298}^o)]_{CO_2} + n_{H_2O}[\bar{h}_f^o + (\bar{h}_T^o - \bar{h}_{298}^o)]_{H_2O(g)}$$

$$+ n_{O_2}[\bar{h}_f^o + (\bar{h}_T^o - \bar{h}_{298}^o)]_{O_2} + n_{N_2}[\bar{h}_f^o + (\bar{h}_T^o - \bar{h}_{298}^o)]_{N_2}$$

$$= 3[-393.52 + (\bar{h}_T^o - \bar{h}_{298}^o)_{CO_2}] + 4[-241.82 + (\bar{h}_T^o - \bar{h}_{298}^o)_{H_2O(g)}]$$

$$+ 5[0 + (\bar{h}_T^o - \bar{h}_{298}^o)_{O_2}] + 37.6[0 + (\bar{h}_T^o - \bar{h}_{298}^o)_{N_2}]$$

$$= -2147.84 + 3(\bar{h}_T^o - \bar{h}_{298}^o)_{CO_2} + 4(\bar{h}_T^o - \bar{h}_{298}^o)_{H_2O(g)} + 5(\bar{h}_T^o - \bar{h}_{298}^o)_{O_2} + 37.6(\bar{h}_T^o - \bar{h}_{298}^o)_{N_2}$$

For adiabatic combustion,

$$H_P = H_R = -103.85 \text{ MJ/kmol fuel}$$

$$\therefore \quad 3(\bar{h}_T^o - \bar{h}_{298}^o)_{CO_2} + 4(\bar{h}_T^o - \bar{h}_{298}^o)_{H_2O(g)} + 5(\bar{h}_T^o - \bar{h}_{298}^o)_{O_2} + 37.6(\bar{h}_T^o - \bar{h}_{298}^o)_{N_2}$$

$$= 2147.84 - 103.85 = 2044 \text{ MJ/kmol fuel}$$

Most of the combustion products are N_2, the first guess for the temperature can be obtained by assuming that the products of combustion contain nitrogen only.

$$\therefore \quad 37.6(\bar{h}_T^o - \bar{h}_{298}^o) = 2044$$

or $$\bar{h}_T^o - \bar{h}_{298}^o = 54.36 \text{ MJ/kmol}$$

Corresponding to this value the temperature from Table 3.3 is approximately equal to 1900 K. Since CO_2 and H_2O in the products have higher specific heats than N_2, they will absorb more heat energy than N_2. Hence, the temperature will be much lower than 1900 K.

For first trial, assume $T = 1500$ K; at this temperature,

$$H_P = -2147.84 + 3(61.714) + 4(48.095) + 5(40.610) + 37.6(38.405)$$

$$= -123.24 \text{ MJ/kmol fuel}$$

This value of H_P is lower than –103.85 MJ/kmol fuel

Try $T = 1600$ K; at this temperatre,

$$H_P = -2147.84 + 3(67.580) + 4(52.844) + 5(44.279) + 37.6(41.903)$$

$$= 63.22 \text{ MJ/kmol fuel}$$

This value of H_P is higher than –103.85 MJ/kmol of fuel. Therefore, the adiabatic flame temperature lies between 1500 K and 1600 K. This can be estimated by linear interpolation as follows:

$$\frac{T-1500}{1600-1500} = \frac{-103.85-(-123.24)}{63.22-(-123.24)}$$

$$\therefore \qquad T = 1500 + \frac{19.39 \times 100}{186.46}$$

$$= 1500 + 10.4$$

$$= \boxed{1510.4 \text{ K}} \quad \textbf{Ans.}$$

EXAMPLE 3.15 Calculate the adiabatic flame temperature, when a chemically correct mixture of methane and air initially at 600 K and 1 atm. burns at (a) constant pressure process and (b) constant volume process.

Specific heat at constant pressure can be obtained from the relation:

$$\frac{c_p}{R} = a + bT + cT^2 + dT^3 + eT^4$$

For methane, $a = 4.503$, $b = -8.965 \times 10^{-3}$, $c = 37.38 \times 10^{-6}$, $d = -36.49 \times 10^{-9}$ and $e = 12.22 \times 10^{-12}$.

Solution: The combustion equation becomes

$$CH_4 + 2O_2 + 2(3.76)N_2 \rightarrow CO_2 + 2H_2O + 7.52N_2$$

$$H_P = \sum_P n_j[\bar{h}_f^\circ + (\bar{h}_T^\circ - \bar{h}_{298}^\circ)]_j = \sum_P n_j[\bar{h}_f^\circ + \Delta\bar{h}^\circ]$$

where $\Delta\bar{h}^\circ = \bar{h}_T^\circ - \bar{h}_{298}^\circ$.

$$\therefore \qquad H_P = 1(-393.52 + \Delta\bar{h}_{CO_2}^\circ) + 2(-241.82 + \Delta\bar{h}_{H_2O}^\circ) + 7.52(0 + \Delta\bar{h}_{N_2}^\circ)$$

$$= \Delta\bar{h}_{CO_2}^\circ + \Delta\bar{h}_{H_2O}^\circ + 7.52\Delta\bar{h}_{N_2}^\circ - 877.16 \text{ MJ/kmol fuel}$$

$$\Delta\bar{h}_{CH_4}^\circ = \int_{298}^{600} cp\,dT = \bar{R}\int_{298}^{600} (a + bT + cT^2 + eT^4)dT$$

$$\therefore \qquad \Delta\bar{h}_{CH_4}^\circ = 8.314\left[a(600-298) + \frac{b}{2}\times 10^4(6^2 - 2.98^2) + \frac{c}{3}\times 10^6(6^3 - 2.98^3) + \frac{d}{4}\times 10^8(6^4 - 2.98^4) + \frac{e}{5}\times 10^{10}(6^5 - 2.98^5)\right]$$

$$= 8.314\left[4.503 \times 302 + \left(-\frac{8.965}{2}\right)10(6^2 - 2.98^2) + \frac{37.38}{3}(6^3 - 2.98^3) - \frac{36.49}{4 \times 10}(6^4 - 2.98^4) + \frac{12.22}{500}(6^5 - 2.98^5)\right]$$

$$= 8.314(1360 - 1215.6 + 2361.6 - 1110.3 + 184.3)$$

$$= 131{,}36 \text{ kJ/kmol} = 13.136 \text{ MJ/kmol}$$

$$H_R = \sum_R n_i [\bar{h}_f^\circ + \Delta\bar{h}^\circ]_i$$

$$= 1(-74.87 + 13.136) + 2(9.247) + 7.52(8.891) = 23.62 \text{ MJ/kmol fuel}$$

(a) For constant pressure process, $H_P = H_R = 23.62$ MJ/kmol fuel

$$\therefore \quad \Delta\bar{h}_{CO_2} + 2\Delta\bar{h}^\circ_{H_2O} + 7.52\Delta\bar{h}^\circ_{N_2} = 877.16 + 23.62 = 900.78 \text{ MJ/kmol fuel}$$

To guess the required temperature, assume

$$7.52\Delta\bar{h}^\circ_{N_2} = 900.78$$

$$\therefore \quad \Delta\bar{h}^\circ_{N_2} = \frac{900.78}{7.52} = 119.8 \text{ MJ/kmol}$$

Corresponding to this $T \simeq 3700$ K (from Table 3.3)
Adiabatic combustion temperature is much below this value.

Try, $T = 2800$ K

$$H_P = \Delta\bar{h}^\circ_{CO_2} + 2\Delta\bar{h}^\circ_{H_2O} + 7.52\,\Delta\bar{h}^\circ_{N_2} - 877.16$$

$$= 140.444 + 2(115.294) + 7.52(85.345) - 877.16 = 135.67 \text{ MJ/kmol fuel}$$

Here, H_p is much higher than H_R, so try with a smaller value of temperature.

Try, $T = 2500$ K

$$H_P = 121.926 + 2(98.964) + 7.52(74.312) - 877.16 = 1.52 \text{ MJ/kmol fuel}$$

Here, H_p is less than H_R, so try with a little higher temperature.

Try, $T = 2600$ K

$$H_P = 128.085 + 2(104.37) + 7.52(77.973) - 877.16 = 46.02 \text{ MJ/kmol fuel}$$

Here, H_p is more than H_R. So, the required temperature lies between 2500 K and 2600 K. By linear interpolation,

$$\frac{T - 2500}{2600 - 2500} = \frac{23.62 - 1.52}{46.02 - 1.52}$$

$$\therefore \quad T = 2500 + \frac{100 \times 22.1}{44.5} = 2549.7 \text{ K} \simeq \boxed{2550 \text{ K}} \quad \textbf{Ans.}$$

(b) For constant volume process:

Internal energy of reactant, $U_R = H_R - (n\bar{R}T)_R$

$$\therefore \quad U_R = 23.62 - 10.52 \times 8.314 \times 10^{-3} \times 600 = -28.86 \text{ MJ/kmol fuel}$$

Internal energy of products, $U_P = H_P - (n\bar{R}T)_p$

$$= H_P - 10.52 \times 8.314 \times 10^{-3}\, T_P = H_P - 0.08746\, T_P$$

$$\therefore \quad U_P = \Delta\bar{h}^\circ_{CO_2} + 2\Delta\bar{h}^\circ_{H_2O} + 7.52\,\Delta\bar{h}^\circ_{N_2} - 877.16 - 0.08746\,T_P$$

The reactants burned at constant pressure transfer work to the surroundings, whereas the reactants burned at constant volume do not transfer work to the surroundings. Therefore, the adiabatic flame temperature at constant volume is higher than that obtained at constant pressure process.

For constant volume process, $U_P = U_R = -28.86$ MJ/kmol fuel

Try, $T = 3000$ K

$$\therefore \quad U_P = 152.862 + 2(126.361) + 7.52(92.738) - 877.16 - 0.08746 \times 3000$$
$$= -36.57 \text{ MJ/kmol fuel}$$

This value of U_P is less than U_R. Therefore, try with an increased value of T.

Try, $T = 3200$ K

$$\therefore \quad U_P = 165.331 + 2(137.553) + 7.52(100.161) - 877.16 - 0.08746 \times 3200$$
$$= 36.62 \text{ MJ/kmol fuel}$$

This value of U_P is more than U_R. Therefore, the required temperature lies between 3000 K and 3200 K.

By linear interpolation,

$$\frac{T - 3000}{3200 - 3000} = \frac{-28.86 + 36.57}{36.62 + 36.57}$$

$$\therefore \quad T = 3000 + \frac{200 \times 7.71}{73.19} = 3000 + 21 = \boxed{3021 \text{ K}} \quad \textbf{Ans.}$$

EXAMPLE 3.16 Calculate the adiabatic flame temperature when methane burns in the presence of air at constant pressure process at 600 K and 1 atm., having (a) 50% excess air and (b) 20% less air, leading to incomplete combustion. Calculate the loss of thermal energy due to incomplete combustion. Take the mean value of $\bar{C}_p$ for methane as 52.234 kJ/kmol K.

Solution: (a) The combustion equation with 50% excess air for 1 kmol of fuel is

$$CH_4 + 3O_2 + 3 \times 3.76N_2 \rightarrow CO_2 + 2H_2O + 11.28N_2 + O_2$$

$$H_R = \sum_R n_i[\bar{h}_f^o + (\bar{h}_T^o - \bar{h}_{298}^o)]_i$$
$$= 1[-74.87 + 52.234(600 - 298) \times 10^{-3}]$$
$$+ 3(0 + 9.247) + 11.28(0 + 8.891)$$
$$= 68.936 \text{ MJ}$$

$$H_P = \sum_P n_j[\bar{h}_f^o + (\bar{h}_T^o - \bar{h}_{298}^o)]_j = \sum_P n_j[\bar{h}_f^o + \Delta\bar{h}^o]_j$$

where $\Delta\bar{h}^o = \bar{h}_T^o - \bar{h}_{298}^o$

$$\therefore \quad H_P = 1(-393.52 + \Delta\bar{h}_{CO_2}^o) + 2(-241.82 + \Delta\bar{h}_{H_2O}^o)$$
$$+ 11.28(0 + \Delta\bar{h}_{N_2}^o) + 1(0 + \Delta\bar{h}_{O_2}^o)$$

$$= \Delta\bar{h}^{\circ}_{CO_2} + 2\Delta\bar{h}^{\circ}_{H_2O} + 11.28\Delta\bar{h}^{\circ}_{N_2} + \Delta\bar{h}^{\circ}_{O_2} - 877.16 \text{ MJ}$$

For constant pressure adiabatic combustion, $H_P = H_R$.

$$\therefore \quad \Delta\bar{h}^{\circ}_{CO_2} + 2\Delta\bar{h}^{\circ}_{H_2O} + 11.28\Delta\bar{h}^{\circ}_{N_2} + \Delta\bar{h}^{\circ}_{O_2} = 877.16 + 68.936 = 946.096 \text{ MJ}$$

To guess the required temperature, assume

$$11.28\Delta\bar{h}^{\circ}_{N_2} = 946.096$$

i.e.
$$\Delta\bar{h}^{\circ}_{N_2} = \frac{946.096}{11.28} = 83.87 \text{ MJ/kmol}$$

Corresponding to this, we have $T \simeq 2800$ K.

The adiabatic combustion temperature is much below this value due to the presence of other constituents in the product.

Try $\quad T = 2000$ K

$$\therefore \quad H_P = 91.45 + 2(72.689) + 11.28(56.141) + 59.199 - 877.16$$
$$= 52.137 \text{ MJ} < H_R$$

Increase the temperature. Try, $T = 2100$ K.

$$\therefore \quad H_P = 97.5 + 2(77.831) + 11.28(59.748) + 62.986 - 877.16$$
$$= 112.95 \text{ MJ} > H_R$$

The required temperature therefore lies between 2000 K and 2100 K.

By linear interpolation,

$$\frac{T - 2000}{2100 - 2000} = \frac{68.936 - 52.137}{112.95 - 52.137}$$

or
$$T = 2000 + \frac{16.799 \times 100}{60.813} = 2000 + 27.6 = \boxed{2027.6 \text{ K}} \quad \textbf{Ans.}$$

(b) For incomplete combustion with 20% less air, i.e. with 80% of chemically correct air. For 1 kmol. of fuel, the combustion equation becomes

$$CH_4 + 0.8 \times 2O_2 + 1.6 \times 3.76N_2 \rightarrow XCO + (1 - X)CO_2 + 2H_2O + 6.016N_2$$

By oxygen balance:
$$1.6 = \frac{X}{2} + 1 - X + 1 \quad \text{or} \quad X = 0.8$$

The combustion equation becomes

$$CH_4 + 1.6O_2 + 6.016N_2 \rightarrow 0.8CO + 0.2CO_2 + 2H_2O + 6.016N_2$$

$$H_R = \sum_R n_i[\bar{h}^{\circ}_f + (\bar{h}^{\circ}_T - \bar{h}^{\circ}_{298})]_i$$

$$= 1[-74.87 + 52.234\,(600 - 298) \times 10^{-3}] + 1.6(0 + 9.247) + 6.016(0 + 8.891)$$

$$= 9.188 \text{ MJ}$$

$$H_P = \sum_P n_j[\bar{h}_f^\circ + (\bar{h}_T^\circ - \bar{h}_{298}^\circ)]_j$$

$$= 0.8(-110.52 + \Delta\bar{h}_{CO}^\circ) + 0.2(-393.52 + \Delta\bar{h}_{CO_2}^\circ)$$

$$+ 2(-241.82 + \Delta\bar{h}_{H_2O}^\circ) + 6.016(0 + \Delta\bar{h}_{N_2}^\circ)$$

$$= 0.8\Delta\bar{h}_{CO}^\circ + 0.2\Delta\bar{h}_{CO_2}^\circ + 2\Delta\bar{h}_{H_2O}^\circ + 6.016\Delta\bar{h}_{N_2}^\circ - 650.76$$

Try $T = 2000$ K

$$H_P = 0.8(56.739) + 0.2(91.45) + 2(72.689) + 6.016(56.141) - 650.76$$

$$= -103.96 \text{ MJ} < H_R$$

Increase the temperature. Try $T = 2400$ K.

$$H_P = 0.8(71.34) + 0.2(115.788) + 2(93.604) + 6.016(70.651) - 650.76$$

$$= 41.71 \text{ MJ} > H_R$$

Reduce the temperature. Try $T = 2300$ K.

$$H_P = 0.8(67.676) + 0.2(109.671) + 2(88.295) + 6.016(67.007) - 650.76$$

$$= 5.019 \text{ MJ} < H_R$$

The required temperature therefore lies between 2300 K and 2400 K.

By linear interpolation,

$$\frac{T - 2300}{2400 - 2300} = \frac{9.188 - 5.019}{41.710 - 5.019}$$

$$\therefore \quad T = 2300 + \frac{4.169 \times 100}{36.691} = 2300 + 11.4 = \boxed{2311.4 \text{ K}}$$ **Ans.**

(Note that the adiabatic flame temperature obtained with excess air and less air are less than those obtained with stoichiometric air as obtained in Example 3.15, $T = 2550$ K.)

The loss of thermal energy due to incomplete combustion is the enthalpy of combustion of the combustible constituents in the products.

The only combustible substance is CO and from Table 3.4, $\Delta\bar{h}_c^\circ$ for CO is -283.022 MJ/kmol.

$$\therefore \quad \text{Thermal energy loss} = 0.8 \times 283.022 = \boxed{226.4 \text{ MJ/kmol fuel}}$$ **Ans.**

3.9 DISSOCIATION

At high temperatures, products dissociate into smaller constituents. For example, CO_2 dissociates into CO and O_2 with absorption of heat energy. When CO and O_2 combine to form CO_2, heat energy is liberated and when CO_2 dissociates into CO and O_2, heat energy is absorbed and therefore the temperature of the products decreases. When chemical equilibrium is reached, the reaction proceeds in both directions, so there is no net change in either the reactants or the products. The equilibrium equation is written as

$$CO + \frac{1}{2}O_2 \rightleftharpoons CO_2$$

A definite proportion of CO, O_2 and CO_2 exists in the equilibrium mixture at each temperature. At higher temperatures, the amount of CO and O_2 increases and that of CO_2 decreases. Because of this, the theoretical temperature calculated on the assumption that all the carbon present in the fuel has been converted into CO_2 (if sufficient oxygen is present) cannot be obtained.

Consider a weak mixture of CO and O_2. The final composition of the mixture at equilibrium can be obtained as follows:

If n is the number of excess moles of oxygen in the mixture, then the original mixture will be $CO + \left(\frac{1}{2} + n\right)O_2$.

Reaction without dissociation can be written as

$$CO + \left(\frac{1}{2} + n\right)O_2 \rightarrow CO_2 + nO_2 \tag{3.29}$$

If α is the degree of dissociation of CO_2, then αCO_2 will dissociate as follows.

$$\alpha CO_2 \rightarrow \alpha CO + \frac{\alpha}{2}O_2$$

$(1 - \alpha)CO_2$ is not dissociated.

The final products will be:

$$(1-\alpha)CO_2 + \alpha CO + \frac{\alpha}{2}O_2 + nO_2$$

The overall reaction can be written as

$$CO + \left(\frac{1}{2} + n\right)O_2 \rightarrow (1-\alpha)CO_2 + \alpha CO + \left(n + \frac{\alpha}{2}\right)O_2 \tag{3.30}$$

Consider a rich mixture of CO and O_2. The final composition of the mixture at equilibrium can be obtained as follows:

Let n be the number of moles of excess CO in the mixture, then the original mixture will be $(1 + n)CO + \frac{1}{2}O_2$.

Reaction without dissociation can be written as

$$(1+n)CO + \frac{1}{2}O_2 \rightarrow nCO + CO_2 \tag{3.31}$$

Let α be the degree of dissociation of CO_2, then αCO_2 will dissociate as follows.

$$\alpha CO_2 \rightarrow \alpha CO + \frac{\alpha}{2}O_2.$$

$(1 - \alpha)CO_2$ is not dissociated.

The final products will be: $(1-\alpha)CO_2 + \alpha CO + \frac{\alpha}{2}O_2 + nCO$

The overall reaction can be written as

$$(1+n)CO + \frac{1}{2}O_2 \rightarrow (1-\alpha)CO_2 + (n+\alpha)CO + \frac{\alpha}{2}O_2 \quad (3.32)$$

EXAMPLE 3.17 Determine the final composition of the mixture at equilibrium and write the overall reaction, when the correct mixture of octane (C_8H_{18}) and oxygen burns and α_1 and α_2 are the degrees of dissociations of CO_2 and H_2O respectively. Assume that CO_2 is dissociated into CO and O_2 and H_2O is dissociated into H_2 and O_2.

Solution: Reaction without dissociation:

$$C_8H_{18} + 12.5O_2 \rightarrow 8CO_2 + 9H_2O$$

Consider the dissociation of CO_2:

$$\alpha_1 CO_2 \rightarrow \alpha_1 CO + \frac{\alpha_1}{2}O_2$$

$\therefore$ $8\alpha_1$ moles of CO_2 will dissociate as

$$8\alpha_1 CO_2 \rightarrow 8\alpha_1 CO + 4\alpha_1 O_2$$

and $8(1-\alpha_1)CO_2$ will not dissociate and will remain in the products.

Consider the dissociation of H_2O,

$$H_2O \rightarrow H_2 + \frac{1}{2}O_2$$

$\therefore$ $9\alpha_2$ moles of H_2O will dissociate as

$$9\alpha_2 H_2O \rightarrow 9\alpha_2 H_2 + 4.5\alpha_2 O_2$$

$9(1-\alpha_2)$ H_2O will not dissociate and remain in the products.

The final composition will be:

$$8(1-\alpha_1)CO_2 + 8\alpha_1 CO + 4\alpha_1 O_2 + 9(1-\alpha_2)H_2O + 9\alpha_2 H_2 + 4.5\alpha_2 O_2$$
$$= 8(1-\alpha_1)CO_2 + 8\alpha_1 CO + 9(1-\alpha_2)H_2O + 9\alpha_2 H_2 + (4\alpha_1 + 4.5\alpha_2)O_2 \quad \textbf{Ans.}$$

The overall reaction will be:

$$C_8H_{18} + 12.5O_2 \rightarrow 8(1-\alpha_1)CO_2 + 8\alpha_1 CO + 9(1-\alpha_2)H_2O + 9\alpha_2 H_2 + (4\alpha_1 + 4.5\alpha_2)O_2 \quad \textbf{Ans.}$$

3.10 CHEMICAL EQUILIBRIUM

To evaluate the degree of dissociation, it is necessary to study chemical equilibrium in the chemical reactions. As a reaction proceeds, some of the products dissociate. When chemical equilibrium is reached, the reaction proceeds in both directions, so there is no net change in either the reactants or the products.

Consider a general case of a stoichiometric chemical equilibrium reaction:

$$\nu_A \cdot A + \nu_B \cdot B \rightleftharpoons \nu_C \cdot C + \nu_D \cdot D \tag{3.33}$$

where, *A*, *B*, *C* and *D* are the individual species in the burned gases that react together, produce and remove each species at equal rates. No net change in species composition results. The ν_A, ν_B, ν_C and ν_D represent the stoichiometric coefficients for the balanced chemical equilibrium reaction.

The equilibrium constant K_p is defined as

$$K_p = \frac{(p_C)^{\nu_C} \cdot (p_D)^{\nu_D}}{(p_A)^{\nu_A} \cdot (p_B)^{\nu_B}} \tag{3.34}$$

where p_A, p_B, p_C and p_D are the partial pressures of the constituents. These could be expressed in terms of mole fractions and total pressure as

$$p_i = x_i \cdot p \tag{3.35}$$

where p_i is the partial pressure of the *i*th constituent, x_i is the mole fraction of the *i*th species and *p* is the total pressure.

$$\therefore \qquad K_p = \frac{(x_C)^{\nu_C} \cdot (x_D)^{\nu_D}}{(x_A)^{\nu_A} \cdot (x_B)^{\nu_B}} \cdot p^{\nu_C + \nu_D - \nu_A - \nu_B} \tag{3.36}$$

In this expression ν_A, ν_B, ν_C and ν_D are the stoichiometric coefficients obtained from the stoichiometric balanced equilibrium equation, whereas the mole fractions x_A, x_B, x_C and x_D are taken from the actual chemical reaction attaining equilibrium, where the reactants may not be the stoichiometric mixture. The equilibrium constant K_p is not really a constant but is only a constant for a given temperature for an ideal gas.

The K_p does not depend on the amount of the various constituents initially present in the mixture. The K_p for a given reaction can be measured at various temperatures by analyzing the equilibrium gas mixture. The values for the natural logarithm of the equilibrium constant K_p for several ideal gas reactions are given in Table 3.5. When the temperature of a gas is increased, there is a tendency for the gas to dissociate. So at any temperature, there is an equilibrium mixture of the gas and the dissociated products. The dissociation process is endothermic, and as such tends to reduce the total energy of a system to a minimum value for a given temperature. The dissociation reduces the adiabatic flame temperature as some of the chemical energy is used in the dissociation process instead of raising the thermal energy.

The vaue of K_p for the equilibrium equation

$$CO + \frac{1}{2}O_2 \rightleftharpoons CO_2$$

will be reciprocal to the value of K_p for the equilibrium equation

$$CO_2 \rightleftharpoons CO + \frac{1}{2}O_2$$

$$\therefore \qquad \ln K_p \text{ for } CO + \frac{1}{2}O_2 \rightleftharpoons CO_2 \text{ will be } - \ln K_p \text{ for } CO_2 \rightleftharpoons CO + \frac{1}{2}O_2.$$

Table 3.5 Natural logarithm of equilibrium constant K_p (ln K_p)

Temperature (K)	$H_2 \rightleftharpoons 2H$	$O_2 \rightleftharpoons 2O$	$N_2 \rightleftharpoons 2N$	$H_2O \rightleftharpoons H_2 + \frac{1}{2}O_2$	$H_2O \rightleftharpoons \frac{1}{2}H_2 + OH$	$CO_2 \rightleftharpoons CO + \frac{1}{2}O_2$	$\frac{1}{2}N_2 + \frac{1}{2}O_2 \rightleftharpoons NO$
298	–164.018	–186.988	–367.493	–92.214	–106.214	–103.768	–35.052
500	–92.840	–105.643	–213.385	–52.697	–60.287	–57.622	–20.295
1000	–39.816	–45.163	–99.140	–23.169	–26.040	–23.535	–9.388
1200	–30.887	–35.018	–80.024	–18.188	–20.289	–17.877	–7.569
1400	–24.476	–27.755	–66.342	–14.615	–16.105	–13.848	–6.270
1600	–19.650	–22.298	–56.068	–11.927	–13.072	–10.836	–5.294
1800	–15.879	–18.043	–48.064	–9.832	–10.663	–8.503	–4.536
2000	–12.853	–14.635	–41.658	–8.151	–8.734	–6.641	–3.931
2200	–10.366	–11.840	–36.404	–6.774	–7.154	–5.126	–3.433
2400	–8.289	–9.510	–32.024	–5.625	–5.838	–3.866	–3.019
2600	–6.530	–7.534	–28.317	–4.654	–4.725	–2.807	–2.671
2800	–5.015	–5.839	–25.130	–3.818	–3.769	–1.900	–2.372
3000	–3.698	–4.370	–22.372	–3.092	–2.943	–1.117	–2.114
3200	–2.547	–3.085	–19.950	–2.457	–2.218	–0.435	–1.888
3400	–1.529	–1.948	–17.813	–1.897	–1.582	0.163	–1.690
3600	–0.622	–0.939	–15.911	–1.398	–1.014	0.695	–1.513
3800	0.189	–0.032	–14.212	–0.951	–0.507	1.170	–1.356
4000	0.921	0.783	–12.673	–0.548	–0.050	1.593	–1.216
4500	2.473	2.500	–9.427	0.306	0.914	2.484	–0.921
5000	3.712	3.882	–6.820	0.990	1.683	3.191	–0.686
5500	4.730	5.010	–4.679	1.554	2.312	3.765	–0.497
6000	5.577	5.950	–2.878	2.026	2.837	4.239	–0.341

Source: JANAF Thermochemical Tables, Document PB 168–370, Clearinghouse for Federal Scientific and Technical Information, 1965.

EXAMPLE 3.18 Determine the per cent dissociation of carbon dioxide into carbon monoxide and oxygen at 3000 K and 4000 K and at 1 atm pressure.

Solution: From Table 3.5, for the chemical equilibrium equation $CO_2 \rightleftharpoons CO + \frac{1}{2}O_2$, , $\ln K_p = -1.117$ at 3000 K. Therefore, $\ln K_p = 1.117$ at 3000 K for $CO + \frac{1}{2}O_2 \rightleftharpoons CO_2$. Let α denote the dissociation of 1 mole of CO_2. The equilibrium equation becomes

$$\alpha CO + \frac{\alpha}{2}O_2 \rightleftharpoons (1-\alpha)CO_2$$

The total number of moles present at equilibrium is

$$\alpha + \frac{\alpha}{2} + (1-\alpha) = \frac{2+\alpha}{2}$$

The partial pressure of each constituent may be expressed in terms of mole fraction and total pressure p as follows:

$$p_{CO} = \frac{2\alpha}{2+\alpha}p$$

$$p_{O2} = \frac{\alpha}{2+\alpha}p$$

$$p_{CO2} = \frac{2(1-\alpha)}{2+\alpha}p$$

$$\therefore \quad K_p = \frac{(p_{CO_2})^1}{(p_{CO})^1 \cdot (p_{O_2})^{1/2}} = \frac{(1-\alpha)(2+\alpha)^{1/2}}{\alpha^{3/2}}p^{-1/2}$$

Since, $p = 1$ atm. and $\ln K_p = 1.117$ at 3000 K,

$$\therefore \quad K_p = \frac{(1-\alpha)(2+\alpha)^{1/2}}{\alpha^{3/2}} = 3.056$$

This equation can be solved for α by a trial-and-error method.

$$\alpha = 0.4, \quad K_p = 3.674$$

$$\alpha = 0.5, \quad K_p = 2.236$$

α can be obtained approximately by linear interpolation as follows:

$$\frac{\alpha - 0.4}{0.5 - 0.4} = \frac{3.674 - 3.056}{3.674 - 2.236}$$

$$\therefore \quad \alpha = 0.443$$

At 3000 K, 44.3% of CO_2 will dissociate **Ans.**

(A more accurate value is 43.4% obtained by trial and error.)

At 4000 K, from Table 3.5, for the equation $CO_2 \rightleftharpoons CO + \frac{1}{2}O_2$, we have $\ln K_p = +1.593$.

∴ For the equation $CO + \frac{1}{2}O_2 \rightleftharpoons CO_2$, $\ln K_p = -1.593$.

∴ $$K_p = 0.2033$$

or $$\frac{(1-\alpha)(2+\alpha)^{1/2}}{\alpha^{3/2}} = 0.2033$$

By trial and error,

$$\alpha = 0.9, \quad K_p = 0.1995$$

$$\alpha = 0.89, \quad K_p = 0.2227$$

∴ $$\alpha = 0.89 + \frac{(0.9-0.89)(0.2033-0.2227)}{0.1995-0.2227}$$

$$= 0.8984$$

At 4000 K, $\boxed{89.84\%}$ of CO_2 will dissociate. **Ans.**

It is thus observed that as the temperature increases, the dissociation of CO_2 increases.

EXAMPLE 3.19 A stoichiometric mixture of carbon monoxide and oxygen is burned in a closed vessel, initially at 1 atm and 300 K. Calculate the mole fraction of the products of combustion at 2400 K and the pressure of the product mixture.

Solution: For $CO_2 \rightleftharpoons CO + \frac{1}{2}O_2$, at 2400 K, $\ln K_p = -3.866$ (from Table 3.5)

∴ For $CO + \frac{1}{2}O_2 \rightleftharpoons CO_2$, at 2400 K, $\ln K_p = 3.866$.

∴ $$K_p = 47.75$$

If α is the degree of dissociation of CO_2, the equilibrium equation becomes,

$$\alpha CO + \frac{\alpha}{2}O_2 \rightleftharpoons (1-\alpha)CO_2; \ (1-\alpha)CO_2; \text{ All constituents are present in the product.}$$

The total number of moles present at equilibrium in the product is $n_P = \alpha + \frac{\alpha}{2} + (1-\alpha)$

$$= \frac{\alpha + 2}{2}$$

Mole fractions:

$$x_{CO2} = \frac{2(1-\alpha)}{\alpha+2}$$

$$x_{CO} = \frac{2\alpha}{\alpha+2}$$

$$x_{O2} = \frac{\alpha}{\alpha+2}$$

For this mixture, the number of moles of reactants, $n_R = 1 + \frac{1}{2} = 1.5$

From the ideal gas law,

$$p_R V = n_R \bar{R} T_R \quad \text{and} \quad p_P V = n_P \bar{R} T_P$$

$$\therefore \qquad \frac{\bar{R}}{V} = \frac{p_R}{n_R T_R} = \frac{p_P}{n_P T_P}$$

$$\therefore \qquad \frac{p_P}{n_P} = \frac{p_R T_P}{n_R T_R} = \frac{1 \times 2400}{1.5 \times 300} = 5.333 \text{ atm/mole}$$

Now,

$$K_p = \frac{(x_{CO_2})^1}{(x_{CO})^1 \cdot (x_{O_2})^{1/2}} \cdot p_P^{-1/2} = 47.75$$

or

$$\frac{2(1-\alpha)/(\alpha+2)}{\{2\alpha/(\alpha+2)\}\{\alpha/(\alpha+2)\}^{1/2}}\{5.333 \times (\alpha+2)/2\}^{-1/2} = 47.75$$

or

$$\frac{1-\alpha}{\alpha^{3/2}} = 77.973$$

The above equation can be solved by trial-and-error method, and thus α is found approximately equal to 0.053. Therefore,

$$n_P = \frac{\alpha + 2}{2} = \frac{0.053 + 2}{2} = 1.0265$$

$$x_{CO2} = \frac{2(1 - 0.053)}{2 + 0.053} = \boxed{0.9226} \quad \textbf{Ans.}$$

$$x_{CO} = \frac{2 \times 0.053}{2.053} = \boxed{0.0516} \quad \textbf{Ans.}$$

$$x_{O2} = \frac{0.053}{2.053} = \boxed{0.0258} \quad \textbf{Ans.}$$

Pressure of the product, $p_P = 5.333 n_P = 5.333 \times 1.0265$

$$= \boxed{5.474 \text{ bar}} \quad \textbf{Ans.}$$

EXAMPLE 3.20 Hydrogen is steadily burned with chemically correct air at 25°C and 1 atm. Determine the adiabatic flame temperature taking dissociation into account.

For H_2 at 2200 K,

$$\bar{h}_T^o - \bar{h}_{298}^o = 59.86 \text{ MJ/kmol}$$

and at 2400 K,

$$\bar{h}_T^o - \bar{h}_{298}^o = 66.915 \text{ MJ/kmol}$$

Solution: The stoichiometric equation without dissociation is

$$H_2 + \frac{1}{2}O_2 + \frac{1}{2}(3.76)N_2 \rightarrow H_2O + 1.88 N_2$$

Let a be the degee of dissociation of H_2O, i.e.

$$\alpha H_2O \rightarrow \alpha H_2 + \frac{\alpha}{2} O_2$$

The chemical equation with dissociation is

$$H_2 + \frac{1}{2} O_2 + 1.88 N_2 \rightarrow (1-\alpha) H_2O + \alpha H_2 + \frac{\alpha}{2} O_2 + 1.88 N_2$$

$$H_R = \sum_R n_i [\bar{h}_f^\circ + (\bar{h}_T^\circ - \bar{h}_{298}^\circ)]_i = 0, \text{ since all the reactants are elements and } T = 298 \text{ K.}$$

$$H_P = \sum_P n_j [\bar{h}_f^\circ + (\bar{h}_T^\circ - \bar{h}_{298}^\circ)]_j$$

$$= (1-\alpha)\,[\bar{h}_f^\circ + \Delta \bar{h}^\circ]_{H_2O} + \alpha\, \Delta \bar{h}_{H_2}^\circ + \frac{\alpha}{2} \Delta \bar{h}_{O_2}^\circ + 1.88\, \Delta \bar{h}_{N_2}^\circ$$

Guess any temperature, and at this temperatue, determine α and then calculate H_P from the above expression.

If $H_P = H_R = 0$, then the solution is obtained. Otherwise we have to guess another temperature till this condition is satisfied.

Guess $T = 2200$ K. At this temperature for the reaction $H_2 + \frac{1}{2} O_2$,

$$\ln K_p = -6.774$$

$$\therefore \quad K_p = 0.001143$$

Now,
$$K_p = \frac{(p_{O_2})^{1/2} \cdot (p_{H_2})^1}{(p_{H_2O})^1}$$

No. of moles of products at equilibrium $= (1-\alpha) + \alpha + \frac{\alpha}{2} + 1.88$

$$= 2.88 + 0.5\alpha$$

Partial pressures:

$$p_{H_2O} = \frac{1-\alpha}{2.88 + 0.5\alpha} p$$

$$p_{H_2} = \frac{\alpha}{2.88 + 0.5\alpha} p$$

$$p_{O_2} = \frac{0.5\alpha}{2.88 + 0.5\alpha} p$$

$$p = 1 \text{ atm.}$$

$$\therefore \quad K_p = \frac{0.7071\alpha^{3/2}}{(1-\alpha)(2.88 + 0.5\alpha)^{1/2}} = 0.001143$$

By trial-and-error method, the above equation can be solved for $\alpha \simeq 0.02$.

The product side of the combustion equation is

$$0.98H_2O + 0.02H_2 + 0.01O_2 + 1.88N_2$$

$$H_P = 0.98(-241.82 + 83.036) + 0.02(59.86) + 0.01(66.802) + 1.88(63.371)$$

$$= -34.6 \text{ MJ/kmol}$$

Guess $T = 2400$ K, $\ln K_p = -5.625$

∴ $K_p = 0.0036$

or

$$\frac{0.7071\,\alpha^{3/2}}{(1-\alpha)\,(2.88 + 0.5\alpha)^{1/2}} = 0.0036$$

By trial-and-error method, we obtain $\alpha \simeq 0.04$.

$$H_P = 0.96(-241.82 + 93.604) + 0.04(66.915) + 0.02(74.492) + 1.88(70.651)$$

$$= -5.3 \text{ MJ/kmole}$$

we have, $H_P - H_R = -34.6$ MJ/kmol at 2200 K

and $H_P - H_R = -5.3$ MJ/kmol at 2400 K, and

we have to find the temperature for which

$$H_P - H_R = 0$$

This can be obtained by linear extrapolation (Figure 3.4) as

$$\frac{T - 2200}{34.6} = \frac{2400 - 2200}{-5.3 + 34.6}$$

∴ $T = 2200 + 236 = \boxed{2436 \text{ K}}$ **Ans.**

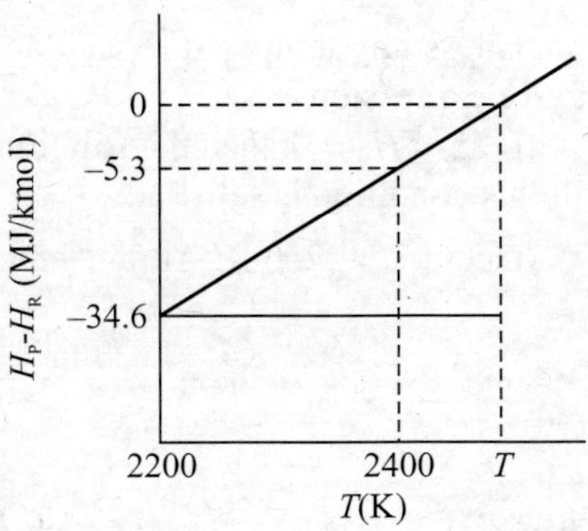

Figure 3.4 Extrapolation: Exercise 3.20.

REVIEW QUESTIONS

1. Define the exothermic and endothermic chemical reactions. What type of chemical reaction is carried out in the combustion of internal combustion engines?
2. Give a brief account of the properties of air in atmosphere. What is the ratio of moles of nitrogen to moles of oxygen in air?
3. Write the stoichiometric equation for combustion of fuel, having the general formula $C_mH_nO_p$, with air.
4. What do you understand by the minimum allowable oxygen content in the reactants per mole of fuel?
5. Determine the number of moles of the product for the combustion of $C_mH_nO_p$ with air under the following conditions: (a) $Y \geq Y_{cc}$ and (b) $Y_{min} \leq Y \leq Y_{cc}$, where Y is the moles of oxygen actually supplied, Y_{cc} is the chemically correct oxygen and Y_{min} is the minimum allowable oxygen per mole of fuel.
6. Define the equivalence ratio and the relative air/fuel ratio.
7. Define the enthalpy of formation. What is the reference state for the enthalpy of formation? Give the justification for taking this as a reference state.

8. In writing enthalpy as $\bar{h}^\circ$, what do the superscript (°) and bar over h represent?
9. Write an expression for the first law of a steady-state steady-flow chemical reaction.
10. Define enthalpy of combustion and write an expression to calculate it. Determine the enthalpy of combustion for methane, when water in the product is liquid and also when it is in gaseous form.
11. Define internal energy of combustion. Write an expression to calculate the internal energy of combustion.
12. Define the higher heating value and the lower heating value of a fuel. Give the relation between them.
13. Explain the term adiabatic flame temperature. How does this temperature change if the combustion is incomplete, when excess air is used and when dissociation takes place?
14. Write an expression to calculate the adiabatic flame temperature. Draw the *H–T* diagram for the steady-flow adiabatic combustion.
15. What do you understand by dissociation of a substance? Explain the term with the help of an example.
16. Explain the term chemical equilibrium. Write expressions for the equilibrium constant K_p in terms of partial pressures and mole fractions of the constituents.

PROBLEMS

3.1 Calculate the theoretical air/fuel and fuel/air ratios for the complete combustion of ethanol (C_2H_5OH). If the air is supplied at 1 bar and 25°C, calculate the volume of air required for complete combustion of fuel per kg and per kg mole of fuel.

[**Ans.** A/F = 8.953, F/A = 0.1117, $V = 7.657$ m³/kg fuel, $\bar{V} = 353.8$ m³/kmol fuel]

3.2 A fuel oil $C_{12}H_{26}$ is burned with 30% excess air. Determine the equivalence ratio, volumetric analysis of the products of combustion in percentage, molecular weight of the reactants and products and the dew point of the products, if the pressure is 1 bar.

[**Ans.** $\phi = 0.7693$, $CO_2 = 9.919\%$, $H_2O = 10.746\%$, $N_2 = 74.747\%$, $O_2 = 4.588\%$, $M_R = 30.06$, $M_P = 28.696$, $T = 47.25$°C]

3.3 Fuel oil $C_{12}H_{26}$ is used in a compression-ignition engine having bore of 100 mm and stroke length of 120 mm. The compression ratio is 16. The percentage composition of dry products of combustion by volume is $CO_2 = 12.5\%$, $O_2 = 3.8\%$, $CO = 0.5\%$ and $N_2 = 83.2\%$. Write the combustion equation. Calculate the air/fuel ratio, the per cent theoretical air and the mass of the residual gases left in the cylinder at the end of the exhaust stroke. The pressure and temperature of the exhaust gas are 1.08 bar and 397°C respectively.

[**Ans.** $C_{12}H_{26} + 20.43O_2 + 76.82N_2 \rightarrow 11.54CO_2 + 0.4617CO + 3.509O_2 + 76.82N_2 + 10.3H_2O + 2.7H_2$, A/F = 16.5, 110.44%, $m = 3.53 \times 10^{-5}$ kg]

3.4 An IC engine fuel has the following composition by weight: Carbon 85%, hydrogen 10%, oxygen 3% and the rest is nitrogen. Determine the chemically correct air/fuel ratio. If 30% excess air is supplied, find the percentage of dry products of combustion by weight and by volume.

[**Ans.** A/F = 13.2; by weight: $N_2 = 76.66\%$, $O_2 = 5.28\%$, $CO_2 = 18.06\%$; by volume: $N_2 = 82.6\%$, $O_2 = 5.0\%$, $CO_2 = 12.4\%$]

3.5 A hydrocarbon fuel has the following composition of dry products of combustion by volume:

$$CO_2 = 11.5\%, \quad CO = 1\%, \quad O_2 = 3.25\%, \quad N_2 = 84.25\%$$

Write the chemical equation for 100 moles of dry products. Determine the air/fuel ratio, the equivalence ratio and the fuel composition on the mass basis.

[**Ans.** $C_{12.5}H_{28.64} + 22.41(O_2 + 3.76N_2) \rightarrow 11.5CO_2 + CO + 3.25O_2 + 84.25N_2 + 14.32H_2O$, A/F = 17.22, $\phi = 0.877$, C = 83.97%, H_2 = 16.03%]

3.6 A petrol has a composition of carbon 86% and hydrogen 14% by weight. The air supplied is 85% of that theoretically required for complete combustion. Assume that all the hydrogen is burnt and that the carbon burns to CO and CO_2, so that there is no free carbon left. Calculate the percentage analysis of dry exhaust gases by volume.

[**Ans.** CO = 7.76%, CO_2 = 9.61%. N_2 = 82.63%]

3.7 A fuel mixture of 50% C_7H_{16} and 50% C_8H_{18} by volume is oxidized with 20% excess air. Determine the mass of air required for 100 kg of fuel and for volumetric analysis of dry products of combustion.

[**Ans.** m = 1809 kg, CO_2 = 11.93%, N_2 = 84.33%, O_2 = 3.74%]

3.8 A gaseous fuel normal octane (C_8H_{18}) undergoes a steady-state steady-flow reaction with air. Determine the heat transfer per mole of fuel when the reactants and products are both at 25°C and 1 atm pressure. [**Ans.** Heat released = 5511.78 MJ/mole fuel]

3.9 An engine produces 750 kW power and uses gaseous $C_{12}H_{26}$ as a fuel at 25°C; 200% theoretical air is used and air enters at 500 K. The products of combustion leave at 800 K. The heat loss from the engine is 175 kW. Determine the fuel consumption for complete combustion. [**Ans.** 98.82 kg/h]

3.10 A mixture of liquid isooctane and oxygen in the proper ratio for the complete combustion and at 25°C and 1 atm reacts in a constant volume bomb calorimeter. Heat is transferred until the products of combustion are at 600 K. Determine the heat transfer per mole of fuel.

[**Ans.** Heat liberated = 4918.8 kJ/mol fuel]

3.11 Determine the amount of heat transfer per kg of fuel during the complete combustion of methane in an open steady-flow system with 50% excess air. Methane enters at 400 K and air at 500 K and the products leave at 1000 K. The specific heat at constant pressure can be obtained from the relation

$$\frac{c_p}{R} = a + bT + cT^2 + dT^3 + eT^4$$

For methane $a = 4.503$, $b = -8.965 \times 10^{-3}$, $c = 37.38 \times 10^{-6}$, $d = -36.49 \times 10^{-9}$, $e = 12.22 \times 10^{-12}$

[**Ans.** Heat liberated 33.81 MJ/kg fuel]

3.12 Propane burns in the presence of theoretical air. Calculate the standard enthalpy of combustion and standard internal energy of combustion, when the water in the products of combustion appears (a) as a liquid and (b) as a gas.

[**Ans.** (a) $\Delta H = -2219.91$ MJ/kmol, $\Delta U = -2212.48$ MJ/kmol
(b) $\Delta H = -2043.99$ MJ/kmol, $\Delta U = -2046.47$ MJ/kmol]

3.13 The lower calorific value of a liquid fuel at constant pressure is 40,000 kJ/kg. The analysis of fuel by mass is 85% carbon and 15% hydrogen. Determine the higher calorific value at constant pressure and the lower and higher calorific values at constant volume.

[**Ans.** $(\text{HHV})_p = 43297$ kJ/kg, $(\text{HHV})_v = 43204$ kJ/kg, $(\text{LHV})_v = 39907$ kJ/kg]

3.14 Gaseous methane is burned with 200% theoretical air. Determine the adiabatic flame temperature for the steady-state, steady-flow process. [**Ans.** $T = 1480.6$ K]

3.15 Calculate the adiabatic flame temperature, when a chemically correct mixture of propane and air initially at 600 K and 1 atm burns at (a) constant pressure process and (b) constant volume process. The specific heat at constant pressure can be obtained from the relation,

$$\frac{c_p}{R} = a + bT + cT^2 + dT^3 + eT^4$$

For propane: $a = -0.4861$, $b = 36.63 \times 10^{-3}$,

$c = -18.91 \times 10^{-6}$, $d = 3.814 \times 10^{-9}$, $e = 0$.

[**Ans.** (a) 2619 K (b) 3112.7 K]

3.16 Calculate the adiabatic flame temperature when propane burns in the presence of air at constant pressure process at 600 K and 1 atm having (a) 50% excess air and (b) 20% less air leading to incomplete combustion. Calculate the loss of thermal energy due to incomplete combustion. The value of c_p can be obtained from the data given in Problem 3.15.

[**Ans.** (a) 2066 K (b) 2382 K, 566 MJ/kmol fuel]

3.17 Determine the percentage dissociation of water into hydrogen and oxygen at 3000 K and at 1 atm pressure. [**Ans.** 14.78%]

4.1 INTRODUCTION

In the analysis of the air-standard cycle as discussed in Chapter 2, it was assumed that the working substance was only air and the air was a perfect gas and had a constant specific heat. The heat was added from the external source and the engine rejected heat to the surrounding. Compression and expansion were assumed to be frictionless adiabatic processes. Since this analysis is based on a highly simplified assumption, the estimated engine performance is on the higher side compared to the actual performance. The actual indicated thermal efficiency for a petrol engine with compression ratio of 8 is nearly 30%, whereas that for the air-standard cycle is 56.47%. This large deviation is due to many reasons, such as non-instantaneous burning, incomplete combustion, dissociation of the products of combustion at high temperature, variation of specific heats of gases with temperature, heat transfer, friction, valve timings, pressure drop across the valve during intake and exhaust, etc. The working substance in the actual cycle is not only air but it is also a mixture of air, fuel and residual gases left in the clearance space from the previous cycle.

The analysis of the actual cycle is thus very complicated and the best method of approach to the real cycle is through several approximations, starting with the air-standard cycle and modifying it step-by-step. The next step is to consider a theoretical cycle, called the fuel-air cycle. The results obtained from the analysis of this cycle are much closer to the results obtained from the actual engine.

The analysis of the air-standard cycle shows the effect of compression ratio on thermal efficiency. The analysis of the fuel-air cycle, on the other hand, is able to predict the effect of the air/fuel ratio on thermal efficiency. This is not possible to be analysed with the air-standard cycle as the working substance is only air. The fuel-air cycle can also predict closely the peak pressure and the peak temperature during the cycle with different air/fuel ratios. Both the peak pressure and the peak temperature affect the engine design and material selection.

4.2 FUEL-AIR CYCLE

The fuel-air cycle calculations can take into consideration the following factors, which are not possible to be considered in the analysis of the air-standard cycle.

1. *The actual composition of the cylinder gases:* The working fluid is considered to be a mixture of air, fuel and residual gases.
2. *The variation of specific heat of the cylinder gases with temperature:* Both the specific heat at constant pressure c_p and the specific heat at constant volume c_v increase with the increase in temperature except in the case of monoatomic gases. Their difference is constant and equal to R but the ratio of the specific heats γ changes with temperature.
3. *The dissociation of gases at high temperature:* At higher temperatures, the products of combustion dissociate into smaller constituents.
4. *The variation in the number of molecules:* The number of molecules present in the cylinder, after combustion, depends upon the fuel/air ratio and upon the temperature and pressure existing in the cylinder.

Besides taking the above factors into consideration, the following assumptions are also made for simplicity:

1. The fuel is completely vaporized and adiabatically mixed with air and residual gases.
2. There is no change in the composition of charge during compression. Compression is assumed to be reversible adiabatic.
3. Combustion is assumed to be complete without any heat loss at constant volume or constant pressure or limited pressure.
4. Expansion of the burned gases is assumed to be reversible adiabatic.
5. The law of mass action (chemical equilibrium) is obeyed at temperatures above 1600 K and species concentration is frozen (fixed) at and below 1600 K.
6. Exhaust blowdown during exhaust is assumed to be ideal adiabatic and the exhaust gases are frozen in composition.
7. Piston displacement is assumed to be frictionless.
8. There is no heat exchange between the gases and the cylinder walls during the complete cycle.
9. Inside the engine cylinder the velocity of gases is negligibly small.

Briefly, the fuel-air cycle consists of compression, combustion, expansion and blowdown processes similar to the air-standard cycle. There is no friction and no heat transfer between the cylinder wall and the gases. Variation of specific heat and dissociation of gases at high temperatures are considered. With these considerations it has been found that the indicated thermal efficiency of the fuel-air cycle is about 15% higher than that obtained for the actual engine.

4.3 FACTORS AFFECTING THE FUEL-AIR CYCLE

The following factors affect the analysis of the fuel-air cycle:

4.3.1 Composition of Cylinder Gases

In an actual operation of an engine, the air/fuel ratio may vary for different operating conditions. This affects the composition of the cylinder gases before and after combustion. Before combustion the cylinder gases are a mixture of air, fuel and residual gases in the clearance space from the previous cycle. The products of combustion may be CO_2, CO, H_2O, O_2, and N_2. The amount of exhaust gases in the clearance volume varies with the speed and the load on the engine. In order to avoid laborious and time-consuming calculations by hand, the fuel-air cycles are analysed with the help of combustion charts. Separate charts are used for unburned and burned mixtures for each fuel/air ratio of interest. However, with the availability of fast digital computers, it is now possible to analyse the fuel-air cycle by means of suitable numerical techniques, which may produce fast and accurate results.

4.3.2 Variation of Specific Heats

All gases, except monatomic gases, show an increase in specific heats with increase in temperature. The specific heat is the amount of heat required to raise the temperature of a unit mass by one degree. As the temperature is raised, larger fractions of the input heat go to produce motion of the atoms within the molecules. Since temperature is the result of motion of the molecules, the energy which goes into moving the atoms does not contribute to the temperature rise, so the final temperature and pressure are lowered. Since at high temperatures, a given amount of heat results in a smaller increase in temperature, therefore the specific heats increase with temperature.

The variation of specific heats with temperature does not follow any specific law. Figure 4.1 shows the variation of specific heat at constant pressure c_p versus temperature for a few ideal gases. Specific heat at constant volume c_v can be obtained from the relation $c_v = c_p - R$, where R is the gas constant.

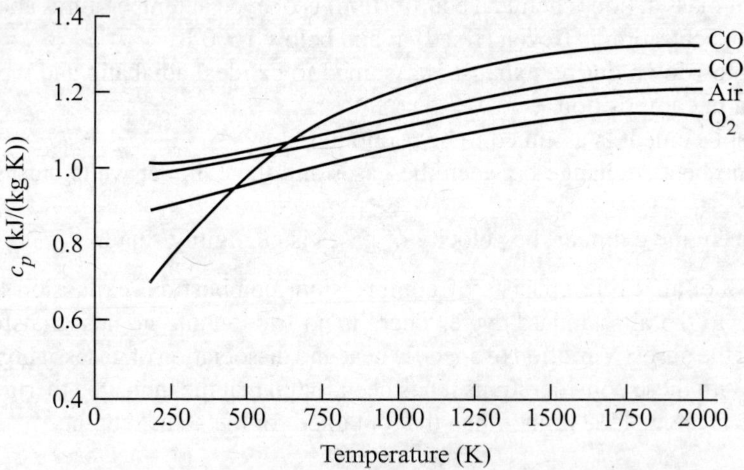

Figure 4.1 c_p for some ideal gases.

Table 4.1 shows the polynomial coefficients for different gases, which may be used to evaluate c_p from the relation

$$\frac{c_p}{R} = \frac{\bar{c}_p}{\bar{R}} = a + bT + cT^2 + dT^3 + eT^4 \tag{4.1}$$

The value of the ratio of specific heats (c_p/c_v) for monatomic gases is constant and for all other gases it decreases with increase in temperature.

The effect of variation of specific heats with temperature on Otto cycle is shown in Figure 4.2.

Table 4.1 Polynomial coefficients to evaluate c_p

Gas	*Temperature Range* (K)	*a*	$b \times 10^3$ (K^{-1})	$c \times 10^6$ (K^{-2})	$d \times 10^9$ (K^{-3})	$e \times 10^{12}$ (K^{-4})
Air	300–1000	3.721	−1.874	4.719	−3.445	0.8531
	1000–3000	2.786	1.925	−0.9465	0.2321	−0.02229
CO_2	300–1000	2.227	9.992	−9.802	5.397	−1.281
	1000–3000	3.247	5.847	−3.412	0.9469	−0.1009
CO	300–1000	3.776	−2.093	4.880	−3.271	0.6984
	1000–3000	2.654	2.226	−1.146	0.2851	−0.02762
C_2H_6	300–1500	0.8293	20.75	−7.704	0.8756	0.0
CH_4	300–1000	4.503	−8.965	37.38	−36.49	12.22
	1000–3000	−0.6992	15.31	−7.695	1.896	−0.1849
C_3H_8	300–1500	−0.4861	36.63	−18.91	3.814	0.0
H_2	300–1000	2.892	3.884	−8.850	8.694	−2.988
	1000–3000	3.717	−0.9220	1.221	−0.4328	0.05202
N_2	300–1000	3.725	−1.562	3.208	−1.554	0.1154
	1000–3000	2.469	2.467	−1.312	0.3401	−0.03454
O_2	300–1000	3.837	−3.420	10.99	−10.96	3.747
	1000–3000	3.156	1.809	−1.052	0.3190	−0.03629
H_2O	300–1000	4.132	−1.559	5.315	−4.209	1.284
	1000–3000	2.798	2.693	− 0.5392	−0.001783	0.009027

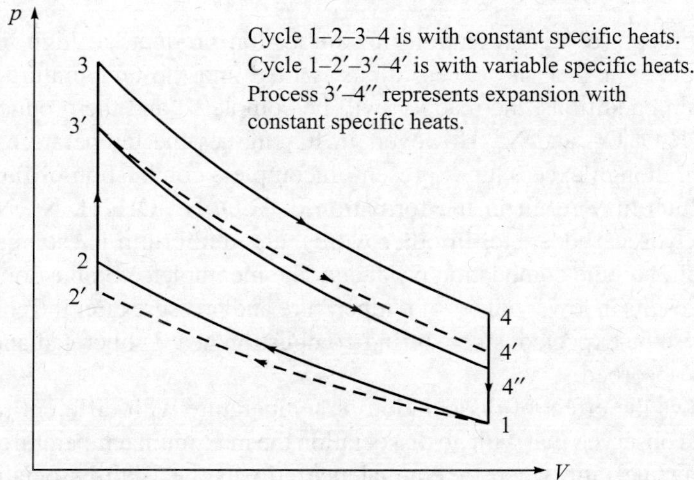

Figure 4.2 Effect of variation of specific heats on Otto cycle.

If the variation of specific heats is considered during the compression stroke, the values of temperature and pressure at the end of compression would be lower than the values obtained by taking the fixed values of specific heats.

With variable specific heats, the state point at the end of compression will be 2′ instead of 2.

With constant specific heats, the temperature T_2 after compression can be obtained from the relation $T_2 = T_1 r^{(\gamma-1)}$, where r is the compression ratio and γ is the ratio of specific heats, c_p/c_v.

With variable specific heats, the temperature T_2' at the end of compression can be obtained only by step-by-step calculations. For this, the compression process is divided into a number of smaller steps and assuming constant specific heats for the smaller step, the temperatures at the end of each step are evaluated and finally T_2' is obtained. The larger the number of steps, the more accurate will be the result.

For the combustion process, with constant specific heats, the process is represented by 2–3, whereas with variable specific heats the process is represented by 2′–3′. As the temperature increases during the combustion process, the specific heats also increase, so the rise in temperature with variable specific heats will be smaller than the rise in temperature without considering the variation in specific heats, i.e.

$$(T_3' - T_2') < (T_3 - T_2)$$

During the expansion process, the temperature decreases, therefore the specific heats also decrease. Due to decrease in specific heats, the temperature at the end of the expansion stroke T_4' would be higher than the temperature T_4'' obtained without considering the variation in specific heats.

It can be observed from Figure 4.2 that by considering the variation in specific heats, the area enclosed by the cycle 1–2′–3′–4′ reduces and hence the work done also reduces in comparison to the area of standard air cycle 1–2–3–4, where constant specific heats are taken.

A step-by-step procedure is also required to evaluate the properties at state points 3′ and 4′ as explained during the compression process.

4.3.3 Effect of Dissociation

Dissociation is a process of disintegration of combustion products at high temperatures. If a mixture of a hydrocarbon fuel and excess air is ignited and the temperature does not exceed 1600 K under certain conditions, the reaction will be completed and the products of combustion will contain CO_2, H_2O, O_2, and N_2. However, in IC engines the temperatures of the products are very high. Even though excess air is present, incomplete combustion of fuel results because of dissociation, which may result in the formation of H_2O, H_2, OH, H, N_2, NO, N, CO_2, CO, O_2, O, A (argon), hydrocarbons, etc. In other words, an equilibrium is established between the uncombined fuel, air, and the combustion products. The incomplete combustion of fuel and air at high temperatures results in lower values of temperature and pressure after the combustion process than would otherwise be expected, since during combustion heat is liberated and whereas during dissociation heat is absorbed.

Figure 4.3 shows the effect of dissociation on temperature with different mixture strengths. From the figure, it is observed that with no dissociation the maximum temperature obtained is at the chemically correct air/fuel ratio, where the equivalence ratio ϕ is one. With dissociation, the maximum temperature is obtained when the mixture is slightly rich. Dissociation reduces the maximum

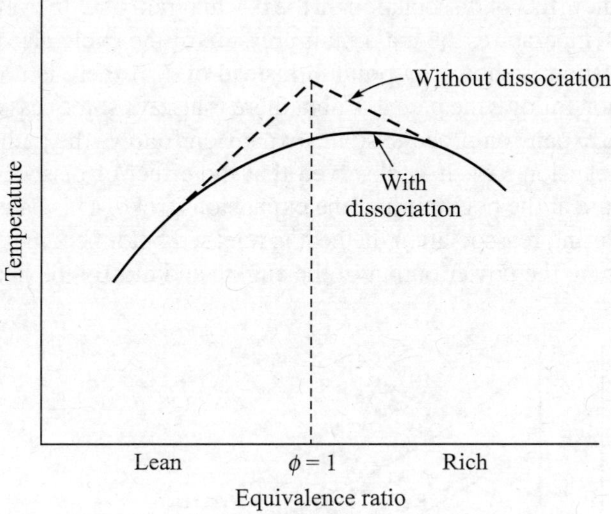

Figure 4.3 Effect of dissociation on temperature.

temperature by about 300°C. The presence of CO in the products of combustion tends to reduce the dissociation of CO_2. This can be noted in a rich mixture which produces more CO and suppresses dissociation of CO_2. On the lean side too, the dissociation is suppressed because of lower temperature.

Figure 4.4 shows the effect of dissociation on power. If there is no dissociation, the brake power (bp) is maximum when the mixture strength is chemically correct. With dissociation, the bp is maximum when the mixture is slightly rich. The difference in the ordinates between the two curves of bp shows the power lost at the given mixture strength.

The effect of dissociation is generally small compared with the effect of variation in specific heats.

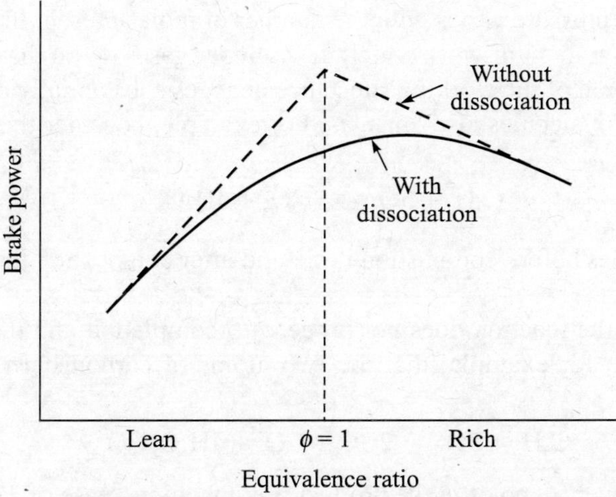

Figure 4.4 Effect of dissociation on power.

Figure 4.5 shows the effect of dissociation on the p-V diagram of Otto cycle. As the dissociation lowers the maximum temperature, the maximum pressure of the cycle also reduces and the state of the gas after combustion is shown by point 3′ instead of 3. If there is no reassociation during expansion, the expansion follows the path 3′4″, but there is always some reassociation due to fall in temperature during the expansion process, so the expansion follows the path 3′–4′. By comparing with the isentropic expansion 3–4, it is observed that the effect of dissociation is to reduce the temperature and pressure at the beginning of the expansion stroke. This causes loss of power and efficiency. Although during reassociation the heat is released but it becomes too late to contribute any significant increase to the power output of the engine and mostly the heat is lost in exhaust.

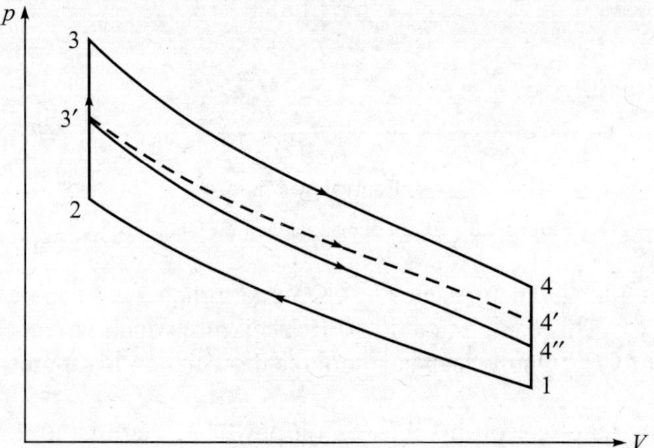

Figure 4.5 Effect of dissociation on Otto cycle.

4.3.4 Effect of Number of Molecules

The number of molecules present in the cylinder after combustion depends upon the fuel/air ratio and upon the pressure and temperature existing in the cylinder. For a given temperature from the gas law $pV = n\bar{R}T$, the pressure varies with the number of moles present in the cylinder. This has a direct effect upon the amount of work which the cylinder gases do on the piston.

Since the composition of the working substance changes substantially during the combustion process, the number of molecules also changes. For example, consider the complete reaction of propane and oxygen,

$$C_3H_8 + 5O_2 \rightarrow 3CO_2 + 4H_2O \tag{4.2}$$

The number of moles before combustion is six and after combustion this number is seven, an increase of 16.7%.

In the actual case, the reaction does not proceed to completion and the change may be still greater. Let us assume, for example, that the two atoms of carbon burn to CO, the chemical reaction then becomes

$$C_3H_8 + 5O_2 \rightarrow 2CO + CO_2 + 4H_2O + O_2 \tag{4.3}$$

Now, the total number of moles in the product is eight, an increase of 33.3% over the mixture before combustion.

It is evident that the change in the number of molecules is dependent on the composition of the fuel, the fuel/air ratio, and on the completeness of the reaction. The number of molecules is greatest at a high temperature where the chemical equilibrium involves considerable amount of OH, NO, O_2, H_2 and CO. As the temperature falls, the number of molecules decreases, but even if combustion were complete most fuels would give a larger total number of molecules after combustion than the number contained in the original mixture of fuel and air.

4.4 EFFECT OF ENGINE VARIABLES ON THE PERFORMANCE OF FUEL-AIR CYCLES

Fuel-air cycles have been calculated for constant inlet pressure and inlet temperature, but for varying fuel/air ratios and compression ratios. These two parameters are engine variables for fuel-air cycle analysis.

Figure 4.6 shows the effect of compression ratio and equivalence ratio on thermal efficiency of the constant volume fuel-air cycle using octene as a fuel with initial pressure, $p_1 = 1$ atm., initial temperature, $T_1 = 388$ K and residual gas fraction, $f = 0.05$. The thermal efficiency is the ratio of the work done to the energy content of the fuel supplied. It is observed from the figure that as the compression ratio increases the thermal efficiency increases too, in the same manner as the air-cycle efficiency. With the increase in equivalence ratio thermal efficiency decreases.

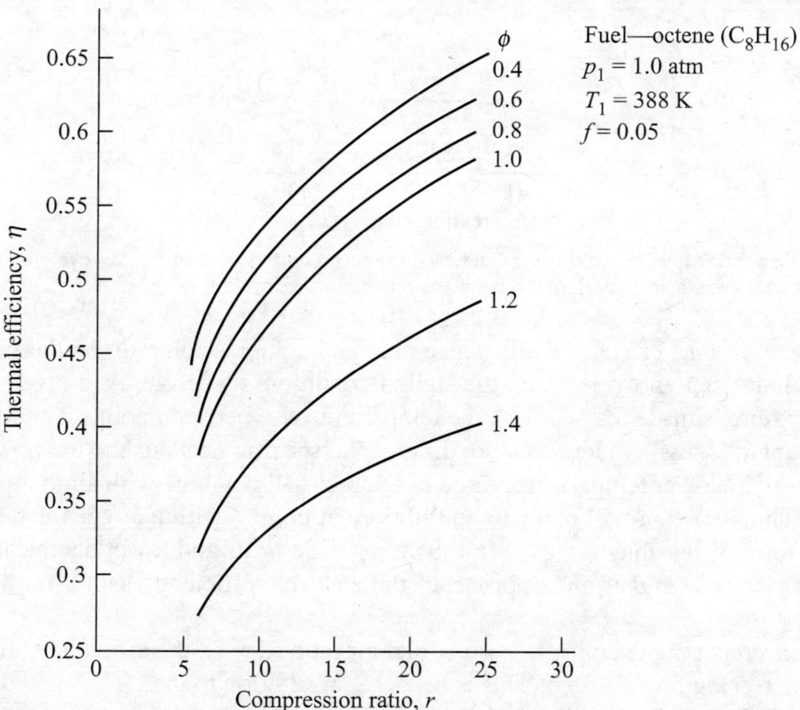

Figure 4.6 Effect of compression ratio r and equivalence ratio ϕ on thermal efficiency η of the constant volume fuel-air cycle.

Figure 4.7 shows the effect of compression ratio and equivalence ratio on the ratio of fuel-air cycle efficiency to air-standard cycle efficiency of the constant volume fuel-air cycle under the similar conditions as before. It is observed that the ratio of these efficiencies slightly increases with increase in compression ratio at a given equivalence ratio but decreases as the amount of fuel is increased.

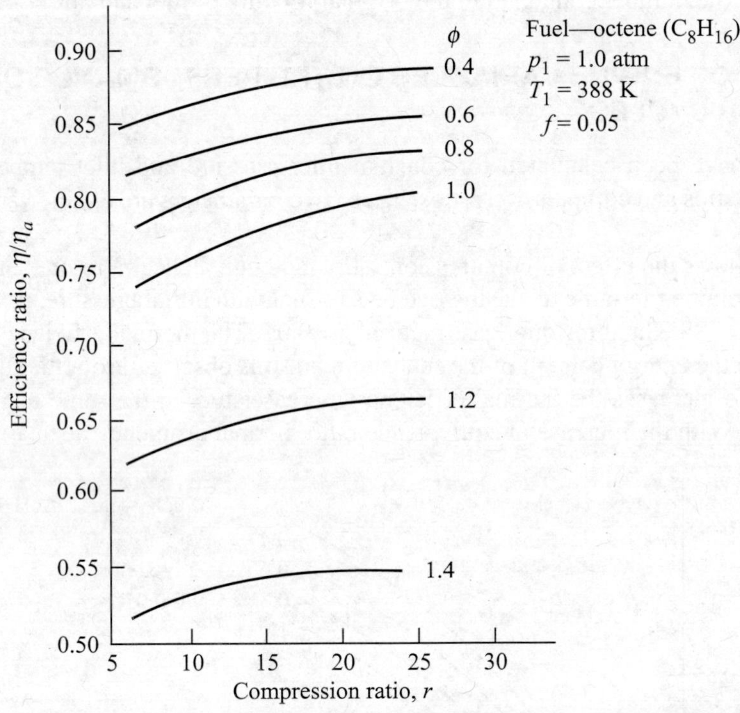

Figure 4.7 Effect of compression ratio r and equivalence ratio ϕ on the ratio of fuel-air cycle efficiency η to air-standard cycle efficiency η_a of the constant volume fuel-air cycle.

Figure 4.8 shows the effect of equivalence ratio and compression ratio on thermal efficiency of constant volume fuel-air cycle under the similar conditions as before. At a given compression ratio, as the mixture is made lean ($\phi < 1$), the temperature rise between points 2 and 3 will be less as the energy input is less. Under these conditions, the specific heats of the gases will be nearly constant. Also at the lower temperature, more of the fuel will combine with air at top dead centre, i.e. the combustion reaction will come to equilibrium at point 3 with a larger fraction of the fuel energy in the form of sensible energy. With lower specific heats and lower chemical equilibrium losses, the efficiency is higher and approaches the air-cycle efficiency as a limit as the fuel/air ratio is reduced.

On the rich side of the chemically correct fuel/air ratio ($\phi > 1$), the efficiency falls even more rapidly with increasing fuel/air ratio. This is because, in addition to the effects noted above, there is insufficient air to completely utilize all the fuel present, regardless of the chemical equilibrium. The exhaust gases of rich mixtures will contain combustibles in the form of CO and H_2, which represent a direct waste of fuel.

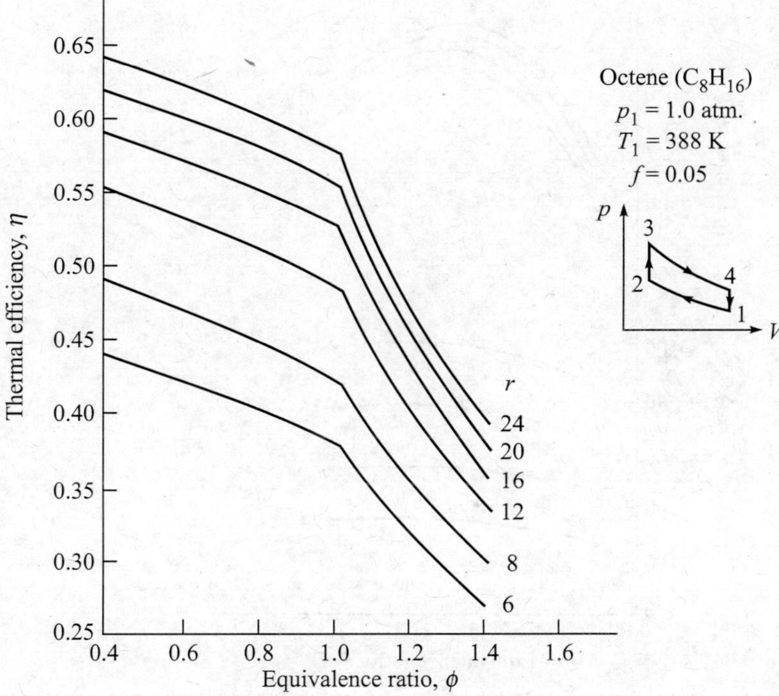

Figure 4.8 Effect of equivalence ratio ϕ and compression ratio r on thermal efficiency η of constant volume fuel-air cycle.

Apart from the effect of compression ratio and equivalence ratio on thermal efficiency, other performances, such as maximum temperature T_3, maximum pressure p_3, temperature T_4 and mean effective pressure are also important for the analysis of fuel-air cycles.

Figure 4.9 shows the effect of equivalence ratio and compression ratio on maximum temperature T_3 and maximum pressure p_3 of the constant volume fuel-air cycle with p_1 = 1 atm, T_1 = 288 K using isooctane as a fuel. In the region of weak mixture ($\phi < 1$), as the equivalence ratio is increased, more sensible energy is produced. This will result in increase in temperature T_3. At the stoichiometric condition ($\phi = 1$), there is just enough fuel present to use up all the oxygen, but because of chemical equilibrium both fuel and oxygen will be present in the engine cylinder. The use of additional fuel ($\phi > 1$) will affect chemical equilibrium in such a way as to cause more complete combustion at point 3, and therefore more fuel will combine with oxygen. T_3 continues to rise because of this effect until an equivalence ratio of about 1.065 is reached. At the richer fuel/air ratios the incomplete combustion causes a decrease in the value of T_3.

From the gas law, $pV = n\bar{R}T$, the pressure of a gas inside the given volume of an engine cylinder depends upon the number of moles and its temperature. The curve of p_3 vs. ϕ tends, therefore, to follow the curve of T_3 vs. ϕ, except that more molecules are formed by the combustion of rich mixtures. Because of increasing number of moles, p_3 does not start to decrease until the mixture is somewhat richer than that for maximum T_3. Maximum p_3 occurs at the equivalence ratio of about 1.2, i.e. about 20% rich.

With the increase in compression ratio, both p_3 and T_3 will increase because p_2 and T_2 are higher at higher compression ratios.

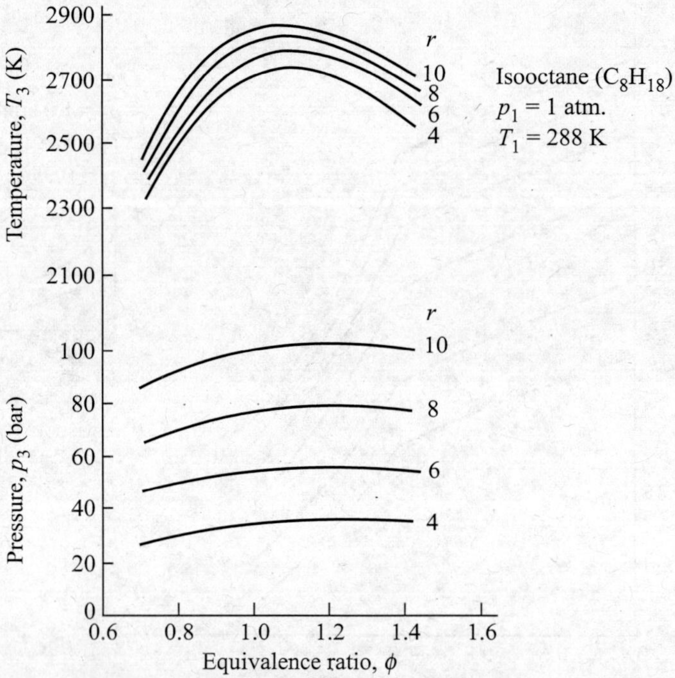

Figure 4.9 Effect of equivalence ratio ϕ and compression ratio r on T_3 and p_3 of the constant volume fuel-air cycle.

Figure 4.10 shows the effect of equivalence ratio and compression ratio on temperature T_4. It is observed from the figure that the maximum value of T_4 is obtained at the chemically correct fuel/air ratio ($\phi = 1$). After expansion, the temperature of the gases is low enough to shift the chemical equilibrium in such a way that both fuel and oxygen will be completely used up at the chemically correct fuel/air ratio.

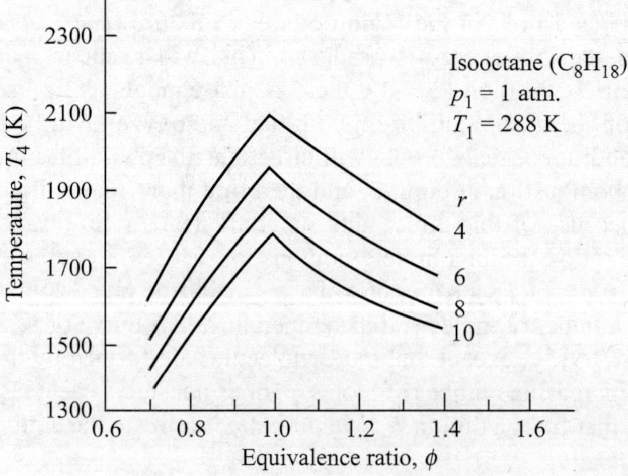

Figure 4.10 Effect of equivalence ratio ϕ and compression ratio r on T_4 of the constant volume fuel-air cycle.

At point 4, the temperature of the gases is low, therefore, the effect of chemical equilibrium is not important at this point. The use of a weak mixture ($\phi < 1$) means that less fuel is burned and therefore the temperature will be smaller. The use of a rich mixture results in the formation of more CO instead of CO_2, with less sensible energy developed and again there will be a lower temperature. As the compression ratio increases, T_4 decreases. This is because by increasing the compression ratio the expansion process is also increased, which causes the gas to do more work on the piston, leaving less heat to be rejected at the end of the expansion stroke.

Figure 4.11 shows the effect of the equivalence ratio and the compression ratio on mean effective pressure (mep) of the constant volume fuel-air cycle with $p_1 = 1$ atm., and $T_1 = 288$ K. It is observed from the figure that the mean effective pressure of the fuel-air cycle is maximum at slightly richer than the chemically correct mixture. This is because the average pressure during the expansion stroke would be affected by the same factors that influence p_3 and p_4. With the increase in compression ratio, the mean effective pressure also increases because of the higher thermal efficiency at high compression ratios.

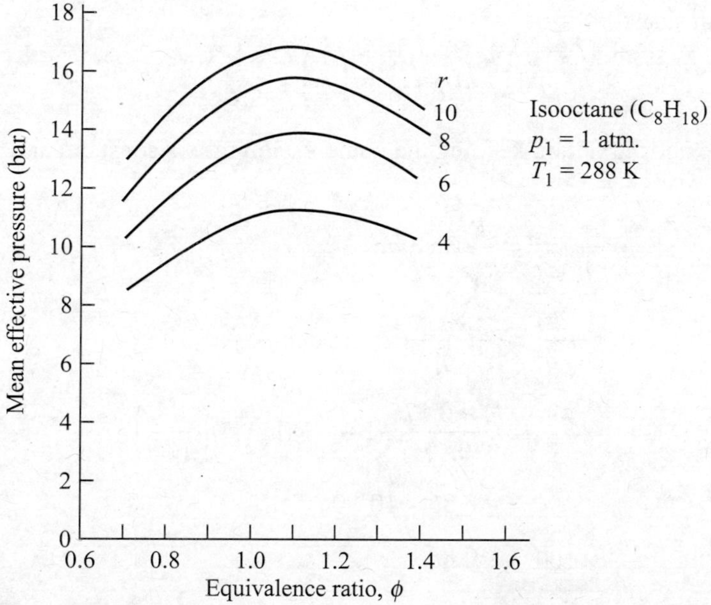

Figure 4.11 Effect of equivalence ratio ϕ and compression ratio r on mep on the constant volume fuel-air cycle.

The results obtained from the analysis of fuel-air cycle are close to the experimental results obtained from the actual engine. The main difference between the two is that in an actual engine, combustion is not instantaneous but it takes some time to complete. The time required for combustion (i.e. the combustion duration) causes the actual efficiency to fall considerably below the fuel-air cycle efficiency.

EXAMPLE 4.1 What will be the percentage change in efficiency of an Otto cycle having a compression ratio of 8.5, when the specific heat at constant volume increases by 1.4%?

Solution: The efficiency of the Otto cycle is given by

$$\eta = 1 - \left(\frac{1}{r}\right)^{\gamma-1} = 1 - \left(\frac{1}{8.5}\right)^{0.4} = 0.575$$

Now, $$c_p - c_v = R \quad \text{and} \quad \frac{c_p}{c_v} = \gamma$$

$$\therefore \quad \gamma - 1 = \frac{R}{c_v}$$

$$\therefore \quad \eta = 1 - \left(\frac{1}{r}\right)^{R/c_v}$$

or $$1 - \eta = \left(\frac{1}{r}\right)^{R/c_v}$$

Taking log on both sides,

$$\ln(1-\eta) = \frac{R}{c_v}\ln\left(\frac{1}{r}\right) = -\frac{R}{c_v}\ln r$$

Differentiating the above relation keeping the compression ratio r constant, and where R is a gas constant, we get

$$-\frac{1}{1-\eta}d\eta = \frac{R}{c_v^2}(\ln r)dc_v$$

$$\therefore \quad \frac{d\eta}{\eta} = -\frac{1-\eta}{\eta}(\gamma - 1)(\ln r)\frac{dc_v}{c_v}$$

$$= \frac{1-0.575}{0.575}(1.4-1)(\ln 8.5)\left(\frac{1.4}{100}\right)$$

$$= -8.86 \times 10^{-3}$$

$$\therefore \quad \frac{d\eta}{\eta} \times 100 = -0.886\%$$

Therefore, the efficiency decreases by 0.886% **Ans.**

EXAMPLE 4.2 What will be the percentage change in efficiency of a Diesel cycle having a compression ratio of 18 and cut-off taking place at 6% of the stroke, when the specific heat at constant volume increases by 2%? Take c_v = 0.717 kJ/(kg K) and R = 0.287 kJ/(kg K).

Solution: The efficiency of the Diesel cycle is given by

$$\eta = 1 - \left(\frac{1}{r}\right)^{\gamma-1}\left[\frac{\beta^{\gamma} - 1}{\gamma(\beta - 1)}\right]$$

where r is the compression ratio and β is the cut-off ratio.

$$\therefore \qquad 1-\eta = \left(\frac{1}{r}\right)^{\gamma-1}\left[\frac{\beta^{\gamma}-1}{\gamma(\beta-1)}\right]$$

Taking log on both sides,

$$\ln(1-\eta) = -(\gamma-1)\ln r + \ln(\beta^{\gamma}-1) - \ln\gamma - \ln(\beta-1)$$

Differentiating the above equation with respect to γ, keeping r and β constant,

$$-\frac{1}{1-\eta}\frac{d\eta}{d\gamma} = \ln r + \frac{\beta^{\gamma}\ln\beta}{\beta^{\gamma}-1} - \frac{1}{\gamma}$$

$$\therefore \qquad \frac{d\eta}{\eta} = \frac{1-\eta}{\eta}d\gamma\left[\ln r - \frac{\beta^{\gamma}\ln\beta}{\beta^{\gamma}-1} + \frac{1}{\gamma}\right] \qquad \text{(i)}$$

But, $\qquad \dfrac{R}{c_v} = \gamma - 1$

Differentiating the above equation,

$$\frac{R}{c_v^2}dc_v = d\gamma$$

$$\therefore \qquad d\gamma = -\frac{R}{c_v}\frac{dc_v}{c_v} = -(\gamma-1)\frac{dc_v}{c_v} \qquad \text{(ii)}$$

Substituting the value of $d\gamma$ from Eq. (ii) into (i),

$$\frac{d\eta}{\eta} = -\left(\frac{1-\eta}{\eta}\right)(\gamma-1)\left[\ln r - \frac{\beta^{\gamma}\ln\beta}{\beta^{\gamma}-1} + \frac{1}{\gamma}\right]\frac{dc_v}{c_v}$$

Compression ratio, $r = \dfrac{V_1}{V_2} = 18$ (Refer to Figure 4.12)

$\therefore \qquad V_1 = 18V_2$

Swept volume, $\qquad V_s = V_1 - V_2$

$\qquad = 18V_2 - V_2 = 17V_2$

$V_3 - V_2 = 0.06V_s$

$\therefore \qquad V_3 = 0.06 \times 17V_2 + V_2 = 2.02V_2$

Cut-off ratio, $\qquad \beta = \dfrac{V_3}{V_2} = 2.02$

$c_p = c_v + R = 0.717 + 0.287 = 1.004$ kJ/(kg K)

$\gamma = \dfrac{c_p}{c_v} = \dfrac{1.004}{0.717} = 1.4$

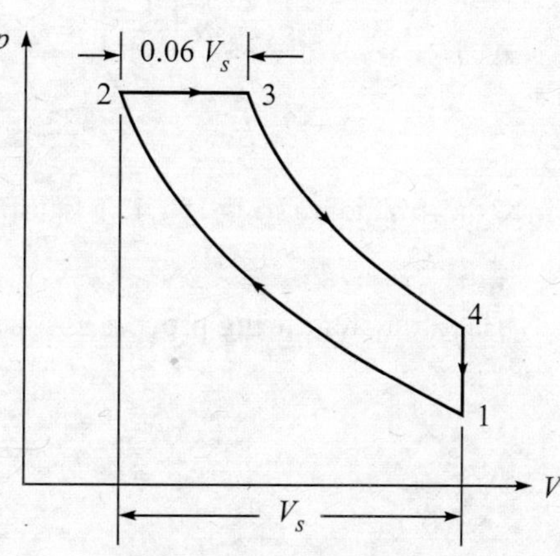

Figure 4.12 Example 4.2.

$$\eta = 1 - \left(\frac{1}{r}\right)^{\gamma-1}\left[\frac{\beta^{\gamma}-1}{\gamma(\beta-1)}\right]$$

$$= 1 - \left(\frac{1}{18}\right)^{0.4}\left[\frac{2.02^{1.4}-1}{1.4(2.02-1)}\right] = 0.6306$$

$$\frac{d\eta}{\eta} = -\left(\frac{1-0.6306}{0.6306}\right)0.4\left[\ln 18 - \frac{2.02^{1.4}\ln 2.02}{2.02^{1.4}-1} + \frac{1}{1.4}\right]0.02 = -0.01163$$

$$\therefore \quad \frac{d\eta}{\eta}\times 100 = -1.163\%$$

Therefore, the efficiency decreases by $\boxed{1.163\%}$ **Ans.**

EXAMPLE 4.3 The compression ratio of an engine working on an Otto cycle is 8 and the air/fuel ratio is 15 : 1, the pressure and temperature at the beginning of a compression stroke being 1 bar and 60°C respectively. The calorific value of the fuel is 44,000 kJ/kg. Determine the maximum temperature and pressure in the cylinder, if the index of compression is 1.32 and the specific heat at constant volume is given by $c_v = (0.678 + 0.00013T)$ kJ/(kg K), where T is the temperature in kelvin. Compare this value with that of constant specific heat $c_v = 0.717$ kJ/(kg K).

Solution: Refer to Figure 4.13.

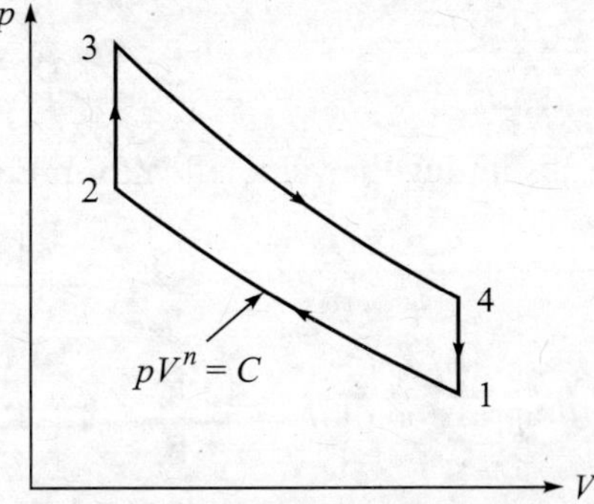

Figure 4.13 Example 4.3.

Compression ratio, $r = \frac{V_1}{V_2} = 8$

$$p_1 = 1 \text{ bar},\ T_1 = 60 + 273 = 333 \text{ K}$$

$$n = 1.32$$

Consider the compression process 1–2.

$$p_2V_2^n = p_1V_1^n$$

$$\therefore \quad p_2 = p_1\left(\frac{V_1}{V_2}\right)^n = 1(8)^{1.32} = 15.56 \text{ bar}$$

$$T_2 = T_1 r^{n-1} = 333\,(8)^{0.32} = 647.8 \text{ K}$$

Since the air/fuel ratio is 15 : 1, 1 kg of mixture has $\frac{1}{16}$ kg of fuel and $\frac{15}{16}$ kg of air.

$\therefore$ Heat transfer during process 2–3 per kg of mixture, $q_{2\text{-}3} = \frac{44{,}000}{16} = 2750$ kJ/kg

Also,
$$q_{2\text{-}3} = \int_{T_2}^{T_3} c_v\,dT$$

$$\therefore \quad \int_{647.8}^{T_3} (0.678 + 0.00013T)dT = 2750$$

or $$0.678(T_3 - 647.8) + 0.000065(T_3^2 - 647.8^2) = 2750$$

or $$0.000065T_3^2 + 0.678T_3 - 3216 = 0$$

or $$T_3^2 + 10{,}430T_3 - 494.8 \times 10^5 = 0$$

$$\therefore \quad T_3 = \frac{-10{,}430 + \sqrt{(10{,}430)^2 + 4 \times 494.8 \times 10^5}}{2} = \boxed{3541\text{ K}} \quad \textbf{Ans.}$$

and $$p_3 = \frac{T_3}{T_1} \times \frac{V_1}{V_3} \times p_1 = \frac{3541}{333} \times 8 \times 1 = \boxed{85\text{ bar}} \quad \textbf{Ans.}$$

With constant value of c_v,

$$c_v(T_3 - T_2) = 2750$$

or $$0.717(T_3 - 647.8) = 2750$$

$$\therefore \quad T_3 = \boxed{4483\text{ K}} \quad \textbf{Ans.}$$

and $$p_3 = \frac{4483}{333} \times 8 \times 1 = \boxed{107.7\text{ bar}} \quad \textbf{Ans.}$$

Maximum pressure and maximum temperature decrease while considering the variation of specific heat with temperature.

EXAMPLE 4.4 The compression ratio of an engine working on Diesel cycle is 21 and the air/fuel ratio is 29 : 1. The temperature at the end of compression is 1000 K. The calorific value of fuel is 42,000 kJ/kg and the specific heat at constant volume of the products of combustion is given by $c_v = (0.71 + 28 \times 10^{-6}T)$ kJ/(kg K) and $R = 0.287$ kJ/(kg K). Determine the percentage of stroke at which combustion is complete.

Solution: Refer to Figure 4.14.

The air/fuel ratio is 29 : 1, therefore, 1 kg mixture has $\frac{1}{30}$ kg of fuel.

$$c_p = c_v + R = 0.71 + 28 \times 10^{-6}T + 0.287$$
$$= 0.997 + 28 \times 10^{-6}T$$

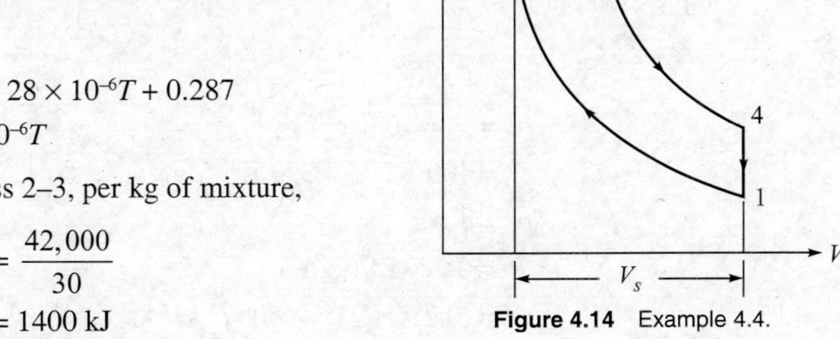

Figure 4.14 Example 4.4.

Heat transfer during process 2–3, per kg of mixture,

$$q_{2\text{-}3} = \frac{42{,}000}{30}$$
$$= 1400 \text{ kJ}$$

Also, $$q_{2\text{-}3} = \int_{T_2}^{T_3} c_p dT$$

$$\therefore \quad \int_{T_2=1000\,\text{K}}^{T_3} (0.997 + 28 \times 10^{-6} T)dT = 1400$$

or $$0.997(T_3 - 1000) + 14 \times 10^{-6}(T_3^2 - 1000^2) = 1400$$

or $$14 \times 10^{-6} T_3^2 + 0.997 T_3 - 2411 = 0$$

$$\therefore \quad T_3 = \frac{-0.997 + \sqrt{0.997^2 + 4 \times 2411 \times 14 \times 10^{-6}}}{28 \times 10^{-6}} = 2341 \text{ K}$$

For constant pressure process 2–3,

$$V_3 = \frac{T_3}{T_2} \times V_2 = \frac{2341}{1000} V_2 = 2.341 V_2$$

Swept volume, $$V_s = V_1 - V_2 = V_2\left(\frac{V_1}{V_2} - 1\right) = V_2(r - 1) = V_2(21 - 1)$$

$$\therefore \quad V_s = 20V_2$$

$$V_3 - V_2 = 2.341V_2 - V_2 = 1.341V_2$$

Percentage stroke during which the combustion is completed,

$$\frac{V_3 - V_2}{V_s} \times 100 = \frac{1.341V_2}{20V_2} \times 100 = \boxed{6.705\%} \quad \textbf{Ans.}$$

EXAMPLE 4.5 An oil engine working on a dual combustion cycle, has a compression ratio of 16 : 1 and the cut-off takes place at 6% of the stroke. The maximum pressure obtained is 70 bar. The pressure and temperature at the beginning of compression are 1 bar and 100°C respectively. Determine the pressure and teperature at all the important points of the cycle. Assume R = 0.287 kJ/(kg K) and $c_v = (0.716 + 125 \times 10^{-6}T)$ kJ/(kg K).

Solution: Refer to Figure 4.15.

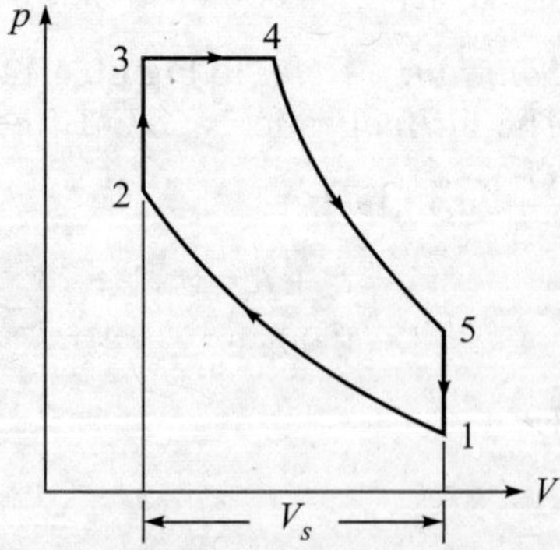

Figure 4.15 Example 4.5.

Given: Compression ratio $r = \dfrac{V_1}{V_2} = 16$

$$V_4 - V_3 = 0.06V_s$$
$$p_3 = p_4 = 70 \text{ bar}$$
$$T_1 = 100 + 273 = 373 \text{ K}$$
$$p_1 = 1 \text{ bar}$$
$$R = 0.287 \text{ kJ/(kg K)}$$
$$c_v = (0.716 + 125 \times 10^{-6}T) \text{ kJ/(kg K)}$$

c_v varies with temperature, so the value of c_p and γ also vary. For compression and expansion processes, the mean value of γ will be different and it is not known.

Consider process 1–2: Assume $\gamma = 1.4$, but its validity has to be checked.

$$T_2 = T_1(r)^{\gamma-1} = 373(16)^{0.4} = 1131 \text{ K}$$

Mean value of c_v for the temperature range 373 K to 1131 K is

$$c_v = \frac{\int_{T_1}^{T_2} c_v dT}{T_2 - T_1} = \frac{1}{1131 - 373} \times \int_{373}^{1131} (0.716 + 125 \times 10^{-6} T) dT$$

$$= \frac{0.716(1131 - 373) + 62.5 \times 10^{-6}(1131^2 - 373^2)}{1131 - 373} = 0.81$$

$$c_p = c_v + R = 0.81 + 0.287 = 1.097$$

and $$\gamma = \frac{c_p}{c_v} = \frac{1.097}{0.81} = 1.354$$

The assumed value of γ is therefore not correct.

Now, assume $\gamma = 1.354$.

$$\therefore \quad T_2 = T_1(r)^{\gamma-1} = 373(16)^{0.354} = 995 \text{ K}$$

$$\text{Mean, } c_v = \frac{1}{995 - 373} \int_{373}^{995} (0.716 + 125 \times 10^{-6} T) dT = 0.8015$$

$$c_p = c_v + R = 0.8015 + 0.287 = 1.0885$$

$$\therefore \quad \gamma = \frac{c_p}{c_v} = \frac{1.0885}{0.8015} = 1.358$$ (This value is very close to the assumed value.)

Thus the mean value of γ is taken as 1.358 during compression.

$$T_2 = T_1(r)^{\gamma-1} = 373(16)^{0.358} = \boxed{1006.4 \text{ K}}$$ **Ans.**

$$p_2 = \frac{T_2}{T_1} \times \frac{V_1}{V_2} \times p_1 = \frac{1006.4}{373} \times 16 \times 1 = \boxed{43.17 \text{ bar}}$$ **Ans.**

Consider process 2–3: It is a constant volume process, i.e. $V_2 = V_3$.

$$\therefore \quad T_3 = \frac{p_3}{p_2} \times T_2 = \frac{70}{43.17} \times 1006 = \boxed{1631 \text{ K}}$$ **Ans.**

$$V_4 - V_3 = 0.06 V_s$$

$$V_4 - V_3 = 0.06(V_1 - V_2) = 0.06 V_2\left(\frac{V_1}{V_2} - 1\right) = 0.06 V_3(r - 1)$$

$$\therefore \quad \frac{V_4}{V_3} - 1 = 0.06(r - 1)$$

or $$\frac{V_4}{V_3} = 0.06(r - 1) + 1 = 0.06(16 - 1) + 1 = 1.9$$

Consider process 3–4: It is a constant pressure process, i.e. $p_3 = p_4$.

$$T_4 = \frac{V_4}{V_3} T_3 = 1.9 \times 1631 = \boxed{3099 \text{ K}} \quad \textbf{Ans.}$$

$$p_4 = p_3 = \boxed{70 \text{ bar}} \quad \textbf{Ans.}$$

Consider process 4–5: It is a reversible adiabatic expansion process. The value of γ is not the same as that for the compression process. At higher temperatures, γ decreases.

Assume γ = 1.3 and check its validity.

$$\frac{V_5}{V_4} = \frac{V_5}{V_3} \times \frac{V_3}{V_4} = \left(\frac{V_1}{V_2}\right)\left(\frac{V_3}{V_4}\right) = \left(\frac{V_1}{V_2}\right)\left(\frac{V_3}{V_4}\right) = \frac{16}{1.9} = 8.421$$

$$\frac{T_5}{T_4} = \left(\frac{V_4}{V_5}\right)^{\gamma-1} = \left(\frac{1}{8.421}\right)^{0.3}$$

or
$$T_5 = \frac{3099}{(8.421)^{0.3}} = 1635 \text{ K}$$

$$\text{Mean, } c_v = \frac{1}{T_5 - T_4}\int_{T_4}^{T_5} c_v\, dT = \frac{1}{T_4 - T_5}\int_{T_5}^{T_4} c_v\, dT$$

$$= \frac{1}{3099 - 1635}\int_{1635}^{3099} (0.716 + 125 \times 10^{-6}\, T)dT$$

$$= \frac{1}{1464}[0.716(3099 - 1635) + 62.5 \times 10^{-6}(3099^2 - 1635^2)]$$

$$= 1.012$$

$$c_p = R + c_v = 0.287 + 1.012 = 1.299$$

$$\gamma = \frac{c_p}{c_v} = \frac{1.299}{1.012} = 1.284 \text{ (The assumed value of } \gamma \text{ is therefore not correct.)}$$

Try γ = 1.284:

$$T_5 = 3099\left(\frac{1}{8.421}\right)^{0.284} = 1692 \text{ K}$$

$$c_v = \frac{1}{3099 - 1692}[0.716(1407) + 62.5(3.099^2 - 1.692^2)] = 1.015$$

$$c_p = c_v + R = 1.015 + 0.287 = 1.302$$

$$\gamma = \frac{c_p}{c_v} = \frac{1.302}{1.015} = 1.283 \text{ (The assumed value of } \gamma \text{ is therefore correct.)}$$

$\therefore$
$$T_5 = 3099\left(\frac{1}{8.421}\right)^{0.283} = \boxed{1696 \text{ K}} \quad \textbf{Ans.}$$

Now, $\frac{p_1 V_1}{T_1} = \frac{p_5 V_5}{T_5}$ and $V_1 = V_5$

$\therefore$ $p_5 = \frac{T_5}{T_1} p_1 = \frac{1696}{373} \times 1 = \boxed{4.55 \text{ bar}}$ **Ans.**

EXAMPLE 4.6 A four-cylinder 2.0*l* SI engine operates on an air-standard Miller cycle with early valve closing. It has a compression ratio of 8 and an expansion ratio of 10. Fuel is isooctane with air/fuel ratio of 15. Heating value of fuel is 44,300 kJ/kg. When the intake valve closes, conditions in the cylinder are 100 kPa and 60°C. Assume that a 3% exhaust residual is left over from the previous cycle. Take $c_v = 0.821$ kJ/(kg K) and ratio of specific heat $\gamma = 1.35$.

Calculate:

(a) Temperature and pressure at all points in the cycle
(b) Work done per cycle
(c) Indicated thermal efficiency
(d) Indicated mean effective pressure

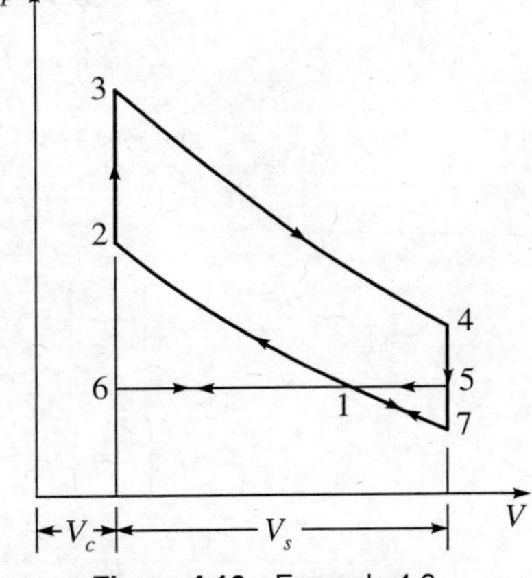

Figure 4.16 Example 4.6.

Solution: Refer to Figure 4.16.

Swept volume, $V_s = \frac{2.0}{4} = 0.5\, l$

Expansion ratio, $e = \frac{V_4}{V_3} = \frac{V_s + V_c}{V_c} = 10$

$$V_c = \frac{V_s}{e-1} = \frac{0.5}{9} = 0.0556\, l = V_2 = V_3 = V_6$$

$$V_4 = 10 V_3 = 10 \times 0.0556 = 0.556\, l = V_5 = V_7$$

Compression ratio,

$$r = \frac{V_1}{V_2} = 8$$

$$V_1 = 8 V_2 = 8 \times 0.0556 = 0.445\, l$$

(a) *Given:*

$$T_1 = 60°C = 333 \text{ K}$$

$$p_1 = p_5 = p_6 = \boxed{100\ kPa}$$ **Ans.**

$$p_7 = p_1 \left(\frac{V_1}{V_7}\right)^{\gamma} = 100 \left(\frac{0.445}{0.556}\right)^{1.35} = \boxed{74 \text{ kPa}}$$ **Ans.**

$$T_7 = T_1 \left(\frac{V_1}{V_7}\right)^{\gamma-1} = 333 \left(\frac{0.445}{0.556}\right)^{0.35} = 308 \text{ K} = \boxed{35°\text{C}}$$ **Ans.**

$$T_2 = T_1 (r)^{\gamma-1} = 333(8)^{0.35} = 689 \text{ K} = \boxed{416°\text{C}}$$ **Ans.**

$$p_2 = p_1(r)^{\gamma} = 100(8)^{1.35} = \boxed{1656 \text{ kPa}} \quad \textbf{Ans.}$$

Mass of the gas in the cylinder,

$$m = \frac{p_1 V_1}{RT_1}$$

$$\therefore \quad m = \frac{100 \times 10^3 \times 0.445 \times 10^{-3}}{287 \times 333} = 0.466 \times 10^{-3} \text{ kg}$$

$$\frac{m_a}{m_f} = 15:1 \quad \text{and exhaust residual} = 3\%$$

$$\therefore \quad m_f = \frac{1}{16} \times 0.97 \times 0.466 \times 10^{-3} = 0.0283 \times 10^{-3} \text{ kg}$$

Heat supplied, $Q_s = m_f \times \text{CV} = 0.0283 \times 10^{-3} \times 44,300 = 1.254$ kJ

Also, $Q_s = mc_v(T_3 - T_2)$

$$\therefore \quad 0.466 \times 10^{-3} \times 0.821(T_3 - 689) = 1.254$$

$$\therefore \quad T_3 = 3967 \text{ K} = \boxed{3694°\text{C}} \quad \textbf{Ans.}$$

$$p_3 = p_2\left(\frac{T_3}{T_2}\right) = 1656 \times \frac{3967}{689} = \boxed{9535 \text{ kPa}} \quad \textbf{Ans.}$$

$$T_4 = T_3\left(\frac{V_3}{V_4}\right)^{\gamma-1} = 3967\left(\frac{1}{10}\right)^{0.35} = 1772 \text{ K} = \boxed{1499°\text{C}} \quad \textbf{Ans.}$$

$$p_4 = p_3\left(\frac{V_3}{V_4}\right)^{\gamma} = 9535\left(\frac{1}{10}\right)^{1.35} = \boxed{426 \text{ kPa}} \quad \textbf{Ans.}$$

$$T_5 = T_4\left(\frac{p_5}{p_4}\right) = 1772 \times \frac{100}{426} = 416 \text{ K} = \boxed{143°\text{C}} \quad \textbf{Ans.}$$

(b)

$$W_{1-2} = \frac{m\,R(T_1 - T_2)}{\gamma - 1} = \frac{0.466 \times 10^{-3} \times 287(333 - 689)}{0.35} = -136.0 \text{ J}$$

$$W_{2-3} = 0$$

$$W_{3-4} = \frac{m\,R(T_3 - T_4)}{\gamma - 1} = \frac{0.466 \times 10^{-3} \times 287(3967 - 1772)}{0.35} = 838.8 \text{ J}$$

$$W_{4-5} = 0$$

$$W_{5-1} = p_1(V_1 - V_5) = 100 \times 10^3(0.445 - 0.556) \times 10^{-3} = -11.1 \text{ J}$$

Work done per cycle, $W_{\text{net}} = -136.0 + 838.8 - 11.1 = 691.7 \text{ J} = \boxed{0.6917 \text{ kJ}}$ **Ans.**

(c) $\eta_{th} = \dfrac{W_{net}}{Q_s} = \dfrac{0.6917}{1.254} = 0.552 = \boxed{55.2\%}$ **Ans.**

(d) $p_{m_i} = \dfrac{W_{net}}{V_s} = \dfrac{0.6917}{0.5 \times 10^{-3}} = 1383 \text{ kPa} = \boxed{13.83 \text{ bar}}$ **Ans.**

EXAMPLE 4.7 An engine working on an Otto cycle, having a compression ratio of 8, uses octane C_8H_{18} as a fuel. The lower heating value of the fuel is 44,000 kJ/kg. The air/fuel ratio is 15:1. Determine the maximum pressure and temperature reached in the cycle (a) without considering the molecular expansion and (b) with molecular expansion. Assume $c_v = 0.71$ kJ/(kg K), compression follows the law $pV^{1.3}$ = constant, the pressure and temperature of the mixture at the beginning of compression being 1 bar and 60°C respectively. Determine the percentage molecular expansion.

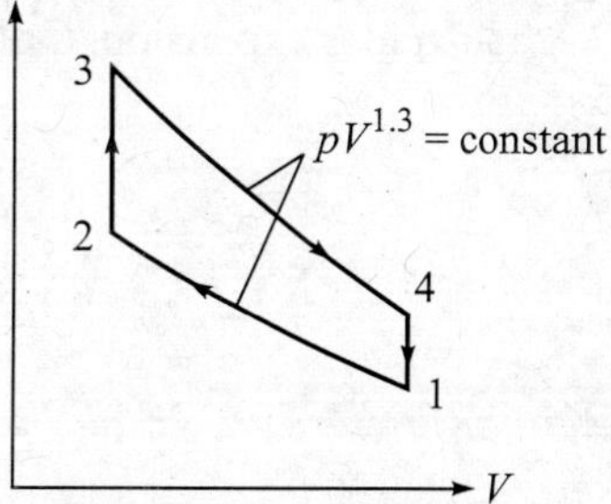

Figure 4.17 Example 4.7.

Solution: Refer to Figure 4.17.

Compression ratio, $r = 8$

The stoichiometric equation can be written as

$$C_8H_{18} + 12.5(O_2 + 3.76N_2) \rightarrow 8CO_2 + 9H_2O + 47N_2$$

$$\text{Stoichiometric air/fuel ratio} = \frac{12.5(32 + 3.76 \times 28.161)}{(12 \times 8) + (1 \times 18)}$$

$$= 15.12$$

With the given air/fuel ratio as 15, the mixture is rich in fuel. Therefore the combustion will be incomplete.

The chemical equation becomes:

$$C_8H_{18} + Y(O_2 + 3.76N_2) \rightarrow xCO + (8 - x)CO_2 + 9H_2O + 3.76YN_2$$

or $$\frac{Y(32 + 3.76 \times 28.161)}{(12 \times 8) + (1 \times 18)} = \frac{\text{air}}{\text{fuel}} = 15$$

$\therefore$ $$Y = 12.4$$

By oxygen balance: $$Y = \frac{x}{2} + (8 - x) + 4.5$$

or $$12.5 - \frac{x}{2} = 12.4$$

$\therefore$ $$x = 0.2$$

The chemical equation now becomes:

$$C_8H_{18} + 12.4(O_2 + 3.76N_2) \rightarrow 0.2CO + 7.8CO_2 + 9H_2O + 46.624N_2$$

No. of moles before combustion = 1 + 12.4 + (12.4 × 3.76) = 60.024

No. of moles after combustion = 0.2 + 7.8 + 9 + 46.624 = 63.624

$\therefore$ $$\text{Molecular expansion} = \frac{63.624 - 60.024}{60.024} \times 100 = \boxed{6\%}$$ **Ans.**

(a) $$p_1 = 1 \text{ bar}, \quad T_1 = 60 + 273 = 333 \text{ K}, \quad n = 1.3$$

$$T_2 = T_1 (r)^{n-1} = 333(8)^{0.3} = 621.4 \text{ K}$$

$$q_{2-3} = c_v (T_3 - T_2)$$

or $$\frac{44{,}000}{16} = 0.71(T_3 - 621.4)$$

∴ $$T_3 = \boxed{4495 \text{ K}} \quad \textbf{Ans.}$$

$$p_3 = \frac{V_1}{V_3} \times \frac{T_3}{T_1} \times p_1 = 8 \times \frac{4495}{333} \times 1 = \boxed{108 \text{ bar}} \quad \textbf{Ans.}$$

(b) Since the mass of the reactants and products is the same and specific heats are assumed same, the temperature of the products with molecular expansion will remain the same as without molecular expansion. Only the pressure will change.

∴ $$T_3 = \boxed{4495 \text{ K}} \quad \textbf{Ans.}$$

$$pV = n\bar{R}T$$

$$p \propto n$$

$$\frac{p_3'}{p_3} = \frac{n'}{n}$$

where n is the number of moles of products without molecular expansion and n' is the number of moles of products with molecular expansion.

∴ $$p_3' = p_3 \times \frac{n'}{n} = 108 \times \frac{63.624}{60.024} = \boxed{114.5 \text{ bar}} \quad \textbf{Ans.}$$

REVIEW QUESTIONS

1. Why is the thermal efficiency of an air-standard cycle much higher than that of an actual engine?
2. Why is the analysis of the fuel-air cycle important?
3. What are the differences between the analysis of the air-standard cycle and that of the fuel-air cycle?
4. What are the assumptions made in the analysis of the fuel-air cycle?
5. How does the composition of cylinder gases affect the fuel-air cycle analysis?
6. How do the specific heats vary with temperature? What is the physical explanation for this variation? What will be its effect on an Otto cycle?
7. Explain the phenomenon of dissociation. Under what conditions is the dissociation of products more pronounced? How does the presence of CO affect dissociation?
8. Show the effect of dissociation on temperature and power of different equivalence ratios.
9. Show the effect of dissociation on the Otto cycle.
10. Explain the effect of change in number of moles during combustion on the maximum pressure in an Otto cycle.

11. How do the compression ratio and the equivalence ratio affect the thermal efficiency and the efficiency ratio (the ratio of fuel-air cycle efficiency to air-standard cycle efficiency) of a constant volume fuel-air cycle?
12. How do the equivalence ratio and the compression ratio affect the maximum pressure and the maximum temperature of a fuel-air cycle?
13. How do the equivalence ratio and the compression ratio affect the temperature at the end of the expansion stroke?
14. How do the equivalence ratio and the compression ratio affect the mean effective pressure of a fuel-air cycle?
15. Prove that the change in thermal efficiency with respect to the change in specific heat at a constant volume of an Otto cycle having compression ratio r is given by

$$d\eta = -(1-\eta)(\gamma - 1)(\ln r)\,\frac{dc_v}{c_v}$$

16. Derive an expression to determine the change in efficiency of a Diesel cycle with respect to change in specific heat at constant volume having compression ratio r and cut-off ratio β.
17. What are the two different types of combustion charts? Describe these charts with the help of simple curves without showing values. Explain the procedure to use these charts for the analysis of a fuel-air cycle.
18. How are the properties of burned gases obtained, starting with the properties of the unburned mixture just before combustion in the case of (a) a constant volume adiabatic combustion and (b) a constant pressure adiabatic combustion?

PROBLEMS

4.1 What will be the percentage change in the efficiency of an Otto cycle having a compression ratio 10, when the specific heat at constant volume increases by 1.5%?

[**Ans.** 0.914% decrease]

4.2 What will be the percentage change in the efficiency of a Diesel cycle having a compression ratio 20 and a cut-off ratio 2.62, when the specific heat at constant volume increases by 1.8%? [**Ans.** 1.06% decrease]

4.3 An engine working on Otto cycle has the air/fuel ratio as 15 : 1. The temperature at the beginning of a compression stroke is 57°C. The calorific value of fuel is 44,000 kJ/kg. The maximum temperature in the cylinder is 3550 K. The index of compression is 1.33 and the specific heat at constant volume in kJ/(kg K) is given by $c_v = 0.678 + 0.00013T$, where T is the temperature in kelvin. Determine the compression ratio of the engine. [**Ans.** 8.18]

4.4 The compression ratio of an engine working on Diesel cycle is 20. The temperature at the end of compression is 1050 K. The calorific value of fuel is 42,000 kJ/kg and the specific heat at constant pressure of the products of combustion is given by $c_p = (0.997 + 28 \times 10^{-6}T)$ kJ/(kg K). Combustion is completed at 6.5% of the stroke. Determine the air/fuel ratio. [**Ans.** 30:1]

4.5 An engine working on Dual combustion cycle, has a compression ratio of 18 and the cut-off takes place at 6.5% of the stroke. The maximum pressure of the cycle is 75 bar.

The pressure and the temperature at the beginning of compression are 1 bar and 67°C respectively. Determine the pressure and the temperature at all the salient points of the cycle. Assume $R = 0.287$ kJ/(kg K) and $c_v = (0.716 + 125 \times 10^{-6}\, T)$ kJ/(kg K).

[**Ans.** $T_2 = 962$ K, $p_2 = 50.93$ bar, $T_3 = 1417$ K, $p_3 = p_4 = 75$ bar, $T_4 = 2983$ K, $T_5 = 1601$ K, $p_5 = 4.71$ bar]

4.6 A single cylinder, 0.3 *l* SI engine operates on an air-standard Miller cycle with late valve closing. It has a compression ratio of 9 and the expansion ratio of 11. Air/fuel ratio is 15:1 and the heating value of fuel is 44,200 kJ/kg. When the intake valve closes, conditions in the cylinder are 1.0 bar and 27°C. Assume that there is a 5% residual gas. Take $c_v = 0.821$ kJ/kg K and $\gamma = 1.35$.
Calculate:

(a) Temperature and pressure at all points in the cycle
(b) Work done per cycle
(c) Work ratio
(d) Indicated thermal efficiency
(e) Indicated mean effective pressure

[**Ans.** (a) $T_2 = 647.3$ K, $p_2 = 19.42$ bar, $T_3 = 3844$ K, $p_3 = 115.3$ bar, $T_4 = 1661$ K, $p_4 = 4.53$ bar, $p_5 = 1.0$ bar, $T_5 = 367$ K, (b) 466.05 J, (c) 0.83 (d) 56.6% (e) 15.54 bar]

4.7 An engine working on Otto cycle with a compression ratio 10, uses C_8H_{18} as a fuel. The calorific value of fuel is 44,000 kJ/kg. The air/fuel ratio is 16 : 1. Determine the percentage molecular expansion. Compute the maximum pressure and the temperature reached in the cycle (a) without considering molecular expansion and (b) with molecular expansion. Assume $c_v = 0.71$ kJ/(kg K), compression follows the law $pV^{1.3}$ = constant. The pressure and the temperature of the mixture at the beginning of compression are 1 bar and 57°C respectively.

[**Ans.** 5.47%, (a) $T_3 = 4304$ K, $p_3 = 130.4$ bar; (b) $T_3 = 4304$ K, $p_3 = 137.5$ bar]

5.1 INTRODUCTION

An engine operating on the air cycle converts input heat into useful work. The remainder heat is rejected at the end of power stroke and represents the energy lost in the hot exhaust gases. Air cycle analysis predicts higher efficiency of the engine than what is actually obtainable, because of a very simplified approach adopted.

When the losses due to chemical equilibrium, variable specific heat and the effect of increase in the number of molecules after combustion are added to the exhaust heat loss, the fuel-air cycle results. The analysis of the fuel-air cycle predicts better results but still not so close as desired.

The actual cycle analysis considers other losses in addition to the above losses. This analysis predicts very close results compared to the results obtained actually by running the engine and taking measurements by sophisticated instruments. An estimate of the losses can be made from previous experience and from simple tests performed on the engines. It is now possible to analyse the actual cycle with the help of computer programs developed for this purpose.

5.2 DIFFERENCE BETWEEN THE ACTUAL CYCLE AND THE FUEL-AIR CYCLE

The possible causes of the observed differences between the actual cycle and the fuel-air cycle include the following losses that are taken into account for the analysis of the actual cycle, and which are not considered in the analysis of the fuel-air cycle.

1. Leakage
2. Imperfect mixing of fuel and air
3. Progressive burning
4. Burning time losses, due to the motion of the piston during combustion.
5. Heat losses to the cylinder walls
6. Exhaust blowdown loss
7. Fluid friction
8. Gas exchange or pumping loss

5.2.1 Leakage

At higher piston speeds, leakage is usually insignificant in a well adjusted engine. However, at low piston speeds and high gas pressure, the gas flows into the regions between the piston, piston rings and cylinder walls and gets cooled by heat transfer through cylinder walls. These regions are called *crevice* regions. The gases flowing into these regions usually remain unburned and some of the gases return to the cylinder during the later part of the expansion stroke and the remaining gas leaks past the piston rings to the crank case. Leakage can be estimated by measuring blow by, that is, the mass of the gases flowing out from the crank case breather. This leakage loss reduces the cylinder pressure during combustion and during the early part of the expansion stroke, thus reducing the net power output of the engine.

5.2.2 Imperfect Mixing of Fuel and Air

In practice, it is not possible to obtain a perfect homogeneous mixture of fuel, air and residual gases in the cylinder before the ignition takes place, because of insufficient turbulence. In one part of the cylinder, there may be excess oxygen and in another part excess fuel may be present. The excess fuel may not find enough oxygen for complete combustion, which may result in the appearance of CO, H_2 and unburned fuel in the exhaust. The efficiency of the engine will decrease because of the wastage of fuel. Figure 5.1 shows the composition of exhaust gases of a typical gasoline engine at various equivalence ratios. The wastage of fuel can be reduced by using a lean mixture. This will ensure the complete utilization of fuel, thus providing maximum economy.

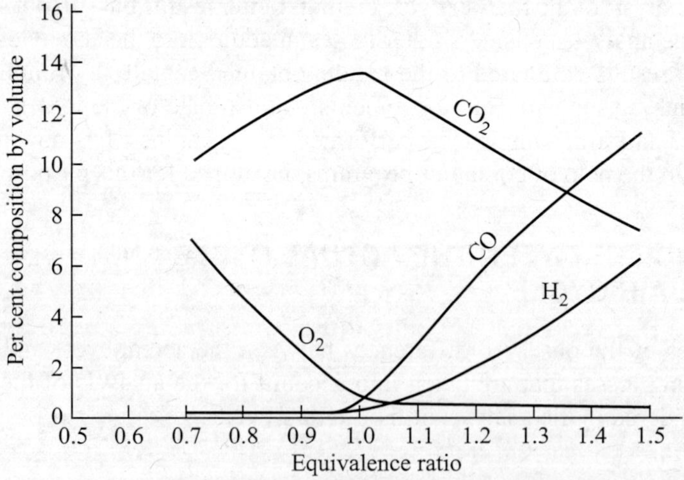

Figure 5.1 Composition of the exhaust gases of a typical gasoline engine at various equivalence ratios.

Even if the unburned fuel and oxygen combine later during the expansion stroke, resulting in no loss of fuel in the exhaust, still there will be a loss in efficiency, since the sensible energy present with the combustion of this part of fuel is not utilized at top dead centre and hence does not contribute to the pressure rise at that point.

A carburettor produces a wet, non-homogeneous mixture and fuel is unevenly distributed between the inlet manifold branches of a multi-cylinder engine. Thus, some cylinders receive a much leaner mixture than others. This maldistribution of fuel in a multi-cylinder engine does not permit to run the engine with the most economical fuel/air ratio. With such mixture, the mixture in some cylinders may be too lean. This may cause slow burning of fuel or the fuel may not burn at all.

The effect of imperfect mixing of fuel and air on engine efficiency is not much, provided the inlet system is designed properly and the inlet manifold is placed over the exhaust manifold to heat the charge for better mixing.

5.2.3 Progressive Burning

In the analysis of constant volume fuel-air cycle it is assumed that the ignition takes place at top dead centre and combustion is instantaneous. In an actual SI engine, combustion starts at a certain point and continues by moving the flame front. Combustion is complete when the flame front has passed through the entire charge. It takes some time in doing so. The time required for this varies with the fuel composition, the combustion chamber shape and size including the number and position of the ignition point and engine operating conditions. Different amounts of charge burn at different times even if the piston is nearly stationery during combustion. The spark ignites a very small portion of the charge immediately adjacent to it. The flame then spreads progressively throughout the mixture. This phenomenon of burning is called *progressive burning*.

The calculations of fuel-air cycles with progressive burning show that the mean effective pressure and efficiency of the cycles are almost the same as those for similar fuel-air cycles with instantaneous burning. For such cycles, the piston is assumed to remain at top dead centre during burning of all parts of the charge.

The losses due to leakage, imperfect mixing and progressive burning are too small and cannot be evident on the *p-V* diagram. The remaining losses, namely the time loss, the heat loss, the exhaust blowdown loss are significant and can be shown on the *p-V* diagram.

5.2.4 Burning Time Losses

The crankshaft normally rotates through 40° or more between the time the spark is produced and the time the charge is completely burned. The time in degree crank angle (°CA) depends upon the flame speed and the distance between the position of spark plug and the farthest side of combustion chamber. The flame travel distance can be reduced by locating the spark plug at the centre of the cylinder head or by using more than one spark plug. A hemispherical combustion chamber often uses two spark plugs mounted on the cylinder head on opposite sides to reduce the flame travel distance. The motion of the flame front depends upon how fast the heat is transferred from the flame front to the unburned mixture just ahead of the flame front. Heat is generated by the chemical reaction at the flame front.

As the crankshaft rotates, the piston moves and if the piston motion during combustion is taken into account the burning time losses are determined, which results in loss of work and efficiency. However, the burning time losses are quite large, if:

(a) The fuel/air ratio is made too lean, or too rich.
(b) The throttle is partially closed, reducing the suction pressure.
(c) The point of ignition is not properly set.

5.2.5 Heat Losses to the Cylinder Walls

Heat transfer between the cylinder gases and the cylinder walls during the compression stroke, up to the point of ignition at which combustion starts, appears to be negligible, owing to the gas temperature being not high. During combustion and early part of the expansion stroke, the temperature of the cylinder gases is high and a considerable amount of heat flows from the hot gases through the cylinder walls and cylinder head into the water jacket in water-cooled engines or cooling fins in air-cooled engines. Some heat enters into the piston head, and from there it flows to the piston rings. As the piston rings are in contact with the cylinder walls, so the heat flows to the cylinder walls. Out of this heat, some heat is taken away by the engine oil present between the piston rings and the cylinder walls for lubrication. The rest of the heat goes to the cooling media.

As a result of loss of heat to the cylinder walls the work and efficiency of the cycle are reduced because some of the heat energy liberated by combustion is not utilized for producing work during expansion. However, the loss of heat during combustion reduces the maximum temperature and the dissociation of CO_2 and H_2O is also reduced, which results in complete combustion, but the improvement in work and efficiency is only marginal.

5.2.6 Exhaust Blowdown Loss

The pressure at the end of the expansion stroke is much higher than the atmospheric pressure. If the exhaust valve is opened at bottom dead centre, the piston would have to do a large amount of work in order to expel the high pressure exhaust gases during the exhaust stroke. If the exhaust valve is opened too early, part of the expansion work is lost. The best position is to open the exhaust valve 40° to 70° before BDC. In this case too, some expansion work is lost but the work spent by the piston during the exhaust stroke is reduced. The net result will be the gain in some work. The early opening of the exhaust valve releases the pressure of the gas before the piston reaches BDC. This process is called *exhaust blowdown*. Figure 5.2 shows the effect of exhaust valve opening time on exhaust blowdown loss. With proper designing and timing of the exhaust valve opening, the blowdown loss is not much.

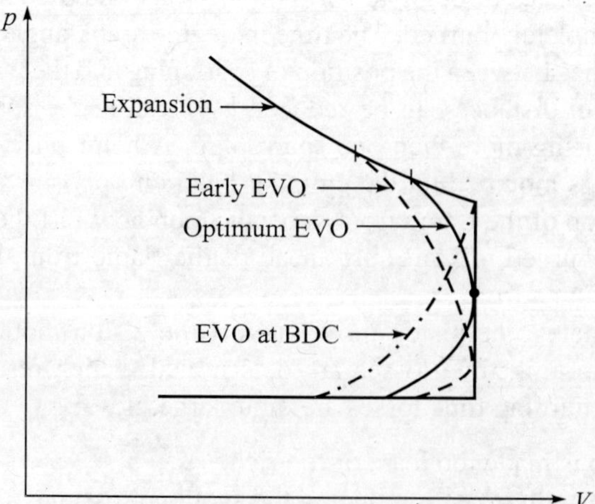

Figure 5.2 Effect of exhaust valve opening (EVO) time on exhaust blowdown loss.

Figure 5.3 shows the comparison of the constant volume fuel-air cycle with the actual cycle, indicating the burning time loss, heat loss and exhaust loss. It is a power cycle and does not include gas exchanges.

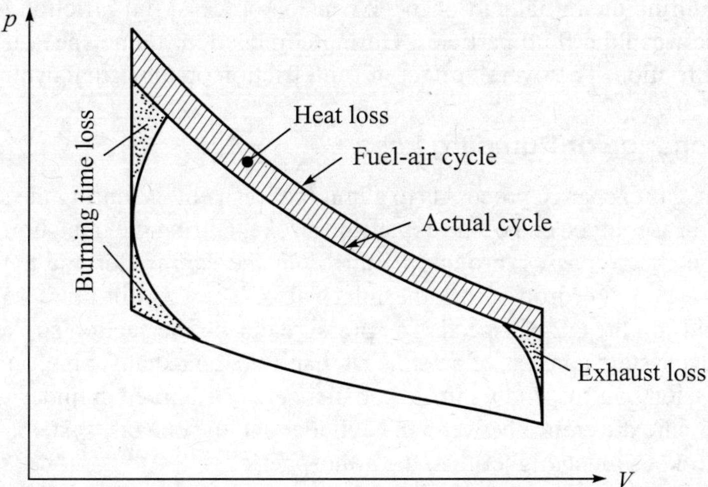

Figure 5.3 Comparison of constant volume fuel-air cycle with the actual cycle showing the different losses.

Loss ratio

It is the ratio of the loss area to the fuel-air cycle area. It is used for the sake of comparison of different losses. The following loss ratios are defined:

(i) *Time loss ratio:* The ratio of the burning time loss area to the area of the fuel-air cycle diagram is called the time loss ratio.

(ii) *Heat loss ratio:* The ratio of the heat loss area to the area of fuel-air cycle diagram is called the *heat loss ratio.*

(iii) *Blowdown loss ratio:* The ratio of the exhaust blowdown area to the area of the fuel-air cycle diagram is called the *blowdown loss ratio.*

(iv) *Lost work ratio:* It can be expressed as the ratio of the difference of area of fuel-air cycle and area of actual cycle to the area of fuel-air cycle, i.e.

$$\text{Lost work ratio} = 1 - \frac{\text{area of actual cycle}}{\text{area of fuel-air cycle}}$$

For a well adjusted and good condition typical SI engine, the ratio of the work of the actual cycle to that of the fuel-air cycle can be safely assumed as 0.8. Total lost work is therefore 20% of the fuel-air cycle work. Let us assume based on experience that out of this lost work 30% goes to the time loss, 60% goes to the heat loss and 10% goes to the exhaust blowdown loss.

For the above example, the time loss comes out to be 6%, heat loss 12% and the blowdown loss comes out to be 2% of the fuel-air cycle work.

5.2.7 Fluid Friction

At high engine speeds, turbulence inside the engine cylinder causes friction between fluid particles. During suction and exhaust processes, fluid friction is appreciable. Near the end of the compression process and during the initial part of the expansion process, fluid friction is high because of higher pressure between the fluid particles. During combustion, flame speed is turbulent, which causes more fluid friction. The overall effect of fluid friction on the actual cycle is very little.

5.2.8 Gas Exchange or Pumping Loss

The purpose of the gas exchange process is to admit the fresh charge during the suction stroke and remove the burned gases at the end of the expansion stroke. During the induction process, pressure losses occur as the charge passes through the air filter, the carburettor and the intake manifold. There is additional pressure drop across the inlet valve. The drop in pressure along the intake system also depends on the engine speed. The pressure during the suction stroke is below atmospheric. The exhaust system consists of an exhaust manifold, an exhaust pipe, and often a catalytic converter for emission control, and a muffler or silencer. The burned cylinder gases get expelled because of the pressure difference between the cylinder and the exhaust system. The pressure during the exhaust stroke is much higher than the atmospheric.

Figure 5.4 shows the gas exchange processes in a conventional SI engine. The difference of work done in expelling the exhaust gases and the work done by the fresh charge during the suction stroke is called the *pumping work* and the loop formed is called the *pumping loop*. The area of the pumping loop indicates negative work and it represents pumping losses. The term pumping is used as the gas from the lower inlet pressure is pumped to the higher exhaust pressure. The pumping loss increases at part throttle because throttling reduces the suction pressure. The pumping loss also increases with speed.

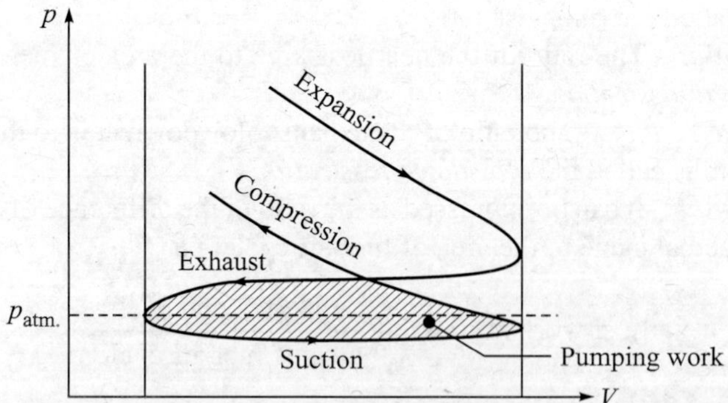

Figure 5.4 Gas exchange processes.

5.3 EFFECT OF ENGINE VARIABLES ON FLAME SPEED

Flame speed is a very important parameter which affects the actual cycle diagram and hence the work output from the engine. The flame speed depends upon certain engine variables as explained in the following subsections.

5.3.1 Fuel/Air Ratio

The flame speed is highly dependent upon temperature. At higher temperatures, the flame speed increases. Figure 4.9 shows that the maximum temperature is obtained with a slightly rich mixture and hence the flame speed is maximum. It is obvious that for a lean mixture, less chemical energy is converted into heat energy, and hence the flame temperature is low. A very rich mixture results in incomplete combustion producing less heat. A slightly rich mixture results in maximum flame temperature during combustion and therefore gives the maximum flame speed.

5.3.2 Inlet Pressure

For a given engine speed, the inlet pressure can be changed either by throttling or by supercharging the engine. When the throttle valve is partially closed, the inlet pressure reduces causing lower maximum temperature, thus resulting in reduction of the flame speed. On the other hand, if the engine is supercharged the inlet pressure increases, causing higher maximum temperature and resulting in increase in the flame speed.

5.3.3 Engine Speed

At higher engine speeds, the mean piston speed increases, which causes a higher inlet velocity of the mixture in the cylinder through the inlet valve. This forms vortices and hence turbulence in the incoming charge. The turbulence persists during burning, causing relative motion between the flame front and the unburned mixture. This motion increases the heat transfer between them and therefore the flame speed increases.

5.3.4 Engine Size

The flame speed does not depend upon the engine size of engines of similar design. Engines of similar design will have the same mean piston speed (S) and the same ratio of area of piston head to the effective area of the inlet valve (A_p/A_v) regardless of their size.

Mean piston speed, $S = 2LN$, where L is the stroke length and N is the engine speed in revolutions per unit time.

In order to keep the mean piston speed almost the same, the larger engines having longer stroke lengths run at lower engine speeds and the smaller engines having shorter stroke lengths run at higher engine speeds.

Assuming the density constant of entering mixture through valve and applying the continuity equation,

$$\text{Inlet velocity of the charge} = S(A_p/A_v).$$

Since for the similar designed engines, S and A_p/A_v are the same, therefore, the inlet velocity of the charge, the turbulence and flame speed will approximately be the same. With the same flame speed, the time taken for completing the combustion process in the larger engine will be more because the distance across the combustion space is greater. The speed (revolutions per minute (rpm)) of the larger engine is low, so the crank takes longer to complete a revolution. The result is that the crank angle required for burning will be almost the same regardless of the size of the engine. The optimum ignition advance will therefore be approximately the same and the burning time losses will be almost the same percentage of the energy input.

5.3.5 Residual Gas

If the residual gas is more, the fresh charge is diluted. It will cause flame speed to reduce and the burning time losses will be more. The amount of residual gas present depends upon the size of the clearance space and the exhaust pressure. The smaller the size of clearance space and lower the exhaust pressure, the less will be the amount of the residual gas present, which can be achieved with a higher compression ratio. The higher the inlet pressure, the more will be the fresh charge. The flame speed increases with more fresh charge and less residual gas.

5.4 EFFECT OF SPARK-TIMING ON THE ACTUAL CYCLE OF SI ENGINES

Suppose the spark timing is set in such a way that as the piston reaches the TDC position, the spark is striked and the combustion of fuel takes place. Combustion is completed only after the piston has travelled a considerable distance from the TDC on expansion stroke. Figure 5.5 shows the actual indicator diagram with ignition at TDC and also shows the corresponding fuel-air cycle indicator diagram. It is observed that the maximum pressure is considerably reduced in comparison to the maximum pressure obtained with the fuel-air cycle. A considerable area of the diagram is lost, and thus the actual power and efficiency are low.

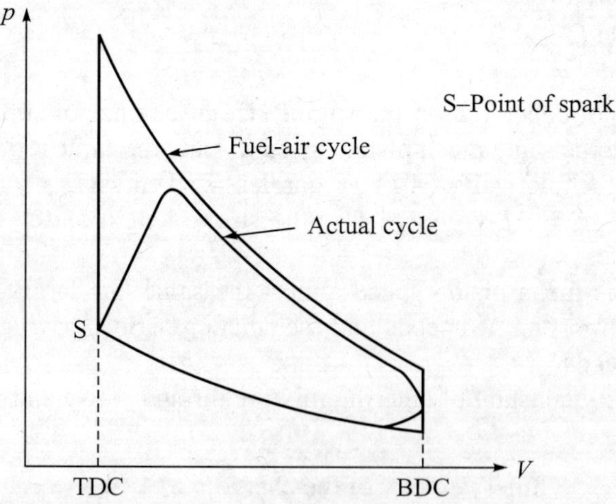

Figure 5.5 Shows the actual indicator diagram with ignition at TDC and also shows the corresponding fuel-air cycle indicator diagram.

If the spark timing is much advanced, the ignition takes place early on the compression stroke. It is possible to get complete combustion in this case just as the piston reaches TDC, but additional work is required to compress the working gases, so that once again the work area is less, and thus the power and efficiency are low. Figure 5.6 shows the actual indicator diagram with early ignition and also shows the corresponding fuel-air cycle indicator diagram.

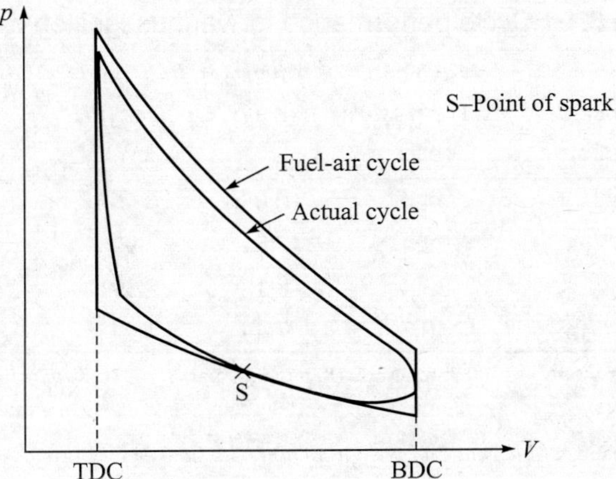

Figure 5.6 Shows the actual indicator diagram with early ignition and also shows the corresponding fuel-air cycle indicator diagram.

A moderate ignition advance gives the best result. There will be small losses on both the compression and expansion strokes, thus giving maximum power and efficiency. Figure 5.7 shows an actual indicator diagram with optimum ignition advance (15° to 30° bTDC) and also shows the corresponding fuel-air cycle indicator diagram.

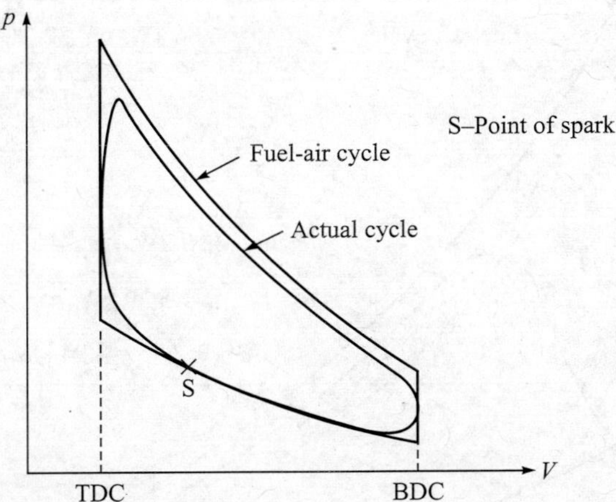

Figure 5.7 Shows the actual indicator diagram with optimum ignition advance and also shows the corresponding fuel-air cycle indicator diagram.

The timing of the spark highly depends upon the speed and the load. The centrifugal and vacuum advance mechanisms are used to provide practically perfect spark timing for all operating conditions. The vacuum unit produces greater advance when the engine is running under part throttle and the centrifugal unit produces greater advance when the engine speed is increased.

Table 5.1 compares the results of the engine performance, such as maximum cycle pressure, mean effective pressure and efficiency, considering the analysis of the actual cycle for various ignition timings with respect to the fuel-air cycle performance. The results are for Cooperative Fuel Research (CFR) engine. The specifications and the operating conditions are mentioned in the table.

Table 5.1 Cycle performance for various ignition timings

Cycle	*Ignition advance,* θ (degree)	*Maximum cycle pressure* (bar)	*mep* (bar)	*Indicated power (ip)* (kW)	*Effciency,* η (%)	*Actual* η / *Fuel cycle* η
Fuel-air cycle	0°	44.0	10.03	6.66	32.2	1.00
Actual cycle	0°	22.4	7.51	4.99	24.1	0.75
	17°	33.4	8.18	5.39	26.3	0.82
	35°	40.7	7.44	4.94	23.9	0.74

(d = 82.6 mm, L = 114.3 mm, r = 6, N = 1300 rpm, ϕ = 1.18, p_i = 0.97 bar, T_i = 310 K, p_e = 1.02 bar, calorific value of fuel = 44,000 kJ/kg)

Source: Rogowski, A.R:, *Elements of Internal Combustion Engines*, McGraw-Hill Book Co., New York, 1953.

Figure 5.8 shows the effect of spark-advance on the p-V diagram. The cycles are superimposed for comparison. The indicated power (ip) and efficiency (η) are marked at different angles of ignition advance. It is clear from the diagram that for given specifications and operating conditions a certain value of ignition advance is required for optimum results. A greater than or less than this value will result in reduced power and poor efficiency.

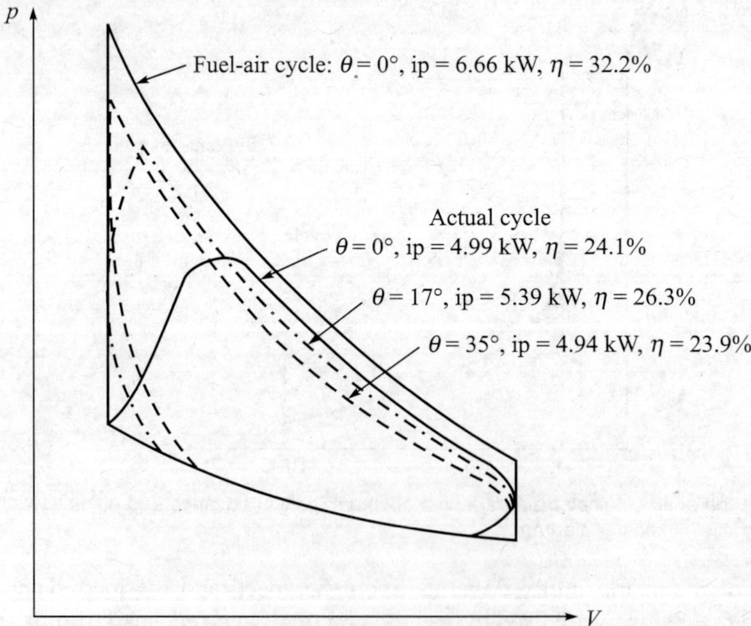

Figure 5.8 The effect of spark-advance on the p–V diagram.

Figure 5.9 shows the effect of deviating spark timing from optimum on the relative indicated mean effective pressure (imep). With optimum spark timing the relative imep is taken as unity. As the ignition advance is increased or decreased from the optimum value, there is a reduction in the imep, and as a result there is a loss of power. In actual practice, sometime a deliberate spark

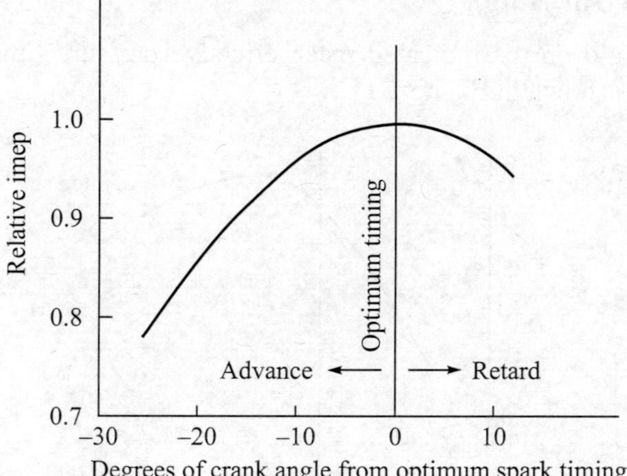

Figure 5.9 Effect of deviating spark timing from optimum on relative indicated mean effective pressure.

retardation from optimum may be necessary in order to avoid knocking and to simultaneously reduce exhaust emission of hydrocarbons and carbon monoxide.

5.5 POWER AND EFFICIENCY OF THE ACTUAL CYCLE

The power and efficiency of the actual engine depend upon many factors. Here, the effects of compression ratio and fuel/air ratio are considered.

5.5.1 Effect of Compression Ratio

The effect of compression ratio on the indicated thermal efficiency of the actual cycle of spark-ignition engine is shown in Figure 5.10. For the sake of comparison, the air-cycle efficiency is also shown. The nature of the two curves is the same, but the efficiency of the actual cycle is lower. The efficiency of the cycle increases with the increase in compression ratio. As the efficiency increases, the indicated power also increases.

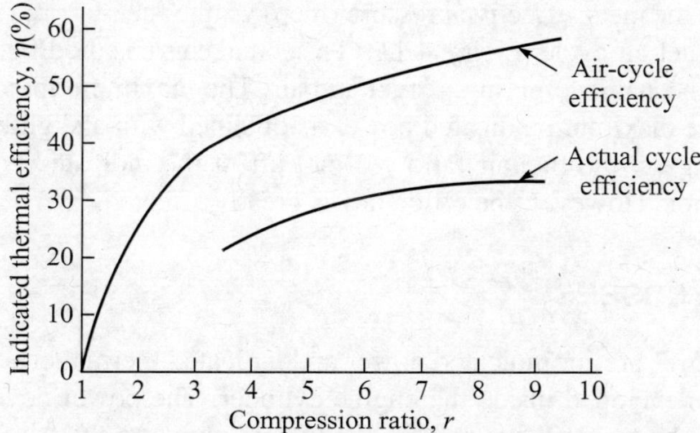

Figure 5.10 The effect of compression ratio on the efficiencies of air-cycle and actual cycle.

5.5.2 Effect of Fuel/Air Ratio

The effect of fuel/air ratio on the indicated mean effective pressure (imep) and the indicated thermal efficiency (η_i) is shown in Figure 5.11.

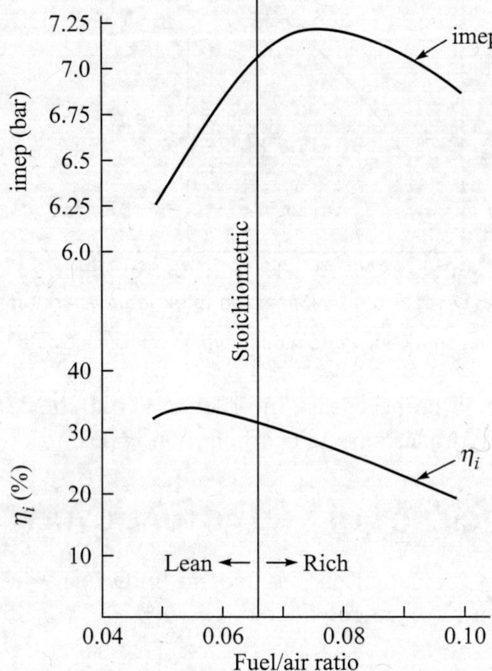

Figure 5.11 Effect of fuel/air ratio on the indicated thermal efficiency (η_i) and the indicated mean effective pressure (imep).

Very rich and very lean mixtures lower the burning rates of the fuel, and thus increase the burning time losses. The effect of these losses is to lower the efficiency at the rich as well as at the lean end. The efficiency of the actual engine is maximum at a fuel/air ratio somewhat leaner than the stoichiometric mixture.

A typical curve of the mean effective pressure (mep) vs. the fuel/air ratio for an actual cycle is similar to that for the fuel-air cycle (Figure 4.11). The actual curve is modified mainly for burning time losses and poor mechanical mixing of fuel and air. The maximum indicated mean effective pressure, and hence the maximum indicated power, is obtained with a slightly rich mixture. Near the peak a small change in the fuel/air ratio will not affect the indicated power as the curve is usually flat in this region. However, the efficiency is greatly affected.

5.6 FRICTIONAL LOSSES

The curves of Figure 5.11 are for indicated power and indicated thermal efficiency. These are the power and efficiency developed inside the engine cylinder. The power developed at the engine shaft will be less than the power developed on the piston, owing to the frictional losses between the various engine parts, such as friction between the piston and cylinder walls, friction in bearings and friction in auxiliary equipment like cooling water pump, lubricating pump, fuel pump, valves, ignition system, fans, etc. The shaft power is called the *brake power* (bp). Similarly, the thermal efficiency of the engine based upon the shaft output will always be less than the indicated thermal efficiency. This efficiency is called the *brake thermal efficiency* η_b, and is calculated as shown in Table 5.2. With the increase in engine speed, the frictional losses increase rapidly.

Table 5.2 Typical losses in SI engine ($r = 8$)

Item	*At load*	
	Full load (%)	*Half load* (%)
Air-standard cycle efficiency, η_a	56.5	56.5
(i) Losses due to variation of specific heat and chemical equilibrium	13.0	13.0
(ii) Loss due to progressive burning	4.0	4.0
(iii) Loss due to burning time	3.0	3.0
(iv) Heat loss	4.0	5.0
(v) Exhaust blowdown loss	0.5	0.5
(vi) Pumping loss	0.5	1.5
(vii) Frictional losses	3.0	6.0
Fuel air cycle efficiency (η_a – (i))	43.5	43.5
Gross indicated thermal efficiency, $(\eta_i)_{gross}$ $= \eta_a - [(i) + (ii) + (iii) + (iv) + (v)]$	32.0	31.0
Net indicated thermal efficiency, $(\eta_i)_{net} = (\eta_i)_{gross}$ – (vi)	31.5	29.5
Brake thermal efficiency, $\eta_b = (\eta_i)_{net}$ – (vii)	28.5	23.5

5.7 THE ACTUAL CYCLE OF COMPRESSION-IGNITION ENGINES

The combustion process of a compression-ignition engine is completed in the following three stages:

1. The first stage is a delay period. It is the time period between the start of fuel injection and the appearance of a flame or measurable pressure rise due to combustion.
2. The second stage of combustion is a rapid rise in pressure. If the delay period is long or longer than the injection period, most of the fuel burns during the period of rapid pressure rise.
3. The third stage is a relatively slow combustion. The remaining unburned fuel finds the limited oxygen to burn slowly. This slow combustion extends over a considerable part of the expansion stroke.

The crank angles during the above three stages of combustion vary with design and operating conditions. At a given speed these timings can be controlled by means of injection timing, spray characteristics and fuel composition.

A longer delay period is undesirable because of the resultant high maximum pressure and high rates of pressure rise causing engine knocking. The limited pressure fuel-air cycle is used as the basis of evaluating the actual compression-ignition engine cycle. The losses of the actual CI engine cycle in addition to those of limited pressure fuel-air cycle are:

1. Leakage
2. Heat losses
3. Time losses
4. Exhaust losses
5. Fluid friction
6. Gas exchange or pumping loss.

In SI engines, incomplete mixing of fuel and air and progressive burning are also included. In CI engines, these losses may be large, but since burning occurs during the mixing process, such losses cannot be separated from the time losses. The descriptions of all other losses mentioned above for CI engines are the same as those described for SI engines.

In spark-ignition engines, the rate of burning starts slowly and accelerates to its highest velocity near the end of the process. In CI engines, the reverse is true. It is because the available supply of oxygen decreases as burning progresses, the mixing process, hence the burning process, tends to slow down in the later stages of combustion.

5.8 ACTUAL AND FUEL-AIR CYCLES OF CI ENGINES

In a Diesel cycle the losses are less in comparison to the losses in an Otto cycle. The main loss is due to incomplete combustion and it is the cause of the main difference between the fuel-air cycle and the actual cycle of a CI engine. Figure 5.12 shows the *p-V* diagram for the actual Diesel cycle and the limited pressure fuel-air cycle. The two cycles are shown on the same diagram for the sake of comparison. A ratio between the actual and fuel-air cycle efficiency of 0.85–0.90 is attainable.

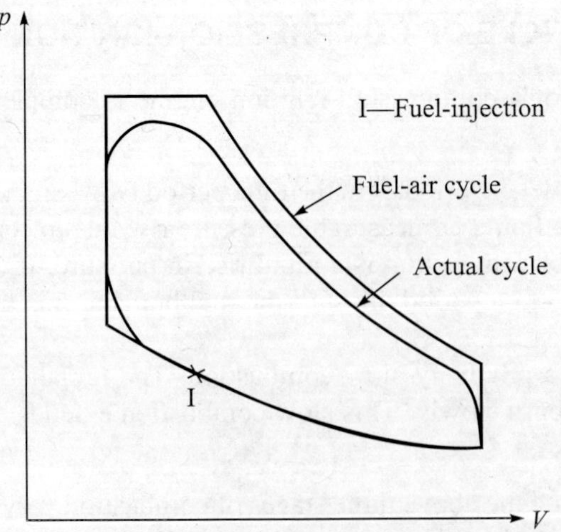

Figure 5.12 Comparison of the actual and limited pressure fuel-air cycles of compression-ignition engines.

REVIEW QUESTIONS

1. What are the losses considered in the analysis of the actual cycle in addition to those considered in the analysis of the fuel-air cycle?
2. What do you understand by crevice regions? Discuss leakage losses in the actual SI engine cycle.
3. Why is perfect mixing of fuel and air not possible? Show the composition of exhaust gases for a typical gasoline engine at various equivalence ratios.
4. What is a maldistribution of fuel in a multi-cylinder engine? What will be its effect?
5. What do you mean by progressive burning? What are the factors on which the time of progressive burning depends?
6. What do you understand by burning time losses? How can such losses be reduced? What are the causes of a large burning time?
7. During which part of the cycle the heat losses are significant? Where does the heat dissipate?
8. What do you understand by exhaust blowdown loss? Show the effect of exhaust valve opening time on exhaust blowdown loss.
9. Define the time loss ratio, the heat loss ratio, the blowdown loss ratio and the lost work ratio. Compare the actual cycle of an SI engine with the constant volume fuel-air cycle on *p-V* diagram, showing the different losses.
10. During which processes of the actual cycle is the fluid friction appreciable?
11. What do you understand by pumping losses during gas exchange? Explain with the help of a *p-V* diagram. What will be the effect of throttling and speed on pumping losses?
12. How does the flame speed vary with respect to the following engine variables?
    (a) Fuel/air ratio, (b) Inlet pressure, (c) Engine speed, (d) Engine size, and (e) Residual gas.
13. Show the actual indicator diagram with (a) ignition at TDC, (b) early ignition advance, and (c) optimum ignition advance. Also, show the corresponding fuel-air cycle in each case. Comment on actual power and efficiency.
14. What are the two important factors on which spark timing depends? How can a practically perfect spark timing be achieved for all operating conditions?
15. Show and compare the effect of spark advance on a single *p-V* diagram.
16. Show the effect of deviating spark timing from the optimum on relative indicated mean effective pressure.
17. What will be the effect of compression ratio on the indicated thermal efficiency of the actual cycle? Compare the efficiency of the actual cycle with that of the air cycle with the help of a diagram.
18. Describe with the help of a diagram the effect of fuel/air ratio on indicated mean effective pressure and indicated thermal efficiency of an actual cycle of a spark-ignition engine.
19. What are the three stages of combustion in a compression-ignition engine? What will be the effect of a longer delay period? What are the losses considered in the evaluation of an actual cycle of a CI engine?
20. For a CI engine, compare with the help of a diagram the actual cycle with the limited pressure fuel-air cycle.

6.1 INTRODUCTION

In a conventional spark-ignition engine a homogeneous mixture of fuel and air is supplied. The combustion in a gaseous fuel-air mixture ignited by a spark is characterized by a rapid development of a flame that starts from the point of ignition and spreads outwards in a continuous manner. When the flame spread continues to the end of the combustion chamber without any abrupt change in its speed and shape, combustion is called *normal*. When the unburned mixture ahead of the flame ignites and burns before the flame reaches it, the phenomenon is called *autoignition*. When there is a sudden increase in the reaction rate, accompanied by a sudden pressure rise forming pressure waves, the phenomenon is called *detonation* which causes engine knock.

When combustion is initiated by a spark, it is called *controlled combustion* and when combustion is initiated by a hot spot, it is called *uncontrolled combustion*. Under normal combustion conditions, the combustion is controlled and this is a designer's objective. Uncontrolled combustion is associated with preignition and running-on. Autoignition and detonation come under the category of abnormal combustion. In this case the fuel-air mixture ignites spontaneously without an ignition source.

The combustion of fuel-air mixture depends on chain reactions. First only a few highly active constituents surrounding the ignition point cause reactions. These in turn generate additional active constituents to cause reactions. Soon a point is reached where the chain breaking reactions dominate the chain forming reactions. In the flame front, the chain forming reactions can only reach a certain distance into a relatively cool, unburned charge before they are broken and thus a definite flame boundary is established. However, if the unburned gases become hot enough to sustain chain reactions, the remaining gas will suddenly auto-ignite.

6.2 NORMAL COMBUSTION

Towards the end of the compression stroke the cylinder contains more or less a homogeneous mixture of vaporized fuel, air and residual gases. A single intense and high temperature spark is produced between the spark plug electrodes and as it passes from one electrode to the other it leaves a thin thread of flame. Combustion spreads to the envelope of the mixture containing the thread at a rate depending primarily on the temperature of the flame front and secondarily on the temperature and density of the surrounding envelope. Thus a small hollow nucleus of flame at first grows up gradually and as the flame front expands with steadily increasing speed it travels across the chamber until finally the whole of the mixture is engulfed. Depending on the degree of turbulence in the cylinder, the flame front wrinkles, thus presenting a greater surface area from which heat is radiated; hence the flame speed is increased enormously, and this speeds up the combustion process.

A theoretical pressure–crank angle (p–θ) diagram in an ideal four-stroke SI engine is shown in Figure 6.1. In an ideal engine, compression and expansion take place during 180° of crank rotation and combustion takes place instantaneously at TDC. During combustion the volume remains constant and there is a sudden pressure rise.

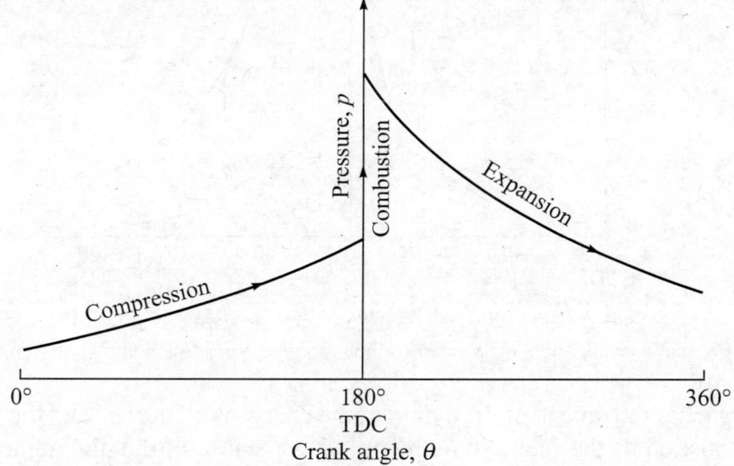

Figure 6.1 Theoretical pressure vs. crank angle diagram.

However, in an actual spark-ignition engine, combustion does not occur instantaneously. It is initiated by a spark produced before TDC at a definite time and the flame takes a finite time to travel across the combustion chamber, burning the charge and raising the cylinder pressure as it proceeds. The thermal efficiency will be higher if the combustion approaches the constant volume process, however, the requirement of smooth and quiet engine operation imposes the restriction.

6.2.1 Stages of Combustion in SI Engine

Figure 6.2 shows the pressure–crank angle (p–θ) diagram indicating three stages of combustion in a spark-ignition engine. The spark is initiated at point A, at point B pressure rise against the motoring curve is observed, point C shows the point of maximum pressure and point D indicates the completion of the combustion process. The p–θ diagram for motoring is obtained when the engine is not firing. Thus AB represents the first stage, BC the second stage and CD the third stage of combustion.

It is important to understand the two terms, i.e. *combustion* and *burning*. Combustion proceeds until the point of chemical equilibrium is reached, on the other hand burning is that part of the combustion process which is associated with the travel of a flame front across the cylinder.

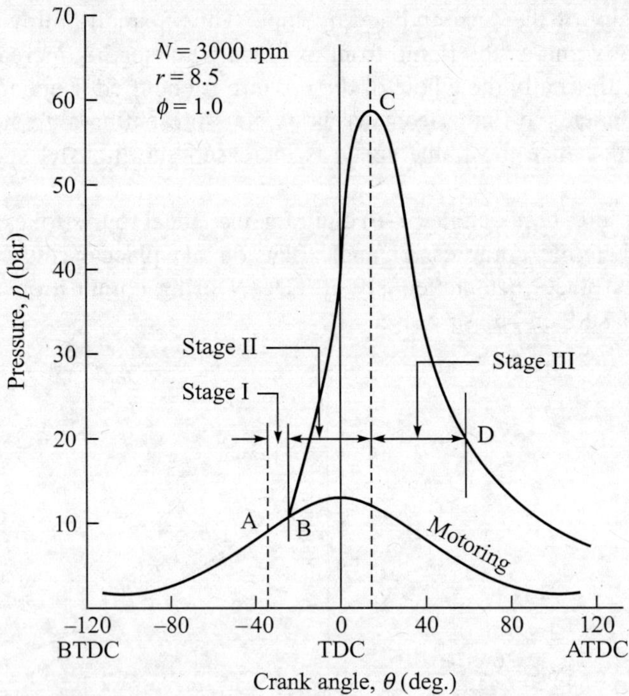

Figure 6.2 Pressure–crank angle diagram showing stages of combustion in a spark-ignition engine.

When the flame reaches the far end of the cylinder, the charge is all burned and the burning process is over. The combustion of the fuel will be incomplete because of the high temperature causing dissociation.

The first stage (AB)

This stage is called *ignition lag* or *preparation phase*. It corresponds to the time for the growth and development of a self-propagating nucleus of the flame. The starting point of the first stage is the point A, where the spark is produced and the end of the stage is marked with point B, where the first measurable pressure rise against the motoring curve is observed. The first stage is mainly a chemical process and depends on the nature of the fuel, temperature and pressure of the fuel-air mixture, the concentration of the residual from the previous cycle present in the cylinder and the chemical reaction rate. It is also influenced by local turbulence.

Although the first stage of combustion is called the ignition lag, as it is analogous to the delay period of compression-ignition engine, and actually the nucleus of combustion appears instantaneously near the spark plug electrodes; initially the flame spreads very slowly and the fraction of the burned mixture is very little, so there is no appreciable pressure rise against the motoring curve.

The second stage (BC)

This stage is called the *main stage*. It corresponds to the propagation of the flame practically at a constant speed. The starting point of the second stage is taken as point B, where the first measurable pressure rise against the motoring curve is observed. The end of the second stage is marked with point C, where the maximum pressure is attained. This stage is both a physical and a chemical process. The heat release depends on the chemical composition and on the prevailing temperatures and pressures and the degree of turbulence in the cylinder. During this stage heat transfer to the cylinder wall is low, since the burning mixture comes in contact with a small part of the cylinder wall. The rate of pressure rise is almost proportional to the rate of heat release because during this stage, the combustion chamber volume does not change much.

The third stage (CD)

This stage is called *afterburning*. Although the point C indicates the completion of the flame travel, it does not follow that the whole of the heat of the fuel has been liberated at this point. Even after the passage of the flame, during expansion some of the constituents re-associate and liberate heat. The starting point of the third stage is usually taken at the instant when maximum pressure is reached on the indicator diagram (like the point C). The end of this stage is marked with point D. This point corresponds to the point where equilibrium is reached and after which the products of combustion are assumed to be frozen. During this stage the flame speed decreases and the rate of combustion is slow. Since the expansion stroke starts before this stage with the piston moving away from TDC, there will be pressure fall during this stage.

6.2.2 Flame Speed Pattern

The flame speed across the chamber follows a pattern similar to that shown in Figure 6.3. It shows the relation between the relative distance travelled by the flame across the chamber with respect to the relative time taken by the flame to travel across the chamber. The slope of the curve indicates the flame speed.

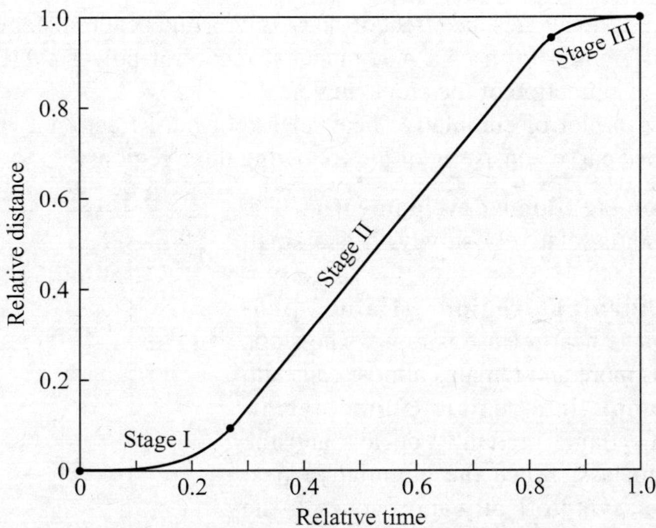

Figure 6.3 Relative distance of flame travel across the chamber vs. relative time of flame travel across the chamber.

The flame travel appears to pass through three distinct stages. During stage I, the flame front progresses slowly because of low transposition rate and low turbulence. Since there is a relatively small mass of charge burned at the start, there is a very little transposition of the flame front. It is thus propagated almost entirely by the reaction rate, resulting in a slower advance. The heat liberated by the burning portion of the flame front prepares the adjacent portion of the unburned charge for the combustion reaction. Also, since the spark plug is located in a quiescent layer of gas, close to the cylinder wall, the lack of turbulence reduces the reaction rate and further lowers the flame speed.

As the flame front proceeds into a more turbulent region leaving the quiescent zone and commences to consume a greater mass of the unburned mixture, it progresses more rapidly and almost at a constant rate as shown in stage II. The average speed of the flame during this period of travel is referred to as the flame speed.

Towards the end of flame travel, i.e. in stage III, the volume of the unburned charge is reduced appreciably, and the transposition rate again becomes negligible, thereby reducing the flame speed. The reaction rate is further reduced since the flame enters a zone of relatively low turbulence.

Thus the flame travel pattern divides the combustion process into four distinct phases:

1. Spark initiation
2. Early flame development
3. Flame propagation
4. Flame termination.

The factors affecting the flame speed have already been discussed in Chapter 5.

6.2.3 Finite Heat Release Model for Combustion Process of SI Engine

Finite heat release model evaluates the effect of spark timing and heat transfer on engine work and efficiency. It is a differential form of a mathematical model of power cycle in which the heat release is expressed as a function of the crank angle.

Figure 6.4 shows a plot of cumulative heat release or burn fraction $x_b(\theta)$ versus the crank angle. It is an S-shaped curve consisting of the following three regions:

1. **Initial ignition lag (flame development):** The heat release starts relatively slowly. It is a small slope region.
2. **A rapid burning region (flame propagation):** The heat release is rapid. The slope of the curve is more and remains almost constant.
3. **Burning completion region (flame termination):** The flame gradually decays and the slope is again less. Since the cumulative heat release curve asymptotically approaches 1, the end of combustion can be considered when x_b = 0.9 to 0.99.

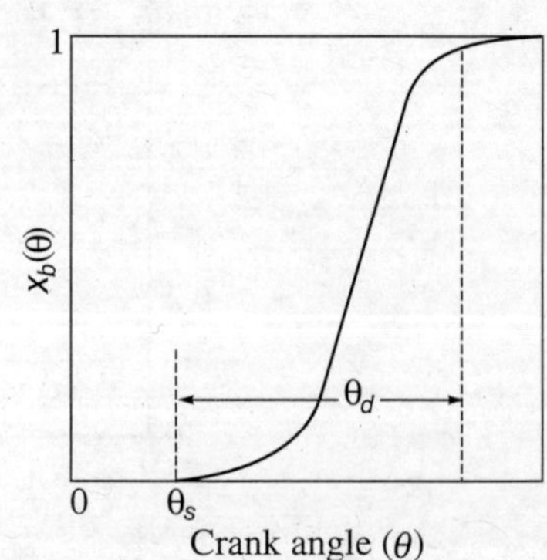

Figure 6.4 Burn fraction versus crank angle.

The S-shaped curve can be represented mathematically by a Weibe function.

$$x_b(\theta) = 1 - \exp\left[-a\left(\frac{\theta - \theta_s}{\theta_d}\right)^n\right] \tag{6.1}$$

where,

θ = crank angle

θ_s = start of heat release (spark timing)

θ_d = duration of heat release

n = Weibe form factor

a = Weibe efficiency factor

Typical values of $a = 5$ and $n = 3$ can be taken for SI engine. However, they depend to some degree on the engine load, speed and type of engine.

$$\frac{dx_b}{d\theta} = -\exp\left[-a\left(\frac{\theta - \theta_s}{\theta_d}\right)^n\right]\left[-an\left(\frac{\theta - \theta_s}{\theta_d}\right)^{n-1}\right]\cdot\frac{1}{\theta_d}$$

$$= \exp\left[-a\left(\frac{\theta - \theta_s}{\theta_d}\right)^n\right]\cdot an\left(\frac{\theta - \theta_s}{\theta_d}\right)^{n-1}\cdot\frac{1}{\theta_d}$$

$$= (1 - x_b)\frac{an}{\theta_d}\left(\frac{\theta - \theta_s}{\theta_d}\right)^{n-1} \tag{6.2}$$

Now, the rate of heat release as a function of crank angle is given by

$$\frac{dQ}{d\theta} = Q_{\text{in}}\frac{dx_b}{d\theta}$$

$$\therefore \quad \frac{dQ}{d\theta} = na\frac{Q_{\text{in}}}{\theta_d}(1 - x_b)\left(\frac{\theta - \theta_s}{\theta_d}\right)^{n-1} \tag{6.3}$$

where Q_{in} is the heat supplied by the fuel.

Pressure–crank angle (p–θ) relation:

First law of thermodynamic for a closed system is

$$\delta Q = \delta W + dU$$

For reversible process,

$$\delta Q = pdV + mc_v dT \tag{6.4}$$

Equation of state,

$$pV = mRT$$

$$pdV + Vdp = mRdT$$

$$m\,dT = \frac{p\,dV + Vdp}{R} \tag{6.5}$$

$$\delta Q = p\,dV + \frac{R}{(\gamma - 1)}\left(\frac{p\,dV + Vdp}{R}\right)$$

$$= p\,dV + \frac{1}{(\gamma - 1)}(p\,dV + Vdp)$$

∴

$$\frac{dQ}{d\theta} = \frac{\gamma}{(\gamma - 1)} \cdot \frac{p\,dV}{d\theta} + \frac{1}{(\gamma - 1)} \cdot \frac{Vdp}{d\theta}$$

∴

$$\frac{dp}{d\theta} = \frac{(\gamma - 1)}{V}\frac{dQ}{d\theta} - \frac{p\,\gamma}{V}\frac{dV}{d\theta} \tag{6.6}$$

Using the relation of $(dQ/d\theta)$ from Eq. (6.3) and $(dV/d\theta)$ from Eq. (1.9), $dp/d\theta$ can be calculated from Eq. (6.6). The differential equation (6.6) is integrated numerically for the pressure at different crank angles by using fourth-order Runge–Kutta method. The integration proceeds degree by degree to top dead center and back to bottom dead center. Once the pressure is computed as a function of crank angle, the work is obtained from $\delta W = pdV$ and temperature from ideal gas law, $T = pV/mR$.

6.2.4 Pressure and Temperature Variation as a Function of Crank Angle

Figure 6.5 shows the variation of pressure and temperature with respect to the crank angle. The flame reaches the cylinder wall farthest from the spark plug about 15° after TDC. At this point the maximum pressure p_{max} is reached, but the combustion is not completed; it continues around parts of the chamber periphery for another few degrees of crank angle, so the maximum temperature T_{max} is obtained about 10° after the maximum pressure is reached. Both pressure and temperature decrease as the cylinder volume continues to increase during the remainder of the expansion stroke.

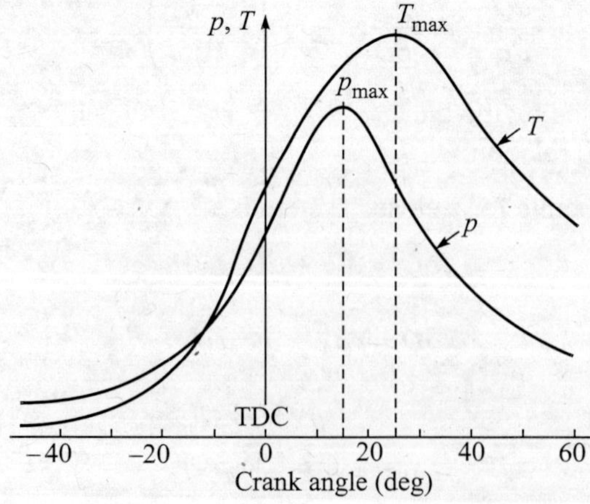

Figure 6.5 Variation of pressure and temperature with respect to the crank angle.

6.2.5 Effect of Spark Timing on Indicator Diagram

Figure 6.6 shows the effect of spark timing on cylinder pressure versus the crank angle. If the spark timing is over-advanced, the combustion process starts while the piston is moving towards TDC, so the compression work (negative work) increases. If the spark timing is too much retarded, the combustion process is progressively delayed, the peak cylinder pressure occurs later in the expansion stroke and its magnitude is reduced. The expansion work (positive work) is also reduced. The optimum spark timing is the timing for which the maximum brake torque is obtained. It is called the MBT timing. The spark timing which is advanced or retarded from MBT timing gives less torque. The MBT timing depends on the rate of flame development, propagation and termination. It also depends on the distance of the flame travel path across the combustion chamber.

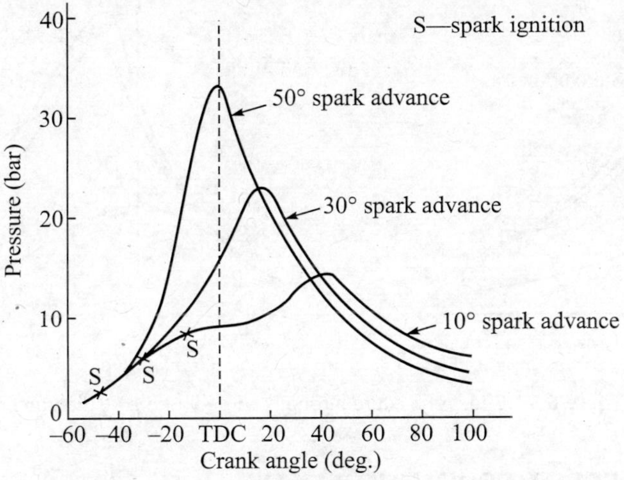

Figure 6.6 Pressure vs. crank angle for over-advanced spark timing (50°), MBT timing (30°) and retarded timing (10°).

Figure 6.7 shows the effect of spark advance on brake torque at constant speed and air/fuel ratio at wide open throttle. With optimum spark timing the maximum pressure occurs at about 15° after TDC and half the charge is burned at about 10° after TDC.

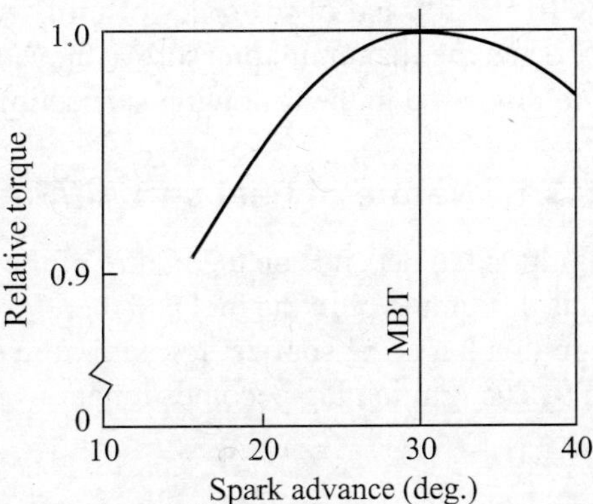

Figure 6.7 Effect of spark advance on brake torque.

6.2.6 Effect of Fuel/Air Ratio on Indicator Diagram

Figure 6.8 shows the effect of mixture strength on indicator diagrams. The fuel/air ratio of the charge influences the rate of combustion and the amount of heat evolved. The maximum flame speed occurs when the mixture strength for hydrocarbon fuels is about 10% rich. When the mixture is made leaner or is further enriched, the flame speed deceases. Lean mixtures release less thermal energy, resulting in lower flame temperature and hence lower flame speed. Very rich mixtures suffer incomplete combustion, hence release less thermal energy resulting in low flame speed. Indicator diagrams for rich, stoichiometric and weak mixtures correspond to equivalence ratio 1.1, 1.0, and 0.9 respectively.

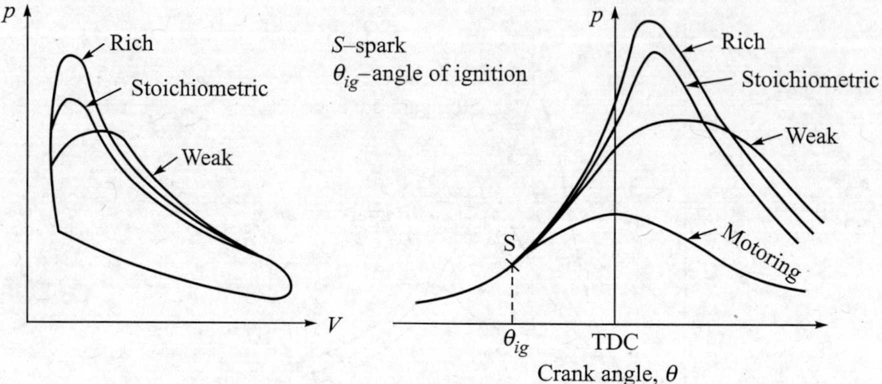

Figure 6.8 Effect of mixture strength on $p-V$ and $p-\theta$ diagrams.

6.3 FACTORS AFFECTING IGNITION LAG

The first phase of a combustion process is called the ignition lag. This is the period between spark initiation and flame development. It is a chemical process. The ignition lag in terms of crank angle is 10° to 20° and of the order of milliseconds. As the ignition lag decreases, the rate of pressure rise and the maximum pressure achieved increase. The factors affecting the period of ignition lag are discussed in the following subsections.

6.3.1 Nature of Fuel and Air/Fuel Ratio

Ignition lag depends on the nature of the fuel. If the self-ignition temperature of the fuel is higher, it is difficult to burn the fuel and therefore the ignition lag will be longer. It has been found that the ignition lag is the shortest for a mixture slightly richer than the stoichiometric as shown in Figure 6.9. The ignition lag becomes longer as the mixture becomes lean or very rich.

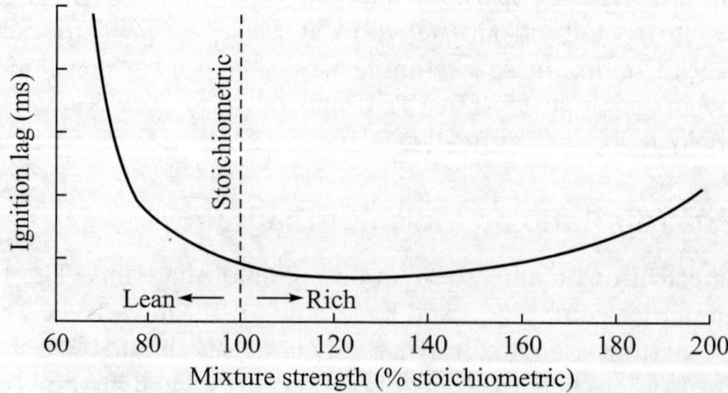

Figure 6.9 Effect of mixture strength on ignition lag.

6.3.2 Initial Temperature and Pressure

The rate of chemical reaction depends to a greater extent on temperature but to a smaller extent on pressure. Increasing the initial temperature and pressure increase the temperature and pressure at the point of ignition, therefore the ignition lag decreases.

6.3.3 Compression Ratio

The temperature and pressure at the point of ignition increase with the increase in compression ratio, therefore the ignition lag reduces. High compression ratios also reduce the concentration of the residual gas and therefore the ignition lag is further reduced.

6.3.4 Spark Timing

As the spark timing is retarded, i.e. when the spark is produced towards the TDC, the compression process at the point of ignition is covered more, resulting in an increase in temperature and pressure, and therefore the ignition lag reduces. Advancing the spark timing increases the ignition lag.

6.3.5 Turbulence and Engine Speed

Ignition lag is not much affected by the intensity of turbulence. Turbulence is directly proportional to the engine speed. Therefore, the engine speed does not affect the ignition lag much in terms of milliseconds. However, as the engine speed increases and keeping the time in milliseconds almost fixed, the crank angle in degrees increases linearly with the speed. For this reason it becomes necessary to advance the spark timing at higher speeds by using the ignition advance mechanisms.

Excessive turbulence of the charge in the vicinity of the spark plug causes flame quenching, since it increases the rate of heat transfer from the combustion zone and leads to unstable development of the flame nucleus. That is why the spark plug is usually located in a small recess in the wall of the combustion chamber.

6.3.6 Electrode Gap of Spark Plug

A suitable spark plug electrode gap is necessary to establish the flame nucleus. If the gap is too small, quenching of the flame nucleus may result and if the gap is too large, the spark intensity is reduced. In both the cases the range of the fuel/air ratio is reduced for the development of the flame nucleus.

Figure 6.10 shows the range of equivalence ratios, which could be used for different electrode gaps and for different compression ratios of the engine. As the compression ratio is increased, the range of the equivalence ratio also increases for a given electrode gap. A higher electrode gap is required for an engine with lower compression ratio. For an engine having compression ratio 9.0, the spark plug electrode gap ranging from 0.8 mm to 0.9 mm is quite satisfactory.

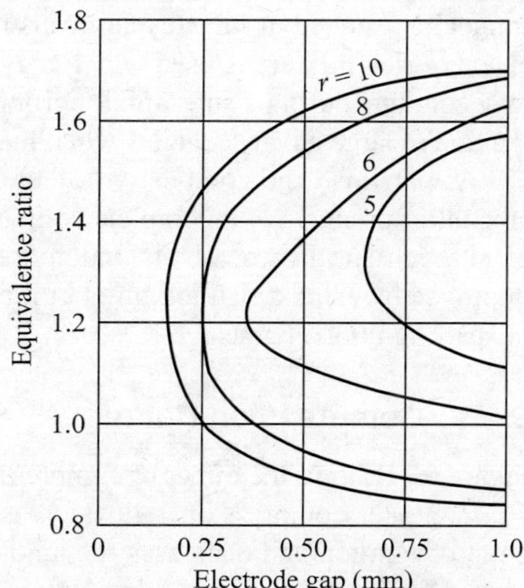

Figure 6.10 Effect of electrode gap on the equivalence ratio for different compression ratios.

6.4 FACTORS AFFECTING COMBUSTION IN SPARK-IGNITION ENGINES

The combustion process of the spark-ignition engine is affected by the factors as discussed in the following subsections.

6.4.1 Composition of the Mixture

The composition of the mixture affects the rate of combustion and the amount of heat evolved, and changes the pressure and temperature of the gases in an engine cylinder. Figure 6.8 shows the effect of mixture strength on indicator diagrams. The angle of ignition θ_{ig} is set for the maximum brake torque (MBT) condition. For a rich mixture with the equivalence ratio between 1.1 and 1.2, the duration of the first stage of combustion, the ignition lag and the duration of the main phase are all minimum, resulting in the maximum rate of pressure rise ($dp/d\theta$). The flame speed, the heat liberation and consequently the power developed by the engine are the maximum. When the equivalence ratio is less than 1.1, the energy content is reduced, hence the duration of the first phase of combustion increases, and the angle of ignition advance must therefore be increased. The duration of the main phase of combustion in the second stage changes slightly, resulting in reduction in maximum pressure and also reduction in the rate of pressure rise ($dp/d\theta$). These could be improved by slightly advancing the spark timing.

For a lean mixture with the equivalence ratio between 0.85 and 0.9, the power output is reduced but this range of equivalence ratio corresponds to the minimum brake specific fuel consumption and represents the most economical range.

6.4.2 Load

When the load is reduced, the power of an engine is reduced by throttling. The initial pressure and the pressure at the point of ignition decrease and the residual gases in the mixture increase. The first phase of combustion prolongs and the combustion process loses its stability and frequently cannot be resumed in some cycles, causing cyclic variations. To overcome this difficulty to some extent, a rich mixture is used which may ensure proper combustion, but the combustion process may continue during a substantial portion of the expansion stroke. This is because of interrupted ignition at large advance angles when the compression pressures are still very low.

At part load the combustion of fuel in the spark-ignition engine is poor, causing a large amount of products of incomplete combustion in the exhaust including carbon monoxide, oxides of nitrogen and hydrocarbons which are responsible for air pollution. Part load combustion is improved by using a rich mixture but it causes wastage of fuel. These are the main shortcomings of spark-ignition engines.

6.4.3 Compression Ratio

Figure 6.11 shows the effect of compression ratio on the indicator diagram.

A higher compression ratio increases the pressure and temperature of the mixture at the point of ignition and decreases the amount of residual gas in the mixture. These are favourable conditions for the ignition of the mixture. The duration of ignition lag in the first phase decreases and the rate of pressure rise in the main phase increases. A high compression ratio increases the surface-to-volume ratio of the combustion chamber, thus increasing the relative amount of mixture near the walls. This part of the mixture afterburns in the third phase. All this retards the MBT timing at higher compression ratios. The combustion duration up to the point of maximum pressure also decreases. The maximum pressure approaches TDC. Heat liberated up to the point of maximum pressure is reduced and the importance of the afterburning process in the third phase increases.

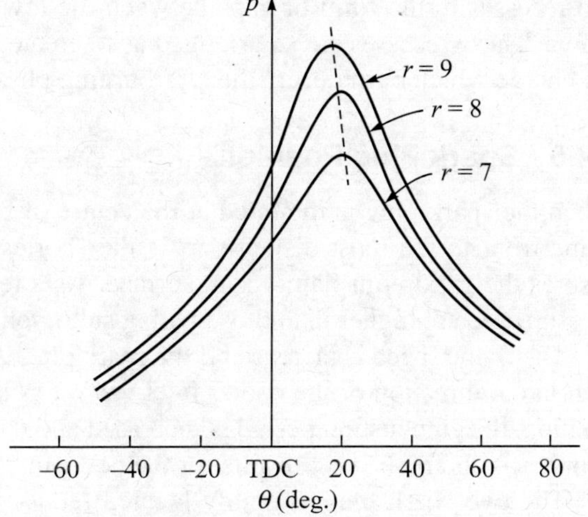

Figure 6.11 Effect of compression ratio r on the p–θ diagram.

6.4.4 Speed

When the speed increases, the time in terms of milliseconds required for the development of the flame in the first phase of combustion is not affected much and the turbulence of the charge increases. The flame speed in the main phase of combustion increases with the increase in speed, while the duration of the main phase expressed in degrees of crank angle (θ_{II}) remains practically the same. The duration of the first phase of combustion (θ_I) in degrees of crank angle increases with the increase in speed.

If the engine speed is increased without changing the angle of ignition θ_{ig}, the duration of the development of flame in the first phase increases as shown in Figure 6.12(a). If the angle of ignition θ_{ig} is advanced at higher speeds, the pressure rise in the main phase of combustion can be practically made to coincide at different speeds as shown in Figure 6.12(b). The duration of the afterburning phase increases with the speed.

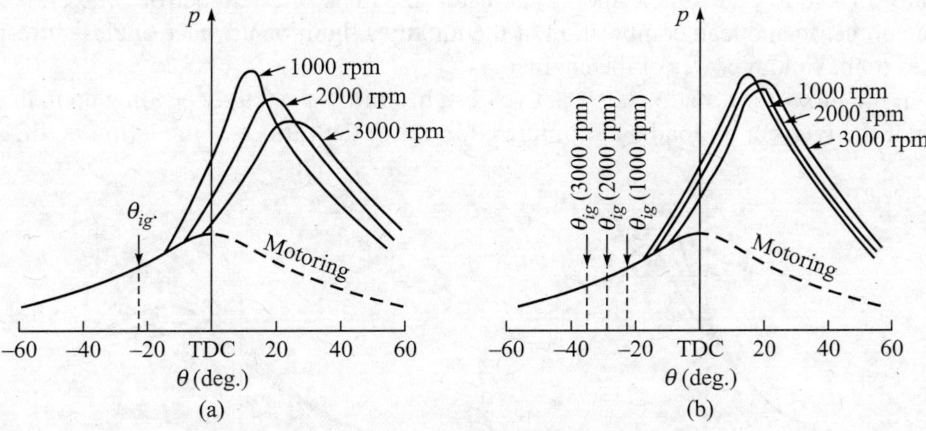

Figure 6.12 Effect of engine speed on indicator diagram.

6.4.5 Turbulence and Shape of Combustion Chamber

Turbulence of the charge starts as it enters into the cylinder through narrow sections of inlet pipes and intake valve. Turbulence can be intensified by using a properly shaped combustion chamber and recesses in the form of gaps between the lower surface of the cylinder head and the piston crown. These recesses are so arranged as to create an additional swirling motion in those parts of the charge which burn during the afterburning phase and thus cause rapid afterburning.

6.4.6 Spark Plug Position

When the spark plug is mounted at the centre of the cylinder head, the distance travelled by the flame front to the most distant part is the shortest. The central position of the spark plug also ensures the maximum flame front surface. As a result, the rate of heat evolution and the rate of pressure rise are higher than those with a side-mounted spark plug.

The flame speed is increased if the spark plug is located more towards the hotter exhaust valve than in the direction of the cooler inlet valve. As the spark plug is moved away from the central position, the combustion period is increased and the ignition requires to be advanced accordingly, in order to obtain the best results for the new plug location.

The two spark plugs suitably located reduce the flame travel paths and give a higher rate of pressure rise. This requires that ignition advance be reduced. The use of two spark plugs with synchronized sparks is standard on aircraft engines. It provides reliability and improved performance. The thermal efficiency is increased and the specific fuel consumption is reduced. With large diameter cylinders the use of two plugs gives better performance results, whereas in small cylinders a single plug will give satisfactory results, owing to the reduced flame travel path.

6.5 CYCLIC VARIATION

One of the prominent characteristics of the spark-ignition engine combustion process is a wide variation from cycle to cycle of the pressure–crank angle diagram. This variation increases greatly as the mixture strength approaches either the weak end or the rich end of the range.

Figure 6.13 shows the superimposed indicator diagrams of a number of consecutive cycles with stoichiometric and lean composition of the mixtures. Ignition advance angles correspond in both cases to maximum power of the engine.

When the mixture is made leaner above certain limits ($\phi < 0.9$) depending upon the design features of the engine, its load and compression ratio, the rate of combustion is different in

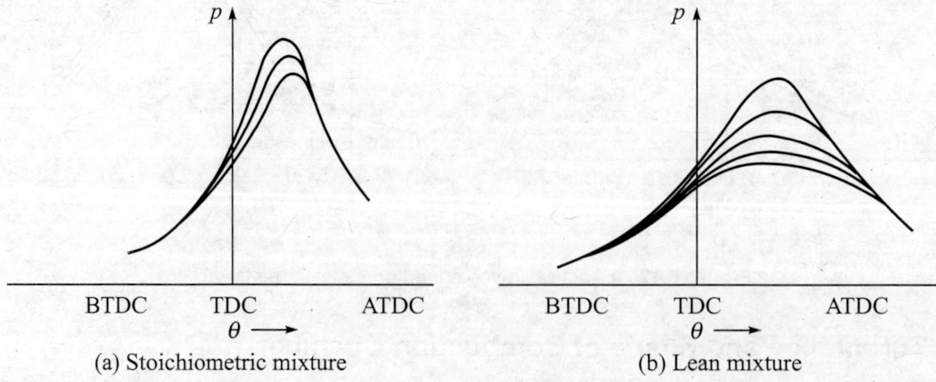

Figure 6.13 Pressure–crank angle diagrams showing cyclic variation for (a) stoichiometric mixture and (b) lean mixture.

consecutive cycles. The reasons for cyclic variation are due to incomplete mixing of fuel, air and residual gas, and the variation of mixture velocity that exists within the cylinder near the spark plug. Since a large amount of random turbulent motion exists inside the cylinder, it is evident that in presence of incomplete mixing, the spark may occur in mixtures of varying fuel/air ratios, resulting in different rates of flame development. There is a possibility of misfiring as well, in some cycles where the flame may not develop at all. The cyclic variation is greater when the residual exhaust gas in the mixture is more. The residual gas will tend to influence the inlet flame temperature and therefore will affect the flame speed. It is observed that the cyclic variation is more at lower compression ratios and at reduced loads for which the residual gas in the mixture is more.

Elimination of cycle-to-cycle variation is important for obtaining improved engine performance. If all cycles were alike and equal to the average cycle, the maximum cylinder pressure would be low, the efficiency would be greater and the detonation limit would be higher.

Cyclic variation can be reduced by using multiple ignition points, by increasing the engine speed and turbulence. Cyclic variation is greatly reduced by a tangentially-oriented swirl created by a shrouded inlet valve. A spark of extended duration also reduces cyclic variation.

6.6 RATE OF PRESSURE RISE

The rate of pressure rise during the combustion process influences the peak pressure, the power produced and the smoothness with which the forces are transmitted from the gas to the piston. The rate of pressure rise depends on the mass rate of combustion of the mixture in the cylinder.

Figure 6.14 shows the pressure–crank angle diagrams for a high, normal and a low rate of combustion. It is observed from the figure that the low rate of combustion requires more ignition advance because of the longer time required to complete combustion. A higher rate of combustion increases the rate of pressure rise and generally produces peak pressures at a point closer to TDC.

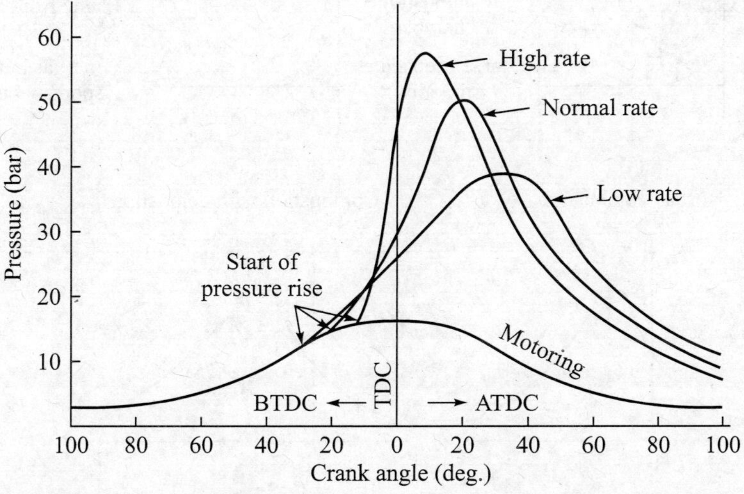

Figure 6.14 Pressure–crank angle diagram for different rates of combustion.

A higher peak pressure closer to TDC is generally desirable because it produces a greater force on the piston acting through a larger portion of the power stroke, and hence increases the power output. There is a practical limit on the rate of pressure rise. The higher rate of pressure rise may result in rough running of the engine because of vibrations and jerks. If the rate of pressure rise is excessively high, it may result in abnormal combustion called *detonation*. A compromise between these two opposing factors is necessary. This can be achieved by designing and operating the engine in such a way that approximately one-half of the pressure rise takes place as the piston reaches the TDC, thus ensuring peak pressure to be reasonably close to the beginning of the power stroke, while maintaining smooth engine operation.

6.7 ABNORMAL COMBUSTION—AUTOIGNITION AND DETONATION

In normal combustion, as shown in Figure 6.15(a), after the flame is initiated by the spark, the flame front travels with a fairly uniform speed across the combustion chamber compressing the unburned gas ahead of it. The gas ahead of the flame front is called the *end-gas*. This is the last part of the charge to burn. The end-gas receives heat due to compression by expanding the burned gases and by radiation from the advancing flame front. If the temperature and pressure are below certain critical values, the flame front moves across the combustion chamber through the unburned charge to the farthest point of the chamber in the normal manner, thus burning the mixture completely. The pressure–crank angle diagram is a smooth curve as shown in Figure 6.15(b).

If the temperature and pressure of the end-gas are high enough, it will ignite spontaneously before the flame front reaches it. Under this abnormal condition, the earlier stages of combustion are normal, but towards the end of combustion, namely at about the last 25 % of the flame travel distance, sudden inflammation of the remaining portion of the end-gas occurs. The end-gas is said to be auto-ignited (Figure 6.15(c)).

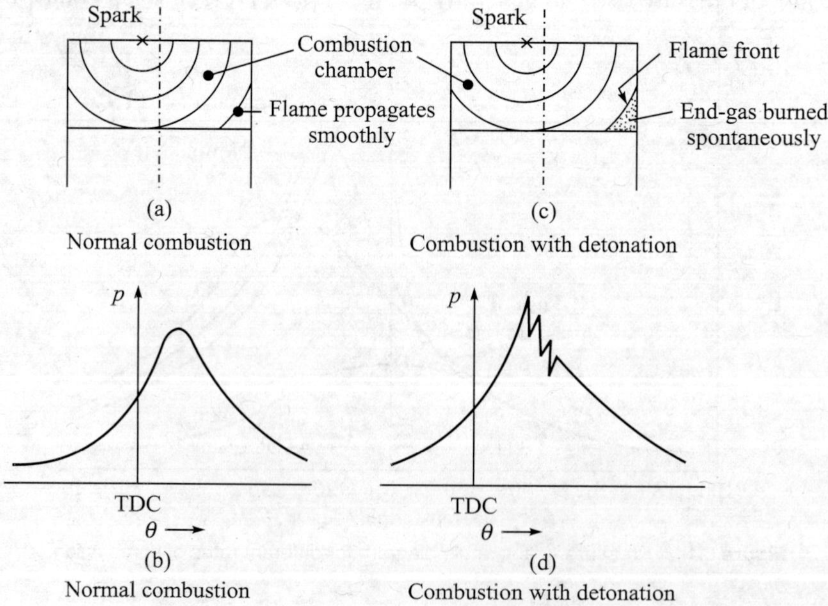

Figure 6.15 Combustion with spark-ignition engine.

In autoignition, the charge must remain above a certain critical temperature for a certain length of time. During this period certain chemical reactions take place which prepare the charge for autoignition. The time required for this preparation phase is called *ignition delay* or *delay period.*

In autoignition the rate of reaction is very high and the burning is almost instantaneous which results in an extremely rapid release of energy causing pressure fluctuations as shown in Figure 6.15(d), and causing pressure of the end-gas to increase almost 3 to 4 times from about 50 bar to 150–200 bar. The pressure rise for most of the charge is around 50 bar but because of autoignition the pressure of the last part of the charge goes to 150–200 bar. This large pressure difference gives rise to severe pressure waves which strike the cylinder wall and set it to vibrate, giving rise to a characteristic high pitched metallic ringing sound. This phenomenon is therefore known as *knocking* or *detonation*. The flame speed during detonation is of the order of 300 to 1000 m/s.

It may be noted that knocking in diesel engines is a different phenomenon. Knocking or detonation in a spark-ignition engine always occurs near the end of combustion, whereas in a diesel engine it occurs in the beginning of combustion.

Knock or detonation is a characteristic audible noise which is transmitted through the engine structure because of spontaneous autoignition of a portion of the end-gas ahead of the propagating flame. When this abnormal combustion takes place, a very high local pressure, far in excess of the average pressure in the cylinder, and pressure waves of substantial amplitude across the combustion chamber are created. A ringing metallic knocking sound clearly indicates detonation caused by shock waves formed in the gases and repeatedly reflected from the walls of the combustion chamber. The frequency of these pressure oscillations is the same as the fundamental frequency of the audible knocks. It depends on the velocity of the shock waves and their path between consecutive reflections from the walls.

Intensive detonation causes loud knocks with a higher frequency that arises in each cycle, the power of the engine drops and black smoke appears intermittently in the exhaust gases. It causes mechanical damage to the engine.

6.8 DETRIMENTAL EFFECTS OF KNOCKING

An engine should never be allowed to operate for a long time with knocking. The impact of knock will depend on its duration and intensity. If the knock duration is short, it is unlikely to cause damage, but a heavy constant knock can easily lead to severe damaging effects as explained below.

Noise and vibration: A violent pressure rise initiates a wave at the point of detonation, which is reflected back and forth across the combustion chamber, causing an audible knock. The pitch of this sound depends upon the size and shape of the combustion space and the velocity of the wave, which is propagated at the speed of sound in the cylinder gases. A knocking engine thus produces a loud and pulsating noise annoying human ears and is very objectionable in automobile engines. Detonation also causes vibration of engine parts.

Increase in heat transfer: Detonation increases the rate of heat transfer to the combustion chamber walls. The increase in the rate of heat transfer is because of higher temperature of the gas in a detonating engine due to rapid completion of combustion. Under detonating conditions the pressure waves scour away the protective layer of inactive stagnant gas on the cylinder walls and this increases the rate of heat transfer. The increased heat transfer is responsible for the damage to the engine.

Mechanical damage: The extremely rapid combustion increases the impact pressure, causing fractures of aluminium alloy pistons, and also increases the temperature, causing overheating and burning of the gaskets between the cylinder and its head, electrodes and insulators of spark plug, piston crowns and the heads of the valves. In severe cases of detonation, increased cylinder and piston temperatures cause collapsing of the piston crowns and sometimes burning of the cylinder head. Prolonged running of the engine under detonating conditions causes loosening of valve insert rings in the cylinder head, burning of the sides of piston due to blow-by of the very hot gases and gumming of the piston rings in their grooves which seizes the piston and the rings in the bore.

Preignition: Ignition of the mixture by some hot surface within the combustion space, before the normal spark ignition occurs is called *preignition*. The detonation wave crossing and recrossing the cylinder causes a flow of hot gases in and out of the spark plug cavity. The hot gases greatly increase the amount of heat picked up by the spark plug. Overheating of the spark plug may be to such an extent that the plug electrodes may become incandescent and ignite the fresh charge long before the ignition is supposed to occur. Thus, overheating of the spark-plug electrodes due to the effects of detonation can lead to preignition. The effect of preignition is similar to the effect of early ignition advance, shown in Figure 5.6. The burning time losses are generally increased, and the power and efficiency are reduced.

Power and efficiency: One of the principal detrimental effects of detonation is that of reduced power output. It is because of the fact that detonation is associated with an increase in the heat losses to the cylinder walls and piston. It also results in the loss of thermal efficiency.

Carbon in the exhaust: Over a long period of running the engine, a large amount of carbon gets deposited in the combustion chamber. As a result of the violent pressure fluctuations under detonation, some of the deposited carbon is not only blown out of the exhaust but also leaves the surface of the combustion chamber, and the top of the piston becomes rough and pitted.

Under detonating conditions, puffs of gray smoke in the exhaust can usually be seen against light background. At night, these puffs are seen as bright yellow flashes. Without detonation, the exhaust is of blue colour with occasional splashes of yellow and orange. Apart from deposited carbon, some free carbon is also present as one of the exhaust products under the detonating condition. Puffs of black exhaust smoke representing free carbon in the exhaust gases also indicate detonation; these puffs occur intermittently and are different from the black smoke of over-rich mixtures.

6.9 THEORIES OF KNOCKING

There is no complete explanation for the origin of knock over the full range of engine operating conditions at which knock occurs. The fundamental explanation of the knocking phenomenon given by high speed cinematography tests led to two general theories of knocking—the autoignition theory and the detonation theory.

Autoignition theory: According to this theory when the fuel-air mixture in the end-gas region is compressed to sufficiently high pressures and above the self-ignition temperatures of the fuel before the flame front reaches it, the preflame reactions take place in the parts or all of the end-gas. During the preflame reactions, extensive decomposition of the mixture takes place producing aldehydes, nitrogen peroxide, hydrogen peroxide and free radicals. The energy released by these reactions and the presence of active chemical species and free radicals greatly accelerate the chemical reaction, producing a very high pressure locally in the end-gas region and leading to autoignition. Strong pressure waves propagate across the combustion chamber, and also knocking sound due to the acoustic vibration of the gases at the appropriate resonant frequency is transmitted through the engine structure.

In autoignition theory it is assumed that the flame front propagates with a normal speed before the start of autoignition.

Detonation theory: According to this theory, under the knocking conditions, the advancing flame front, called *detonating waves*, accelerates to sonic velocity, and consumes the end-gas at a rate much faster than would occur with normal flame speeds. Here also, there is a rapid release of chemical energy in the end-gas which creates a high pressure in the end-gas region, propagates strong pressure waves and produces knocking.

There is much less evidence to support the detonation theory compared to the evidence to support the autoignition theory as the knock initiating process. Most recent evidence indicates that the knock originates with the autoignition of one or more local regions within the end-gas. The rest of the regions then ignite until the end-gas is completely reacted. This sequence of operations occurs extremely rapidly. Thus, the autoignition theory is most widely accepted. Therefore, the more general term ‘knock’ is preferred to detonation to describe this phenomenon.

6.10 EFFECT OF ENGINE VARIABLES ON KNOCK

To prevent knock in the SI engine the end-gas should have low temperature, pressure and density, a long ignition delay and a non-reactive composition. Thus the major factors that appear to be involved in producing and preventing knocking are: temperature, pressure, density, time, the composition of the unburned charge and engine design.

When engine conditions are changed, the effect of the change may be reflected by more than one of the variables mentioned above. Since the effects of temperature, pressure and density are closely related with one another, these are grouped together in subsection 6.10.1.

6.10.1 Temperature, Pressure and Density Factors

As the temperature of the charge is increased, the flame speed increases and the possibility of the end-gas to reach its critical temperature for autoignition also increases. This increases the tendency to knock. An increase in pressure reduces the delay period of the last part of the charge as this part of the charge is subjected to a high pressure. Increase in density of the charge tends to increase the possibility of knocking by increasing the preflame reactions in the end-gas, thus releasing higher energy. The following factors too tend to affect the tendency to knock:

The compression ratio: As the compression ratio is increased, the pressure, the temperature and the overall density of the charge increase. Therefore, an increase in compression ratio increases the knocking tendency of the engine. For a given engine setting and a fuel, there is a critical compression ratio above which knock would occur. This compression ratio is called the highest useful compression ratio (HUCR). Materials with high conductivity such as aluminium alloys are used for cylinder head, since a cool combustion chamber wall is required for an engine to have a high compression ratio without knock.

The mass of inducted charge: The mass of the induced charge can be increased by supercharging the engine. The charge density will be increased which in turn will increase the tendency to knock. The mass of the inducted charge can also be increased by opening the throttle from the idling to the full power output position, so that both the charge pressure and temperature are increased during compression and subsequent combustion. The knock tendency will therefore increase with the opening of the throttle and reach a maximum at full throttle position.

The inlet temperature of the mixture: An increase in the inlet temperature of the mixture increases the temperature at the end of compression, which in turn increases the temperature of the last part of the charge to burn, thus shortening the delay period and greatly increasing the tendency to knock. The volumetric efficiency of the engine is also reduced as the inlet temperature is raised. Therefore, it is preferred to keep the inlet temperature low but not too low so as to cause starting and vaporization problems.

The temperature of the combustion chamber walls: An increase in the temperature of the combustion chamber walls increases the tendency to knock. To avoid knocking, the end-gas should not be compressed against the spark plug and exhaust valve as these are the two hottest parts in the combustion chamber.

Spark timing: An increase in spark advance from the optimized timing increases the peak pressure of the cycle and therefore increases the pressure and temperature to which the last part of the charge is subjected. This shortens the delay period and increases the tendency to knock.

The coolant temperature: An increase in the coolant temperature increases the temperature of the end-gas. This increase in temperature is due to reduced heat transfer rate from the gas to the coolant. It results in shortening the delay period of the end-gas and therefore knocking is increased.

Power output: An increase in the power output of the engine increases the temperatures of the cylinder and combustion chamber walls and also increases the temperature and pressure of the end-gas, thus, reducing the delay period. Therefore, the tendency to knock increases.

Exhaust back pressure: Increasing the exhaust back pressure, increases the compression temperature, increases the residual fraction and lowers the maximum pressure. The first effect tends to increase the knocking, while the others tend to reduce it. Increasing the exhaust back pressure has only a small effect on the knocking tendency, the opposing factors tend to balance it out in this case.

Cycle-to-cycle variation: Due to cyclic variation, the cycles with lower peak pressure may not knock, while the cycles with higher peak pressure will knock. A series of cycles with little cyclic variation will show a higher knock-limited indicated mean effective pressure than a series with of large cyclic variation, the average peak pressure being the same in both the cases.

Carbon deposits: Incomplete combustion of fuel causes carbon deposits on combustion chamber walls, valve head and piston crown. A part of the heat released by combustion of fuel is absorbed by these deposits due to the relatively poor thermal conductivity and this heat is transferred back to the fresh charge which is relatively cool. Thus the temperature of the fresh charge is increased and the tendency to knock increases. Apart from this, the clearance volume is also reduced because of the deposits on the combustion chamber walls and therefore the compression ratio is increased, causing an increase in the knocking tendency.

Cylinder deposits can be reduced by using a proper grade of lubricating oil. It is found that naphthenic oil is much superior to a paraffinic oil in keeping down the cylinder deposits.

6.10.2 Time Factors

In general, any action which tends to decrease the normal flame speed or shortens the ignition delay period, will tend to increase knocking. Such an action will autoignite the end-gas before the flame front reaches it. The following factors tend to affect the tendency to knock:

Turbulence: Turbulence depends on the engine speed and the design of the combustion chamber. Decreasing turbulence decreases the flame speed and increases the time available for the end-gas to attain autoignition conditions easily. Thus, knocking increases with the decrease in turbulence.

Engine speed: A decrease in engine speed decreases the turbulence within the cylinder and therefore decreases the flame speed and increases the time available for preflame reactions. The length of the delay period is not greatly affected by engine speed. Therefore, the knocking increases with the decrease in engine speed.

Flame travel distance: The flame travel distance can be increased by increasing the size of the engine and the combustion chamber. It can also be increased by locating the spark plug away from the centre. Increasing the flame travel distance increases the time taken by the flame front to travel across the combustion chamber. This gives more time for the end-gas to autoignite and, therefore, knocking increases. The following factors affect the flame travel distance.

Combustion chamber shape: In a compact combustion chamber, the normal flame can be made to reach the last part of the charge more quickly, so the combustion time will be shorter. Further, if the combustion chamber is highly turbulent, the combustion rate is high and the combustion time is further reduced. Thus a compact combustion chamber reduces knocking.

Engine size: Large engines operate at low rpm, while the small engines operate at high rpm. Thus the piston speed, turbulence and the flame speed are almost the same in similar engines, regardless of the size. Therefore, the time required for the flame to travel across the combustion space would be longer in the larger engines. The delay period is not much affected by size. The larger cylinders will therefore be more likely to knock. SI engine is generally limited to the size of 150 mm bore.

Location of spark plug: A spark plug which is centrally located in the head of the combustion chamber has the minimum tendency to knock, since the flame travel distance is minimum. The flame travel distance can further be reduced by using two or more spark plugs.

Location of exhaust valve: The exhaust valve should be located close to the spark plug. The flame starts from the spark plug, therefore the end-gas is far away from it. Locating the exhaust valve near the spark plug means that the exhaust valve is also not situated near the end-gas region. So, the temperature of the end-gas will not increase due to hot exhaust valve, and the delay period will also remain long. A long delay period means that there is a less chance of knocking.

6.10.3 Composition Factors

Once the compression ratio and the engine dimensions are selected, the fuel/air ratio and the properties of the fuel play an important role in controlling the engine knock. The following composition factors affect the knock.

Octane rating of the fuel: The tendency of an engine to knock depends on the properties of the fuel used. A lower self-ignition temperature of the fuel and a high preflame reactivity would increase the tendency of knocking. The octane number is the measure of resistance to knock. A higher octane number of the fuel reduces the tendency to knock. In general, the hydrocarbons of the paraffin series have the maximum tendency to knock while those in the aromatic series have the minimum tendency to knock. The naphthene series comes in between these two.

The knock tendency of paraffins depends on the molecular structure indicated by the following general relationships:

(a) Increasing the length of the carbon chain increases the knocking tendency.
(b) Centralizing the carbon atoms decreases the knocking tendency.
(c) Adding methyl groups (CH_3) to the side of the carbon chain in the 2, or centre position decreases the knocking tendency.

The unsaturated aliphatic hydrocarbons show less knocking tendency than the corresponding saturated hydrocarbons with exceptions of ethylene, acetylene and propylene. Thus acetylene knocks much more easily than ethane.

Naphthenes and aromatics show the following relationships between the molecular structure and the knocking tendency:

(a) The naphthenes have greater knocking tendency than the corresponding aromatics. Thus, cyclohexane (C_6H_{12}) knocks much more easily than benzene (C_6H_6).
(b) Two or three double bonds have less knocking tendency, whereas one double bond knocks easily.
(c) Lengthening the side chain increases the knocking tendency in both groups of fuel, whereas branching of the side chain decreases the knocking tendency.

In general, for most hydrocarbons the increased compactness of the molecular structure reduces the knocking tendency.

Fuel/air ratio: For a slightly rich fuel-air mixture for which the best power is obtained, the flame temperature is maximum resulting in maximum flame speed and minimum delay period. These two effects counteract each other regarding knocking. However, the short delay is the predominating factor and therefore even by using the best power fuel/air ratio ($\phi = 1.1–1.2$) the knock is maximum. The length of the delay period increases as the mixture is made richer or leaner than this and the tendency of knocking will be reduced. For weaker mixtures the octane requirement will decrease continuously since the reaction rate and temperature are continuously reduced as the fuel proportions are reduced. For richer mixtures the knock is reduced by the cooling due to the latent heat of the fuel and the lower flame temperature.

Humidity of air: Increasing the humidity of the entering air tends to reduce knocking by increasing the reaction time of the end gas. The ignition delay is prolonged and autoignition of the end gas is suppressed.

Stratifying the mixture: The probability of knock is decreased by stratifying the mixture, which makes the end-gas less reactive.

Maldistribution: The unequal distribution of air and fuel between the various cylinders in a multi-cylinder engine is called maldistribution, which may result in different knocking tendencies in different cylinders because of change in the air/fuel ratio locally. Usually some retardation of ignition timing or enrichment of the mixture is required to reduce the knocking tendency in those cylinders which have the greater tendency to knock. Such adjustments may reduce the power output and will increase the specific fuel consumption.

Dilution of the charge: The dilution of the charge with the inert gases increases the reaction time and reduces the flame speed. The increase in reaction time of the end gas reduces knocking and decrease in the flame speed increases knocking. The first factor dominates. Therefore, by introducing cooled exhaust gas with the inlet air, the tendency to knock can be reduced.

Water or water-alcohol injection: Injection of water or water-alcohol mixtures into the inlet system of the engine reduces knocking by increasing the reaction time of the end gas. It increases the length of ignition delay and autoignition of the end gas is suppressed.

Fuel additives: Several substances have been found which have a pronounced anti-knock effect and increase the octane ratings when added to petrol in a very small proportion, called *dope*. Typical examples of these include benzole, ethanol, methanol, acetone, nitrobenzene and tetra ethyl lead, etc. The most important of these is tetra ethyl lead $Pb[C_2H_5]_4$, which is soluble in petrol and enables high compression ratios to be used compared to those with the petrol alone. A proportion of ethylene dibromide is added to tetra ethyl lead in order to prevent the deposition of lead inside the engine. The lead and bromine (expelled with the exhaust gas as lead bromide) combine and greatly reduce the amount of the deposits but some part of the deposit may still be found in the cooler part of the exhaust system.

In recent years, the use of leaded fuel has been restricted, since it pollutes the atmosphere and destroys the effectiveness of the noble metal catalysts of catalytic converters, used for controlling the air pollution from the exhaust of the engine. The other drawbacks, associated with the prolonged use of leaded fuels, are the deposition of lead salts upon the spark plugs, exhaust valves and combustion chambers.

6.10.4 Effect of Design

The knocking tendency of the engine is affected by the following design considerations:

Effect of shrouded inlet valve: Plain valves and shrouded inlet valves are shown in Figure 6.16(a) and Figure 6.16(b) respectively. The use of a shrouded inlet valve provides the flow of charge in a definite direction, so that the combustion time is reduced. This will reduce the tendency to knock. The shrouded valve also tends to reduce the cycle-to-cycle variation, especially when oriented so as to give tangential flow into the cylinder..

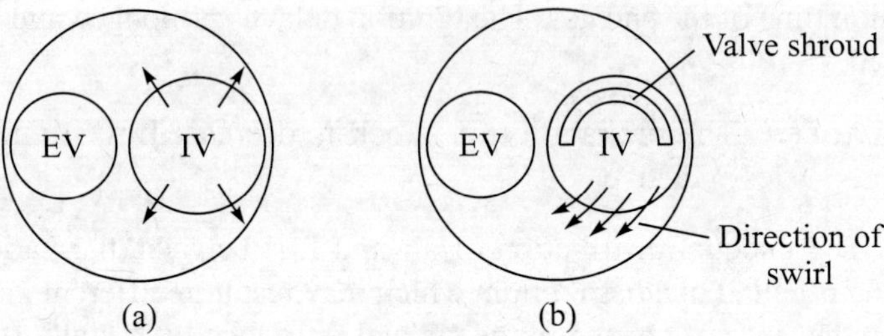

Figure 6.16 (a) Plain valves and (b) shrouded inlet valves (IV-Inlet valve, EV-Exhaust valve).

Effect of piston shape: Plain piston with flat top and squish piston are shown in Figure 6.17(a) and Figure 6.17(b) respectively. In squish piston, the charge is squeezed radially inwards, near the top dead centre and the tendency to knock is less. The thin space above the piston in the combustion chamber is called the *quench space*. Near the end of the flame travel the end-gas is located in a thin space where it makes good contact with the cylinder walls, which are at a lower temperature than the end-gas. Thus the quench space is cooled which reduces the possibility of autoignition and hence the knocking. Because of the reduced space above the squish piston, the combustion chamber becomes effectively more compact and the possibility of turbulence increases. Both of these factors tend to decrease the knocking tendency.

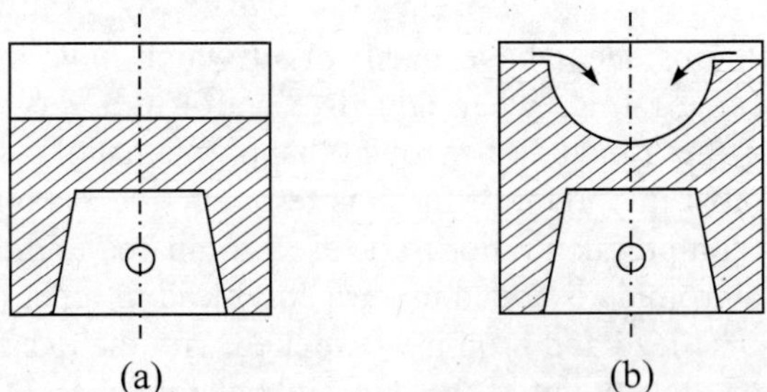

Figure 6.17 (a) Flat Piston and (b) squish piston.

Effect of cylinder bore: When engines of similar design but of increased bore run at the same piston speed and with the same inlet conditions, fuel/air ratio and exhaust back pressure, the combustion time increases and the temperature of the inner surface of the cylinder also increases. Both of these factors tend to increase the knocking tendency. The octane requirement of the fuel increases with the increase in bore even though the compression ratio and the engine rpm remain the same.

Table 6.1 provides a summary of the factors that influence the tendency to knock in SI engines.

Table 6.1 Summary of the factors influencing tendency to knock in SI engines

Variable	*Major effect on unburned charge*	*Action to be taken to reduce knocking*	*Can the operator control?*
Compression ratio	Temperature, pressure and density	Reduce	No Can be controlled in a VCR engine
Mass of inducted charge	Pressure and density	Reduce	Yes, by throttling
Inlet temperature	Temperature	Reduce	Yes, in some cases
Combustion chamber wall temperature	Temperature	Reduce	Not ordinarily
Spark timing	Pressure and temperature	Retard	Yes
Coolant temperature	Temperature	Reduce	Yes
Load (power output)	Temperature and density	Reduce	Yes
Cyclic variation	Pressure	Reduce	No
Carbon deposits	Temperature	Reduce	In some cases
Turbulence	Time	Increase	Yes, by engine speed
Engine speed	Time	Increase	Yes
Flame travel distance	Time	Reduce	No
Octane rating of fuel	Self-ignition temperature, preflame reactivity	High	In some cases by additives.
Fuel/air ratio	Temperature and time	Use very rich or lean mixture	Yes
Humidity	Reaction time	Increase	No
Mal-distribution	Temperature and time	Reduce	No
Dilution of charge	Reaction time	More	Yes, in some cases

6.11 DETECTION OF KNOCKING

Detection of knocking is very important. Once knocking is recognized, control measures can be applied before the damage is done. The following simple methods can be used to detect knocking:

1. Knocking sound can be heard in engines fitted with the silencer in the exhaust pipe. Under loud exhaust or propeller noise, it is often impossible to detect knocking.
2. The temperature measurement of a spark-plug gasket by a thermocouple embedded in it can indicate the knocking. A sudden or abnormal temperature rise under steady operating conditions shows the presence of possible knock. Steady operating conditions are necessary since the spark plug gasket temperature may also be affected by changing the air/fuel ratio, the engine speed, the manifold pressure, and the rate of cooling, etc.

3. Knock intensity can be detected by a pressure transducer which is flush mounted in the combustion chamber. It is a pressure sensitive unit in which the diaphragm is exposed to the gases in the cylinder and the pressure signals are converted to electrical signals. This electrical signal is amplified and recorded on a knock meter. With increasing amplitude of the signal, the scale reading of the knock meter increases and a relative measure of knock intensity is obtained. This unit can be used to apply knocking control measures automatically.
4. It is often possible to detect knocking by the presence of intermittent puffs of gray smoke in the exhaust, which appear bright yellow flashes when the test is carried out in dark.

6.12 UNCONTROLLED COMBUSTION

Under certain conditions, the fuel-air mixture is ignited by a hot spot in the cylinder. Initiation of a flame front by a hot surface other than the spark is called *surface ignition*. It comes under the category of uncontrolled combustion. The hot surface might be the spark plug insulator or electrode, the exhaust valve head, the combustion deposits on the combustion chamber surfaces, etc. Surface ignition occurring before the spark is called *preignition* and that occurring after the spark is called *postignition*. Run-on, run-away, wild ping and rumble are caused by surface ignition which are harmful.

6.12.1 Preignition

It is uncontrolled inflammation of the combustible mixture in an engine by a hot surface before the spark occurs. This is the most severe form of uncontrolled combustion. Under severe operating conditions, some part of the cylinder surface may be hot enough (nearly 1100°C) to ignite the charge before the spark does so. This is equivalent to advancing the ignition, but since the hot spot surface is larger than the spark, the combustion rate would be faster than that of the normal combustion, creating very high cylinder pressures and temperatures and thus resulting in excessive negative compression work and increased heat loss to the walls. The overall effect will be the loss in power.

Preignition will also cause higher temperatures and pressures in the end-gas than those caused by normal ignition because of its earlier occurrence on the compression stroke. Thus preignition leads to autoignition and hence knock, and autoignition encourages preignition. Knock and preignition are different phenomena. Knock is due to the rapid combustion of the last part of the mixture following the initiation of flame by the spark, whereas preignition is the ignition of the charge by a hot body before the spark occurs.

The results of preignition are to increase the work of the compression stroke, decrease the net work of the cycle, increase the engine pressures, increase the heat loss from the engine and decrease the efficiency. Preignition if not checked gets progressively worse, culminating in severe engine damage.

Preignition can be detected by switching off the ignition when irregular firing might occur for a few strokes before the engine speed drops. The sudden loss of power with no evidence of mechanical malfunctioning may also indicate preignition. The best proof of preignition is to take the indicator diagram with a high speed indicator and analyse for preignition.

Water is a very effective inhibitor of preignition. A small proportion of water is always added in commercial alcohol to avoid preignition.

6.12.2 Run-on Surface Ignition

When the ignition is switched off and the throttle is closed (fuel-air mixture is supplied through the idling jet), the condition in which the engine continues to fire is called run-on. It might be due to a hot surface in the cylinder, but the major cause is spontaneous ignition of the fuel-air mixture. The physical factors influencing spontaneous ignition are (a) an elevated temperature of the inlet mixture, (b) poor cooling of the combustion chamber surface, (c) duration of the valve overlap, and (d) a high compression ratio. The inlet temperature is elevated at the low speed condition by the low rate of air flow through the induction system, often in close proximity to the hot exhaust. At idling speed, the combustion chamber surface is not properly cooled due to poor coolant circulation.

6.12.3 Run-away Surface Ignition

In severe cases of surface ignition, the run-away surface ignition develops. Surface ignition in one cycle heats the surface ignition source to still higher temperatures in consecutive cycles and a series of earlier preignitions is set up. The run-away surface ignition results in considerable damage to pistons and other engine parts. Engine may catch fire as the preigntion advances to the time when the intake valve is open and fuel-air mixture is entering.

The run-away surface ignition is usually caused by an overheated spark plug, exhaust valve or piston head.

6.12.4 Wild Ping

Wild ping is one or several irregular, but very sharp, combustion knocks caused by early surface ignition from deposit particles after the inlet valve is closed. Knock occurs in an erratic way. It produces sharp cracking sound. A probable reason for wild ping is that some glowing carbon particles attached lightly to the combustion chamber surface break free, and then floating erratically through the chamber ignite the charge until they are finally carried away past the exhaust valve.

6.12.5 Rumble

Rumble is the name assigned to intermittent roughness caused by combustion chamber deposits which create secondary flame fronts. It is a low pitched noise distinctly different from spark knock. It follows that the rate of pressure rise and the maximum pressure become very high. Rumble develops early and at multiple points.

Rumble is avoided or minimized by eliminating deposits usually by fuel additives. The type of lubricating oil and gasoline without tetra ethyl lead can also reduce deposits and therefore rumble.

Rumble causes vibrations of the crank shaft arising from a high rate of pressure rise with consequent deflection of mechanical parts.

The combustion phenomena in a spark-ignition engine are summarized in Figure 6.18.

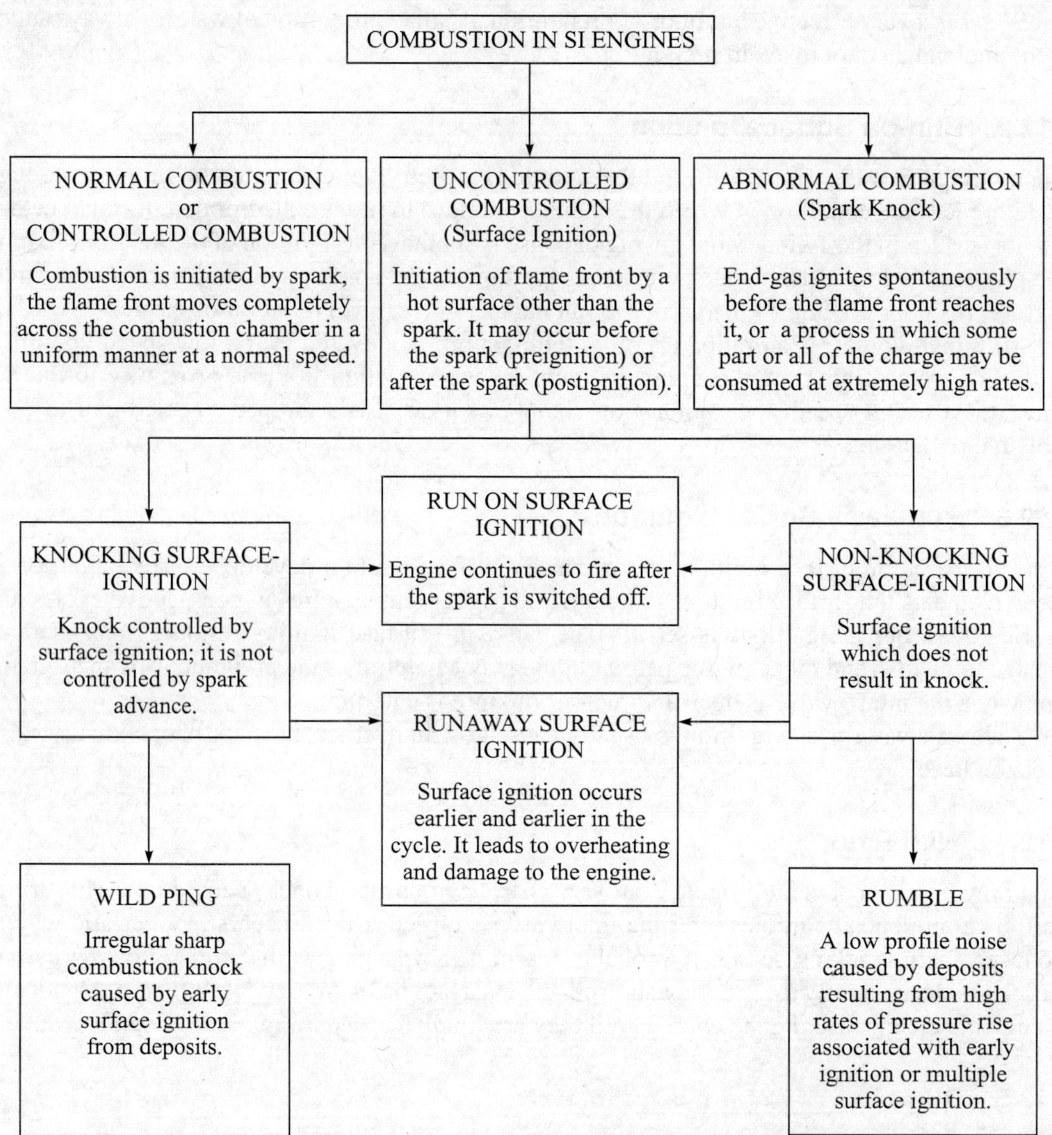

Figure 6.18 Summary of combustion phenomena in a spark-ignition engine.

6.13 COMBUSTION CHAMBERS FOR SPARK-IGNITION ENGINES

A proper design of the combustion chamber for the spark-ignition engine is important as it affects the engine performance, its knocking tendencies and exhaust pollutants. The design involves the shape of the cylinder head and piston crown, the location of the spark plug, and location, size and number of inlet and exhaust valves. The design of the intake port also influences the turbulence and flow pattern of the charge in the combustion chamber. The optimum design of a combustion chamber is a subject of research and development. During the days of development of the engine, the compression ratio used was only 4 and now it is possible to raise the compression ratio to above 10 without knock.

6.13.1 Basic Requirements of a Good Combustion Chamber

The basic requirements of a good combustion chamber are to provide:

1. High power output
2. High thermal efficiency and low specific fuel consumption
3. Smooth engine operation
4. Reduced exhaust pollutants.

High power output: For producing a high power output, a combustion chamber requires the following:

(a) A high compression ratio
(b) A little rich mixture
(c) Good turbulence
(d) Large inlet valve to obtain a higher volumetric efficiency
(e) Streamline flow in order to reduce the pressure drop and to increase further volumetric efficiency.

High thermal efficiency and low specific fuel consumption: In order to achieve high thermal efficiency and low specific fuel consumption, the following are the requirements:

(a) A high compression ratio
(b) A small heat loss during combustion, which means a small surface-to-volume ratio and a compact shape.
(c) Faster fuel burning process
(d) A little lean mixture.

Smooth engine operation: Smooth engine operation, with the selection of the highest compression ratio to use for a fuel of given octane rating, requires the following:

(a) A moderate rate of pressure rise during combustion.
(b) Absence of knock, which in turn means:
    (i) A compact combustion chamber to reduce the flame travel distance.
    (ii) Proper location of spark plug and exhaust valve, and their cooling.

Reduced exhaust pollutants: Exhaust pollutants can be reduced by designing a combustion chamber that produces a faster burning rate of fuel. A faster burning chamber with its shorter burning time permits operation with substantially higher amounts of Exhaust Gas Recirculation (EGR), which reduces the oxides of nitrogen (NO_x) in the exhaust gas without substantial increase in the hydrocarbon emissions. It can also burn very lean mixtures within the normal constraints of engine smoothness and response. A faster burning chamber exhibits much less cyclic variations, permitting the normal combustion at part load to have greater dilution of the charge.

Methods of using a 'fast burn' combustion chamber include the following:

(a) Locating the spark plug to a more central position within a compact combustion chamber.
(b) Using two spark plugs.
(c) Increasing the in-cylinder gas motion by creating swirl during the induction process or during the latter stages of compression.

6.14 COMBUSTION CHAMBER DESIGN PRINCIPLES

The general principles governing an efficient combustion chamber design may be enumerated as follows:

High volumetric efficiency: High volumetric efficiency means more charge per stroke and a proportionate increase in the power output. Effective valve open area, which depends on valve diameter and lift, directly affects the volumetric efficiency. To obtain maximum performance and to reduce pumping losses, the size of the valve heads should be large. The valve sizes that can be accommodated depend on the cylinder head geometry.

Minimum path of flame travel: The flame travel path is determined by the location of the spark plug and by the shape of the combustion chamber. A compact combustion chamber reduces the flame travel distance to a minimum. With a given turbulence, this reduces the time of combustion and minimizes the burning time loss. The minimum path of flame travel reduces the possibility of knock. SI engines are generally limited up to 150 mm cylinder bore because of short flame travel distance. There are no such limits on CI engines.

Provision of minimum heat loss zone around the spark plug: It ensures good initial combustion conditions. The spark plug is placed near the exhaust valve to prevent heat loss in the first phase of combustion. The surface-to-volume ratio should be minimum here.

Reduced rate of pressure rise: The second phase of the combustion zone or shock zone should be designed to give a reduced rate of pressure rise to avoid knocking and to avoid excessive shocks on the crankshaft.

Provision of a suitable quench region: The quench region in a combustion chamber is provided at the farthest distance of the flame travel, so that there will be increased cooling of the combustion gases during the most likely knocking period. This condition is obtained by making the surface-to-volume ratio maximum in this region. If the relatively cool inlet valve is located in this region, then cooling of the end-gases will be improved.

Maximum thermal efficiency: Maximum thermal efficiency for the given grade of fuel can be obtained by employing the highest compression ratio for smooth engine operation, without knocking tendency, under all operating conditions.

Short combustion time or fast burn: It is an important consideration in combustion chamber design of a spark-ignition engine. Fast burn results from properly creating turbulence. It improves the possibility of 'lean burn', thus reducing air pollution. Proper turbulence may be created by positioning the inlet valve suitably, designing the inlet passage to create swirl in the induction process and streamlining the combustion chamber. Turbulence may also be created by 'squish', which is better as it does not adversely affect the volumetric efficiency.

Exhaust valve location: To reduce the possibility of knock, there should not be a hot surface in the end-gas region. Exhaust valve being a very hot surface should not be placed in this region. The exhaust valve should be located near the spark plug. In order to reduce the hot surface area, the exhaust valve head diameter is kept small and to avoid flow restrictions, a high lift is employed. The exhaust valve head should be cooled to the desired extent.

Maximum output: For maximum output, two inlet and two exhaust valves are used per cylinder. Engines of high output usually have intake valves in one line and exhaust valves in another. The valve-in-head type of combustion chamber is usually domed. The spark plug is located in the top of the dome, which makes a compact combustion chamber. The flame front area increases rapidly with this type of chamber, the length of flame travel is short and the combustion rate is high.

Cooling of spark plug points: Sufficient cooling of the spark-plug points is required to avoid preignition effects, at the wider throttle openings.

Scavenging of the exhaust gas: A good scavenging of the exhaust gas is required.

Materials for cylinder head: Aluminium-alloy heads are mostly used. Some heads have been built of copper or copper alloys, and others had copper inserts located to contact the last part of the charge to be burned. These are materials of high heat conductivity, which conduct heat away from the hot spots.

6.15 COMBUSTION CHAMBER OPTIMIZATION PROCEDURE

The sequence of steps described in the following subsections constitutes the process of development of a good combustion chamber.

6.15.1 Geometric Considerations

These involve the considerations of the shape of cylinder head and piston crown, the spark plug location, and the inlet and exhaust valves location and size. Open chambers, such as the hemispherical with almost central spark plug location, give close to the maximum flame front surface area ensuring faster burn and provide the lowest chamber surface area in contact with the burned gases ensuring the lowest heat transfer. More compact combustion chamber shapes other than the open chambers, such as the bowl-in-piston, do produce a somewhat faster burn but suffer from lower volumetric efficiency and higher heat losses.

With almost central positioning of the spark plug and providing some squish area the burn rate is improved.

6.15.2 Considerations for Cyclic Variations

Cyclic variations can be reduced by taking care of the following:

1. Improving the uniformity of the flow of air, fuel and residual gases mixture during the intake process.
2. Delivery of equal amounts of constituents to each cylinder to avoid maldistribution.

3. Provision for good mixing between constituents in the intake manifold.
4. Accurate control of mixture composition during engine transients.
5. Achieving nearly similar flow patterns within each engine cylinder to obtain equal burn rates in all cylinders.

6.15.3 Consideration for Proper Turbulence

The higher turbulence levels during combustion are required which can usually be best done by creating swirl during the induction process. Higher than necessary gas velocities within the cylinder result in excessive heat losses and low volumetric efficiency.

6.16 TYPES OF COMBUSTION CHAMBER

Brief descriptions of a few important types of combustion chambers, with locations of valves and spark plug, showing their developments are given below:

6.16.1 T-head Type Combustion Chamber

Figure 6.19 shows the T-head type of combustion chamber. This was the earliest type used by Ford Motor Corporation during early stages of engine development in 1908. The T-head design suffers from the following disadvantages:

(a) The distance across the combustion chamber is very long. The spark plug is located near the exhaust valve, so the flame travel distance from the spark plug to the end-gas (near the inlet valve) increases. Therefore, knocking tendency is increased.
(b) The configuration provides two valves on either side of the combustion chamber, requiring two camshafts.

There was a violent knocking even at compression ratio 4 : 1. This was also because of the low octane number of petrol available at that time, which varied from 45 to 50.

6.16.2 L-head Type or Side Valve Combustion Chamber

Figure 6.20 shows the L-head type of combustion chamber. In this, the combustion chamber is in the form of a more or less flat slab, extending over the piston. The disadvantages of the T-head type of combustion chamber forced the development of L-head, in which the two valves are placed on the same side of the combustion chamber, thus reducing the flame travel distance, and the valves are operated by a single camshaft.

During the period 1910–1930, the side valve combustion chambers were commonly used in SI engines. In such an engine the valves are placed side by side in a detachable block. Manufacturing and maintenance of this type of combustion chamber are both easy. The valve mechanism can easily be enclosed and lubricated. The detachable head can easily be removed for decarbonising without disturbing either the valve gear or the main pipe work.

In its original form, this type of engine gave poor performance because of the following limitations:

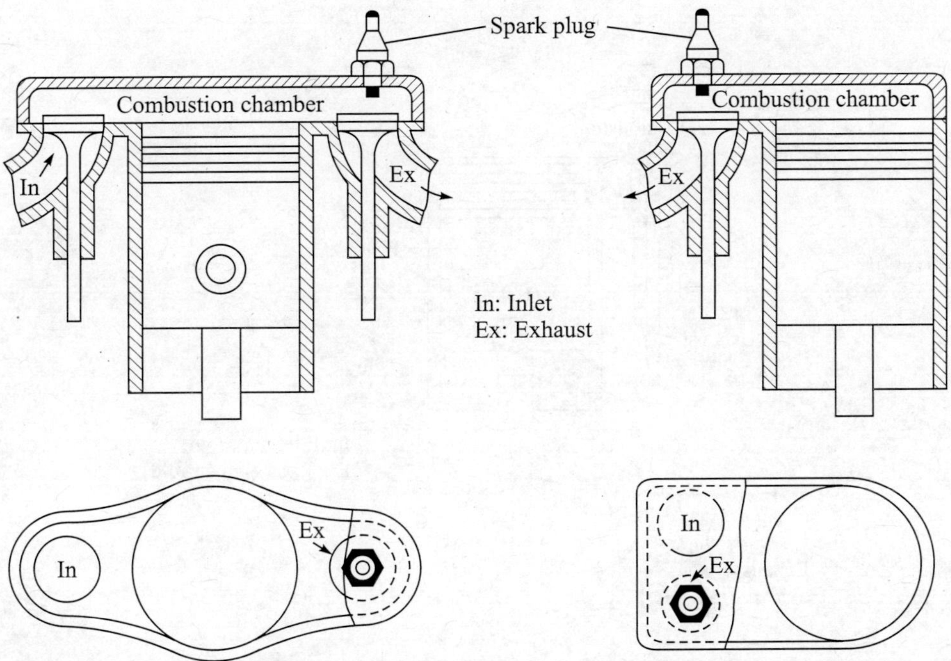

Figure 6.19 T-head combustion chamber

Figure 6.20 L-head or side valve combustion chamber.

(a) Lack of turbulence as the charge had to take two right angle turns to enter into the cylinder and in doing so the initial velocity of the charge got reduced.
(b) Extremely prone to knock due to lack of turbulence, resulting in a low flame speed. The flame travel distance was also large, and it therefore caused knock.
(c) Extremely sensitive to ignition timing due to lower rate of burning and slow combustion.

The side valve engines are not preferred for higher compression ratios on account of inadequate volumetric efficiency, noncompactness and additional requirement of cooling. These engines do not compare well with the overhead valve engines which were developed later to give more power and higher efficiency.

6.16.3 Ricardo Turbulent Head Side Valve Combustion Chamber

Figure 6.21 shows the arrangement of this type of combustion chamber developed in 1919. The main objective of this design was to increase turbulence in order to obtain a higher flame speed and to reduce the knocking tendency of the engine.

A greater volume of the space of combustion chamber was available over the valves and was called the main body of the combustion chamber. A slightly restricted passage-way was provided over the cylinder. During the compression stroke the gases were forced back to the main body through the restricted passage-way that created additional turbulence. By varying the throat area of the restricted passage it was possible to achieve any desired degree of turbulence within the main body of the combustion chamber. The rate of combustion during the second stage of combustion was improved considerably and this resulted in improved performance.

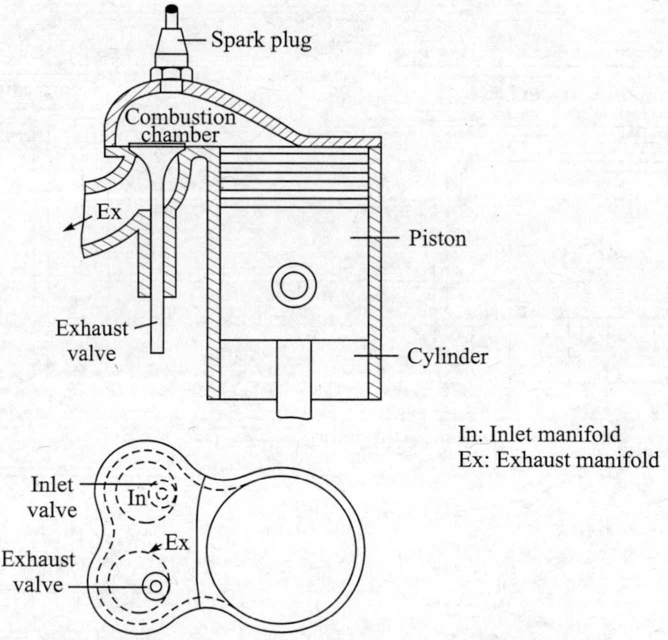

Figure 6.21 Ricardo turbulent combustion chamber (1919).

In order to reduce the knocking tendency to a minimum, the distance of effective flame travel was shortened by forming a very thin layer of entrapped gas between the piston crown and the portion of combustion chamber at the end-gas region, when the piston was at TDC. The surface-to-volume ratio in this region was high, which helped in rejecting enough heat from the end-gas to prevent knocking. The flame travel distance was further reduced by placing the spark plug in the centre of the effective combustion space with a slight shift towards the exhaust valve.

At the time the turbulent head was developed, the octane number of the petrol was in the region of 45–50. With this fuel the compression ratio of the engine was raised from 4 : 1 to 4.8:1 and the power and efficiency of the engine got substantially improved.

As the octane number of petrol improved with time, by about 1935, it was possible to run the engine with a compression ratio 6 : 1. At this ratio the flame speed was increased to the extent that the rate of pressure rise was observed more than 2.5 bar per degree of crank angle exceeding the optimum rate of pressure rise, namely about 2–2.5 bar per degree of crank angle. To combat this problem, the area of the passageway was increased progressively.

With the relatively high octane petrol available today, the compression ratio has been increased, resulting in lack of space to accommodate the valves in L-head or side-valve design. It appears that the side-valve engine can no longer compete with the overhead design. Nowadays a compression ratio of 8 : 1 is normal and ratios up to 10.5:1 or even 11:1 are used in very high output engines and sports car engines.

At compression ratios over 8, the normal turbulence created by the entry of the gases through the inlet valves is usually sufficient. A small increase in turbulence can always be obtained by providing ‘squish’. Squish is the rapid ejection of gas trapped between the piston and some flat or corresponding surface in the cylinder head.

6.16.4 Overhead-valve or I-head Type Combustion Chamber

The overhead-valve combustion chamber is also called the I-head type in which both the inlet and the exhaust valves are located in the cylinder head. The overhead-valve engine is better than a side-valve engine at higher compression ratios. Following are the advantages of the overhead-valve type of arrangement:

(a) The volumetric efficiency is higher because of the better breathing of the engine from larger valves or valve lifts, and the pumping losses are less because of the more direct passage-ways provided for the gas exchange processes with less pressure drops through valves.
(b) The average flame travel distance is reduced and therefore the engine is less prone to knock, resulting in reduced octane requirements.
(c) The surface-to-volume ratio is decreased, therefore, the heat losses through the combustion chamber walls from the burned gas to the coolant are reduced. It results in an increase in the thermal efficiency of the engine. It also provides more complete combustion of fuel, thus producing more power and reducing air pollution.
(d) Hot exhaust valve is placed over the head instead of in the cylinder block, thus confining thermal failures only to the cylinder head which can easily be removed and replaced.
(e) The possibility of leakage of compression gases and jacket water is reduced as in this type of arrangement the cylinder head bolts are subjected to less force.
(f) The casting process is easier, thus leading to reduction in costs.

The two important arrangements of the overhead-valve combustion chamber are described below:

Bath-tub type of combustion chamber

The bath-tub type of combustion chamber is the most simple and convenient form. It consists of an oval-shaped combustion chamber bolted over the main cylinder in such a way that some part of the oval portion overhangs the cylinder. This part may be used for 'squish'. Both the valves are mounted vertically overhead, with the spark plug at the side, as shown in Figure 6.22(a).

The main disadvantage of this type of combustion chamber is that the valves are placed in a single row along the cylinder block, resulting in less space to locate valves of larger diameter. It reduces the volumetric efficiency. More space for valves within the bore diameter can be made available if the stroke/bore ratio is kept unity or less.

Wedge-shaped combustion chamber

It is shown in Figure 6.22(b). The combustion chamber is wedge shaped with slightly inclined valves. This type has also given a very satisfactory performance.

6.16.5 F-head Type Combustion Chamber

The type of combustion chamber in which one valve is located in the head and the other in the block is known as the F-head combustion chamber. Figure 6.23 shows one of the most perfect F-head Wedge-type combustion chambers used by the Rover company. It has a well shaped piston

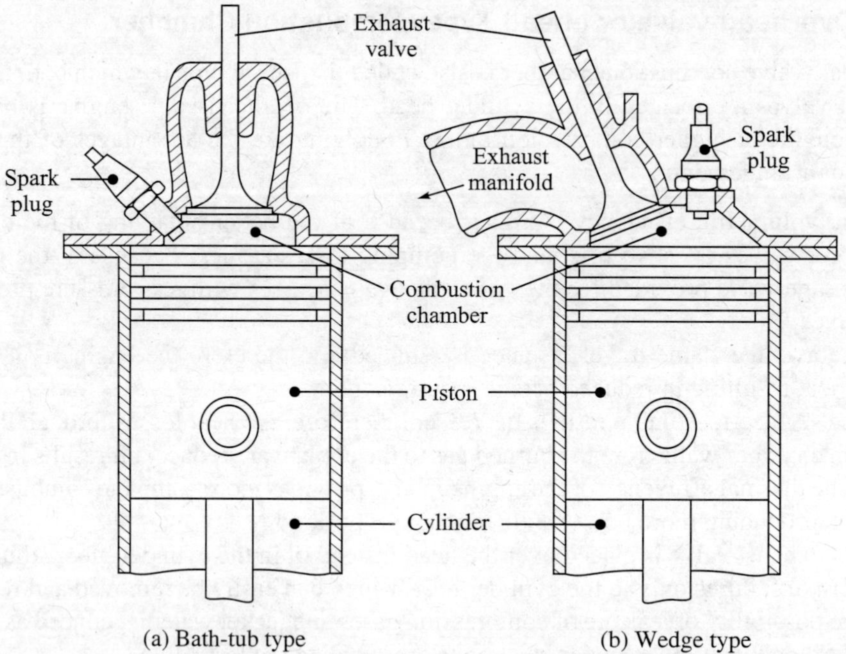

Figure 6.22 Overhead-valve combustion chamber designs.

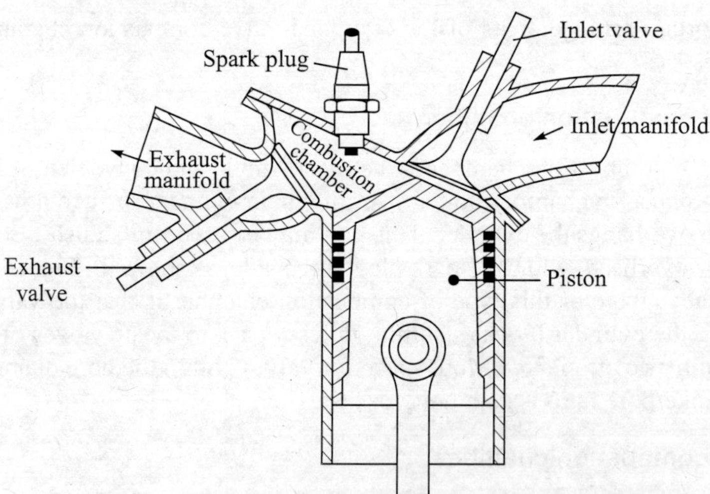

Figure 6.23 F-Head type—the Rover design combustion chamber.

crown with a correspondingly matched sloping cylinder head. The inlet and exhaust valves are inclined. The inlet valve is located in the head and the exhaust valve in the block. The plug is in an excellent position in the flat roof of the chamber. The flat roof allows the use of a larger size of inlet valve than the exhaust valve. The cooling of the spark plug and the exhaust valve is efficient. The flame travel distance is short and the end-gas is reduced to a thin layer, so the knocking tendency is reduced. The operation of the valves involves a complex mechanism.

6.16.6 Hemispherical Combustion Chamber

Figure 6.24 shows the hemispherical head with domed piston used in Jaguar racing car engines.

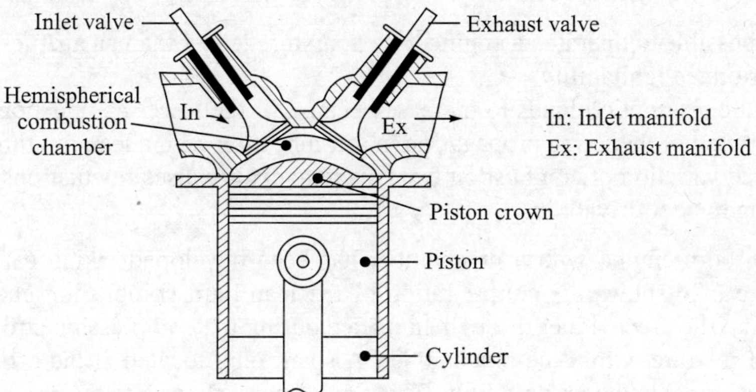

Figure 6.24 Hemispherical type Jaguar combustion chamber

A hemispherical chamber with inclined valves is the best design to use where the maximum specific output is required, involving piston speeds exceeding 15 m/s. Nearly all racing cars have used the hemispherical cylinder head with domed piston.

Following are the advantages of the hemispherical head:

(a) The combustion chamber is very compact.
(b) The surface-to-volume ratio is small which reduces the heat loss to the cylinder wall during combustion, thus providing a higher thermal efficiency.
(c) The larger diameter valves can be employed which may increase the volumetric efficiency.

However, the operation of the valves and placing of the spark plugs in a multi-cylinder engine present difficulties unless a twin overhead camshaft mechanism is used.

6.16.7 Piston Cavity Combustion Chamber

The combustion chamber comprises a bowl in the piston crown in conjunction with a flat cylinder head. The rover piston cavity type combustion chamber is shown in Figure 6.25. This resembles the combustion chamber of the normal direct-injection compression ignition engine. Here an almost ideal chamber shape is provided with all surfaces machined to give an accurately defined volume. Such a design was not possible in the past when long strokes and low compression ratios were used, but now with the use of higher compression ratios and stroke/bore ratios near one or less, this configuration has become practical and likely to appear more in the future.

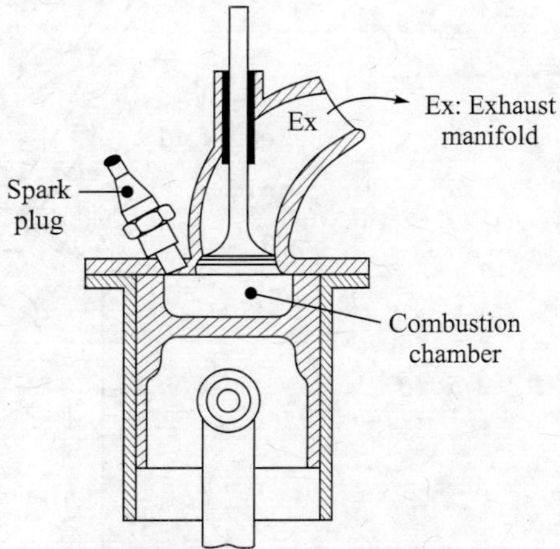

Figure 6.25 Piston cavity type Rover combustion chamber.

6.16.8 Combustion Chamber with a Pre-chamber for Lean Burn Engine

Lean combustion in an engine is one of the most promising methods for reducing the exhaust emission and improving the fuel economy. The problems associated with lean burn are:

(a) It is impossible to operate an engine with a mixture leaner than an air/fuel ratio of 19 due to deteriorated ignitability.
(b) The fuel consumption tends to increase because of the lower combustion speed and the deterioration of the flame propagation in a combustion with a lean mixture.
(c) Increased variation of combustion from cycle to cycle causes fluctuations in torque, thus resulting in poor drivability.

Toyota lean burn engine with a prechamber has been developed taking care of the above problems. Figure 6.26 shows the configuration of the lean burn combustion chamber. A fresh mixture flows into the prechamber through an orifice during the compression stroke, resulting in strong eddies of mixture within the prechamber. A spark plug located at the orifice ignites and produces a flame kernel in the mixture flow. This flame kernel flows into the prechamber with the mixture flow and causes rapid combustion in the prechamber. As a result, an optimum jet flame spouts into the main combustion chamber. This jet flame increases the combustion speed of the lean mixture thus improving the combustion. The prechamber is named the Turbulence Generating Pot (TGP), since its function is to generate turbulence in the main combustion chamber. In addition to the TGP, the carburettor and the exhaust manifold are also modified for use in the lean burn engine. The TGP improves the combustion of the lean mixture. The fuel consumption and torque fluctuations of the lean mixture combustion are reduced. Lean misfire limit is extended remarkably by locating the spark plug in the orifice of the TGP. Both the ignition lag and combustion noise are reduced due to this location of the spark plug.

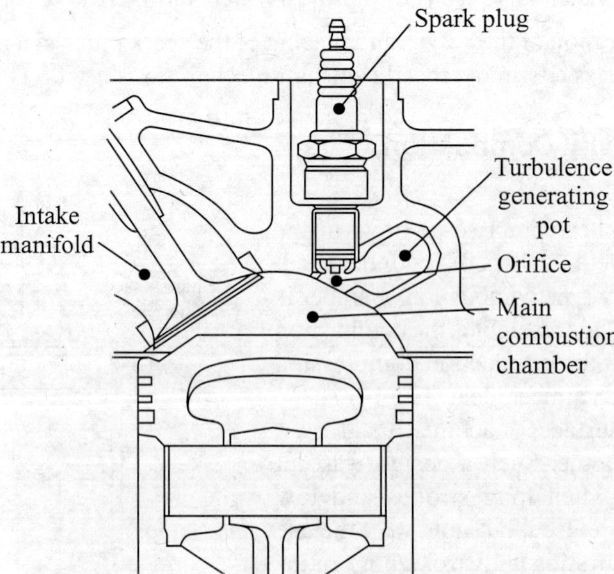

Figure 6.26 Toyota lean burn engine with a prechamber.

6.16.9 Future Trends

The design of the future combustion chambers will be influenced more by energy crisis and increasing amounts of air pollution. Both of these factors require lean burn engines for which fast burning is important and in order to complete the burning fast, the flame travel distance should be short. It requires a smaller bore than the stroke, i.e. a reversal of the modern trend and a little or no quench area in the combustion chamber.

6.17 OCTANE REQUIREMENT

The compression ratio, the performance and the efficiency of an engine are limited by knock. Knock in an engine depends on the antiknock quality of fuel, called the octane number. It determines whether or not a fuel will knock in a given engine under the given operating conditions. A higher octane number will have a higher resistance to knock. The octane number requirement of an engine is defined as the minimum fuel octane number that will resist knock throughout the engine's speed and load range. The following factors affect the octane requirement of an engine:

(a) Composition of the fuel
(b) Combustion chamber geometry
(c) Charge motion
(d) Spark timing
(e) Inlet air, intake manifold and water jacket temperatures
(f) Air/fuel ratio
(g) Ambient conditions—pressure, temperature and relative humidity.

The octane number requirement tends to go down when

(a) the ignition timing is retarded.
(b) the engine is operated at higher altitudes or smaller throttle openings or lower ambient pressures.
(c) the humidity of the air increases.
(d) the inlet air temperature is decreased.
(e) the fuel/air ratio is richer or leaner than that required for producing maximum knock.
(f) the exhaust gas recycle (EGR) system operates at part throttle.
(g) the engine load is reduced.

EXAMPLE 6.1 A spark-ignition engine runs at 2100 rpm. The compression ratio is 9.0, the cylinder bore is 10 cm and the engine is square. Connecting rod length is 17.5 cm. The spark plug is offset by 2 cm from the centre. The spark plug is fired at 20° bTDC. It takes 8° of engine rotation for combustion to develop and get into flame propagation mode. Flame termination occurs at 13° aTDC.

Calculate

(a) Piston displacement from TDC position at the time of flame termination
(b) Flame travel distance
(c) Effective flame speed

Solution:

Given:

(a) Bore $d = 10$ cm

Stroke $L = d = 10$ cm (square engine)

Crank radius, $a = \dfrac{L}{2} = 5$ cm

Connecting rod length, $l = 17.5$ cm

Equation (1.7) gives the piston displacement from TDC position,

$$x = l + a - a\cos\theta - \sqrt{l^2 - a^2\sin^2\theta}$$

$$= 17.5 + 5 - 5\cos 13° - \sqrt{(17.5)^2 - (5)^2\sin^2 13°} = \boxed{0.164 \text{ cm}} \quad \textbf{Ans.}$$

(b)

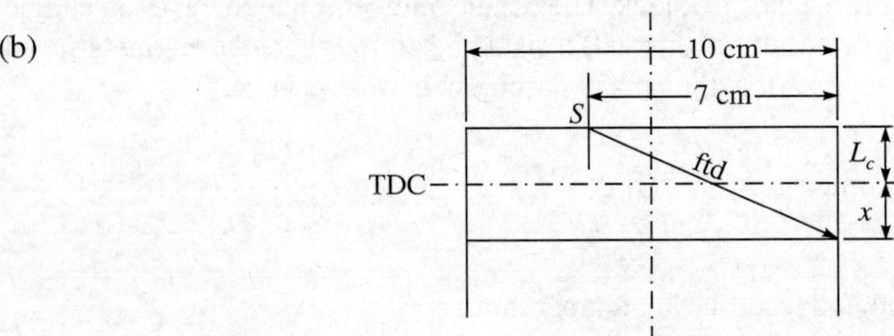

Figure 6.27 Example 6.1

Refer to Figure 6.27.

As clearance volume, $$V_c = \frac{V_s}{r-1}$$

$$\therefore \quad L_c = \frac{L}{r-1} = \frac{10}{8} = 1.25 \text{ cm}$$

$$L_c + x = 1.25 + 0.164 = 1.414 \text{ cm}$$

Flame travel distance, $$\text{ftd} = \sqrt{(7)^2 + (1.414)^2} = \boxed{7.14 \text{ cm}} \quad \textbf{Ans.}$$

(**Note:** ftd can be taken as 7 cm in the absence of enough data to calculate x and L_c)

(c) Ignition lag = 8°

Combustion starts at an angle = 20 – 8 = 12° bTDC

Rotational angle during flame propagation from

$$12°\text{bTDC to } 13°\text{aTDC} = 12 + 13 = 25°$$

$$N = \frac{2100}{60} = 35 \text{ rps} = 35 \times 360° = 12{,}600 \text{ degree/s}$$

Time of flame propagation, $$t = \frac{25}{12{,}600} = 0.001984 \text{ s}$$

Effective flame speed, $v_f = \frac{\text{ftd}}{t} = \frac{7.14}{0.001984} = 3599 \text{ cm/s} = \boxed{35.99 \text{ m/s}}$ **Ans.**

EXAMPLE 6.2 (a) A spark-ignition engine running at 1500 rpm has a 10 cm bore with the spark plug offset by 8 mm from the centre. The spark plug is fired at 18° bTDC. It takes 5° of engine rotation for combustion to develop and get into flame propagation mode, where the average flame speed is 20 m/s. Calculate time for flame front to reach the farthest cylinder wall and crank angle position at the end of combustion.

(b) Now the engine speed is increased to 3000 rpm. The flame termination crank angle position remains the same as in part (a), flame development takes 75% of real time in comparison to part (a) and flame speed is related to engine speed by the relation $v_f \propto N^{0.85}$. Calculate flame speed at 3000 rpm, crank angle position when flame propagation starts and crank angle position when the spark plug should be fired.

Solution:

(a) $N = 1500$ rpm, $d = 10$ cm

$$\text{Flame travel distance, ftd} = 5 + 0.8 = 5.8 \text{ cm}$$

$$\text{Spark position} = 18° \text{ bTDC}$$

$$\text{Ignition lag} = 5°$$

$$\text{Combustion starts at an angle} = 18 - 5 = 13° \text{ bTDC}$$

Flame speed, $v_f = 20$ m/s

$$\text{Time} = \frac{\text{ftd}}{v_f} = \frac{5.8}{20 \times 100} = \boxed{0.0029 \text{ s}} \quad \textbf{Ans.}$$

$$N = 1500 \text{ rpm} = \frac{1500}{60} = 25 \text{ rps} = 25 \times 360° = 9000 \text{ degree/s}$$

Rotational angle during flame propagation = 9000 × 0.0029 = 26.1°

Crank angle position at the end of combustion = 26.1 – 13

$$= \boxed{13.1° \text{aTDC}} \quad \textbf{Ans.}$$

(b) Now, $v_f \propto N^{0.85}$

Let at 3000 rpm flame speed is v'_f.

$$\frac{v'_f}{v_f} = \left(\frac{3000}{1500}\right)^{0.85}$$

$$v'_f = 20 \times (2)^{0.85} = \boxed{36.05 \text{ m/s}} \quad \textbf{Ans.}$$

Time for flame propagation, $t = \frac{\text{ftd}}{v'_f}$

$$= \frac{5.8}{36.05 \times 100} = 0.001609 \text{ s}$$

$$N = 3000 \text{ rpm} = \frac{3000}{60} = 50 \text{ rps} = 50 \times 360 = 18{,}000 \text{ degree/s}$$

$$\text{Rotational angle during flame propagation} = 18{,}000 \times 0.001609$$

$$= 28.96^\circ$$

$$\text{Combustion starts at an angle} = 28.96 - 13.1 = 15.86^\circ \text{ bTDC}$$

$$\text{Real time for ignition lag in part (a)} = \frac{5}{9000} = 0.0005556 \text{ s}$$

$$\text{Real time for ignition lag in part (b)} = 0.0005556 \times 0.75$$

$$= 0.0004167 \text{ s}$$

$$\text{Ignition lag} = 18{,}000 \times 0.0004167 = 7.5^\circ$$

$$\text{Spark at an angle} = 15.86 + 7.5 = \boxed{23.36^\circ \text{ bTDC}}$$ **Ans.**

EXAMPLE 6.3 Using Weibe function for finite heat release model, calculate at what crank angle 50% of the heat is released. Spark is initiated at an angle 20° bTDC and the duration of heat addition is 40°. Assume form factor $n = 3$ and efficiency factor $a = 5$.

Solution:

$$x_b = 1 - \exp\left[-a\left(\frac{\theta - \theta_s}{\theta_d}\right)^n\right]$$

Given $x_b = 0.5$, $a = 5$, $n = 3$, $\theta_s = -20^\circ$, $\theta_d = 40^\circ$. Determine θ.

$$\exp\left[-a\left(\frac{\theta - \theta_s}{\theta_d}\right)^n\right] = 1 - x_b$$

$$-a\left(\frac{\theta - \theta_s}{\theta_d}\right)^n = \ln(1 - x_b)$$

$$\left(\frac{\theta - \theta_s}{\theta_d}\right)^n = \frac{1}{a}\ln\frac{1}{(1 - x_b)}$$

$$\theta = \theta_s + \theta_d\left[\frac{1}{a}\ln\cdot\frac{1}{(1 - x_b)}\right]^{1/n}$$

$$= -20 + 40\left[\frac{1}{5}\ln\cdot\frac{1}{(1 - 0.5)}\right]^{1/3} = 0.7^\circ$$

$$= \boxed{0.7^\circ \text{ aTDC}}$$ **Ans.**

REVIEW QUESTIONS

1. What do you understand by normal combustion?
2. Explain the terms controlled, uncontrolled, and abnormal combustion.
3. How do the combustion reactions proceed and how does a definite flame boundary get established?
4. How does the nucleus of a flame grow and how does the combustion proceed?
5. Draw and explain a theoretical p–θ diagram.
6. Distinguish between combustion and burning.
7. Explain the stages of combustion in the SI engine with the help of a p–θ diagram.
8. Define the terms: ignition lag and afterburning.
9. Show the flame speed pattern with the help of a diagram. How does the flame travel pattern divide the combustion process into distinct phases?
10. Describe, with the help of a diagram, the pattern of the burned mass fraction in a typical SI engine as a function of the crank angle.
11. Describe finite heat release model for combustion process of SI engine to evaluate rate of heat release as a function of crank angle using Weibe function. Also, evaluate an expression to find out rate of pressure rise with respect to crank angle.
12. Show and explain with reasons the variation of pressure and temperature in the SI engine as a function of the crank angle.
13. Show and explain the effect of spark timing on an indicator diagram. What do you understand by the terms MBT, overadvanced and retarded timing?
14. Show the effect of spark advance on brake torque.
15. Show and explain the effect of mixture strengths on p–v and p–θ diagrams.
16. Discuss the effect of the following variables on ignition lag
    (a) Nature of fuel and air/fuel ratio
    (b) Initial temperature and pressure
    (c) Compression ratio
    (d) Spark timing
    (e) Turbulence and engine speed
    (f) Gap between electrodes of the spark plug.
17. Discuss the effect of the following factors on the combustion process of the SI engine.
    (a) Composition of the mixture
    (b) Load
    (c) Compression ratio
    (d) Speed
    (e) Turbulence and shape of the combustion chamber
    (f) Spark plug location.
18. What do you understand by cyclic variation? What are the reasons for it? How does it depend on the mixture strength, the compression ratio and the load? How can it be reduced?

19. On the p–θ diagram, show the effect of rate of pressure rise during the combustion process of an SI engine. How does the ignition timing vary with the rate of combustion? At what point in the cycle is it desirable to locate the peak pressure?
20. How does a normal combustion take place in the SI engine? What do you understand by the term 'end-gas'? Show the p–θ diagram with normal combustion.
21. What do you understand by the term autoignition? Define the ignition delay period.
22. What do you understand by knock or detonation in SI engines? Explain this phenomenon. How does the knock in SI engines differ from the knock in CI engines?
23. Explain the autoignition theory and detonation theory of knocking in SI engines.
24. What are the detrimental effects of detonation?
25. What are the major factors involved in preventing knock in SI engines?
26. Discuss the effect of the following engine variables on knock in SI engines:
    (a) Compression ratio
    (b) The mass of the inducted charge
    (c) Inlet temperature
    (d) Temperature of the combustion chamber wall
    (e) Spark timing
    (f) Coolant temperature
    (g) Power output
    (h) Exhaust back pressure
    (i) Cycle-to-cycle variation
    (j) Carbon deposits.
27. Briefly explain the time factors that affect knock in SI engines.
28. Enumerate the factors that affect the flame travel distance.
29. Briefly explain the composition factors that affect knock in SI engines.
30. Discuss the influence of the following design considerations on knock in SI engines:
    (a) Effect of shrouded inlet valve
    (b) Effect of piston shape
    (c) Effect of cylinder bore.
31. What are the methods of detecting knock?
32. What is surface ignition? How does it occur?
33. Explain the following terms related to surface ignition:
    (a) Pre-ignition
    (b) Run-on
    (c) Run-away
    (d) Wild ping
    (e) Rumble.
34. What are the basic requirements of a good combustion chamber? How can these be achieved?
35. Briefly explain the design principles underlying a good combustion chamber of a spark-ignition engine.

36. Discuss the sequence of steps generally followed in the optimization of a combustion chamber of SI engines?

37. Describe with the help of simple diagrams the T-type, the L-type and the I-type of combustion chamber heads.

38. What are the shortcomings of the T-head type of combustion chamber? What are the advantages of the L-head type over the T-head type? What are the limitations of the L-head type combustion chamber?

39. How is the desired degree of turbulence obtained in a Ricardo turbulent combustion chamber? How is the effective flame travel distance reduced in this type of combustion chamber?

40. What are the advantages of the overhead valve combustion chamber? Describe the bath-tub type and the wedge-type of combustion chambers with the help of diagrams.

41. Describe the F-type of combustion chamber with the help of a simple diagram. How is the knocking tendency reduced in this type of combustion chamber?

42. Describe the hemispherical type combustion chamber. What are its advantages?

43. Describe the piston cavity type combustion chamber. What are its advantages?

44. With the help of a simple diagram, describe the combustion chamber with a prechamber used for the Toyota lean burn engine.

45. Define the octane number requirement. Name the factors that affect the octane requirement of an engine? How does the octane number requirement go down in SI engines?

PROBLEMS

6.1 The spark plug is fired at 20° bTDC in a spark ignition engine running at 2000 rpm. It takes 8.5° of engine rotation to start comubustion and get into flame propagation mode. Flame termination occurs at 15° aTDC. Bore diameter is 9.0 cm and the spark plug is offset 8 mm from the centre line of the cylinder. Calculate the effective flame front speed during flame propagation. [**Ans.** 24 m/s]

6.2 The engine in Problem 6.1 is now run at 3000 rpm. As speed increases, the flame front speed increases at a rate such that $v_f \propto N^{0.8}$. Flame development after spark plug firing still takes 8.5° of engine rotation. Calculate how much ignition timing must be advanced such that flame termination again occurs at 15° aTDC. [**Ans.** 2.23°]

6.3 A spark-ignition engine runs at 2000 rpm. The compression ratio is 10.0, the cylinder bore is 10 cm and stroke length is 12 cm. Connecting rod length is 20 cm. The spark plug is offset by 1.0 cm from centre. The spark plug is fired at 20° bTDC. It takes 9° of engine rotation for combustion to develop and get into flame propagation mode. Flame termination occurs at 15° aTDC.

Calculate:

(a) Piston displacement from TDC position at the time of flame termination
(b) Flame travel distance
(c) Effective flame speed [**Ans.** (a) 0.2648 cm, (b) 6.209 cm, (c) 28.65 m/s]

6.4 The speed of engine in problem 6.3 is now increased to 2800 rpm. The flame termination crank angle position remains the same, flame development takes the same amount of real time and flame speed is related to engine speed as $v_f \propto N^{0.85}$.
Calculate:

(a) Flame speed at 2800 rpm
(b) Crank angle position when flame propagation starts
(c) Crank angle position when the spark plug should be fired

[Ans. (a) 38.14 m/s, (b) 12.35° bTDC, (c) 24.95° bTDC]

6.5 Using Weibe function for finite heat release model, calculate at what crank angles 10%, 50% and 90% of heat is released. Spark is initiated at 15° bTDC and the duration of heat addition is 40°. Assume form factor = 3 and efficiency factor = 5.

[Ans. 3.95° bTDC, 5.7° aTDC, 15.89° aTDC]

Combustion in Compression-Ignition Engines

7.1 INTRODUCTION

The diesel engine is a compression ignition (CI) engine and the typical compression ratios are in the range of 14:1 to 22:1. In order to achieve spontaneous ignition, the compression stroke must raise the air to a much higher temperature than that in the SI engine, and this requires a high compression ratio. Diesel engines have considerable advantages in commercial applications, for example, for continuous heavy-duty operation such as that required in locomotives, heavy road transport and marine engines, and where robustness is at a premium as in tractors and earth-moving vehicles. These engines are also used for stationary industrial applications, such as in pumping sets, small and medium electric power generators, etc.

The thermal efficiency of the CI engine is higher than that of the SI engine because of the higher compression ratio. The CI engine fuels (diesel or crude oil) are less expensive than the SI engine fuels (petrol or gasoline). Moreover, the CI engine fuels have higher specific gravity than that of petrol and since the fuel is sold on volume basis, more kilograms of fuel per litre are obtained in purchasing CI engine fuels. Most of the time engines are run at part load and the supply of fuel through the fuel injectors can be controlled in CI engines, thus reducing the fuel/air ratio at part load. In SI engines, the fuel/air ratio prepared in the carburettor remains almost constant at all loads. These factors reduce the running cost of CI engines. However, the CI engines are not favoured in passenger cars, where the use is limited, because of certain drawbacks compared to SI engines, such as heavier weight for the same power output. It is because of the fact that heterogeneous mixture is used in CI engines, and the compression ratio is high as well. It increases the initial cost of the engine. The CI engines produce more noise and vibrations because of heavier parts. The exhaust emissions from CI engines with heterogeneous mixture are unburned hydrocarbons, oxides of nitrogen, aldehydes, carbon monoxide, smoke particulates and the odour constituents. The major pollutants from SI engines are unburned hydrocarbons, oxides of nitrogen and carbon monoxide.

Smoke and odour are the most noticeable emissions of diesel engines. White smoke is emitted when the engine is cold and consists largely of unburned or partially burned fuel droplets. White smoke may be accompanied by carbonyl compounds and other partial oxidation products. Black smoke is emitted when the engine is operating at a high load. This smoke consists of carbonaceous particles produced by the pyrolytic reactions that occur in the core of the fuel spray where large amounts of fuel are injected into the combustion chamber.

Diesel engines come in many configurations: two stroke and four stroke engines, open and closed (prechamber) combustion chamber types and with especially shaped pistons and inlet valve configurations to provide chamber turbulence and swirl.

The combustion process in a CI engine is very complex and its detailed mechanisms are not well understood. Autoignition occurs at several locations in the combustion chamber where there is a combustible mixture. Meanwhile in some other locations the fuel may still be in the liquid phase. Under most engine operating conditions, ignition starts while some portion of the fuel may still not have been injected. The proportion of fuel present in the combustion chamber at the start of combustion to the total amount of fuel injected greatly affects the combustion process. Also, the distribution of the fuel within the combustion chamber has a great effect on the mechanisms of combustion and emission formation. This distribution depends mainly on the injection process and the air motion.

7.2 AIR MOTION IN CI ENGINES

In open chamber CI engines the fuel is injected in the bulk of the air, so the turbulence of air in the combustion chamber at the time of fuel injection and during the combustion process is very important. The air velocity in the combustion chamber has two components: (a) air swirl (spiral flow), and (b) air radial flow (squish).

Air swirl is created during the intake stroke by inducing the air into the cylinder through a shaped intake port which is tangential to the piston as shown in Figure 7.1. The induced swirl can be increased during compression by transferring the air to the recess in the piston or in the cylinder head. As the piston approaches TDC, the air flows radially inwards, i.e. towards the combustion chamber recess. It is shown in Figure 7.2. The radial streams coming from the opposite sides meet

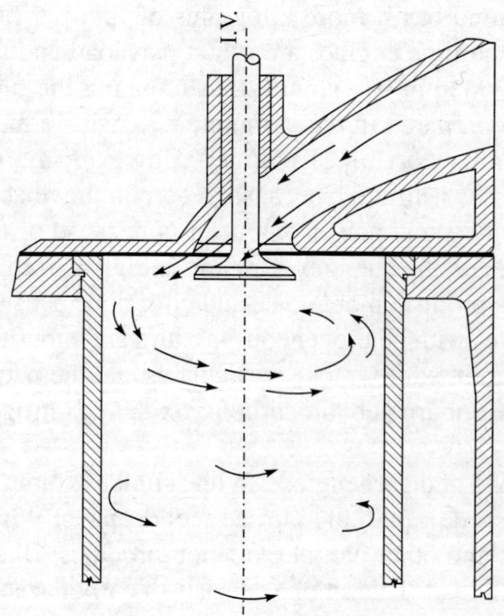

Figure 7.1 Induction induced swirl.

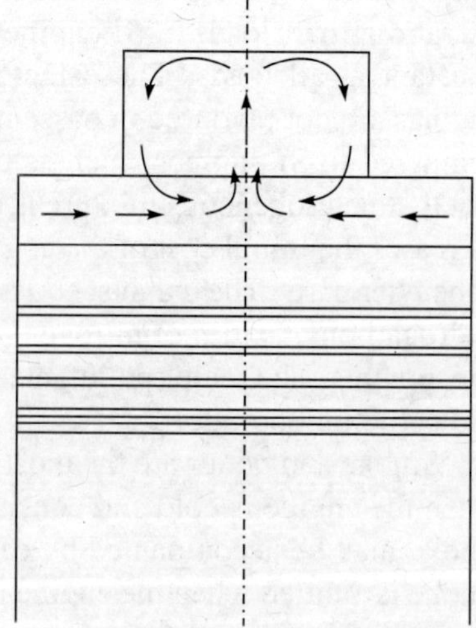

Figure 7.2 Air radial flow (squish).

and get deflected upwards into the chamber. After reaching the end of the chamber, the air flows radially outwards, i.e. towards the outer walls, then downwards, i.e. towards the open end. Here the air is met by air flowing radially inwards from between the cylinder and piston, and is carried around again, producing a toroidal movement within the combustion chamber.

The effect of the squish component on the spray formation is very small compared to the similar effect of the swirl component.

7.3 SPRAY STRUCTURE

A schematic of the spray pattern which results when a fuel jet is injected radially outwards into a swirling flow is shown in Figure 7.3. The structure of the jet is complex because there is relative motion in both the radial and tangential directions between the initial jet and the air. The small droplets are carried away with the air and form the leading edge of the spray. The relatively large droplets are concentrated in the core and the trailing edge of the spray. The average distance between the droplets changes with the location in the spray, being the greatest in the leading edge of the spray.

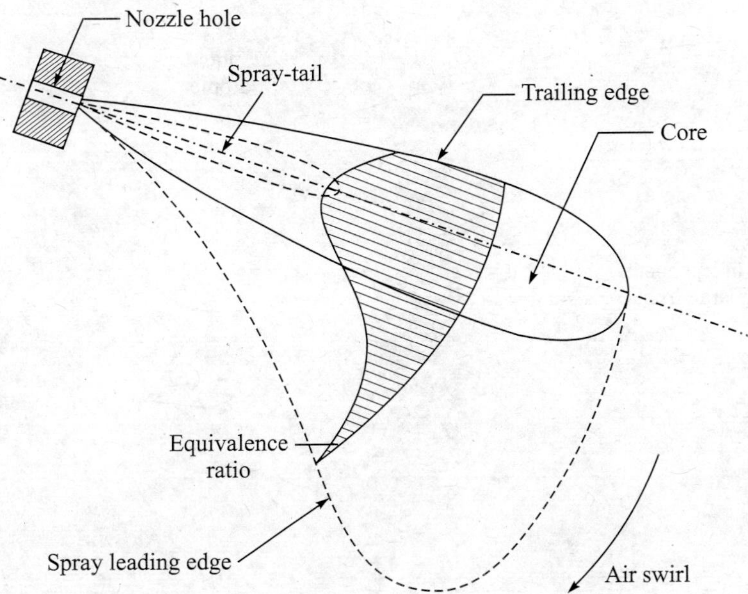

Figure 7.3 Fuel spray injected into swirling air.

A plot of the local fuel/air ratio on a mass basis versus the angle from the centre line of the injector hole is also shown in Figure 7.3. The fuel-air distribution varies with the radial distance from the nozzle hole. The leading edge of the spray always contains the smallest droplets, which are the first to evaporate.

The smaller the droplet diameter the farther it will be carried away from the core, by the swirling air. Thus the mixture near the leading edge of the spray may be assumed to consist of premixed fuel-vapour and air before ignition. In the core the big droplets are concentrated, and they are expected to be in the liquid phase at the start of ignition.

The spray may be divided into several regions depending upon the fuel-air distribution and mechanism of combustion in each region. These regions are: (a) the lean flame region, (b) the lean flameout region, (c) the spray core, (d) the spray tail, (e) the after-injection, and (f) the fuel deposited on the walls.

Lean flame region (LFR)

The concentration of the fuel in the air between the core and the leading edge of the spray is heterogeneous and the local fuel/air ratio may vary from zero to infinity. Ignition nuclei will be formed at several locations where the mixture is most suitable for autoignition. Ignition starts near the leading edge of the spray as shown in Figure 7.4. Once the ignition starts, small independent nonluminous flame fronts propagate from the ignition nuclei and ignite the combustible mixture around them. This mixture is leaner than the stoichiometric mixture and the region is called the 'lean flame region'. In this region combustion is complete and the nitric oxide may be formed at high local concentrations. Under very light loads, the temperature may not be high enough to produce high nitric oxide concentrations at this early stage of the combustion process.

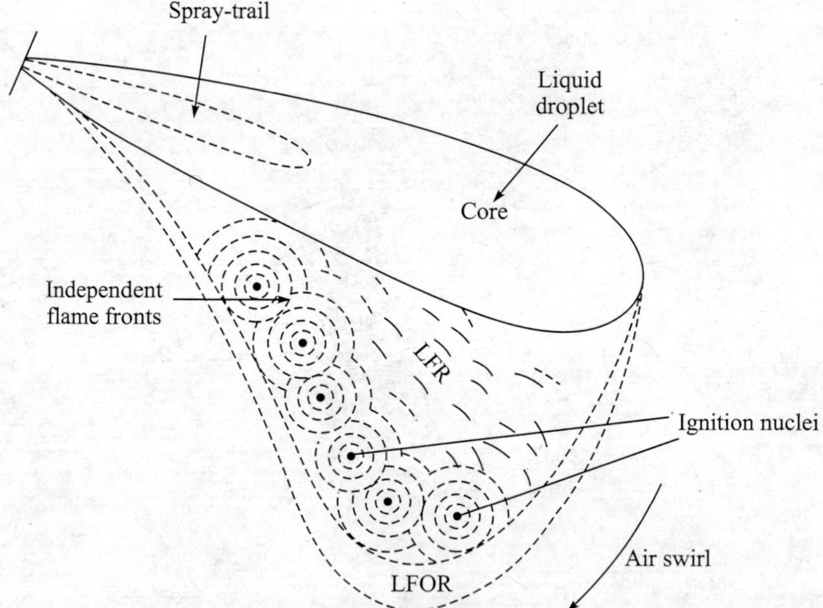

Figure 7.4 Mechanisms of combustion of a fuel spray injected in swirling air.

Lean flame-out region (LFOR)

Near the far leading edge of the spray, the mixture is too lean to ignite or support combustion. This region is called the 'lean flame-out region'. Within this region some fuel decomposition and partial oxidation take place. The decomposition products consist of lighter hydrocarbon molecules and this region is believed to be the main contributor to the unburned hydrocarbons in the exhaust. The partial oxidation products may contain aldehydes and other oxygenates.

Spray core

Following ignition and combustion in the lean flame region, the flame propagates towards the core of the spray. Between these two regions the fuel droplets are larger. They give heat by radiation from the already established flames and evaporate at a higher rate. The increase in temperature also increases the rate of vapour diffusion. These droplets may get completely or partially evaporated. If they are completely evaporated, the flame will burn all the mixture with the rich ignition limit. The droplets which do not get completely evaporated will be surrounded by a diffusion type of flame. The combustion in the core of the jet depends mainly upon the local fuel/air ratio. Under part load operation, this region contains adequate oxygen and combustion is expected to be complete and result in the production of high oxides of nitrogen (NO_x). Near the full-load conditions, incomplete combustion occurs in many locations in the fuel-rich core resulting in unburned hydrocarbons, carbon monoxide, oxygenated compounds and carbon in the exhaust. NO_x are formed at low concentration under these conditions.

Spray tail

The last part of the fuel to be injected usually forms large droplets because of the relatively less pressure difference acting on the fuel near the end of the injection process. This is caused by a decreased fuel injection pressure and increased cylinder gas pressure. The penetration of this part of the fuel is usually poor. This portion is known as the spray tail. Under high load conditions, the spray tail has little chance to get into regions with adequate oxygen concentration. However, the temperature of the surrounding gases is high and the rate of heat transfer to these droplets is fairly high. These droplets therefore, tend to evaporate quickly and decompose into unburned hydrocarbons and a high percentage of carbon molecules. Partial oxidation products include carbon monoxide and aldehydes.

After-injection

When the injector valve remains open for a short time after the end of the main injection, a small amount of fuel is further injected, called 'after-injection', especially under medium and high load conditions. It is injected late in the expansion stroke with very little atomization and penetration. This fuel is quickly evaporated and gets decomposed, resulting in the formation of carbon monoxide, carbon particles (smoke), unburned hydrocarbons and oxygenated hydrocarbons.

Fuel deposited on the walls

Some fuel sprays impinge on the walls. Because of the shorter spray path and the limited number of sprays, this is especially the case in small, high speed direct injected CI engines. If the surrounding gas has a high relative velocity and contains enough oxygen, the flame will propagate from wall to a small distance within the chamber. Combustion of the rest of the fuel on the walls will depend upon the rate of evaporation and mixing of fuel and oxygen. If the surrounding gas has a low oxygen concentration or the mixing is not appropriate, evaporation will occur without complete combustion. Under this condition, the fuel vapour will decompose and form unburned hydrocarbons, partial oxidation products and carbon particles.

7.4 STAGES OF COMBUSTION

The combustion process which continually takes place in an operating diesel engine is basically represented by the pressure vs. crank angle diagram shown in Figure 7.5.

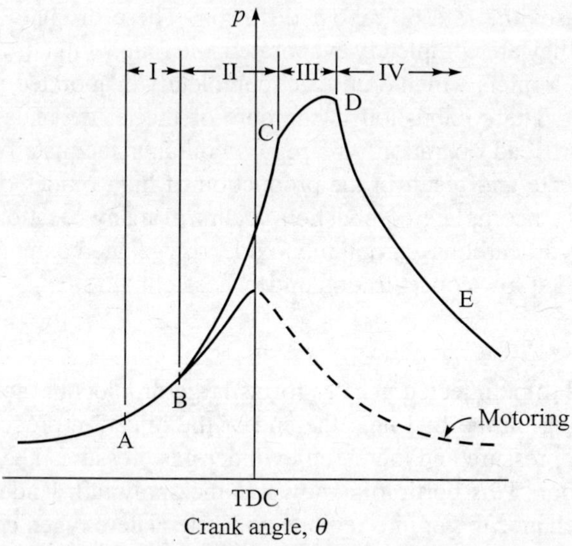

Figure 7.5 Pressure vs. crank angle diagram of a CI engine showing the different stages of combustion. (On an actual p–θ diagram taken from an engine, the dividing points are not pronounced, and the four stages merge one into another gradually.)

At point A, the injector starts to inject fuel into the combustion chamber. A finite time elapses, during process AB, before the fuel droplets reach the ignition temperature, although when this happens most of the fuel injected during the first stage ignites spontaneously, causing an abrupt pressure rise at B and also during the process BC (second stage). The rest of the fuel injected during stage 3 is burnt as soon as it enters the combustion chamber.

Power output is controlled by metering the amount of injected fuel resulting in very good part load efficiency.

The classical diesel compression and combustion pressure diagram shown in Figure 7.5 has been proved to be an acceptable model and can be roughly divided into four broad stages:

(a) Ignition delay period
(b) Rapid or premixed uncontrolled combustion phase
(c) Mixing-controlled combustion phase
(d) Late combustion phase or afterburning.

The first stage—ignition delay period (Process A to B)

During this stage no noticeable deviation of the pressure diagram from the pure air compression curve is observed. The processes that take place during the ignition delay can be divided into the following physical and chemical processes.

The physical processes are:

1. Spray disintegration and droplet formation
2. Heating of the liquid fuel and evaporation
3. Diffusion of the vaporized fuel into the air to form an ignitable mixture.

The chemical processes are:

1. Decomposition of the heavy hydrocarbons into lighter components
2. Preignition chemical reactions between the decomposed components and oxygen.

During this period, the chemical reactions proceed so slowly that no effect is discernible. It is incorrect to think of these two delays—the physical and the chemical—as additive. They overlap, and the rate of chemical reaction increases as the fuel and air become intimately mixed. Therefore, the first part of the ignition delay is considered to be dominated by the physical processes which result in the formation of a combustible mixture. The second part of the ignition delay is considered to be dominated by the chemical changes which lead to autoignition.

The second stage—rapid or premixed uncontrolled combustion phase (Process B to C)

In this phase, combustion of the fuel, which has mixed with air to within flammability limits during the ignition delay period, occurs rapidly in a few crank angle degrees. Ignition in one place is followed by ignition elsewhere, so rapid combustion of the prepared fuel follows the first ignition. The rate and quantity of combustion in the second phase is thus dependent on the duration of the delay period and the rate of preparation of fuel during this period. The speed of this reaction determines the rate of pressure rise ($dp/d\theta$) in the cylinder. A high rate of pressure rise means a sudden application of load to the engine structure, which often results in fatigue damage to the parts. A high rate of pressure rise also produces a violent pounding noise which is known as 'diesel knock'.

The magnitude of the pressure rise during the second period may determine the value of the peak pressure of the cycle. For structural reasons, to reduce mechanical stresses it is important to limit the peak pressure as well as the rate of pressure rise. As the peak pressure increases, the peak temperature also increases, which increases (a) the NO_x emission, (b) the thermal loads on the cooling system, and (c) the temperature and thermal stressing of the combustion chamber walls. Therefore, to limit the pressure and temperature rise during the second period, it is important to keep the delay period as short as possible.

The third stage—mixing-controlled combustion phase (Process C to D)

Once the fuel and air which premixed during the ignition delay have been consumed at the end of the period of rapid combustion, the temperatures within the cylinder are so high that any fuel injected after this time will burn as soon as it finds oxygen, and any further rise in pressure is controlled by the injection rate as well as by the mixing and diffusion processes. Engines running at low rpm should be designed to secure rapid mixing of fuel and air during the third stage in order to complete the combustion process as early as possible in the expansion stroke.

The fourth stage—late combustion phase or afterburning (Process D to E)

A very low rate of combustion, sometimes referred to as the tail of combustion, occurs well down the expansion stroke of the engine. There are several reasons for this. A small fraction of fuel may not yet have burned. The cylinder charge is nonuniform and mixing during this period promotes more complete combustion and less dissociated product gases. The kinetics of the final burnout processes become slower as the temperature of the cylinder gases fall during expansion.

The late combustion phase is undesirable because it reduces the power output and produces smoky exhaust. This can be eliminated by supplying more excess air and creating more turbulence. A flow diagram of the entire process of combustion in CI engines is shown in Figure 7.6.

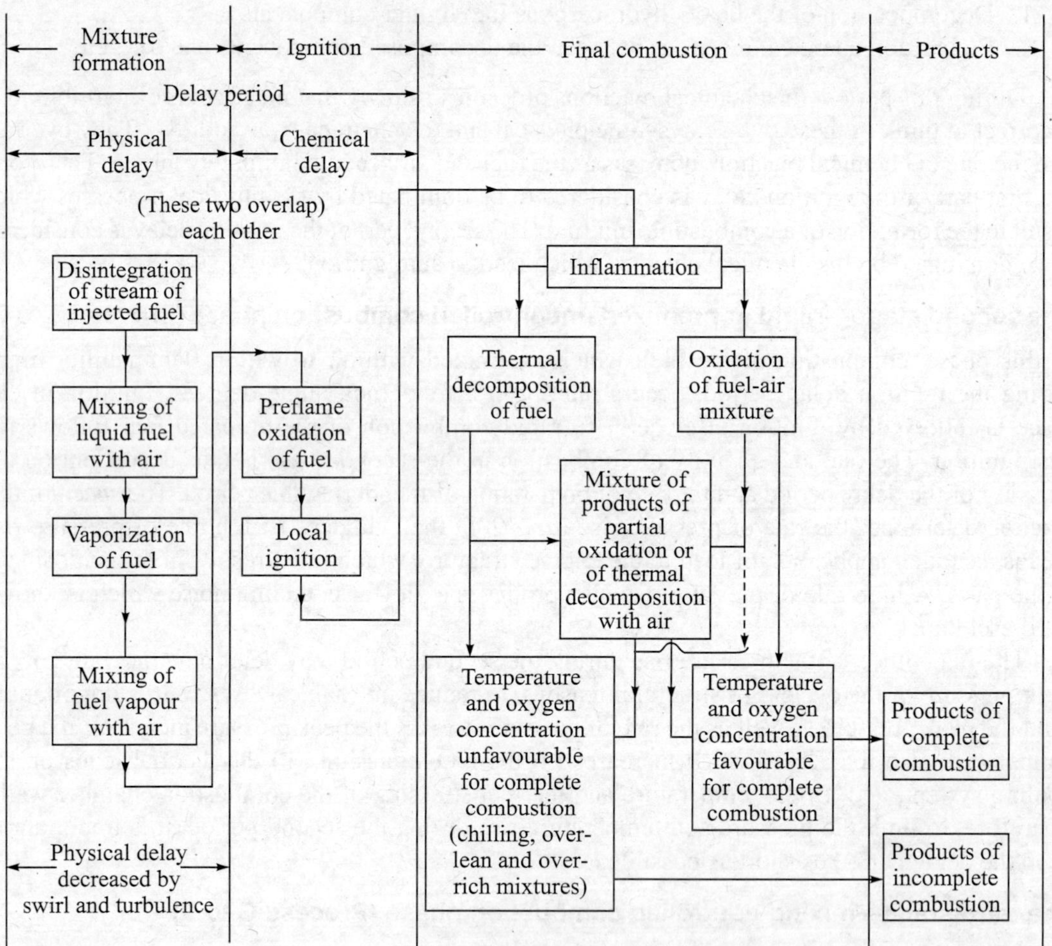

Figure 7.6 A flow diagram of the combustion process in CI engines.

7.5 HEAT RELEASE RATE

A typical heat release rate diagram of a direct injection compression ignition engine, identifying the different diesel combustion phases is shown in Figure 7.7.

The period a–b represents the ignition delay. No heat is released during this period. The period b–c represents the rapid premixed uncontrolled combustion phase. The high heat release rate is the characteristic of this phase. The first peak occurs and results from the rapid combustion of the

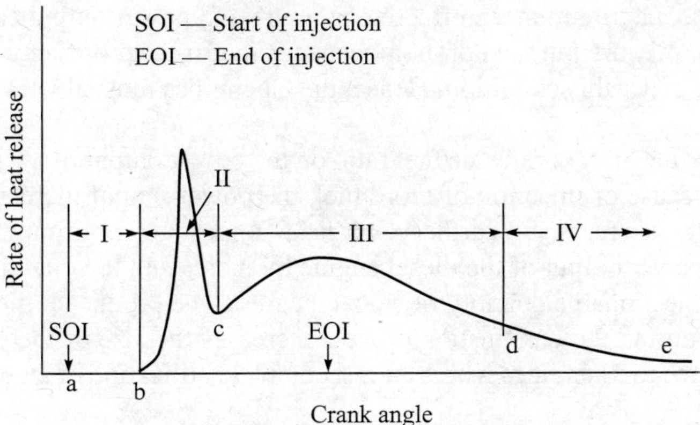

Figure 7.7 Heat release rate diagram.

portion of the injected fuel that has vaporized and mixed with the air during the ignition delay period. The heat release rate in this phase is relatively independent of the load, since the initial mixing is independent of the injection duration. The period c–d represents the mixing-controlled combustion phase. During this period the heat release rate curve reaches a second peak. The heat release during this phase depends on the injection duration. As the injection duration is increased at high load condition, the amount of fuel injected increases, thus increasing the magnitude and duration of the mixing-controlled heat release. It decreases as this phase progresses. The period d–e represents the late combustion phase, the heat release rate decreases further and continues at a lower rate well into the expansion stroke.

7.6 AIR/FUEL RATIO IN CI ENGINES

For a given engine speed, the supply of air in a CI engine is almost constant and does not depend upon the engine load. This engine may be termed a constant air supply engine. With the change in load the quantity of fuel is changed, which alters the air/fuel ratio. As the load increases, more fuel is used. The overall air/fuel ratio may very from 100 : 1 at no-load to 20 : 1 at full-load. In the SI engine, the air/fuel ratio remains almost constant near to stoichiometric from no-load to full-load conditions. The inflammability limit of hydrocarbon fuel ranges from 8 : 1 to 30 : 1. The question arises that how does the combustion of fuel occur in CI engines when the overall mixture is much leaner than this limit? From the study of spray structure it is clear that there is a heterogeneous mixture in the combustion chamber. The local air/fuel ratio may vary from zero to infinity in different parts. In the regions of very rich or very lean mixtures, no definite flame front can be established. There would also be some regions where the local air/fuel ratio is within the combustible limits, and ignition would occur under favourable conditions of temperature. Ignition nuclei are formed at several locations, and multiple flame fronts travel to burn the remaining charge. It is in contrast to the SI engine where a definite flame front travels.

At full-load, the SI engines run with a slightly rich mixture (say 13.5 to 14 : 1) but the CI engines still run with a lean mixture (say 20 : 1).

With a leaner mixture the indicated thermal efficiency of the CI engines will be higher but the mean effective pressure and power output will be decreased. Therefore, the size of the CI engine compared to the SI engine will be larger for a given power output.

Theoretically, the maximum mean effective pressure and power output can be obtained with nearly stoichiometric mixture but the non-homogeneous intermixing of fuel with the air results in objectionable smoke under this condition. Hence, the CI engines must always operate with a lean mixture.

Figure 7.8 shows the effect of the air/fuel ratio on the power output of a CI engine. As the air/fuel ratio reduces because of injection of more fuel, the power output increases, but then comes a limit of reducing the air/fuel ratio further, as it leads to production of an undesirable quantity of smoke. Thus the power output of the diesel engine by increasing the quantity of fuel is limited to the point where objectionable quantity of smoke begins to be produced. Incomplete utilization of air in CI engines during the combustion process increases the size of the engine for the same power output compared to SI engines where almost complete utilization of air due to homogeneous mixing takes place.

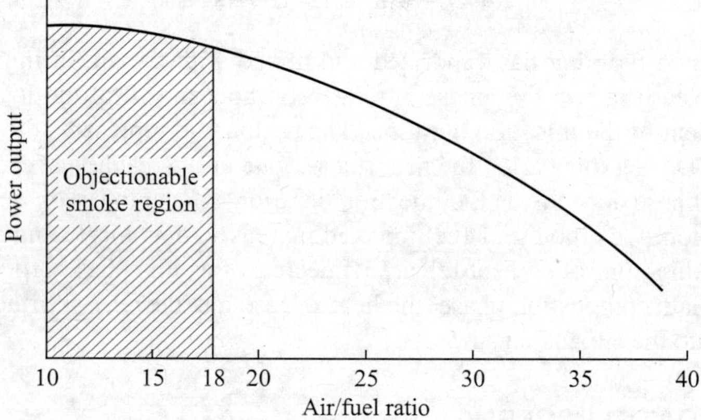

Figure 7.8 Effect of air/fuel ratio on the power output of a CI engine.

7.7 INFLUENCE OF VARIOUS FACTORS ON DELAY PERIOD

The factors described in the following subsections affect the delay period.

Ignition quality of fuel

The self-ignition temperature of a fuel is a very important ignition quality. If the self-ignition temperature of a fuel is low, the delay period is reduced. Cetane number is a scale used for comparing the ignition quality of fuels. By using a fuel with a higher cetane number, all other conditions being the same, the delay period can be reduced and the engine operation made smoother. The dependence of cetane number on fuel molecular structure is as follows: Straight-chain paraffinic compounds (normal alkanes) have the highest ignition quality, which improves as the chain length increases. Aromatic compounds have poor ignition quality as do the alcohols. Hence, difficulties are associated with the use of methanol and ethanol as possible alternative fuels in CI engines. The effect of cetane number on the indicator diagram when the ignition timing is the same is shown in Figure 7.9.

The ignition quality of the fuel may be affected by volatility, latent heat, viscosity and surface tension. Volatility and latent heat affect the time taken to form an envelop of the vapour. The viscosity and surface tension influence the fineness of atomization of fuel.

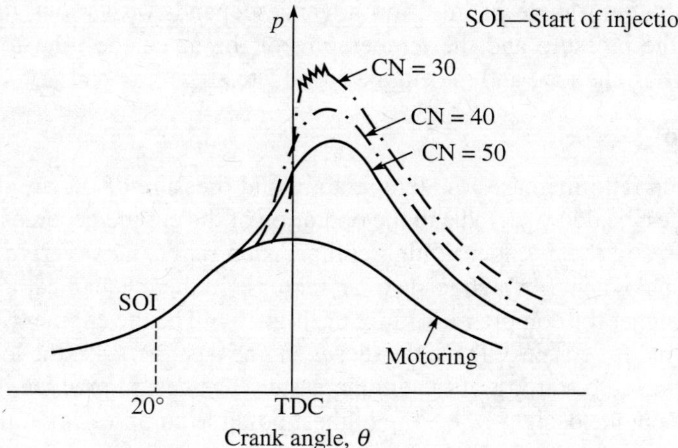

Figure 7.9 Effect of Cetane number (CN) on the p–θ diagram.

The higher the cetane number, the shorter the delay period and the greater the resistance to diesel knock. Conversely, a low cetane number extends the delay period and makes the engine more prone to diesel knock.

An approximate guide to cetane number requirements for different speed range engines is as follows:

Class	*Speed range (rpm)*	*Cetane number*
Slow speed	100 – 500	25–35
Medium speed	300–1500	35–45
High speed	500–5000	45–55

Injection timing

A large injection advance increases the delay period, since the pressure and temperature of the air are lower when the injection begins. If the injection starts later, i.e. closer to TDC, the temperature and pressure are initially slightly higher but then decrease due to the increased heat loss as the delay proceeds. Both advancing and retarding of the injection increase the delay period. The most favourable conditions for ignition lie in between these two conditions.

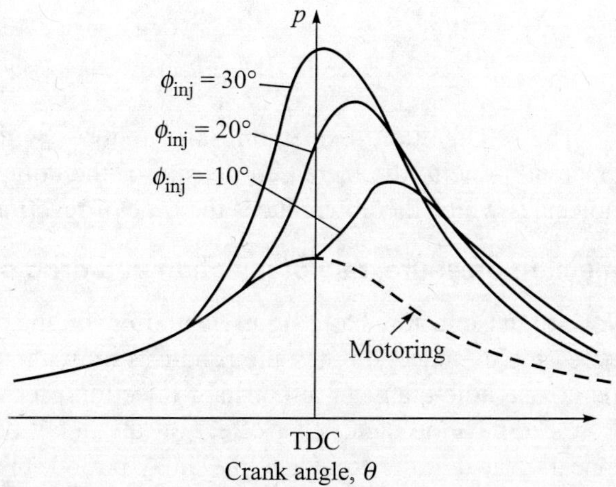

Figure 7.10 Indicator diagrams at various injection advance angles, ϕ_{inj}.

Figure 7.10 shows the indicator diagrams taken at various angles of injection advance at a constant duration of injection, the most advantageous angle at which the injection begins is about 20°

before TDC. The optimum angle of injection advance depends on various factors such as the compression ratio, the pressure and the temperature of the air at the inlet to the cylinder, the injection characteristics, the load and the engine speed, etc.

Compression ratio

A higher compression ratio increases the temperature and pressure of the air at the beginning of injection. This reduces the delay period and the operation of the engine becomes smoother. Owing to the higher pressures in the cylinder at high compression ratios, however, the crank gear parts have to be stronger and heavier, which leads to an increase in the mechanical losses.

In practice, the higher the compression ratio, the higher will be the engine friction, the leakage, and the torque required for starting. Thus, the diesel engine designers use the lowest compression ratio. It also makes the cold starting of the engine easier. The design practice is just the opposite of that adopted in SI engine design, where the highest possible compression ratio is used which is limited only by knocking.

Figure 7.11 shows the typical effect of compression ratio on the p–θ diagram when the injection timing, the speed and the fuel quality are held constant. The delay period reduces by increasing the compression ratio.

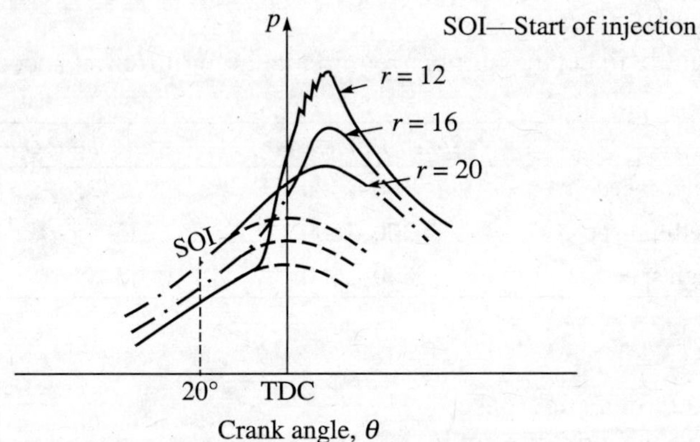

Figure 7.11 Effect of compression ratio on the p–θ diagram.

The self-ignition temperature also reduces with the increase in compression ratio. This is apparently due to the increased density of the compressed air, resulting in closer contact of the molecules which thereby reduces the time of reaction when the fuel is injected.

Injection pressure, rate of injection and drop size

None of these factors has a significant effect on the delay period. At normal operating conditions, increasing the injection pressure produces only a modest decrease in the delay period. Increasing the nozzle hole diameter at constant injection pressure to increase the rate of fuel injection and drop size have no significant effect on the delay period. These can be explained as follows: It appears that in order to reduce the delay period, the fuel should be injected in small droplets to increase the surface area of inflammation but the rate of combustion mainly depends upon the relative motion between the fuel droplets and the swirling air. The smaller droplets will have less momentum and travel in air with less relative velocity resulting in partial suffocation by its own products of combustion, and thus giving rise to a larger delay period. As these two reasons contradict each other, practically, the injection pressure, the rate of injection and the drop size do not significantly affect the delay period.

Intake, jacket water and fuel temperatures

Any increase in the inlet temperature, the cooling water temperature or the fuel temperature results in an increase in the temperature of the charge at the moment of injection and this causes a decrease in the delay period.

Increasing the inlet temperature by preheating the air is undesirable as it reduces the density of air and hence reduces the volumetric efficiency and power output.

Intake pressure

Any increase in the intake air pressure or supercharging reduces the delay period. The rate of pressure rise during combustion also decreases and gives smooth operation of the engine. Since the compression pressure will increase with the inlet pressure, the peak pressure will be higher. As the cylinder will contain more air, it will be possible to inject more fuel to obtain more power. The supercharged CI engines run smoother compared to the unsupercharged engines.

Engine speed

As the engine speed increases the loss of heat during compression decreases, resulting in increased temperature and pressure of the compressed air. Thus the delay period in milliseconds reduces but its value in degrees of crank travel increases. The amount of fuel injected during the delay period depends on crank degrees and not on the absolute time because the fuel pump is geared to the engine. At high speeds, there will be more fuel present in the cylinder to take part in the second stage of uncontrolled combustion, resulting in higher rates of pressure rise and higher maximum pressure. This factor has caused difficulties in the development of high speed CI engines. Very high speed CI engines, therefore, require a greater angle of injection advance, high ignition quality and a special design for the combustion chamber.

Air/fuel ratio

As the air/fuel ratio is increased, the cylinder wall temperature and combustion temperatures are lowered and hence the delay period is increased. There is very little reduction in the maximum rate of pressure rise except at the very high air/fuel ratios. On the other hand, the maximum pressure falls steadily with increasing air/fuel ratios. A practical limit on the minimum air/fuel ratio, however, is set by incomplete combustion accompanied by a smoky exhaust—called the smoke limit.

Load

As the load is increased, the residual gas temperature and the wall temperature increase. This results in a higher charge temperature at injection, thus shortening the delay period.

Engine size

Large engines operate at low rpm because of inertia stress limitations. Large engines have smaller surface-to-volume ratios which causes less heat loss during compression, resulting in higher temperatures of the compressed air. It shortens the delay period and improves the combustion conditions. The design of a CI engine of larger size operating at a lower speed is therefore less difficult.

Combustion chamber wall effects

The impingement of the fuel jet on the combustion chamber wall affects the fuel evaporation and mixing processes. The impingement occurs almost in all smaller size and higher speed engines. If the pressure and temperature inside the cylinder are low, the presence of wall reduces the delay period, but most diesel engines running under normal operating conditions have higher pressure and temperature inside the cylinder and the fuel jet impingement on the wall surface has no significant effect on the delay period. However, the initial rate of burning increases because of increased evaporation and fuel-air mixing rates.

Swirl rate

The air swirl rate affects fuel evaporation, the fuel-air mixing processes and the wall heat transfer during compression and hence the charge temperature at injection. Under engine starting conditions, the engine speed and compression temperature are low and a good air swirl would reduce the ignition delay. At normal operating engine speeds, the air swirl rate has very little effect on delay period.

Exhaust gas recirculation (EGR)

The oxygen concentration in the charge into which the fuel is injected, influences the delay period. The oxygen concentration is changed when the exhaust gas is recycled to the intake for the control of oxides of nitrogen emissions. As the oxygen concentration is decreased, the ignition delay period becomes longer.

Type of combustion chamber

The pre-combustion chamber because of its compactness, produces a shorter delay period compared to an open type of combustion chamber.

A summary of the influence of various factors on delay period is presented in Table 7.1.

7.8 COMBUSTION KNOCK IN CI ENGINES

Combustion knock in CI engines is associated with an extremely high rate of pressure rise during the second phase of combustion (rapid or uncontrolled combustion) and also with heavy vibration accompanied by a knocking sound, thus causing overheating of the piston and the cylinder head, drop in power, damage to bearings and possible piston seizure.

The injection process of a fuel takes place over a definite period of time in terms of degree crank angle. As a result, the first few drops which are injected into the chamber pass through the ignition delay while the additional droplets are being injected into the chamber. Normally, the fuel injected period is more than the delay period. If the delay period of the injected fuel is short, the first few droplets will commence the burning phase in a relatively short time after injection, and a relatively small amount of fuel will be accumulated in the chamber when actual burning commences.

Table 7.1 Effect of various factors on delay period

Increase in variable	*Effect on delay period*	*Reason*
Ignition quality of fuel:		
(i) Self-ignition temperature	Increases	Difficult to ignite.
(ii) Cetane number	Decreases	Reduces the self-ignition temperature.
Injection timing:		
(i) advancing	Increases	At injection point the pressure and temperature of air are lower.
(ii) retarding	Increases	Increased heat loss.
Compression ratio	Decreases	Increases the temperature and pressure of air. Reduces the self-ignition temperature.
Injection pressure, rate of injection and drop size	Decreases a little, but no significant change	Surface area of small droplets increases, this tends to reduce delay. Momentum reduces, this tends to increase delay.
Temperature:		
(i) Intake	Decreases	Increases the air temperature
(ii) Jacket water	Decreases	Increases the wall temperature and hence that of air.
(iii) Fuel	Decreases	Better vaporization, increases reaction.
Intake pressure	Decreases	Compression pressure increases.
Engine speed	Decreases in milliseconds Increases in °CA.	Heat loss decreases during compression.
Air/fuel ratio	Increases	Wall and combustion temperatures are lowered.
Load	Decreases	Residual gas and wall temperature increase
Engine size	Decreases	Smaller surface-to-volume ratio, less heat loss resulting in higher temperature of compressed air.
Fuel jet impingement on the wall	Decreases a little, but no significant effect.	Increase in fuel evaporation and mixing process.
Swirl rate	Decreases.	Increases evaporation and mixing, reduces heat transfer, increases the charge temperature.
Exhaust Gas Recycle (EGR)	Increases	Less concentration of oxygen.
Type of combustion chamber	Decreases for engines with pre-combustion chamber	Compactness of the chamber causing increased temperature and pressure of the charge.

As a result, the rate of burned mass of fuel will be such as to produce a rate of pressure rise that will exert a smooth force on the piston, as shown in Figure 7.12(a). If, on the other hand the delay period is longer, the burning of the first few droplets is delayed and therefore a greater quantity of fuel droplets will accumulate in the chamber. When the actual burning commences, the additional fuel may cause rapid rate of pressure rise (Figure 7.12(b)) resulting in rough engine operation. If the delay period is too long, much fuel will be accumulated resulting in instantaneous rise in pressure (Figure 7.12(c)). Such a situation produces pressure waves striking on cylinder walls, piston crown and cylinder head, producing knock and vibrations. In fact, the combustion mechanism of diesel engine is based on the autoignition of the charge and hence, mild knock may always be present. When it exceeds a certain limit, the engine is said to be knocking.

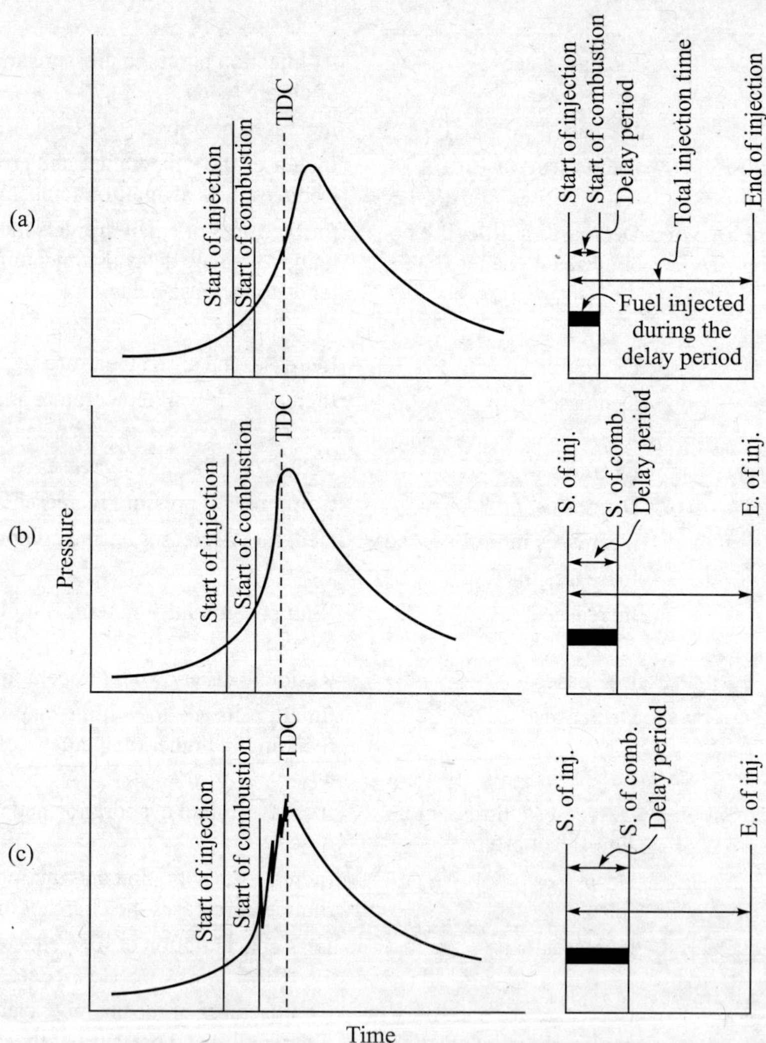

Figure 7.12 Diagrams illustrating the effect of ignition delay on the rate of pressure rise in a CI engine.

Though the long ignition delay improves the mixing process of fuel and air and makes the mixture more homogeneous, it helps the process of autoignition and makes the engine more prone to knock.

7.9 COMPARISON OF KNOCK IN SI AND CI ENGINES

It is now possible to make an interesting comparison between knocking in SI and CI engines. While knocking in the SI engine and in CI engine have essentially the same basic cause, i.e. autoignition followed by a rapid pressure rise, it is important to note the following differences:

(a) In the SI engine, knocking occurs due to autoignition of the last part of the charge (end-gas), i.e. at the end of combustion, while in the CI engine knocking occurs in the first part, i.e. at the start of combustion (Figure 7.13).

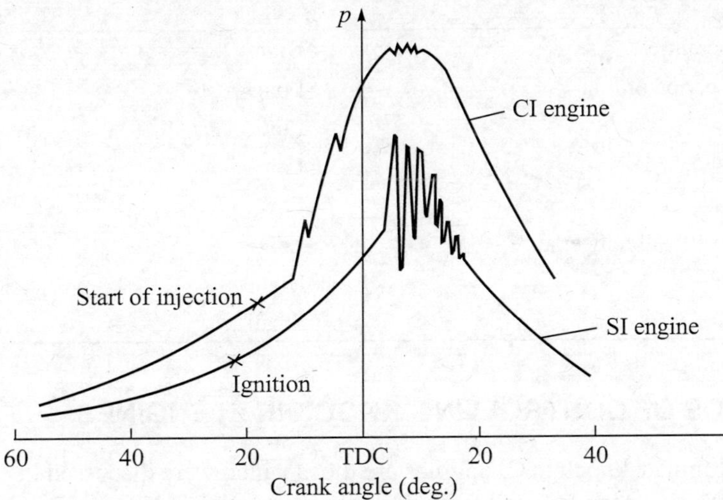

Figure 7.13 Comparison of knock in SI and CI engines.

(b) In the SI engine, it is the homogeneous charge that autoignites and causes knocking, resulting in a very high rate of pressure rise and high peak pressure. In the CI engine the fuel and air are not homogeneously mixed and hence the rate of pressure rise is normally lower than that in the knocking part of the charge in the SI engine. However, the peak pressure is higher due to a high compression ratio.

(c) In the CI engine only the air is compressed during the compression stroke and the ignition can take place only after the fuel is injected just before TDC. Therefore, there is no question of preignition occurring in the CI engine, whereas preignition may occur in the SI engine due to the presence of both fuel and air during compression.

(d) In the SI engine, knocking can easily be detected by human ear but in the CI engine there is no clear distinction between knocking and normal combustion, since normal combustion in the CI engine is itself by autoignition and a mild knock may always be present. It is therefore a personal judgment that is involved here. When such noise becomes excessive and there is excessive vibration in the engine structure, in the opinion of the observer, the engine is said to be knocking.

Those factors which tend to prevent knock in SI engines, the same very factors promote knock in CI engines. To prevent knock in SI engines, the autoignition of the last part of the charge should not take place; this requires a long delay period and a high self-ignition temperature. To prevent knock in CI engines, the autoignition of the first part of the charge should be achieved as early as possible and therefore it requires a short delay period and a low self-ignition temperature. It may also be noted that a good SI engine fuel is a bad CI engine fuel and vice versa.

Table 7.2 presents a comparative statement of the various factors to be varied in order to reduce knock in SI and CI engines.

Table 7.2 Factors tending to reduce knock in SI and CI engines

Factor	*SI engines*	*CI engines*
Self-ignition temperature	High	Low
Time lag or delay period of fuel	Long	Short
Compression ratio	Low	High
Inlet temperature	Low	High
Inlet pressure	Low	High
Combustion chamber wall temperature	Low	High
Engine speed, rpm	High	Low
Cylinder size	Small	Large

7.10 METHODS OF CONTROLLING KNOCK IN CI ENGINES

The methods to eliminate knock in CI engines are those which were discussed in conjunction with the decrease of the delay period. Apart from the operating variables discussed above, the knock in CI engines can be controlled by the following methods:

1. Certain fuels cause knocking in a given CI engine, others do not. Thus, fuels of high cetane number are obtained by adding chemical dopes, called ignition accelerators. The two common chemical dopes are ethyl nitrate and amyl nitrate in concentrations of 8.8 g/l and 7.7 g/l respectively. The chemical dopes increase the preflame reactions and reduce the flash point. The cetane number of the diesel fuel is increased and the fuel autoignites at lower temperatures. However, these dopes are expensive and produce more oxides of nitrogen emissions in the exhaust as they contain nitrogen.
2. The turbulent air-cell method of combustion generally gives knockless running of the engine and if the amount of fuel injected is not excessive, a smokeless exhaust is emitted.
3. Two-stage injection is used to avoid knocking in CI engines. In order to reduce diesel knock or avoid it altogether, a small quantity of fuel is injected just before the main amount is delivered to the injector. It is known as *pilot injection*. The use of two-stage injection gives a better control of the rate of pressure rise after the main delivery of fuel has taken place and avoids a sudden pressure rise in the cylinder. Special types of fuel injection pumps have been developed to give this initial delivery of a small fuel charge before the main one.

7.11 COMBUSTION CHAMBER FOR CI ENGINES

7.11.1 Combustion Chamber Characteristics

The proper design of a combustion chamber is very important. In a CI engine the fuel is injected during a period of some 20 to 35 degrees of crank angle. In this short period of time an efficient preparation of the fuel-air charge is required, which means:

(a) An even distribution of the injected fuel throughout the combustion space, for which it requires a directed flow or swirl of the air.
(b) A thorough mixing of the fuel with the air to ensure complete combustion with the minimum excess air, for which it requires an air swirl or squish of high intensity.

An efficient smooth combustion depends upon:

(a) A sufficiently high temperature to initiate ignition; it is controlled by the selection of the proper compression ratio.
(b) A small delay period or ignition lag.
(c) A moderate rate of pressure rise during the second stage of combustion.
(d) A controlled, even burning during the third stage; it is governed by the rate of injection.
(e) A minimum of afterburning.
(f) Minimum heat losses to the walls. These losses can be controlled by reducing the surface-to-volume ratio.

The main characteristics of an injection system that link it with a given combustion chamber are atomization, penetration, fuel distribution, and the shape of the fuel spray.

7.11.2 Classification of CI Engine Combustion Chambers

In order to attain the above objectives, the CI engines are divided into two basic categories according to their combustion chamber design: (a) direct-injection (DI) engines, which have a single open combustion chamber into which fuel is injected directly; (b) indirect-injection (IDI) engines, where the chamber is divided into two regions and the fuel is injected into the prechamber which is situated above the piston crown and is connected to the main chamber via a nozzle or one or more orifices. The IDI engine designs are only used in the smallest engine sizes. Within each category there are different chamber geometries, air flow, and fuel injection arrangements.

Figure 7.14 shows the classification of CI engine combustion chambers.

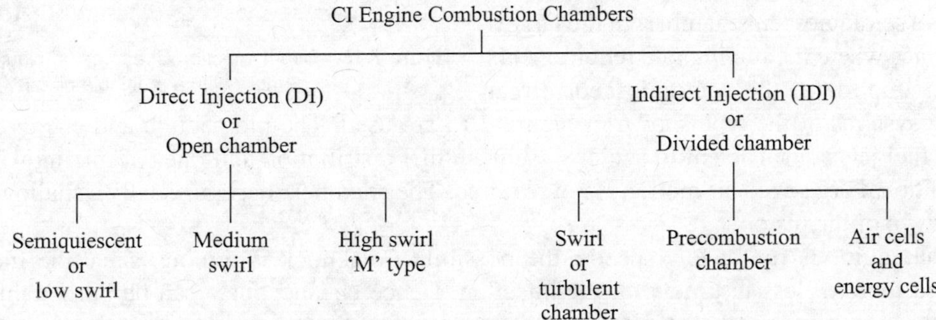

Figure 7.14 Classification of CI engine combustion chambers.

7.12 DIRECT INJECTION (DI) ENGINES OR OPEN COMBUSTION CHAMBER ENGINES

An open chamber has the entire compression volume in which the combustion takes place in one chamber formed between the piston and the cylinder head. The shape of the combustion chamber may create swirl or turbulence to assist fuel and air. Swirl denotes a rotary motion of the gases in the chamber more or less about the chamber axis. Turbulence denotes a haphazard motion of the gases.

In this combustion chamber, the mixing of fuel and air depends entirely on the spray characteristics and on air motion, and it is not essentially affected by the combustion process. In this type of engine, the spray characteristics must be carefully arranged to obtain rapid mixing. Fuel is injected at high injection pressure and mixing is usually assissted by a swirl, induced by directing the inlet air tangentially, or by a squish which is the air motion caused by a small clearance space over part of the piston. The open chamber design can be classified as follows:

(a) Semiquiescent or low swirl open chamber
(b) Medium swirl open chamber
(c) High swirl open chamber ('M' type).

7.12.1 Semiquiescent or Low Swirl Open Chamber

In this type of engine, mixing the fuel and air and controlling the rate of combustion mainly depend upon the injection system. The nozzle is usually located at the centre of the chamber. It has a number of orifices, usually six or more, which provide a multiple-spray pattern. Each jet or spray pattern covers most of the combustion chamber without impinging on the walls or piston. The contour of the inlet passage way does not encourage or induce a swirl or turbulence, so the chamber is called quiescent chamber. However, the air movement in the chamber is never quiescent, so it is better to call the chamber a semiquiescent chamber. In the largest size engines where the mixing rate requirements are least important, the semiquiescent direct injection systems of the type shown in Figure 7.15 are used. The momentum and energy of the injected fuel jets are sufficient to achieve adequate fuel distribution and rates of mixing with air. Any additional organized air motion is not required. The chamber shape is usually a shallow bowl in the crown of the piston.

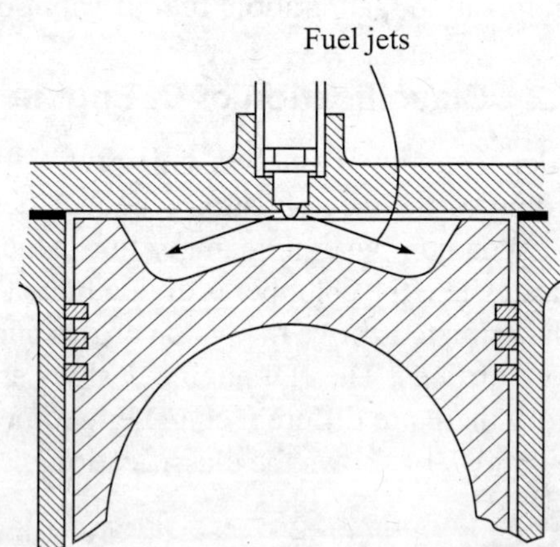

Figure 7.15 Direct injection CI engine semiquiescent chamber with a multihole nozzle.

If the engine is run at low speeds, the possibility of knock is remote, since the fuel can be burned more or less in time with the injection. Hence cheaper fuels can be burned and low combustion pressures can be held. Low combustion temperatures, and low turbulence and swirl reduce heat loss to the coolant.

The advantages of the open chamber design with a slow speed engine are as follows:

1. The specific fuel consumption is less, because of the following reasons:
   (a) The fuel is burned close to TDC because the time in degree crank angle is long. It is an approach towards achieving the Otto cycle efficiency.
   (b) The air/fuel ratio is high, therefore, combustion should be relatively complete with an approach towards the air standard efficiency.
   (c) Percentage heat loss is reduced. It is an approach towards adiabatic combustion. Following are the possible reasons:
      (i) Either low swirl or low turbulence
      (ii) Low surface-to-volume ratio of an undivided chamber
      (iii) Low overall combustion temperatures.
2. Starting is relatively easier. It is because of low heat losses.
3. Less heat is rejected to the coolant and to the exhaust gases. It requires smaller radiator and pumps. The life of the exhaust valve is increased.
4. The engine is quiet and provides relative freedom from combustion noise.
5. The residual fuels can be burned. It favours the operation of two-stroke engines.

7.12.2 Medium Swirl Open Chamber

As the engine size decreases and the speed increases, the quantity of fuel injected per cycle is reduced and the number of holes in the nozzle is necessarily less (usually 4). As a result, the injected fuel needs help in finding sufficient air in a short time. Faster fuel-air mixing rates can be achieved by increasing the amount of air swirl. Air swirl is generated by a suitable design of the inlet port. The air swirl rate can be increased as the piston approaches TDC by forcing the air towards the cylinder axis. Figure 7.16 shows a bowl-in-piston type of medium swirl open chamber with a centrally located multihole injector nozzle. The amount of liquid fuel which impinges on the piston cup walls is kept minimum. This type is used in medium size (10 to 15 cm bore) diesel engines.

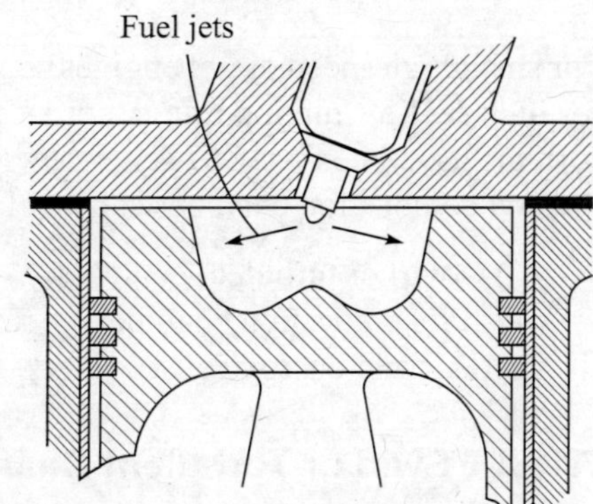

Figure 7.16 Direct injection CI engine bowl-in-piston chamber with medium swirl and a multihole nozzle.

7.12.3 High Swirl Open Chamber ('M' type)

Spiral intake ports produce a high speed rotary air motion in the cylinder during the induction stroke. Here, a single coarse spray is injected from a pintle nozzle in the direction of the air swirl, and tangential to the spherical wall of the combustion chamber in the piston. The fuel strikes against the wall of the spherical combustion chamber where it spreads to form a thin film which will evaporate under controlled conditions. The air swirl in the spherically shaped combustion chamber is quite high which sweeps over the fuel film, peeling it from the wall layer by layer for progressive and complete combustion. The flame spirals slowly inwards and around the bowl, with the rate of combustion controlled by the rate of vaporization.

Figure 7.17 shows the 'M', 'MAN' Meurer wall burning combustion system, optimized to give greater swirl and mixing for complete combustion. This combustion system was developed around 1954 by the Maschin-enfabrik Augsburg Nürnburg (MAN) AG of Germany for small, high speed engines. It has single-hole fuel-injection, so oriented that most of the fuel is deposited on the piston bowl walls. In practice, this engine gives good performance even with fuels of exceedingly poor ignition quality. Its fuel economy appears to be extremely good for an engine of small size. Because of the vaporization and mixing processes, the 'M' engine is ideally suited as a multifuel engine.

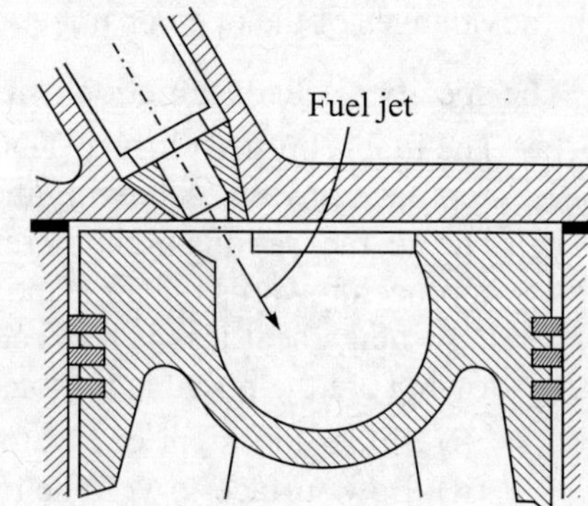

Figure 7.17 Direct injection bowl-in-piston chamber with high swirl and a single hole nozzle, M-type.

7.13 INDIRECT-INJECTION (IDI) ENGINES OR DIVIDED COMBUSTION CHAMBER ENGINES

For small high speed diesel engines such as those used in automobiles, the inlet generated air swirl for high fuel-air mixing rate is not sufficient. Indirect injection (IDI) or divided chamber engine systems have been used to generate vigorous charge motion during the compression stroke. The divided combustion chamber can be classified as:

(a) Swirl or turbulent chamber
(b) Precombustion chamber
(c) Air and energy cells.

7.13.1 Swirl or Turbulent Chamber

The swirl chamber design is shown in Figure 7.18. The spherically shaped swirl chamber contains about 50 per cent of the clearance volume and is connected to the main chamber by a tangential throat offering mild restriction. Because of the tangential passageway, the air flowing into the chamber on the compression stroke sets up a high swirl.

During compression the upward moving piston forces a flow of air from the main chamber above the piston into the small antechamber, called the swirl chamber, through the nozzle or orifice. Thus, towards the end of compression, a vigorous flow in the antechamber is set up. The connecting passage and chamber are shaped so that the air flow within the antechamber rotates rapidly. Fuel is usually injected into the antechamber through a pintle nozzle as a single spray. In some cases sufficient air may be present in the antechamber to burn completely all but the overload quantities of the fuel injected. The pressure built up in the antechamber by the expanding burning gases forces the burning and the unburned fuel and air mixtures back into the main chamber, where the jet issuing from the nozzle entrains and mixes with the main chamber air, imparting high turbulence and therefore further assisting combustion. The glow plug shown on the right of the antechamber in Figure 7.18 is a cold starting aid.

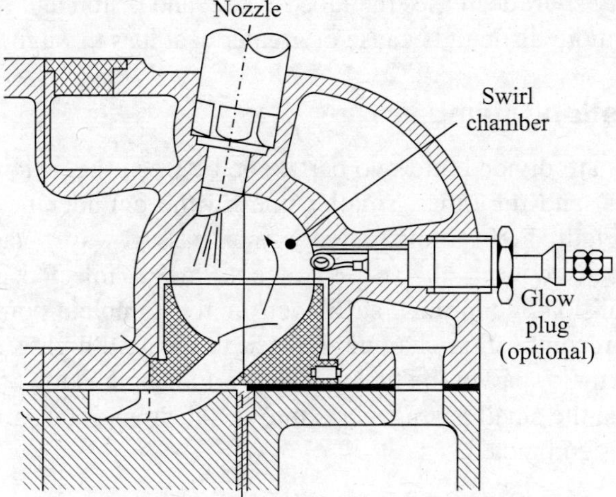

Figure 7.18 Ricardo swirl chamber (IDI).

Since the antechamber is small, deep penetration of the spray is not required. Since the swirl is high, a single hole nozzle is sufficient, although a well atomized fuel spray is desirable. A pintle type nozzle offers these qualities.

The advantages of the indirect injection swirl chamber over the open chamber are as follows:

1. Higher speed, brake mean effective pressure and power with less smoke are feasible. It is because of:
   (a) Higher volumetric efficiency—since the nozzle is at the side, there is more room for the larger intake and exhaust valves.
   (b) Shorter delay period—since the antechamber is compact and the air swirl in the chamber is very high.
2. Less mechanical stress and noise. It is because of the lower rate of pressure rise and the lower maximum pressure in the main chamber due to the throttling effect of throat.
3. Less maintenance—since the pintle nozzle is self-cleaning and the mechanical stress is less.
4. Wider range of fuels can be used. It can serve as a multi-fuel engine with minimal changes.
5. Smoother and quieter idling—since matching of small air supply with the small fuel supply is possible.
6. Cleaner exhaust resulting in less air pollution.

The disadvantages of the IDI swirl chamber over the open chamber are as follows:

1. Higher specific fuel consumption resulting in poorer fuel economy. It is because of greater heat losses and pressure losses through the throat which result in lower thermal efficiency and higher pumping losses.
2. The flow of combustion gases through the throat leads to thermal cracks in the cylinder head and creates sealing problems.
3. Cylinder construction is more expensive.
4. More thermal energy is lost to the exhaust gases. It may decrease the life of the exhaust valve which will run hotter and increase cracking and sealing problems of the exhaust manifold.

5. Higher secondary swirl decreases the lube-oil life, and piston and ring life.
6. Cold starting is more difficult because of greater heat loss through the throat.

7.13.2 Precombustion Chamber

Here also the chambers are divided into two parts, one between the piston and the cylinder head (i.e. the main chamber) and the other, smaller one, in the cylinder head (i.e. precombustion chamber) as shown in Figure 7.19. Comparatively small passageways, made more restricted than those in a swirl chamber, connects the two chambers. Fuel is injected into the precombustion chamber, and under full-load conditions sufficient air for complete combustion is not present in this chamber. The precombustion chamber is used to create a high secondary turbulence for mixing and burning the major part of the fuel and air. Partial combustion of the fuel discharges the burning mixture through the small passageways into the air in various parts of the main chamber where the combustion is completed.

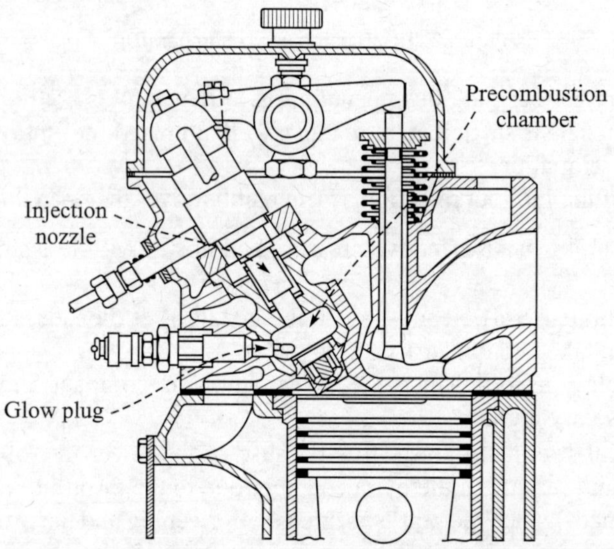

Figure 7.19 Precombustion chamber.

The prechamber contains 20 to 30 per cent of the clearance volume (versus 50 per cent and higher in the swirl chamber), with one or more outlets leading to the main chamber. The passageways may be oriented to create primary turbulence in the prechamber. Fuel is injected by a single open nozzle with one large orifice to obtain a jet with a concentrated core.

This type of combustion chamber produces a smooth combustion process but has high fluid friction and heat transfer losses. The advantages and disadvantages of the precombustion chamber relative to open chamber type are, in general, the same as those described for the swirl chamber.

7.13.3 Air Cells

The air cell type of combustion chamber does not depend upon the organized air-swirl like the precombustion chamber. The air cell is a separate chamber used to communicate with the main chamber through a narrow restricted neck. The air cell contains 5 to 15 per cent of the clearance volume. Fuel is injected into the main combustion space and ejects in a jet across this space to the open neck of the air cell, as shown in Figure 7.20. Some fuel enters and ignites in the air cell. This raises the pressure in the air cell and the burning mixture is discharged into the main chamber. Some combustion also takes place in the main combustion chamber. Combustion is completed on the downstroke of the piston while the air is discharged from the air cell into the partly burned mixture. The effective expansion ratio is curtailed and both the efficiency and the power output are reduced, but easy starting and reasonably smooth running are obtained with fairly low maximum pressures. It is best suited for small engines of medium duty where a relatively high fuel consumption can be tolerated. The air cell was considered unnecessary after it became possible to generate swirl in open chamber and therefore the air cell types of combustion chambers are now obsolete.

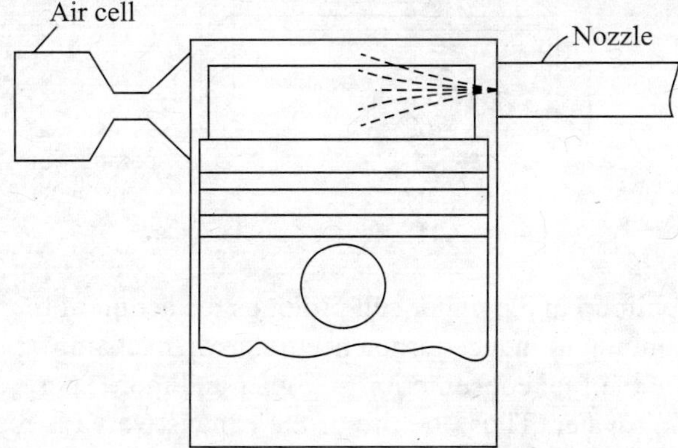

Figure 7.20 Air-cell CI engine.

7.13.4 Energy Cells

In the precombustion chamber, fuel is injected into the air-stream entering the prechamber during the compression stroke. As a result, it is not possible to inject the main body of the fuel spray into the most important place for burning. In the air cells, unburned fuel in the main chamber may not find enough turbulence. These drawbacks can be overcome in the energy cells. It is a hybrid design between the precombustion chamber and the air cell. The energy cell contains about 10–15 % of the clearance volume. It has two cells, major and minor, which are separated from each other by a restrictive orifice. The energy cell is separated from the main chamber by a narrow restricted neck (Figure 7.21).

As the piston moves up on the compression stroke, some of the air is forced into the major and minor chambers of the energy cell. When the fuel is injected through the pintle type nozzle, part of the fuel passes across the main combustion chamber and enters the minor cell, where it is mixed with the air entering during compression from the main chamber. Combustion first commences in the main chamber where the temperature is higher, but the rate of burning is slower in this location due to insufficient mixing of the fuel and air. The burning in the minor cell is slower at the start, but due to better mixing, progresses at a more rapid rate. The pressures built up in the minor cell, create added turbulence and produce better combustion in this chamber. In the

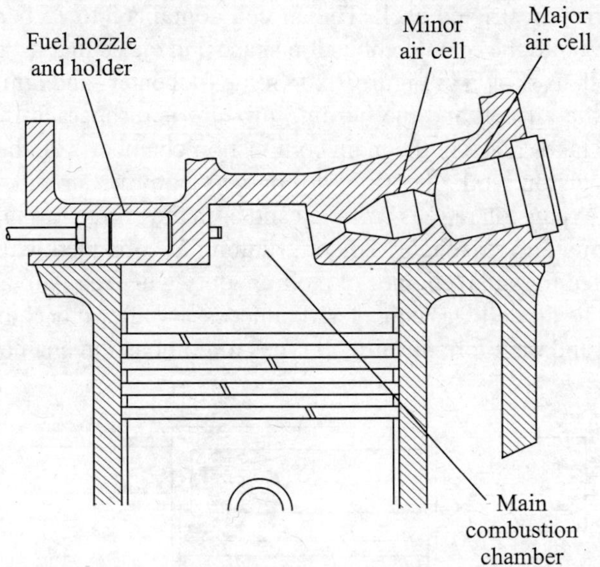

Figure 7.21 Energy cell CI engine

meantime, the pressure built up in the major cell prolongs the action of the jet stream entering the main chamber, thus continuing to induce turbulence in the main chamber.

In this type of engine the fuel consumption is higher and more energy is carried away in the exhaust and in the cooling water. Thus, the life of the exhaust valve is reduced and this type of engine requires a larger radiator and fan.

7.14 COMPARISON OF CHARACTERISTICS OF COMBUSTION CHAMBERS OF CI ENGINE

The characteristics of CI engine combustion chambers including the advantages and disadvantages of all different types of combustion chambers may be deduced from the various remarks made above and from general thermodynamic and gas dynamic principles. These characteristics are summarized in Table 7.3. The remarks made for different combustion chambers are only the generalizations and may not all be true for any particular engine. The tabulated dimensions and operating characteristics are indicative of only the typical ranges for each different type of CI engine and its combustion system.

7.15 STARTING METHODS AND AIDS

Before the CI engine can be started, some external means must be provided to rotate the crankshaft of the engine until combustion starts. The engine can be rotated by any one of the following means:

1. ***Hand starting:*** Single cylinder small engines can be cranked manually by rotating the crankshaft with the help of a handle. For easy rotation of the crank, a decompression valve is provided. Initially this valve is kept open, so that as the long as rotating air is not

Table 7.3 Comparison of characteristics of combustion chambers of CI engine

Basis for comparison	*Direct injection*			*Indirect injection*		
	Semiquiescent	*Medium swirl*	*High swirl 'M'*	*Swirl chamber*	*Pre-combustion*	*Energy cell*
Chamber	Open or shallow dish	Bowl-in-piston	Deep bowl-in-piston	Swirl pre-chamber	Prechamber	Air cells
Size	Largest	Medium	Smaller	Small	Small	Small
Cycle	2/4-stroke	4-stroke	4-stroke	4-stroke	4-stroke	4-stroke
Speed (rpm)	120–2100	1800–3500	2500–5000	3500–5000	4500	3500
Bore (mm)	900–150	150–100	130–80	95–70	95–70	95–70
Stroke/bore	3.5–1.2	1.3–1.0	1.2–0.9	1.1–0.9	1.1–0.9	1.1–0.9
Compression ratio	12–15	15–16	16–18	20–24	22–24	13–16
Air flow pattern	Low swirl	Medium swirl	High swirl	Very high swirl in swirl chamber	Very high turbulence in prechamber	High secondary turbulence
Air utilization	Fairly moderate	Moderate	Moderate	Very good	Very good	Good
No. of nozzle holes	Multi (6–8)	Multi (4)	Single	Single	Single	Single
Injection pressure (bar)	Very high (1300)	High (540)	Medium (270)	Medium (340)	Low (135)	Medium (270)
Rate of pressure rise in the main chamber (bar/degree)	High (6.8–10.2)	Medium (6.8–8.5)	Low (4.0–5.4)	Low 4.0	Very low 2.7	Very low 2.7
Heat loss	Low	Moderate	Moderate	High	High	High
Pumping loss	Low	Lower	Medium	High	High	High
Throttling loss	No	No	No	Some	More	More
bmep potential	Highest	High	Medium	Medium	Good	Medium
Volumetric Efficiency	Low	Lower	Lowest	High	High	High
Cooling load	Low	Low/Medium	Medium	Highest	High	High
Combustion noise	High	Medium	Lowest	Low	Low	Low
Cold starting	Excellent	Very good	Good	Needs aid	Needs aid	Needs aid

compressed, the free rotation is possible. When the engine gets momentum, this valve is closed and the air is compressed to ignite the fuel by compression.

2. ***Electric motor:*** It is used for cranking small engines. It needs a 12 volt battery.
3. ***An auxiliary SI engine:*** It can be cranked by hand or else by an electric motor.
4. ***Air pressure:*** Large CI engines are normally started by air pressure. It requires a storage tank for the compressed air and a compressor. On the expansion stroke of the engine, compressed air enters the cylinder through the air valve and cranks the engine.

Cold starting of a CI engine in some cases is a very serious problem. The problem is aggravated further if the weather is extremely cold, the cylinder liner is worn or the valves are leaky. In order to start ignition, the temperature must sufficiently exceed the self-ignition temperature of the fuel. The pressure must also be high enough to ensure good contact and hence rapid heat transfer from hot air to the surface of the liquid fuel.

At very low speed during compression the heat loss to the walls will be more. The temperature and pressure of the air may not be enough to ignite the fuel. At very high speed although the temperature and pressure will be high enough but the time for vaporization and mixing will not be enough to ignite the fuel. Therefore, for a CI engine there will be an optimum starting speed in between these two limits. The optimum speed will depend upon:

1. ***Surface-to-volume ratio:*** The larger the size of the cylinder, the less will be the surface-to-volume ratio and hence less will be the rate of loss of heat during compression.
2. ***The intensity of air-swirl:*** It determines the rate of loss of heat by convection during compression.
3. ***The physical conditions:*** Such as weather, leakage, etc.

It is evident that the open-chamber direct-injection engine will give the easiest cold starting, since it will have both the smallest surface-to-volume ratio and the lowest intensity of air swirl. However, sometime physical conditions may not be favourable and cold starting may not be possible without the use of some aid to ignition.

The majority of high speed engines require some starting aids. A few of them are mentioned below:

1. ***Electric glow plugs in the combustion chamber:*** A glow plug can get damaged by combustion over a period of time.
2. ***Pintaux nozzle:*** It is a combination of pintle with the auxiliary spray. At lower rpm, the needle valve lift is small and hence less fuel is supplied. To increase the supply of fuel, the auxiliary orifice delivers more fuel which helps in cold starting. As the speed picks up, the increased lift of the needle valve supplies more fuel through the main jet. The auxiliary orifice then delivers less fuel.
3. ***Intake manifold heater:*** It ignites a small feed of fuel.
4. ***Injection of diethyl ether into the intake in a controlled amount:*** Ether has a very low self-ignition temperature.
5. ***Injection of a small amount of lubricating oil or fuel oil:*** It raises temporarily the compression ratio and seals both the piston rings and valves.

REVIEW QUESTIONS

1. What is the typical range of the compression ratio in CI engines?
2. What are the advantages and disadvantages of the CI engine over the SI engine? Where does the CI engine find its application?
3. Describe with the help of diagrams the air-swirl and squish in the CI engine combustion chamber.
4. Draw and describe the spray pattern when fuel is injected into swirling air. Show a plot of equivalence ratio on the spray pattern.
5. Describe the mechanisms of combustion of a fuel spray injected in swirling air.
6. Describe the different stages of combustion in a CI engine.
7. What is meant by a delay period? Describe it, indicating its importance.
8. Describe the heat release rate pattern of a CI engine during the different stages of combustion.
9. What is the inflammability limit of a diesel fuel? How does the combustion of a fuel occur in a CI engine, when the overall mixture is much leaner than this limit? How does the air/fuel ratio in a CI engine change with the change in load?
10. Show the effect of the air/fuel ratio on power output from a CI engine. What is the limit of using rich fuel?
11. Describe the influence of the following factors on delay period:
    (a) Ignition quality of fuel and cetane number: show the effect of cetane number on the p–θ diagram.
    (b) Injection timing: show the effect of injection timing on the p–θ diagram.
    (c) Compression ratio: show the effect of compression ratio on the p–θ diagram.
    (d) Injection pressure, rate of injection and drop size
    (e) Intake, jacket water and fuel temperatures
    (f) Intake pressure
    (g) Engine speed
    (h) Air/fuel ratio
    (i) Load
    (j) Engine size
    (k) Combustion chamber wall effect
    (l) Swirl rate
    (m) EGR
    (n) Type of combustion chamber.
12. Summarize in a tabular form the effect of various factors on the CI engine delay period.
13. Describe the phenomenon of knock in CI engines.
14. Illustrate and describe the effect of ignition delay on the rate of pressure rise in a CI engine.
15. Compare the knocking phenomena in SI and CI engines. Explain clearly that the factors which tend to prevent knock in SI engines, in fact promote knock in CI engines.
16. Give a comparative statement of the various factors to be increased or decreased, which tend to reduce knock in SI and CI engines.

17. Briefly describe the different methods used to control knock in CI engines.
18. What are the requirements of a good combustion chamber of a CI engine?
19. Give the classifications of different types of combustion chambers of a CI engine.
20. What is direct-injection (DI) type combustion chamber for a CI engine? What are the different types of this combustion chamber?
21. Briefly describe with the help of simple diagrams the following types of direct-injection combustion chambers for CI engines:
(a) Semiquiescent (b) Medium Swirl (c) 'M' type.
22. What are the advantages of the open chamber design of combustion chamber for a CI engine?
23. What is indirect-injection (IDI) type combustion chamber for a CI engine? What are the different types of this combustion chamber?
24. Describe with the help of a diagram the swirl chamber IDI type combustion chamber for a CI engine.
25. What are the advantages and disadvantages of the IDI swirl chamber over the open chamber?
26. Describe with the help of a diagram the IDI precombustion chamber for a CI engine. What are the advantages and disadvantages of this type of combustion chamber compared to the open chamber type.
27. Describe the air cell type of combustion chamber with the help of a diagram.
28. Describe the energy cell type of combustion chamber with the help of a diagram. What are its advantages over the air cell type?
29. Briefly compare the characteristics of different types of combustion chambers in a tabular form.
30. What are the different methods of starting a diesel engine?
31. Why is it not possible to start an engine at a very low or at a very high speed? On what factors does the optimum starting speed depend?
32. Why are the open chamber direct injection engines easy to cold start?
33. What are cold starting aids? Describe them briefly.

8.1 INTRODUCTION

Fuel is a substance which participates easily with oxygen in a self-sustaining exothermic reaction. The subject of fuels for IC engines has been studied ever since the engines have been in existence. Engine performance depends upon the fuel characteristics. It is therefore essential to study the characteristics of different types of fuels to understand the combustion phenomenon. The character of the fuel used may have considerable influence on the design, the output, the efficiency, the fuel consumption, the air pollution and in many cases on the reliability and durability of the engine.

The characteristics of the IC engine fuels must be such that the following requirements are met:

1. The fuel should get effectively atomized, vaporized and well mixed with the air.
2. The combustion process must be fast.
3. Starting of the engine should be quick and reliable at any ambient condition.
4. The surface of the combustion chamber should remain free from carbon and other deposits.
5. The cylinder face, the piston and the piston rings should not get subjected to excessive wear and corrosion.
6. The basic elements of the engine should remain free from thermal stresses due to temperature gradient developed during combustion.
7. Combustion should be complete without the evolution of harmful exhaust gases.

The IC engines can be operated on many different kinds of fuels, including liquid, gaseous, and even solid. The selection of a fuel for a particular use in engines is mainly governed by (a) the type of the equipment required to store, supply and burn the fuel in the engine, (b) the heating value per unit volume of the fuel, and (c) the availability and cost of the fuel at the site of the engine.

8.2 CLASSIFICATION OF FUELS

The principal constituents of any fuel are carbon and hydrogen. Fuels may be classified as primary or secondary according to whether they occur in nature or are prepared. Fuels can be in solid, liquid or gaseous state. The classification of important fuels can be summarized as follows:

1. ***Solid fuels***
   Natural (*Primary*): Wood; Peat; Coal, the major ranks of coal from the lowest to the highest are: (i) Lignite, (ii) Subbituminous, (iii) Bituminous, (iv) Semianthracite, and (v) Anthracite.

   Prepared (*Secondary*): Charcoal; Coke; Briquetted coal; Pulverized coal.

2. ***Liquid fuels***
   Natural: Petroleum

   Prepared: (i) Petroleum based: gasoline, kerosene, diesel oil, fuel oil, lubricating oil. (ii) Non-petroleum based: benzol, alcohols.

3. ***Gaseous fuels***
   Natural: Natural gas

   Prepared: Liquefied petroleum gas (LPG), producer gas, coal gas, hydrogen.

8.3 SOLID FUELS

8.3.1 Brief Description of Solid Fuels

Wood: It is a naturally available solid fuel. It is not generally used as a commercial fuel except in industries where a large amount of waste wood is available.

Peat: It is a mixture of decayed vegetable matter with water. It is dried with air. It contains a high percentage of moisture. It is used as a fuel in a gas producer plant.

Coal: This includes all types of natural solid fuels from lignite to anthracite. Peat is not a coal but represents the first stage in the formation of coal. In the early stages of transformation, varying amounts of water, methane and carbon dioxide got gradually eliminated by increased pressure and temperature. Peat was changed to lignite and then to bituminous coal. The volatile matter of the fuel was reduced further to form the anthracite coals. Lignites have a wood or clay like appearance associated with a high amount of moisture and it has low heat content. Lignites are usually amorphous in character and pose transportation difficulties as they break easily. Anthracite is very hard and has a shining black lustre. The moisture content is very low. The heating value of this fuel is high. The quality of coal improves gradually from lignite to anthracite.

Charcoal: It is prepared by destructive distillation of wood. The by-products resulting from distillation are methyl alcohol, acetic acid, acetone, gaseous compounds, and tar. It absorbs 12 to 15 per cent moisture from atmosphere and this lowers its heating value.

Coke: It is the solid residue left after the destructive distillation of certain soft coals. It consists of carbon, mineral matter with about 2% sulphur and small quantities of hydrogen, nitrogen and phosphorus. It is mainly used in blast furnaces.

Briquetted coal: It is a block of compressed coal dust. The blocks are prepared from fine coal and slacks of all types produced in mining by compressing the material under high pressure. It gives satisfactory strong briquettes (bricks) without the addition of a binder.

Pulverized coal: Reducing the coal to powder or dust is called pulverized coal. Low grade fuel is efficiently burnt by pulverizing.

8.3.2 Use of Solid Fuels in IC Engines

The use of solid fuels leads to problems of complicated injection systems as well as difficulties associated with solid residue or ash. Attempts have been made to use pulverized coal in CI engines, but there has been very little development of the use of fuel in this form. Solid fuels consequently find little practical application at present. The problem of using coal directly by pulverizing it and burning in a CI engine is being brought nearer to a practical solution but the main obstacle is the excessive wear of the cylinder liners and the piston rings.

Since coal is our most abundant fuel, considerable research is underway on methods that in the future will produce either gasoline or gas from coal. A number of processes have been developed for converting coal into liquid fuels. Basically, the process involves the forming of a synthesis gas by partially burning coal with insufficient air. This yields a product high in CO and H_2 content. The synthesis gas is then passed through a synthesis reactor where, in the presence of a catalyst and the proper temperature and pressure, the CO and H_2 are reformed to produce hydrocarbons of the paraffin, olefin, and alcohol families. These products are further refined to form suitable liquid fuels.

8.4 LIQUID FUELS

8.4.1 Petroleum Fuels (Petra = rock + oleum = oil)

The most widely accepted theory is that the oil was formed by the decomposition of dead animals and plants at high temperature and under high pressure which brought about a type of semi-destructive distillation. This along with other metamorphic processes resulted in the formation of petroleum. After a period of many centuries, the oil was forced through layers of porous rock strata until it finally became entrapped under a domed-shaped hard rock (Figure 8.1). This made it impossible for the gas and oil to escape, and the water kept the pool of oil under pressure.

In drilling for oil, the dome-shaped hard rock is penetrated, and the existing pressure forces the oil, gas, and some water to the surface. As the petroleum is removed from the underground

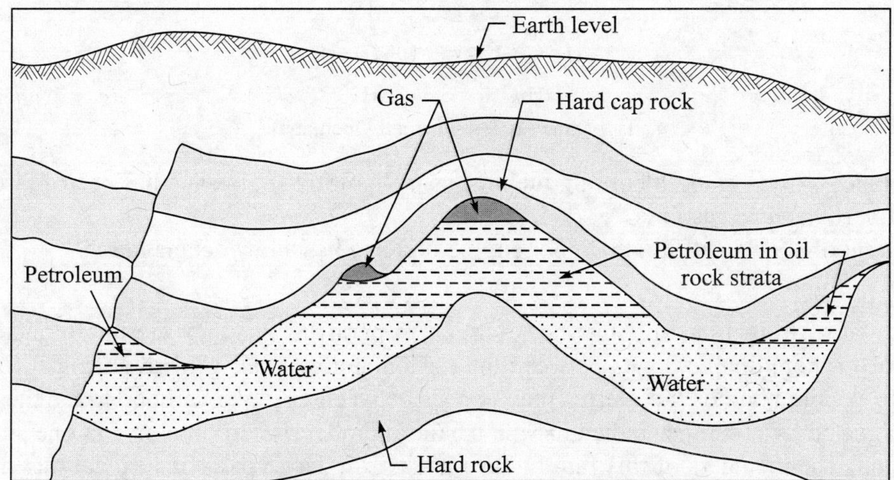

Figure 8.1 Typical oil-pool formations.

deposit, the water moves in to take its place. Eventually, the natural pressure decreases to the point where it will not be possible to force the oil and gas to the surface. When this occurs, the gas or water may be pumped back into the well to increase the pressure and hence force the crude oil from the pool to the surface.

Petroleum is a mixture of many different hydrocarbons, with some sulphur and other impurities. Hydrocarbons present in fuel may be classified into four different groups: (a) Paraffins, (b) Olefins, (c) Naphthenes, and (d) Aromatics.

Paraffins (Alkanes)

Normal or straight-chain paraffins are stable, saturated compounds having the general chemical formula C_nH_{2n+2}. The compounds in this group have a name ending in –ane. The first four members of this series (n = 1 to 4) are gases. The members from n = 6 to 18 are liquids and then gradually the transformation takes place from n = 18 to 21, after which they are solids. The molecular structures of three of the members of the normal paraffin family with n = 1, 3 and 8 are shown below:

```
    H              H  H  H              H  H  H  H  H  H  H  H
    |              |  |  |              |  |  |  |  |  |  |  |
 H—C—H          H—C—C—C—H           H—C—C—C—C—C—C—C—C—H
    |              |  |  |              |  |  |  |  |  |  |  |
    H              H  H  H              H  H  H  H  H  H  H  H
```

Methane (CH_4) Propane (C_3H_8) Octane (C_8H_{18})

The hydrocarbons which have the same chemical formula but different structural formula are known as *isomers*. Any compound with four or more carbon atoms may possess isomers. Isooctane has the same general chemical formula and molecular weight as octane but a different structure and different physical characteristics. Isoparaffins are also stable compounds.

```
              H
              |
           H—C—H
              |
    H         |        H        H        H
    |         |        |        |        |
 H—C————C————C————C————C—H
    |         |        |        |        |
    H         |        H        |        H
              |                 |
           H—C—H           H—C—H
              |                 |
              H                 H
```

Isooctane (2, 2, 4 – trimethylpentane)

Isooctane is a very smooth burning fuel in a spark-ignition engine and has been chosen as the standard for 100 octane gasoline.

In contrast, normal octane is a rather poor engine fuel and has a low octane rating.

Olefins (Alkenes)

Olefins are also straight chain compounds similar to paraffins but they are unsaturated as they contain one or more double bonds between the carbon atoms. Their general chemical formula is C_nH_{2n}. The double bond of the olefins may appear between any two carbon atoms, the position being designated by a number indicating the smaller number of carbon atoms at one side of the double bond. The members of this family have names similar to paraffins except that the suffix -ene is used. The structure of 3-Heptene (C_7H_{14}) is shown below:

```
     H   H        H   H   H
     |   |        |   |   |
H—C—C—C=C—C—C—C—H
     |   |   |   |   |   |   |
     H   H   H   H   H   H   H
```

3-Heptene (C_7H_{14})

The diolefins (C_nH_{2n-2}) have a structure similar to that of olefins, but have two double bonds in the open-chain structure, and the suffix "diene" is used

```
          H   H        H
          |   |        |
H—C=C—C—C—C=C—C—H
     |   |   |   |   |   |   |
     H   H   H   H   H   H   H
```

1, 5-Heptadiene (C_7H_{12})

The olefins and the still more unsaturated diolefins are unstable compounds and are thought to be the cause of much gum formation in gasoline. However, since they possess very good ignition characteristics, substantial quantities of these compounds appear in modern high octane gasoline.

Naphthenes (Cyclanes)

Naphthenes are saturated, stable compounds with a ring structure. The molecular structure of cyclopentane is shown below:

Cyclopentane (C_5H_{10})

The naphthenes are saturated compounds even though the general chemical formula is C_nH_{2n}, the same as for the unsaturated olefins. The names of the members of this family are the same as those of the corresponding paraffin except that the prefix cyclo– is added to denote the ring structure of the naphthenes.

The physical properties of this group are similar to those of the normal paraffins, but the combustion properties are more like those of isoparaffins.

Aromatics (Benzene derivatives)

The aromatics are unsaturated but stable ring compounds. The general chemical formula for this group is C_nH_{2n-6}. Benzene (C_6H_6) is the most characteristic member of the group, and all other

Benzene (C_6H_6)

Tolune ($C_6H_5CH_3$)

aromatics consist of some variation of the benzene ring. Various aromatic compounds are formed by replacing one or more of the hydrogen atoms of the benzene molecules with an organic radical. By adding a methyl group (CH_3), tolune is formed.

This group has very desirable combustion characteristics for use in SI engines, and the members of this group are often added to gasoline to produce high octane fuels.

8.4.2 Refining Process of Petroleum

Crude oil coming out from the oil wells is a liquid consisting mainly of hydrocarbons with traces of sulphur, nitrogen, oxygen and a few impurities such as water and sediment. Crude petroleum is rarely used as a fuel. It is refined so as to produce desirable commercial products. Crude petroleum has different hydrocarbon constituents that have different boiling points. The boiling points of the various constituents increase more or less regularly with the increase in molar mass. This fact is used in the refining process to separate crude oil into more desirable products by fractional distillation.

Fractional distillation

The distillation of petroleum is carried out in a tubular furnace with a tall steel fractionating column. The crude oil is pumped continuously through a heated pipe and flashed into the fractionating column. The vapours of the oil as they rise up the fractionating column become cooler and condense on the shelves at various heights. The heavier compounds have a high boiling point and, as the vapours rise in the tower and are cooled, the heavy fractions precipitate first and are removed from the tower. The light hydrocarbons rise to the top of the column before condensing, while the compounds with intermediate boiling points are removed at the intermediate stages. Outlets are provided in the side of the column at suitable heights for withdrawal of several fractions.

Figure 8.2 shows the primary fractions, obtained from a fractionating tower, and the order in which they occur. From top to bottom of the fractionating tower, the fractions are gases—

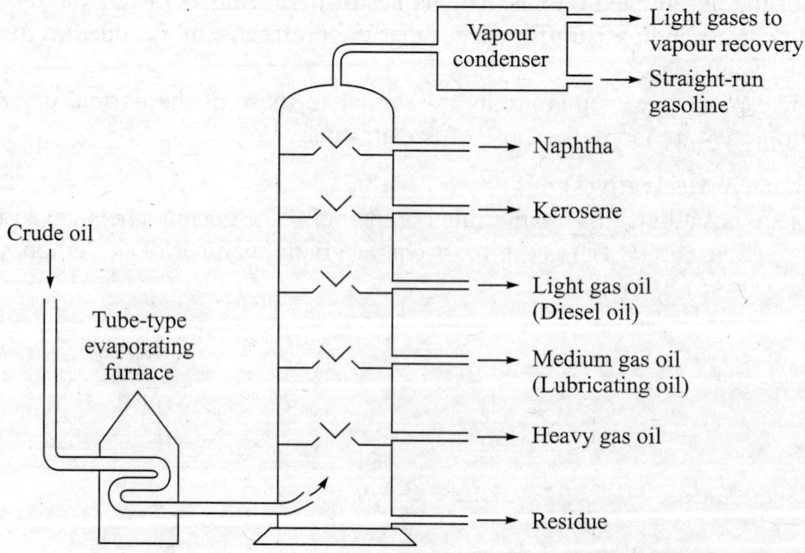

Figure 8.2 Fractions obtained from the fractional distillation process.

principally butane, propane and ethane. Next come the gasolines, naphthas, kerosene, diesel oil, lubricating oils, heavy gas oil, in that order. After distillation, a residue of paraffin wax or asphalt is left, depending on the base of the oil, i.e. whether it is paraffin or naphthene.

The demand ratio of certain components, such as gasoline to other petroleum components, far exceeds the naturally occurring ratio of gasoline found in crude oil. This has led to the development of refinery processes to 'crack' the larger molecules into smaller molecules and 'polymerize' or 'alkylate' small molecules into larger molecules to get the boiling points of the resulting components in the gasoline range. Some of such processes are briefly described here.

Cracking: It is the process of breaking down the large and complex molecules into lighter and simpler compounds. Cracking reactions may be thermal or catalytic. In thermal cracking the heavy hydrocarbons are subjected to high temperature and pressure. At high temperature, the kinetic energy of the molecules increases and as a result they move faster and strike harder. Thus, some of the chemical bonds holding the carbon atoms together break away and the molecules split into lighter and smaller compounds. In catalytic cracking, somewhat lower pressure and temperature are required and the smaller molecules are found to be isomeric rather than normal hydrocarbons. This is desirable because the isomeric compounds have better antiknock characteristics.

An example of a cracking reaction is the breaking off the chain-like structure of paraffin, n-tetradecane into two smaller constituents, one being a paraffin, C_7H_{16}, and the other an olefin, C_7H_{14}, i.e.

$$C_{14}H_{30} \rightarrow C_7H_{16} + C_7H_{14} \tag{8.1}$$

Hydrogenation: It may be described as a cracking process which occurs in an atmosphere of hydrogen. Here, certain unsaturated compounds pick up more hydrogen to become saturated, and the yield and quality of gasoline are increased. This process gives a high-octane product with a high boiling point and high flash point. It is especially suited for aviation gasoline. The hydrogenation process is also used to produce liquid fuel from coal and other similar materials.

Absorption: The gases leaving the refinery processes may contain some heavier hydrocarbons in the vapour state, which may fall into the gasoline range. The absorption process is used to recover these vapours. In this process the gases come into contact with a kerosene or light oil which absorbs the heavy hydrocarbon vapours. The vapours are then driven out from the oil by heating. Finally, the vapours are cooled to obtain gasoline.

Polymerization: It is the process of combining the unsaturated products of one family (two or more olefinic molecules) to form heavier and stable compounds that have a high antiknock rating. A typical chemical reaction is given below:

$$\underset{\text{(Two moles of isobutylene)}}{(CH_3)_2C{=}CH_2 + (CH_3)_2C{=}CH_2} \longrightarrow \underset{\text{(One mole of di-isobutylene)}}{CH_3{-}C(CH_3)_2{-}CH_2{-}C(CH_3){=}CH_2} \tag{8.2}$$

Alkylation: It is the process of combining light undesirable hydrocarbons of one chemical family with another family to form a larger molecule. An isoparaffin (usually isobutane) combines with an olefin (usually butene or propene) to form a larger isoparaffin (usually isooctane or isoheptane) that has a very high octane number.

Isomerization: It is the process of changing the relative position of atoms within the molecule of a hydrocarbon without changing its molecular formula. It produces isomers of the original hydrocarbons.

Cyclization: It is a process of joining together the ends of a straight chain molecule to form a ring compound of the naphthene family.

Aromatization: It is a process of joining together the ends of a straight chain molecule to form an aromatic compound.

Reforming: It is used to convert low antiknock quality gasoline into high antiknock quality. It does not increase the total gasoline volume as in polymerization and alkylation.

Finished blended products: It is a process of mixing certain products to obtain a commercial product of desired quality. The products from the various refinery processes are blended together to form liquid fuels with the proper physical characteristics to give high quality gasoline, fuel oil, kerosene, diesel fuel, etc. These liquid products are then put through the various finishing processes to reduce the sulphur and wax contents.

8.4.3 Petroleum-based Liquid Fuels

Some petroleum-based liquid fuels of importance are discussed below:

Gasoline

It consists of a mixture of liquid hydrocarbons having four to ten carbon atoms, small amounts of lighter and heavier hydrocarbons, minute quantities of crude petroleum impurities such as sulphur and nitrogen. Some additives in very low percentages are used to impart performance benefits. Gasoline is the lightest petroleum fraction in the liquid form. The boiling range of gasoline lies between 30°C and 200°C. The specific gravity lies between 0.70 and 0.78. The chemical composition of its constituents varies widely, depending on the base crude and the process of refining. The heating value of a typical gasoline is 44,000 kJ/kg. It is widely used in spark-ignition engines.

Kerosene

It is the next fraction heavier than gasoline. It is widely used in lamps, heaters, stoves and similar appliances. It may also be used in CI engines and gas turbines. The specific gravity of kerosene lies between 0.78 and 0.85.

Diesel oils

These are petroleum fractions heavier than kerosene. These oils cover a wide range of specific gravity and a very wide distillation range. Their composition is controlled to make them suitable for use in various types of CI engines. The heating value of typical diesel oil is 42,000 kJ/kg.

Fuel oils

The range of specific gravity and the distillation range are similar to those of diesel oils. Their composition does not require accurate control as is required in the case of diesel oils. These fuels are used in continuous burners.

Lubricating oils

These are made up in part from heavy distillation of petroleum and in part from residual oils that remain after distillation. These are used for lubricating purposes.

8.5 LIQUID ALTERNATIVE FUELS

With increasing demand and escalating prices of the conventional petroleum fuels such as gasoline and diesel, it becomes a necessity to find alternative fuels for the internal combustion engines. Also, currently the world is facing the problem of fast depleting natural oil resources. Fuels derived from renewable sources are called alternative fuels. Various liquid fuels are used as substitutes for petroleum-based fuels.

8.5.1 Benzol

It is a distillate of coal tar. It contains about 70% benzene (C_6H_6), 20% tolune (C_7H_8), 10% xylene (C_8H_{10}) and traces of sulphur compounds. It has the high antiknock characteristic. It has a freezing point of 5.6°C and therefore it is not a suitable fuel to be used in a cold climate. Because of its high antiknock characteristic, it can be blended with gasoline to be used as a fuel in SI engines.

8.5.2 Alcohol

Alcohol is of organic origin and can be produced from a wide range of abundantly available raw materials. Ethanol (C_2H_5OH) can be produced by fermentation of carbohydrates which occur naturally and abundantly in some plants like sugarcane and can also be produced from starchy materials like corn, potatoes, maize and barley. The starchy material is first converted into sugar which is then fermented by yeast.

For the large-scale production of methanol (CH_3OH), the following methods are commonly employed: (a) destructive distillation of wood, (b) synthesis from water gas, (c) from natural gas but it is petroleum based, (d) from coal, a relatively abundant fossil fuel.

Alcohols have high antiknock characteristics which permit spark-ignition engines to run at higher compression ratios. A lean mixture will burn and the exhaust gas temperature will be lower. Alcohols, therefore, will reduce CO and NO_x in the exhaust. The alcohol fuelled SI engines can produce a slightly higher power output.

Methanol is considered to be one of the most likely alternative automotive fuels. However, several major technical difficulties must be resolved before 100% methanol can become a commercially acceptable fuel for use in vehicles. The most commonly mentioned difficulties are cold start (no start below 15°C), safety (explosive mixture in the fuel tanks, invisible flame), and corrosion and wear of engine and fuel system materials. In addition, the vehicle range (distance covered) will also be reduced substantially, unless the size of the fuel tank is greatly increased, because the volumetric energy density of methanol is only about one-half of that of gasoline. Most of these problems may be resolved by using a medium concentration (30–70% by volume) blend of methanol and gasoline. However, the use of such blends may compromise some of methanol's key technical advantages, namely increased engine efficiency and power, and decreased NO_x emissions.

The physical properties of methanol, ethanol and gasoline are compared in Table 8.1.

Table 8.1 Important properties of methanol, ethanol and gasoline

Property	*Methanol*	*Ethanol*	*Gasoline*
Chemical formula	CH_3OH	C_2H_5OH	C_mH_n
Molecular weight	32.04	46.06	107
Composition by weight (per cent)			
Carbon	37.5	52.2	84.0
Hydrogen	12.5	13.0	16.00
Oxygen	50.0	34.8	NIL
Specfic gravity	0.792	0.785	0.7 – 0.78
Boiling point (°C)	65.0	77.8	38 – 205
Latent heat of vaporiztion (kJ/kg)	1168	921	290 – 420
Vapour pressure at 311 K (bar)	0.31	0.21	0.48 – 1.03
Lower calorific value (MJ/kg)	20.1	27.0	44.0
Stoichiometric air/fuel ratio	6.4	9.0	14.7
Ignition limit, air/fuel ratio	3.15 – 12.8	3.5 – 17.0	7.0 – 22.0
Self-ignition temperature (°C)	574	537	300 – 450
Octane number:			
(a) Research	114	111	91
(b) Motor	94	94	82
Cetane number	3	8	8.14

8.5.3 Biodiesel

Biodiesel is made from animal fats or vegetable oils, which are renewable resources derived from plants such as soybean, sunflower, corn, olive, peanut, palm, coconut, sesame and cottonseed. Biodiesel is typically prepared by chemically reacting vegetable oil or animal fat with an alcohol producing fatty acid esters.

Biodiesel is a liquid which varies in colour between golden and dark brown depending on the production feedstock. It is immiscible with water and has a high boiling point and low vapour pressure. It has a density of 0.88 g/cm^3, which is higher than diesel (0.85 g/cm^3). The calorific value of biodiesel is about 37.27 MJ/kg. This is 9% lower than regular diesel. Variations in biodiesel energy density depend on the feedstock used.

Biodiesel has better lubricating properties and much higher cetane ratings than today's lower sulphur diesel fuels. Biodiesel reduces fuel system wear. It gives complete combustion, thus increasing the engine energy output and partially compensating for the higher energy density of diesel.

A variety of oils can be used to produce biodiesel. These include the following:

1. Virgin oil feedstock—rapeseed and soybean oils are most commonly used. It can also be obtained from jatropha and other crops, such as mustard, jojoba, flax, sunflower, palm oil and coconut.

2. Waste vegetable oil
3. Animal fats including tallow, lard and the by-products resulting from the production of omega-3 fatty acids from fish oil
4. Algae which can be grown by using waste materials such as sewage and without displacing land currently used for food production.

8.5.4 Emulsified Diesel Fuel

Emulsified fuels are composed of water, petroleum and additives that bind water and petroleum together. Emulsified fuels are manufactured from base fuels such as diesel, heavy oil or biodiesel and water through the chemical bonding of additives and then passing all three fluids through a blending unit.

The water content of emulsion turns to steam during combustion, atomising the oil into smaller droplets. The contact surface between fuel and air is increased. As a result, combustion is more efficient. This improved process reduces soot residue and particulate matters, while the presence of water lowers the combustion temperature and oxides of nitrogen (NO_x).

The addition of water in emulsified diesel reduces the energy content of the fuel, and so some reduction in power and fuel economy can be expected.

8.5.5 Acetone

Acetone (CH_3COCH_3) can be used as an additive for boosting octane levels in gasoline. It can be added to the fuel tank in small amounts. It improves fuel's ability to vaporize completely by reducing the surface tension that inhibits vaporization of some fuel droplets. Acetone aids in the vaporization of the gasoline or diesel, increasing fuel efficiency and performance, as well as reducing hydrocarbons emissions. However, a high amount of acetone in fuel can affect the rubber seals of vehicle engines.

8.5.6 Diethyl Ether

Diethyl ether is an organic compound in the ether class with the formula $(C_2H_5)_2O$. It is a colourless, highly volatile and flammable liquid with a characteristic odour. It has a high cetane number of 85–96 and is used as a starting fluid for diesel and gasoline engines because of its high volatility and low auto-ignition temperature.

8.5.7 Vegetable Oils

Vegetable oil can be used as diesel fuel without being converted to biodiesel. The downside is that straight vegetable oil (SVO) is much more viscous than conventional diesel fuel or biodiesel, and it does not burn the same in the engine. It can damage the engine. Preheating the oil is not enough to ensure that it will combust properly inside the engine. It needs a complete system, including specially made injector nozzles and glow plugs optimised for vegetable oils.

The older IDI diesel engines are generally more suitable for SVO use. Mechanical injection is better for SVO use than computerized injection. Inline pump is also suitable. Rotary pumps should not be used with SVO system. SVO systems containing copper parts should be avoided. Copper will catalyse the oil, causing it to decompose.

8.6 GASEOUS ALTERNATIVE FUELS

Gaseous fuels present no difficulty regarding mixing with air, distributing homogeneous mixture to various cylinders and cold starting. However, gaseous fuels create problems of storage and handling large volumes especially in automobiles. Consequently, gaseous fuels find applications in stationary power plants located near an abundant supply of the fuel. Some gaseous fuels can be liquefied under pressure to increase the density and thus reduce the storage tank volume. Use of natural gas and liquefied petroleum gas (LPG) is increasing and attempts to use hydrogen as a fuel in IC engines are in progress.

8.6.1 Natural Gas

Natural gas is a mixture of several different gases. The primary constituent is methane, which typically makes up 85–99 % of the total volume. The other constituents include other hydrocarbons, inert gases such as nitrogen, helium and carbon dioxide, and traces of hydrogen sulphide and water. The non-methane hydrocarbons present in natural gas consist primarily of ethane. The remainder is made up mostly of propane and butane, with some traces of C_5 and higher species.

Natural gas is an excellent fuel for SI engines. As a gas under normal conditions, it mixes readily with air in any proportion. Unlike liquid fuels it does not need to vaporize before burning. Thus cold engine starting is easier especially at low temperatures, and cold-start enrichment is not required. Cold-start enrichment is a major source of CO emissions and emissions-related problems in gasoline-fuelled SI engines.

Natural gas has a high ignition temperature, and is resistant to self-ignition. It has excellent antiknock properties. Pure methane has an equivalent research octane number (RON) of 130, the highest of any commonly used fuel. Because of its antiknock properties, natural gas can safely be used with engine compression ratios as high as 15:1 (compared to 8–10 : 1 for 91 octane gasoline). Natural gas engines using these higher compression ratios can reach significantly higher efficiencies than are possible with gasoline.

8.6.2 Liquefied Petroleum Gas (LPG)

Liquefied Petroleum Gas (LPG) is a product of petroleum gases, principally propane (C_3H_8), propylene (C_3H_6) and butane (C_4H_{10}). These gases can be liquefied at normal temperatures by subjecting them to a moderate pressure. Owing to the demand from industry for butane derivatives, LPG sold as a fuel is made up largely of propane. Liquefied petroleum gases are used as fuels for stoves, trucks, buses and tractors in many parts of the world. The LPG has higher heating value compared to gasoline. Since propane and butane are heavier than air, the escaping gas will tend to settle and collect in pockets thus creating an explosion hazard. The LPG is suitable for IC engines because of its availability and low carbon content, thus resulting in drastic reduction in exhaust emissions. The LPG has a high self-ignition temperature and a high octane number, which makes it more suitable for SI engines.

Engines with natural gas and LPG can run lean because of their better distribution and higher misfire limits. Also, their higher octane numbers, allow an increase in the compression ratio in SI engines, which consequently improves the thermal efficiency and reduces the exhaust emissions.

8.6.3 Producer Gas

It is made by burning carbonaceous material (coal, wood, charcoal, coke, etc.) with a large deficiency of air and treating with steam. The products of this partial combustion contain CO and H_2 in sufficient quantities, so that they can be used in an engine as a fuel. The producer gas has a high percentage of N_2, since air is used. Thus, it has a low heat value.

8.6.4 Coal Gas

Coal is heated to temperatures up to 1500°C in the presence of very little air. The complex organic compounds of the coal decompose at the high temperature and form simpler, volatile products and coke.

8.6.5 Hydrogen

With the imposition of stringent emission standards along with the decreasing availability of petroleum products, it is imperative that a search for low polluting alternative fuels be made. Hydrogen, as an SI engine fuel acquires special significance in view of its unlimited supply potential and almost non-polluting characteristics. Even though with current economics hydrogen would be a costly automotive fuel, based on long-term considerations, its relative cost standing may improve considerably. As and when a cheaper method of hydrogen production becomes available, it could be used for aircraft, marine vessels, railways and automotive vehicles.

Hydrogen has emerged as a potential fuel for internal combustion engines. It is generally considered to be non-polluting because hydrogen contains no carbon. Species such as carbon monoxide and unburned hydrocarbons, which are normally found in gasoline fuelled engines, would be virtually eliminated in the exhaust. Hydrogen is found in abundant quantities in various forms and can be considered to be an almost inexhaustible fuel. It can be adapted as a fuel to engines without major design changes.

The problems generally experienced in a hydrogen-fuelled engine are the backfiring, preignition, knocking and rapid rate of pressure rise during the combustion process because of the higher flame speed. Backfiring is mainly due to less ignition energy for a hydrogen-air mixture. Localized hot points in the chamber and the temperature of the residual gas are sometimes sufficient to cause backfiring.

Hydrogen-fuelled engines can be run at a much leaner equivalence ratio than a gasoline-fuelled engine, although lean operations of hydrogen-fuelled engine generally reduce NO_x emissions and show improved thermal efficiency. Problems associated with too lean mixtures increase the ignition delay and cause severe cyclic variations. The hydrogen peroxide is present in the exhaust products of a hydrogen-fuelled engine operating with very lean mixtures.

Compared with hydrocarbon fuels, hydrogen has certain advantages due to its high chemical reactivity: (a) a higher flame propagation speed, (b) wider ignition limits, and (c) lower ignition energy. The high flame propagation speed of hydrogen certainly benefits the thermodynamic efficiency of the engine as the combustion process is closer to the optimum theoretical combustion at constant volume. The wider ignition limits provide the possibility to run with extremely lean mixtures. A lower ignition energy is favourable for ignition of lean mixtures in SI engines, but has the disadvantage of resulting in abnormal combustion (knock and surface ignition), especially near stoichiometry.

Because backfire must be avoided at all costs, it is necessary to avoid knock and surface ignition. This can be done by leaning the mixture. Since this reduces the available power, its use should be limited. Knocking and the rate of pressure rise can also be controlled by increasing the flame travel distance. This can be done by locating the spark plug near the periphery, away from the centre of the cylinder head.

On-board storage of hydrogen remains a major technical challenge. As a gas, hydrogen has a very low energy density. This leads to a large tank size even with high pressure storage and short vehicle range. Storing hydrogen in liquid form is also problematic as it liquifies at –235°C. Storage of hydrogen may be achieved by solid state hydride storage materials, liquid hydrides, microglass sphere storage, storage in carbon, storage in zeolites, and similar such storage.

The properties of hydrogen, natural gas (methane) and gasoline are compared in Table 8.2.

Table 8.2 Properties of hydrogen, methane and gasoline

Property	*Hydrogen*	*Methane*	*Gasoline*
Molecular weight	2.016	16.043	107.0
Heat of combustion (low), MJ/kg	119.93	50.02	44.0
Specific heat, c_p at NTP, kJ/(kg K)	14.89	2.22	1.62
Viscosity at NTP, g cm^{-1} s^{-1}	0.0000875	0.00011	0.000052
Specific heat ratio at NTP	1.383	1.308	1.05
Gas constant R, kJ/(kg K)	4.124	0.518	0.0777
Diffusion coefficient in air, cm^2/s	0.61	0.16	0.005
Limits of flammability in air, vol %	4.0 to 75.0	5.3 to 15.0	1.0 to 7.6
Stoichiometric composition in air, vol %	29.53	9.48	1.76
Minimum energy for ignition in air, mJ	0.02	0.29	0.24
Autoignition temperature, K	858	813	573-723
Flame temperature in air, K	2318	2148	2470
Burning velocity in air at NTP, cm/s	265 to 325	37 to 45	33 to 47

* If CNG (compressed natural gas) has a significant amount of non-methane components, the burning velocity will be lower than that for pure methane. None of the other properties are significantly dependent on the exact formulation of the natural gas. LNG (liquefied natural gas) is almost exclusively methane.

8.6.6 Hythane

Hythane is a mixture of natural gas and hydrogen, usually 5–7% hydrogen by energy. Hydrogen and methane are complementary vehicle fuels in many ways. Methane has a relatively narrow flammability range that limits both the fuel efficiency and the reduction of oxides of nitrogen (NO_x) that are possible at lean air/fuel ratios. The addition of even a small amount of hydrogen, however, increases the lean flammability range significantly. Methane has a slow flame speed, especially in lean air/fuel mixtures, while hydrogen has a flame speed about eight times faster. Methane is a fairly stable molecule that can be difficult to ignite, but hydrogen has an ignition energy requirement about 25 times lower than that of methane. Hydrogen is a powerful combustion stimulant for accelerating the methane combustion within an engine. A large number of buses in India will be converted to run on hythane in near future.

8.6.7 Dimethyl Ether

Dimethyl ether (DME), which has the chemical formula CH_3OCH_3, is the simplest ether compound and is known to be both non-toxic and environmentally benign. DME can be prepared from a wide variety of resources, including coal, methane, natural gas and biomass.

DME has high vapour pressure and low boiling temperature, and it is a gas fuel at room temperature and atmospheric pressure. It is regarded as an ultra-clean, renewable and low-carbon fuel. The short carbon chain compound with its high oxygen content leads to very low emissions of particulate matter, oxides of nitrogen (NO_x) and carbon monoxide (CO) during combustion. It promises a smoke-free combustion. For these features and being sulfur-free, DME meets the most stringent emission regulations.

DME is a promising fuel in CI engines, SI engines (30% DME/70% LPG) and gas turbines owing to its high cetane number, which is 55, compared to diesel, which has the cetane number 40–53. The heat value of DME is significantly lower than that of the conventional diesel fuel. Therefore, the fuel supply and ignition system and the combustion system of the engine should be redesigned or modified.

8.6.8 Biogas

Biogas is a renewable energy fuel. It refers to a gas produced by the biological breakdown of organic matter in the absence of oxygen. Organic waste such as dead plant and animal materials, animal faeces and kitchen waste can be converted into a gaseous fuel called biogas.

Biogas is produced by the anaerobic digestion or fermentation of biodegradable materials, such as biomass, manure, sewage, municipal waste, plant materials and crops.

India is the largest cattle breeding country. Abundant raw materials are available for producing biogas. Cow dung and municipal sewage can be used for this purpose. Biogas consists of approximately 50–65% methane, 25–35% carbon dioxide and traces of hydrogen, nitrogen, oxygen and hydrogen sulphide.

Carbon dioxide does not help in combustion process but it reduces the calorific value of biogas. The calorific value of biogas is around 35 MJ/m^3. The quality of biogas is not high enough to be used as a fuel gas in engines. The corrosive nature of H_2S alone is enough to destroy the internal parts. This problem is overcome by biogas upgrading or purification process whereby contaminants in the raw biogas stream are absorbed or scrubbed, leaving more methane per unit volume of gas.

The air to biogas stoichiometric ratio by volume for complete combustion is from 9.5:1 to 10:1. Biogas has very slow flame velocity, only 0.29 m/s at its highest. Biogas has higher octane number compared to gasoline and alcohol. This means a higher compression ratio can be used with biogas. It is highly knock resistant and reduces pollution.

8.7 CHARACTERISTICS OF SI ENGINE FUELS

The performance of the spark-ignition engine depends upon the following fuel characteristics, which are of importance: (a) volatility, (b) sulphur contents, (c) gum deposits, (d) carburettor detergent additives, and (e) antiknock quality.

8.7.1 Volatility of Liquid Fuels

Volatility is the tendency of a liquid to evaporate at given conditions. Petroleum fuels consist of a large number of different hydrocarbons, each having a different boiling point. Hence, upon heating a gasoline of fuel oil, it is observed that all of the liquid does not pass to the vapour phase at one temperature, as is the case when water is heated at constant pressure. Instead, a small percentage of the fuel will vaporize at a low temperature, and a much higher temperature must be reached before the fuel completely vaporizes. Volatility characteristics of fuels can be obtained by the following tests.

ASTM distillation test

The relative volatility of fuels is ordinarily measured by means of the ASTM (American Society for Testing Materials) distillation test. The apparatus used is shown in Figure 8.3. Heat is supplied to the flask, which contains 100 ml of the fuel. The vapours pass through the condenser tube, which is surrounded by cracked ice, and the condensate drips into the graduated receiver where the quantity of condensed fuel is measured.

A reading on the thermometer is taken at the time the first drop of the condensate appears. This temperature is known as the *initial boiling point* of the fuel. Subsequent temperature readings are taken as each additional 10 % of the fuel is evaporated. The final temperature required to evaporate the liquid completely is called the *end point temperature*.

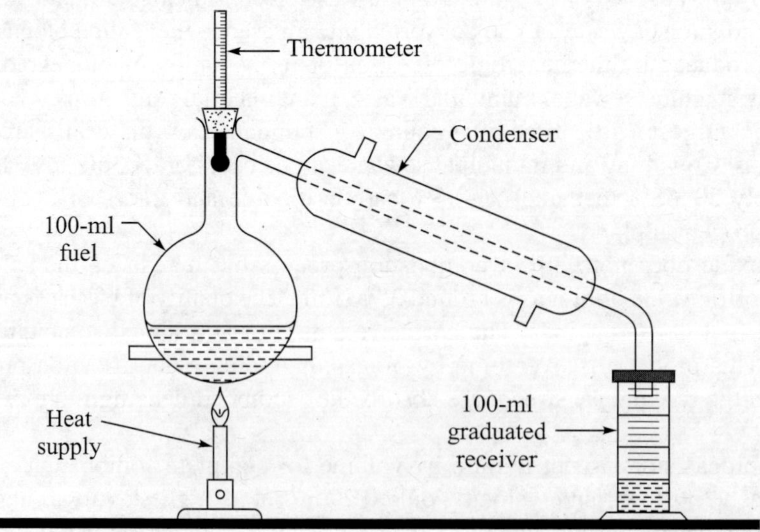

Figure 8.3 ASTM distillation test apparatus

However, it is found that not all of the fuel is really driven from the flask, even though the flask appears to be empty, and a residue will condense upon cooling. It is also found that a portion of the fuel has unavoidably escaped. Since the sum of the condensed portion and the residue will not total the original amount, the portion of the fuel that has unavoidably escaped is called *loss*. It is arbitrarily assumed that the loss represents the most volatile part of the fuel and therefore occurs at the very start of the distillation. Typical results of the ASTM distillation tests of several products of petroleum refining are given in Figure 8.4.

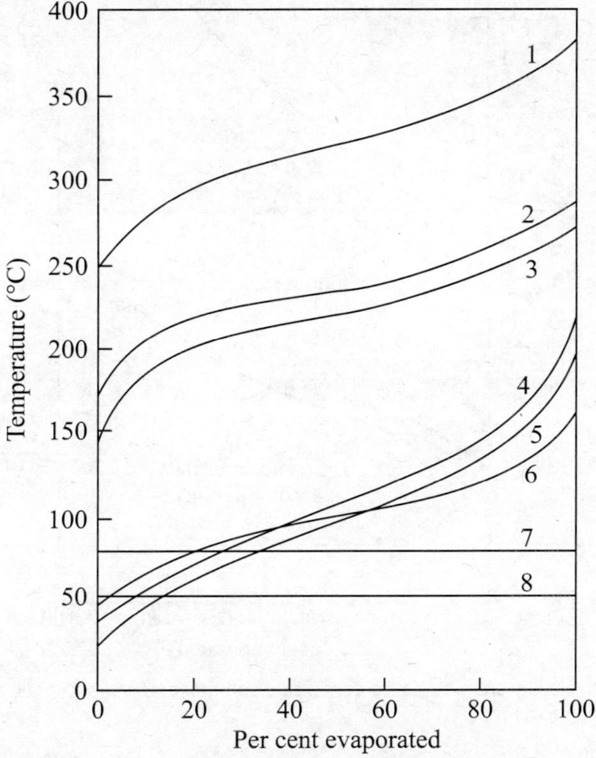

Figure 8.4 Typical ASTM distillation curves: 1, heavy diesel oil; 2, light diesel oil; 3, kerosene; 4, summer gasoline; 5, winter gasoline; 6, aviation gasoline; 7, ethyl alcohol; 8, benzene.

Equilibrium air distillation (EAD) test

In the ASTM distillation test the liquid fuel evaporates in the presence of fuel vapour. In IC engines, the fuel is evaporated in the pressure of air, and a greater percentage of the fuel will be evaporated at a given temperature than is indicated by the standard ASTM results. The EAD test may be used to obtain more accurate information concerning the actual volatility characteristics of the fuels in use.

Comparisons of characteristics of average summer and winter premium gasoline, as determined by the ASTM distillation test and the EAD test for 16:1 air/fuel ratios, are made in Figure 8.5. It is easier and quicker to conduct the ASTM test, hence it is performed more often and the EAD temperatures are computed from the ASTM data with the help of suitable charts.

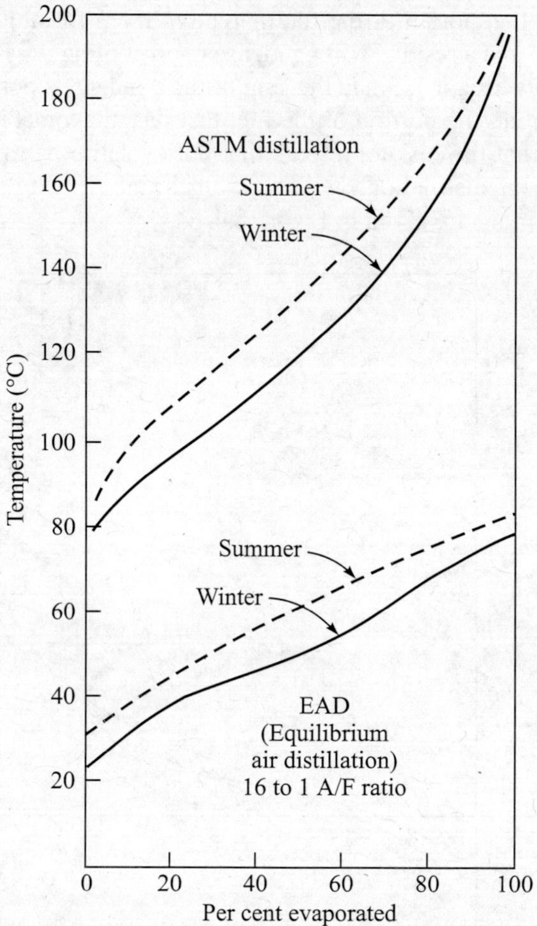

Figure 8.5 Comparison of volatility characteristics of average summer and winter premium gasolines as determined by the ASTM and EAD tests.

Reid vapour pressure bomb

It affords another method for measuring the volatility characteristic of a fuel. Figure 8.6 shows the Reid vapour pressure bomb. In conducting this test, a sample of 100 ml of fuel is sealed in the bomb and the instrument is immersed in a 38°C water bath. A portion of the fuel, which vaporizes, rises into the air chamber above the fuel. The increase in pressure in the air chamber is measured by a pressure gauge. The vapour pressure in the air chamber is reported as the Reid vapour pressure of the fuel.

8.7.2 Effect of Volatility on the Performance of SI Engines

Distillation curves shown in Figure 8.5 can be divided into three portions: (a) Front-end volatility: it covers 0 to 20% of volatility of fuel, (b) mid-range: it covers 20 to 80% of volatility, and (c) tail-end: it covers 80 to 100% of volatility.

Front-end volatility

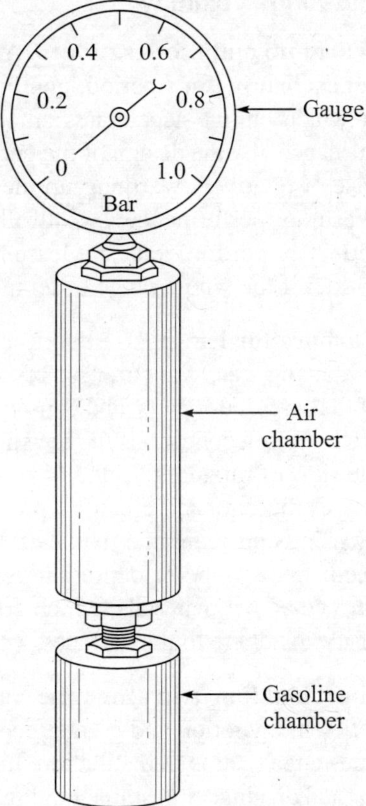

Figure 8.6 Reid vapour-pressure bomb.

The following important performance characteristics are affected by the first part of volatility.

Cold starting: In the winter and in the cold climate, an air-vapour mixture of about 13:1 is required for easy starting (in less than 10 revolutions). This ratio can be obtained in the engine by choking the carburettor and so restricting the inlet flow of air. If the carburettor delivers a 1.3:1 air/fuel ratio and 10% of the fuel vaporizes, the desired 13:1 air/vapour ratio will be obtained. It is therefore desirable to have a relatively low evaporation temperature of the front-end of the fuel, so that almost 10% of the total fuel evaporates as quickly as possible.

Hot starting: If the engine is started immediately after a hot shutdown, the amount of fuel vapours entering the intake manifold will be high, and the mixture formed in the combustion chamber will be too rich to ignite. It creates the problem of hot starting. In order to avoid this hot starting problem especially on hot days, volatility temperature of the first portion of the fuel should be high. This requirement is opposite to the requirement of cold starting. Therefore, it is a normal practice to use less volatile gasoline in summer than that used in winter. The problem of hot starting can be avoided by proper design and proper placement of the fuel system. The fuel system should be placed away from the hot engine parts. The use of an electrical pump helps in placing the pump and the fuel lines away from the hot engine parts and thus in reducing the problem of hot starting.

Vapour lock: Vapour lock is the restriction of the fuel supply to the engine caused by excessively rapid formation of vapour in the fuel supply system or carburettor. The vapour will occupy a greater volume than the liquid and therefore, the amount of fuel flow to the engine will be reduced, causing loss in power or else complete stoppage of the engine. The possibility of vapour lock increases with highly volatile fuels related to the front-end volatility and by the exposure of the fuel to either high temperatures or low pressures in the fuel system.

In order to avoid vapour lock, the fuel lines, the pump and the carburettor should be placed in relatively cool regions and not exposed to radiated heat from the engine and exhaust pipes. These parts should be cooled by the flow of air from the fan.

Evaporation loss: Vaporization and loss of the lighter fractions of gasoline from the fuel tank and the carburettor occur at all times. The evaporation loss depends on the vapour pressure of the fuel at the storage temperature. It decreases the fuel economy and the anti-knock quality of the fuel, since the lighter fractions have higher anti-knock properties. The evaporation loss is related to 10% ASTM distillation temperature. The front-end volatility temperature should be higher to reduce evaporation in order to reduce evaporation loss and vapour lock.

Mid-range volatility

Warm-up and acceleration: After the engine has been started, a warm-up period of the engine begins. During this period, engine temperatures gradually increase to those of normal operation at which engine accelerates smoothly from a given speed. The warm-up period is, of course, influenced by the design of the engine in quickly securing a minimum mixture temperature but the ease of relatively warming-up the given designed engine depends upon the volatility of the fuel. Warm-up performance is controlled to a large extent by the mid-range of the distillation curve (50–70 % portion) and to a lesser extent by the front and tail ends. The warm-up period will be shorter if the whole range of temperatures on the ASTM curve is lower.

Carburettor icing: Carburettor icing is formed due to the vaporization of gasoline into the air containing water vapour. This results in a rapid drop in temperature of the air-fuel mixture and that of the carburettor parts, and consequently, under some conditions, ice is formed on the throttle blade. Under idling conditions the ice slides down the throttle blade and restricts the passage, preventing the flow of mixture past the throttle, thereby causing the engine to run slower and to stall.

Carburettor icing can be prevented by the use of less volatile fuels. It can also be reduced by using anti-icing additives with volatile gasolines. Two types of anti-icing additives have been used: freezing-point depressants, such as isopropyl alcohol or methyl alcohol (1 to 2 %), and surface-active materials, which coat the metal surface with a film, thus minimizing the tendency of ice to adhere to the surfaces.

Short and long trip economy: In short-trip driving the warm-up period is quite significant. For efficient operation and greater economy, it requires a fuel having relatively more volatility in the mid-range section of distillation. In long-trip driving the warm-up period is insignificant compared to total driving. A gasoline having higher density will give more kilometres per litre in a warm-up engine.

Tail-end volatility

Crankcase dilution: Liquid gasoline in the cylinder is undesirable, since it washes away the lubricating oil from the cylinder walls. It reduces lubrication and tends to increase the friction between the piston rings and the cylinder, thus causing damage to the engine. The degree of crankcase oil dilution is directly related to the tail-end volatility temperatures of the mixture. The 90 % temperatures of the ASTM and EAD distillations evaluate the dilution tendency of the fuel; the lower the temperature of the 90% point (the more volatile the tail-end portion of fuel), the less will be the dilution of the crankcase oil, assuming that almost all of the gasoline in the cylinder will be vaporized.

Engines using heavy fuels, such as kerosene, being less volatile, may suffer from poor lubrication because of excessive dilution.

Sludge deposits: Certain types of hydrocarbons present in gasoline may not evaporate even during the tail-end evaporation, due to very high evaporation temperature and may leave sludge deposits inside the engine. These deposits may cause sticking of piston rings and valves, thus resulting in poor operation. It may also cause spark-plug fouling. The lower the tail-end volatility temperature, the less will be the chance of sludge deposits.

8.7.3 Sulphur Content

High sulphur content in gasoline in the form of free sulphur, hydrogen sulphide and other sulphur compounds is undesirable because of the formation of SO_3 whose combination with water vapour forms H_2SO_4, which is a very corrosive substance that may attack various parts of the engine, thus affecting engine performance and life. Since sulphur has a low ignition temperature, the presence of sulphur can reduce the self-ignition temperature of the fuel, and thus promote knock in SI engines. Consequently, the gasoline specifications limit the permissible quantity of sulphur which may be present. Sulphur contents less than 0.1% are demanded for gasolines used in SI engines.

8.7.4 Gum Deposits

Reactive hydrocarbons and impurities in the fuel have a tendency to oxidize and form viscous liquids and solids called *gum*. It deteriorates the gasoline during the long period of storage at high ambient temperatures. The pure stable hydrocarbons of the paraffin, naphthene, and aromatic families form little gum, while cracked gasolines form considerable amount of gum. A gasoline with high gum content will cause operating difficulties, such as sticking valves and piston rings, carbon deposits in the engine, gum deposits in the manifold, clogging of carburettor jets and lacquering (varnish appearing residue) of the valve stems, the cylinders and pistons. Sticking of the inlet valve and formation of gum deposits in the intake manifold reduce volumetric efficiency greatly.

The amount of gum increases with increased concentrations of oxygen, with the rise in temperature, with exposure to sunlight and also on contact with metals. In storing fuels, these factors should be kept in mind.

Inhibitors of gum deposits are almost invariably added to thermally cracked gasoline in order to ensure stability. Certain dyes can be added to colour the gasoline, and also to inhibit the formation of gum. Such inhibitors have preference for oxidization over gasoline and the activity of the inhibitor decreases. This fades the colour of the gasoline. Thus, the loss of colour of the gasoline may be an indication of the age or exposure of the fuel to gum-forming conditions. Gasoline specifications therefore limit both the gum content of the fuel and its tendency to form gum during storage.

8.7.5 Carburettor Detergent Additives

The intake manifold and carburettor deposits may result from airborne contaminants, from gums content in the gasoline, from incomplete combustion products and crankcase vapours. These deposits restrict the flow of charge past the throttle plate, especially in the idle position, thus causing rough idling and starting.

Many gasolines now contain detergent additives to prevent the formation of these deposits and to remove the existing deposits. Alphanaphthol is used as antioxidant additive to control the oxidation of fuels.

8.7.6 Anti-knock Quality

Detonation in SI engine causes a very rapid and uncontrolled burning of the fuel and air mixture in a cylinder, and this results in an abnormally rapid pressure rise. This sets up vibrations of the gases, the cylinder walls, and other metallic surfaces giving a distinct knock or noise. Therefore, the characteristics of the fuel should be such that the knocking tendency is resisted, and this property of fuel is called the anti-knock quality. The anti-knock quality of a fuel depends on the self-ignition temperature of the fuel and the chain reaction mechanism by which the fuel burns. A liquid such as iso-octane exhibits very smooth burning characteristics. In contrast with this, the longer chain compound, normal-heptane displays a very strong tendency towards detonation. In general, the best SI engine fuel will be that which has the highest anti-knock property, since this permits the use of higher compression ratios, resulting in higher thermal efficiency and power output.

8.8 CHARACTERISTICS OF CI ENGINE FUELS

Most CI engine fuels are obtained in the fractions of crude petroleum near kerosene and gas oil. These fuels are heavier and more viscous than the gasoline used in the SI engine. Some of the important characteristics of CI engine fuels are: (a) ignition quality, (b) volatility (c) viscosity, (d) gravity, (e) corrosion and wear, (f) handling ease, (g) safety, and (h) cleanliness.

8.8.1 Ignition Quality

The ignition quality of a fuel is one of the most important characteristics. It is a measure of ability of a fuel to ignite promptly after injection, thus ensuring a progressive smooth burning and easy starting. The ignition quality is measured in terms of the delay period, which is the lapse of time between the beginning of injection and the ignition of the fuel showing the appearance of an appreciable pressure rise. The ignition quality is better with a shorter delay period. In larger, low-speed engines the delay is not so noticeable and the usual fuels give a satisfactory performance. In smaller, high-speed engines, a short delay is very important.

A fuel with a lower self-ignition temperature will ignite more quickly when injected into the combustion chamber than the one with a higher self-ignition temperature. The desired chemical structure for CI engine fuels is opposite to that desirable for SI engines. The best fuels for the CI engine are straight-chain paraffins with average molecular weights greater than those of the gasolines. The rating of diesel fuels is given by the cetane number. A fuel with a higher cetane number gives better ignition quality in CI engines. The ignition quality of CI engine fuels can be improved by certain additives. The ignition quality of a CI engine fuel has a marked influence on cold starting, engine roughness, and compression ratio.

Cold starting

Diesel fuels are less volatile and more viscous than gasoline. Because of these properties the formation of a combustible mixture during cold starting becomes difficult. In order that the fuel should start the cold engine easily, a high cetane rating of the fuel is required. It reduces the self-ignition temperature.

Volatile fuels, such as ether, have been found to reduce the delay period and cold starting becomes easier when a small amount of it is introduced with the intake air.

Engine roughness

The intensity of vibration of various engine parts is the measure of engine roughness, which is caused by high rates of pressure rise in the combustion chamber. As the cetane number increases, the maximum rate of pressure rise decreases, resulting in reduced engines roughness. A high cetane number is usually less important to the divided combustion chamber type of CI engine than to the open chamber type. The cetane number is also less important to low-speed large cylinders, since only a small fraction of the fuel is burned during the rapid pressure rise period compared to that in the high-speed small cylinders.

Compression ratio

The compression ratio is normally kept low in order to avoid excessively high cylinder pressure, which will facilitate easy starting. With the increase in cetane number of the fuel, the ignition quality improves and the compression ratio can be lowered without facing the knocking problem in CI engines.

8.8.2 Volatility

Petroleum fuels used in CI engines cover a wide range of volatility. The range of volatility of CI engine fuels lies at lower volatility compared to the range of volatility of SI engine fuels. High volatility fuels are generally not used in CI engines, partly because of the high demand for high volatility fuels for SI engines used in automobiles and partly because of poor ignition quality.

Volatility affects the spray characteristics and may affect both power and efficiency. Increase in volatility increases the rate of evaporation of fuel and hence the rate of mixing of fuel and air. The fuel should be sufficiently volatile in the operating temperature range to produce good mixing and combustion, and thus reduce objectionable smoke and odour in the exhaust gases. The smaller, high-speed engines require rapid evaporation of the fuel and consequently should have fuels with more of the low boiling and less of the high boiling constituents compared with the same constituents required by the larger, low-speed engines. The ASTM distillation curve of a typical diesel fuel is presented in Figure 8.7.

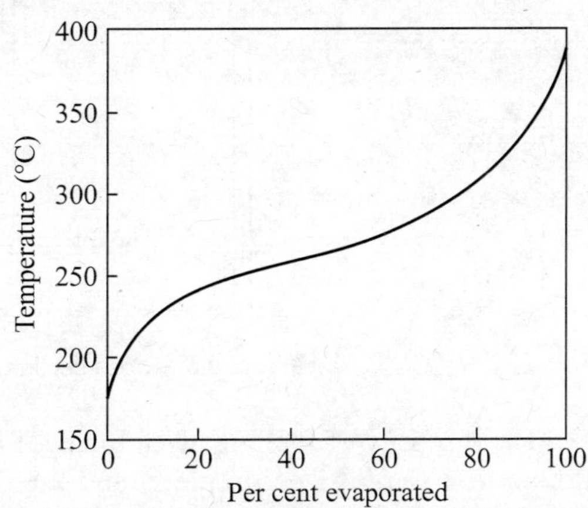

Figure 8.7 ASTM distillation curve of a typical diesel fuel.

8.8.3 Viscosity

Viscosity is defined as the ratio of shearing stress in a fluid to the rate of shear (angular deformation), and is a measure of the resistance of fluid flow. It is an important characteristic, as it affects the atomization of fuel and operation of the high pressure fuel pumps.

The viscosity of oil increases with the increase in the number of carbon atoms in it. If there are same number of carbon atoms in two hydrocarbons, the one with the lower number of hydrogen content will have a higher viscosity. The viscosity of oil decreases rapidly with the increase in temperature.

The time taken for the flow of a given quantity of oil through a standard orifice under the action of gravity is the measure of the viscosity of that oil. The standard test apparatus used to determine the viscosity of oil is the Saybolt Universal (SU) viscosimeter. A cross-section of Saybolt viscosity test apparatus is shown in Figure 8.8. The viscosity is reported as Saybolt Universal Seconds (SUS) at the temperature of the test. The viscosity in SUS may be converted to absolute or kinematic viscosity by tables available for that purpose.

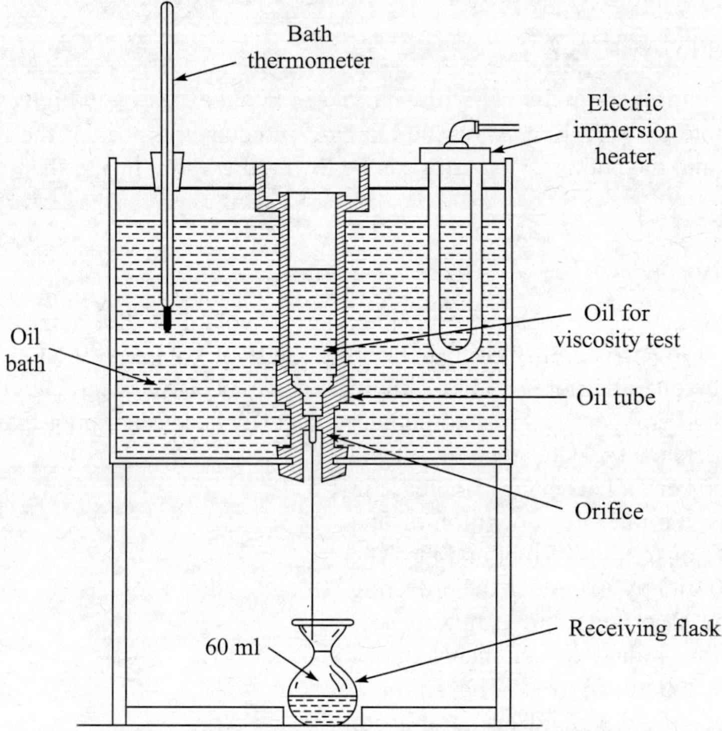

Figure 8.8 Cross-section of Saybolt viscosity test apparatus.

The viscosity of fuel greatly influences the spray characteristics. High viscosity causes low atomization (large-sized droplets) and high penetration of the spray jet. In small combustion chambers, the effect of viscosity is critical, hence the maximum and minimum values suitable for the engine should be specified. In cold engines, a high viscous oil may cause starting problems and also a smoky exhaust may appear. If the viscosity of the fuel is low, leakage past the piston in the pump will be increased. The lubricating qualities of low-viscosity fuels are poor, thus resulting in wear.

The SU viscosity required for most high-speed engines ranges between 35 and 70 s at 37.8°C (100°F). As viscosity changes rapidly with temperature, just a numerical value of viscosity has no significance unless the temperature is specified.

8.8.4 Specific Gravity

The specific gravity is defined as the mass of a unit volume of fluid to that of the same volume of water preferably at the same temperature (say 15.6°C). It is commonly designated as 'sp.gr.

15.6/15.6°C', indicating that both the oil and the water are weighed and measured at a temperature of 15.6°C (60°F).

The oil industry uses a scale adopted by the American Petroleum Institute (API) for measuring the relative density of fuels, giving readings in degrees API. The relation between the specific gravity and the API gravity is given by

$$\text{API gravity (in degrees)} = \frac{141.5}{\text{sp.gr.at } 15.6°\text{C}} - 131.5$$

Thus a light fuel, which has a low specific gravity, has a higher API gravity. Limitations imposed by viscosity limits on CI engines more or less confine the limits of specific gravity to about 0.83 to 0.90 or 39° to 26° API.

8.8.5 Corrosion and Wear

The fuel should be such that it should not cause corrosion and wear before and after combustion. In order to avoid corrosion and wear, the fuel should not contain much sulphur, ash and carbon residue.

Sulphur

The percentage of sulphur in diesel oil is higher than that in gasoline. The wear and fouling that arise from sulphur in the fuel result from the formation of SO_3, during the combustion process, due to combination of sulphur with large amount of excess air. The SO_3 may attack the lubricating oil on the cylinder walls to form resinous materials which harden to form varnish and carbon. It may also react with water to form sulphuric acid. Wear is caused due to acidic corrosion and due to abrasion with the carbonaceous material. This condition is especially bad when the combustion products are cooled enough to condense some of the water vapour. In this case excessive corrosion of the combustion equipment and exhaust passages will result.

The amount of sulphur in the diesel oil can be greatly reduced by expensive refining processes. Sulphur contents over 1.0% are harmful, while amounts of 0.5% are economically feasible.

Carbon residue

When a fuel is burned with a limited amount of oxygen, carbon residue is usually left. The heavier ends of the liquid fuel suffer from the incomplete combustion and therefore yield carbon in the combustion chamber. High carbon residues increase the deposits in the combustion chamber and around the nozzle tips, thus adversely affecting the spray characteristics.

Ash

The ash content of a fuel is the solid material which remains after complete combustion of the fuel. It is a measure of abrasiveness of the products of combustion that could cause wear in the engine. The ash content should not exceed 0.12 % by weight for the heaviest fuel and should be 0.01% for light fuels used in high speed engines.

8.8.6 Handling Ease

The fuel oil used in CI engines is a liquid that will readily flow under all conditions. This requirement is measured by the cloud point and the pour point of the fuel.

Cloud point

It is the temperature below which the wax content of the petroleum oil separates out in the form of a solid. The waxy solid may clog the fuel lines and fuel filters.

Pour point

The pour point of an oil is found by cooling a sample in a test tube until no movement of the oil occurs for 5 seconds after the tube is tilted from the vertical to the horizontal position. The pour point is important only when the engine has to run at low temperatures. In such cases, the oil should have a pour point 5 to 10°C below the operating temperature. The pour point indicates that it may not be possible to have gravity feeding of fuel from the reservoir to the engine below this temperature.

8.8.7 Safety

The safety of a diesel oil is measured by its flash point and fire point.

Flash point

The flash point is the lowest temperature at which a fuel will vaporize sufficiently to form a combustible mixture of fuel vapour and air above the fuel. It can be determined by heating a quantity of the fuel in a special container while passing a flame above the liquid to ignite the vapours. A distinct flash of flame occurs when the flash point temperature has been reached. The flash point is important for safety purposes and serves as a measure of the fire hazard. A minimum flash point of 65°C is specified for safety.

Fire point

The fire point is the temperature at which enough vapours will rise to produce a continuous flame above the liquid fuel. The flame must sustain at least for five seconds.

The fire hazard increases with increase in volatility. As the volatility of diesel oil is less than that of gasoline, it is safer under most circumstances.

8.8.8 Cleanliness

The cleanliness factor is very important is CI engines, because of the precisely fitted parts in the fuel pump and nozzle. Dirt and water in the oil may damage engines. Since diesel oil is more viscous than gasoline, so it has a tendency to hold more solid particles in suspension. It is therefore necessary to pass diesel oil through an elaborate filtering process before it enters the pipe lines, fuel pumps and nozzles.

8.9 KNOCK RATING OF FUELS

It determines whether or not a fuel will knock in a given engine under the given operating conditions.

8.9.1 Knock Rating of SI Engine Fuels

A practical measure of a fuel's resistance to knock in SI engines is the fuel's octane number. The higher octane number (ON) indicates higher resistance to knock and the higher compression ratio may be used without knocking. The octane number used depends on the engine design and the operating conditions during the test. The octane number (ON) scale is based on two hydrocarbons which define the ends of the scale. The scale has been set up in which isooctane (C_8H_{18}; 2-2-4-trimethyl pentane) being a very good antiknock fuel is arbitrarily assigned a rating of 100 octane number, and normal heptane (n-C_7H_{16}), on the other hand, has very poor antiknock qualities and is given a rating of zero octane number. These hydrocarbons were selected because of the great difference in their ability to resist knock and the fact that isooctane had a higher resistance to knock than any of the fuels available at the time the scale was established. A gasoline is rated as a 90 octane number, if its tendency to detonate in the test engine is the same as that of a mixture of 90% isooctane and 10% normal heptane by volume. Thus, the octane number rating is an expression which indicates the ability of a fuel to resist knock in SI engines.

There are two common procedures for determining the octane rating of fuels—the research method (Testing code: ASTM D – 2699) and the motor method (Testing Code: ASTM D – 2700). In the motor method, the engine operating conditions are more severe and thus there are more chances of knock being produced. The engine operating conditions of the two methods are presented in Table 8.3.

Table 8.3 Engine operating conditions for research and motor methods

Variable	*Research method*	*Motor method*
Inlet temperature	52°C (125°F)	149°C (300°F)
Inlet pressure	Atmospheric	Atmospheric
Humidity (kg/kg dry air)	0.0036 – 0.0072	0.0036 – 0.0072
Coolant temperature	100°C (212°F)	100°C (212°F)
Engine speed	600 rpm	900 rpm
Spark advance	13° BTDC (constant)	19°–26° BTDC (varies with compression ratio)
Air/fuel ratio	Adjusted for maximum knock	Adjusted for maximum knock

The engine used in the research and motor methods is a special CFR engine developed by the Cooperative Fuel Research Committee (now the Coordinating Research Council, Inc.) in 1931. It has a single cylinder, overhead valves, a three-bowl carburettor and a variable compression ratio. The compression ratio can be changed (even with the engine running) by raising or lowering the entire cylinder and head assembly relative to the crankshaft and crankcase through a worm gear turned by the hand crank.

Fuel sensitivity

Since the motor method of determining the octane number used more severe operating conditions than the research method, the motor octane number (MON) is lower than the research octane number (RON). The difference between these octane numbers is called the *fuel sensitivity*, i.e.

$$\text{Fuel sensitivity} = \text{RON} - \text{MON} \tag{8.3}$$

Sensitivity is a measure of the extent to which a gasoline is downgraded under severe conditions. The higher sensitivity indicates poor performance under severe conditions.

The primary reference fuels—isooctane and *n*-heptane—are paraffins having the same RON and MON. In general, the paraffins are the least sensitive, while olefins, naphthenes and aromatics are more sensitive. Therefore, the straight-run gasolines containing a high percentage of saturated hydrocarbons have low sensitivity, while the cracked gasolines containing a large percentage of unsaturated hydrocarbons have high sensitivity.

Road octane number

Automobile engines run on the road under variable speed, load and weather conditions. The spark-timing also changes with speed. On the other hand, the CFR engines, to determine research octane number and motor octane number, are run at constant speed, full throttle, and fixed spark-timing.

Therefore, the road octane number requirement differs from RON and MON. The road ratings of fuels usually lie between the research and motor ratings and can be expressed as

$$\text{Road ON} = a(\text{RON}) + b(\text{MON}) + c \tag{8.4}$$

where a, b, c are experimentally determined constants. For most gasolines used in automobiles, $a \simeq b \simeq 0.5$ and $c \simeq 0$ give good agreement with the practical results.

The antiknock index of a fuel can be expressed as the mean of the RON and MON. It characterizes the antiknock quality.

$$\text{Antiknock index} = \frac{\text{RON} + \text{MON}}{2} \tag{8.5}$$

Additives to improve octane number

The octane number of a gasoline can be improved by adding certain antiknock agents, such as tetraethyl lead, $(C_2H_5)_4$ Pb (TEL), tetramethyl lead, $(CH_3)_4$ Pb (TML), and methyl-cyclopentadienyl manganese tricarbonyl (MMT).

The introduction of TML permits better distribution of octane amongst the cylinders of an engine because it boils in the mid-range of a gasoline (110°C), whereas TEL boils at the high end (200°C). MMT is a supplementary antiknock agent for TEL. Engines fitted with catalytic converters for the reduction of exhaust emissions require unleaded fuel, because the catalysts used will be poisoned by the presence of lead in the fuel and will soon loose their activities. Low concentrations of MMT may be used as an antiknock additive in unleaded gasoline. This also produces deposits in the catalytic converter, which may increase the resistance to flow, developing back pressure on the engine and thus reducing the performance of the engine. Nowadays, it is more common to use alcohols and ethers to increase the octane rating of the gasoline. Alcohols and ethers possess excellent antiknock qualities.

Fuels superior to isooctane in antiknock qualities have become increasingly important. For fuels having octane numbers greater than 100, a mixture of isooctane and tetraethyl lead is used, and the octane number scale according to ASTM standards is defined as follows:

$$\text{ON} = 100 + [28.28\ \text{TEL}/\{1 + 0.736\ \text{TEL} + \sqrt{(1 + 1.472\ \text{TEL} - 0.035216\ \text{TEL}^2)}\}] \tag{8.6}$$

where, TEL is in milliliters of TEL per US gallon in isooctane.

Performance number

Performance number (PN) is the measure of antiknock effectiveness. Figure 8.9 shows the performance number related to octane number and tetraethyl lead in isooctane. It is observed that as the amount of tetraethyl lead in isooctane increases, the antiknock effect also increases, but the increase is not linear. The antiknock effect first increases rapidly and then with further addition of

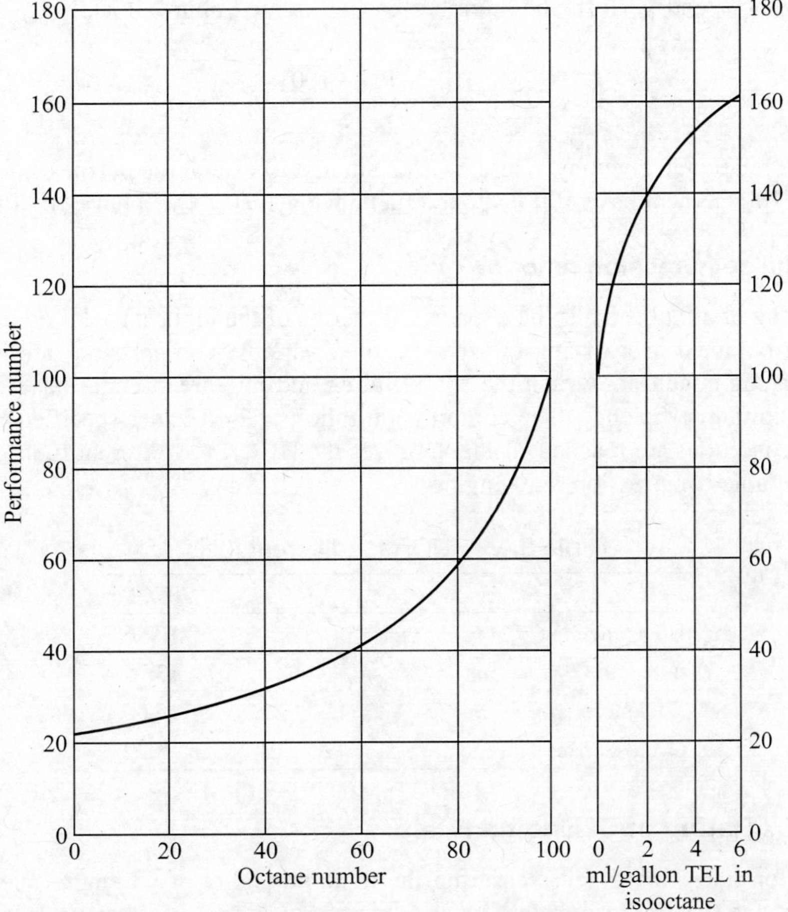

Figure 8.9 Performance number equivalents for octane number and tetraethyl lead in isooctane.

TEL the antiknock effect increases slowly. Also, as the octane number increases from 0 to 100, the antiknock effect is not linear. An increase in octane number at the higher range of octane scale produces a greater antiknock effect compared to the same increase in octane number at the lower end of the scale. For example, the increase in octane number from 80 to 90 will produce a greater antiknock effect than the increase from 20 to 30. Because of this nonlinear variation, another scale is used, which is known as *Performance Number* (PN). It expresses almost the relative engine performance.

The performance number is the ratio of the knock-limited indicated mean effective pressure (klimep) of the test fuel to the knock-limited indicated mean effective pressure (klimep) of isooctane. That is,

$$\text{PN} = \frac{\text{klimep of test fuel}}{\text{klimep of isooctane}} \tag{8.7}$$

Isooctane is arbitrarily assigned a PN of 100. A fuel rated at 120 PN can produce approximately 1.2 times the power (without knock) that it can develop with a 100 PN fuel (without knock).

Octane number and performance number can be related approximately by the following relation:

$$\text{ON} = 100 + \frac{\text{PN} - 100}{3} \tag{8.8}$$

The above relation is given by Wiese; it is an attempt to extend the octane scale beyond 100. A fuel having 100 ON will have 100 PN, but a fuel having 120 ON will have 160 PN.

Highest useful compression ratio

The knock rating of a fuel can also be expressed in terms of the highest useful compression ratio (HUCR). It is obtained by carrying out the test on a variable compression ratio engine under specified operating conditions, when the spark timing and mixture strength have been adjusted to give the maximum efficiency. The compression ratio is raised under specified conditions till the knocking conditions are reached. Table 8.4 gives the HUCR for different fuels determined in Ricardo E6 variable compression ratio engine.

Table 8.4 HUCR for different fuels

Fuel	*HUCR*
Isooctane	10.96
n-heptane	3.75
Toluene	15.00
Cyclo-hexane	8.20

8.9.2 Knock Rating of CI Engine Fuels

The methods for determining and measuring the ignition quality of CI engine fuels are (a) the cetane number, and (b) the diesel index.

Cetane number

The cetane number determines the ignition quality of a diesel fuel. It is the most used method. An increase in cetane number reduces the ignition delay period and thus reduces the tendency to knock. The cetane number scale is defined by blending cetane (n-hexadecane $C_{16}H_{34}$), a hydrocarbon of high ignition quality that represents the top of the scale with a cetane number 100. An isocetane (Hepta-methylnonane, HMN) having low ignition quality, represents the bottom of the scale with a cetane number 15. In the original procedure α-methylnapthalene ($C_{11}H_{10}$) with a cetane number of zero represented the bottom of the scale, but HMN, a more stable compound, has replaced it.

The cetane number (CN) of a fuel is given by

$$\text{CN} = \text{per cent } n\text{-cetane} + 0.15 \times \text{per cent HMN} \tag{8.9}$$

The ASTM method for rating the cetane number of a given fuel is determined by performing the test in a CFR engine. It is a single cylinder, variable compression ratio engine. The operating conditions are: engine speed 900 rpm, intake air temperature 65.6°C (150°F); coolant temperature 100°C; injection timing 13°bTDC; injection pressure 10.3 MPa (1500 lb/in^2). The cetane number of the fuel to be determined is tested in the above engine with the specified operating conditions, the compression ratio of the engine is varied until the combustion starts at TDC, i.e. an ignition delay period of 13° is obtained. This compression ratio is now kept fixed and the test engine with the specified operating conditions is run with the different blends of the reference fuels until combustion starts once again at TDC. Knowing the percentage of both the reference fuels in the blend, the cetane number is calculated from Eq. (8.9). The recommended cetane numbers are 25 to 35 for low speed, 35 to 45 for medium speed and 45 to 60 for high speed engines. Experiments have shown that, with a given fuel, if the engine has a good start and operates satisfactorily, raising the cetane number may not increase the engine performance, and it is quite possible that the performance on the other hand may somewhat deteriorate, resulting in decreased power and increased fuel consumption.

The ignition quality of a diesel fuel has a nonlinear relationship with the cetane number, but it is not a serious problem, since CI engines burn fuels in a narrow range of cetane scale.

The cracked fuels and fuels of other than paraffinic base have low cetane number. Certain additives are required to raise the cetane number to desirable values for high speed engines. Such additives reduce the self-ignition temperature of the fuel. Most additives for CI engine fuels are mild explosives. Some of the CI fuels additives are: Isopropyl nitrate, ethyl nitrate, amyl nitrate, ethyl nitrite, butyl peroxide and methyl acetate.

TEL is not a suitable additive for CI engine fuels. It increases the ignition delay and hence the knocking tendency increases too.

Diesel index

The diesel index depends on the fact that aromatic hydrocarbons mix completely with aniline at comparatively low temperatures, whereas the paraffins require considerably higher temperatures before becoming completely miscible. First, the 'aniline point' is determined. It is the lowest temperature at which equal volumes of the fuel and aniline become just miscible. The aniline point is determined by heating a mixture consisting of equal volumes of the test sample and freshly distilled, water free aniline, C_6H_7N, until a clear solution is obtained. Then, while the solution is cooling, the temperature at which turbidity appears is noted.

The diesel index is computed from the following expression:

$$\text{Diesel index} = \frac{\text{Aniline point (°F)} \times \text{API gravity}}{100} \tag{8.10}$$

$$= \left(\text{Aniline point (°C)} \times \frac{9}{5} + 32\right) \times \frac{\text{API gravity}}{100} \tag{8.11}$$

The diesel index usually gives values slightly above the cetane number. Figure 8.10 shows the relationship between the cetane number and the diesel index number.

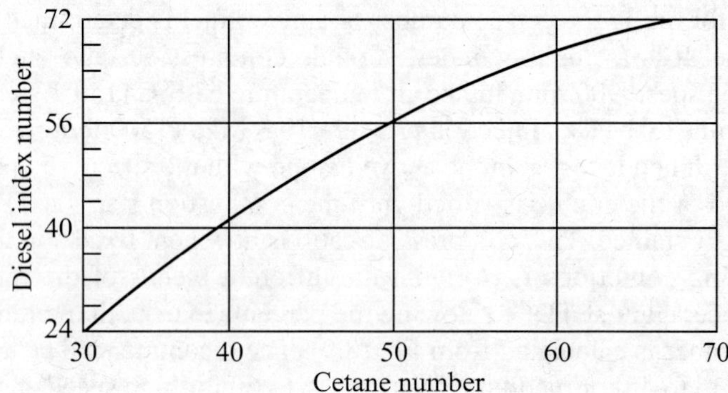

Figure 8.10 Relation between the cetane number and the diesel index number.

REVIEW QUESTIONS

1. Define a fuel. What are the requirements of IC engine fuels?
2. What are the criteria for the selection of IC engine fuels?
3. How are the fuels classified?
4. Briefly describe the following types of solid fuels: wood, peat, coal (lignite to anthracite), charcoal, coke, briquetted coal, and pulverized coal.
5. Comment on the use of solid fuels in IC engines.
6. Describe, with the help of a diagram, a typical oil-pool formation. How is the crude oil forced out from the pool to the surface?
7. How are the hydrocarbons present in the fuel classified?. Briefly describe them with their structural formulae.
8. Describe the refining process of petroleum by the method of fractional distillation.
9. Briefly describe the following processes used after the fractional distillation of crude oil. Cracking, Hydrogenation, Absorption, Polymerization, Alkylation, Isomerization, Cyclization, Aromatization, Reforming, and Finished blended products.
10. Briefly describe the different petroleum based liquid fuels of importance in IC engines.

11. Briefly describe the following non-petroleum based liquid fuels used in IC engines:
    Benzol, Methyl alcohol, Ethyl alcohol, Biodiesel, Emulsified diesel fuel, Acetone, Diethyl ether and Vegetable oils.
12. What are the advantages and disadvantages of using gaseous fuels in IC engines?
13. Describe the composition, properties and suitability to use the following gaseous fuels in IC engines: Natural gas, liquefied petroleum gas, hydrogen, hythane, dimethyl ether, biogas.
14. What are the different fuel characteristics on which the performance of SI engines depends?
15. Describe the volatility characteristics of petroleum fuels.
16. Describe the ASTM distillation test procedure for the measurement of relative volatility of fuels.
17. How does the EAD test differ from the ASTM distillation test? Enumerate the importance of each of these tests?
18. Describe the Reid vapour pressure bomb for measuring the volatility characteristic of a fuel.
19. Briefly describe the effect of volatility on the following performance characteristics of SI engines:
    Cold starting, Hot starting, Vapour lock, Evaporation loss, Warm-up and acceleration, Carburettor icing, Short and long trip economy, Crankcase dilution, and Sludge deposits.
20. Briefly describe the effect of the following fuel characteristics on the performance of SI engines:
    Sulphur content, Gum deposits, Carburettor detergent additives, and Antiknock quality.
21. Mention the important characteristics of CI engine fuels.
22. Briefly describe the effect of the following fuel characteristics on the performance of CI engines:
    Ignition quality, Volatility, Viscosity, Gravity, Corrosion and wear, Handling ease, Safety and cleanliness.
23. Describe the influence of ignition quality of CI engine fuels on cold starting, engine roughness and compression ratio.
24. How is Saybolt Universal Viscosimeter used to determine the viscosity of oil?
25. Define API gravity.
26. Define the following terms: Cloud point, Pour point, Flash point, and Fire point of a liquid fuel.
27. How are the SI engine fuels rated in terms of octane number? Define the octane number scale and give the reasons for selecting this scale.
28. What are the two common procedures for determining the octane ratings of fuels? Briefly describe them.
29. Define fuel sensitivity. Give examples of hydrocarbons having low sensitivity and high sensitivity.

30. Define Road Octane Number and Antiknock index.
31. Mention some of the additives to increase the octane number of fuels. What are the advantages and disadvantages of using leaded gasolines?
32. How is the fuel superior to isooctane in antiknock quality rated?
33. Define the performance number of a fuel. Give the approximate relation between the octane number and the performance number.
34. How is the rating of a fuel expressed in terms of HUCR?
35. Describe the following methods for determining and measuring the ignition quality of CI engine fuels: (a) Cetane number (b) Diesel index.

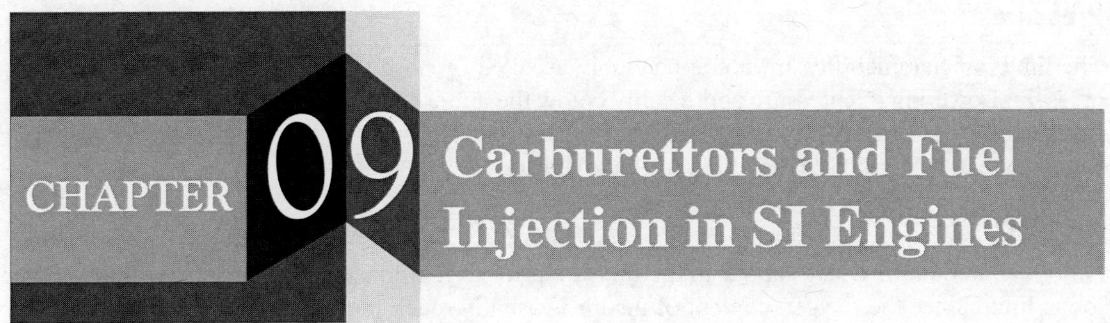

9.1 INTRODUCTION

Carburettor is a device for atomizing and vaporizing the fuel and mixing it with the air in varying proportions to suit the entire operating range of SI engines. The process of breaking up and mixing the fuel with the air is called *carburetion*. The carburettor supplies a mixture of vaporized fuel and air in the proper proportion to the intake manifold and finally to the cylinders. The intake manifold is normally kept over the exhaust manifold to heat the atomized fuel, which helps the vaporization process. The optimum air/fuel ratio for SI engines is that which gives the required power output with the minimum fuel consumption. The air/fuel ratio that provides the minimum fuel consumption, smooth and reliable engine operation and satisfies the emission requirements at the required power output, depends on engine speed and load. The air/fuel ratio for maximum power is not the same as the air/fuel ratio for maximum economy. The mixture requirements for starting, warm-up and acceleration are also different from those for steady operation.

9.2 LIMITS OF FLAMMABILITY

There are two limits of flammability for each fuel. The lower limit of flammability is the leanest mixture (the least amount of fuel in the mixture), which will support combustion. The upper limit is the richest mixture which can propagate a flame. At both ends of the flammable limits, the rate of flame propagation is the lowest. Since combustion engines run at various speeds and provide various air velocities in the combustion chamber, the flame will be stabilized with air/fuel ratio well within the flammability limits. For isooctane, the rich limit, the stoichiometric and the lean limit for engines are 8, 15.2 and 18 air/fuel ratios, by weight respectively.

Various other terms, such as inflammation limit, explosion limit, limits of inflammability are often used interchangeably with the terms 'limits of flammability'. The limits of flammability are affected by the following factors in combustion engines.

Temperature

Temperature changes of a few degrees at the time of ignition have little effect on the flammability limits. At temperatures considerably above the room temperature, the range of flammability usually increases by extending both the limits.

Pressure

The limits of flammability are not appreciably affected by normal variations of the atmospheric pressure. Lowering the pressure appreciably below the atmospheric usually decreases the range of flammability. The effect of pressure above the atmospheric varies with each fuel, and no general statement can be applied to the effect of flammability.

Humidity

A normal amount of water vapour in the mixture prior to ignition seems to have no effect on the lower limit, since the oxygen content of the air is usually much more than the fuel content of the mixture. However, the upper limit is lowered, because the water vapour decreases the oxygen content of the air.

9.3 STEADY-RUNNING MIXTURE REQUIREMENTS

Steady running means the continuous operation of the engine at a given speed and power output with normal engine temperatures. Under the steady-running condition the mixture requirements for maximum power and for minimum specific fuel consumption are different.

9.3.1 Mixture Requirements for Maximum Power

Figure 9.1 shows the effect of the fuel/air ratio on the indicated mean effective pressure at full throttle and part throttle positions. The maximum indicated mean effective pressure occurs at a fuel/air ratio of about 0.08 for gasoline, which is a little more than the chemically correct amount of fuel. At full throttle and with a maximum indicated mean effective pressure mixture, the flame speed is high, and the burning time losses are very small, since the piston is near the TDC and moves very little during the burning process. The effect of closing the throttle reduces the indicated mean effective pressure and the indicated power by reducing both the intake manifold pressure and the volumetric efficiency. The entire indicated mean effective pressure curve, therefore, lowers

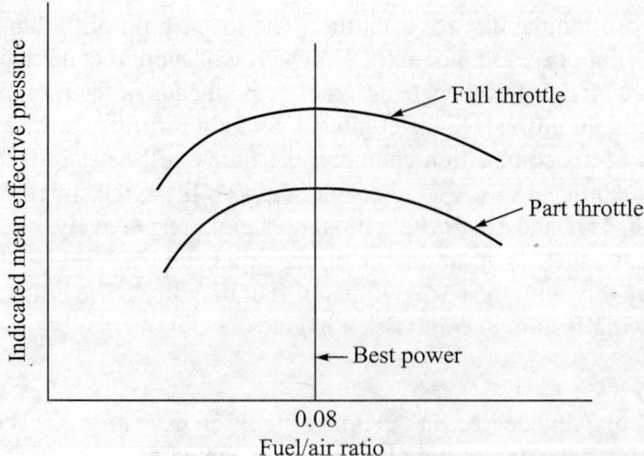

Figure 9.1 Effect of fuel/air ratio on the indicated mean effective pressure at full throttle and part throttle positions.

itself at part throttle, however the intake manifold pressure also reduces the flame speed, which increases the burning time losses and which further reduces the indicated mean effective pressure. However, the burning time losses at fuel/air ratio of 0.08 are almost negligible. For any given throttle position, the indicated mean effective pressure will be maximum at a fuel/air ratio of about 0.08.

9.3.2 Mixture Requirements for Minimum Specific Fuel Consumption

The maximum indicated thermal efficiency occurs at a fuel/air ratio of about 0.06 for gasoline, which is slightly leaner than the chemically correct fuel/air ratio because excess air ensures complete combustion of the fuel. A lean mixture lowers the maximum temperature, which favours the chemical equilibrium and specific heat of gases. If the mixture is made too lean, the flame speed reduces, which increases the burning time losses and lowers the efficiency. The indicated specific fuel consumption (isfc) reduces as the indicated thermal efficiency increases. The important effect of throttling on isfc is the change in time losses. The time losses are slightly affected at fuel/air ratio 0.08, but as the mixture is made richer or leaner, the effect of throttling on time losses becomes much larger. It is observed from Figure 9.2 that the isfc curves for part-throttle operation are very close to the full throttle curve at fuel/air ratio of 0.08 and deviate at leaner and richer mixtures. The point of minimum isfc occurs at progressively richer fuel/air ratios as the throttle is reduced.

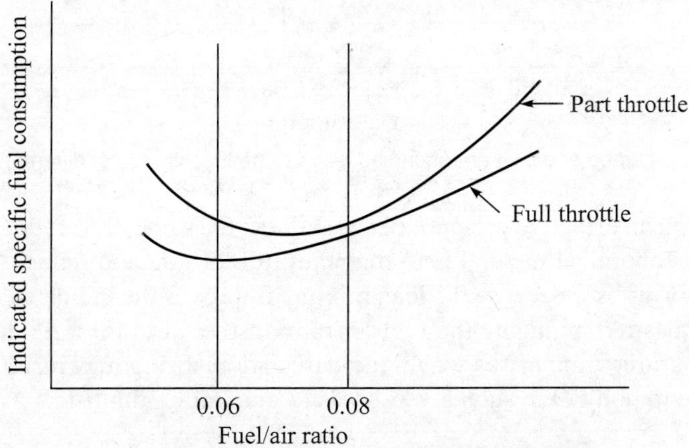

Figure 9.2 Effect of throttling on indicated specific fuel consumption at various fuel/air ratios and constant rpm.

9.3.3 Mixture Requirements for Various Outputs

Figure 9.3 shows the typical curves of brake mean effective pressure (bmep) and the corresponding brake specific fuel consumption (bsfc) versus the fuel/air ratios for an engine operating at different throttle positions. From the figure it is observed that there is only one fuel/air ratio and the throttle position which will result in maximum power. This corresponds to point A at full throttle with fuel/air ratio of 0.08. If it is required to reduce power by 80% of maximum, it can be done by reducing the throttle with fuel/air ratio of 0.08 corresponding to point B. This will result in an increase in bsfc from A to B on the bsfc curves. The increase is due to the reduced throttle causing less imep,

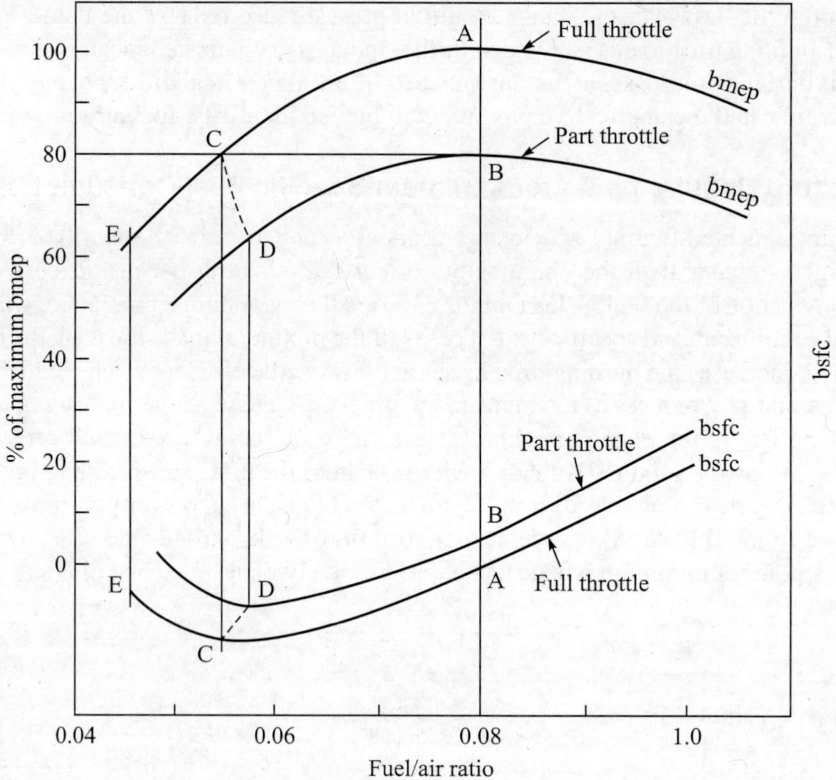

Figure 9.3 Curves of bmep and the corresponding bsfc vs. the fuel/air ratio at different throttle positions.

keeping the friction mean effective pressure (fmep) almost the same, thus reducing the mechanical efficiency. A more economical method is to maintain full throttle and reduce the fuel/air ratio to point C. This gives lower bsfc because the lean mixture improves the indicated thermal efficiency. If still less power is desired, reducing the fuel-air ratio further at full throttle to point E would be undesirable, as the burning time losses would increase and misfiring may result. It would be better to reduce the throttle to point D in such a way so as to give minimum bsfc.

9.3.4 Mixture Requirements for Idling, Cruising and High Power

The mixture requirements for idling, cruising, and high power ranges are shown in Figure 9.4.

Idling range

The engine is said to idle when it is operating at no external load with the throttle almost closed. An idling engine requires a rich mixture, as shown by point A in Figure 9.4. It is because of the following reasons:

The mass of the exhaust gas remaining as residual in the clearance space at the end of the exhaust stroke remains fairly constant throughout the throttle range. On the other hand, the mass of the fresh charge induced on each intake stroke depends upon the manifold pressure and therefore

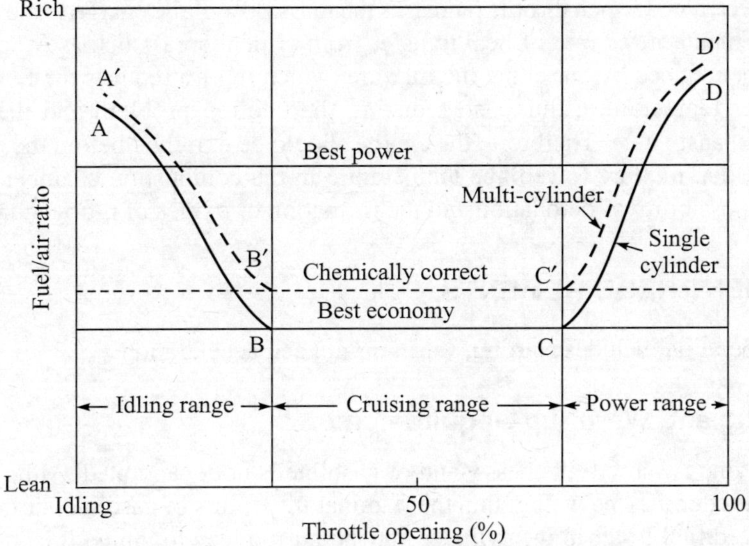

Figure 9.4 Mixture requirements for idling, cruising and high power range.

it depends upon the throttle position. The manifold pressure and hence the mass of the fresh charge inducted during idling are much less than those during the full throttle operation, due to the restrictions imposed by throttle. The result is a much larger proportion of exhaust gas being mixed with the fresh charge, under idling conditions causing dilution of the fresh charge. The presence of the exhaust gas reduces the intimate contact of fuel and air particles, resulting in poor combustion and thus causing power loss. It is, therefore, necessary to provide more fuel particles for enriching the charge. This increases the probability of contact of fuel and air particles, and thus improves combustion.

As the throttle is gradually opened from A to B (Figure 9.4), the exhaust gas dilution of the fresh charge diminishes. Mixture requirements then proceed along the curve AB to a leaner fuel/air ratio.

Cruising range

In the cruising range from B to C (Figure 9.4), the engine is operating at part throttle. The main objective is to obtain the maximum fuel economy. Consequently, in this range, it is desirable that the carburettor should provide the engine with the best economy mixture, which is slightly leaner than the stoichiometric.

With lean mixtures, the flame speed is relatively slow, and even slower when the mixture is diluted with exhaust gas. Hence the spark is advanced as the manifold vacuum increases.

High power range

During high power operation, when the engine is operating with the fully opened throttle, the engine requires a richer mixture, as indicated by the curve CD (Figure 9.4). It has already been observed that the mixture requirement for maximum power is with a slightly richer mixture than the stoichiometric. The carburettor has to be set in the vicinity of the best power mixture.

At high power, a wide open throttle increases the mass flow of the charge to the cylinder, which increases the demand of the rate of heat transfer from critical areas such as exhaust valve. This demand could be reduced by enriching the mixture, which in turn reduces the flame temperature and the cylinder temperature, thus also reducing the cooling problem and the possibility of damaging the exhaust valve. Therefore, the charge should be enriched before the throttle is made wide open. The rich mixture lowers the temperature in the combustion chamber and thus helps in reducing the possibility of detonation and the formation of oxides of nitrogen in the exhaust.

9.4 TRANSIENT REQUIREMENTS

The transient operation includes starting, warming-up, and acceleration.

9.4.1 Starting and Warm-up Requirements

When a cold engine is started, the heavy end of gasoline is not evaporated. Although the fuel/air ratio at the carburettor may be well within the flammability limits of gasoline-air mixtures, but the ratio of evaporated fuel-to-air in the cylinder may be far too lean to ignite. It is, therefore, necessary to supply 5 to 10 times richer fuel at the carburettor to obtain enough evaporated light ends to ignite, until the manifold and cylinder parts become warm.

As the engine warms up, the fuel/air ratio requirement at the carburettor must be reduced to refrain the evaporated fuel/air ratio from becoming too rich.

9.4.2 Acceleration Requirement

Under steady-running conditions, there is a tendency for some non-vaporized liquid droplets to form a thin liquid film and move along the inner wall of the intake manifold to the cylinders. The air and evaporated fuel mixture take much less time than the liquid streams along the wall to reach the cylinder from the carburettor. When a sudden acceleration is required and the throttle is suddenly opened, the gaseous charge of air and fuel moves rapidly into the cylinders. The liquid film, due to its greater inertia lags behind. It causes lean mixture to move to the engine cylinder for a short time. This temporary lean mixture prevents the engine from developing full power just at the time when it is required the most.

In order to compensate for this tendency of the carburettor, during acceleration, to fail momentarily to supply a sufficiently rich mixture, a mechanical accelerating device is provided, which is directly connected to the throttle mechanism.

9.5 MIXTURE REQUIREMENTS IN A MULTI-CYLINDER ENGINE

Air-fuel maldistribution between the cylinders is inherent in a conventional multi-cylinder spark-ignition engine. Complete atomization and vaporization of the fuel by the carburettor is difficult to obtain. Even under the best conditions, the intake manifold contains an appreciable amount of liquid as well as large fuel droplets. A thin film of liquid fuel also adheres to the inner walls of the intake manifold. These droplets have greater inertia than the gaseous mixtures. Consequently, whenever the direction of flow is changed abruptly, the droplets tend to continue in their original direction of movement. As a result, there is a variation in the air/fuel ratio between the cylinders, depending upon the cylinder location and the manifold design.

For good distribution, it is important to heat the mixture in the intake manifold, causing the less volatile parts of the fuel to vaporize. Poor distribution of fuel to the cylinders of a multi-cylinder engine results in a loss of power and efficiency. A slight enrichment of the overall mixture will improve the engine performance, so that the leanest cylinder receives the required air/fuel ratio. It is indicated by the curve A′ B′ C′ D′ in Figure 9.4.

9.6 CARBURETTOR REQUIREMENTS

The following are the requirements for an ideal carburettor:

1. It should provide easy starting of the engine from the cold.
2. It should provide properly atomized fuel with the correct fuel-air mixture at each speed corresponding to the throttle position.
3. It should provide the correct fuel-air mixture at each throttle opening under different loads and speeds.
4. It should enable the engine to run slowly during idling without hunting or missfiring the engine and thus eliminating undue wastage of fuel.
5. It should generate maximum acceleration when the throttle is suddenly or slowly opened. There should not be any flat spots (hesitation to pick up speed) throughout the throttle opening range.
6. It should be so designed that when the throttle is fully opened the maximum quantity of the correct mixture flows into the engine. Sudden bends and restrictions must be avoided.
7. It should function correctly under different climatic conditions, such as temperature, barometric or altitude and atmospheric moisture changes.
8. It should enable maximum distance per litre under the above conditions.

9.7 A SIMPLE CARBURETTOR

A simple jet type of carburettor is shown in Figure 9.5. The modern carburettor in its various forms has been evolved from this elementary carburettor. An explanation of its construction and operation will aid in understanding the basic principles underlying all carburettors.

It consists of a fuel jet of small diameter placed in a constricted tube called venturi or choke tube, and a float chamber having a hollow thin metal float provided with a conical needle valve.

The fuel pump delivers fuel from the fuel tank to the float chamber. When sufficient fuel enters the float chamber, the float is lifted due to buoyancy and the conical needle valve engages with a similarly shaped seating in the petrol pipe union, and thus shuts off the fuel. In this way the fuel level in the float chamber is always maintained constant. If the fuel level tends to fall, the float drops and thereby opens the needle valve, thus admitting more fuel. The height at which the fuel is maintained in the float chamber is governed by the level required in the discharge jet. This level should stand a little below the orifice at the tip of the jet to prevent spilling. The float chamber is vented to atmosphere through a small hole in the cover, hence the pressure on the surface of the fuel remains constant and equal to that of the atmosphere.

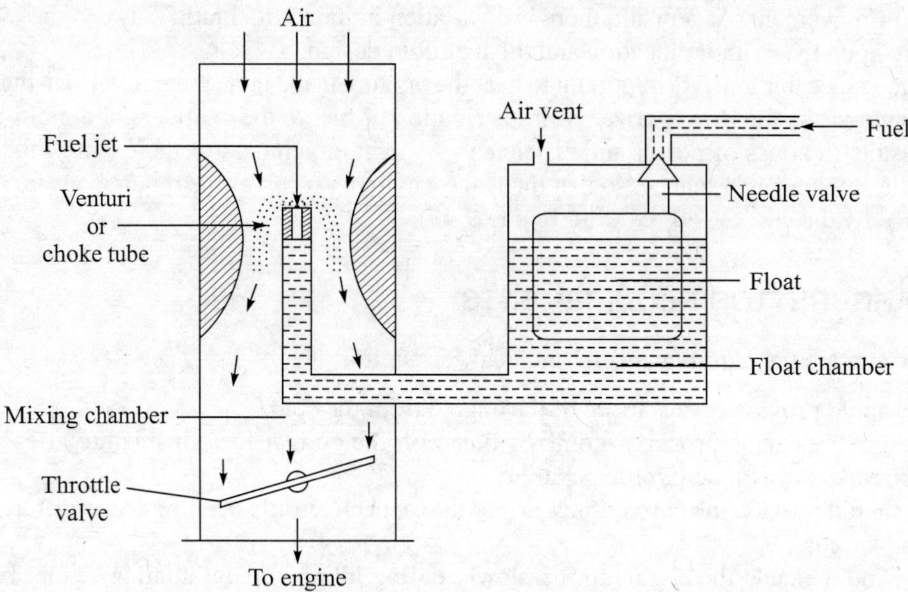

Figure 9.5 A simple carburettor.

The air is induced in the venturi tube by the suction created by the descending piston of the engine cylinder. The SI engine is quantity governed which means that when less power is required at a particular speed the amount of charge delivered to the cylinders is reduced. This is achieved by means of a throttle valve of the butterfly type which is situated at the exit of the venturi tube. At the throat of the venturi tube, the area of cross-section is minimum, which is shaped to give the minimum resistance to the air flow. The fuel discharge jet is situated at the throat. The pressure at the throat is below atmospheric, since the air velocity has been increased from that at the inlet to the carburettor to a maximum at the throat. The surface of the fuel in the float chamber is exposed to atmospheric pressure, while that at the jet opening in the venturi tube is less than this; it thus follows that the fuel is forced out of the jet due to this pressure difference, where it mixes with the high velocity of air, being atomized in the process, and passes with it into the engine via the intake manifold and intake valve. The rate of fuel flow is controlled or metered by the size of the smallest section in the petrol passage. This is provided by the main jet and the size of this jet is chosen to give the required engine performance. The pressure at the throat at the fully open throttle condition lies usually between 38 and 50 mm Hg below atmospheric.

In the elementary carburettor described above, the choke or throat has a constant area and the pressure changes with the throttle opening and engine speed. It is referred to as a fixed choke type of carburettor. It is evident that with this type of carburettor the correct fuel/air ratio is obtained only at one particular speed. As the engine speed increases the mixture is enriched and as the speed falls the mixture is weakened. It is usual to indicate the size of the carburettor by quoting the diameter of the venturi tube in millimetres, and the jet size in hundredths of a millimetre, i.e. the jet number 50 has a diameter of 0.050 mm.

The mathematical analysis of the performance of a simple carburettor follows in the next section.

9.8 CALCULATION OF THE AIR/FUEL RATIO FOR A SIMPLE CARBURETTOR

Figure 9.6 shows a simple carburettor where the section AA (plane 1) is taken at the entry to the carburettor and the section BB (plane 2) at the venturi throat.

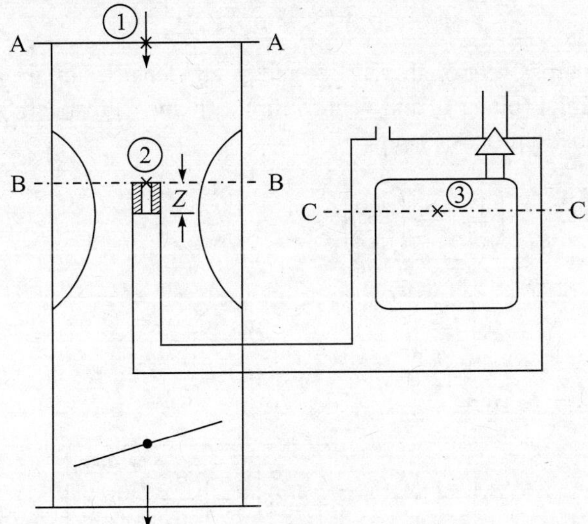

Figure 9.6 A simple carburettor showing sections AA, BB and CC.

Applying the steady-flow energy equation between the sections AA and BB and considering unit mass of air flow, we can write

$$h_1 + \frac{C_{a1}^2}{2} + q = h_2 + \frac{C_{a2}^2}{2} + w \tag{9.1}$$

Here, q and w are the heat and work transfers per unit mass of air flow between planes 1 and 2 and h and C_a denote the enthalpy and velocity of air respectively.

The flow is assumed to be reversible adiabatic (isentropic) and there is no work transfer between planes 1 and 2; therefore, $q = 0$ and $w = 0$. The approach velocity of air C_{a1} is negligible compared to velocity C_{a2}, therefore C_{a1} may be taken as zero. Substituting these values in Eq. (9.1), the velocity

$$C_{a2} = \sqrt{2(h_1 - h_2)} \tag{9.2}$$

Assuming air to behave like an ideal gas, therefore, taking $h = c_p T$, Eq. (9.2) becomes

$$C_{a2} = \sqrt{2c_p (T_1 - T_2)} = \sqrt{2c_p T_1 \left(1 - \frac{T_2}{T_1}\right)} \tag{9.3}$$

For an isentropic process,

$$\frac{T_2}{T_1} = \left(\frac{p_2}{p_1}\right)^{(\gamma-1)/\gamma} \tag{9.4}$$

$$\therefore \quad C_{a2} = \sqrt{2c_p T_1 \left[1 - \left(\frac{p_2}{p_1}\right)^{(\gamma-1)/\gamma}\right]} \tag{9.5}$$

From the continuity equation, the mass flow of air is

$$\dot{m}_a = \rho_1 A_1 C_{a1} = \rho_2 A_2 C_{a2} \tag{9.6}$$

where A_1 and A_2 are the cross-sectional areas, ρ_1 and ρ_2 are densities of air, and C_{a1} and C_{a2} are the air velocities at the air inlet (plane 1) and venturi throat (plane 2) respectively.

For reversible adiabatic process, $p_1 v_1^\gamma = p_2 v_2^\gamma$

$$\therefore \quad \frac{p_2}{p_1} = \left(\frac{v_1}{v_2}\right)^\gamma = \left(\frac{\rho_2}{\rho_1}\right)^\gamma$$

$$\therefore \quad \rho_2 = \rho_1 \left(\frac{p_2}{p_1}\right)^{1/\gamma} \tag{9.7}$$

Now,

$$\dot{m}_a = \rho_2 A_2 C_{a2}$$

$$= \rho_1 \left(\frac{p_2}{p_1}\right)^{1/\gamma} A_2 \sqrt{2c_p T_1 \left[1 - \left(\frac{p_2}{p_1}\right)^{(\gamma-1)/\gamma}\right]}$$

$$= \rho_1 A_2 \sqrt{2c_p T_1 \left[\left(\frac{p_2}{p_1}\right)^{2/\gamma} - \left(\frac{p_2}{p_1}\right)^{(\gamma+1)/\gamma}\right]} \tag{9.8}$$

From equation of state, $\rho_1 = \dfrac{p_1}{RT_1}$

$$\therefore \quad \dot{m}_a = \frac{p_1}{R\sqrt{T_1}} \cdot A_2 \sqrt{2c_p \left[\left(\frac{p_2}{p_1}\right)^{2/\gamma} - \left(\frac{p_2}{p_1}\right)^{(\gamma+1)/\gamma}\right]} \tag{9.9}$$

The actual rate of mass flow of air is given by

$$(\dot{m}_a)_{\text{actual}} = Cd_a \dot{m}_a,$$

where Cd_a is the coefficient of discharge for the venturi.

$$\therefore \quad (\dot{m}_a)_{\text{actual}} = Cd_a \frac{p_1}{R\sqrt{T_1}} A_2 \sqrt{2c_p \left[\left(\frac{p_2}{p_1}\right)^{2/\gamma} - \left(\frac{p_2}{p_1}\right)^{(\gamma+1)/\gamma}\right]} \tag{9.10}$$

For the calculation of the mass rate of flow of fuel, the Bernoulli's theorem can be used as the fuel is considered to be incompressible. Applying the Bernoulli's theorem between sections CC (plane 3) and BB (plane 2),

$$\frac{p_3}{\rho_f} + \frac{C_{f3}^2}{2} = \frac{p_2}{\rho_f} + \frac{C_{f2}^2}{2} + gZ \tag{9.11}$$

where, ρ_f is the density of fuel, C_f is the fuel velocity and Z is the height of the nozzle exit above the level of fuel in the float chamber.

Here, the velocity of fuel C_{f3} at section CC is negligible, since the level of fuel does not drop in the reservoir.

The fuel velocity at the nozzle exit C_{f2} can be obtained from Eq. (9.11), which is given by

$$C_{f2} = \sqrt{2\left(\frac{p_3 - p_2}{\rho_f} - gZ\right)} \tag{9.12}$$

Pressures at plane 1 and plane 3 are both atmospheric, therefore, $p_3 = p_1$.

$$\therefore \quad C_{f2} = \sqrt{2\left(\frac{p_1 - p_2}{\rho_f} - gZ\right)} \tag{9.13}$$

$$= \sqrt{\frac{2}{\rho_f}(\Delta p - \rho_f gZ)} \tag{9.13a}$$

From the continuity equation, the mass rate of fuel is given by

$$\dot{m}_f = A_j C_{f2} \rho_f = A_j \sqrt{2\rho_f(\Delta p - \rho_f gZ)} \tag{9.14}$$

where A_j is the area of cross-section of the fuel jet at the exit from the nozzle.

The actual rate of mass flow of fuel is

$$(\dot{m}_f)_{\text{actual}} = Cd_f \cdot A_j \sqrt{2\rho_f(\Delta p - \rho_f gZ)} \tag{9.15}$$

where Cd_f is the coefficient of discharge for fuel nozzle.

$$\text{Air/fuel ratio,} \quad \frac{A}{F} = \frac{(\dot{m}_a)_{\text{actual}}}{(\dot{m}_f)_{\text{actual}}}$$

$$\therefore \quad \frac{A}{F} = \frac{Cd_a p_1}{Cd_f \cdot R\sqrt{T_1}} \cdot \frac{A_2}{A_j} \sqrt{\frac{c_p\left[\left(\frac{p_2}{p_1}\right)^{2/\gamma} - \left(\frac{p_2}{p_1}\right)^{(\gamma+1)/\gamma}\right]}{\rho_f(\Delta p - \rho_f gZ)}} \tag{9.16}$$

9.9 AIR/FUEL RATIO NEGLECTING THE COMPRESSIBILITY OF AIR

Neglecting the effect of compressibility of air, the Bernoulli's theorem can be used for flow of air as well. Therefore, applying the Bernoulli's theorem between the section AA (plane 1) and section BB (plane 2) and neglecting the change in potential energy (for air being very light in weight) compared to the change in pressure and kinetic energy, the equation becomes,

$$\frac{p_1}{\rho_a} + \frac{C_{a1}^2}{2} = \frac{p_2}{\rho_a} + \frac{C_{a2}^2}{2} \tag{9.17}$$

The approach velocity C_{a1} may be neglected.

$$\therefore \quad C_{a2} = \sqrt{\frac{2(p_1 - p_2)}{\rho_a}} = \sqrt{\frac{2\Delta p}{\rho_a}} \tag{9.18}$$

Now,

$$\dot{m}_a = A_2 C_{a2} \rho_a$$

$$= A_2 \sqrt{2\rho_a \Delta p} \tag{9.19}$$

$$(\dot{m}_a)_{\text{actual}} = Cd_a A_2 \sqrt{2\rho_a \Delta p} \tag{9.20}$$

$$\frac{A}{F} = \frac{(\dot{m}_a)_{\text{actual}}}{(\dot{m}_f)_{\text{actual}}}$$

$$\therefore \quad \frac{A}{F} = \frac{Cd_a}{Cd_f} \cdot \frac{A_2}{A_j} \sqrt{\frac{\rho_a \Delta p}{\rho_f(\Delta p - \rho_f gZ)}} \tag{9.21}$$

If $Z = 0$;

$$\frac{A}{F} = \frac{Cd_a}{Cd_f} \cdot \frac{A_2}{A_j} \sqrt{\frac{\rho_a}{\rho_f}} \tag{9.22}$$

9.10 COMMENTS ON AIR/FUEL RATIO SUPPLIED BY A SIMPLE CARBURETTOR

The following points relevant to the air/fuel ratio supplied by a simple carburettor are worth noting:

1. From Eq. (9.14), it is clear that when $\Delta p \leq \rho_f gZ$, there will be no flow of fuel. The fuel flow will take place only when $\Delta p > \rho_f gZ$. As this pressure difference increases the rate of mass flow of fuel increases and the mixture becomes progressively richer.
2. Minimum air velocity at the throat which may cause fuel flow can be estimated approximately from Eq. (9.18) as

$$C_{a2} = \sqrt{\frac{2\Delta p}{\rho_a}} = \sqrt{\frac{2\rho_f gZ}{\rho_a}} \tag{9.23}$$

3. At high rate of flow of air, Δp is very large compared to $\rho_f gZ$. Hence $\rho_f gZ$ can be neglected compared to Δp in Eq. (9.21), and the air/fuel ratio approaches

$$\frac{A}{F} = \frac{Cd_a}{Cd_f} \cdot \frac{A_2}{A_j} \sqrt{\frac{\rho_a}{\rho_f}} \tag{9.24}$$

4. Equation (9.21) also reveals that as the density of air reduces the air/fuel ratio also decreases, i.e. the mixture becomes richer. At high altitudes, the density of air is low. The density of air at the throat also reduces for a high rate of air flow through the carburettor.

9.11 DEFICIENCIES OF THE ELEMENTARY CARBURETTOR

The deficiencies of the elementary carburettor can be listed as follows:

1. At low loads the throttle valve is partially open, so the mixture becomes lean whereas the engine requires the rich mixture at low loads.
2. At intermediate loads, the equivalence ratio increases slightly because the proportionate increase in fuel flow is more than the increase in air flow. However, the engine requires an almost constant equivalence ratio.
3. Near to wide open throttle the elementary carburettor provides the maximum air flow and the equivalence ratio remains constant. However, the engine requires the rich mixture with equivalence ratio 1.1 or more to develop maximum power.
4. The elementary carburettor cannot enrich the mixture during engine starting and warm-up.
5. The elementary carburettor cannot adjust to changes in altitude.

9.12 ESSENTIAL PARTS OF A MODERN CARBURETTOR

Modern carburettors vary considerably in design and in the means adopted for mixture compensation for speed and throttle opening. The essential parts of all such modern automobile carburettors, in addition to float chamber, venturi tube, fuel nozzle and throttle, are choke, main metering system, idling system, accelerating system, economizer system and power system.

9.12.1 Choke

All modern carburettors are provided with a choke valve in the air intake passage of the carburettor. It is a butterfly type of valve and is shown in Figure 9.7.

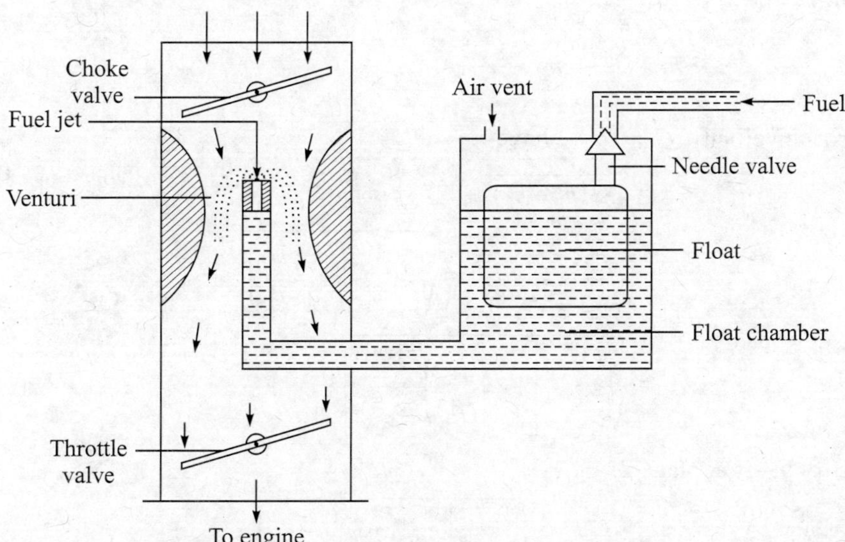

Figure 9.7 A simple carburettor with choke valve.

When a cold engine is started, especially at low ambient temperatures, the starting motor cranks the engine slowly (70 to 150 rpm). This produces low manifold vacuum, which draws less fuel from the jet causing the too lean fuel to ignite. The method usually employed for starting the engine from cold is to shut off most of the main air supply to the main jets and thus produce rich mixture necessary for cold starting. The choke valve is held in a partially closed position by the thermostat when the engine is being started, and is opened automatically as the engine warms up, thus gradually supplying the leaner mixture. Some popular cars still use a hand-operated choke controlled through linkage by a knob on the instrument panel. Pulling the knob out closes the choke valve partially, providing a rich mixture for starting. After the engine has started, the knob is pushed back to open the valve which increases the air flow through the carburettor, thus leaning the mixture.

EXAMPLE 9.1 A simple jet carburettor is required to supply 5 kg of air per minute and 0.4 kg per minute of fuel of density 780 kg/m³. The air is initially at 1.013 bar and 27°C. Calculate the throat diameter of the choke for an air flow velocity of 90 m/s. Take the velocity coefficient for the venturi to be 0.80 and the coefficient of discharge of the main fuel jet to be 0.6. Assume isentropic flow and the flow to be compressible. If the pressure drop across the fuel metering orifice is 0.75 of that at the choke, calculate the orifice diameter.

Solution: Refer to Figure 9.8.

Given: $m_a = 5$ kg/min

$\dot{m}_f = 0.4$ kg/min

$p_1 = 1.013$ bar, $T_1 = 300$ K

$C_2 = 90$ m/s, $Cv_a = 0.8$, $Cd_f = 0.6$

$d_2 = ?$, $d_j = ?$

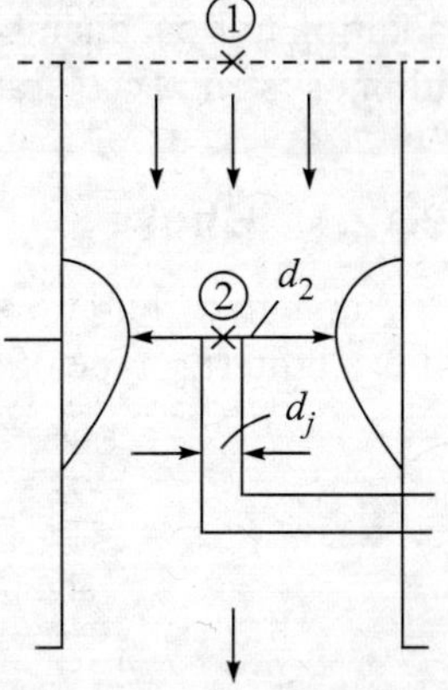

Figure 9.8 Example 9.1.

Applying the flow energy equation between planes 1 and 2, for isentropic flow,

$$h_1 + \frac{C_1^2}{2} = h_2 + \frac{C_2^2}{2}$$

Taking approach velocity, $C_1 = 0$, we have

$$C_2 = \sqrt{2(h_1 - h_2)}$$

$$= \sqrt{2c_p(T_1 - T_2)} = \sqrt{2c_pT_1\left(1 - \frac{T_2}{T_1}\right)}$$

$$= \sqrt{2c_pT_1\left[1 - \left(\frac{p_2}{p_1}\right)^{(\gamma-1)/\gamma}\right]}$$

Actual velocity of air at plane 2,

$$C_2 = C_{v_a}\sqrt{2c_pT_1\left[1 - \left(\frac{p_2}{p_1}\right)^{(\gamma-1)/\gamma}\right]}$$

or $$1-\left(\frac{p_2}{p_1}\right)^{(\gamma-1)/\gamma} = \frac{C_2^2}{C_{v_a}^2 \cdot 2c_p T_1}$$

or $$p_2 = p_1\left[1-\frac{C_2^2}{C_{v_a}^2 \cdot 2c_p T_1}\right]^{\gamma/(\gamma-1)}$$

$\therefore$ $$p_2 = 1.013\left[1-\frac{8100}{0.8\times 0.8\times 2\times 1005\times 300}\right]^{1.4/0.4}$$

$$= 1.013(1-0.02099)^{3.5}$$

$$= 0.9405 \text{ bar}$$

$\therefore$ Throat pressure, $p_2 = 0.9405$ bar

Density of air at inlet, $$\rho_1 = \frac{p_1}{RT_1} = \frac{1.013\times 10^5}{287\times 300} = 1.177 \text{ kg/m}^3$$

For isentropic flow, $$p_1 v_1^{\gamma} = p_2 v_2^{\gamma}$$

$\therefore$ $$\frac{p_1}{\rho_1^{\gamma}} = \frac{p_2}{\rho_2^{\gamma}}$$

Density of air at the throat, $$\rho_2 = \rho_1\left(\frac{p_2}{p_1}\right)^{1/\gamma} = 1.177\left(\frac{0.9405}{1.013}\right)^{1/1.4}$$

$$= 1.116 \text{ kg/m}^3$$

To determine the throat area, apply the continuity equation, i.e.

$$\dot{m}_a = \rho_2 A_2 C_2$$

$\therefore$ $$A_2 = \frac{\dot{m}_a}{\rho_2 C_2} = \frac{5/60}{1.116\times 90} = (8.297\times 10^{-4})\text{ m}^2 = 8.297\text{ cm}^2$$

$\therefore$ $$\frac{\pi}{4} d_2^2 = 8.297$$

$\therefore$ $$d_2 = \sqrt{\frac{4\times 8.297}{\pi}} = \boxed{3.25 \text{ cm}}$$ **Ans.**

Pressure drop at venturi = 1.013 – 0.9405 = 0.0725 bar

$\therefore$ Pressure drop at jet = 0.75 × 0.0725 = 0.0544 bar

Now, $$(\dot{m}_f)_{\text{actual}} = Cd_f A_j \sqrt{2\rho_f(\Delta p - gZ\rho_f)}$$

Here, $$Z = 0$$

$\therefore$ $$A_j = \frac{\dot{m}_f}{Cd_f\sqrt{2\rho_f \Delta p}} = \frac{0.4/60}{0.6\sqrt{2\times 780\times 0.0544\times 10^5}}$$

$$= (3.814 \times 10^{-6}) \text{ m}^2$$

$$= 3.814 \text{ mm}^2$$

$$\therefore \quad \frac{\pi}{4} d_j^2 = 3.814$$

$$\therefore \quad d_j = \sqrt{\frac{4 \times 3.814}{\pi}} = \boxed{2.2 \text{ mm}} \quad \textbf{Ans.}$$

EXAMPLE 9.2 A four-stroke petrol engine of 2 litre capacity is required to develop maximum power at 4500 rpm. The volumetric efficiency at this speed is assumed to be 75% and the air/fuel ratio is 14:1. Two carburettors are to be fitted and it is expected that at peak power the air speed at the choke will be 100 m/s. The coefficient of discharge for the venturi is assumed to be 0.85 and that of the main petrol jet is 0.66. An allowance should be made for the emulsion tube, the diameter of which can be taken as 0.4 of the choke diameter. The petrol surface is 6 mm below the choke at this engine condition. Calculate the size of a suitable choke and that of the main jet. The specific gravity of petrol is 0.75.

The atmospheric pressure and temperature are 1.013 bar and 15°C respectively.

Solution: Swept volume, V_s = 2 litre = 0.002 m^3. Volume of air inducted for four-stroke engine per second = $\eta_V \cdot V_s \cdot \dfrac{N}{2 \times 60}$

where η_V is the volumetric efficiency and N is the engine rpm.

$$\therefore \quad \text{Volume of air inducted} = \frac{0.75 \times 0.002 \times 4500}{2 \times 60} = 0.05625 \text{ m}^3\text{/s}$$

$$\text{Each carburettor delivers an air flow of } \frac{0.05625}{2} = 0.028125 \text{ m}^3\text{/s}$$

$$\therefore \quad \dot{V}_1 = 0.028125 \text{ m}^3\text{/s}$$

$$\dot{m}_a = \frac{p_1 \dot{V}_1}{RT_1} = \frac{1.013 \times 10^5 \times 0.028125}{287 \times 288} = 0.03447$$

kg/s

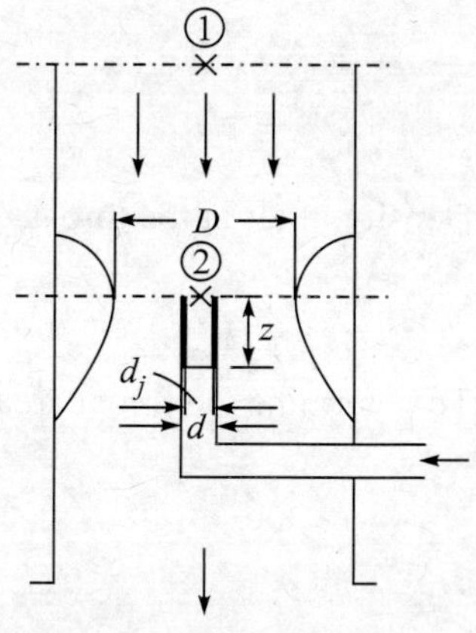

Figure 9.9 Example 9.2.

$$\text{Velocity at throat, } C_2 = \sqrt{2 c_p T_1 \left[1 - \left(\frac{p_2}{p_1}\right)^{(\gamma-1)/\gamma}\right]}$$

$$\therefore \quad \frac{p_2}{p_1} = \left[1 - \frac{C_2^2}{2 c_p T_1}\right]^{\gamma/(\gamma-1)}$$

$$= \left(1 - \frac{100 \times 100}{2 \times 1005 \times 288}\right)^{3.5} = 0.9408$$

$$\therefore \quad p_2 = 1.013 \times 0.9408 = 0.953 \text{ bar}$$

Density of air,

$$\rho_1 = \frac{p_1}{RT_1} = \frac{1.013 \times 10^5}{287 \times 288} = 1.2256 \text{ kg/m}^3$$

$$\therefore \qquad \rho_2 = \rho_1 \left(\frac{p_2}{p_1}\right)^{1/\gamma} = 1.2256(0.9408)^{1/1.4} = 1.1733 \text{ kg/m}^3$$

Throat area,

$$A_2 = \frac{\dot{m}_a}{\rho_2 C_2 Cd_a} = \frac{0.03447}{1.1733 \times 100 \times 0.85} = (3.456 \times 10^{-4}) \text{ m}^2 = 345.6 \text{ mm}^2$$

$$\therefore \qquad \frac{\pi}{4}(D^2 - d^2) = 345.6$$

Now,

$$d = 0.4D$$

$$\therefore \qquad \frac{\pi}{4}(D^2 - 0.16D^2) = 345.6$$

or

$$0.84 \times \frac{\pi}{4} D^2 = 345.6$$

or

$$D = \sqrt{\frac{345.6 \times 4}{\pi \times 0.84}} = \boxed{22.89 \text{ mm}} \quad \textbf{Ans.}$$

$$\dot{m}_f = Cd_f A_j \sqrt{2\rho_f (p_1 - p_2 - gZ\rho_f)}$$

$$\dot{m}_f = \frac{\dot{m}_a}{14} = \frac{0.03447}{14} = 0.002462 \text{ kg/s}$$

For the petrol, the pressure difference across the main jet is given by

$$p_1 - p_2 - gZ\rho_f = 1.013 - 0953 - \frac{9.81 \times 0.006 \times 750}{10^5}$$

$$= 1.013 - 0.953 - 0.00044$$

$$= 0.05956 \text{ bar} = (0.05956 \times 10^5) \text{ N/m}^2$$

Area of the jet,

$$A_j = \frac{\dot{m}_f}{Cd_f \sqrt{2\rho_f (p_1 - p_2 - gZ\rho_f)}}$$

$$= \frac{0.002462}{0.66\sqrt{(2 \times 750 \times 0.05956 \times 10^5)}}$$

$$= (1.248 \times 10^{-6}) \text{ m}^2 = 1.248 \text{ mm}^2$$

$$\therefore \qquad d_j = \sqrt{\left(\frac{4 \times 1.248}{\pi}\right)} = \boxed{1.26 \text{ mm}} \quad \textbf{Ans.}$$

EXAMPLE 9.3 A four-cylinder four-stroke spark ignition engine with 80 mm bore and 90 mm stroke runs at 4000 rpm and uses a fuel having 84% carbon and 16% hydrogen by mass. The volumetric efficiency of the engine at that speed is 80%. The ambient conditions are: pressure = 1.0 bar, temperature = 25°C. The depression at the venturi throat is 0.06 bar. The actual quantity of air supplied is 0.95 of the stoichiometric value. Calculate the fuel flow rate, the air velocity at the throat and the throat diameter.

Take R(air) = 287 J/(kg K); R(fuel vapour) = 98 J/(kg K).

Solution: As the gas constant R for fuel vapour is given, instead of only volume of air supplied to the engine, the volume of the mixture can be considered, which is more accurate.

∴ Volume of mixture supplied to the engine,

$$V = \frac{\pi}{4} d^2 L \eta_V \cdot \frac{N}{2 \times 60} \times \text{No. of cylinders}$$

$$= \frac{\pi}{4} (0.08)^2 (0.09)(0.8) \cdot \frac{4000}{2 \times 60} \times 4$$

$$= 0.04825 \text{ m}^3/\text{s}$$

$$\text{Stoichiometric air/fuel ratio} = \frac{100}{23}\left(\text{C} \times \frac{32}{12} + \text{H} \times 8\right)$$

$$= \frac{100}{23}\left(0.84 \times \frac{32}{12} + 0.16 \times 8\right) = 15.3:1$$

The actual mass of air supplied per kg of fuel

$$= 0.95 \times 15.3 = 14.535 \text{ kg/kg fuel}$$

∴ Actual air/fuel ratio = 14.535 : 1

The density of air at 1 bar and 298 K,

$$\rho_a = \frac{p}{RT} = \frac{10^5}{287 \times 298} = 1.169 \text{ kg/m}^3$$

The density of fuel vapour, $\rho_v = \dfrac{10^5}{98 \times 298} = 3.424 \text{ kg/m}^3$

The distribution of air and fuel is usually approximated by

Volume flow rate of air + Volume flow rate of fuel = Volume flow rate of mixture

i.e.

$$\frac{\dot{m}_a}{\rho_a} + \frac{\dot{m}_f}{\rho_v} = 0.04825$$

Also,

$$\dot{m}_f = \frac{\dot{m}_a}{14.535}$$

$$\therefore \quad \dot{m}_a\left(\frac{1}{1.169} + \frac{1}{14.535 \times 3.424}\right) = 0.04825$$

$\therefore \quad \dot{m}_a = 0.05511 \text{ kg/s}$

and $\quad \dot{m}_f = \dfrac{0.05511}{14.535} = (3.792 \times 10^{-3}) \text{ kg/s} = \boxed{0.003792 \text{ kg/s}}$ **Ans.**

The depression at the venturi throat = 0.06 bar

i.e. $\quad \Delta p = p_1 - p_2 = 0.06 \text{ bar}$

$\therefore \quad p_2 = p_1 - 0.06 = 1.00 - 0.06 = 0.94 \text{ bar}$

or $\quad \dfrac{p_2}{p_1} = 0.94$

Velocity of air at the throat, $\quad C_2 = \sqrt{2c_pT_1\left[1 - \left(\dfrac{p_2}{p_1}\right)^{(\gamma-1)/\gamma}\right]}$

$$= \sqrt{2 \times 1005 \times 298\,[1 - (0.94)^{0.286}]}$$

$$= \boxed{102.5 \text{ m/s}} \quad \textbf{Ans.}$$

The density of air at the throat, $\quad \rho_2 = \dfrac{p_2}{RT_2}$

$$T_2 = T_1\left(\frac{p_2}{p_1}\right)^{(\gamma-1)/\gamma} = 298\,(0.94)^{0.286} = 292.8 \text{ K}$$

$$\therefore \quad \rho_2 = \frac{0.94 \times 10^5}{287 \times 292.8} = 1.119 \text{ kg/m}^3$$

If the coefficient of discharge for the throat is assumed as unity, the cross-sectional area of the venturi throat,

$$A_2 = \frac{\dot{m}_a}{\rho_2 C_2} = \frac{0.05511}{1.119 \times 102.5} = (4.805 \times 10^{-4}) \text{ m}^2$$

$$= 4.805 \text{ cm}^2$$

$$\therefore \quad d_2 = \sqrt{\frac{4.805 \times 4}{\pi}} = 2.47 \text{ cm} = \boxed{24.7 \text{ mm}} \quad \textbf{Ans.}$$

EXAMPLE 9.4 A four-stroke four-cylinder spark-ignition engine having a bore of 100 mm and stroke of 120 mm and running at 3000 rpm has a carburettor venturi with a 35 mm throat diameter. The volumetric efficiency of the engine at this speed is 80%, the coefficient of discharge of air flow is 0.82. The ambient pressure and temperature are 1.013 bar and 25°C respectively. The air/fuel ratio is 15. The top of the jet is 5 mm above the petrol level in the float chamber. The coefficient of discharge for fuel flow is 0.7. Determine the depression at the throat and the diameter of the fuel jet of a simple carburettor. The specific gravity of petrol is 0.75.

Solution: Volume of air inducted per second $= \frac{\pi}{4} d^2 \times L \times \eta_V \cdot \frac{N}{2 \times 60} \times$ no. of cylinders

$$= \frac{\pi}{4} (0.1)^2 \times (0.12) \times 0.8 \times \frac{3000}{2 \times 60} \times 4$$

$$= 0.0754 \text{ m}^3/\text{s}$$

Density of air, $\rho_a = \frac{p}{RT} = \frac{1.013 \times 10^5}{287 \times 298} = 1.184 \text{ kg/m}^3$

$\therefore$ Mass of air/s $= \dot{m}_a = 0.0754 \times 1.184 = 0.08927$ kg/s

As the throat velocity and depression are both not given, the problem is solved by the approximate method, i.e. by assuming incompressible flow of air.

$$\dot{m}_a = Cd_a A_2 \sqrt{2\rho_a (p_1 - p_2)}$$

$$A_2 = \frac{\pi}{4} d_2^2 = \frac{\pi}{4} (0.035)^2 = (9.621 \times 10^{-4}) \text{m}^2$$

$$\therefore \quad p_1 - p_2 = \frac{\dot{m}_a^2}{Cd_a^2 \, A_2^2 \, 2\rho_a}$$

$$= \frac{(0.08927)^2}{(0.82)^2 (9.621 \times 10^{-4})^2 (2 \times 1.184)} = 5407 \text{ N/m}^2$$

$$= \boxed{0.05407 \text{ bar}} \quad \textbf{Ans.}$$

$$\dot{m}_f = \frac{\dot{m}_a}{15}$$

or $\dot{m}_f = \frac{0.08927}{15} = 5.951 \times 10^{-3}$ kg/s

Now, $\dot{m}_f = Cd_f A_j \sqrt{2\rho_f (p_1 - p_2 - gZ\rho_f)}$

$$\therefore \quad A_j = \frac{\dot{m}_f}{Cd_f \sqrt{2\rho_f (p_1 - p_2 - gZ\rho_f)}}$$

$$= \frac{5.951 \times 10^{-3}}{0.7 \sqrt{2 \times 750 \, (5407 - 9.81 \times 0.005 \times 750)}}$$

$$= 3 \times 10^{-6} \text{ m}^2 = 3 \text{ mm}^2$$

$$\therefore \quad d_j = \sqrt{\frac{3 \times 4}{\pi}} = \boxed{1.95 \text{ mm}} \quad \textbf{Ans.}$$

EXAMPLE 9.5 An engine having a single jet carburettor consumes 6.0 kg/h of fuel. The density of fuel is 750 kg/m³. The level in the float chamber is 3 mm below the top of the jet when the engine is not running. Ambient conditions are 1.013 bar and 21°C. The jet diameter is 1.2 mm and its discharge coefficient is 0.65. The discharge coefficient of air is 0.80. The air/fuel ratio is 15.3:1. Determine the critical air velocity, the depression at the throat in mm of H_2O and the effective throat diameter. Neglect the compressibility of air.

Solution: Density of air, $\rho_a = \dfrac{p}{RT} = \dfrac{1.013\times10^5}{287\times294} = 1.2\ \text{kg/m}^3$

Neglecting the compressibility of air, the critical air velocity at the throat,

$$C_{a2} = Cd_a\sqrt{\frac{2gZ\rho_f}{\rho_a}} = 0.8\sqrt{\frac{2\times9.81\times0.003\times750}{1.2}}$$

$$= \boxed{4.852\ \text{m/s}} \quad \textbf{Ans.}$$

Fuel flow per second,

$$\dot{m}_f = Cd_f A_j\sqrt{2\rho_f(p_1 - p_2 - gZ\rho_f)}$$

Given: $\dot{m}_f = \dfrac{6}{3600} = \dfrac{1}{600}\ \text{kg/s};\qquad Cd_f = 0.65;\qquad Z = 0.003\ \text{m}$

$$A_j = \frac{\pi}{4}d_j^2 = \frac{\pi}{4}\times(1.2)^2 = 1.131\ \text{mm}^2 = (1.131\times10^{-6})\ \text{m}^2$$

$$2\rho_f(p_1 - p_2 - gZ\rho_f) = \frac{\dot{m}_f^2}{Cd_f^2\,A_j^2}$$

$$\therefore\qquad p_1 - p_2 = \frac{\dot{m}_f^2}{Cd_f^2\,A_j^2\,2\rho_f} + gZ\rho_f$$

$$= \frac{1}{(600)^2(0.65)^2(1.131\times10^{-6})^2(2\times750)} + 9.81\times0.003\times750$$

$$= 3427 + 22 = 3449\ \text{N/m}^2 = 0.03449\ \text{bar}$$

In metre of water, $h = \dfrac{3449}{\rho_{H_2O}\times g} = \dfrac{3449}{1000\times9.81}$

In mm of H_2O, $h = \dfrac{3449}{9.81} = \boxed{351.6\ \text{mm of}\ H_2O}$ **Ans.**

Air flow per second,

$$\dot{m}_a = Cd_a A_2\sqrt{2\rho_a(p_1 - p_2)}$$

Also, $\dot{m}_a = \dfrac{1}{600}\times15.3 = 0.0255\ \text{kg/s}$

Here, $Cd_a = 0.8$, $\rho_a = 1.2$ kg/m³, $p_1 - p_2 = 3449$ N/m²

$$\therefore \quad A_2 = \frac{\dot{m}_a}{Cd_a\sqrt{2\rho_a(p_1 - p_2)}}$$

$$= \frac{0.0255}{0.8\sqrt{(2 \times 1.2 \times 3449)}} = (3.5035 \times 10^{-4}) \text{ m}^2$$

$$= 3.5035 \text{ cm}^2$$

$$\therefore \quad d_2 = \sqrt{\frac{3.5035 \times 4}{\pi}} = 2.112 \text{ cm}$$

$$= \boxed{21.12 \text{ mm}} \quad \textbf{Ans.}$$

EXAMPLE 9.6 A simple carburettor has a venturi throat diameter of 22 mm and the coefficient of air flow is 0.82. The fuel orifice has a diameter of 1.2 mm and the coefficient of fuel flow is 0.70. The petrol surface is 4 mm below the throat. Find:

(a) The air/fuel ratio when the nozzle lip is neglected.
(b) The air/fuel ratio for a pressure drop of 0.075 bar when the nozzle lip is taken into account.
(c) The minimum velocity of air or critical air velocity required to start the fuel flow when the nozzle lip is provided.

Take density of air and fuel as 1.2 kg/m³ and 750 kg/m³ respectively.

Solution: (a) When the nozzle lip Z is neglected:

Rate of mass flow of air, $\dot{m}_a = Cd_a A_2 \sqrt{2\rho_a \, \Delta p}$

Rate of mass flow of fuel, $\dot{m}_f = Cd_f A_j \sqrt{2\rho_f \, \Delta p}$

Air/fuel ratio, $$\frac{\dot{m}_a}{\dot{m}_f} = \frac{Cd_a \, A_2}{Cd_f \, A_j}\sqrt{\frac{\rho_a}{\rho_f}}$$

$$= \frac{Cd_a \, d_2^2}{Cd_f \, d_j^2}\sqrt{\frac{\rho_a}{\rho_f}}$$

$$= \frac{0.82}{0.70} \times \left(\frac{22}{1.2}\right)^2 \times \sqrt{\frac{1.2}{750}} = \boxed{15.75} \quad \textbf{Ans.}$$

(b) When the nozzle lip Z is considered:

$\dot{m}_a$ will remain the same.

But $\dot{m}_f = Cd_f A_j \sqrt{2\rho_f(\Delta p - gZ\rho_f)}$

Now, $gZ\rho_f = 9.81 \times 0.004 \times 750 = 29.43\ \text{N/m}^2$

$= 0.0002943$ bar

and $\Delta p - gZ\rho_f = 0.075 - 0.0002943 = 0.0747$ bar

Air/fuel ratio,
$$\frac{\dot{m}_a}{\dot{m}_f} = \frac{Cd_a}{Cd_f} \cdot \frac{A_2}{A_j} \sqrt{\frac{\rho_a \, \Delta p}{\rho_f (\Delta p - gZ\rho_f)}}$$

$$= \frac{0.82}{0.7} \left(\frac{22}{1.2}\right)^2 \sqrt{\frac{1.2 \times 0.075}{750 \times 0.0747}} = \boxed{15.78} \quad \textbf{Ans.}$$

(c) When the nozzle lip is provided, the flow of fuel will start only when there is sufficient depression to overcome the nozzle lip effect. The depression is created by the critical velocity of air.

∴ Minimum velocity of air or critical velocity at the throat,

$$C_{a2} = \sqrt{\frac{2gZ\rho_f}{\rho_a}}$$

$$= \sqrt{\frac{2 \times 9.81 \times 0.004 \times 750}{1.2}} = \boxed{7\ \text{m/s}} \quad \textbf{Ans.}$$

EXAMPLE 9.7 Determine the air/fuel ratio in an airplane engine carburettor at 4000 m altitude. The carburettor is adjusted to give an air/fuel ratio of 14.7 : 1 at sea level where the air temperature is 22°C and the pressure 1.013 bar. The temperature variation with altitude is given by the expression

$$T = T_{\text{sea level}} - 0.0065h,$$

where h is the height in metres and $T_{\text{sea level}}$ is the sea level temperature in °C. The air pressure decreases with altitude as per the relation, $h = 19200 \log_{10} (1.013/p)$, where p is in bar.

Solution: *Given:*

$$T = T_{\text{sea level}} - 0.0065h$$
$$= 22 - (0.0065 \times 4000)$$
$$= 22 - 26 = -4\text{°C} = 269\ \text{K}$$

Now, $h = 19200 \log_{10}(1.013/p)$

or $4000 = 19200 \log_{10}(1.013/p)$

or $$\log_{10}\left(\frac{1.013}{p}\right) = \frac{4000}{19200} = 0.2083$$

or $$\frac{1.013}{p} = (10)^{0.2083} = 1.6155$$

∴ $$p = \frac{1.013}{1.6155} = 0.627\ \text{bar}$$

$$\frac{\text{air/fuel ratio at altitude}}{\text{air/fuel ratio at sea level}} = \sqrt{\frac{\rho_{\text{altitude}}}{\rho_{\text{sea level}}}}$$

where $$\rho_{\text{altitude}} = \frac{p}{RT} = \frac{0.627 \times 10^5}{287 \times 269} = 0.81214 \text{ kg/m}^3$$

and $$\rho_{\text{sea level}} = \frac{1.013 \times 10^5}{287 \times 295} = 1.196 \text{ kg/m}^3$$

$$\therefore \quad \frac{\text{air/fuel ratio at altitude}}{14.7} = \sqrt{\frac{0.81214}{1.196}}$$

$$\therefore \quad \text{Air/fuel ratio at altitude} = 14.7\sqrt{\frac{0.81214}{1.196}} = \boxed{12.11} \quad \textbf{Ans.}$$

It is observed that with the increase in altitude the mixture becomes rich. Therefore, some altitude compensating device should be incorporated for proper combustion of fuel.

EXAMPLE 9.8 A carburettor is tested without an air-cleaner. The main metering system is adjusted to give air/fuel ratio of 14.5:1. The pressure at the venturi throat is 0.825 bar. The atmospheric pressure is 1.013 bar. The same carburettor is tested again with an air-cleaner. The pressure drop to the air-cleaner is found to be 37.5 mm of Hg. The air flow is 260 kg per hour. Assuming zero nozzle lip and constant coefficient of flow, determine (a) the throat pressure when the air-cleaner is fitted and (b) the air/fuel ratio with the air-cleaner fitted.

Solution: (a) Without the air-cleaner, the depression at the throat,

$$\Delta p_a = 1.013 - 0.825 = 0.188 \text{ bar}$$

When the air-cleaner is fitted, let p_2 be the throat pressure,

then, $$\Delta p'_a = \left(1.013 - \frac{37.5}{750} - p_2\right) \text{bar}$$

$$= (0.963 - p_2) \text{ bar}$$

For the same air flow and constant coefficient of flow,

$$\Delta p_a = \Delta p'_a$$

or $$0.188 = 0.963 - p_2$$

$$\therefore \quad p_2 = 0.963 - 0.188 = \boxed{0.775 \text{ bar}} \quad \textbf{Ans.}$$

(b) Without the air-cleaner, $\Delta p_f = \Delta p_a = 0.188$ bar

With the air-cleaner, $\Delta p_f = 1.013 - 0.775 = 0.238$ bar

As Δp_f has increased, more fuel will flow through, thus making the mixture richer.

The air flow is given by

$$\dot{m}_a = Cd_a A_2 \sqrt{2\rho_a \Delta p_a}$$

Since Δp_a and all other terms are same in both the cases, i.e. with and without the air-cleaner, $\dot{m}_f$ remains the same.

The fuel flow without lip ($Z = 0$),

$$\dot{m}_f = Cd_f A_j \sqrt{2\rho_f \, \Delta p_f}$$

Cd_f, A_j and ρ_f are the same in both the cases, but Δp_f is different.

$$\therefore \quad \frac{(\dot{m}_f)_{\text{without air-cleaner}}}{(\dot{m}_f)_{\text{with air-cleaner}}} = \sqrt{\frac{(\Delta p_f)_{\text{without air-cleaner}}}{(\Delta p_f)_{\text{with air-cleaner}}}}$$

or

$$\frac{(\dot{m}_a/\dot{m}_f)_{\text{with air-cleaner}}}{(\dot{m}_a/\dot{m}_f)_{\text{without air-cleaner}}} = \sqrt{\frac{(\Delta p_f)_{\text{without air-cleaner}}}{(\Delta p_f)_{\text{with air-cleaner}}}}$$

$$\therefore \quad \text{Air/fuel ratio with the air-cleaner} = 14.5\sqrt{\frac{0.188}{0.238}}$$

$$= \boxed{12.89} \quad \textbf{Ans.}$$

EXAMPLE 9.9 A petrol engine develops 8 kW brake power having brake thermal efficiency of 30% when working at the full-load condition. The calorific value of the fuel is 44,000 kJ/kg. The suction conditions of the engine are 1.013 bar and 300 K. The carburettor fitted on this engine has a single jet of 2.5 mm^2 and the nozzle lip is 8 mm. Determine the venturi throat diameter of the carburettor to provide an air/fuel ratio of 15:1. Take the following data:

$$Cd_a = 0.9, \qquad \rho_f = 750 \text{ kg/m}^3, \qquad Cd_f = 0.7$$

and

$$v_a(\text{at NTP}) = 0.8 \text{ m}^3/\text{kg}$$

Solution: Specific volume of air at atmospheric pressure and 300 K

$$v_a = \frac{v_a(\text{at NTP}) \times 300}{273} = \frac{0.8 \times 300}{273} = 0.8791 \text{ m}^3/\text{kg}$$

$\therefore$ Density of air at the inlet condition,

$$\rho_a = \frac{1}{v_a} = \frac{1}{0.8791} = 1.1375 \text{ kg/m}^3$$

Brake thermal efficiency,

$$\eta_b = \frac{\text{bp(kW)}}{\dot{m}_f\,(\text{kg/s}) \times \text{CV (kJ/kg)}}$$

or

$$0.3 = \frac{8}{\dot{m}_f \times 44{,}000}$$

$$\therefore \quad \dot{m}_f = \frac{8}{0.3 \times 44{,}000} = 0.000606 \text{ kg/s}$$

From the continuity equation,

$$\dot{m}_f = Cd_f \rho_f A_j C_f$$

∴ Velocity of fuel,
$$C_f = \frac{\dot{m}_f}{Cd_f \rho_f A_j}$$

$$= \frac{0.000606}{0.7 \times 750 \times 2.5 \times 10^{-6}} = 0.4617 \text{ m/s}$$

Now,
$$C_f = \sqrt{\frac{2}{\rho_f}(\Delta p - \rho_f g Z)}$$

or
$$\Delta p - \rho_f g Z = \frac{C_f^2 \rho_f}{2}$$

or
$$\Delta p = \frac{C_f^2 \rho_f}{2} + \rho_f g Z$$

$$= \frac{(0.4617)^2\, 750}{2} + 750 \times 9.81 \times 0.008$$

$$= 79.94 + 58.86 = 138.8 \text{ N/m}^2$$

Now, velocity of air at the throat,

$$C_a = \sqrt{\frac{2\Delta p}{\rho_a}} = \sqrt{\frac{2 \times 138.8}{1.1375}} = 15.62 \text{ m/s}$$

Air supplied to the engine per second,

$$\dot{m}_a = \dot{m}_f \times \frac{\text{A}}{\text{F}} = 0.000606 \times 15 = 0.00909 \text{ kg/s}$$

and
$$\dot{m}_a = Cd_a \rho_a A_2 C_a$$

∴
$$A_2 = \frac{\dot{m}_a}{Cd_a \rho_a C_a} = \frac{0.00909}{0.9 \times 1.1375 \times 15.62} = (5.684 \times 10^{-4}) \text{ m}^2 = 5.684 \text{ cm}^2$$

∴
$$d_2 = \sqrt{\frac{4 \times 5.684}{\pi}} = \boxed{2.69 \text{ cm}} \quad \textbf{Ans.}$$

9.13 PROBLEMS ASSOCIATED WITH CARBURETTORS

Some of the problems associated with satisfactory running of the carburettor are discussed in the following subsections.

9.13.1 Ice Formation

When a volatile liquid evaporates, the temperature around is lowered owing to the effect of latent heat of evaporation. Applied to the fuel jet of a carburettor this evaporation lowers the temperature of the intake air and if the air contains any moisture, some of it will be condensed. If the outside temperature is low, as in winter or at higher altitudes as in the case of aircraft engines, the moisture in the air will tend to form ice on all internal surfaces exposed to the mixture stream. Moisture in the intake air tends to form ice on the throttle blade and on the parts of the carburettor barrel near to the blade. In the course of time the whole of the top edges become coated and the air flow is restricted, therefore at low throttle openings the engine may cease to operate.

As this poses a serious problem, steps must be taken to overcome this difficulty. The following methods may be employed:

1. Heating the intake air.
2. Heating the metal surface, i.e. the choke and mixture chamber by coolant from the water jackets or exhaust gases, on which ice is liable to form.
3. The use of ice preventives or inhibitors. If alcohol such as ethanol or methanol is added to the fuel, freezing can be avoided. An ice detector and an automatic servomechanism can be used to deliver alcohol on to the surfaces liable to be covered with ice until the ice is dispersed.

9.13.2 Vapour Lock in Fuel Systems

The improved volatility of modern fuels, and the necessity of providing heat to prevent icing troubles, has led to the occasional occurrence of carburetion difficulties, due to vaporization of the fuel in the petrol pipes, fuel feed pump, carburettor float chamber and the jet walls. Apart from the fuel volatility, excessive heating of the fuel can be due to petrol pipes being too near the engine and due to heat conducted from the engine metal to the fuel pump and carburettor inlet manifold flanges.

After a car engine has been used for a road journey and is left to idle, the cooling system becomes less efficient due to lack of cooling air, and therefore the bonnet temperature rises. A few minutes after stopping the car, but with the engine idling, the carburettor and the cooling water temperatures usually rise considerably. Vapour lock can occur under these conditions.

The formation of fuel vapour in the carburettor may result in a weak mixture. The vapour will occupy a greater volume than the liquid and therefore the amount of fuel flow will be reduced. The reduction will cause either a loss in power or else complete stoppage of the engine.

The heat insulation of the fuel pump and the carburettor flanges by non-conducting type gaskets will often prevent vapour lock.

9.13.3 Backfiring or Popping in the Carburettor

Backfiring or popping in the carburettor is an occasional weak explosion in the inlet pipe and carburettor. It occurs due to too weak a mixture or insufficient heating. The mixture is so weak that the explosion flame travels very slowly through it with the result that inflammation occurs not only during the firing and exhaust strokes, but also continues when the inlet valve opens again. It often happens in cold weather at starting. The remedy is to increase the fuel supply or reduce the air supply, and to look for extraneous sources of air leakage.

9.14 CARBURETTOR DRAWBACKS

The following are the main drawbacks of the carburettor which prompted the use of fuel-injection system in SI engines:

1. The carburettor has certain wearing parts. When wear occurs, it usually operates less efficiently.
2. There is a maldistribution of mixture quantity and quality in multi-cylinder engines, since the induction passages are of unequal lengths and offer different resistances to mixture flow.
3. There is a loss of volumetric efficiency on account of restrictions of free flow passages for mixture through the venturi tube, the jet, the throttle valve, the inlet pipe bend, etc.
4. Freezing of moisture of air at low temperatures may take place unless some means are provided to eliminate it.
5. When the carburettor is tilted, while the vehicle is negotiating a grade, surging of the fuel is caused in the float chamber.
6. There is a possibility of backfire in the intake manifold and popping in the carburettor in cold weather, unless the flame traps are fitted. It involves an additional complication which tends to reduce the volumetric efficiency.
7. Vapour lock in the fuel systems may result in hot weather.

9.15 FUEL-INJECTION SYSTEMS IN SI ENGINES

In view of the several disadvantages of the carburettor, the fuel-injection system in SI engines is the right solution. This system is getting more popular on modern vehicles with multi-cylinder engines. Table 9.1 shows the comparison between the carburettor-mounted SI engine and the fuel-injection SI engine.

Table 9.1 Carburettor-mounted vs. fuel-injection SI engines

Description	*Carburettor-mounted SI engine*	*Injector-mounted SI engine*
Applications	Small engines	Larger engines having multi-cylinders.
Precision in manufacturing	To a high degree	To a much higher degree
Power output	Usual	Produces 10 to 20% more power
Necessity of choke	Yes	No
Fuel distribution	Good	Very good
Detonation	Possible	Reduced
Compression ratio	High	Much higher
Fuel consumption	Usual	Comparatively less
Engine response to throttle	Quick	Very fast
Time lag between throttle movement and fuel injection	Short	Much shorter
Maintenance	Less	More
Cost	Less	High
Exhaust emissions	More	Less

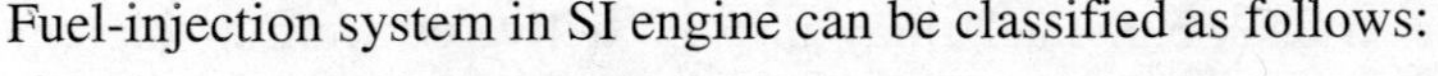

Fuel-injection system in SI engine can be classified as follows:

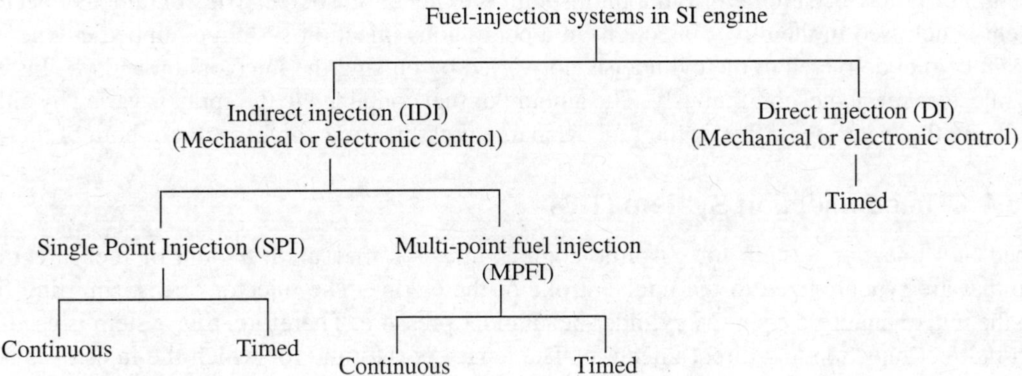

In indirect injection, fuel is injected into the air stream prior to entering the combustion chamber. In direct injection, fuel is injected directly inside the combustion chamber.

9.15.1 Single-point Injection (SPI)

It is also called single-point throttle body injection or simply throttle body injection (TBI). The injector is usually placed just upstream of the throttle valve in the intake manifold. It uses a single injector. It has low cost. Many of the carburettor's supporting components, such as the air filter, throttle valve and intake manifold could be reused. There is no venturi restriction, so higher volumetric efficiency is obtained. Injection pressure is higher compared to carburettor discharge pressure, which speeds up and improves the atomization of the liquid fuel. However, maldistribution of fuel cannot be avoided. To overcome maldistribution of fuel multi-point fuel injection system can be used.

Single-point injection was extensively used on American-made passenger cars and light trucks during 1980–1995, as well as in some European cars in the early and mid-1990s.

9.15.2 Multi-point Fuel Injection

It is also called multi-point port fuel injection or indirect multi-point injection (IMPI) or simply multi-port injection (MPI). The injectors are positioned in the intake ports just upstream of each cylinder's intake valve. It requires one injector per cylinder and in some systems, one or more injectors are used to supplement the fuel flow during starting and warm-up periods. The advantages of port fuel injection are increased power and torque through improved volumetric efficiency and more uniform fuel distribution to each cylinder, more rapid engine response to changes in throttle position and more precise control of the equivalence ratio during cold start and engine warm-up.

The above two schemes fall under the category of indirect injection. Indirect injection can be discharged at relatively low pressure (2–6 bar). With indirect injection arrangement, two methods are used in discharging the fuel into the air stream. They are continuous injection and timed injection.

9.15.3 Continuous Injection System

In a continuous injection system (CIS), the injection nozzle and its valve are permanently open while the engine is running. Fuel is injected into the airstream at all times from the fuel injectors prior to entering the combustion chamber. Fuel spray may be delivered from a single-point injection or it may be supplied from a multi-point fuel injection source. The continuous injection system is not used in the direct injection. In a continuous injection system equipped engine, the amount of fuel delivered to the cylinder is not varied by pulsing the injectors on and off. Instead CIS injectors spray fuel continuously. The amount of fuel contained in the spray is varied by either regulating the metering orifice or the fuel discharge pressure, or a combination of both of these.

9.15.4 Timed Injection System (TIS)

Timed fuel injection system for gasoline engines injects a measured amount of fuel in certain time that are synchronized to the intake stroke of the engine. The injector opens, squirting fuel into the intake manifold or in the cylinder head under pressure. Therefore, this system is used for both indirect injection and direct injection. The length of the time for which the injector is held open depends on the input signals seen by the engine management system from its various engine sensors.

Many modern electronic fuel injection (EFI) systems utilize multi-point fuel injection in which injection is timed to coincide with each cylinder's intake stroke. However, in newer gasoline engines, direct injection systems are beginning to replace MPFI system.

9.15.5 Direct Injection

In a direct injection engine, fuel is injected directly into the combustion chamber. Modern gasoline engines may utilize direct injection using electronic control, which is referred to as gasoline direct injection (GDI). In IC engines, GDI is also known as petrol direct injection, direct petrol injection, spark ignition direct ignition (SIDI) or fuel stratified injection (FSI). This is the next step in evolution from multi-point injection. It reduces emissions and fuel consumption.

9.16 MECHANICAL FUEL-INJECTION SYSTEMS

Some important fuel-injection system in SI engine using mechanical control are described in the following subsections.

9.16.1 Continuous Injection System with Mechanical Control

The principle of the continuous injection system is to introduce a steady flow of fuel at low pressure into the air supply. The principle of this system is illustrated in Figure 9.10. The fuel is drawn from a fuel tank (1) by a fuel pump (2) and delivered to a speed sensing mechanism (3), which is driven by the engine. The fuel at a pressure around 2 bar is delivered to the main metering system (4), which is a density bellows chamber unit for the purpose of regulating the amount of fuel according to the inlet manifold and atmospheric pressures. The fuel then passes through an engine idling unit (5). The acceleration pump (6) delivers an extra supply of fuel for quick acceleration purposes. It consists of a diaphragm which is forced upwards by the fuel pressure so as to store the extra fuel for acceleration. When the accelerator pedal is depressed sharply, this fuel is forced into the delivery side of the fuel nozzle injection system. Automatic starting requirements for the engine are taken care of by a bimetallic regulator device (7), which when actuated controls a separate starting or idling air valve in the body of the throttle. It also regulates the fuel flow for starting and the fuel flow during the engine warming-up period. The throttle plate (8) is as usual. The injection nozzle (9) can be arranged under the throttle body or at each of the cylinder inlet ports. The nozzle has a spring-actuated pressure regulator which controls the fuel flow in accordance with the fuel line pressure.

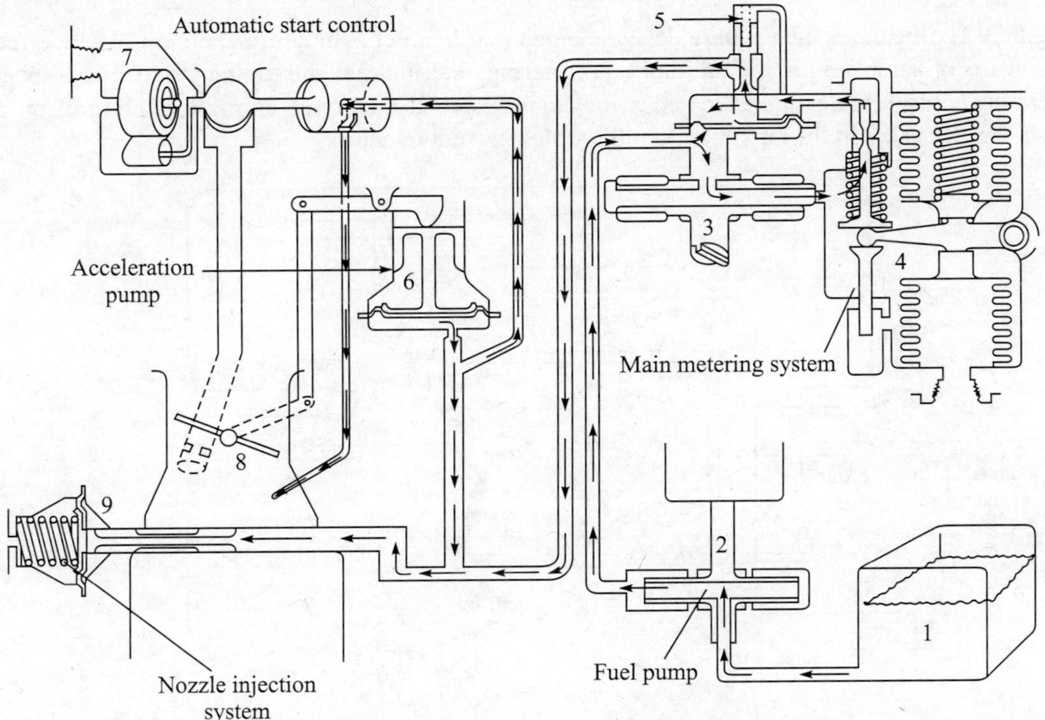

Figure 9.10 Continuous injection system

9.16.2 Timed Injection System with Mechanical Control

In this system there is a single pump unit which delivers fuel under pressure to a rotating distributor, the purpose of which is to deliver the metered quantity of fuel at exactly the correct time to each cylinder in turn in relation to the piston's position in its compression stroke. The fuel supply is regulated by engine speed, inlet manifold vacuum, atmospheric pressure and temperature, warm and cold starting requirements, idling conditions and rapid acceleration. In place of the rotating distributor, a single pump unit containing as many fuel pump plungers and barrels as there are cylinders in the engine can also be used. This is similar to the diesel engine fuel-injection pump. The individual plunger controls the amount of fuel and its time of injection. The controls of the SI engine fuel-injection system differ considerably from the CI engine fuel-injection system.

In the method described above the fuel is directly injected inside the engine cylinder during the compression stroke; this method is called the direct cylinder injection method using the timed injection system.

An alternative to the above method is to use port injection, which delivers a measured small quantity of fuel into each cylinder during the induction stroke at low pressure, but at a definite time and over a definite period of this stroke. The two methods of timed injection system are described below.

Direct cylinder injection

Figure 9.11 illustrates the primary design elements of a direct cylinder injection of gasoline fuel. It consists of a transfer pump T, a filter F, a metering, distributing, and timing pump P with speed-metering control (this pump is geared to the engine), a load-metering control R, a mixture control M (a throttle valve in the air stream), and nozzles N (fuel injectors).

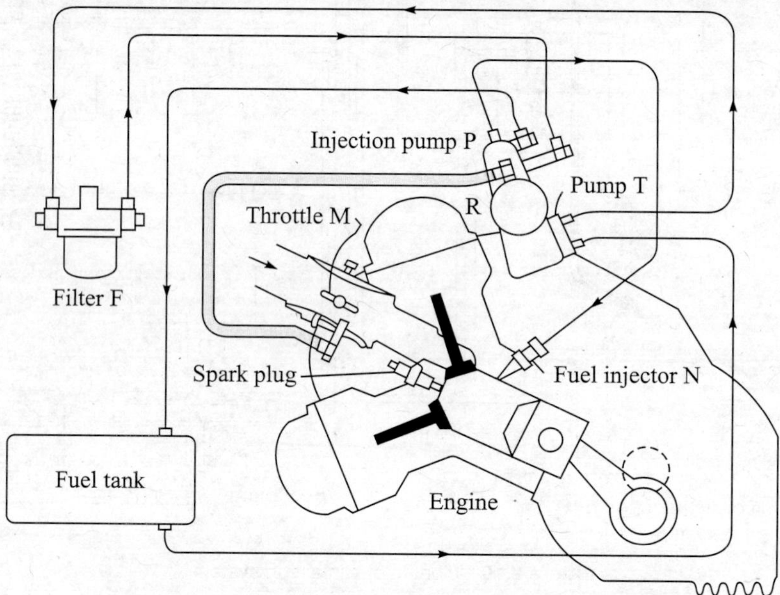

Figure 9.11 Direct cylinder injection of gasoline fuel.

The pump plunger raises the pressure and meters the correct amount of fuel for the load on the engine, and also delivers the fuel into the cylinder over a particular interval of the cycle. The control R is a device used to exert the manifold pressure on a diaphragm attached to the rack of the pump. Thus at wide open throttle, the maximum quantity of fuel per stroke will be delivered, and the fuel quantity will decrease with decrease in pressure in the throttled manifold.

Direct-injection of gasoline into the cylinder using mechanical control is rarely done because of (a) the difficulty of finding space in the head for an injector, (b) the added cooling and casting complications, (c) the added cost, (d) the refinements that are necessary for idling, and (e) the problems of exact metering at light loads from cylinder to cylinder owing to the individual plungers.

Lucas petrol-injection system

Figure 9.12 shows the schematic layout of the Lucas petrol-injection system. The timed port injection of fuel is used in this system. It contains a high pressure (7 bar) gear pump P, a metering

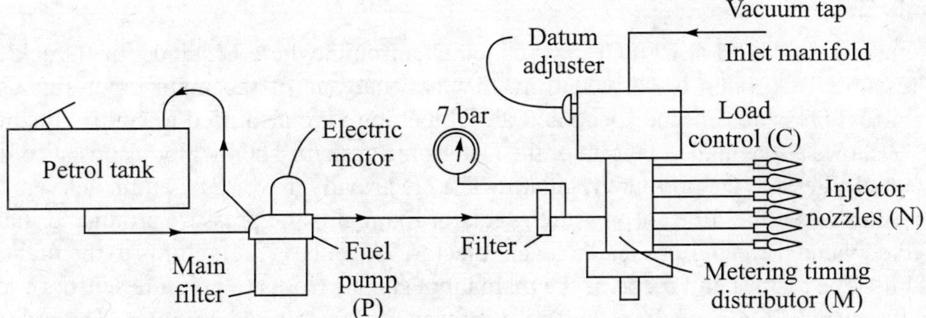

Figure 9.12 Schematic diagram of the Lucas fuel-injection system.

and timing distributor M (geared to the engine), a load control C, and the atomizing injector nozzles N (3.5 bar).

An electrically driven fuel pump supplies filtered fuel at a pressure of nearly 7 bar to the combined metering distributor and mixture control unit mounted on, and driven by, the engine. From the metering distributor, the accurately timed and metered quantities of fuel are delivered at each injector in turn. A relief valve returns excess fuel to the tank and maintains the line pressure at 7 bar.

9.17 ELECTRONIC FUEL-INJECTION SYSTEMS (EFIs)

Early fuel-injection systems were mechanical and used complex designs. They have now been superseded by electronic fuel-injection systems (EFIs). Some important fuel-injection system in SI engines using electronic control are described in the following subsections.

9.17.1 Single-point Throttle Body Injection with Electronic Control

In this system an electronically controlled injector meters the fuel and injects it into the air flow directly above the throttle body. It provides electronic control of fuel metering at a cost lower than the multipoint port injection systems. The injector meters the fuel in response to calibrations of air flow based on intake manifold pressure, air temperature, and engine speed. This system creates problems in obtaining equal distribution of fuel to all cylinders, and the fuel particles have a tendency to deposit on the walls of the manifold when cold. It produces about 10 per cent less power than a multi-point port injection.

9.17.2 Multi-point Port Injection with Electronic Control

In this system, fuel is injected into the intake port of each engine cylinder. Injectors are located immediately before the inlet valve of each cylinder. The combination of emission control and fuel economy gives the impact for using EFI. With this system it is possible to operate the engine at the stoichiometric air/fuel ratio and to apply closed-loop control. These features result in an appreciable fuel economy. Two basic approaches are described below.

D-Jetronic EFI system

The first generation of EFI at BOSCH was called D-Jetronic, where D stands for 'Druck', which means pressure. This name is derived from the fact that one of the main input signals is the intake manifold pressure. Engine speed and air temperature are also used to control the air flow. Figure 9.13 shows a schematic diagram of the D-Jetronic system. The fuel loop consists of the fuel pump, the fuel filter and the pressure regulator. The electrically driven fuel pump delivers the fuel through a filter to the fuel line. A pressure regulator maintains the pressure around 2.7 bar in the line at a fixed value. Branch lines lead to each injector. The excess fuel returns to the fuel tank via a second line. The control unit receives the main input signals from the pressure sensors connected to the intake manifold and receives the speed information from the distributor. The inducted air flows through the air filter, past the throttle plate to the intake manifold. The other components, i.e. the cold start valve, the thermal timing switch, the temperature sensor and auxiliary air valve are used for cold start and warm-up control.

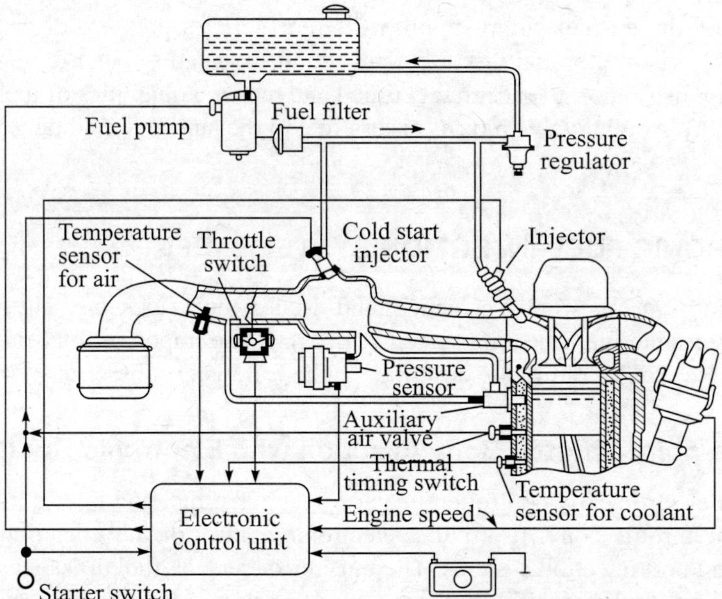

Figure 9.13 D-Jetronic EFI system with multi-point port injection.

L-Jetronic EFI system

This is the second generation EFI system developed at BOSCH. The L-Jetronic uses the air flow rate as one of the basic input signals. Its name is due to this fact: L stands for 'Luftmengenmessung', which means air flow measurement. A schematic diagram of the L-Jetronic multi-point port fuel injection (MPFI) system is shown in Figure 9.14.

The fuel loop is basically the same as in the D-Jetronic system, except that the pressure regulator is connected through a hose to the intake manifold. Thus the fuel pressure is a function of the manifold pressure, and consequently the pressure drop across the injectors is kept constant.

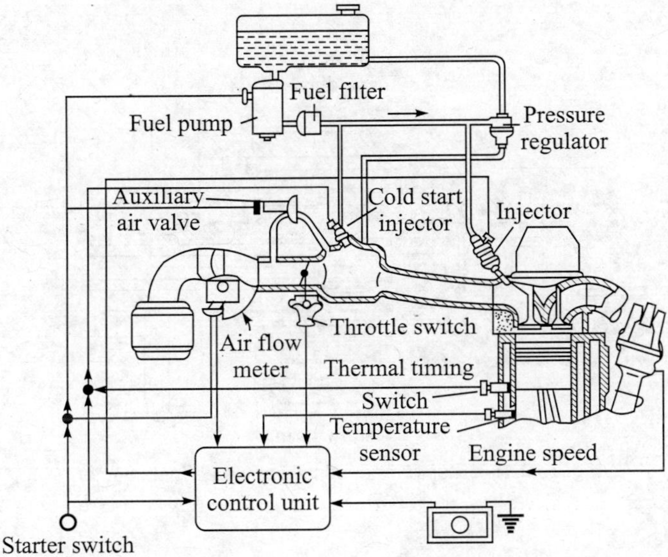

Figure 9.14 L-Jetronic EFI system with multi-point port fuel injection.

The air flow rate is measured by the airflow meter whose movable measuring plate is opened by the air stream against the force of a spring. The position of the measuring plate is sensed by a potentiometer. Its voltage represents the air flow rate, and is one of the main input signals going into the control unit. The second one is the engine speed taken from the distributor. For cold starting, additional fuel in the highly vapourized form is injected by the cold start injector which is mounted on the intake manifold. The cold start valve is controlled by the starter switch and the thermo time switch. During the warm-up period, additional air bypassing the throttle plate is controlled by the auxiliary air valve. The throttle position switch is used to adjust the air/fuel ratio at full-load and at idling and overrun.

In this system, the cost is reduced by using integrated circuits in the electronic control unit, which at the same time results in higher reliability due to the reduced number of components.

Fuel-injector

An electromagnetically actuated fuel injection valve is shown in Figure 9.15. It is located either in the intake port or in the intake manifold tube of each cylinder. It consists of valve housing, the injector spindle, the magnetic plunger to which the spindle is connected, the helical spring and the solenoid coil. When the solenoid is not energized, the solenoid plunger of the magnetic circuit is forced with its seal against the valve seat by the helical spring and closes the fuel passage. When the solenoid coil is energized, the plunger is attracted and lifts the spindle by about 0.15 mm, so that the fuel can flow through the calibrated annular passage around the valve stem. The front end of the injector spindle is shaped as an atomizing pintle with a ground top to atomize the injected fuel. The mass of the fuel injected per injection is controlled by varying the duration of the current pulse that excites the solenoid coil. Typical injection times for automobile applications range from about 1.5 to 10 ms. The appropriate coil excitation pulse duration or width is set by the electronic control unit (ECU). The control unit also initiates mixture enrichment during cold-engine operation and during acceleration that are detected by the throttle sensor.

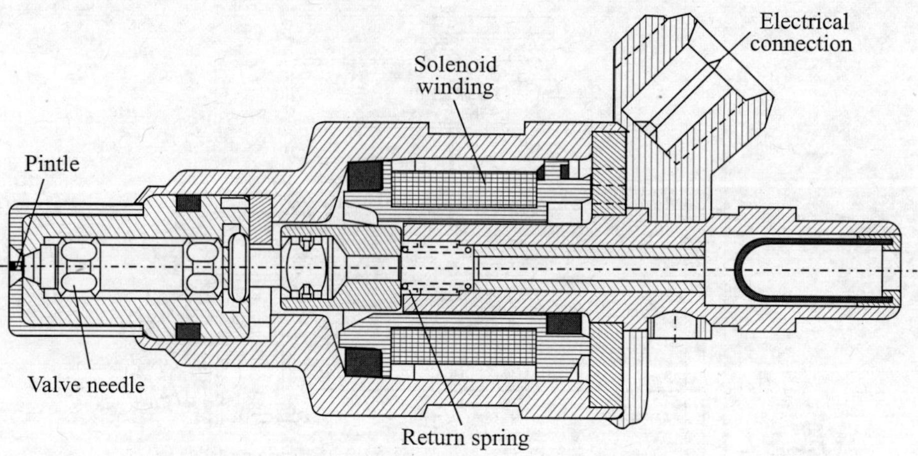

Figure 9.15 Electromagnetically actuated fuel-injector.

Feedback control

Figure 9.16 shows the closed-loop EFI feedback control. With this control it is possible to obtain very low exhaust emissions. The λ-sensor which is a detector for oxygen helps to run the engine with stoichiometric air/fuel ratio ($\lambda = 1$). In this case the emission standards can be met with one catalyst.

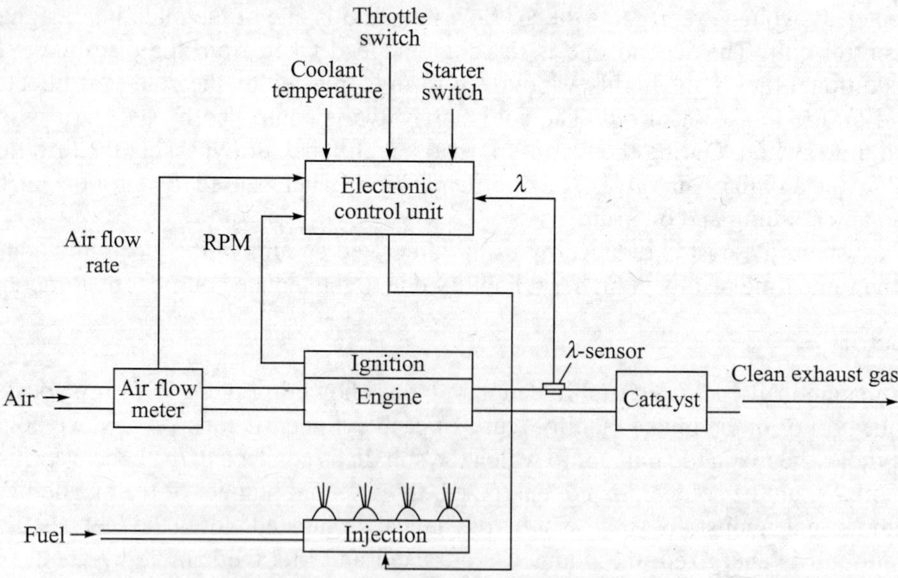

Figure 9.16 Closed-loop EFI feedback control.

9.18 ADVANTAGES OF THE MULTI-POINT FUEL-INJECTION SYSTEM

An SI engine with the multi-point fuel injection system has the following advantages over the carburetted engine:

1. Increase in volumetric efficiency resulting in higher power output and torque. The volumetric efficiency is increased because of elimination of intake manifold heating and carburettor pressure loss.
2. Lower specific fuel consumption since the amount of fuel injected is reduced at part throttle and during decceleration.
3. Faster acceleration, since the fuel is injected into or close to the cylinder and need not flow through the manifold.
4. Less maldistribution to each cylinder.
5. Elimination of throttle plate icing, since the fuel is not vaporized before the throttle.
6. Easier starting, since atomization of fuel does not depend on the cranking speed.
7. Less knocking tendency, since heat need not be supplied for better distribution. The temperature of the mixture, therefore, is lowered. Thus, a fuel having a lower octane number or a higher compression ratio can be used.
8. The tendency of backfiring or popping in the carburettor is reduced, since the combustible mixture is not in the intake manifold.
9. Less exhaust emissions are produced. For emission control, fuel-injection exhibits the higher accuracy of fuel metering. This allows a leaner adjustment of the engine with acceptable driveability.
10. Fuel response is practically instantaneous, so the flat spot is eliminated.
11. Engine bonnet can be appreciably lower owing to the absence of the usual down-draught carburettor and owing to the fact that the position of the injection unit is not critical.
12. Engines fitted with injection systems can be used in high angles of tilt; the tilt may produce fuel surge problems in the float chamber of the carburettor.

Figure 9.17 shows the typical performance curves for the fuel-injection and carburettor systems. It is observed that the power output with the fuel-injection system is higher than that with

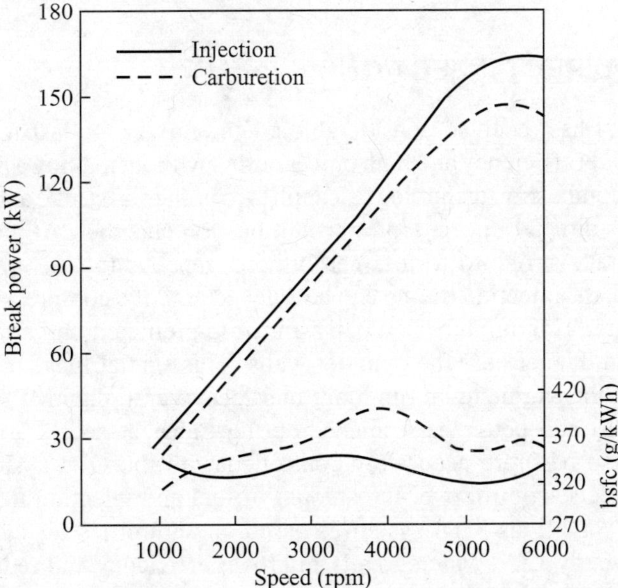

Figure 9.17 Comparison of performance of the carburettor system with that of the fuel-injection system.

the carburettor system. There is also a marked reduction in fuel consumption over a wide range of rpm with the fuel-injection system.

9.19 DISADVANTAGES OF THE MULTI-POINT FUEL-INJECTION SYSTEM

The disadvantages of the SI engine multi-point fuel-injection system compared to the carburetted engine are as follows:

1. The working mechanism of the fuel-injection equipment is much more complicated. The components of the injection system are of high precision workmanship and finish. Therefore, it has much higher cost.
2. The fuel injection pump continuously runs at half the speed of the crankshaft and thus causes wear of plungers and barrels and valves and their seatings. There are no similar wearing components in the carburettor.
3. The individual fuel pipes are required from the pump to each cylinder. It increases the risk of breakdown.
4. There is a possibility of a fuel nozzle carbonization, similar to the high speed diesel engines. It requires increased attention to maintenance and to training of technical personnel.
5. Weight and bulk are usually more compared to the carburettor. The carburettor needs only one feed supply pipe which can supply mixture to many cylinders, whereas the fuel-injection system needs separate fuel-feed pipe lines.
6. Additional screwed holes are required in the cylinders for individual injectors.

Earlier, owing to their appreciably higher cost, the fuel injection systems were mainly used on racing car engines. Now, due to increased pressure to reduce air pollution from motor vehicles and because of the advantages of better power and fuel economy, the SI fuel-injection system is being used in a large number of cars.

9.20 GASOLINE DIRECT INJECTION

GDI engine technology has received considerable attention over the last few years as a way to significantly improve fuel efficiency and high power output without making a major shift away from conventional internal combustion technology. It applies to both four-stroke and two-stroke engines.

In GDI, the fuel is directly injected into the combustion chamber. At higher load, the fuel is injected during the intake stroke to form a nearly homogeneous fuel-air mixture at the time of ignition. At lower load, the injection timing can be delayed until the compression stroke to produce a stratified fuel mixture. This mixture is ideally uniform, premixed and stoichiometric near the centre, and it is devoid of fuel near the cylinder walls. This spatial localization translates into a faster burn and allows the engine to be run more fuel-lean overall than MPFI engines, providing improved fuel economy and better performance during transient acceleration or decceleration. Emission level can also be more accurately controlled with the GDI system. The cited gains are achieved by the precise control over the amount of fuel and injection timings that are varied according to the load conditions. GDI system is based on unthrottled air induction and a higher compression ratio of nearly 12:1, which greatly improves efficiency and reduces pumping losses in engines without a throttle plate.

Basic technical features of GDI engines are the:

- Cavity on the piston's surface
- Upright straight intake port
- High-pressure fuel pump
- High-pressure swirl injector
- Higher compression ratio
- Electronic control unit (ECU) and the engine management system (EMS)

The cavity on the piston's surface

The combustion takes place in a toroidal cavity on piston's surface. The cavity is displaced to one side of the piston, the side that has the fuel injector. It controls both the air-fuel mixture and the air flow inside the combustion chamber and has an important role in maintaining ultra-lean air-fuel mixtures.

The upright straight intake port

The GDI engine has upright straight intake ports rather than horizontal intake ports used in the conventional engines. The upright straight intake ports efficiently direct the in-cylinder airflow down at the curved-top piston, which redirects the airflow into a strong reverse tumble for optimal fuel injection.

The high-pressure fuel pump

The gasoline is highly pressurized by a high-pressure pump and injected via a common rail fuel line directly into the combustion chamber of each cylinder.

The high-pressure swirl injector

High-pressure swirl injectors provide the ideal spray pattern to match each engine operational modes. These operational modes are described next. By applying highly swirling motion to the entire fuel spray, injectors enable sufficient fuel atomization that is mandatory for the GDI.

Many of the engines are using multiple injector sprays per stroke. The first spray occurs as the air starts flowing into the intake stroke. The second spray occurs right before the spark-plug fires. This creates a stratified charge of fuel for a better burn pattern. The stratified charge allows less fuel to be injected while still getting a good ignition pattern. It allows a much leaner mixture in the cylinder. It can be incorporated by using piezoelectronic fuel injectors. With their extremely fast response time, multiple injection events can occur during each cycle of each cylinder of the engine.

The high compression ratio

By virtue of better dispersion and homogeneity of the directly injected fuel, the cylinder and piston are cooled, thereby permitting higher compression ratios and more aggressive ignition timing, with resultant enhanced power output and efficiency.

The electronic control unit and the engine management system

The electronic control unit receives signals from all the electronic sensors and interprets the engine's needs. The ECU makes continuous adjustments to engine's fuel delivery circuits as well as the ignition timing to provide proper air and fuel mixing being ignited at the optimum time in the combustion chamber. This ensures that the engine is operating at the utmost peak power and economy level.

Engine management system regulates fuel injection and ignition timing. More precise management of the fuel injection event also enables better control of emissions. The engine management system continually chooses among three combustion modes: ultra-lean burn, stoichiometric and full power output. Each mode is characterized by the air/fuel ratio. The stoichiometric air/fuel ratio for gasoline is 14.7 by weight, but ultra-lean mode can involve ratios as high as 65:1. These mixtures are much leaner than those in a conventional engine and considerably reduce fuel consumption.

Ultra-lean burn mode

Ultra-lean burn mode is used for light-load running conditions. In this mode, fuel injection occurs at the latter half of the compression stroke, so the small amount of air-fuel mixture is optimally placed near the spark plug and ignition occurs at an ultra-lean air/fuel ratio of 30–40. This stratified charge is mostly surrounded by air, which keeps the fuel and the flame away from the cylinder walls for lowest emissions and heat losses.

Stoichiometric mode

Stoichiometric mode is used for moderate load conditions. Fuel is injected during the intake stroke, creating a homogeneous fuel-air mixture in the cylinder. From the stoichiometric ratio, an optimum burn results in a clean exhaust emission, which is further cleaned by the catalytic converter.

Full power mode

When the GDI engine operates with higher loads or at higher speeds, fuel injection takes place during the intake stroke. The air-fuel mixture is homogeneous and the ratio is slightly richer than stoichiometric, which helps prevent knock.

Advantages of GDI engines are as follows:

- Lower fuel consumption and more power output
- Improved volumetric efficiency
- Very efficient intake and relatively high compression ratio
- Lower octane requirement of fuel
- More precise air-fuel ratio control
- More rapid starting
- Reduced emissions
- Improved transient response
- In two-stroke engines, the benefits of direct injection are even more pronounced, because it eliminates much of the pollution they cause

Disadvantages of GDI engines

Direct-injection systems are more expensive to build because their components should be more rugged. They handle fuel at significantly higher pressures than indirect injection systems and the injectors themselves must be able to withstand the heat and pressure of combustion inside the cylinder.

REVIEW QUESTIONS

1. What is the function of a carburettor? What is carburetion?
2. What are the limits of flammability? What are the factors on which the flammability limits depend?
3. Explain the mixture requirements for maximum power and minimum specific fuel consumption under steady running of the engine.
4. Describe the mixture requirements for various power outputs with the help of bmep and the corresponding bsfc curves versus the fuel/air ratio at different throttle positions.
5. Show and explain with reasons the mixture requirements for idling, cruising and high power range at various throttle openings.
6. Explain the mixture requirements for starting, warm-up and acceleration.
7. Compare and explain with reasons the mixture requirements for a multi-cylinder engine with a single cylinder for idling, cruising and high power range at various throttle openings.
8. What are the requirements for an ideal carburettor?
9. Explain the construction and operation of a simple carburettor with the help of a diagram.
10. Derive an expression to calculate the air/fuel ratio for a simple carburettor taking into account the compressibility of air.
11. Derive an expression to calculate the air/fuel ratio for a simple carburettor neglecting the compressibility of air.
12. What are the deficiencies of an elementary carburettor?
13. What are the essential parts of a modern carburettor?
14. Describe the function, the location and the working of a choke valve with the help of a simple diagram.
15. How does the ice formation take place in a carburettor? What are the possible steps required to overcome this difficulty?
16. How does the vapour lock take place in the fuel system? What are its detrimental effects? How can they be avoided?
17. What is backfiring in an intake system? What are the causes of this backfiring? Suggest methods for its remedy.
18. What are the main drawbacks of the carburettor which prompted the use of the fuel-injection system in SI engines?
19. Show the comparison between the carburettor-mounted SI engines and the injector-mounted SI engines.
20. What are the different fuel-injection systems used in SI engines? Describe them briefly.
21. Describe the principle and working of a continuous injection system with the help of a diagram.
22. Describe the principle of a timed injection system. What are the two methods of fuel injection using timed injection?
23. Describe the direct cylinder injection method in SI engines using mechanical control with the help of a diagram. Why is this system rarely used in SI engines?

24. Describe the Lucas petrol-injection system with the help of a diagram.
25. Describe the principle of single-point throttle body EFI system.
26. Describe the principle of multi-point port EFI system.
27. Describe the working of D-Jetronic Bosch EFI system with the help of a diagram.
28. Describe the working of L-Jetronic Bosch EFI system with the help of a diagram.
29. Describe an electromagnetically actuated fuel injector with the help of a diagram.
30. Draw a schematic diagram of a closed-loop EFI feedback control. What is its purpose?
31. What are the advantages of the SI engine multi-point fuel-injection system? Compare the performance curves of the multi-point fuel-injection system in SI engines with those of the carburettor system.
32. What are the disadvantages of the SI engine multi-point fuel-injection system?
33. Describe technical features of a GDI engine. What are its advantages and disadvantages?

PROBLEMS

9.1 A gasoline engine has a carburettor of 35 mm throat diameter. The fuel jet diameter is 2.2 mm. The depression at the throat is 60 mm of Hg. The ambient pressure is 1 bar and temperature 27°C. The coefficient of discharge for venturi is 0.8 and for fuel jet 0.6. The density of gasoline is 750 kg/m^3. Neglect the nozzle lip. Determine the air velocity and air flow, the fuel velocity and fuel flow and the air/fuel ratio in the following cases:
(a) Considering the compressibility of air
(b) Neglecting the compressibility of air.

[**Ans.** (a) 119.2 m/s, 0.1004 kg/s, 4.618 m/s, 0.0079 kg/s, 12.7,
(b) 117.4 m/s, 0.1049 kg/s, 4.618 m/s, 0.0079 kg/s, 13.3]

9.2 A venturi of a simple carburettor has a throat diameter of 30 mm. The fuel jet diameter is 2.0 mm. The coefficient of air flow is 0.88 and the coefficient of fuel flow is 0.65. The fuel surface is 6 mm below the jet. The depression of pressure at the venturi throat is 0.075 bar, the densities of air and fuel are 1.2 kg/m^3 and 750 kg/m^3 respectively. Determine:
(a) The air/fuel ratio when the nozzle lip is neglected.
(b) The air/fuel ratio when the nozzle lip is considered.
(c) The minimum velocity of air required to start the fuel flow when the nozzle lip is provided. [**Ans.** (a) 12.18, (b) 12.22, (c) 8.58 m/s]

9.3 What will be the percentage change in the fuel/air ratio if the velocity of air at the throat is increased three times. Initially, the depression at the throat is $\frac{1}{15}$th of the ambient pressure. [**Ans.** 39.8%]

9.4 A four-stroke SI engine of 1100 cc capacity is required to develop maximum power at 4000 rpm. The volumetric efficiency at this speed is 80%. The air speed at the choke is 100 m/s. The coefficient of discharge for the venturi is 0.86 and that of the fuel jet is 0.65. The fuel surface is 5 mm below the top of the jet. Determine the throat diameter of the choke tube and the air/fuel ratio, if the jet diameter is 1.2 mm. The density of the fuel is 750 kg/m^3, and ambient conditions are 1.0 bar and 27°C. [**Ans.** 21.28 mm, 15.9]

9.5 A four-stroke four-cylinder SI engine having 70 mm bore and 84 mm stroke runs at 3000 rpm and uses a fuel having 84% carbon and 16% hydrogen by mass. The volumetric efficiency of the engine is 85%. The ambient conditions are 1.0 bar and 27°C. The depression at the venturi throat is 0.065 bar. The actual quantity of air supplied is 1.1 times that of the stoichiometric value. Calculate the fuel flow rate, the air velocity at the throat and the throat diameter. Take R(air) = 287 J/(kg K), and R(fuel vapour) = 98 J/(kg K).

[**Ans.** 0.001859 kg/s, 107.1 m/s, 18.33 mm]

9.6 A four-stroke four-cylinder SI engine having a bore of 90 mm and a stroke of 100 mm running at 3600 rpm has a venturi throat with 30 mm diameter. The volumetric efficiency of the engine at this speed is 85%. The coefficient of discharge of air flow is 0.85. The ambient pressure and temperature are 1.0 bar and 27°C. The top of the jet is 5.5 mm above the fuel level in the float chamber. The coefficient of discharge for fuel flow is 0.68. The diameter of the fuel jet is 2 mm and the specific gravity of the fuel is 0.76. Determine the depression at the throat and the air/fuel ratio. [**Ans.** 0.06773 bar, 11.03]

9.7 Determine the air/fuel ratio in an aircraft engine carburettor at 5000 m altitude. The air/fuel ratio at sea level is 15:1. The ambient conditions are 1.013 bar and 27°C. The temperature variation with altitude is given by: $T = T_{\text{sea level}} - 0.0065\,h$, where h is the height in metres and $T_{\text{sea level}}$ is the sea level temperature in °C. The air pressure decreases with altitude as per the relation.

$$h = 19200 \log_{10} (1.013/p), \text{ where } p \text{ is in bar.}$$

[**Ans.** 11.77]

10.1 INTRODUCTION

In compression-ignition engines, a fuel-injection system is required to inject a definite quantity of fuel at the desired time and at a definite rate into the combustion chamber. The system must also atomize the fuel and distribute it to the various parts of the combustion chamber. The injection system is responsible for initiating and controlling the combustion process. Consequently, the performance of CI engines largely depends upon the good design of the fuel-injection system.

In CI engines, during the suction stroke, air is drawn into the engine cylinder, and during the compression stroke it is compressed to a high pressure, thus raising the temperature of the compressed air higher than the self-ignition temperature of the injected fuel. Fuel is injected into the combustion chamber near the end of the compression stroke, and at that time the pressure in the cylinder is usually between 20 bar and 35 bar. As the combustion of the injected fuel proceeds, the maximum cylinder pressure reaches up to 70 bar. The fuel is thus injected into a high pressure combustion chamber. It requires the injection pressure of the fuel to be much higher than this pressure. Also, the good atomization of the fuel depends upon the high injection velocities which require a high pressure difference. Thus high pressures, usually between 100 and 200 bar, are required for the fuel-injection system.

10.2 REQUIREMENTS OF INJECTION SYSTEMS

The fuel-injection system of the CI engine has to fulfil the following requirements for a proper running and good performance of the engine:

1. Metering the quantity of fuel as required by the engine speed and load. Metering means the supply of the desired quantity of fuel in each injection, at an acceptable rate and over an acceptable crank angle. It is very critical as even a small variation may affect the performance of the engine drastically.
2. The distribution of the metered fuel in a multi-cylinder engine should be equal among the cylinders.
3. The injection of the fuel should be at the correct time in the cycle so that more power is obtained with the least amount of fuel, thus ensuring clean burning.

4. The fuel should be injected at the correct rate, so that it results in the desired heat-release pattern during combustion.
5. The fuel should be injected with the proper spray pattern and atomization depending upon the design of the combustion chamber in order to ensure proper mixing of fuel and air, both in time and space.
6. Injection should start and stop sharply to eliminate dribbling of fuel droplets into the cylinder.
7. The injection system must have the minimum weight and overall dimensions. It should be inexpensive to manufacture, and easy to maintain and repair.
8. The service life of all the elements of the fuel-injection system should be compatible with that of the engine itself.

In order to accomplish the above requirements, the following components are required in a fuel-injection system:

1. *Pumping elements:* Pumping elements pump the fuel from the fuel tank to the cylinder through pipe lines and injectors.
2. *Metering elements:* Metering elements measure and supply the fuel at the rate required by the engine speed and load.
3. *Metering controls:* Metering controls adjust the rate of the metering elements for changes in engine speed and load.
4. *Distributing elements:* Distributing elements distribute the metered fuel equally among the cylinders.
5. *Timing controls:* Timing controls adjust the start and the stop of injection.
6. *Mixing elements:* Mixing elements atomize and distribute the fuel in the combustion chamber.

10.3 INJECTION SYSTEMS

Based on the methods used to produce the required pressure for atomization of the fuel, there are basically two distinct methods of fuel injection in CI engines:

1. Air-injection system
2. Airless-or solid-injection system.

10.3.1 Air-Injection System

Air-injection applies to systems injecting air along with the liquid fuel. This method was first successfully developed by Rudolf Diesel and is shown in Figure 10.1. It requires an air compressor for supplying air at 70 bar or even a higher pressure. A camshaft driven fuel pump meters and discharges a definite quantity of fuel into the injection valve. The injection valve is mechanically opened, and the high-pressure air drives the fuel charge and some air into the combustion chamber.

The following are the advantages of the air-injection system:

1. Very good atomization and distribution of the fuel, resulting in comparatively high mean effective pressures.
2. High viscous fuels, which are less expensive than those used by the engine having the solid injection system, can be utilized without any problem.

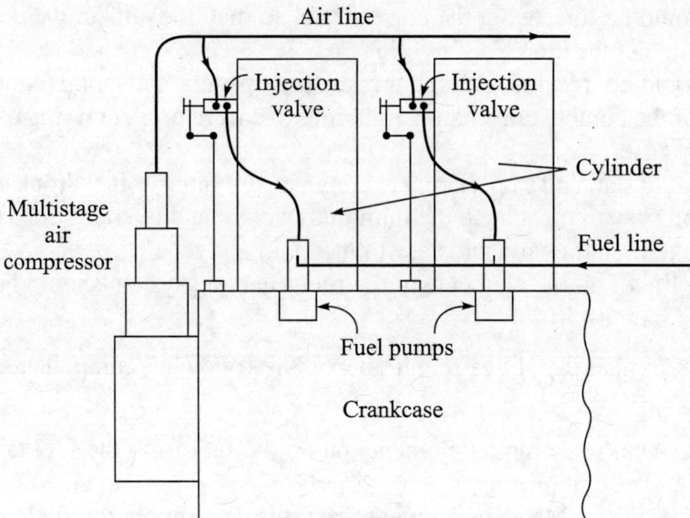

Figure 10.1 Air-injection system.

3. The fuel pump is required to develop less pressure than that required by the engine having the solid injection system.

The following are the disadvantages of the air-injection system.

1. Complexity of the engine due to a multistage air compressor, which often is a source of trouble and requires expert attention.
2. A separate mechanical linkage is required to operate the fuel valve at the proper time.
3. Due to the compressor and the linkage the engine weight increases and as some power is absorbed in operating the compressor and the linkage, the brake power reduces.
4. The fuel in the combustion chamber ignites very near to the injection nozzle, which may result in overheating and burning of the fuel valve and the valve seat.
5. The fuel valve seating requires regular attention to guard against any leakage.
6. If the fuel valve sticks, the system becomes very dangerous because of the presence of high-pressure air.

The size and cost of the air compressor, along with the power required for its operation, has made the air-injection system obsolete.

10.3.2 Airless- or Solid-Injection System

In this system the fuel is injected at a very high pressure into the combustion chamber without the aid of the compressed air. Hence, it is called the airless injection system. It is also called the solid injection system. The main parts of this system are the fuel pump and the injector. Depending upon the location of the fuel pumps, the grouping, the method of actuating the pumps and the methods used to meter the fuel, the solid injection systems can be classified as follows:

1. Individual pump system or the divided fuel-feed device
2. Unit injector system or the undivided fuel-feed device

3. Distributor system
4. Common rail system.

All of the above systems consist of the following components:

1. Fuel tank to store the fuel
2. Filters:
   (a) A primary stage filter to remove course particles (larger than 0.025 mm). It is a metal-edge filter.
   (b) A secondary stage filter to remove fine particles from about 4 microns to 0.025 mm. It is a replaceable cloth, paper, or felt element.

   The primary and secondary filters are placed between the fuel tank and the transfer pump.
3. A low pressure (3 bar) transfer pump (gear or vane type) is used to lift the fuel from the tank, to overcome the pressure drops in the filters and to charge the metering and pressurizing unit.
4. Final stage filter to remove fine particles that might have escaped the secondary stage. It is a sealed non-replaceable element. It is used to guard the high-pressure unit.
5. High-pressure fuel injection pump to meter and pressurize the fuel (100 to 200 bar) for injection.
6. Governor to ensure that the amount of fuel injected is in accordance with change in load.
7. Injector to receive the high-pressure fuel from the injection pump and to inject the fuel in the combustion chamber, and atomize it into fine droplets.

10.3.3 Individual Pump System or the Divided Fuel-feed Device

Figure 10.2 shows the schematic diagram of an individual pump system. In this system, each cylinder is provided with one pump and one injector. The injector is located on the cylinder, while the pump is on the side of the engine. Each pump may be placed close to the cylinder to which it delivers the fuel as shown in Figure 10.3(a) or pumps may be arranged in a cluster, as shown in Figure 10.3(b). The smaller engines combine the individual pumps into one assembly. The high pressure pump plunger is actuated by a cam and produces the fuel pressure necessary to open the spring-loaded injector valve at the correct time. This valve is a hydraulically operated automatic plunger valve, i.e. no mechanism is required to operate it. The amount of fuel injected depends on the effective stroke of the plunger. The spray pattern from the injector nozzle depends upon the type of the orifice attached to the nozzle. This system is used for large slow-speed engines (output about 150 kW and above).

The fuel pump used in this system is of reciprocating type. This requires robust and heavy components causing the pump to be always accompanied by a jerking noise. The pump is therefore called a *jerk* pump.

10.3.4 Unit Injector System or the Undivided Fuel-feed Device

The high-pressure pipe line connecting the individual pump and the associated injector can be avoided by the design of a unit injector. In this system the pump and the injector nozzle are combined in one housing. Each cylinder has one of these injector units. Fuel is brought up to the injector by a low-pressure pump. This system requires a push rod and a rocker arm to actuate the

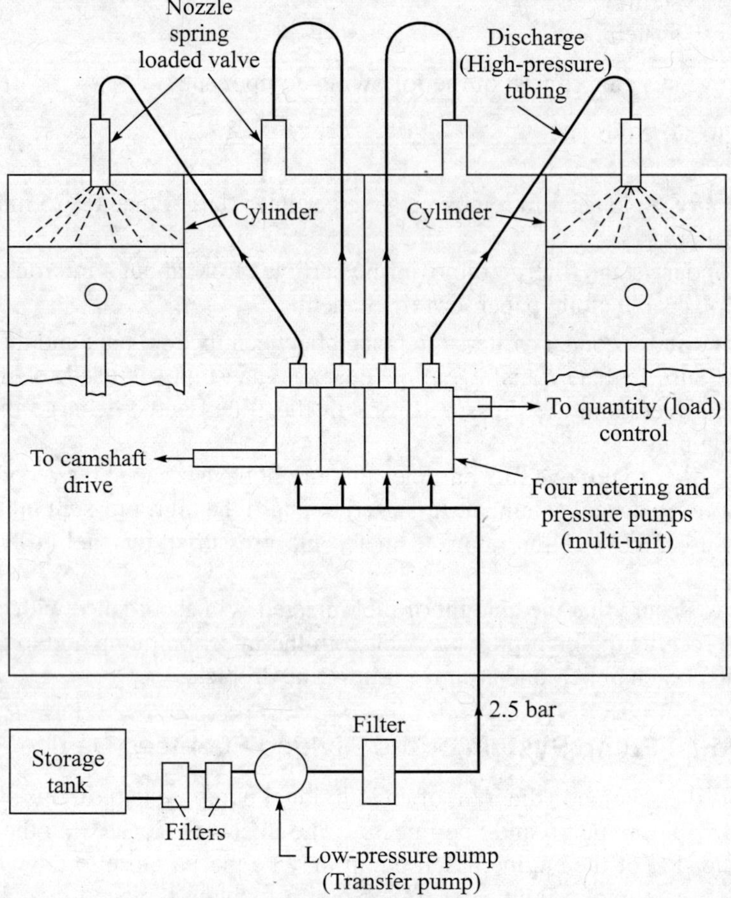

Figure 10.2 Individual pump system.

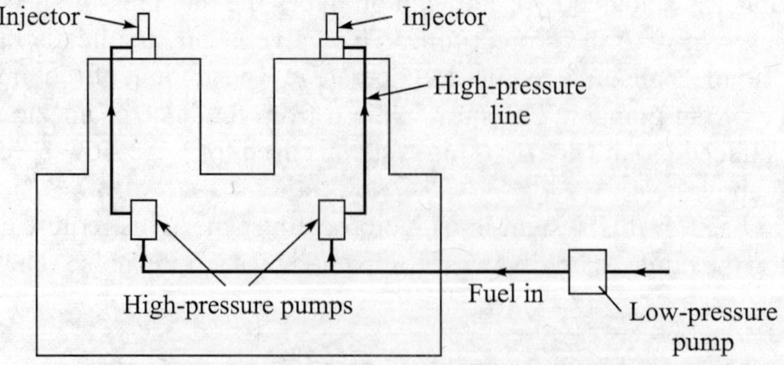

Figure 10.3(a) Individual pump and nozzle with separated pumps.

plunger and injects the fuel into the cylinder at the proper time. The quantity of fuel injected is regulated by the effective stroke of the plunger. The unit injector system is shown in Figure 10.4. It is used extensively on large two-stroke cycle diesel engines.

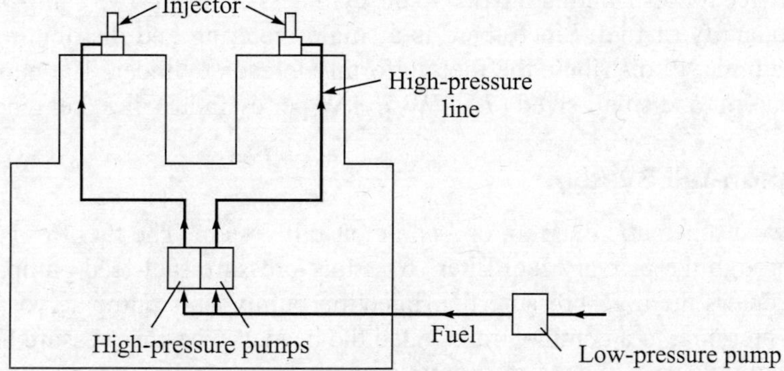

Figure 10.3(b) Individual pump and nozzle with pumps in cluster.

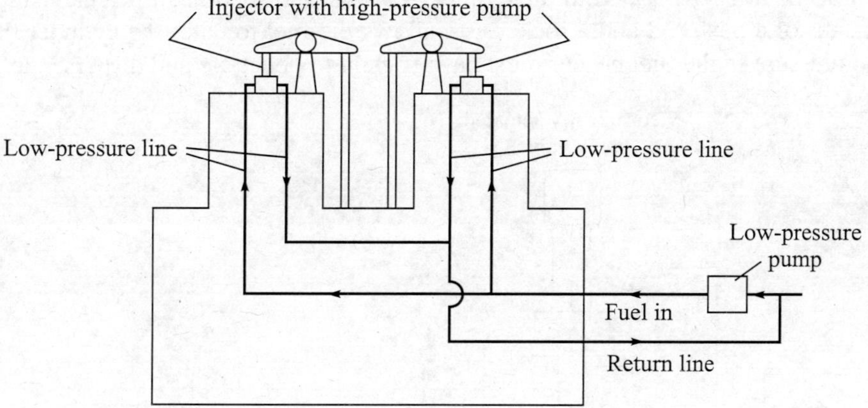

Figure 10.4 Unit injector system.

10.3.5 The Distributor System

Figure 10.5 shows a schematic diagram of the distributor system. The individual pump system, described earlier, requires a separate metering and compression pump for each cylinder, which increases the cost of the system. In the distributor system, a single pump for compressing the fuel

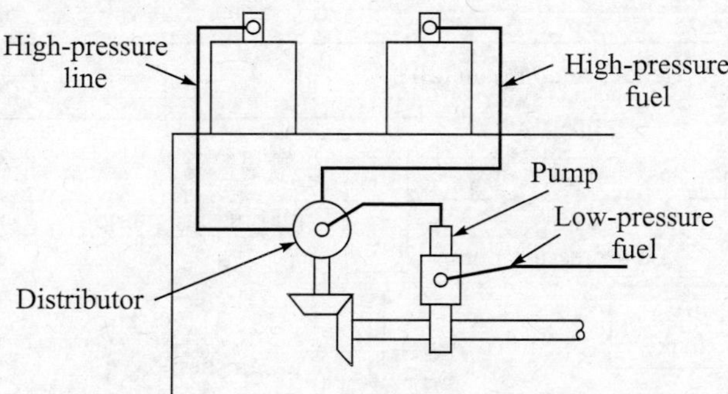

Figure 10.5 Schematic diagram of the distributor system.

and a dividing device for distributing the fuel to the cylinders are used. The pump or the distributor may meter the quantity of fuel. Since there is a single metering and distributing element, it is expected that the unit will distribute the fuel uniformly to each cylinder. The pumps of this type are found on medium- and small-sized (7.5 kW–75 kW per cylinder) diesel engines.

10.3.6 Common-rail System

Figure 10.6 shows a schematic diagram of a common-rail system. The fuel from the fuel storage tank is drawn through the primary fuel filters by a low-pressure fuel-feed pump. The discharge from this pump enters the high-pressure fuel-injection pump. This pump serves only to deliver fuel, under high pressure, to a common rail, called the header, with the pressure held constant by a pressure regulating valve. Thus the maximum pressure is under direct control and the metering problem is not handled by the high-pressure pump. The high pressure in the header forces the fuel to each of the nozzles located in the cylinders. At the proper time, a mechanically operated valve by means of a push rod and a rocker arm allows the fuel to enter the cylinder through the nozzle. The pressure in the fuel header must be sufficient to penetrate and disperse the fuel in the

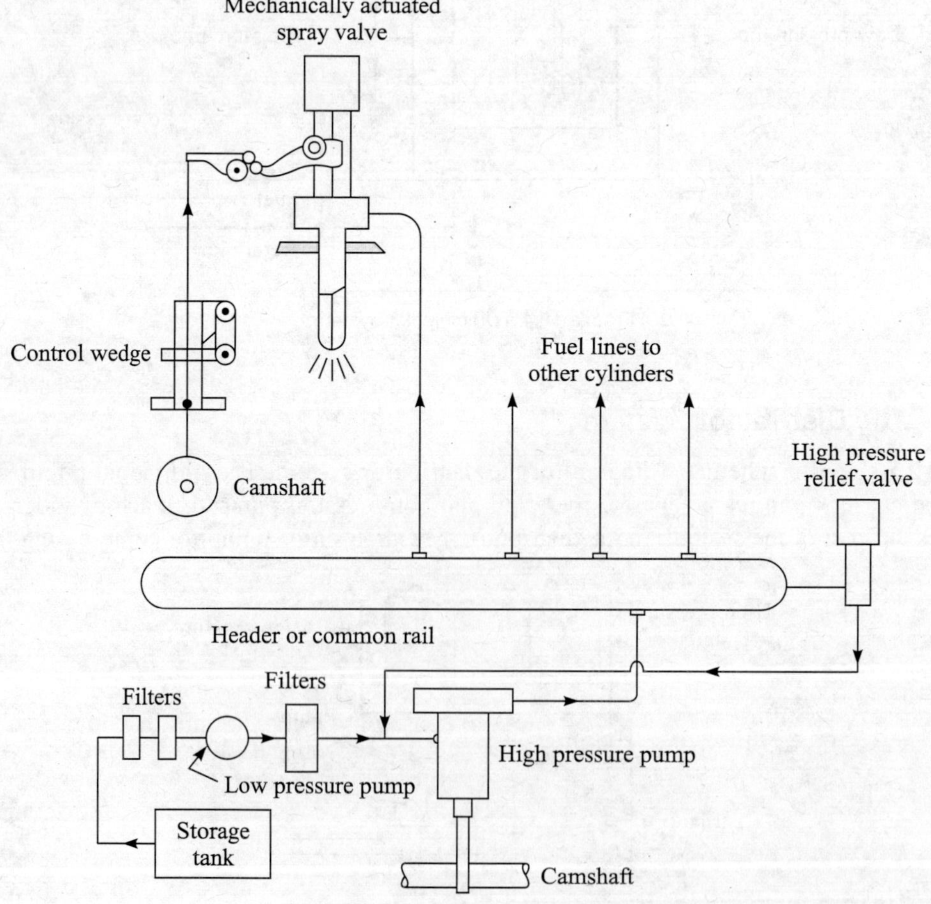

Figure 10.6 Common-rail system.

combustion chamber and must be in accordance with the injector system design. The amount of fuel supplied to the combustion chamber is regulated by varying the length of the push rod stroke.

The common-rail system tends to be self-governing. If the engine speed falls, an increased quantity of fuel is automatically injected, since the time taken for the same crank angle rotation, during which period the fuel is supplied, is increased. As the fuel pressure is maintained constant, with increased time, the fuel supply will also be increased.

Very accurate design and workmanship are required in this type of fuel-injection mechanism. The lift of the valve spindle is very small, it requires absolute rigidity of the operating mechanism and freedom from vibration and temperature effects, moreover, the wear must be negligible. The common-rail systems were, once, quite popular for large, slow-speed engines, but over the years, have been replaced by the jerk-pump injection.

10.4 FUEL-INJECTION PUMPS

The high-pressure fuel-injection pump is one of the main and most complicated units of the fuel-injection system in a CI engine. It meters out the fuel in conformity with the engine duty and supplies fuel to the injector at the proper time. Fuel pumps in which every working section delivers fuel to individual cylinders of the engine are known as multi-sectional pumps. These are generally of jerk type. The other types are single- or double-plunger distributing pumps in which one working section feeds fuel to several cylinders.

10.4.1 In-line or Jerk Type Fuel-Injection Pump

Figure 10.7 shows the in-line or jerk type Bosch high-pressure fuel-injection pump. It consists of a barrel in which a plunger reciprocates. The pump plunger is lifted by a cam on a camshaft driven by the engine. There is a very small clearance between the barrel and the plunger, which is of the order of 2 to 3 thousandths of a millimetre. It provides perfect sealing even at very high pressures and low speeds.

The pump barrel has two radially opposing ports. These are the inlet port and the spill or bypass port. The plunger moves vertically in the barrel with a constant stroke. To enable the pump to vary the quantity of fuel delivered per stroke, the plunger is provided with a vertical channel extending from its top edge to an annular groove, the upper edge of which is formed as a helix. The helix runs a little way down the plunger length. The effective stroke is varied by means of the helix on the plunger which permits the by-passing of fuel at any position of the delivery stroke, thereby controlling the quantity delivered. This is accomplished by means of a control rod or rack, which turns a toothed control sleeve that turns the plunger to the desired position.

Operation

When the plunger is at the bottom of its stroke, the inlet port and the spill port are uncovered. Fuel enters the barrel and as the plunger rises, both the ports are covered. The trapped fuel is compressed and lifts the delivery valve and then the injection begins. As the plunger continues to rise, the spill port is uncovered by the helical groove on the plunger and the high-pressure oil above the plunger returns to the sump. When the pressure above the plunger falls, the delivery valve is closed by the spring and then the injection stops.

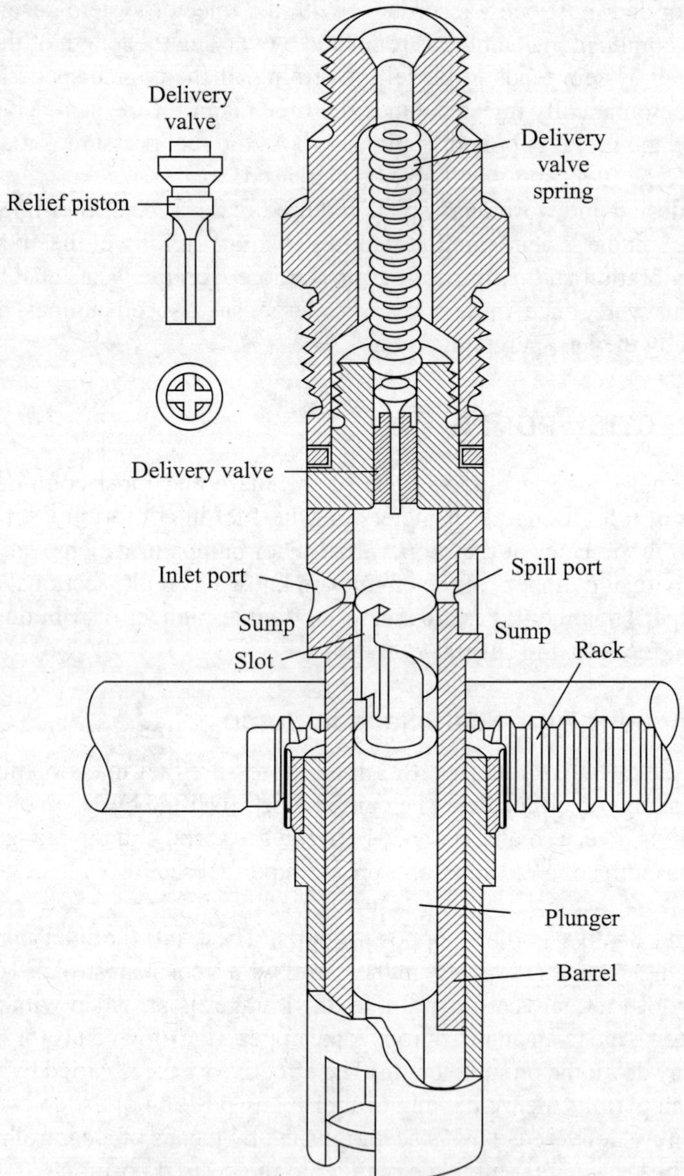

Figure 10.7 Sectional view of Bosch pump elements.

By moving the rack, the quantity of fuel injected can be varied in accordance with the load. Figure 10.8(a) shows the position of the plunger for full load. Here, the effective travel is maximum before the spill port is uncovered by the helix. It ensures the maximum delivery of fuel required for full load. Figure 10.8(b) shows a shorter effective travel obtained by rotating the plunger with the help of the rack, It reduces the delivery of fuel which is suitable for part load. Figure 10.8(c) shows that the slot in the plunger is in line with the spill port. This position is obtained by further

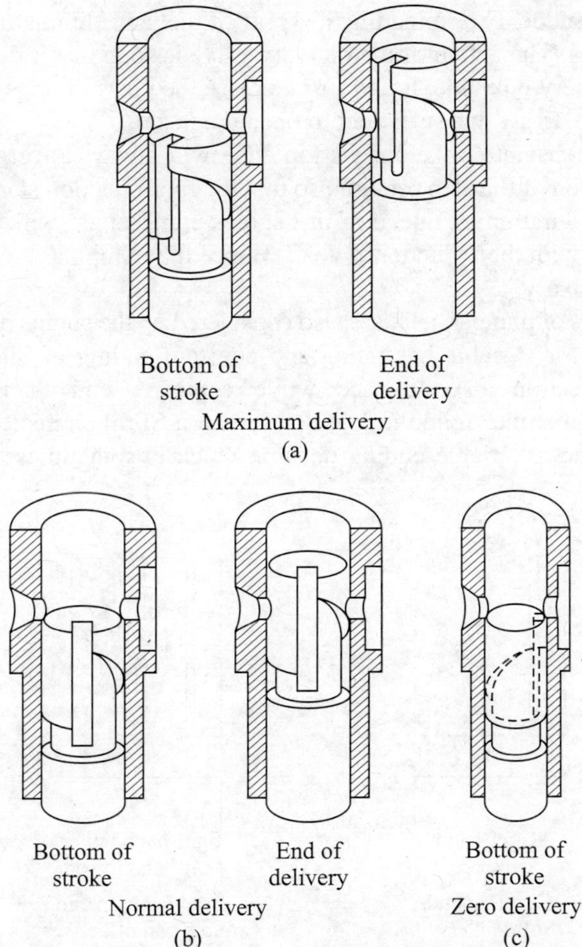

Figure 10.8 Position of the helix for various load conditions.

rotating the plunger with the help of the rack. At this position, fuel is not trapped by the upward movement of the plunger. All the fuel is returned to the sump via the spill port and the fuel is not delivered. This is the stop position for shutting down the engine. Thus, the overall displacement of the plunger remains constant at all speeds and loads but the effective travel is varied by the helix in accordance with the load on the engine.

The delivery valve maintains the high pressure in the delivery pipe and also stops the injection through nozzles abruptly. When the pump is on its delivery stroke, the pressure of the fuel above the plunger increases, the delivery valve is forced upwards but the flow does not begin until the relief piston of the delivery valve leaves the passage. The fuel can now be delivered through the longitudinal grooves over the valve face to the nozzle. By the movement of the plunger, when the pressure falls in the barrel, the delivery valve is closed by the spring and the relief piston returns into its housing. The return of the relief piston increases the volume of the delivery line and reduces the pressure. The effect of this is to cause the spring-loaded nozzle valve in the fuel nozzle to snap on its seat, thus suddenly terminating the spray of fuel and eliminating dribbling of the fuel through the nozzle holes. The action of the delivery valve is such that it does not relieve entirely the pressure in the delivery pipe line. It helps to increase the pressure in the pipe line quickly, on the next injection stroke, to a value sufficient to open the nozzle.

The plunger is of constant stroke and its top edge will always cover the ports in the pump barrel at the same position of the cam rotation, so that the fuel-injection starts at the same position of the crankshaft. The duration of injection in degrees crank angle will be a maximum at full load, and will decrease with the reduction in load. Hence the pump has a constant beginning and a variable ending of delivery.

Two more variations of plunger helix are also considered by the engine designer. Figure 10.9(a) shows a plunger helix for variable beginning and constant ending of injection. As the load is increased the start of injection can be advanced while keeping the end of delivery at constant crank angle. This design is sometimes found in low compression SI oil engines. Figure 10.9(b) shows a variable beginning and a variable ending design, which is sometimes specified for small CI automotive engines.

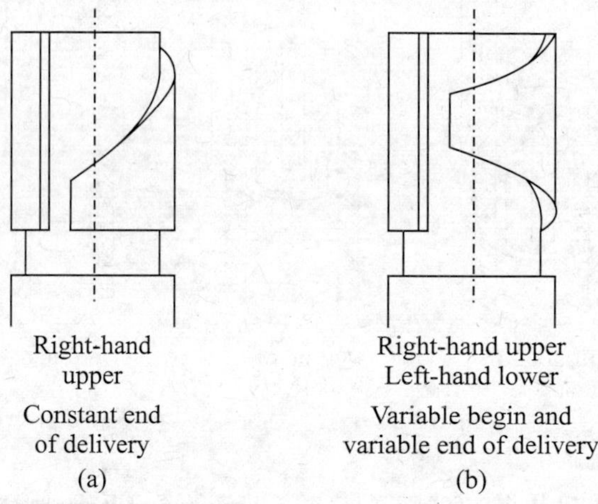

Figure 10.9 Plunger helix.

Delivery characteristics

The characteristic of a fuel pump related to the speed is the relation between the delivery percycle and the camshaft speed with the governing member (rack) in a constant position. Figure 10.10 shows the delivery characteristics of a port-controlled displacement pump. The curve A of Figure 10.10(a) shows the actual fuel delivery per cycle versus the pump speed. It has a rising delivery characteristics.

A fuel-injection pump tends to cause an increase in fuel discharge per stroke with rising speed. This is basically because at low pump speed, there is more time for fuel leakage to take place between the barrel and the plunger. In addition, as plunger moves upward, a pressure is built up in the space above the plunger, but throttling loss (pressure drop) appears when the intake port is partially closed by the plunger edge and the cut-off port is partially opened by the helical edge. This throttling loss is more at lower speed as more time is available for the loss to take place.

At low speed, the compression of the fuel will occur only after the ports are completely covered by the plunger. However, at higher speeds the pressure will build up above the plunger faster than the fuel can bypass through the ports, and the delivery valve, leading the fuel to the injector, will open before the bypass port is completely covered. Thus the fuel injected at higher speeds increases, and the rising delivery characteristics as shown by curve A in Figure 10.10(a) is obtained.

In CI engines, the ideal value of fuel delivery should be in proportion to mass of air inducted by the engine per intake stroke. Therefore, it should follow a pattern of volumetric efficiency with speed which is a slightly rising and then falling characteristic, as shown by curve B of Figure 10.10 (a).

Therefore, it follows that the air-fuel ratio will tend to become richer at higher speed. Consequently, there will be an excess of fuel delivered in the upper speed range of the engine.

Apart from leakage and throttling, the other factors that affect the delivery of the fuel are compressibility of the fuel and resilience of the elements in the fuel-feed devices like plunger barrel and delivery pipe. These factors may cause some of the supplied fuel to accumulate in the high-pressure region between the delivery valve and the injector. This part of the fuel does not enter into the engine cylinder and the delivery per cycle diminishes.

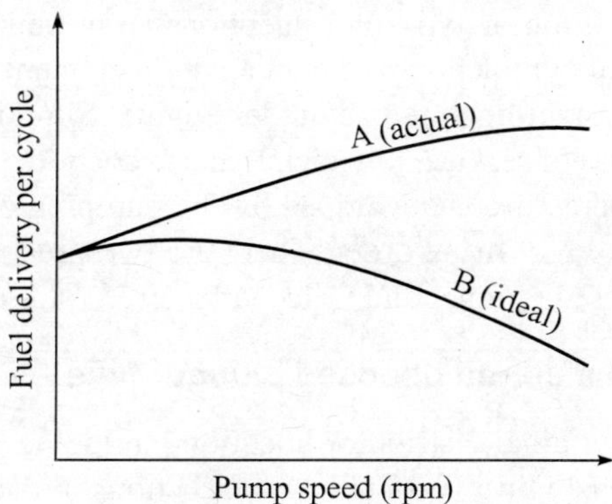

Figure 10.10(a) Delivery characteristics.

To obtain a satisfactory air-fuel ratio match throughout the engine's speed range, the injection pump output characteristics can be changed slightly to produce a progressively decreasing fuel delivery per plunger stroke, as the speed increases, by modifying the design of the delivery valves which may progressively increase the accumulated volume between the delivery valve and the injector.

Figure 10.10(b) shows the delivery characteristics at three positions of the fuel pump rack. Curve 1 is for rack position corresponding to the full delivery. At this position the delivery per cycle increases slightly with speed. Curves 2 and 3 are for rack positions corresponding to part delivery, the delivery per cycle increases more intensively with the speed. This indicates the growth in the influence of fuel throttling in the transition from full to part delivery adjusted by rack position.

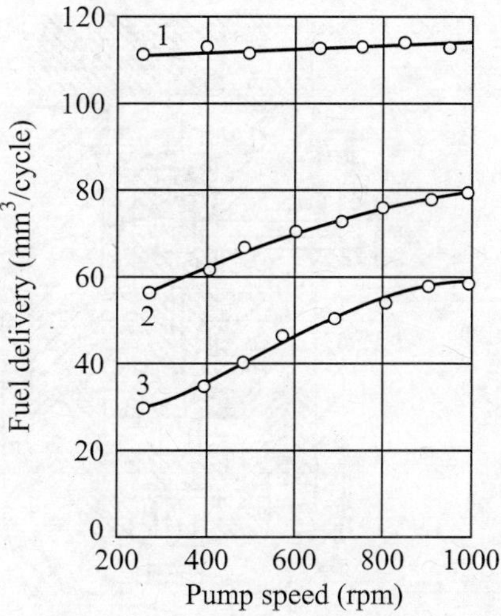

Figure 10.10(b) Delivery per cycle versus pump speed.

A constant-speed fuel pump characteristic, for the same type of pump, can also be drawn, which gives the fuel delivery per cycle with respect to the rack position. This characteristic is shown in Figure 10.11. Direct proportionality between the fuel delivery per cycle and the rack stroke exists above a certain rack stroke. At shorter rack strokes, the direct proportionality between the fuel delivery and the rack stroke gets more and more violated owing to the greater influence of throttling in the barrel ports.

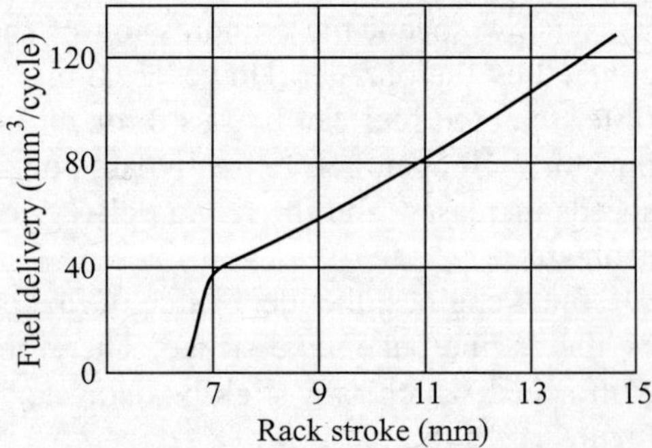

Figure 10.11 Delivery of pump per cycle versus rack stroke

10.4.2 Distributor Type Fuel-Injection Pump

A distributor type fuel-injection pump generally uses one pumping element to pressurize fuel for all cylinders rather than a single element for each cylinder. A single–plunger pump is used to inject fuel in a multi-cylinder engine. Simplicity, the major advantage of this design, contributes to ease of service, low maintenance costs, dependability and ability to operate at high speeds. High fuel pressure is developed by the pumping element and distributed by a rotor to each cylinder in the correct firing order. There are two predominant types of distributor pumps: the inlet metered opposed plunger and a sleeve metered rotating and reciprocating pump plunger.

Inlet metered opposed plunger type

The original distributor injection pump, the Roosa Master, was an opposed plunger inlet metered pump. Figure 10.12 shows the sections of the Roosa Master injection pump. This type of injection pump used only one metering valve to control the fuel and opposed plungers to pump the fuel.

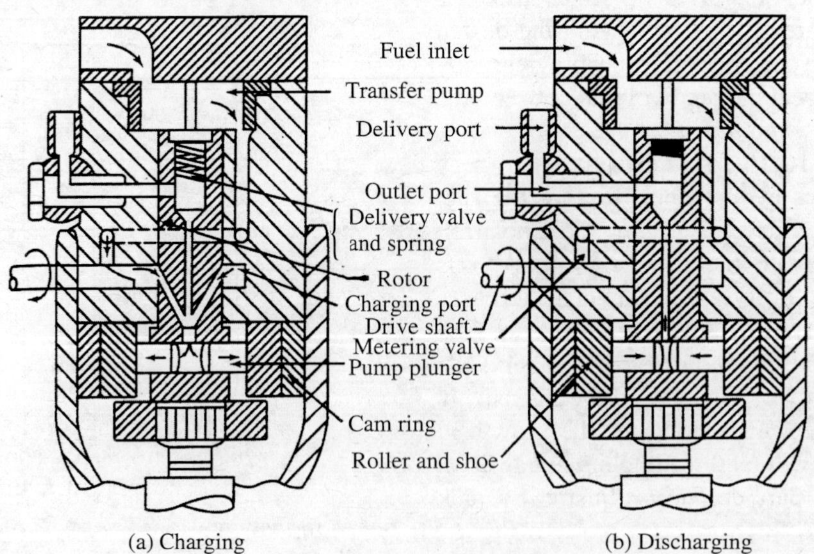

Figure 10.12 Sections of the Roosa Master distributor pump.

The driveshaft, pumping and distributing rotor, and sliding vane type transfer pump are an integral unit. The rotor is driven by the driveshaft. It is used to distribute the metered fuel to the injectors. The rotor incorporates two charging passages and a single axial bore. One discharge port serves all the outlet ports to the injection lines.

As the rotor revolves, the charging passages in the rotor connect with one of the charging ports in the charging ring. The transfer pump pressure forces the plungers apart a distance proportional to the amount of fuel required for injection. The plungers will move to the most outward position only at full load. While the charging passages in the rotor are in contact with one of the charging ports, the discharge passage is out of contact with the outlet ports.

When the plungers are forced together by the cam ring profile, the charging passages are no longer in contact with the charging ports and fuel is forced to travel to the rotor central drilling through the delivery valve to one of the distributor ports and on to its attached injector.

Sleeve metered rotating and reciprocating pump plunger

The Bosch VE pump has a rotating and reciprocating plunger, so the rotor is still used to distribute the fuel. The functioning and metering of this pump is different from Roosa Master pump. Metering is accomplished by sleeve metering. The positioning of this sleeve will vary the effective stroke of the plunger, thereby varying the amount of fuel injected per cycle.

Each time the plunger reciprocates in its bore, it will create an injection pulse, depending on the position of the sleeve, and as it also rotates simultaneously, it will deliver that pulse to each cylinder in sequence according to the firing order.

Figure 10.13 shows the sections of the Bosch, a single-plunger distributor pump of sleeve metered type. The plunger has four inlet slots in a four-cylinder application. The plunger is also centre drilled and cross-drilled.

Fuel enters the barrel at the end of the downward stroke, and the compression of the fuel begins as the intake ports are closed by the upward motion of the plunger. Delivery begins when the upper annulus in the plunger uncovers the high pressure duct from the delivery valve. At this time the distributing slot connects this annulus with the duct to the nozzle. Delivery ends when the second annulus uncovers the upper edge of the control sleeve. The amount delivered is controlled by varying the vertical position of the control sleeve. Thus, all cylinders are supplied with fuel from the same plunger, the same delivery valve, and ducts of the same shape and length.

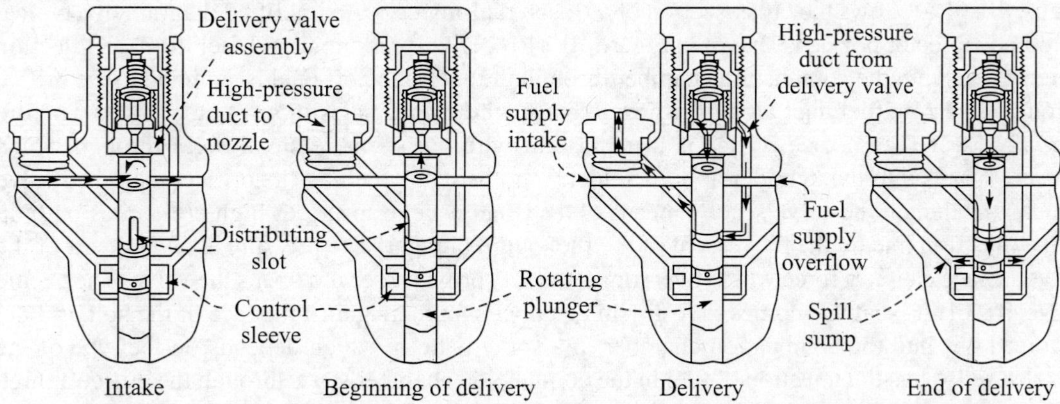

Figure 10.13 Sections of the American Bosch single-plunger multi-cylinder engine pump head.

The cost of the Bosch pump is about 60 % of an individual pump assembly with comparable delivery characteristics and life.

10.5 FUEL INJECTOR

The successful operation of a CI engine depends on the functional efficiency of its injector assembly. The selection of the injection holder and nozzle depends on the construction of the combustion chamber, the location of the holder and the nozzle assembly.

10.5.1 Fuel-Injection Holder

The main purpose of the fuel-injection holder is to position and to hold the fuel-injection nozzle in the cylinder head. The fuel duct within the body connects the fuel inlet with the injection nozzle. The leak-off fuel, used to lubricate the nozzle valve, fills the spindle and the adjusting device area and returns via the leak-off connection to the fuel tank.

10.5.2 Fuel-Injection Nozzle

The fuel-injection nozzle has a nozzle body and a needle valve. The design of the injector nozzle must be such that the liquid fuel forced through the nozzle will get broken up into fine droplets, or get atomized, as it enters into the combustion chamber. This will ensure proper mixing of the fuel and air in the combustion chamber during the first phase. The fuel must then be properly distributed in the desired areas of the combustion chamber. For this, the injection pressure, the density of air in the cylinder, the physical qualities of fuel such as viscosity, the surface tension etc. and the nozzle design are the important factors. Higher injection pressure results in better distribution and greater penetration of the fuel into the desired locations. It also produces finer droplets which tend to mix more readily with the air. Fuel spray must reach the air in the combustion space farthest from the nozzle, but should not impinge the surrounding wall, as it may form gummy deposits which will cause the piston rings to stick in their grooves and produce carbon deposits, unpleasant odour, smoky exhaust and cause an increase in fuel consumption.

Injector action

Figure 10.14(a) shows the cross-section of a Bosch fuel-injector. A simplified diagram of a typical automatic injector nozzle is shown in Figure 10.14(b). High-pressure liquid fuel from the injection pump passes into the combustion chamber through the fuel-injector. Fuel is supplied to the orifice through a narrow fuel duct along the nozzle body which terminates in an annular gallery before the valve seat. The nozzle orifice is usually built with one or more small holes through which the fuel sprays into the cylinder at high velocity. Immediately behind the orifice, there is a valve which is seated on the valve seat by means of a stiff adjustable spring. A high pressure is built up in the injection line by the movement of the fuel pump plunger. The pressure acting on part of the valve surface causes a force which overcomes the set spring force, and opens the valve. The needle valve lifts off its seat and comes to rest with its upper shoulder against the face of the holder. Fuel is forced out into the combustion chamber in a spray pattern, which depends on the type of the nozzle used. Thus the injection of fuel in the combustion chamber starts through the orifice at high velocity, which ensures good atomization and penetration of the fuel in the combustion chamber.

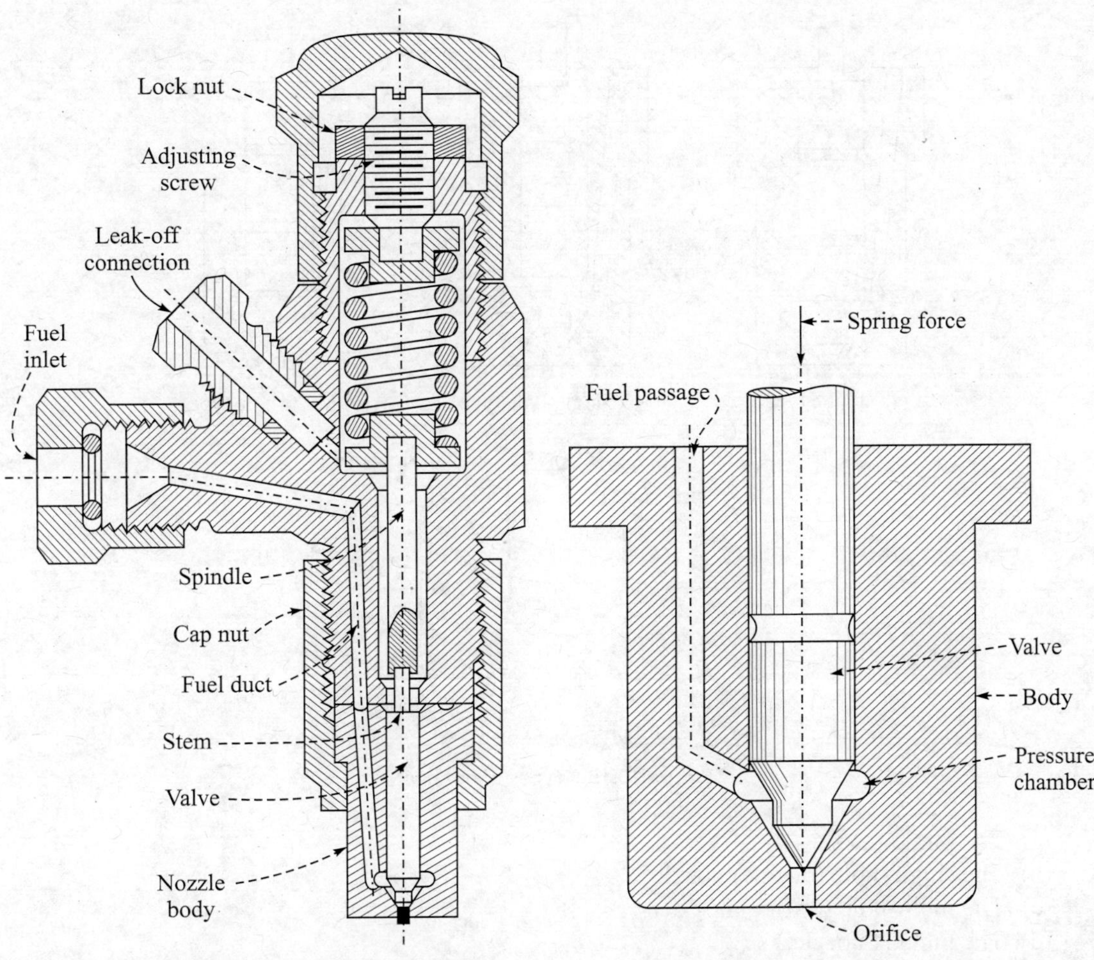

Figure 10.14(a) Typical spring-loaded nozzle with differential valve. (American Bosch Corp.)

Figure 10.14(b) Automatic injector nozzle.

When the fuel-injection pump spills back by the plunger movement, the delivery from the pump ceases and the delivery valve of the pump closes, which increases the volume of the injection line and reduces pressure. The resulting sudden drop of pressure at the nozzle causes the pressure spring to snap the needle valve into its seat and the nozzle valve closes instantly. The fuel injection thus stops suddenly without dribbling.

10.6 TYPES OF NOZZLES

The combustion chamber design dictates the type of nozzle, the droplet size and the spray required to achieve complete combustion within a given time and space. There is a wide variety of nozzle designs to provide the spray characteristics required for each type of combustion chamber. The most common types illustrated in Figure 10.15 are:

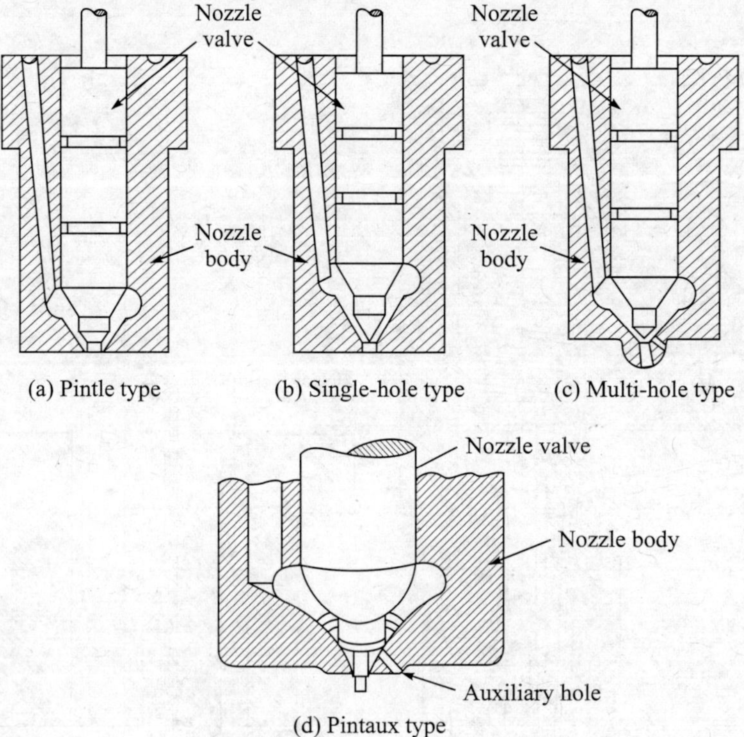

Figure 10.15 Types of nozzles.

(a) The pintle nozzle
(b) The single-hole (orifice) nozzle
(c) The multi-hole (orifice) nozzle
(d) The pintaux nozzle.

10.6.1 Pintle Nozzle

The pintle type nozzle is shown in Figure 10.15(a). The stem of the nozzle valve has a profiled extension to form a pintle that protrudes into the valve body orifice to form an annular spray hole. The pintle shape is controlled to vary the annular orifice area as the valve lifts. Two types of pintle nozzles are shown in Figure 10.16. Figure 10.16(a) shows a standard pintle nozzle and Figure 10.16(b) shows a throttling pintle nozzle. They are quite similar except for the increased

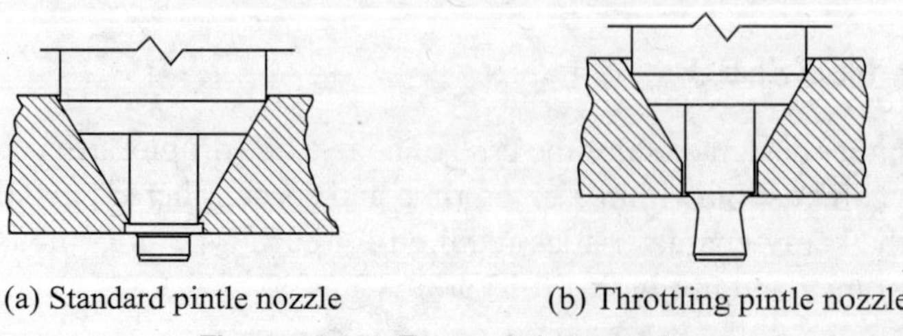

Figure 10.16 Types of pintle nozzles.

orifice length and greater pintle protrusion of the throttling type. Depending upon the pintle profile, the spray can be made to issue in the form of a hollow cone with an included angle as high as 60 degrees for standard pintle nozzles, or as a relatively heavy core-spray with an included angle of only a few degrees of throttling nozzles.

The throttling nozzle gives smoother combustion by reducing the initial rate of injection during the ignition lag period. Figure 10.17(a) shows a comparison of typical rates of discharge between the standard pintle nozzle and the throttling nozzle. Both the beginning and ending of the throttle nozzle are slower.

The relation of the orifice area to valve lift for both nozzles is shown in Figure 10.17(b). This shows that the orifice area for the throttle nozzle is very small up to 0.48 mm valve lift, whereas with the standard pintle nozzle the orifice area increases rapidly above 0.10 mm valve lift.

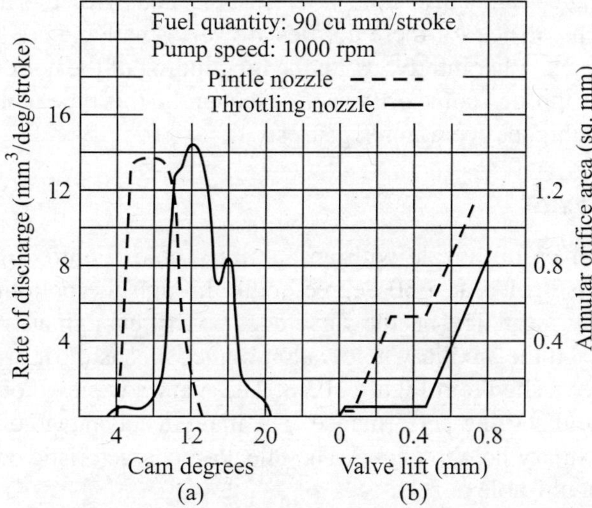

Figure 10.17 Comparative rates of discharge and orifice areas of pintle and throttling type nozzles.

Standard pintle or throttling types are used in engines with precombustion chambers, turbulence chambers, air or energy cells.

10.6.2 Single-hole (Orifice) Nozzle

The single-hole nozzle (Figure 10.15(b)) consists of a single hole in the end of the nozzle, through which the fuel passes into the combustion chamber. The size of the hole is usually of the order of 0.2 mm. Injection pressure is of the order of 80 to 100 bar. The hole may be drilled centrally or at an angle to the centre line of the nozzle, such that the spray cone angle is about 15°.

The main disadvantages of the single-hole nozzle are:

(a) The spray angle is very narrow which does not facilitate good mixing unless higher air velocities are provided.
(b) It requires high injection pressure.
(c) It has a tendency to dribble.

10.6.3 Multi-hole Nozzle

The multi-hole nozzles (Figure 10.15(c)) are most suitable for open chamber engines because of the versatility with which changes can be made to provide any desired penetration, dispersion, and duration of injection. The orifices can be drilled as small as 0.125 mm diameter or as large as 0.85 mm, and the number of orifices varies from 4 to 10 depending upon the cylinder bore and air swirl. The use of more than 10 orifices generally results in interference between the adjacent sprays. The nozzle opening pressure for these nozzles varies from 165 bar to 200 bar for small engines and 240 to 300 bar for large engines. The nozzle opening and closing pressures should be sufficiently high to ensure that the valve is fully seated against high combustion chamber pressures after the injection has ended, otherwise the combustion gases will enter the nozzle and foul the orifices and the valve with carbon.

In general, the single- and multi-hole nozzles are used with the non-turbulent type of combustion chamber. The orifices of these nozzles are very small and are subject to clogging by carbon particles which may either interfere with the functioning of the nozzle stream, or may even completely stop the flow through some orifices. Consequently, this type of nozzle usually requires greater maintenance and higher operating expenses.

10.6.4 Pintaux Nozzle

It is a throttling type of pintle nozzle with an auxiliary hole (Figure 10.15(d)). An auxiliary 0.2 mm diameter hole is drilled at a 30 degree angle through the bottom of the nozzle into a space just below the valve seat. The needle valve does not lift enough at low speeds and most of the fuel is injected through the auxiliary hole, but at higher speeds most of the fuel is discharged normally past the pintle as shown in Figure 10.18. The main advantage of the pintaux nozzle is that it provides better cold starting performance. The main disadvantage of this type of nozzle is the tendency for the auxiliary hole to choke. The injection characteristic of the pintaux nozzle is poorer than that of the multi-hole nozzle.

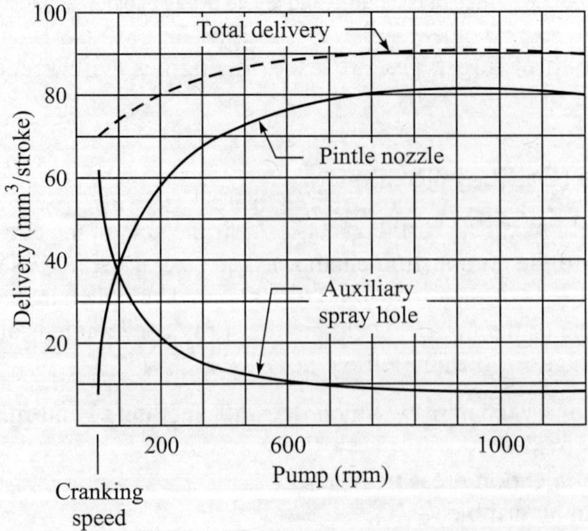

Figure 10.18 Fuel delivery characteristics of a pintaux nozzle.

10.7 UNIT INJECTOR

Unit injector combines the fuel pump and injection nozzle in one unit and thus eliminates the high-pressure pipeline connection which creates pressure waves and compressibility of fuel and disturbs the fuel discharge. Consequently, the reduction in compressibility of fuel and increased rigidity of the unit injector enable much higher injection pressure and higher rates of injection with more spray holes to be utilized. As a result, proportionally more energy for the fuel-air mixing comes from the injected fuel spray, and therefore lower rates of air swirl can be tolerated by medium-speed engines so that shallow quiescent chambers can be used effectively, as these tend to reduce exhaust emissions and consume less fuel. Conversely, higher engine speeds can be tolerated with direct-injection small diesel engines, since the thoroughness of mixing the fuel and air with the higher injection rates and pressures can be sustained at high engine speeds, even though injection and mixing time are considerably reduced.

10.7.1 Mechanically Operated Unit Injector

Figure 10.19 shows a mechanically operated unit injector and its driving mechanism. It consists of a plunger which reciprocates and rotates in its barrel, a plunger type follower with its return spring which imparts its reciprocating motion to the plunger via a rocker arm, a push rod, a roller follower and a cam.

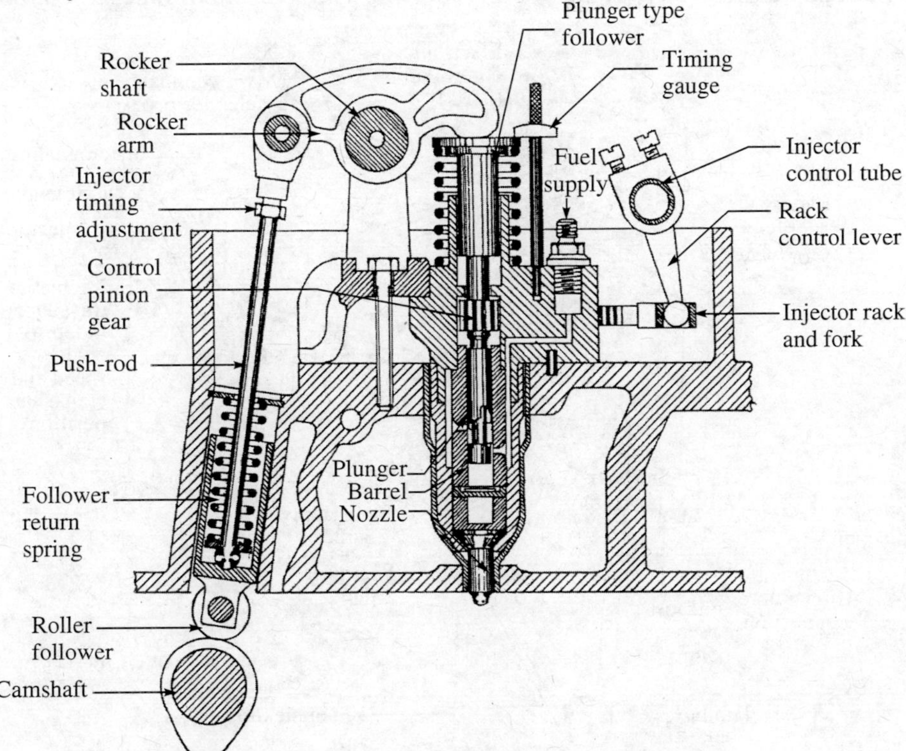

Figure 10.19 Unit Injector and its driving mechanism.

The lower end of the plunger has an annular groove with a helix control edge machined partially around it. A cross-hole drilled in the grooved portion of the plunger intersects a central hole drilled at the base of the plunger. The barrel has two ports, upper and lower ports. In addition to the reciprocating motion of the plunger, the plunger can be rotated about its axis by a control pinion gear which meshes with the control-rack.

The upward motion of the plunger allows fuel to flow from the fuel cavity through the upper port and plunger passage, as well as through the lower port into the chamber ahead of the plunger. During the first part of the downward stroke of the plunger, the fuel returns to the cavity until plunger covers the lower port and later the upper helix covers the upper port. The plunger motion builds up a fuel pressure that opens the check valve and forces the fuel through the small orifices. The injection ends when the lower helix uncovers the lower port and thus relieves the fuel pressure.

Turning of plunger by horizontal movement of the rack changes the position of the helix, thus varying the quantity of fuel injected and the timing of the injection period.

10.7.2 Electronically Controlled Unit Injector System

Figure 10.20 shows an electronically controlled unit injector system. The microprocessor in the electronic control unit (ECU) receives information from various sensors and compares it with the pre-programmed data stored in its memory. It then computes appropriate fuel delivery and timing requirements and transmits these signals to the solenoid. The solenoid operates a control valve in

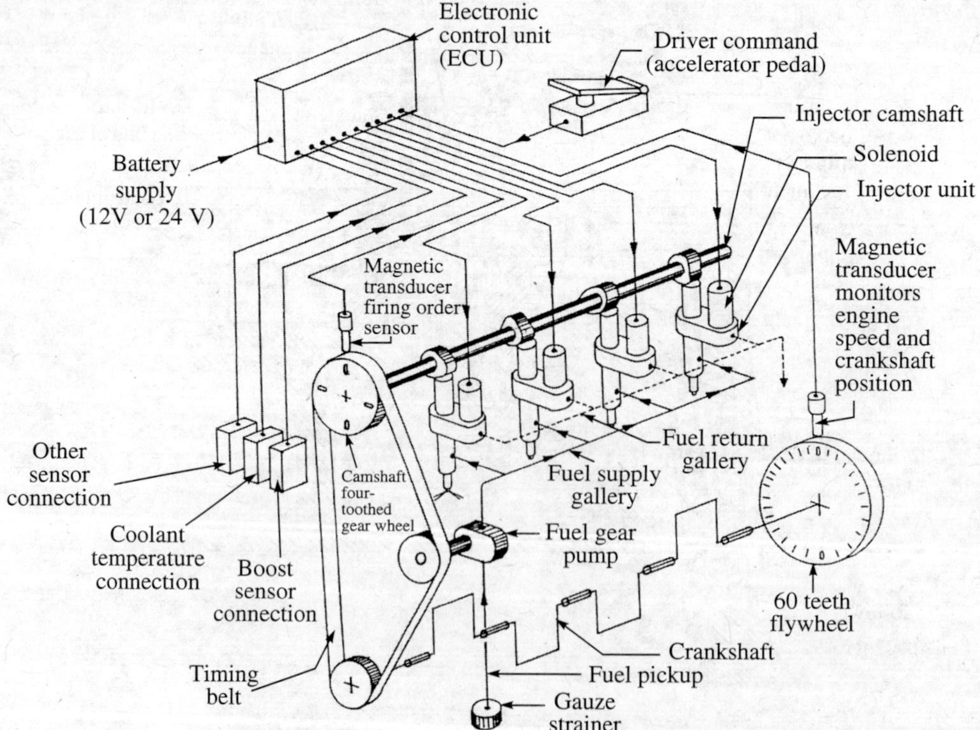

Figure 10.20 Electronically controlled unit injector system.

relation to the requirements. Electronically controlled fuel injector is shown in Figure 10.21. Fuel injection commences when energized solenoid closes the control valve. The duration of the valve closure determines the quantity of the fuel injected. The injection terminates as soon as the control valve opens. The injection pressure is generated by the camshaft-driven plunger, and the needle valve nozzle is employed as usual. The electronically controlled unit fuel injection system provides increased flexibility in fuel metering and injection timing.

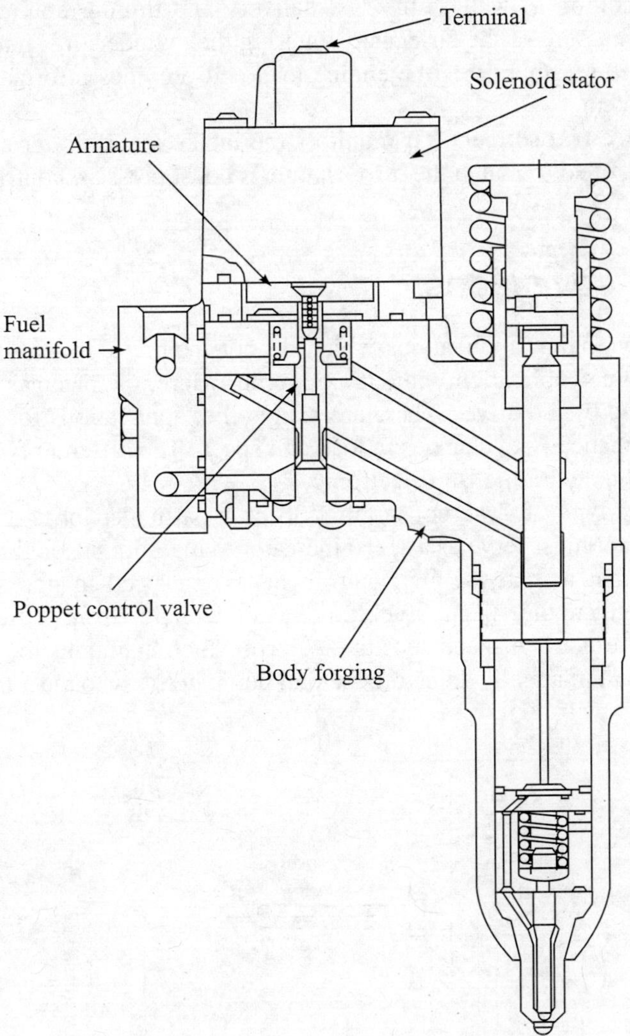

Figure 10.21 Electronically controlled unit injector.

Main sensors (transducers) which provide input signals to the electronic control unit are as follows:

1. **Accelerator pedal position:** It monitors the demand of the drivers.
2. **A crankshaft-driven 60-teeth flywheel with a magnetic transducer:** It monitors engine speed. A unique tooth indicates TDC position of cylinder 1 on its compression stroke

for injection timing purposes. Engine crankshaft angular data are used to actuate solenoid for required fuel delivery and injection timing.

3. **Firing order sensor:** It is an induction type magnetic transducer. It gives signals for the sequence of the engine's firing order to ECU.
4. **Temperature transducers:** These transducers monitor coolant temperature, air temperature and oil temperature. The coolant temperature indicates the cold start and normal operating condition as well as the fuel delivery and timing requirements. The air temperature is a measure of the air density entering the cylinder. It signals the fuel need. The oil temperature sensor relates the engine load and provides information to the engine's protection system.
5. **Boost pressure transducer:** It monitors the intake manifold air charge pressure. Fuel supply is controlled to match the turbocharger's boost pressure contribution over engine's speed range.

10.8 CI ENGINE GOVERNORS

There are only a few engine applications which do not require a governor. The passenger car is an example of engine application without a governor. Here, the changes in engine speed and engine load are sensed by the driver. There are many other applications for which governors are used to readjust automatically to changes in load and speed. The different types of governors are: mechanical, pneumatic, hydraulic, and electronic.

Regardless of the type of governor required to do a particular job, each must perform two major tasks. A governor must serve as a speed indicator to measure the engine speed and must act as a power mechanism to actuate the fuel control whenever a speed change occurs. Further more, the governor must regulate the injected fuel to prevent the engine from stalling or overspeeding. The conventional governor is a speed sensitive control which maintains the engine speed within the desired limits by automatically adjusting the fuel pump delivery to meet the variations in load.

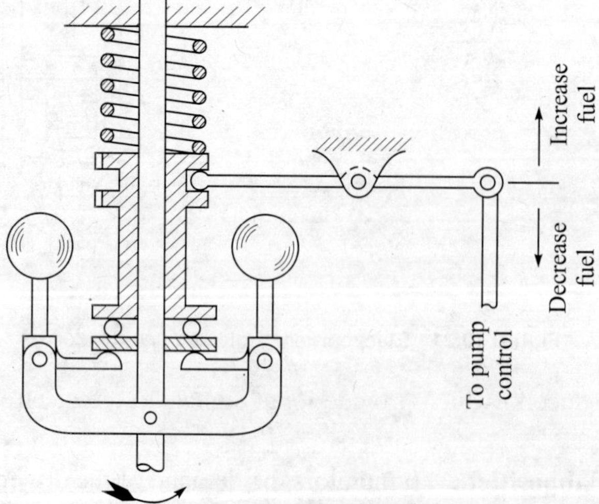

Figure 10.22 Simple mechanical governor.

Only the mechanical governor is described here. Figure 10.22 shows a schematic diagram of a simple mechanical governor. It has a pair of engine driven flyweights, working against a spring, for speed sensing. When the engine speed increases due to decreased load, the centrifugal force of the rotating flyweights forces the weight outwards against the spring force until both the forces are equal. At the same time the fuel control rack of the injection pump is moved towards the low-fuel position and the governor maintains a constant engine speed. The reverse action takes place when the engine speed decreases due to increased load. Thus the engine speed is maintained fairly constant irrespective of the engine load.

10.9 SPRAY CHARACTERISTICS

The efficiency and power output of CI engines are highly dependent on the characteristics of the fuel spray injected from the nozzle into the combustion chamber. The process of spraying has two objectives. It divides the liquid fuel into a large number of fine droplets to increase its surface area for rapid heat transfer and combustion, and it distributes the fuel through the combustion chamber for intimate mixing of the fuel and air.

10.9.1 Spray Formation

Depending upon the injection pressure and viscosity of the liquid fuel, the first portion of fuel coming out from the nozzle usually appears as a cylindrical jet. At a certain distance from the nozzle the jet breaks up into fine droplets to form a conical-shaped spray.

The most widely used method for spraying liquid fuels in CI engine is by solid or pressure injection. Fuel is forced through an orifice under pressure to form an unstable jet which disintegrates as it comes into contact with the air in the combustion chamber.

10.9.2 Atomization

The uniformity and fineness of the droplets in a spray defines its degree of atomization. The fuel velocity is the most important factor that affects the degree of atomization. This depends primarily upon the injection pressure, being a function of the square root of the difference between the injection pressure and the compression pressure. Increasing the injection pressure reduces the mean diameter of the particles as well as varies their size. The percentage of particles with the large diameters decreases rapidly with an increase in the injection pressure. A small orifice diameter results in a large surface-to-volume ratio for the fuel stream. This should result in better atomization.

The density of the air into which the fuel is injected affects atomization because of the important part played by the air in the fuel jet disintegration process. As the air pressure is increased, there is a consistent decrease in droplet size.

The fuel factors that affect atomization are viscosity, surface tension and density.

10.9.3 Penetration

The distance to which the tip of a fuel spray will penetrate the air in the combustion chamber in a given time depends primarily upon the jet velocity, the combustion chamber air density and the orifice size. The fuel viscosity has a small effect on penetration.

An increase in injection pressure increases the spray-tip penetration. An increase in combustion chamber air density decreases the penetration. An increase in orifice diameter increases the penetration of the spray tip. The orifice should have a high coefficient of discharge and should be of such length that the fuel leaves the orifice as a stream flowing along the axis of the orifice if high penetration is desired. An orifice's length-to-diameter ratio between 4 and 7 gives the highest spray penetration.

For a given orifice and combustion chamber pressure, it is found that

$$s = f(t\sqrt{\Delta p}) \tag{10.1}$$

where t is the time required for the spray to penetrate the distance s and Δp is the pressure difference between the injection pressure and the combustion chamber pressure. The functional relation (10.1) can be represented by a curve shown in Figure 10.23(a). With the variation of injection pressure the results lie on a single curve. The two points 1 and 2 on the curve satisfy the functional relation, therefore,

$$s_1 = f(t_1\sqrt{\Delta p_1}) \quad \text{and} \quad s_2 = f(t_2\sqrt{\Delta p_2}) \tag{10.2}$$

If the penetration $s_1 = s_2$, then

$$t_1\sqrt{\Delta p_1} = t_2\sqrt{\Delta p_2} \tag{10.3}$$

For the same value of penetration the above relation is satisfied. This relation may be used to determine the time required for the spray to penetrate the same distance when one injection pressure is known.

It is also found that

$$\frac{s}{d} = f\left(\frac{t}{d}\right) \tag{10.4}$$

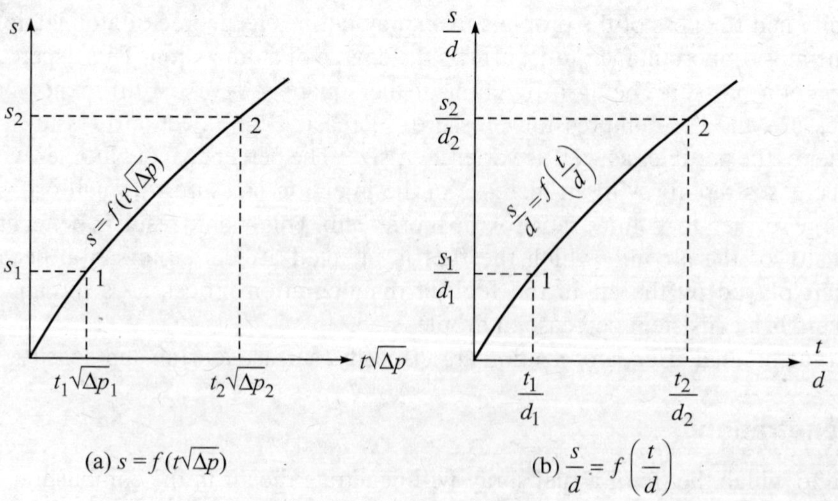

Figure 10.23 Spray penetration curves.

where d is the orifice diameter. This functional relation is represented by a curve shown in Figure 10.23(b). The two points 1 and 2 satisfy the functional relation, therefore

$$\frac{s_1}{d_1} = f\left(\frac{t_1}{d_1}\right) \quad \text{and} \quad \frac{s_2}{d_2} = f\left(\frac{t_2}{d_2}\right) \tag{10.5}$$

For the similar points on the penetration curve,

$$\frac{s_1}{d_1} = \frac{s_2}{d_2} \quad \text{and} \quad \frac{t_1}{d_1} = \frac{t_2}{d_2} \tag{10.6}$$

10.9.4 Dispersion

The primary objective in fuel injection is to distribute the fuel throughout the inducted air in the combustion chamber. This is achieved by the spray having the required penetration and dispersion, which involves the atomization and spreading of the fuel. Spray dispersion has been defined as the ratio of spray volume to fuel volume at a given phase of the injection interval.

The dispersion of the fuel spray depends upon the injection-nozzle design. The type of orifice that gives a large spray cone produces better dispersion of the fuel than does a small spray cone. This indicates a small length-to-diameter ratio for the orifice. However, an increase in spray dispersion decreases the penetration.

Spray dispersion is affected by the following factors:

1. The process of dispersion takes time, and so it becomes more uniform as the distance from the orifice is increased.
2. The dispersion becomes more even as the air density increases.
3. As the oil viscosity is decreased the dispersion becomes more uniform.
4. The dispersion improves as the injection pressure is increased.

10.10 RATE OF FUEL INJECTION IN CI ENGINES

Consider Figure 10.24, showing fuel-injection into a cylinder through an injection nozzle. Take two sections 1–1′ and 2–2′ as shown in the diagram.

Let

p_1 = fuel-injection pressure

p_2 = pressure in the cylinder at the time of fuel injection.

ρ_f = density of fuel (assumed incompressible)

c_1 = velocity at section 1–1′

c_2 = velocity at section 2–2′

From Bernoulli's equation (neglecting change in potential energy),

$$\frac{c_1^2}{2} + \frac{p_1}{\rho_f} = \frac{c_2^2}{2} + \frac{p_2}{\rho_f} \tag{10.7}$$

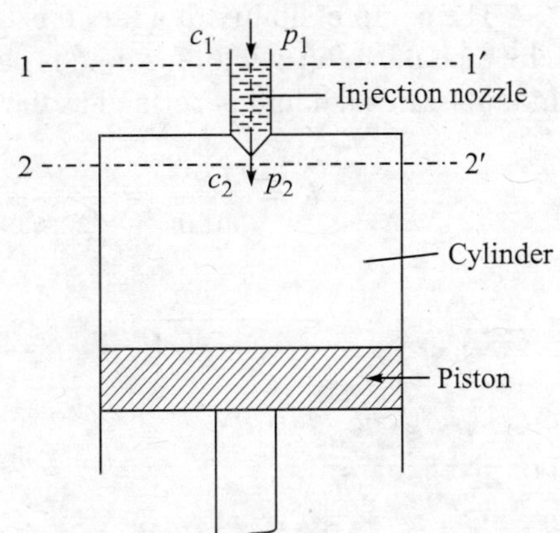

Figure 10.24 Fuel injection through an injection nozzle.

Neglecting the initial velocity of fuel, c_1 being very small compared to c_2, we have

$$c_2 = \sqrt{\frac{2(p_1 - p_2)}{\rho_f}} \tag{10.8}$$

If c_f is the actual velocity of fuel, C_v is the velocity coefficient for the orifice and $\Delta p = p_1 - p_2$,

$$c_f = C_v \sqrt{\frac{2\Delta p}{\rho_f}} \tag{10.9}$$

Rate of mass flow of fuel,

$$\dot{m}_f = \rho_f a_f c_f = C_d A_f \sqrt{2\rho_f \Delta p} \tag{10.10}$$

where

A_f = area of cross-section of a fuel nozzle,
a_f = area of cross-section of the fuel jet at venna-contracta,
$a_f = A_f C_c$, with C_c as the coefficient of contraction
C_d = coefficient of discharge = $C_c C_v$.

10.11 FUEL-LINE HYDRAULICS

Efficient combustion requires that the fuel be injected at the proper time and rate, and the injection pressure must be sufficiently high for adequate atomization and penetration. This requires not only the mechanical characteristics of the pump, the discharge tubing and the injector, but also the compressibility and dynamics of the fuel flow between the pump and the nozzle.

10.11.1 Fuel Compressibility

Since fuel is compressible at high pressure, there is a time lag between the beginning of delivery by the pump and the beginning of discharge from the nozzle, and the rate of delivery from the pump is not identical with the rate of discharge from the nozzle.

The compressibility of a liquid is the reciprocal of the modulus of elasticity or bulk modulus. The modulus of elasticity for liquids can be expressed in the same way as for the solids. The bulk modulus K of a liquid is defined as the pressure required to produce unit volumetric strain.

$$K = \frac{\text{stress}}{\text{strain}} = \frac{\text{increase in pressure}}{\text{decrease in volume per unit original volume}}$$

$$\therefore \quad K = \frac{\Delta p}{-\Delta V/V} \tag{10.11}$$

mass, $m = \rho V$

For given mass,

$$\rho V = \text{constant.}$$

Taking log and differentiating,

$$\frac{\Delta\rho}{\rho} = -\frac{\Delta V}{V}$$

The bulk modulus is a function of temperature and pressure but here it will be considered to be a constant. Then with adequate accuracy, we can write

$$K = \frac{\Delta p}{\Delta\rho/\rho} \tag{10.12}$$

where $\Delta\rho$ is the increase in density and ρ is the density of the fluid.

The coefficient of compressibility k is the reciprocal of the bulk modulus K, i.e.

$$k = \frac{1}{K} = \frac{-\Delta V/V}{\Delta p} \tag{10.13}$$

10.11.2 Pressure Waves in Fuel Lines

As fuel oil is a compressible fluid, the movement of the pump plunger initiates a pressure wave and propagates it through the discharge tube at the speed of sound in the fuel oil. The pressure wave is caused by the pressure that is built up by the plunger in compressing the fuel at the pump, while accelerating the fuel column towards the nozzle. These waves not only travel from the pump to the nozzle, but are also reflected towards the pump from the nozzle. The pressure waves affect the designed rates of injection and pressures.

Consider Figure 10.25 showing a simplified representation of plunger and barrel of injection pump. Assume that the plunger and all fluid particles are moving in the cylinder at a velocity v, and that the plunger is being accelerated and the velocity is instantly increased by Δv. The fluid adjacent to the plunger will also increase its velocity at the same instant by Δv, but at section X-X a finite time will pass before the pressure increase is evident. The additional mass of fluid displaced in time Δt by the plunger because of the velocity increase Δv would be $(\rho \cdot \Delta v \cdot \Delta t)A$, where ρ is the density of the fluid and A is the area of cross-section of the barrel. This displacement causes a pressure wave to travel down the pipe at a sonic velocity v_S and increases the density of the fluid. The additional mass of the fluid displaced in the barrel by the plunger because of the increased velocity must be equal to the gain in mass of the fluid resulting from the increased density. This gain in mass would be $(\Delta\rho v_S \Delta t)A$. Equating these two expressions,

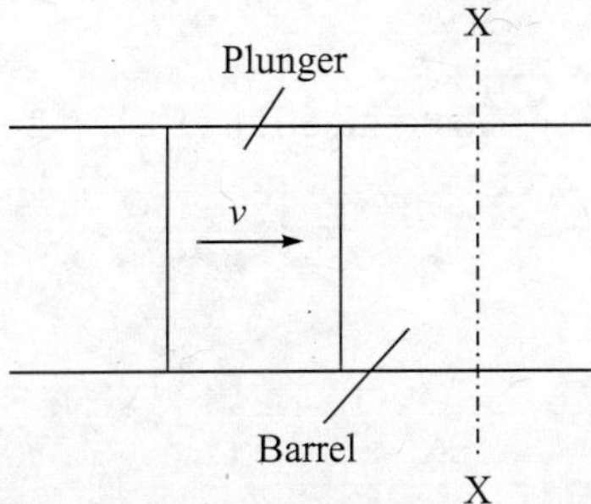

Figure 10.25 Simplified representation of plunger and barrel of injection pump.

$$(\rho\Delta v\Delta t)A = (\Delta\rho v_S \Delta t)A$$

$$\therefore \quad v_S = \rho\frac{\Delta v}{\Delta\rho} \tag{10.14}$$

where

v_s = velocity of propagation of pressure disturbance in the fluid (sonic velocity), in m/s

ρ = density of fluid, in kg/m3

Δv = change in velocity of plunger, in m/s

$\Delta\rho$ = increase in density resulting from increased pressure, in kg/m^3.

Another expression for sonic velocity v_s can be obtained from Newton's law,

$$F = ma$$

or

$$\Delta p A = m\frac{\Delta v}{\Delta t}$$

where Δp is the increase in pressure in N/m^2 corresponding to the acceleration $\Delta v/\Delta t$ and m is the mass of fluid accelerated.

Now,

$$m = \rho v_s \Delta t A$$

$\therefore$

$$\Delta p A = \rho v_S \Delta t A \frac{\Delta v}{\Delta t}$$

or

$$\Delta p = \rho v_s \Delta v \tag{10.15}$$

$\therefore$

$$v_S = \frac{1}{\rho}\frac{\Delta p}{\Delta v} \tag{10.16}$$

Upon multiplying Eqs. (10.14) and (10.16),

$$v_S^2 = \frac{\Delta p}{\Delta \rho} \tag{10.17}$$

Upon substituting Eq. (10.12) in Eq. (10.17) and simplifying,

$$v_S^2 = \frac{K}{\rho}$$

$\therefore$

$$v_S = \sqrt{\frac{K}{\rho}} \tag{10.18}$$

By substituting this value of v_s in Eq. (10.16),

$$\sqrt{\frac{K}{\rho}} = \frac{1}{\rho}\frac{\Delta p}{\Delta v}$$

or

$$\frac{\Delta p}{\Delta v} = \sqrt{K\rho} \tag{10.19}$$

From Eq. (10.18), $\sqrt{\rho} = \dfrac{\sqrt{K}}{v_S}$. Substituting it in Eq. (10.19),

$$\frac{\Delta p}{\Delta v} = \frac{K}{v_S} \tag{10.20}$$

EXAMPLE 10.1 A six-cylinder four-stroke diesel engine develops a power of 250 kW at 1500 rpm. The brake specific fuel consumption is 0.3 kg/kWh. The pressures of air in the cylinder at the beginning of injection and at the end of injection are 30 bar and 60 bar respectively. The fuel injection pressures at the beginning and end of injection are 220 bar and 550 bar respectively. Assume the coefficient of discharge for the injector to be 0.65, specific gravity of fuel to be 0.85 and the atmospheric pressure to be 1.013 bar. Assume the effective pressure difference to be the average pressure difference over the injection period.

Determine the nozzle area required per injection if the injection takes place over 15° crank angle. If the number of orifices used in the nozzle are 4, find the diameter of the orifice.

Solution: Brake specific fuel consumption, bsfc (kg/kWh) $= \dfrac{\dot{m}_f\,(\text{kg/h})}{\text{bp}\,(\text{kW})}$

$$\therefore \quad \dot{m}_f = \text{bsfc} \times \text{bp} = 0.3 \times 250 = 75 \text{ kg/h} = 1.25 \text{ kg/min}$$

$$\text{Fuel injected per cycle per cylinder} = \frac{\text{fuel consumed per min}}{\text{no. of cycles per min}} \times \frac{1}{\text{no. of cylinders}}$$

$$= \frac{1.25}{(1500/2)} \times \frac{1}{6} = 0.0002778 \text{ kg}$$

$$15°\text{CA (crank angle)} = \frac{15}{360} \text{ revolution}$$

$$\text{Duration of injection (s)} = \frac{\text{no. of revolutions}}{\text{revolution per second}} = \frac{15/360}{1500/60}$$

$$= 1.667 \times 10^{-3} \text{ s} = 0.001667 \text{ s}$$

$$\therefore \quad \text{Mass of fuel injected per second, } \dot{m}'_f = \frac{0.0002778}{0.001667} = 0.1666 \text{ kg/s}$$

$$\text{Pressure difference at the beginning} = 220 - 30 = 190 \text{ bar}$$

$$\text{Pressure difference at the end} = 550 - 60 = 490 \text{ bar}$$

$$\text{Average pressure difference, } \Delta p = \frac{190 + 490}{2} = 340 \text{ bar}$$

The mass of fuel injected per second,

$$\dot{m}'_f = C_d A_f \sqrt{2\rho_f \cdot \Delta p}$$

$\therefore$ Area of cross-section of the nozzle,

$$A_f = \frac{\dot{m}'_f}{C_d \sqrt{2\rho_f \cdot \Delta p}}$$

$$= \frac{0.1666}{0.65\sqrt{2 \times 850 \times 340 \times 10^5}}$$

$$= 1.066 \times 10^{-6} \text{ m}^2 = \boxed{1.066 \text{ mm}^2} \quad \textbf{Ans.}$$

Now, the diameter of orifice, d_o can be calculated from the relation,

$$\left(\frac{\pi}{4} d_o^2\right) \times \text{no. of orifices} = A_f$$

or
$$\frac{\pi}{4} d_o^2 \times 4 = 1.066$$

∴
$$d_o = \sqrt{\frac{1.066}{\pi}} = \boxed{0.58 \text{ mm}} \quad \textbf{Ans.}$$

EXAMPLE 10.2 A single-cylinder four-stroke diesel engine develops a power of 30 kW at 3000 rpm. The specific fuel consumption is 0.28 kg/kWh fuel of 35° API. The fuel is injected at an average pressure of 160 bar. The duration of injection is 28° of crank travel. The pressure in the combustion chamber is 35 bar. The coefficient of velocity is 0.92. Determine the velocity of injection of the fuel and diameter of the fuel orifice.

Solution:
$$\text{Specific gravity} = \frac{141.5}{131.5 + °\text{API}} = \frac{141.5}{131.5 + 35} = 0.85$$

∴
$$\text{Density of fuel, } \rho_f = 0.85 \times 1000 = 850 \text{ kg/m}^3$$

$$\text{Duration of injection} = \frac{\text{no. of revolutions}}{\text{revolution per second}}$$

$$= \frac{28/360}{3000/60} = 1.556 \times 10^{-3} \text{ s}$$

$$\text{Fuel consumption per cycle} = \frac{\text{bsfc} \times \text{kW}}{\text{cycles per hour}}$$

$$= \frac{0.28 \times 30}{(3000/2) \times 60} = 9.333 \times 10^{-5} \text{ kg}$$

∴ Mass of fuel flow per second,

$$\dot{m}_f = \frac{9.333 \times 10^{-5}}{1.556 \times 10^{-3}} = 6 \times 10^{-2} = 0.06 \text{ kg/s}$$

Velocity of injection of fuel,

$$c_f = C_v \sqrt{\frac{2\Delta p}{\rho_f}}$$

$$= 0.92 \sqrt{\frac{2(160 - 35) \times 10^5}{850}} = 157.8 \text{ m/s}$$

$$\dot{m}_f = \rho_f \times A_f \times c_f \text{ (neglecting contraction of the jet)}$$

∴
$$A_f = \frac{\dot{m}_f}{\rho_f c_f} = \frac{0.06}{850 \times 157.8} = 4.473 \times 10^{-7} \text{ m}^2$$

$$= 0.4473 \text{ mm}^2$$

$$\therefore \quad d = \sqrt{\frac{4}{\pi} \times 0.4473} = \boxed{0.755 \text{ mm}} \quad \textbf{Ans.}$$

EXAMPLE 10.3 A closed injection nozzle has an orifice diameter of 0.8 mm and the maximum cross-sectional area of the passage between the needle cone and the seat is 1.65 mm². The discharge coefficient for the orifice is 0.9 and that for the passage is 0.85. The injection pressure is 170 bar and the compression pressure of the charge during injection is 25 bar, when the needle valve is fully open. Determine the discharge of fuel through the injector and the jet velocity at that instant. The density of the fuel is 850 kg/m³.

Solution: Velocity of flow, $c_f = C_v \sqrt{\frac{2\Delta p}{\rho_f}}$

Neglecting contraction of the jet, the coefficient of contraction, $C_c = 1.0$

∴ Coefficient of velocity, C_v = coefficient of discharge, C_d.

$$\therefore \quad c_f = C_d \sqrt{\frac{2\Delta p}{\rho_f}}$$

Discharge,

$$Q = A \times c_f$$

where A is the area of cross-section of the orifice.

$$\therefore \quad Q = AC_d \sqrt{\frac{2\Delta p}{\rho_f}}$$

$$\therefore \quad \Delta p = \left(\frac{Q}{AC_d}\right)^2 \frac{\rho_f}{2} [\text{N/m}^2]$$

$$= \frac{1}{2} \rho_f \left(\frac{Q}{AC_d}\right)^2 \frac{1}{10^5} [\text{bar}]$$

Let p be the pressure immediately before the orifice in bar. We now have the following two equations:

Flow through passage:

$$170 - p = \frac{1}{2} \times 850 \frac{Q^2}{(1.65 \times 10^{-6} \times 0.85)^2 \times 10^5} [\text{bar}]$$

$$\therefore \quad 170 - p = 2.161 \times 10^9 Q^2 \qquad \text{(i)}$$

Flow through orifice:

$$p - 25 = \frac{1}{2} \times 850 \frac{Q^2}{\left(\frac{\pi}{4}(0.8 \times 10^{-3})^2 \times 0.9\right)^2} \times \frac{1}{10^5}$$

$$\therefore \quad p - 25 = 20.77 \times 10^9 Q^2 \qquad \text{(ii)}$$

Adding Eqs. (i) and (ii), we get

$$145 = 22.931 \times 10^9 Q^2$$

$$\therefore \quad Q = \sqrt{\frac{145}{22.931 \times 10^9}} = 0.795 \times 10^{-4}\ \text{m}^3/\text{s}$$

$$= \boxed{79.5\ \text{cm}^3/\text{s}} \quad \textbf{Ans.}$$

From Eq. (i),

$$170 - p = 2.161 \times 10^9 \times (0.795)^2 \times 10^{-8}$$

$$\therefore \quad p = 170 - 13.66 = 156.34\ \text{bar}$$

Velocity of fuel flow through the orifice,

$$c_f = C_d \sqrt{\frac{2\Delta p}{\rho_f}}$$

$$= 0.9\sqrt{\frac{2(156.34 - 25) \times 10^5}{850}}$$

$$= \boxed{158.2\ \text{m/s}} \quad \textbf{Ans.}$$

EXAMPLE 10.4 (a) A spray penetration of 20 cm in 15.7 ms is obtained with an injection pressure of 150 bar. Determine the time required for the spray to penetrate this distance when an injection pressure of 450 bar is used. The orifice and the combustion chamber density remain constant. The combustion chamber pressure is 15 bar.

(b) The penetration for an orifice of 0.34 mm diameter is 20 cm in 12 ms for a certain injection pressure. Determine a similar point for the penetration curve of an orifice having a diameter of 0.17 mm.

Solution: (a) $s_1 = f(t_1\sqrt{\Delta p_1})$ and $s_2 = f(t_2\sqrt{\Delta p_2})$

These two results lie on a single curve.

$\therefore$ If $s_1 = s_2$, then $t_1\sqrt{\Delta p_1} = t_2\sqrt{\Delta p_2}$

$$\therefore \quad 15.7\sqrt{150 - 15} = t_2\sqrt{450 - 15}$$

$$\therefore \quad t_2 = \frac{15.7\sqrt{135}}{\sqrt{435}} = \boxed{8.75\ \text{ms}} \quad \textbf{Ans.}$$

(b) $$\frac{s_1}{d_1} = f\left(\frac{t_1}{d_1}\right) \quad \text{and} \quad \frac{s_2}{d_2} = f\left(\frac{t_2}{d_2}\right)$$

These two results also lie on a single curve. Therefore, for a similar point,

$$\frac{s_1}{d_1} = \frac{s_2}{d_2} \quad \text{and} \quad \frac{t_1}{d_1} = \frac{t_2}{d_2}.$$

$$\therefore \qquad s_2 = d_2\left(\frac{s_1}{d_1}\right) = \frac{0.17 \times 20}{0.34} = \boxed{10 \text{ cm}} \quad \textbf{Ans.}$$

and
$$t_2 = t_1\left(\frac{d_2}{d_1}\right) = \frac{12 \times 0.17}{0.34} = \boxed{6 \text{ ms}} \quad \textbf{Ans.}$$

EXAMPLE 10.5 In a fuel-injection pump of a diesel engine, the volume of the fuel in the pump barrel just before the commencement of the effective stroke is 6.5 cc. The fuel pipe line is 3 mm in diameter and 650 mm in length. The fuel in the injection valve is 2.5 cc. Assume the coefficient of compressibility of the oil to be 78.5×10^{-6} per bar. Take the atmospheric pressure as 1 bar.

Determine:

(a) The pump displacement necessary to deliver 0.1 cc of fuel at a pressure of 180 bar
(b) The effective stroke of the plunger which is 7.5 mm in diameter.

Solution: The coefficient of compressibility,

$$k = \frac{\text{change in volume per unit volume}}{\text{difference in pressure causing compression}}$$

$$= \frac{V_1 - V_2}{V_1(p_2 - p_1)}$$

Total initial volume, V_1 = Vol. of fuel in the barrel + Vol. of fuel in the pipe line + Vol. of fuel in the injection valve.

$$= 6.5 + \frac{\pi}{4}(0.3)^2 \times 65 + 2.5 = 13.59 \text{ cc}$$

Change in volume due to compression,

$$V_1 - V_2 = kV_1(p_2 - p_1)$$

$$= 78.5 \times 10^{-6} \times 13.59\,(180 - 1)$$

$$= 0.191 \text{ cc}$$

Total displacement of the plunger $= (V_1 - V_2) + 0.1$

$$= 0.191 + 0.1$$

$$= \boxed{0.291 \text{ cc}} \quad \textbf{Ans.}$$

$$\therefore \qquad \frac{\pi}{4}d^2 l = 0.291$$

$\therefore$ Effective stroke of the plunger,

$$l = 0.291 \times \frac{4}{\pi} \times \frac{1}{(0.75)^2} = \boxed{0.659 \text{ cm}} \quad \textbf{Ans.}$$

EXAMPLE 10.6 A four-cylinder four-stroke diesel engine running at 2500 rpm, produces 90 kW power. The specific fuel consumption is 0.28 kg/kWh. Each cylinder is provided with a separate fuel pump, pipe line and injector. At the beginning of the effective plunger stroke of one fuel pump, the fuel in the barrel is 3.5 cc, the fuel in the pipe line is 2.5 cc and the fuel inside the injector is 2 cc. The average injection pressure is 280 bar and the compression pressure of air during injection is 30 bar. The density of fuel is 850 kg/m^3, the coefficient of compressibility of fuel is 80×10^{-6} per bar. Fuel enters the pump barrel at 1 bar. Determine the displacement volume of one plunger per cycle and the power lost in pumping the fuel.

Solution: Fuel consumption per cycle $= \dfrac{\text{bsfc} \times \text{kW}}{\text{no. of cycles per hour}}$

$$= \frac{0.28 \times 90}{(2500/2) \times 60} = 3.36 \times 10^{-4} \text{ kg}$$

Fuel consumption per cylinder per cycle $= \dfrac{3.36 \times 10^{-4}}{4} = 0.84 \times 10^{-4}$ kg/cycle

Volume of fuel injected per cylinder per cycle, $V_f = \dfrac{0.84 \times 10^{-4}}{850}$

$$= 0.0988 \times 10^{-6} \text{ m}^3$$

$$= 0.0988 \text{ cm}^3$$

Coefficient of compressibility, $k = \dfrac{V_1 - V_2}{V_1 \cdot \Delta p}$

Total initial volume, V_1 = Fuel in pump barrel + fuel in pipeline + fuel in injector

$$= 3.5 + 2.5 + 2 = 8 \text{ cc}$$

$$V_1 - V_2 = kV_1 \cdot \Delta p$$

$$= 80 \times 10^{-6} \times 8 \times (280 - 1) = 0.1786 \text{ cc}$$

Volume displaced by plunger $= V_f + (V_1 - V_2)$

$$= 0.0988 + 0.1786$$

$$= \boxed{0.2774 \text{ cc}}$$ **Ans.**

Pump work per cycle, W = compression work + injection work.

The compression of fuel may be assumed to be represented by a straight line on the *p-V* diagram, as shown in Figure 10.26(a). The area enclosed represents the compression work. Injection is a constant-volume flow process. The work during this process is shown in Figure 10.26(b).

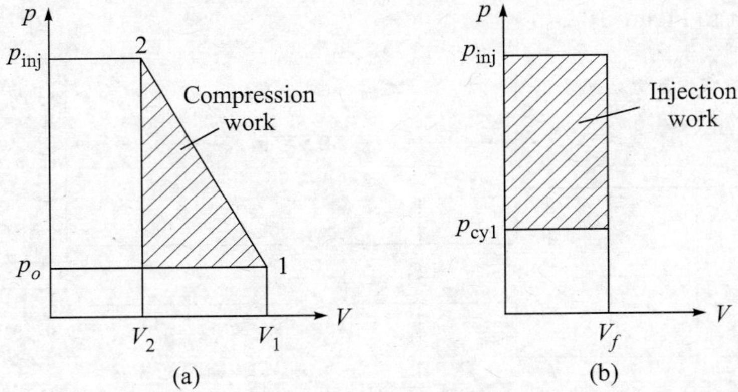

Figure 10.26 *p-V* diagram showing (a) the compression work and (b) the injection work.

∴ Pump work per cycle,

$$W = \frac{1}{2}(p_{inj} - p_o)(V_1 - V_2) + (p_{inj} - p_{cyl})V_f$$

$$= \frac{1}{2}(280 - 1) \times 10^5 \times 0.1786 \times 10^{-6} + (280 - 30) \times 10^5 \times 0.0988 \times 10^{-6}$$

$$= 2.49 + 2.47 = 4.96 \text{ J}$$

$$\text{Power lost per cylinder} = \frac{W \times N}{2 \times 60 \times 1000} \text{ kW}$$

$$= \frac{4.96 \times 2500}{2 \times 60 \times 1000} = 0.103 \text{ kW}$$

Total power lost for pumping the fuel $= 0.103 \times 4 = \boxed{0.412 \text{ kW}}$ **Ans.**

EXAMPLE 10.7 An injection system consists of a pump plunger moving with a velocity of 0.3 m/s and connected to a fuel pipe which is 0.575 m long and has a cross-sectional area $\frac{1}{20}$th of that of the plunger cylinder. The end of the pipe is provided with an open nozzle having a hole whose area is $\frac{1}{40}$th of that of the pipe. The initial pressure in the line is 27.6 bar, and the compression pressure of the engine is 27.6 bar. If the bulk modulus of fuel, K, is 17830×10^5 N/m^2 and the specific gravity of the oil is 0.86, determine the following:

(a) The velocity of pressure disturbances.
(b) The time taken by the disturbance to travel through the pipe line.
(c) The pressure and the velocity at the pump end of the pipe line as the plunger moves.
(d) The magnitude of the first reflected pressure and velocity wave.
(e) The pressure and the velocity at the orifice end of the pipe line after the first reflection.

Solution: Refer to Figure 10.27.

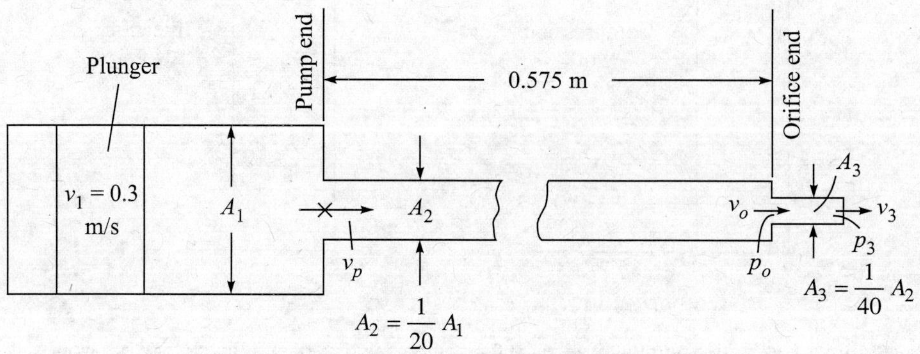

Figure 10.27 Simplified injection system.

(a) The velocity of pressure disturbances in oil of specific gravity 0.86 is

$$v_S = \sqrt{\frac{K}{\rho_f}} = \sqrt{\frac{17{,}830 \times 10^5}{860}} = \boxed{1440 \text{ m/s}} \quad \textbf{Ans.}$$

(b) The time taken by the disturbance to travel through the pipeline,

$$\Delta t = \frac{\text{length of the pipeline}}{v_S} = \frac{0.575}{1440} = \boxed{0.0004 \text{ s}} \quad \textbf{Ans.}$$

(c) Let the velocity of the fuel at the inlet of the pipeline be v_p.

$$\therefore \quad A_1 v_1 = A_2 v_p$$

$$\therefore \quad v_p = \frac{A_1}{A_2} v_1 = 20 \times 0.3 = \boxed{6 \text{ m/s}} \quad \textbf{Ans.}$$

This is the velocity of the fuel at the pump end of the pipeline.
Assume the initial velocity of the fuel at this location to be zero.

$\therefore$ Change in velocity at the inlet to the pipeline, $\Delta v = 6$ m/s

Now, $$\Delta p = \frac{K}{v_S} \Delta v = \frac{17{,}830 \times 10^5 \times 6}{1440} = 74.3 \times 10^5 \text{ N/m}^2 = 74.3 \text{ bar}$$

Initial pressure in the pipeline is 27.6 bar.
The fuel is assumed to increase in velocity from zero to 6 m/s instantaneously;

$$\text{Pressure at this instant} = 74.3 + 27.6 = \boxed{101.9 \text{ bar}} \quad \textbf{Ans.}$$

(d) A pressure disturbance of 74.3 bar and a velocity disturbance of 6 m/s now move down the pipe at a speed of 1440 m/s.

The velocity v_3 of the fluid after passing the orifice will be

$$v_3 = \sqrt{\frac{2\Delta p}{\rho_f}}$$

Velocity in the pipe just before the orifice, v_o, can be obtained from, $A_2 v_o = A_3 v_3$.

$$\therefore \quad v_o = \frac{A_3}{A_2} v_3 = \frac{1}{40}\sqrt{\frac{2\Delta p}{\rho_f}} = \frac{1}{40}\sqrt{\frac{2(p_0 - p_3)}{\rho_f}}$$

where p_o = 27.6 bar (initial) + 74.3 bar (arriving pressure disturbance) + p_r (reflected pressure disturbance).

and p_3 = compression pressure of engine = 27.6 bar.

The velocity of 6 m/s is propagated through the pipe from pump end of the line. When this velocity reaches the orifice end of the line, a reduction of velocity will take place.

Change in this velocity,

$$\Delta v = 6 - \frac{1}{40}\sqrt{\frac{2(p_o - p_3)}{\rho_f}}$$

$$\therefore \quad \Delta v = 6 - \frac{1}{40}\sqrt{\frac{2}{860}(27.6 + 74.3 + p_r - 27.6)10^5}$$

$$= 6 - \frac{1}{40}\sqrt{\frac{2}{860}(74.3 + p_r)10^5}$$

$$= 6 - 0.3812\sqrt{74.3 + p_r}$$

The first reflected pressure wave resulting from Δv will be

$$\Delta p = p_r = \frac{K}{v_S} \cdot \Delta v = \frac{17{,}830}{1440}(6 - 0.3812\sqrt{74.3 + p_r})$$

$$p_r = 74.3 - 4.72\sqrt{74.3 + p_r}$$

By trial, $p_r = \boxed{26.8 \text{ bar}}$ **Ans.**

$$\Delta v = v_r = p_r \times \frac{v_S}{K} = \frac{26.8 \times 1440}{17830} = \boxed{2.16 \text{ m/s}} \quad \textbf{Ans.}$$

(e) $p_o = 27.6 + 74.3 + 26.8 = \boxed{128.7 \text{ bar}}$ **Ans.**

$$v_o = \frac{1}{40}\sqrt{\frac{2(128.7 - 27.6) \times 10^5}{860}}$$

$$= \boxed{3.83 \text{ m/s}} \quad \textbf{Ans.}$$

REVIEW QUESTIONS

1. What are the requirements of a fuel-injection system of a CI engine?
2. What are the main components required in a fuel-injection system?
3. Describe an air-injection system with the help of a diagram. What are the advantages and disadvantages of an air-injection system?
4. What do you mean by a solid injection system? Name the different types of solid-injection systems. What are the main components of this system?
5. Describe the construction and working of an individual pump fuel-injection system with the help of a diagram. Draw the schematic diagrams of two arrangements of the pump in this system.
6. Describe the construction and working of a unit injector system with the help of a diagram.
7. Describe the construction and working of a distributor fuel injection system with the help of a diagram.
8. Describe the common-rail fuel-injection system with the help of a diagram. Why is this system not so popular now?
9. Describe the construction of a Jerk type fuel-injection pump.
10. Describe the working principle of a Jerk type of injection pump. Show the position of the helix for various load conditions.
11. Show the variation of plunger helix indicating the beginning and ending of injection.
12. Draw and describe the actual and theoretical delivery characteristics of a port-controlled displacement pump.
13. Describe and draw the injection pump delivery characteristics corresponding to full and part delivery with respect to pump speed. Also draw the fuel pump delivery characteristics at constant speed versus the rack stroke.
14. Describe the construction and working of inlet metered and sleeve metered distributor types of fuel-injection pumps with the help of suitable diagrams.
15. What is the function of a leak-off fuel in the fuel-injection holder?
16. Describe the construction and working of a fuel injector nozzle with the help of a diagram.
17. What are the different types of nozzles? What is the difference between the pintle type and the pintaux type nozzles?
18. Describe the pintle nozzle with the help of a diagram. What is the main difference between the standard pintle nozzle and the throttling pintle nozzle?
19. Compare the variation of rate of discharge with respect to cam degrees and orifice areas with respect to valve lift of a simple pintle nozzle and a throttling type nozzle.
20. Describe a single-hole nozzle with the help of a diagram.
21. Describe a multi-hole nozzle with the help of a diagram.
22. Describe the pintaux nozzle with the help of a diagram. Show the fuel delivery characteristics of a pintaux nozzle.
23. Mention the benefit of a unit injector. Describe the working principle of a mechanically operated unit injector.

24. Describe the working principle of an electronically controlled fuel-injection system.
25. Describe the simple mechanical governor used for CI engines.
26. Define the degree of atomization. What are the important factors that affect the degree of atomization?
27. What factors affect the penetration of fuel in the combustion chamber of a CI engine?
28. Define spray dispersion. What are the important factors that affect the dispersion of fuel in the combustion chamber of a CI engine?
29. Derive an expression to evaluate the rate of mass flow of fuel from the injector of a CI engine.
30. Define the coefficient of compressibility of a liquid fuel. Derive an expression to evaluate the velocity of propagation of pressure disturbance in the pipe line. Prove that:

$$\frac{\Delta p}{\Delta v} = \frac{K}{v_S}$$

where
Δp = increase in pressure
Δv = change in velocity of the pump plunger
K = modulus of rigidity
v_S = velocity of propagation of pressure disturbance in fluid.

PROBLEMS

10.1 A four-stroke four-cylinder CI engine develops 100 kW at 3000 rpm. The specific fuel consumption is 0.225 kg/kWh. Determine the mass of the fuel injected by the nozzle per cycle per cylinder, in the combustion chamber. [**Ans.** 0.0625 g]

10.2 A four-stroke six-cylinder CI engine develops 240 kW when running at 1000 rpm. The specific fuel consumption is 0.24 kg/kWh. The values of pressure of air in the cylinder at the beginning of injection and at the end of injection are 40 bar and 60 bar respectively. The fuel injection pressures at the beginning and end of injection are 200 and 600 bar respectively. Assume the effective pressure difference to be the average pressure difference during the injection period. Take the coefficient of discharge for the injector to be 0.6 and the density of the fuel to be 850 kg/m^3. Determine the nozzle area required per injection if the injection is carried out during the 12° rotation of the crank. If the number of orifices used in a nozzle are two, find the diameter of an orifice. [**Ans.** 1.093 mm^2, 0.834 mm]

10.3 Evaluate the diameter of the fuel orifice of a four-stroke single-cylinder CI engine which develops a power of 20 kW at 2000 rpm. The specific fuel consumption is 0.25 kg/kWh fuel of 32° API . The fuel is injected at an average pressure of 180 bar over a crank travel of 20°. The pressure in the combustion chamber is 30 bar. The coefficient of discharge is 0.85. [**Ans.** 0.68 mm]

10.4 A four-stroke four-cylinder CI engine operates on air/fuel ratio of 20. The bore and stroke of the cylinder are 12 cm and 16 cm respectively. The volumetric efficiency is 0.86. The condition of air at the beginning of compression is 1 bar and 300 K. Determine the mass of the fuel injected in each cylinder per second. If the speed of the engine is 1500 rpm, the injection pressure is 180 bar, the compression pressure of air is 35 bar and the fuel injection is carried out during the 20° of crank rotation, determine the diameter of the fuel orifice having single-hole nozzle. Take the density of the fuel to be 850 kg/m^3 and the coefficient of discharge for fuel nozzle to be 0.65. **[Ans.** 0.04066 kg/s, 0.712 mm]

10.5 A closed injection nozzle of a CI engine with a single-hole injector, has an orifice diameter of 0.75 mm, and the maximum cross-sectional area of the passage between the needle cone and the seat is 1.6 mm^2. The fuel is injected at an average pressure of 180 bar and the average compression pressure of air during injection is 30 bar. The discharge coefficient for the orifice is 0.85 and that for the passage is 0.80. The density of the fuel is 850 kg/m^3. Determine the volume rate of flow per second of fuel through the injector and the velocity of the jet at that instant. **[Ans.** 67.7 cm^3/s, 153 m/s]

10.6 From the injector nozzle of a CI engine, a spray penetration of 25 cm is obtained in 18 ms from an orifice of 0.6 mm diameter with an injection pressure of 170 bar. The combustion chamber pressure is 20 bar. Determine:

(a) The time required for the spray to penetrate the same distance when an injection pressure of 250 bar is used.

(b) The penetration and time for a similar point on the penetration curve of an orifice having a diameter of 0.4 mm. **[Ans.** (a) 14.54 ms (b) 16.7 cm, 12 ms]

10.7 A four-cylinder four-stroke CI engine running at 2200 rpm, produces 85 kW power. The specific fuel consumption is 0.26 kg/kWh. The volume of the fuel in the pump barrel just before the commencement of the effective stroke is 5cc. The fuel pipeline is 3.2 mm in diameter and 700 mm in length. The fuel inside the injector is 2.5 cc. The average injection pressure is 200 bar and the compression pressure of air during injection is 35 bar. The density of fuel is 860 kg/m^3, and the coefficient of compressibility of fuel is 75×10^{-6} per bar. Fuel enters the pump barrel at 1 bar. Determine the plunger displacement per cycle per cylinder and the effective stroke of the plunger which is 8 mm in diameter. Also calculate the power lost in pumping the fuel. **[Ans.** 0.293 cm^3, 0.583 cm, 0.2612 kW]

10.8 In an injection system of a CI engine, the pump plunger moves with a velocity of 0.25 m/s. The length of the fuel pipe is 0.6 m. The cross-sectional area of pipe is 1/25th of that of plunger barrel. The end of the pipe has an open nozzle having a hole of area which is 1/50th of that of the pipe. The initial pressure in the line is 25 bar and the compression pressure of the engine is 30 bar. If the bulk modulus of the fuel is 18×10^8 N/m^2 and the specific gravity of the oil is 0.85, determine the following:

(a) The velocity of pressure disturbance

(b) The time taken by the disturbance to travel through the pipe line

(c) The pressure and velocity at the pump end of the pipe line as the plunger moves

(d) The magnitude of the first reflected pressure and velocity wave

(e) The pressure and velocity at the orifice end of the pipeline after reflection.

[Ans. (a) 1455 m/s, (b) 0.0004124 s, (c) 102.3 bar, 6.25 m/s
(d) 37.5 bar, 3.03 m/s, (e) 139.8 bar, 3.21 m/s]

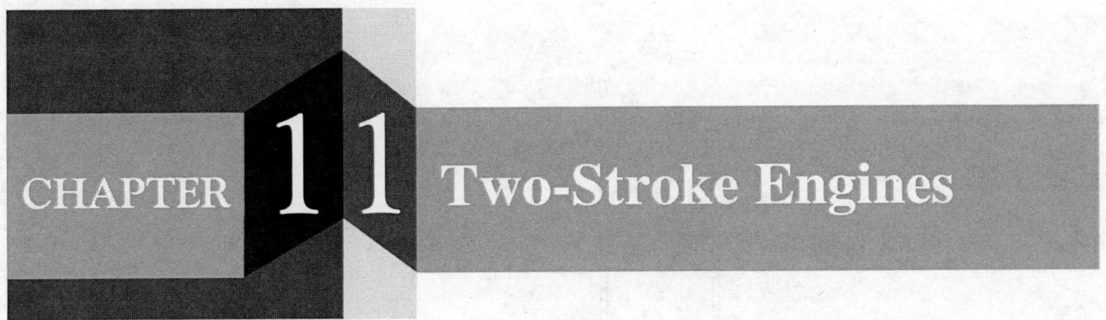

11.1 INTRODUCTION

In two-stroke cycle engines the cycle of operation is completed in two strokes only and each outward stroke of the piston is a power or expansion stroke. The engine piston needs only to compress the fresh charge and to expand the products of combustion. Such operation is made possible by the fact that the pumping function is not carried out in the working cylinders but is accomplished either in a separate mechanism called the scavenging pump or in an enclosed crankcase with the back of the engine piston being used as a scavenging pump. The fresh charge is supplied to the engine cylinder at a high enough pressure to displace the burned gases from the previous cycle. The operation of clearing the exhaust gases from the cylinder and filling it more or less completely with fresh charge is called *scavenging*. The process of scavenging includes both the intake and the exhaust processes.

Many two-stroke engines use the piston as a slide valve in conjunction with the inlet and exhaust ports on the side of the cylinder. This arrangement greatly simplifies the mechanical construction of the engine. Very large marine engines and very small reciprocating piston engines are two-stroke engines.

11.2 CLASSIFICATION OF TWO-STROKE ENGINES

Depending upon the methods of producing the scavenge charge for the scavenging process, the two-stroke engines are classified as follows:

1. Crankcase scavenged engines
2. Separately scavenged engines.

The crankcase scavenged engine is the simplest type and has already been discussed in Section 1.13. The engines of this type seldom have mean effective pressures of over 4 bar. This type of engine is not satisfactory as the crankcase pumping has a very low volumetric efficiency and, instead of an excess air, the engine receives less air than is theoretically necessary.

The separately scavenged engines are widely used in all larger and some small engines. In this type, a separate scavenging pump such as roots blower, as shown in Figure 11.1, is used, which

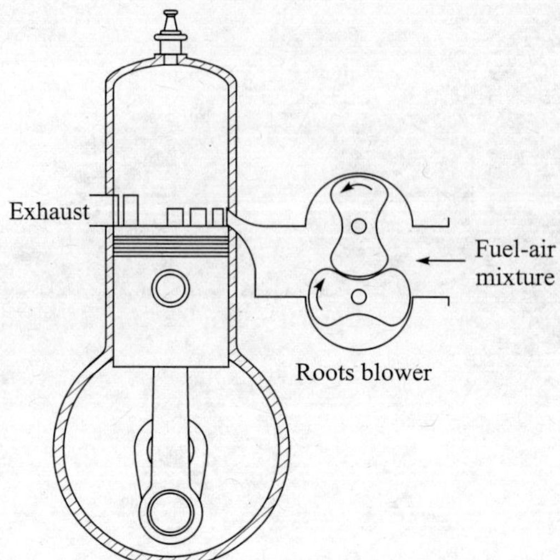

Figure 11.1 Separately scavenged two-stroke engine.

is either driven from the engine crankshaft or driven using outside power. An external blower is used for charge to enter into the cylinder through an intake port. As the piston moves down on the expansion stroke, it uncovers the exhaust ports at approximately 65° crank angle before the bottom dead centre. About 10° later, when the cylinder pressure is considerably lowered, the inlet ports open and the scavenging process takes place. The inlet ports are shaped in such a way that most of the charge flows to the top of the cylinder on the inlet side and back down on the exhaust side, thus expelling the burned gases out through the exhaust port. This ensures scavenging of the upper part of the cylinder as well. Piston deflectors are not suitable as they are heavy and tend to become overheated at high output. The scavenging process is more efficient in properly designed separately scavenged engines than in the usual crankcase compression engines.

11.3 SCAVENGING ARRANGEMENTS

Based on the scavenging arrangements, the scavenging process can be classified as follows:

1. Return-flow scavenging
2. Uniflow scavenging.

The return-flow scavenging is applicable to single-piston engines, while uniflow engines are of side-by-side cylinder, twin piston type. No uniflow petrol engine with valve gear or opposed piston has achieved success.

11.3.1 Return-flow Scavenging

In return-flow scavenging the scavenge air is directed towards the other end of the cylinder by a slant of the inlet ports or by the shape of the piston head or by both and then the air is returned to the piston head pushing the burned gas out through the exhaust port. In doing so, some of the air mixes with the burned gases and escapes with them through the exhaust ports. The scavenging efficiency is thus lowered. The exhaust and inlet ports are opened and closed by the sliding motion of the piston.

Depending on the shape and relative position of the exhaust and scavenge ports, the return flow scavengings are divided into:

(a) Cross-flow scavenging, Figure 11.2(a)
(b) Full-loop scavenging (MAN type), Figure 11.2(b)
(c) Tangential-loop scavenging (Schnuerle type), Figure 11.2(c)
(d) Combination of loop- and cross-scavenging (Curtiss type), Figure 11.2(d).

Most small engines are cross-scavenged. This type of engine is the simplest but it is the least efficient. The cross-flow system requires a piston with shaped crown, and it is necessary for the transfer and exhaust ports to be situated on a diametrical line across the cylinder bore. Thus there is restricted scope for port positioning as a whole, though the shape, the number and the dimensions of the ports can be varied to quite a large extent. The odd-shaped deflector piston has the disadvantage of uneven heat stressing, which makes it liable to distortion. The success of this type largely depends on the design of the deflector, not only in its function with respect to gas flow, but also in its mechanical strength and ability to transfer heat.

For small engines the ports are drilled radially to reduce the cost; in large engines the ports are made rectangular for better breathing. The incoming flow charge is directed upwards by the deflector on the pistons, and then the cylinder walls and head reverse the direction of flow and push the exhaust gases out through the exhaust port.

The cross-flow scavenge engine is the worst type, in that it is susceptible to short-circuiting the charge and is unable to displace the exhaust gases adequately. The alternative loop-scavenged system in its various forms is designed to direct the incoming charge by means of inclined and aimed ports, in such a way that both charge loss and mixing are minimized. Three representative arrangements, as shown in Figures 11.2(b), (c) and (d) will serve as examples, these are MAN-loop scavenging (Maschinenfabrik Augsburg Nuernberg, a German firm), Schnuerle-loop scavenging and Curtiss-loop scavenging arrangements respectively.

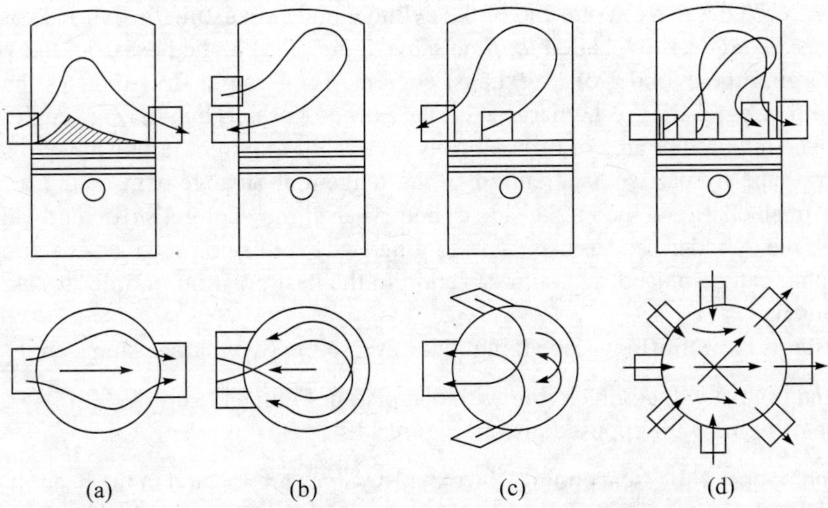

Figure 11.2 Return-flow systems: (a) Cross-flow, (b) MAN-loop scavenging, (c) Schnuerle-loop scavenging, and (d) Curtiss-loop scavenging.

No piston deflector is required with loop-scavenging, the piston being of normal shape, flat, slightly domed, or a combination of both. This is of great benefit, as the troubles associated with an assymmetrical shape are eliminated.

In the MAN-loop scavenge system the entry ports are located immediately above the exhaust ports on the same side of the cylinder (Figure 11.2(b)). In some designs the exhaust port is placed above the inlet port on the same side. The incoming flow of the charge is directed across the chamber. Even when using ports of maximum width and minimum height, however if the ports are of adequate cross-sectional area, a large proportion of the stroke length is taken up by the port operation. Thus, while acceptable on medium-speed diesels, this arrangement is obviously not suitable for smaller high-speed petrol engines.

In the Schnuerle-loop scavenging system (Figure 11.2(c)) two or more transfer ports are often employed, though a similar system can be worked with only one transfer port. The flow is directed tangentially and upwards. The location of the transfer ports at the sides instead of directly opposite the exhaust port, facilitates the use of very short passages having entry from the crankcase via slots or ports in the piston skirt, which of course does not carry much thrust load at these areas.

The Curtiss design (Figure 11.2(d)) makes use of a considerable number of entry ports to make up a large total area. The ports are also individually angled in such a manner that the converging streams meet some way off the central axis of the cylinder and rise up to the head after merging. This promotes a displacement of the exhaust gases from the top and their expulsion from the exhaust ports. Here again, though successful on diesels having sufficient excess air, multiple transfer ports have not shown the same advantage on the high-speed petrol engines. Apart from the extra charge volume contained in the multiple transfer passages, which increases the free space in the crankcase and lowers the pumping volumetric efficiency, it is found that the many transfer streams tend to mix excessively with the exhaust instead of displacing it.

11.3.2 Uniflow Scavenging

Admitting the fresh charge from one end of the cylinder and exhausting the burned gases from the other end gives a straight flow, called *uniflow scavenging*. This is the best from the viewpoint of thoroughly purging the cylinder of its exhaust content. The straight flow reduces the turbulence and hence the mixing of the fresh charge with the burned gases. Thus the scavenging efficiency is increased. The required degree of turbulence is invariably required, it may have to be promoted on the uniflow type by particular attention to the tangential arrangement of the transfer ports, otherwise the fresh charge in the intake side cylinder barrel may not get sufficiently agitated. The passage across the cylinder head assists in promoting swirl at this point, which is a good condition as the spark plug can be placed without restriction in the best position in order to take advantage of this phenomena.

Many systems have uniflow scavenging, out of these the two main systems are:

1. Port and poppet valve scavenging with one piston, Figure 11.3
2. Port scavenging with opposed piston, Figure 11.4.

In port and poppet valve scavenging, the exhaust valves are located in the head. It is shown in Figure 11.3. The admission of fresh charge through the inlet ports is controlled by the piston, while the products of combustion are removed through the poppet valves provided in the cylinder head.

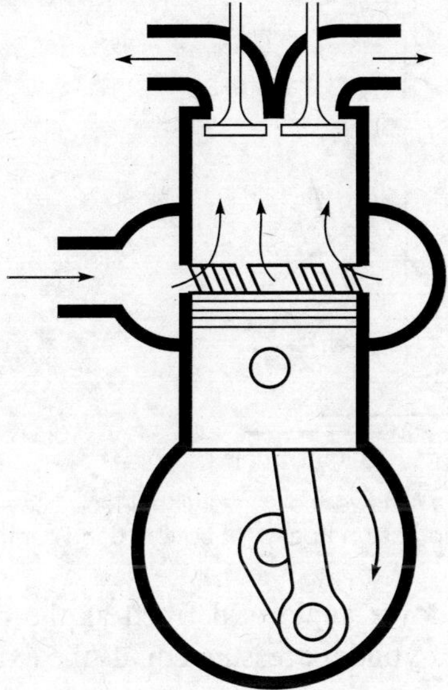

Figure 11.3 Port and poppet valve scavenging.

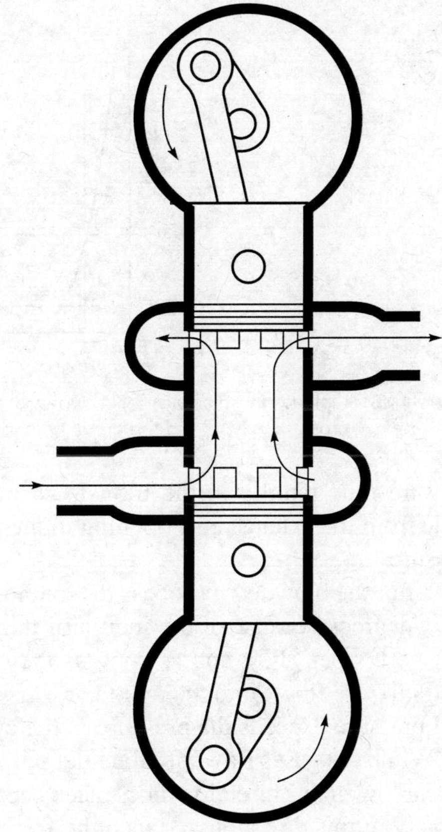

Figure 11.4 Port scavenging with opposed piston.

Provisions for adjusting the timing of the opening and closing of the exhaust valves to suit engine speed and scavenge charge pressure are the main assets of this type of engine.

Port scavenging with opposed pistons (Junker's design) has a very effective scavenging process. This type is also called the end-to-end scavenged engine. It is shown in Figure. 11.4. The intake ports are covered by the lower piston and the exhaust ports are covered by the upper piston. All ports are distributed uniformly over the whole circumference. The piston moves in the opposite direction and a spiral motion of the charge perpendicular to the cylinder axis is created by the tangential arrangement of the inlet ports. Thus, this system produces very good scavenging and charging conditions. The swirl helps to prevent mixing of fresh charge and combustion products during the scavenging process. In this type the exhaust ports are opened before the inlet ports open and also the exhaust ports are closed before the inlet ports close. This timing helps this type of engine in filling its cylinder at full inlet pressure. In this type of engine, the counter-flow within the cylinder is eliminated and there is less opportunity for mixing of fresh charge with the burned gases. It develops high mean effective pressure. The combustion chamber is simple and efficient. The reciprocating masses are well balanced even with one piston slightly leading the other. However, this engine has mechanical disadvantages. It requires a complicated running gear mechanism, there are difficulties in cooling the pistons and there are higher costs of manufacturing and maintenance.

11.4 SCAVENGING PROCESS

Figure 11.5 shows a 'light-spring pressure' vs. crank angle diagram taken of a conventional-loop scavenged cross-flow type two-stroke engine. After the exhaust ports open (EO), the cylinder

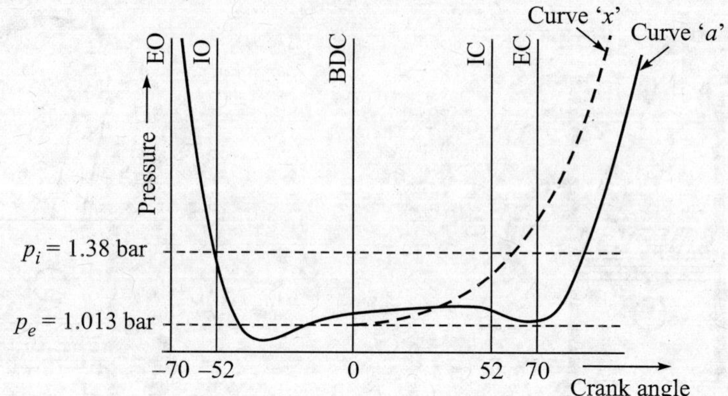

Figure 11.5 Light-spring diagram for a two-stroke engine: the curve '*a*' shows the conventional-loop scavenged cross-flow type and the curve '*x*' shows the adiabatic compression from p_e at bottom dead centre.

pressure falls rapidly in the blowdown process. The blowdown angle is defined as the crank angle from the exhaust port opening to the point at which the cylinder pressure equals the exhaust pressure.

After the blowdown process the cylinder pressure usually falls below the exhaust pressure for a few degrees because of the inertia of the gases. Soon after the exhaust ports begin to open, the inlet ports open (IO); and as soon as the cylinder pressure falls below the scavenging pressure, fresh mixture flows into the cylinder. This flow continues as the inlet ports are open and the inlet total pressure exceeds the pressure in the cylinder.

While the gases flow into the inlet ports, the exhaust gases continue to flow out of the exhaust ports. Flow in this direction takes place because during the blowdown period the gases flow at high velocity in this direction and also the fresh mixture flowing in through the inlet ports eventually builds up a pressure in the cylinder which exceeds the exhaust system pressure.

The crank angle during which both the inlet and exhaust ports are open is called the *scavenging angle*, and the corresponding time is called the *scavenging period.*

Whether the inlet ports close (IC) first or the exhaust ports close (EC) first, depends on the cylinder design. After all the ports are closed, the cycle proceeds through compression, combustion and expansion.

The exhaust ports must open well before BDC in order that the cylinder pressure be substantially equal to the exhaust system pressure before the piston reaches BDC and so that the excessive blow-back of unburned gases into the inlet system can be avoided.

11.5 SCAVENGING PARAMETERS

In order to discuss the gas exchange performance of the two-stroke engines, it is essential to define several scavenging parameters.

Delivery ratio

It compares the actual scavenging air mass or mixture mass (air mass is used in CI engines and air-fuel mixture mass is used in SI engines) to that required in an ideal charging process.

$$\text{The delivery ratio, } R_d = \frac{\text{mass of fresh charge delivered per cycle } (M_{\text{supp}})}{\text{reference mass } (M_{\text{ref}})} \tag{11.1}$$

The reference mass is defined as the product of swept volume and the ambient fresh charge density.

Scavenging ratio

It is the ratio of the 'mass flow rate of the fresh charge supplied to the engine' to the 'mass flow rate of the fresh charge supplied during the scavenging process' which would just fill the cylinder at BDC at inlet temperature and pressure.

$$\therefore \quad \text{Scavenging ratio, } R_{sc} = \frac{\dot{M}_{\text{supp}}}{\dot{M}_{\text{ideal}}} \tag{11.2}$$

where

$\dot{M}_{\text{supp}}$ = mass flow rate of the fresh charge supplied to the engine

$\dot{M}_{\text{ideal}}$ = mass flow rate of the fresh charge supplied during the ideal scavenging process which would just fill the cylinder at BDC at inlet temperature and exhaust pressure.

The suffix i is used for the inlet condition and the suffix e is used for the exhaust condition.

Now,

$$\dot{M}_{\text{ideal}} = n\,V\,\rho \tag{11.3}$$

where

n = engine speed in rps
V = total cylinder volume = $V_c + V_s$
V_c = clearance volume
V_s = swept volume
ρ = density of fresh mixture.

Compression ratio,
$$r = \frac{V}{V_c} = \frac{V_s + V_c}{V_c} = \frac{V_s}{V_c} + 1$$

$$\therefore \quad V_c = \frac{V_s}{r-1}$$

and
$$V = V_c + V_s = \frac{r}{r-1} V_s = \frac{r}{r-1} A_p L \tag{11.4}$$

where

A_p = area of cross-section of the piston
L = stroke.

Density,
$$\rho = \frac{p_e}{RT_i} = \frac{p_e \mu_i}{R_m T_i} \tag{11.5}$$

where

μ_i = molecular weight of fresh charge
R_m = Universal gas constant.

Substituting the values of V and ρ from Eq. (11.4) and Eq. (11.5) into Eq. (11.3),

$$\dot{M}_{\text{ideal}} = n\frac{r}{r-1}A_p L\frac{p_e\mu_i}{R_m T_i}$$

$$= \frac{S}{2}A_p\frac{r}{r-1}\frac{p_e\mu_i}{R_m T_i} \tag{11.6}$$

where $S = 2Ln$ (mean piston speed).

$\therefore$ Scavenging ratio,
$$R_{sc} = \frac{2\dot{M}_{\text{supp}}(r-1)R_m T_i}{S A_p r p_e \mu_i} \tag{11.7}$$

Fresh charge may consist of air, fuel and water vapour.

Fresh charge, $\dot{M}_{\text{supp}} = \dot{M}_i = \dot{M}_a + \dot{M}_f + \dot{M}_v$ and $M_i = M_a + M_f + M_v$

where

$\dot{M}$ = rate of mass flow

M = mass per cycle

Suffix i = fresh charge, a = air, f = fuel, and v = water vapour.

$$\therefore \quad \frac{\dot{M}_i}{\dot{M}_a} = 1+\frac{\dot{M}_f}{\dot{M}_a}+\frac{\dot{M}_v}{\dot{M}_a} = 1 + F + h$$

where

F = fuel/air ratio

h = water vapour/air ratio.

(Both are present in the inlet air from the inlet ports.)

$$\therefore \quad \frac{M_i}{M_a} = \frac{\dot{M}_i}{\dot{M}_a} = 1 + F + h$$

Characteristic gas constant, $R_i = \dfrac{R_m}{\mu_i}$

Number of moles, $N_i = N_a + N_f + N_v$

$$= \frac{M_a}{\mu_a}+\frac{M_f}{\mu_f}+\frac{M_v}{\mu_v}$$

$$= \frac{M_a}{\mu_a}\left(1+\frac{M_f}{M_a}\frac{\mu_a}{\mu_f}+\frac{M_v}{M_a}\frac{\mu_a}{\mu_v}\right)$$

$$= \frac{M_a}{29}\left(1+F\frac{29}{\mu_f}+h\frac{29}{18}\right)$$

$$\therefore \quad N_i = \frac{M_a}{29}\left(1+F\frac{29}{\mu_f}+1.6h\right) \tag{11.8}$$

Now,
$$\mu_i = \frac{M_i}{N_i} = \frac{M_a + M_f + M_v}{\frac{M_a}{29}\left(1 + F\frac{29}{\mu_f} + 1.6h\right)}$$

$$= \frac{29(1 + F + h)}{1 + F\frac{29}{\mu_f} + 1.6h} \quad (11.9)$$

$$R_i = \frac{R_m}{\mu_i} = \frac{R_m\left(1 + F\frac{29}{\mu_f} + 1.6h\right)}{29(1 + F + h)} \quad (11.10)$$

$$\rho = \frac{p_e}{R_i T_i} = \frac{p_e \cdot 29(1 + F + h)}{T_i R_m\left(1 + F\frac{29}{\mu_f} + 1.6h\right)} \quad (11.11)$$

Substituting the value of μ_i from Eq. (11.9) into Eq. (11.7),

$$R_{sc} = \frac{2\dot{M}_i(r-1)R_m T_i\left(1 + F\frac{29}{\mu_f} + 1.6h\right)}{S A_p r p_e 29(1 + F + h)}$$

We know that

$$\dot{M}_i = \dot{M}_a + \dot{M}_f + \dot{M}_v$$

$$= \dot{M}_a(1 + F + h)$$

$$\therefore \quad R_{sc} = \frac{2\dot{M}_a(r-1)R_m T_i\left(1 + F\frac{29}{\mu_f} + 1.6h\right)}{(S A_p r p_e)(29)}$$

$$= \frac{2\dot{M}_a(r-1)}{S A_p r \rho_s} \quad (11.12)$$

where
$$\rho_s = \frac{29\,p_e}{T_i R_m\left(1 + F\frac{29}{\mu_f} + 1.6h\right)} \quad (11.13)$$

Here ρ_s is the density of dry air in the inlet mixture at T_i and p_e. For CI engines, F is zero, since F is the mass ratio of gaseous fuel to dry air upstream from the inlet ports.

Trapping efficiency

It is the ratio of the mass of fresh charge retained in the cylinder to the mass of fresh charge supplied. It indicates what fraction of fresh charge supplied is retained in the cylinder, i.e.

Trapping efficiency,
$$\eta_{tr} = \frac{\dot{M}_{ret}}{\dot{M}_{supp}} \tag{11.14}$$
where

$\dot{M}_{ret}$ = mass of fresh charge retained per unit time in the cylinder after the ports close

$\dot{M}_{supp}$ = mass flow of fresh charge per unit time supplied to the engine.

Scavenging efficiency

It is the ratio of mass of fresh charge retained in the cylinder to the ideal mass retained during scavenging. It indicates the extent to which the residual gases in the cylinder have been replaced with the fresh charge.

Scavenging efficiency,
$$\eta_{sc} = \frac{\dot{M}_{ret}}{\dot{M}_{ideal}} \tag{11.15}$$

where $\dot{M}_{ret}$ and $\dot{M}_{ideal}$ are already defined above.

Hence,
$$\eta_{sc} = \frac{\dot{M}_{ret}}{\dot{M}_{ideal}} = \frac{\dot{M}_{ret}}{\dot{M}_{supp}} \times \frac{\dot{M}_{supp}}{\dot{M}_{ideal}} = \eta_{tr} R_{sc}$$

$$\therefore \quad \eta_{sc} = \eta_{tr} R_{sc} \tag{11.16}$$

Charging efficiency

It is the ratio of the mass of fresh charge retained to the reference mass.
$$\eta_{ch} = \text{charging efficiency} = \frac{\text{mass of fresh charge retained } (M_{ret}) \text{ per cycle}}{\text{reference mass } (M_{ref})} \tag{11.17}$$

It indicates how effectively the cylinder volume has been filled with fresh charge.
$$\eta_{ch} = \frac{M_{ret}}{M_{ref}} = \frac{M_{ret}}{M_{supp}} \frac{M_{supp}}{M_{ref}} = \frac{\dot{M}_{ret}}{\dot{M}_{supp}} \frac{M_{supp}}{M_{ref}} = \eta_{tr} R_d$$

$$\therefore \quad \eta_{ch} = R_d \eta_{tr} \tag{11.18}$$

Purity

The purity of the charge is the ratio of the mass of fresh charge trapped to the mass of trapped cylinder charge. It indicates the degree of dilution of the fresh charge in the cylinder with the burned gases. Thus,
$$\text{Purity} = \frac{\text{mass of fresh charge trapped}}{\text{mass of trapped cylinder charge}} \tag{11.19}$$

11.6 IDEAL MODELS FOR SCAVENGING PROCESS

In a real scavenging process, mixing occurs as the fresh charge displaces the burned gases and some of the fresh charge is expelled. There are three hypothetical models for scavenging processes:

1. Perfect displacement model
2. Complete mixing model
3. Short-circuiting.

11.6.1 Perfect Displacement Model

The perfect displacement model is based on the following assumptions:

1. The fresh charge would push the residual gases without mixing. Thus there is no mixing of the fresh incoming charge with the burned gases.
2. There is no exchange of heat between the fresh charge and the burned gases.

There are two cases that arise with this model.

Case 1: The scavenging ratio is less than or equal to 1 ($R_{sc} \le 1$)

The scavenging ratio less than 1 means that the mass of the fresh charge delivered is less than the mass of the fresh charge which could be present in the total cylinder volume at ambient condition. It indicates that all the burned gases have not been expelled. All the incoming charge has been trapped and some burned gases are also present in the trapped mass. In this case the mass of fresh charge delivered is equal to the mass of fresh charge retained. Under this condition, the following results are obtained:

$$\eta_{sc} = R_{sc} \quad \text{and} \quad \eta_{tr} = 1 \quad \text{for} \quad R_{sc} \le 1 \tag{11.20}$$

Case 2: The scavenging ratio is greater than 1 ($R_{sc} > 1$)

The scavenging ratio greater than 1 means that all the burned gases have been expelled and some of the incoming fresh charge will also go out with the burned gases. The trapped cylinder mass will be full of fresh charge. The following results are obtained under this condition:

$$\eta_{sc} = 1 \quad \text{and} \quad \eta_{tr} = \frac{1}{R_{sc}} \quad \text{for} \quad R_{sc} > 1 \tag{11.21}$$

11.6.2 Complete Mixing Model

The complete mixing model is based on the following assumptions:

1. The fresh charge entering the cylinder mixes instantaneously and uniformly with the residual gases. Thus complete mixing occurs.
2. The scavenging process is considered to be a quasi-steady flow process.
3. The density of the incoming charge and the burned gases is the same.
4. The residual gases are at the same temperature and have the same molar mass as the fresh mixture.
5. The piston remains at BDC during the scavenging process.
6. The cylinder pressure is assumed to be constant during scavenging.

Let x be the volumetric fraction of the fresh charge retained in the cylinder at any time, and V_{supp} be the volume of the fresh charge supplied that enters the cylinder up to that time. Let V be the total volume of the cylinder.

$$x = \frac{V_{\text{fresh charge retained}}}{V}$$

∴ Volume of fresh charge retained, $V_{\text{fresh charge retained}} = xV$

and in differential form, $dV_{\text{fresh charge retained}} = Vdx$

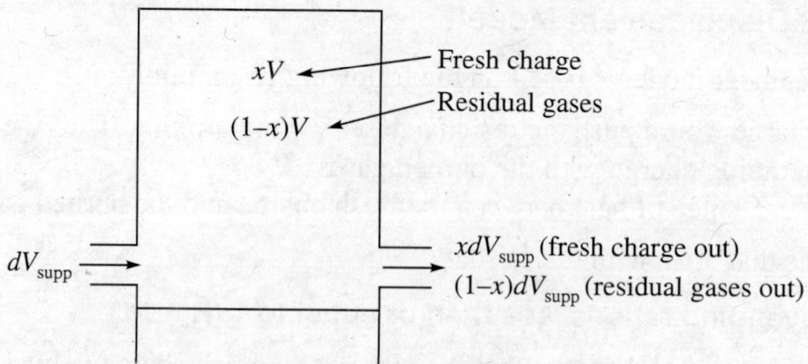

Figure 11.6 Complete mixing model.

Refer to Figure 11.6. Let dV_{supp} be the fresh charge supplied further in the cylinder. Fresh charge and residual gases are in the ratio x:$(1-x)$ in the cylinder. In the same ratio, fresh charge and residual gases move out.

i.e.,

$$\text{fresh charge out} = x\, dV_{supp}$$

$$\text{residual gases out} = (1-x)\, dV_{supp}$$

Applying mass conservation equation,

Mass of fresh charge in – mass of fresh charge out = mass of fresh charge retained

$$\rho dV_{supp} - \rho x dV_{supp} = \rho V dx$$

where density, ρ, is assumed to be the same for fresh charge and residual gas.

$$(1-x)\, dV_{supp} = V\, dx$$

$$\frac{dV_{supp}}{V} = \frac{dx}{1-x}$$

Integration of both sides will give

$$\frac{V_{supp}}{V} = -\ln(1-x)$$

Now,

$$\frac{V_{supp}}{V} = \frac{\rho V_{supp}}{\rho V} = \frac{M_{supp}}{M_{ideal}} = \frac{\dot{M}_{supp}}{\dot{M}_{ideal}} = R_{sc} = \text{scavenging ratio}$$

$$x = \frac{V_{\text{fresh charge retained}}}{V} = \frac{\rho V_{\text{fc ret}}}{\rho V} = \frac{M_{ret}}{M_{ideal}} = \frac{\dot{M}_{ret}}{\dot{M}_{ideal}} = \eta_{sc} = \text{scavenging efficiency}$$

So above expression reduces to

$$R_{SC} = -\ln(1-\eta_{sc})$$

or

$$1 - \eta_{sc} = e^{-R_{sc}}$$

∴

$$\eta_{sc} = 1 - e^{-R_{sc}} \tag{11.22}$$

and

$$\eta_{tr} = \frac{\eta_{sc}}{R_{sc}} = \frac{1 - e^{-R_{sc}}}{R_{sc}} \tag{11.23}$$

11.6.3 Short-circuiting

In this model the fresh mixture might flow through the cylinder and out of the exhaust ports in a separate stream without mixing with the residual gases or without pushing them out. In this case only a little fresh mixture would be trapped, and the scavenging efficiency at any value of the scavenging ratio would be very low.

In real engines, all three of the hypothetical processes described above occur simultaneously during the scavenging process. Another phenomenon which reduces the scavenging efficiency is the formation of pockets or dead zones in the cylinder volume where the burned gases are trapped and have no displacement or entrainment by the fresh scavenging flow. These unscavenged zones are most likely to occur in regions of the cylinder that remain secluded from the main fresh mixture flow path.

11.7 RELATIONSHIP OF SCAVENGING RATIO AND SCAVENGING EFFICIENCY

Figure 11.7 shows the scavenging efficiency η_{sc} and the trapping efficiency η_{tr} for the perfect displacement and complete mixing assumptions as a function of the scavenging ratio R_{sc}. The curve '*a*' shows the relationship between the scavenging efficiency η_{sc} and the scavenging ratio R_{sc}, for the perfect displacement model, when the trapping efficiency η_{tr} is unity. In this case the scavenging efficiency equals the scavenging ratio at all points. The curve '*b*' shows the relationship between the scavenging efficiency η_{sc} and the scavenging ratio R_{sc}. The curve '*c*' shows the relationship between the trapping efficiency η_{tr} and the scavenging ratio R_{sc}, for the perfect mixing model. The curve '*d*' shows the relationship between the scavenging efficiency η_{sc} and the scavenging ratio R_{sc}, with the complete short-circuiting.

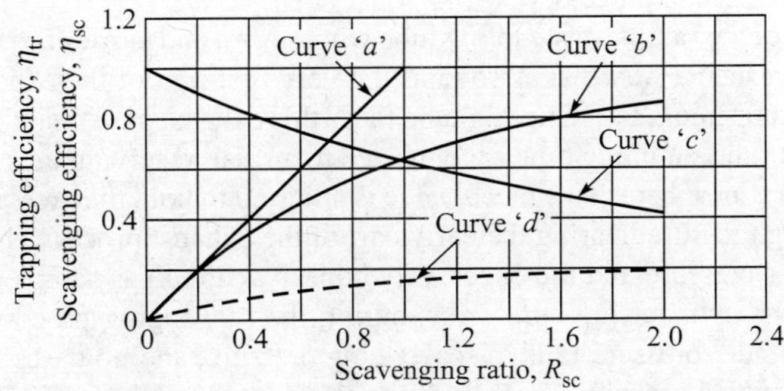

Figure 11.7 Theoretical relationship between scavenging efficiency, trapping efficiency, and scavenging ratio: (a) η_{sc} with perfect scavenging, (b) η_{sc} with perfect mixing, $\eta_{sc} = 1 - e^{-R_{sc}}$, (c) η_{tr} with perfect mixing, and (d) η_{sc} with complete short-circuiting.

11.8 MEASUREMENT OF SCAVENGING EFFICIENCY

Measurement of the scavenging efficiency and the results obtained from the flow through ports are useful to evaluate the performance of the engine and to enable the calibration of hypothetical models of the scavenging process. Many methods of measuring the scavenging efficiency are available but only a few of them which yield reliable results are described here and these are reasonably easy to apply.

11.8.1 Tracer Gas Method

In this method, a small quantity of gas, like mono-methylamine (CH_3 NH_2) and carbon monoxide (CO) is continuously introduced with the fresh charge. Thus the inlet charge will have a fixed percentage of the tracer gas. The tracer gas should burn completely at combustion temperatures but not burn at temperatures below those prevailing in the cylinder when the exhaust ports open. The tracer gas must also mix with the fresh mixture thoroughly. It must be easy to identify qualitatively and quantitatively by chemical analysis.

Then the scavenging efficiency,

$$\eta_{sc} = \eta_{tr} R_{sc} = \frac{x - y}{x} R_{sc} \tag{11.24}$$

where x is the mass fraction of the tracer gas in the inlet charge and y is the mass fraction of the unburned tracer gas in the exhaust gases. By measuring the inflow of the tracer gas and the mass flow of air and fuel through the engine, x can be obtained and by chemical analysis of the exhaust gases, y can be obtained. R_{sc} can also be evaluated by the measured value of mass flow of air and fuel through the engine.

11.8.2 Gas-sampling Method

This method is applicable for Diesel engines and also for SI engines in which the fuel is injected after the ports are closed. The scavenging efficiency is determined from the measurement of mass of fuel supplied and from the chemical analysis of a sample of the gases which were in the cylinder near the end of the expansion process.

For the analysis of the sample, a sampling tube whose open end carries a check valve is placed in the exhaust port in such a way that as soon as the very first part of the exhaust gas comes out from the exhaust port opening, the sampling tube faces this exhaust gas stream as shown in Figure 11.8. The other end of the sampling tube is connected to a small receiving tank. This tank is fitted with a bleed valve to atmosphere. This bleed valve is adjusted to hold the pressure in the tank well above the scavenging pressure. During the early part of the exhaust process, the sampling tube is exposed to a cylinder pressure and the check valve opens and the gas sample is collected in the receiving tank. As the cylinder pressure drops during the scavenging process, the check valve closes before the cylinder pressure falls to scavenging pressure, and no gases enter the sampling tube during the scavenging process. Analysis of the exhaust gas can be made with the help of gas analyzers and the air/fuel ratio can be estimated. Since the quantity of fuel injected is known, it is possible to calculate the mass of the trapped air. The mass of the air inducted into the engine can be measured directly and the delivery ratio, the scavenging efficiency, the trapping efficiency and the charging efficiency can be calculated.

An accurate measurement of the scavenging efficiency is difficult due to the problem of measuring the mass of the trapped air. Estimation of scavenging efficiency using the above methods is reliable. However, flow visualization experiments have proved useful in providing a qualitative picture of the scavenging flow field and in identifying problems such as short-circuiting and dead zones.

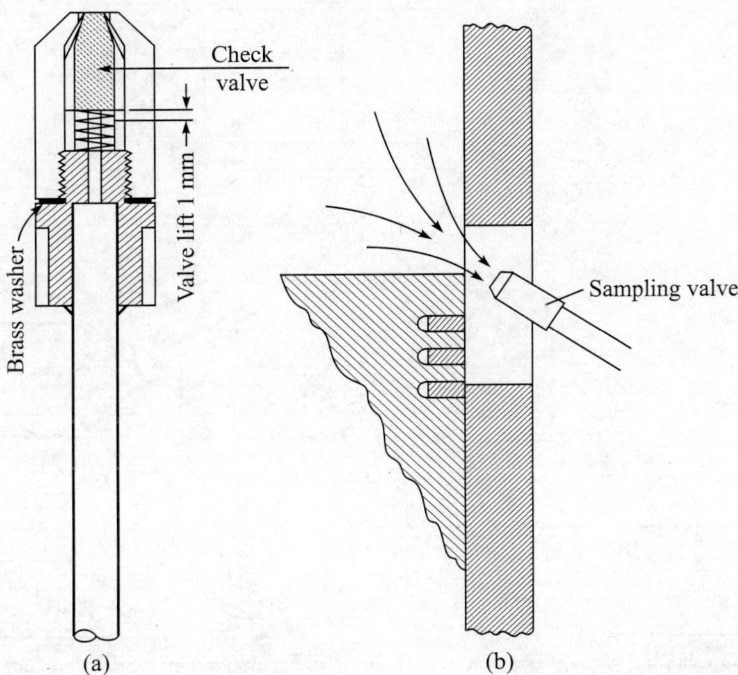

Figure 11.8 (a) Exhaust-blowdown sampling valve and (b) positioning of the sampling valve in the exhaust port.

11.9 FLOW-THROUGH PORTS

The scavenging process of a two-stroke cycle engine is greatly affected by the shape, the size, the number, the location around the circumference, ports timing and the direction and velocity of the jets issuing from the ports into the cylinder. Rectangular ports make the best use of the cylinder wall area and give precise timing control. Ports can be tapered, and may have axial and tangential inclination.

There are several possible designs of ports as shown in Figure 11.9. The conventional square or rectangular type gives the largest flow area and can be accurately dimensioned without trouble, but tends to wear off the piston rings rather quickly. Both wear and scoring can be avoided by the use of a rhomboidal shape, but of course this adds to manufacturing complications. However a sensible compromise is found in the rectangular or square port with its corners radiused. In this case the flow area is reduced slightly compared to that with the sharp-cornered port, but not to any really significant extent.

If the mechanical design allows wide ports to be incorporated, it is possible to obtain an adequate size of flow area. However, a port width exceeding 15% of the cylinder barrel circumference is undesirable, and if this is exceeded the piston rings are liable to spring into the port opening. Wider ports should therefore be bridged. In the case of exhaust ports particularly, the bridges should be of adequate width to avoid excessive heating. In cases of excessive overheating the bridge may distort inwards, with serious consequences to the whole area of the piston surface at the ring grooves, thus resulting in the rings becoming locked solid.

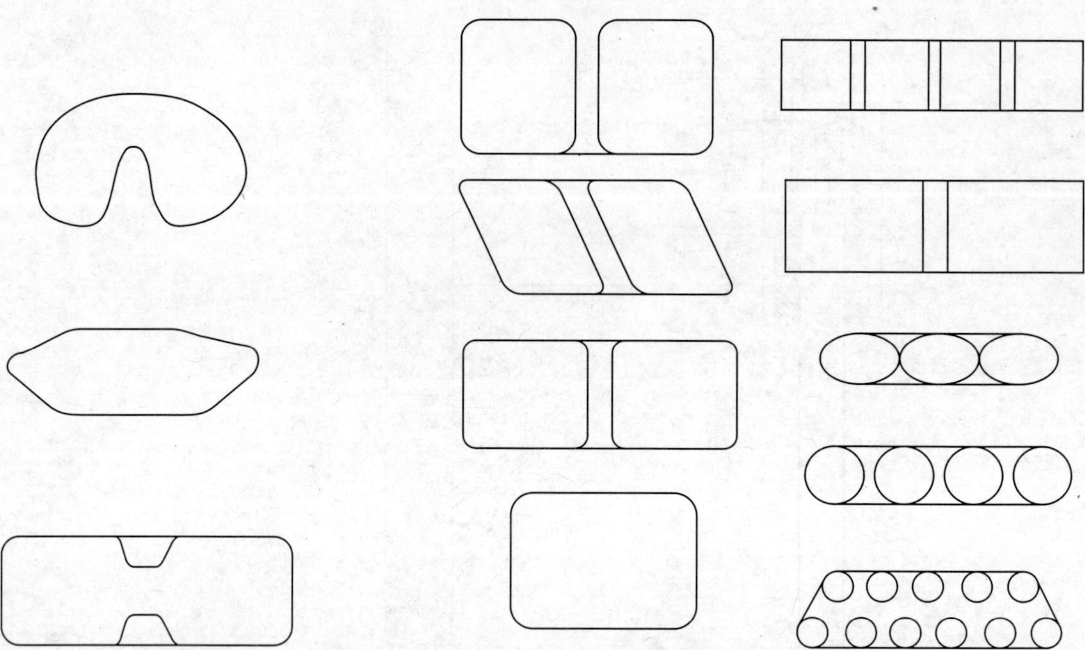

Figure 11.9 Several possible designs of port shapes used on modern engines.

Figure 11.10 illustrates the flow pattern through the piston-controlled inlet ports. For small openings, the flow remains attached to the port walls [Figure 11.10(a) for the port axis perpendicular to the wall, and Figure 11.10(d) for the port axis inclined to the wall]. For fully open ports with sharp corners the flow detaches at the upstream corners [Figure 11.10(b) for the port axis perpendicular to the wall, and Figure 11.10(e) for the port axis inclined to the wall]. For fully open ports with rounded entry and converging taper, there is no flow detachment within the port [Figure 11.10(c) for the port axis perpendicular to the wall].

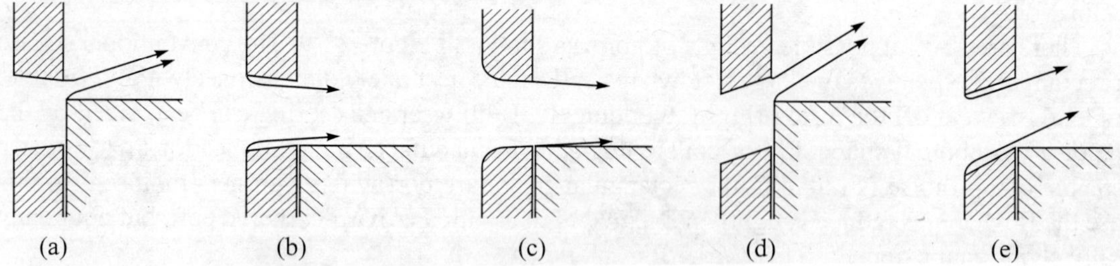

Figure 11.10 Flow pattern through the piston-controlled inlet ports.

Figure 11.11 shows the discharge coefficient as a function of the open fraction of the port for different inlet port designs. It may be observed from the figure that the geometry effects are most significant at small and large open fractions. The coefficient of discharge C_d also varies with pressure ratio across the port. As the pressure ratio increases the C_d also increases.

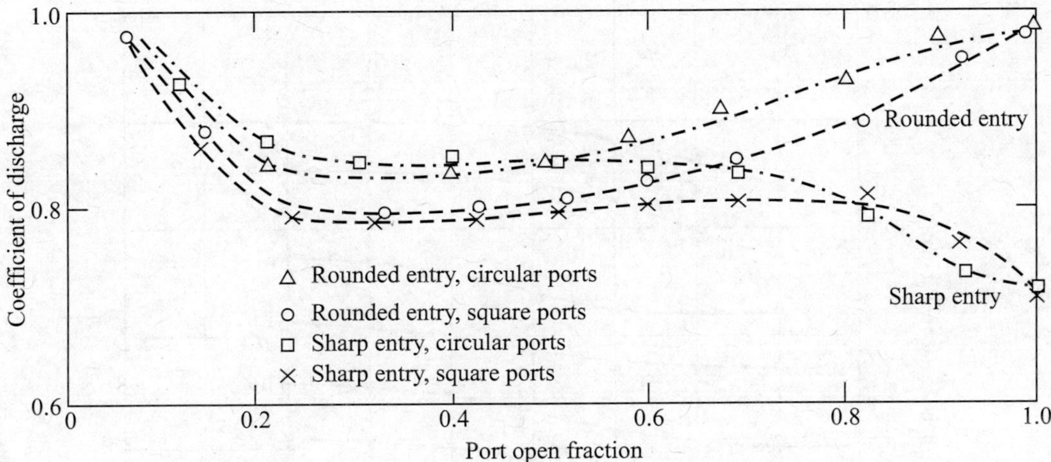

Figure 11.11 Discharge coefficients as a function of port open fraction (uncovered height/port height) for different inlet port designs. Pressure ratio across the port = 2.35.

The tangentially inclined inlet ports are used when swirl is desired to improve scavenging. The discharge coefficient C_d decreases as the tangential inclination of the jet increases.

In the piston-controlled exhaust ports, the angle of the jet α, exiting from the exhaust port increases with increased port height as shown in Figure 11.12. The pressure ratio across the exhaust port varies substantially during the exhaust process. The pressure ratio has a significant effect on the exhaust port discharge coefficient, as shown in Figure 11.13.

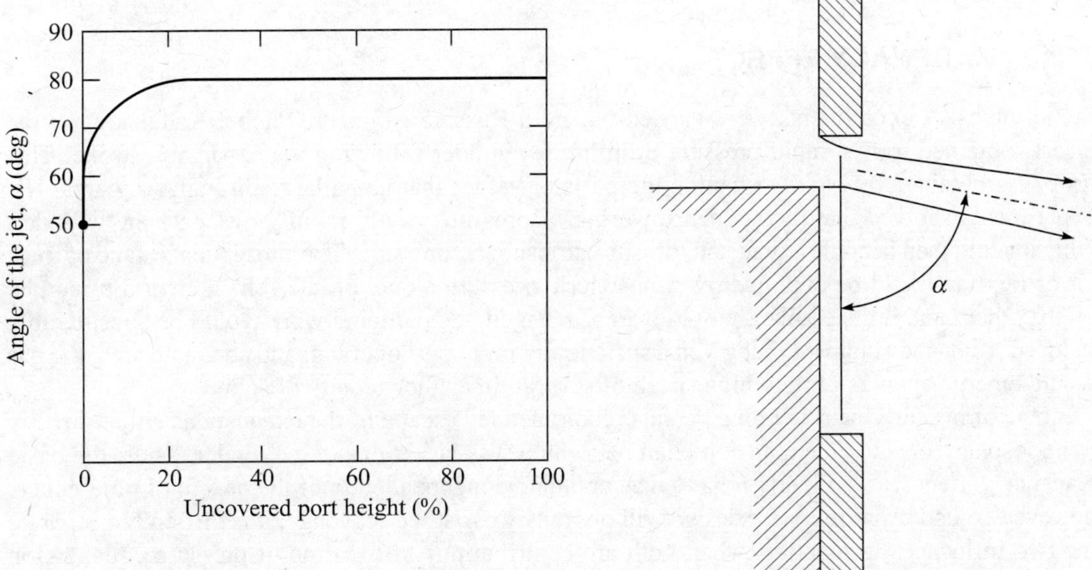

Figure 11.12 Angle of the jet exiting the exhaust port as a function of open port height.

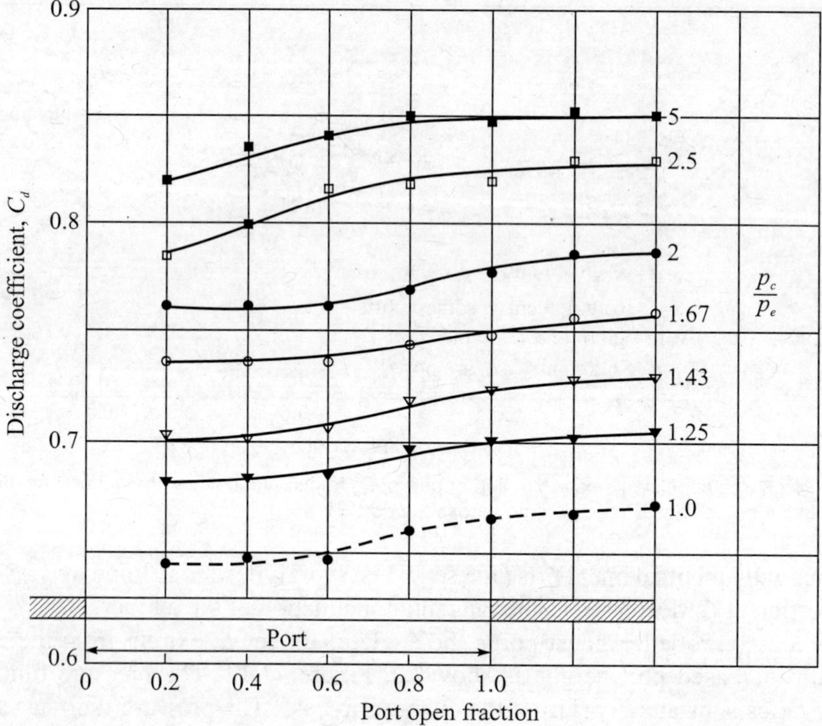

Figure 11.13 Discharge coefficient of a single rectangular exhaust port. p_c = cylinder pressure, p_e = exhaust system pressure.

11.10 KADENACY EFFECT

Some intensive experiments were carried out by a Russian engineer, Michel Kadenacy, on the effects obtained from a rapid pressure drop in the cylinder following the expansion stroke. This drop was obtained by rapidly opening the ports or valves that had a large through-way area. His experiments showed that the sudden lowering of pressure would rapidly evacuate the cylinder, which would then become exhausted to a sub-atmospheric pressure. The subsequent intake of fresh air or mixture could be obtained by atmospheric pressure alone, on the self-induction principle, on the other hand if the crankcase or blower scavenged, the pumping work would be considerably reduced. Kadenacy claimed that given a sufficiently rapid port opening, and adequate area, the gas would emerge at an extremely high speed of several times the velocity of sound.

One of his early petrol engine patents recommended escape of the exhaust gas either directly to atmosphere or into an expansion chamber which was larger than the cylinder. Thus, the basic Kadenacy effect was evidently regarded as obtainable independently of the design of pipe. Since, however, a good exhaust system design will obviously assist the scavenge, it is difficult to separate the two influences in practice, when both are contributing. At the time of pressure equalization in the cylinder, the kinetic energy of the gas in the pipe will be greater than that remaining in the cylinder. Also, in practice, it is difficult to obtain the extreme rapidity of port action for the full Kadenacy effect, the latter becoming negligible at low speeds and with small ports.

11.11 SCAVENGING PUMPS

The performance of a two-stroke engine largely depends on the characteristics of the compressor used as a scavenging pump. Figure 11.14 shows the crankcase, the roots, the centrifugal and the piston types of scavenging pumps.

The crankcase compression is illustrated in Figure 11.14(a). It is used in many smaller (up to about 50 kW per cylinder) SI gas engines and CI oil engines. The volumetric efficiency of such a pump is low (55 to 65%).

The high-speed engines use rotary pumps of the roots type with three lobes, as shown in Figure 11.14(b). The sides of the lobes are helical surfaces in order to give a uniform air motion and to reduce the air noise. The roots blowers are also used in some medium-speed large engines.

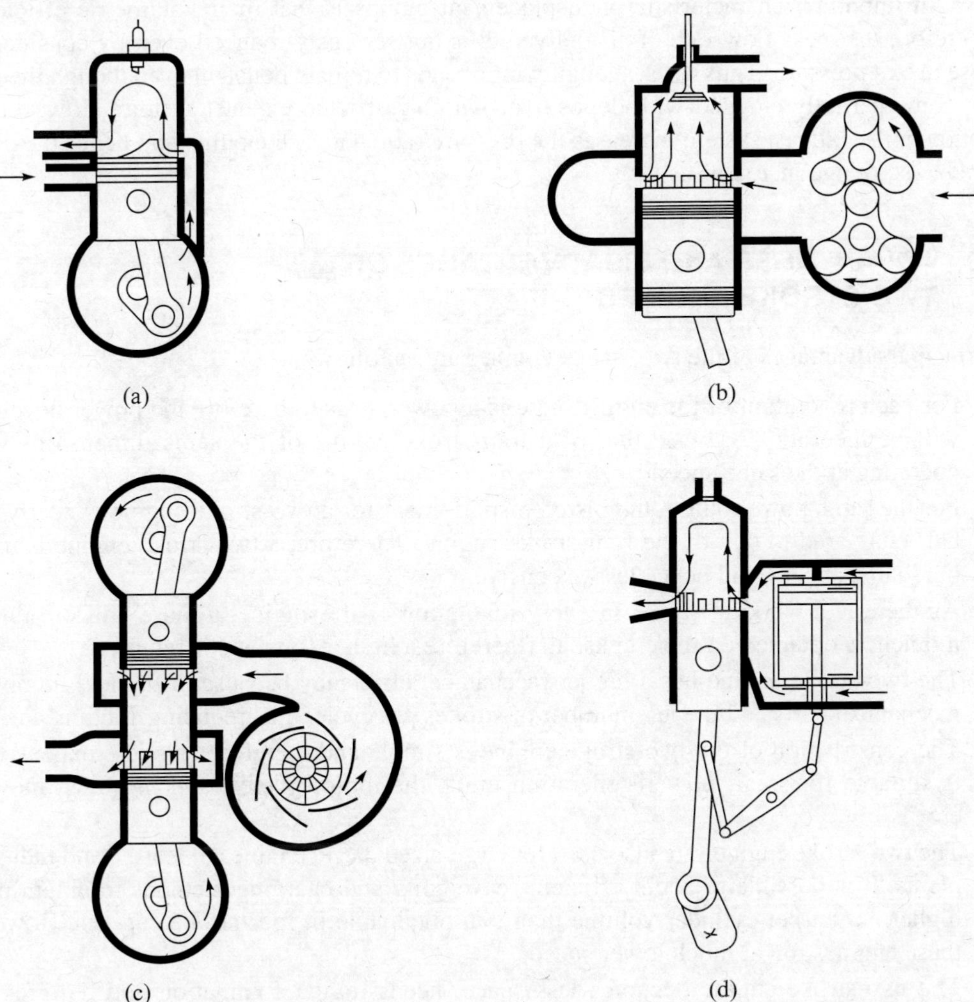

Figure 11.14 Scavenging-pump types: (a) crankcase, (b) roots, (c) centrifugal, and (d) piston.

The centrifugal blower is shown in Figure 11.14(c). It is preferred for large and high output engines. With the centrifugal scavenging pump, the mass flow rate is heavily influenced by the engine and by the exhaust system resistance. Thus the centrifugal type scavenging pump is very satisfactory, provided the resistance of the flow system is not sharply affected by carbon deposits in the ports, or a restricted exhaust, or by long exhaust piping which may develop adverse pressure waves at certain speeds. If the deposits accumulate, an engine having a centrifugal blower will start smoking earlier than that having a roots blower. Therefore the roots blower is preferred due to its lower sensitivity to flow resistance changes for systems where the space for the exhaust ports is limited.

The piston type scavenging pump is shown in Figure 11.14(d). It is generally used for low-speed and single-cylinder engines.

The crankcase, the roots blower and the piston type scavenging pumps are displacement type pumps. An important characteristic of displacement pumps is that their volumetric efficiency, and therefore the mass flow rate of air delivered, is not seriously reduced even by considerable increase in exit pressure. Thus the scavenging ratio tends to remain constant even though the ports may become partially clogged with deposits or with a restricted exhaust system. However, the restriction in the exhaust system increases the pressure ratio across the pump and hence the power required to scavenge increases as well.

11.12 ADVANTAGES AND DISADVANTAGES OF TWO-STROKE ENGINES

The principal advantages of the two-stroke engines are as follows:

1. For each revolution of the engine there is a power stroke, therefore the power developed will be theoretically twice that of a four-stroke engine of the same dimensions when operating at the same speed.
2. For the same power output the piston displacement for a two-stroke engine is nearly one-half compared to that of the four-stroke engine. Therefore, a two-stroke engine is nearly one-half as heavy and hence less expensive.
3. As there is a working stroke in every revolution of the engine, a more uniform turning moment is obtained on the crankshaft, therefore a lighter flywheel is required.
4. The two-stroke engine has a higher mechanical efficiency because of the less number of mechanical parts and the less number of strokes per cycle, thus reducing friction.
5. The construction of the two-stroke engine is simple and therefore it is less expensive. It is suitable for small power generating units, the motor cycles, and the lawn mowers, etc.
6. The two-stroke engines are also used for large-sized diesel engines in marine and industrial plants. It is possible to adopt efficient scavenging and charging methods to obtain much higher output per cylinder volume than that obtainable in four-stroke engines. However, these engines run at much lower speeds.
7. The two-stroke engine occupies less space, needs lighter foundation and requires less maintenance.

The chief disadvantages of the two-stroke engines are as follows:

1. It is difficult to obtain good scavenging owing to the much shorter period of time available. Poor scavenging means a reduction of the oxygen content in the charge, and consequently of the amount of fuel that can be burned. As a result, considerably lower mean indicated pressures are obtained than those that are theoretically expected. With crankcase scavenging, they are even lower than those in four-stroke engines.
2. The brake specific fuel consumption is slightly higher than that in four-stroke engines, except in larger units.
3. The cooling of the piston is not efficient because the piston does not have the benefit of the second revolution when no heat is being generated as in a four-stroke engine. The danger of overheating the piston and some other parts limits the maximum power of a two-stroke engine.
4. The two-stroke engine has limited speed range and also limited mixture strength range compared to the four-stroke engine.
5. Part of the piston stroke is lost because of the provisions for the ports. Thus the effective stroke is reduced, which reduces the power output. The effective compression ratio is also reduced, which reduces the thermal efficiency of the engine. The thermal efficiency is further reduced as some of the unburned fuel escapes during scavenging.
6. To ensure adequate lubrication of the working members in simple two-stroke engines, oil is mixed with petrol. This is an inefficient method compared with high pressure lubrication. It causes a heavier consumption of the lubricating oil.
7. The piston-controlled exhaust ports are liable to form an appreciable amount of carbon deposits around the edges of ports, owing to the fact that lubricating oil is scraped off the piston by these ports. Periodical decarbonizing of these ports therefore becomes essential.
8. Due to sudden release of the burned gases, the exhaust is noisy.

11.13 COMPARISON OF TWO-STROKE SI AND CI ENGINES

The two-stroke SI engines mainly suffer from fuel loss and idling difficulty. The CI engines have no such problems.

In the two-stroke SI engine the air-fuel mixture enters into the engine cylinder while both the inlet and exhaust ports remain open. So, some fuel is liable to go out through the exhaust port before the ports are closed. Thus there is a loss of fuel. In the CI engine, only the air enters into the cylinder while the ports are open and fuel is injected well after the ports are closed, so all the fuel injected takes part in the process of combustion.

At low engine speed during idling the engine may run irregularly and may even stop. This is because of the dilution of fresh charge with the large amount of residual gases that result due to poor scavenging. At low speeds the burning rate is slow which may cause backfiring in the intake system.

The problems of the two-stroke SI engine can be eliminated if the fuel is injected as soon as the exhaust port closes. It will prevent fuel loss and also backfiring as there is no fuel in the intake system.

EXAMPLE 11.1 The scavenging efficiency of a two-stroke engine is 75%. If the scavenging efficiency is increased by 20%, what would be the percentage change in the scavenging ratio?

Solution:
$$\eta_{sc} = 1 - e^{-R_{sc}} = 1 - \frac{1}{e^{R_{sc}}}$$

or
$$\frac{1}{e^{R_{sc}}} = 1 - \eta_{sc} = 1 - 0.75 = 0.25$$

∴
$$e^{R_{sc}} = \frac{1}{0.25} = 4$$

or
$$R_{sc} = \log_e 4 = 1.386$$

Now, a 20% increase in scavenging efficiency means that

$$\eta_{sc} = 0.75 + 0.75 \times 0.2 = 0.90$$

∴
$$\frac{1}{e^{R_{sc}}} = 1 - \eta_{sc} = 1 - 0.9 = 0.1$$

or
$$e^{R_{sc}} = 10$$

or
$$R_{sc} = \log_e 10 = 2.303$$

∴
$$\text{Percentage increase in } R_{sc} = \frac{2.303 - 1.386}{1.386} \times 100$$

$$= \boxed{66.2\%} \quad \textbf{Ans.}$$

EXAMPLE 11.2 A single-cylinder two-stroke CI engine of 120 mm bore and 150 mm stroke with compression ratio 16 is running at 2000 rpm when the following measurements are made:

Actual air flow per hour = 240 kg

Air inlet temperature = 300 K

Exhaust pressure = 1.025 bar

Determine:

(a) The scavenging ratio
(b) The scavenging efficiency
(c) The trapping efficiency.

Solution: (a) Density of air, $\rho = \frac{p_e}{RT_i} = \frac{1.025 \times 10^5}{287 \times 300} = 1.19 \text{ kg/m}^3$

Swept volume,
$$V_s = \frac{\pi}{4} d^2 L = \frac{\pi}{4}(0.12)^2 \times 0.15 = (1.696 \times 10^{-3}) \text{ m}^3$$

Total cylinder volume,
$$V = \left(\frac{r}{r-1}\right) V_s$$

$$= \frac{16}{15} \times 1.696 \times 10^{-3} = (1.809 \times 10^{-3}) \text{ m}^3$$

$\therefore$ Ideal mass in total cylinder volume $= \rho V$

$$= 1.19 \times 1.809 \times 10^{-3}$$
$$= (2.153 \times 10^{-3}) \text{ kg per cycle}$$

Ideal mass per unit time $= 2.153 \times 10^{-3} \times (2000 \text{ rpm})$

$$= 4.306 \text{ kg/min}$$

Actual mass of air supplied per min $= \dfrac{240}{60} = 4.0$ kg/min

$\therefore$ Scavenging ratio,
$$R_{sc} = \frac{\text{actual mass of air supplied}}{\text{ideal mass}}$$
$$= \frac{4.0}{4.306} = \boxed{0.929} \quad \textbf{Ans.}$$

(b) Scavenging efficiency,
$$\eta_{sc} = 1 - e^{-Rsc} = 1 - e^{-0.929}$$
$$= 1 - 0.395 = 0.605$$
$$= \boxed{60.5\%} \quad \textbf{Ans.}$$

(c) Trapping efficiency,
$$\eta_{tr} = \frac{\eta_{sc}}{R_{sc}} = \frac{0.605}{0.929} = 0.651$$
$$= \boxed{65.1\%} \quad \textbf{Ans.}$$

EXAMPLE 11.3 A two-stroke single-cylinder SI engine consumes 6.5 kg/h of fuel when running at 3000 rpm. The air/fuel ratio is 15:1; the calorific value of fuel is 44,000 kJ/kg. The mean piston speed is 9 m/s and the imep is 4.8 bar. The scavenging efficiency of the engine is 85%; the mechanical efficiency is 80%. Determine the bore, the stroke, the brake power and the brake thermal efficiency of the engine. Take R for the mixture as 290 J/(kg K). The pressure and temperature of the mixture are 1.03 bar and 15°C, respectively. Assume perfect displacement model.

Solution: Given, $\dot{m}_f = 6.5$ kg/h, $N = 3000$ rpm, $\dfrac{\dot{m}_a}{\dot{m}_f} = 15$

$\therefore$ $\dot{m}_a = 15 \times 6.5 = 97.5$ kg/h

Calorific value, CV = 44,000 kJ/kg

$$\frac{2LN}{60} = 9 \text{ m/s}; \quad \therefore \; L = \frac{9 \times 60}{2 \times 3000} = 0.09 \text{ m} = \boxed{9 \text{ cm}} \quad \textbf{Ans.}$$

$$p_{mi} = 4.8 \text{ bar}, \quad \eta_{sc} = 0.85, \quad \eta_{mech} = 0.8.$$

Since perfect displacement model is assumed and $\eta_{sc} < 1$,

$$\eta_{tr} = 1$$

and
$$\eta_{sc} = R_{sc} = \frac{\dot{m}_{supp}}{\dot{m}_{ideal}}$$

$$\dot{m}_{\text{supp}} = \dot{m}_f + \dot{m}_a = 6.5 + 97.5 = 104 \text{ kg/h}$$

$$\dot{m}_{\text{ideal}} = \frac{\dot{m}_{\text{supp}}}{\eta_{\text{sc}}} = \frac{104}{0.85} = 122.4 \text{ kg/h}$$

Swept volume, $V_s = \dfrac{\pi}{4} d^2 L$

Volume swept per hour, $\dot{V}_s = \dfrac{\pi}{4} d^2 L(60 \times N)$

$$= \frac{\pi}{4} \times d^2 \times 0.09 \times 60 \times 3000$$

$$= 12723 d^2 \text{ m}^3\text{/h}$$

$$\rho = \frac{p}{RT} = \frac{1.03 \times 10^5}{290 \times 288} = 1.233 \text{ kg/m}^3$$

Since clearane volume is not given, it can be ignored,

$$\dot{m}_{\text{ideal}} = \rho \times \dot{V} \simeq \rho \times \dot{V}_s$$

$$\therefore \quad \dot{m}_{\text{ideal}} = \rho \times \dot{V}_s = 12723 \times 1.233 d^2$$

$$= 15687 d^2 \text{ kg/h}$$

$$\therefore \quad 15687 d^2 = 122.4$$

$$\therefore \quad d = \sqrt{\frac{122.4}{15687}} = 0.0883 \text{ m} = \boxed{8.83 \text{ cm}} \quad \textbf{Ans.}$$

$$\text{ip} = \frac{p_{m_i} LAN}{60 \times 1000} \text{kW}$$

$$= 4.8 \times 10^5 \times 0.09 \times \frac{\pi}{4}(0.0883)^2 \times \frac{3000}{60 \times 1000} = 13.23 \text{ kW}$$

$$\text{bp} = \text{ip} \times \eta_{\text{mech}} = 13.23 \times 0.8 = \boxed{10.58 \text{ kW}} \quad \textbf{Ans.}$$

$$\eta_{\text{th,b}} = \frac{\text{bp}}{\text{heat supplied by fuel per second}} = \frac{\text{bp}}{\dot{m}_f \times \text{CV}}$$

$$= \frac{10.58}{(6.5/3600) \times 44{,}000} = 0.133 = \boxed{13.3\%} \quad \textbf{Ans.}$$

EXAMPLE 11.4 A single-cylinder two-stroke SI engine has 8 cm diameter bore and 10 cm stroke. The compression ratio is 8. The exhaust port opens 60° before BDC and closes 60° after BDC. The air/fuel ratio is 15.1. The temperature of the mixture entering into the engine is 300 K and the pressure in the cylinder at the time of closing the exhaust port is 1.05 bar. Take R for the mixture = 290 J/(kg K). Air supplied to the engine is 150 kg/h. The speed of the engine is 4000 rpm. Considering the effective stroke, calculate the scavenging ratio, the scavenging efficiency and the trapping efficiency.

Solution: *Given:* $d = 8$ cm = 0.08 m , $L = 10$ cm = 0.1 m, compression ratio, $r = 8$, $\frac{\dot{m}_a}{\dot{m}_f} = 15$, $T_1 = 300$ K, $p_e = 1.05$ bar, $\dot{m}_a = 150$ kg/h, $R = 290$ J/(kg k), N = 4000 rpm.

$\therefore$ Scavenging ratio, $$R_{sc} = \frac{\dot{m}_{supp}}{\dot{m}_{ideal}}$$

$$\dot{m}_f = \frac{\dot{m}_a}{15} = \frac{150}{15} = 10 \text{ kg/h}$$

$$\dot{m}_{supp} = \dot{m}_a + \dot{m}_f = 150 + 10 = 160 \text{ kg/h}$$

Calculations based on effective stroke L_e, as shown in Figure 11.15, give better result.

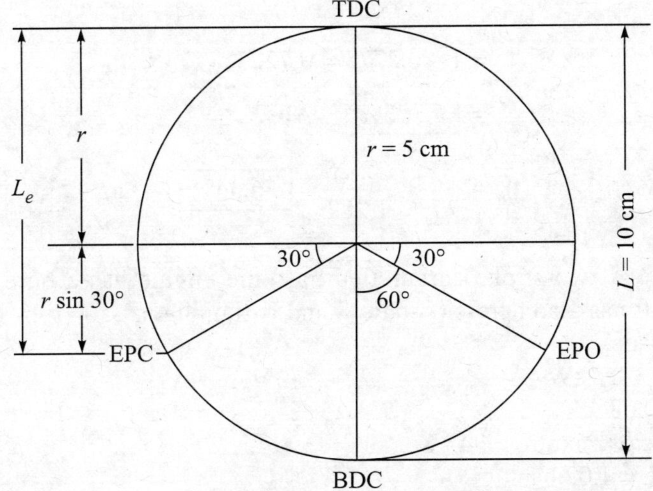

Figure 11.15 Example 11.4.

$$r = \frac{L}{2} = \frac{10}{2} = 5 \text{ cm}$$

Effective stroke, $$L_e = r + r\sin 30° = 5 + 5\sin 30° = 5 + 2.5 = 7.5 \text{ cm}$$
$$= 0.075 \text{ m}$$

Swept volume corresponding to L_e, $$V_{se} = \frac{\pi}{4} d^2 L_e$$

$$= \frac{\pi}{4} \times (0.08)^2 \times 0.075$$

$$= (3.77 \times 10^{-4}) \text{ m}^3$$

Total cylinder volume corresponding to L_e,

$$V = \left(\frac{r}{r-1}\right) V_{se} = \frac{8}{7} \times 3.77 \times 10^{-4} = 4.31 \times 10^{-4} \text{ m}^3$$

Density, $\rho = \dfrac{p_e}{RT_1} = \dfrac{1.05 \times 10^5}{290 \times 300} = 1.207 \text{ kg/m}^3$

Mass of mixture per cycle, $m = 4.31 \times 10^{-4} \times 1.207$

$= 5.202 \times 10^{-4} \text{ kg/cycle}$

Ideal rate of mass flow, $\dot{m}_{\text{ideal}} = m \times 60N$

$= 5.202 \times 10^{-4} \times 60 \times 4000$

$= 124.8 \text{ kg/h}$

Scavenging ratio, $R_{sc} = \dfrac{160}{124.8} = \boxed{1.282}$ **Ans.**

Scavenging efficiency, $\eta_{sc} = 1 - e^{-Rsc}$

$= 1 - e^{-1.282}$

$= 1 - 0.2775 = 0.7225$

$= \boxed{72.25\%}$ **Ans.**

Trapping efficiency, $\eta_{tr} = \dfrac{\eta_{sc}}{R_{se}} = \dfrac{72.25}{1.282} = \boxed{56.36\%}$ **Ans.**

EXAMPLE 11.5 A two-stroke carburetted gasoline engine has a bore of 8.25 cm and a stroke of 11.25 cm. It has a compression ratio 8 and is running at 2500 rpm when the following measurements are made:

Indicated power = 17 kW
Fuel/air ratio = 0.08
Inlet temperature of mixture = 345 K
Exhaust pressure = 1.02 bar
Calorific value of fuel = 44,000 kJ/kg
Indicated thermal efficiency based on fuel retained = 0.29
Molar mass of fuel = 114
Determine the scavenging efficiency.

Solution: Displacement volume, $V_s = \dfrac{\pi}{4} d^2 L = \dfrac{\pi}{4}(8.25)^2 \times 11.25$

$= 601.4 \text{ cm}^3$

Total cylinder volume, $V = \left(\dfrac{r}{r-1}\right) V_s = \dfrac{8}{7} \times 601.4 = 687.3 \text{ cm}^3$

If the air is dry, the density of dry air is

$$\rho_s = \frac{29\, p_e}{R_m T_i} \left(\frac{1}{1 + F_i (29/\mu_f) + 1.6\, h} \right)$$

Given: $F_i = 0.08$, $h = 0$, $\mu_f = 114$, $p_e = 1.02$ bar, $T_i = 345$ K, $R_m = 8314$ J/(kg K)

$$\therefore \qquad \rho_s = \frac{29 \times 1.02 \times 10^5}{8314 \times 345}\left(\frac{1}{1 + 0.08(29/114)}\right) = 1.01 \text{ kg/m}^3$$

$$\dot{V}_{\text{ret}} = \eta_{\text{sc}} \cdot V \cdot N$$

$$(\dot{m}_{\text{ret}})_{\text{air}} = \eta_{sc} \cdot V \cdot N \cdot \rho_s$$

$$(\dot{m}_{\text{ret}})_{\text{fuel}} = \eta_{\text{sc}} \cdot V \cdot N \cdot \rho_s \cdot F_i$$

$$\eta_i = \frac{ip}{(\dot{m}_{\text{ret}})_{\text{fuel}} \cdot Q_c}$$

where, Q_c is the calorific value of fuel.

$$\text{ip} = \eta_{\text{sc}} N V \rho_s F_i Q_c \eta_i$$

Given: $\text{ip} = (17 \times 10^3)$ W, $N = \dfrac{2500}{60}$ rps, $V = (687.3 \times 10^{-6})$ m^3, $\rho_s = 1.01$ kg/m^3

$F_i = 0.08$, $Q_c = 44{,}000$ kJ/kg, $\eta_i = 0.29$

$$\therefore \qquad \eta_{\text{sc}} = \frac{17 \times 10^3}{(2500/60) \times 687.3 \times 10^{-6} \times 1.01 \times 0.08 \times 44000 \times 10^3 \times 0.29}$$

$$= \boxed{57.6\%} \quad \textbf{Ans.}$$

REVIEW QUESTIONS

1. Define scavenging. How are the two-stroke engines classified depending upon the methods of producing the scavenge charge?
2. Distinguish between the crankcase scavenged engines and separately scavenged two-stroke engines.
3. Describe the working principle of separately scavenged two-stroke engines with the help of a diagram.
4. What are the two types of scavenging processes based on the scavenging arrangements? Briefly describe them.
5. What are the different types of return flow scavenging engines? Briefly describe them with the help of diagrams.
6. Describe two main systems of uniflow scavenging with the help of diagrams.
7. Explain the scavenging process for a cross-flow two-stroke engine with the help of pressure–crank angle diagram.
8. Define the terms blow-down angle and scavenging angle.
9. Define the following terms in connection with the two-stroke engine: delivery ratio, scavenging ratio, trapping efficiency, scavenging efficiency, charging efficiency, and purity of charge.

10. Derive an expression to evaluate the scavenging ratio in terms of the rate of mass flow of air, the compression ratio, the piston speed, the piston area and the density of dry air, when the fresh charge consists of air, fuel and water vapour.
11. Describe the perfect displacement model for the scavenging process giving suitable assumptions for the different conditions of the scavenging ratio.
12. Describe the complete mixing model for the scavenging process giving suitable assumptions. Prove that for this model,

$$\eta_{sc} = 1 - e^{-Rsc}.$$

13. What do you understand by short-circuiting in the scavenging process?
14. What processes are involved in real engines during scavenging? What are the factors responsible for reducing the scavenging efficiency?
15. Show the theoretical relationships between the scavenging efficiency and the scavenging ratio for the perfect displacement, complete mixing and complete short-circuiting cases. Also show the relationship between the trapping efficiency and the scavenging ratio for the complete mixing model.
16. Describe the tracer gas method for the measurement of scavenging efficiency.
17. Describe the gas-sampling method for the measurement of scavenging efficiency, giving the sampling valve diagram, and represent the position of the sampling valve in the exhaust port.
18. Show the various possible designs of ports. Which type of design is more suitable and why?
19. Show the flow pattern through the piston-controlled inlet ports for various openings and for different orientations of the port axis.
20. Show the variation of discharge coefficient as a function of the port open fraction for different inlet port designs.
21. Show the variation of the jet angle exiting from the exhaust port as a function of the open port height.
22. Show the effect of pressure ratio (p_c/p_e) on the variation of exhaust port discharge coefficient as a function of the port open fraction.
23. Describe the Kadenacy effect in the process of scavenging of a two-stroke engine.
24. Describe the different types of scavenging pumps with the help of diagrams, mentioning their suitability for particular types of engines.
25. Compare the advantages and disadvantages of the two-stroke engine with the four-stroke engine.
26. Compare the two-stroke SI engines with the two-stroke CI engines.

PROBLEMS

11.1 The scavenging ratio of a loop-scavenged two-stroke engine is 1.2. If this ratio is increased by 30%, find the percentage increase in the scavenging efficiency. **[Ans.** 13%]

11.2 The scavenging efficiency and the scavenging ratio of a loop-scavenged two-stroke engine are 80% and 0.9 respectively. Find the trapping efficiency of the engine. If the trapping efficiency is increased by 10%, find the percentage decrease in the scavenging ratio. Assume that the scavenging efficiency remains constant. **[Ans.** 88.9%, 9.1%]

11.3 A two-stroke single-cylinder SI engine of 100 mm bore having compression ratio 8.5 consumes 15.75 kg/h of fuel when running at 3500 rpm. The piston speed is 14 m/s and the indicated mean effective pressure is 5.0 bar. The air/fuel ratio is 15 : 1, the calorific value of fuel is 44,000 kJ/kg. Take R for the mixture as 290 J/(kg K), the pressure and temperature of the mixture as 1.05 bar and 27°C. Determine the scavenging ratio, the scavenging efficiency and the trapping efficiency. Also determine the ip, the bp and the brake thermal efficiency of the engine, if the mechanical efficiency is 85%.

[Ans. 0.931, 60.6%, 65.1%, 27.5 kW, 23.4 kW, 12.2%]

11.4 A single-cylinder two-stroke CI engine has 10 cm diameter and 12 cm stroke. The compression ratio is 16, the exhaust port opens 60° before BDC and closes 60° after BDC. The temperature of the air entering into the engine is 27°C and the pressure in the cylinder at the time of closing the exhaust port is 1.03 bar. Take R for the air as 287 J/(kg K). The air supplied to the engine is 2.5 kg/min. The speed of the engine is 3000 rpm. Considering the effective stroke, calculate the scavenging ratio, the scavenging efficiency and the trapping efficiency. **[Ans.** 0.9242, 60.3%, 65.3%]

11.5 A two-stroke SI engine has 8 cm bore. The compression ratio is 9. The engine runs at 3000 rpm with a mean piston speed of 10 m/s. The air/fuel ratio is 12.5, the inlet temperature of the mixture is 320 K, the exhaust pressure is 1.04 bar, the calorific value of the fuel is 44,000 kJ/kg, the indicated thermal efficiency is 0.28, the molar mass of the fuel is 114, and the scavenging efficiency is 60%. Determine the indicated power.

[Ans. 18.58 kW]

CHAPTER 12 Ignition Systems

12.1 INTRODUCTION

The ignition system carries the electrical current to the spark plug where the spark carries sufficient energy to increase the temperature of the surrounding charge to the ignition point at which combustion becomes self-sustaining. The spark appears at the plug gap in SI engines just as the piston approaches the top dead centre on the compression stroke, when the engine is idling. At higher speeds or during increased throttle operation, the spark is advanced. It occurs somewhat earlier in the cycle. The mixture thus has ample time to burn and deliver its power. Many means are employed to produce the necessary high voltage required to jump a set gap of the spark plug to produce spark in the combustion chamber for the ignition of the combustible charge at the correct time. When such a spark is produced to ignite a homogeneous air-fuel mixture in the combustion chamber of an engine, it is called the spark-ignition system. The ignition systems are classified as follows:

1. Battery-ignition system
2. Magneto-ignition system
3. Electronic-ignition system.

12.2 IGNITION SYSTEM REQUIREMENTS

The requirements of an ignition system for a spark-ignition engine are detailed in the following subsections.

Intensity of spark

The spark should be sufficiently intense to ignite the mixture, whether rich or weak, under full throttle conditions. On reduced loads and also for very weak mixtures the intensity of spark becomes important as it determines the range and also the rate of burning of the weaker range of mixture. Any further increase in intensity has no effect on the power output. Too high a spark intensity may burn the electrodes of the spark plug.

Voltage

There must be sufficient voltage at the spark-plug electrodes to jump the gap and produce spark. The voltage required to jump the gap increases with the increase in gap. The fuel/air ratio within the engine cylinder also affects the required voltage. Lean mixture near the spark plug will require a higher voltage. A higher voltage is also required when the engine is cold, when the fuel vaporization is poor and when the charge density due to high manifold pressure or compression ratio is more. The difference between the voltage available from the ignition system and the required voltage is called the ignition reserve. Misfiring occurs when there is no ignition reserve available. However, an excessively high voltage may cause erosion of the spark-plug electrodes and insulation failure of the spark plug. The voltage necessary to overcome the resistance of the spark-plug gap and to release energy to initiate the self-propagating flame front in the combustible mixture depends upon the size of the gap. Figure. 12.1 shows the relationship between the sparking voltage and the gap. The curve shows that the sparking voltage increases with the increase in spark-plug gap.

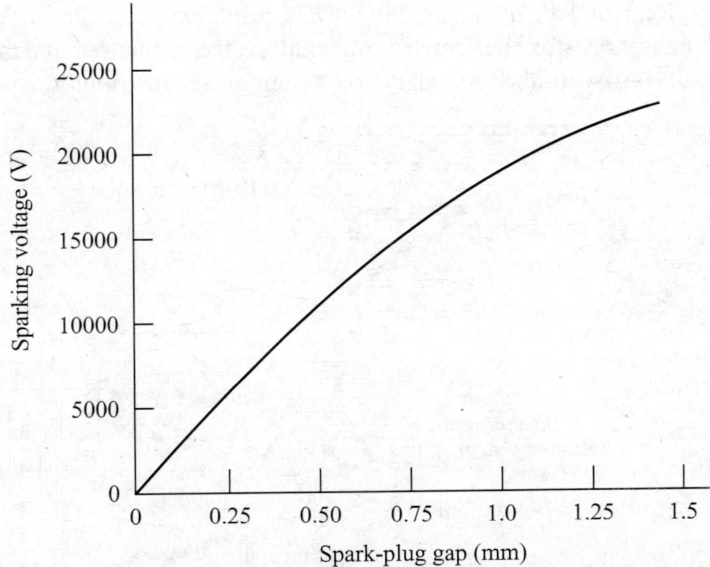

Figure 12.1 Sparking voltage and electrode gap relationship.

Number of molecules

In order to initiate a self-sustaining flame front it is necessary to have some minimum number of molecules between the spark-plug electrodes, which depends upon the fuel/air ratio, the amount of residual gas, the pressure, the temperature and other engine operating conditions.

With decreased manifold pressure, the number of molecules in the gap reduces and this reduction may cause irregular firing even though the voltage requirement is less. If the electrode gap is now increased to increase the number of molecules in the gap, firing will become regular. The minimum spark-plug gap that is satisfactory for engine idle condition is 0.625 mm. Most ignition systems use gap specifications over 0.75 mm.

Protection of ignition system

The complete ignition system must be protected from ingress of moisture to avoid current leakage. It is an essential requirement to obtain a good and consistent spark.

Ignition timing

Ignition timing should automatically be changed with engine speed and load after the initial timing is set.

12.3 BATTERY-IGNITION SYSTEM

The components of a battery-ignition system, shown in Figure 12.2 are (a) a battery, (b) an ignition switch, (c) an ignition coil with or without an added ballast resistor, (d) a distributor which houses the contact breaker points, the cam, the condenser, the rotor, and the advance mechanisms, (e) a spark plug, and (f) the low and high tension wirings. There are two circuits of the ignition system: the primary circuit and the secondary circuit. The primary circuit consists of the battery, the ignition switch, the ballast resistor, the primary coil winding, the condenser, and the breaker points. The secondary circuit consists of the secondary coil windings, the distributor, and the spark plugs.

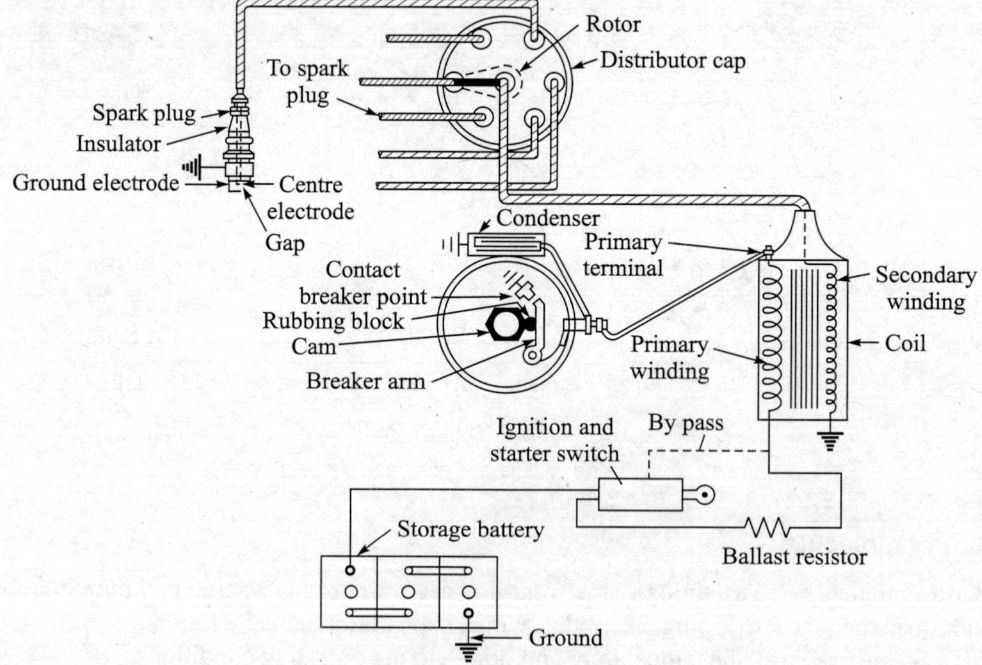

Figure 12.2 Battery-ignition system.

12.3.1 Battery

Electrical energy for ignition is provided by a storage battery. It is an electrochemical device which converts chemical energy into electrical energy. The six-plate 12-volt battery supplies a steady current for ignition, starter motor, lighting and other electrical circuits, and provides a reserve of electricity when the current consumed by the electrical equipment exceeds that being produced by the dynamo. The fully charged battery has an electrolyte of 1.26 to 1.28 specific gravity and contains 23% sulphuric acid. The battery is charged by a dynamo.

12.3.2 Ignition Switch

The ignition switch is placed between the battery and the primary winding of the ignition coil. With the help of this switch the ignition system is turned on or off. The ignition switch has an extra set of contacts that are used when the switch is turned past ON to START. The contacts connect the starting motor solenoid to the battery, so that the starting motor can operate when the engine is started, and when the switch is released, it return to ON. The starting motor is then disconnected from the battery.

12.3.3 Ballast Resistor

A ballast resistor may be placed in series with the primary winding of the ignition coil to regulate the primary current. At low engine speeds the average current is high due to longer closer of the contact breaker point. The ballast resistor is heated up and produces more resistance which cuts down the current. At high speeds the average current is low, the resistor runs cooler and its resistance decreases which in turn increases the flow of current. This resistor is shorted by the ignition switch when turned to START. Now, the full battery voltage is applied to the ignition coil for good performance during cranking. After the engine is started and the ignition switch is turned on, the resistance is in the ignition primary circuit. It thus protects the contact points from excessive current.

12.3.4 Ignition Coil

The ignition coil is used to step up 12 volts of the battery to a very high voltage of 10,000 to 20,000 volts to induce an electric spark across the electrodes of the spark plug. The typical induction coil of metal clad type is shown in Figure 12.3. It consists of a primary winding of 200 to 300 turns of thick wire of about 20 SWG (standard wire gauge) to provide a resistance of about 1.15 Ω and the secondary winding is made up of a large number of turns about 21,000 of fine enameled wire of 38–40 SWG sufficiently insulated to withstand high voltages. These windings are wound upon a cylindrical soft-iron core, and enclosed by a soft iron shell. The secondary coil is wound close to the core and the primary winding is located on the outside of the secondary coil. Each layer of winding is insulated from the next by a sheet of oiled paper, while the entire assembly is sealed within an oil-filled case for cooling and for protecting the coil from moist air. A heavily insulated terminal block, which supports three terminals, is provided on the top of the ignition coil assembly. The terminals are usually marked as SW (switch wire), CB (contact breaker) and HT (high tension). The SW terminal is connected to the ignition switch through a resistor; the CB terminal is connected to the contact breaker; and the HT terminal is connected to the centre of the distributor cap to the rotor arm.

Another type of induction (ignition) coil is a core type coil. It has a U-shaped core surrounded by the primary winding. The secondary winding is outside the core. The size of the primary

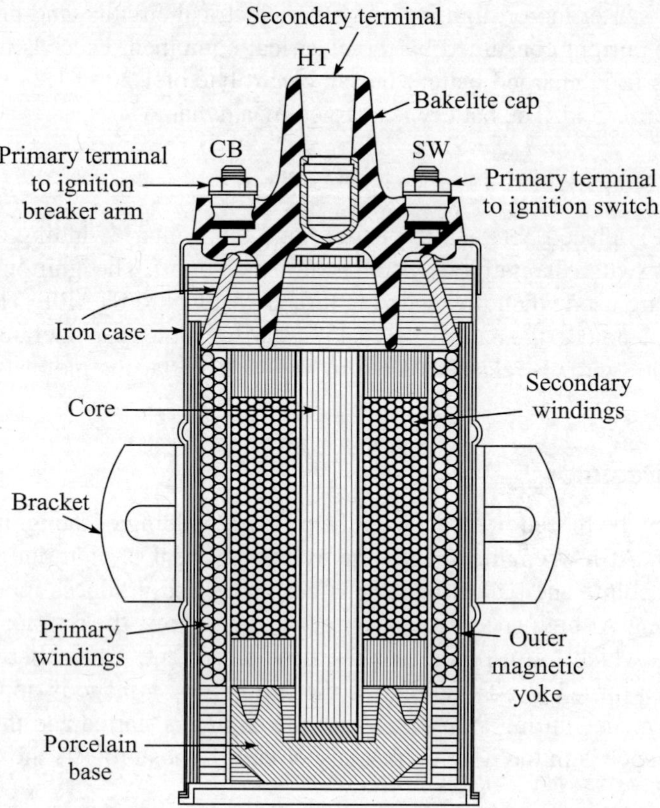

Figure 12.3 Section through an ignition coil.

winding is reduced which reduces the resistance of the primary circuit compared to that of the metal-clad type. It is therefore necessary to use an external ballast resistance in series with the primary coil.

12.3.5 Contact Breaker Points

It is a mechanical device for making and breaking the primary circuit of the ignition system. It consists of a fixed metal point which is grounded and another metal point attached to a movable, spring-loaded insulated breaker arm. The metal used is invariably one of the hardest metals, usually tungsten and each point has a circular flat face of about 3 mm diameter. The breaker arm is connected to the primary circuit and separated from the breaker cam by a non-conducting rubbing block. Each time the cam lobe passes over the rubbing block it forces the breaker arm to separate the points, thus breaking the primary circuit. When the flat part of the cam comes in contact with the rubbing block the points are not separated as they remain held together by the spring pressure, thus closing the primary circuit. The distributor shaft is so meshed with the engine camshaft that the lobes of the distributor cam open the point when the pistons are nearly at the upper end of each compression stroke. The distributor shaft turns at the same speed as the engine camshaft through gears.

If the contact breaker points are burned, pitted and badly worn, they must be removed and either replaced, or their faces must be filed smooth and the gap should be set accurately.

12.3.6 Condenser

As the contact breaker points open, the current from the battery through the primary winding of the coil is stopped. The magnetic field therefore begins to collapse. The collapsing magnetic field induces current which continues to flow in the same direction in the primary circuit and thus charges the condenser plates. This absorbed current surges back out of the condenser and towards the ignition coil, thus helping to bring about the complete collapse of the magnetic field in the coil which in turn induces a high voltage of necessary magnitude in the secondary winding. The collapse of the magnetic field also re-establishes the flow of current in the primary circuit, causing a heavy electric arc across the separating contact points. The points might burn, causing the energy stored in the ignition coil as magnetism to be consumed by the arc. Because of the absorbing qualities of the condenser, the arcing and pitting of the points are prevented.

The condenser is made up of two thin metallic plates separated by an insulator. The plates are thin strips of lead or aluminum foil. They are insulated from each other by a special type condenser paper and wrapped to form a winding. The winding is then installed in a container to prevent it from damage against moisture and external physical contact.

12.3.7 Distributor

It consists of a housing, a drive shaft with breaker cam, an advance mechanism, a breaker plate with contact points and a condenser, a rotor, and a cap. The shaft is driven by the camshaft of the engine through spiral gears. It rotates at one-half of the speed of crankshaft. This shaft is usually coupled to another shaft which drives the oil pump for lubrication.

In the lower part of the distributor housing there is a speed sensitive device, called the centrifugal advance mechanism, whose function is to advance the spark with the increase in engine speed. It also carries the vacuum advance mechanism, which serves to retard the spark as the load on the engine increases. The contact breaker assembly is placed above the advance mechanism. Figure 12.4 shows the top and sectional views of an ignition distributor using the contact points. In the top view, the cap has been removed so that the breaker plate can be seen.

The distributor has several functions. First, it closes and opens the contact points to complete and interrupt the primary circuit between the battery and the ignition coil. When the primary circuit is completed through the closed contact points, the current flows in the ignition coil and builds up a magnetic field. When the points open by the cam rotation, the primary circuit is opened and the current stops flowing. The magnetic field collapses and this produces a high voltage in the secondary winding of the ignition coil.

The second function of the distributor is to distribute each high voltage surge to the correct spark plug at the correct instant. It does this with the distributor rotor. The rotor rotates with the breaker cam and connects the central terminal of the cap with each outside terminal in turn. Each terminal is connected electrically to a spark plug by a spark-plug wire. The high tension secondary current passes from the coil into the centre of the distributor cap where it is transmitted by a carbon brush to the centre of the rotor. The current then passes along a conductor within the rotor and each

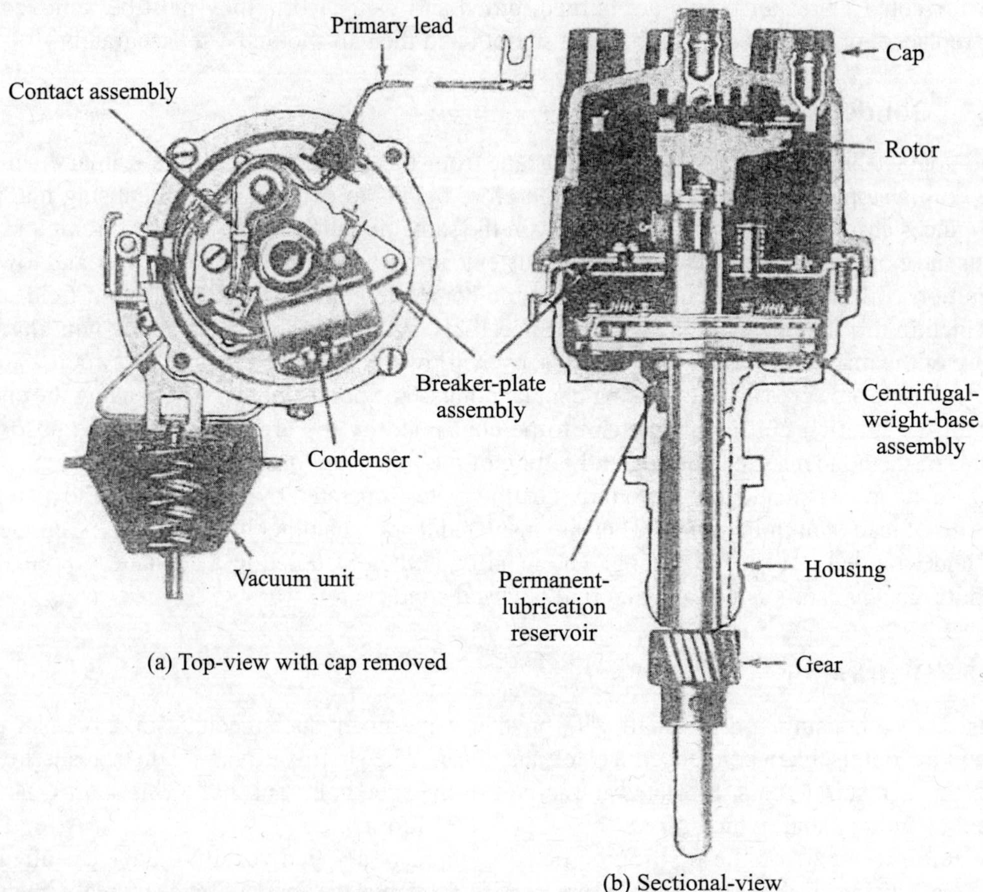

Figure 12.4 (a) Top and (b) sectional views of an ignition distributor.

time the end of the rotor passes a brass terminal in the distributor cap, the current jumps the short gap and passes from the terminal to the spark plug by way of the spark-plug wire. Thus the high voltage surges from the coil are directed to each spark plug in turn according to the firing order.

12.4 FIRING ORDER

The term firing order is used in a multi-cylinder engine. The sequence in which the power impulses occur in an engine is called the firing order. As the number of cylinders is increased, the power impulses for each revolution of the crankshaft increase in frequency, giving a more uniform torque and smoother operation. The more the cylinders in an engine, the more continuous is the flow of power if the power impulses are spaced equally, the less is the vibration, and also less work has to be done by the flywheel in storing and releasing the energy. The flywheels for the multi-cylinder engines, therefore, can be lighter than those used in engines with fewer cylinders.

Figure 12.5 shows the position of crankshaft for a four-cylinder in-line engine. The crank arms for No. 1 and 4 cylinders project in the same direction, and the crank arms for No. 2 and 3

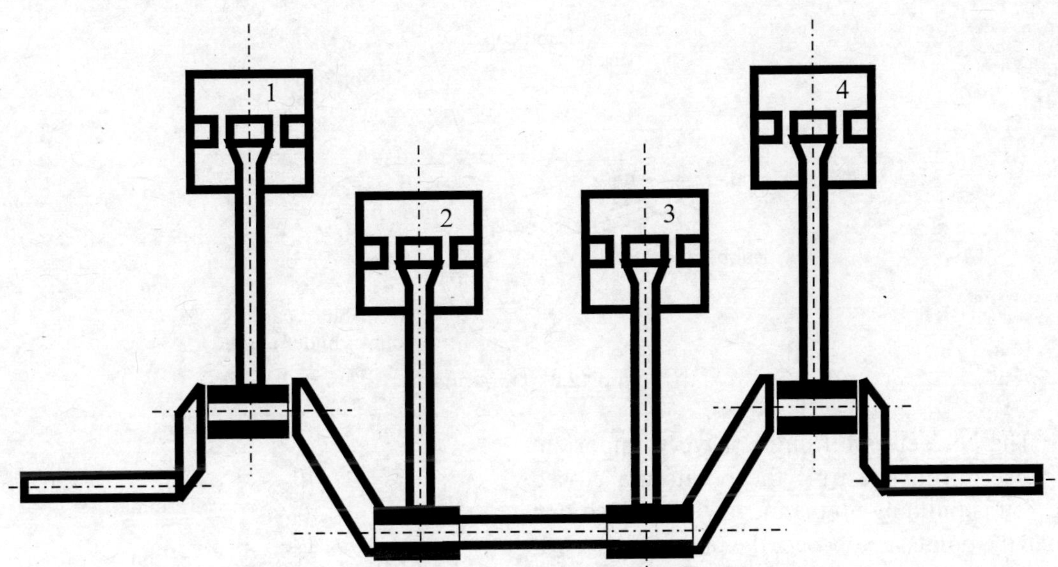

Figure 12.5 Position of crankshaft for a four-cylinder engine.

cylinders project from the opposite side of the crankshaft. Hence the angle between the pair of throws is 180 degrees.

In the four-cylinder engines, the No. 1 and 4 pistons are always moving in the opposite direction from positions 2 and 3. As the No. 1 piston moves downwards on the power stroke, the No. 4 piston must move downwards on the suction stroke; No. 2 piston can be moving upwards on exhaust or compression stroke and No. 3 will be moving upwards on compression or exhaust stroke. With either arrangement, the power impulses are evenly distributed and they are 180° apart. Therefore, for four-cylinder engines the possible firing orders are: 1–3–4–2 or 1–2–4–3. The former is more commonly used. For a three-cylinder in-line engine, the firing order is 1–3–2. For a six-cylinder in-line engine, the firing orders which can be used are 1–5–3–6–2–4, 1–5–4–6–2–3, 1–2–4–6–5–3, and 1–2–3–6–5–4. The first one is commonly used.

12.5 DWELL ANGLE

The dwell angle or the cam angle is the number of degrees that the distributor cam travels during the period when the ignition points are closed (Figure 12.6). In a four-cylinder engine the average cam angle is about 50 degrees. This means that during the 90 degrees of cam rotation the points are closed for 50 degrees and opened for the remaining 40 degrees. If the contact breaker point gap is increased, the dwell angle will obviously be decreased. A decrease in gap will produce a larger dwell angle. The magnitude of the dwell angle also depends upon the angle between the cam lobes. The angle between the cam lobes depends upon the number of engine cylinders. If a single contact breaker is used, the number of lobes on the cam will be the same as the number of cylinders.

As the number of cylinders is increased, the dwell angle is reduced because more and more closing and opening operations of the points must be accomplished during every rotation of the distributor cam.

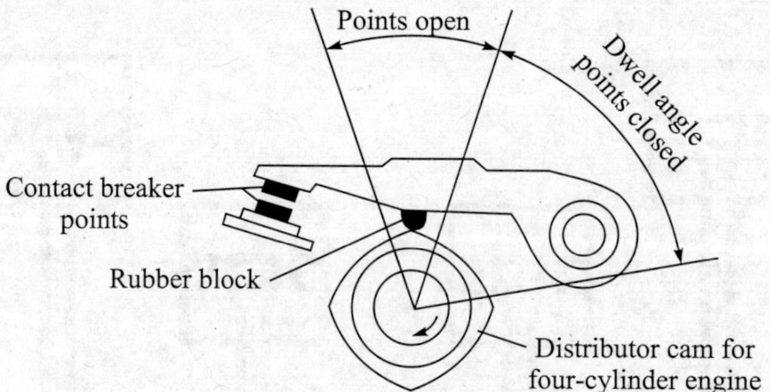

Figure 12.6 Dwell angle.

The correct dwell angle is very important because during the time the points are closed the coil is building up the magnetic field, so that when the points are opened, the proper amount of high tension current will be available at the spark plugs. Too small a dwell angle will result in less secondary voltage, and hence poor sparks or even misfiring. Too large a dwell angle will lead to burning of the condenser and the contact points due to oversaturation of windings.

12.6 SPARK PLUGS

A spark plug consists of three major parts: the shell, the insulator and the two electrodes as shown in Figure 12.7.

The shell supports the insulator and has threads that screw into the cylinder head, sealing the combustion chamber spark-plug hole. The threaded portion must be long enough to allow the electrodes to reach the combustion chamber. This length is called the spark-plug reach. The ceramic insulator is sealed inside the shell, so that it makes a pressure and thermal seal. It must withstand high thermal and mechanical stresses, and insulate the high secondary voltage. There are two electrodes, one usually grounded through the shell of the plug and the other is the central electrode which is well insulated with porcelain or mica.

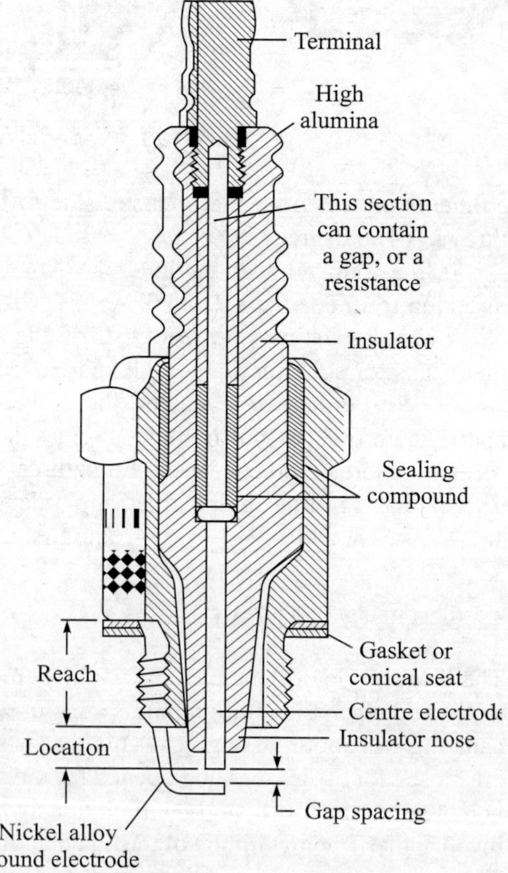

Figure 12.7 Spark plug.

Some spark plugs have a built-in resistor, which is a part of the centre electrode. This resistor reduces the interference to radio and TV receivers from the ignition system. It also reduces the electrode erosion caused by over-long sparking.

12.6.1 Requirements of a Good Spark Plug

The following characteristics are required for a good spark plug:

1. It should be capable of resisting leakage.
2. It should be gas-tight, so that the combustion gases do not leak out. A sealing gasket is provided for this purpose.
3. The temperature at electrodes should be maintained in the range of 400–800°C. Pre-ignition and oxidation of electrodes at high temperatures, as well as carbon and oil deposits at lower temperatures must be avoided.
4. It must not interfere with the radio and television receivers in its vicinity.
5. Erosion of electrodes due to high voltage sparks should be minimum.
6. It has to withstand widely fluctuating temperatures and pressures in the engine cylinder. The maximum pressure may be as high as 50 bar. This requires that the insulator material must possess high electrical resistance, good thermal conductivity and sufficient mechanical strength. Mostly, special nickel alloys having one or more of the elements like manganese, tungsten, silicon and chromium are used.

12.6.2 Factors Affecting the Establishment of Spark

The problem of establishing an electric spark across the gap of the electrodes in a spark plug is affected by many factors:

1. *The length of the gap*: The larger the gap, the more is the required breakdown voltage.
2. *The geometry of the gap*: Small pointed electrodes require lower breakdown voltages.
3. *The temperature of the electrodes*: The higher the temperature of the electrodes, the lower is the required breakdown voltage.
4. The density of the mixture: The higher densities require higher breakdown voltages.
5. *Carbon deposits on the insulator*: The carbon and metallic oxides form electrically conductive coatings on the insulator to increase the leakage. These coatings reduce the impressed voltage across the spark-plug gap, therefore a higher secondary voltage is required.
6. *The rate of increase of the voltage at the gap*: If the ignition system builds up the voltage at a rapid rate, the effect of leakage is minimized, and a higher voltage is made available across the gap.
7. *The air/fuel ratio*: Lean and very rich mixtures require a higher breakdown voltage than that required by the slightly rich mixtures.

12.6.3 Spark-Plug Heat Range

The spark plug should not run too cold or too hot. In other words, it must have a suitable heat range. Too cold a spark plug may form a sooty carbon deposit on the insulator around the centre electrode, which is electrically conducting and may short-circuit the plug, thus preventing the occurrence of spark across the gap. Carbon deposits can also be formed by too-rich mixtures or by too much oil in the cylinder. A hotter running plug burns this carbon away and prevents its formation. If the plug runs too hot the insulator may be damaged. A plug that runs hot, wears out more rapidly. The higher temperatures cause the electrodes to burn away more quickly. In addition, with a hot running plug, there is always a danger of pre-ignition which may occur before the passage of the normal spark.

The operating temperature of the plug depends upon the area of insulation and electrode exposed to the hot gases and the length of the heat path from electrode points and insulation back to the cooled part of the cylinder wall into which the plug is screwed. A plug with a short, centre electrode and insulator will run cooler than the one with a long one. The former is called a cold plug and the latter a hot plug.

When an engine runs under a heavy load, a cold spark plug is required to prevent overheating of the plug. Engines having tendency to run cold under light duty operation require hot spark plugs to keep the plug temperature high enough to reduce fouling of the plug. A well designed plug will perform satisfactorily over a fairly wide range of engine speeds and loads. Figure 12.8 illustrates hot and cold spark plugs, showing the relative paths of heat travel. A hot plug runs at a higher temperature than a cold one because it has a longer path through which the heat that it receives must travel to reach the cooling cylinder head water. The arrows in Figure 12.8 show the direction of heat flow.

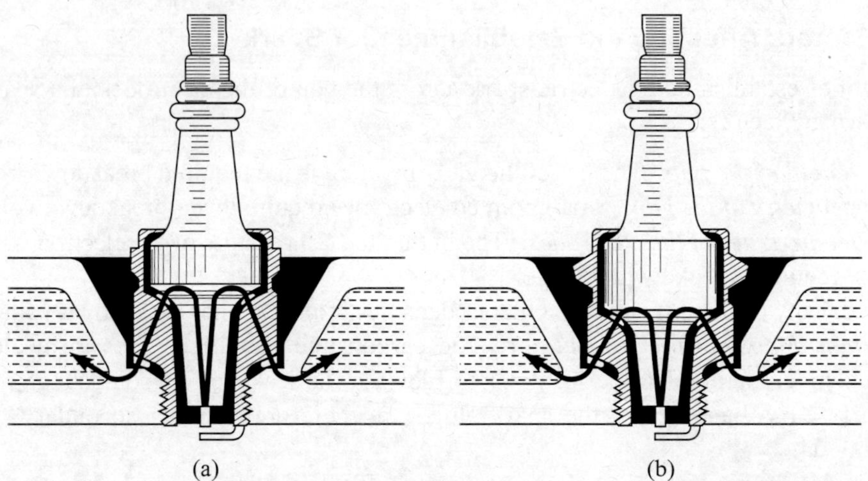

Figure 12.8 Paths of heat travel in a (a) hot spark plug and (b) cold spark plug.

12.7 MAGNETO-IGNITION SYSTEM

The magneto-ignition system is extensively used in mopeds, scooters, three wheelers, motor cycles, sports and racing cars and reciprocating aircraft engines. A schematic diagram of a rotating magnet type for a four-cylinder engine is shown in Figure 12.9. It is similar in principle to the battery-ignition system except that the magnetic field in the core of the primary and secondary windings is produced by a rotating permanent magnet. The magneto has got its own current generating unit without making use of battery and ignition coil. The magneto consists of a fixed armature having primary and secondary windings and a rotating assembly of magnets driven from the engine. It is called a high tension magneto.

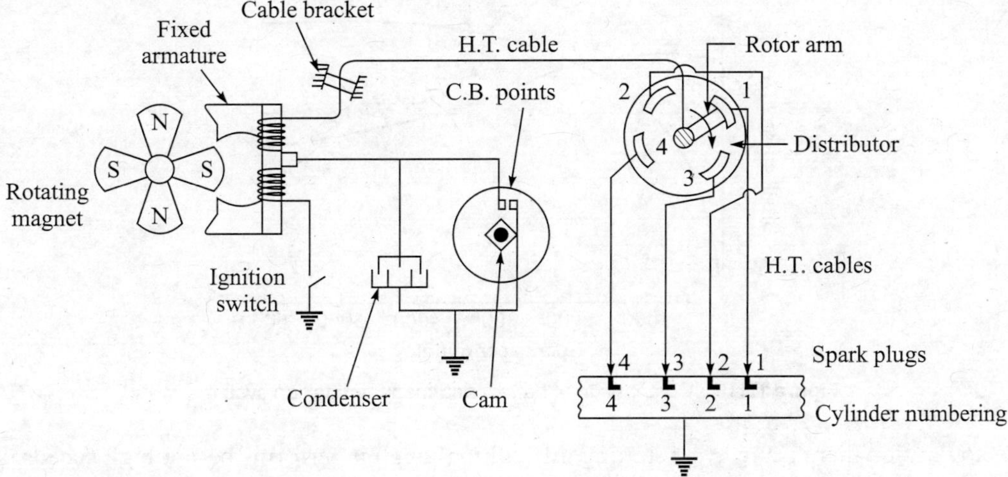

Figure 12.9 Layout of magneto-ignition system showing various parts. The sequence 1–3–4–2 marked on distributor indicates the firing order.

As the magnet turns, the magnetic field produced varies. The change in the magnetic flux induces voltage and current in the primary winding. The primary current produces a magnetic field of its own and the primary and secondary windings are surrounded by the magnetic field. When the contact breaker points are opened by the rotation of the distributor cam, the highly charged condenser discharges itself into the primary circuit and this produces a rapid change in the magnetic flux, which induces a very high voltage in the secondary winding sufficient to produce spark and ignite the combustible mixture.

Another type of magneto-ignition system is that of a rotating armature type. In this type the rotating member consists of an armature carrying both primary and secondary windings, and the armature rotates between the poles of a stationary magnet. A third type of magneto called the polar inductor type is also in use. It consists of three main parts: a magnet, an armature and an inductor. Out of these, the first two parts remain stationary while the inductor rotates with the magneto shaft.

A magneto having no separate secondary windings on its armature is called a low-tension magneto. For the spark to occur, the required voltage is produced on stepping up the low tension supply by means of an ignition coil.

The magneto-ignition system is cheap, reliable and requires little maintenance. The starting of engine is difficult as the magneto does not furnish enough voltage for ignition at low speeds. The efficiency of the system improves with the increasing speed.

12.8 COMPARISON OF BATTERY- AND MAGNETO-IGNITION SYSTEMS

Figure 12.10 shows the relationship of break current and sparks per minute for both battery and high-tension magneto ignition systems. The number of sparks produced per minute depends upon the engine speed and the number of cylinders. The current is zero when the magneto is at rest and the current increases as the speed increases, until at a certain designed speed the current attains a maximum value, which is higher than that of the equivalent battery ignition unit. Thus with the

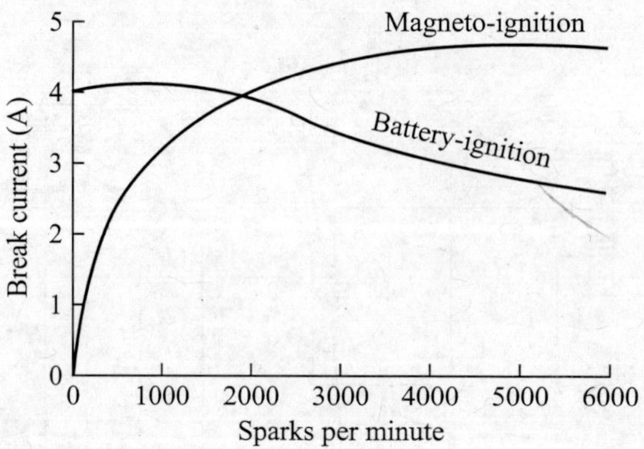

Figure 12.10 Comparison of battery and magneto-ignition systems.

magneto ignition there is always a starting difficulty, though it performs best at high speeds. The magneto ignition is therefore most suitable for high speed engines such as sports and racing cars and aircraft engines.

Table 12.1 gives the comparison of battery- and magneto-ignition systems.

Table 12.1 Comparison of battery- and magneto-ignition systems

	Battery-ignition system		*Magneto-ignition system*
1.	A battery is required. When the battery is discharged, the engine is difficult to start.	1.	It does not require any battery.
2.	More maintenance is required because of the battery.	2.	As the battery is not required, maintenance is less.
3.	Current for the primary circuit is obtained from the battery.	3.	The required current is generated by the magneto.
4.	A good spark is available at the spark plug even at low speeds such as during starting.	4.	During starting, the spark at the spark plug is poor due to cranking at low speeds.
5.	The spark intensity decreases as the engine speed increases. It lowers the efficiency of the system at higher speeds.	5.	The intensity of spark keeps on improving as the speed goes on increasing. The efficiency of the system thus improves as the engine speed increases.
6.	The space occupied by the system is more.	6.	The space occupied by the system is less.
7.	The ignition timing can be easily varied with the ignition advance units.	7.	Varying the ignition timing is more difficult since the breaker points must be opened when the rotating magnets are in the most favourable position.
8.	Commonly used in petrol cars and light commercial vehicles.	8.	Used in scooters, motor cycles, racing cars, etc.

12.9 PROBLEMS ASSOCIATED WITH CONVENTIONAL IGNITION SYSTEMS

The conventional ignition systems described earlier suffer from the following major problems:

1. Arcing, burning and excessive wear of contact breaker points are very common due to interruption of high inductive current during operation.
2. Contact breaker arm tends to bounce at high speeds. This leads to weaker sparks.
3. A decrease in available voltage occurs as the engine speed increases. This is due to limitations in the current switching capability of the breaker system and the decreasing time available to build up the primary coil stored energy.
4. Because of the high source impedance (about 500 kΩ) the system is sensitive to side-tracking across the spark-plug insulator.
5. The breaker points are subjected to electrical wear in addition to mechanical wear due to high current load. It requires short maintenance intervals. Increased currents rapidly reduce the breaker point life and system reliability.

12.10 ELECTRONIC-IGNITION SYSTEMS

In order to overcome the problems associated with conventional ignition systems, the modern ignition systems use electronic components to open and close the circuit between the battery and the ignition coil. All other parts of the electronic-ignition system are practically the same as in the contact point system. The ignition switch, the ignition coil, the wiring and cables, the centrifugal and vacuum advance, and the cap and rotor are the same for both systems. There are some new ignition systems which do not use mechanical, or centrifugal, or vacuum advance controls. Instead, these systems use electronic controls and sensing devices to produce an accurate control of advance. They advance or retard the spark as engine temperature, manifold vacuum, engine speed, atmospheric pressure, and other conditions change.

The following two types of electronic-ignition systems are getting popular:

1. Transistorized-coil ignition (TCI) system
2. Capacitive-discharge ignition (CDI) system.

12.10.1 Transistorized-coil Ignition (TCI) System

A transistor gives very good amplification with an extremely clean cut-off, and therefore is particularly suited to efficient switching operations. It can be used in the primary circuit of an ignition coil; it can act as a make-and-break, to produce the maximum inductive effect in the secondary circuit.

Figure 12.11 shows a circuit diagram of a transistorized-coil ignition system. It is very much similar to the conventional battery-ignition system except for the addition of an electronic control unit. The distributor points and cam assembly of the conventional ignition system are replaced by a magnetic pulse generating system which detects the distributor shaft position and sends electrical pulses to an electronic control unit. There are many types of pulse generators that could trigger the electronic circuit of the ignition system. A magnetic pulse generator, where a gear-shaped armature driven by the distributor shaft rotates past the stationary pole piece of the magnetic pick-up, is usually used. The top view of an electronic distributor with cap and rotor removed is shown in Figure 12.12. As the shaft and armature rotate, the teeth on the armature swing past the magnetic pick-up (Sensor) assembly. A coil in the sensor magnetically detects the movement of each tooth

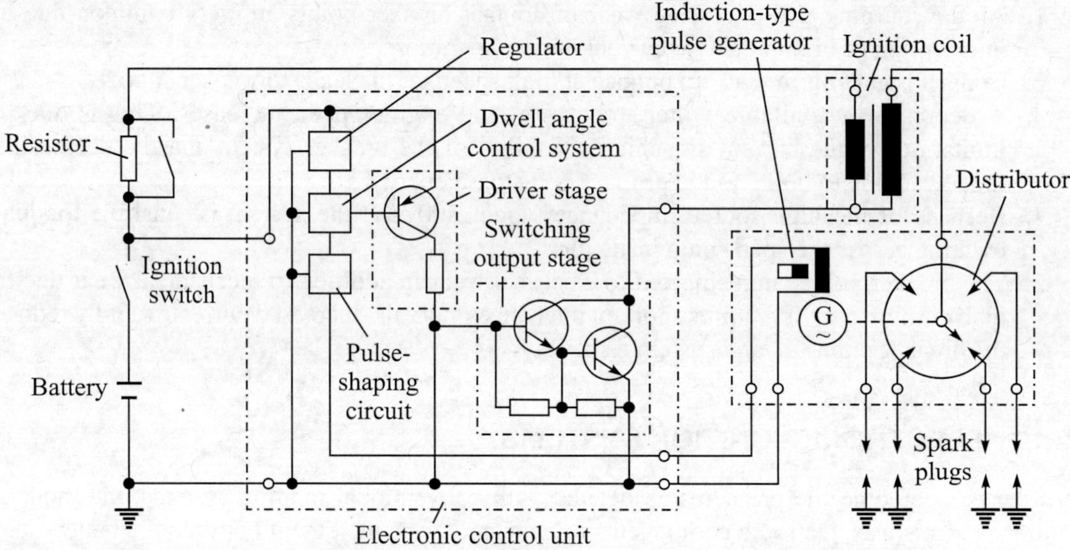

Figure 12.11 Circuit diagram of a transistorized-coil ignition system.

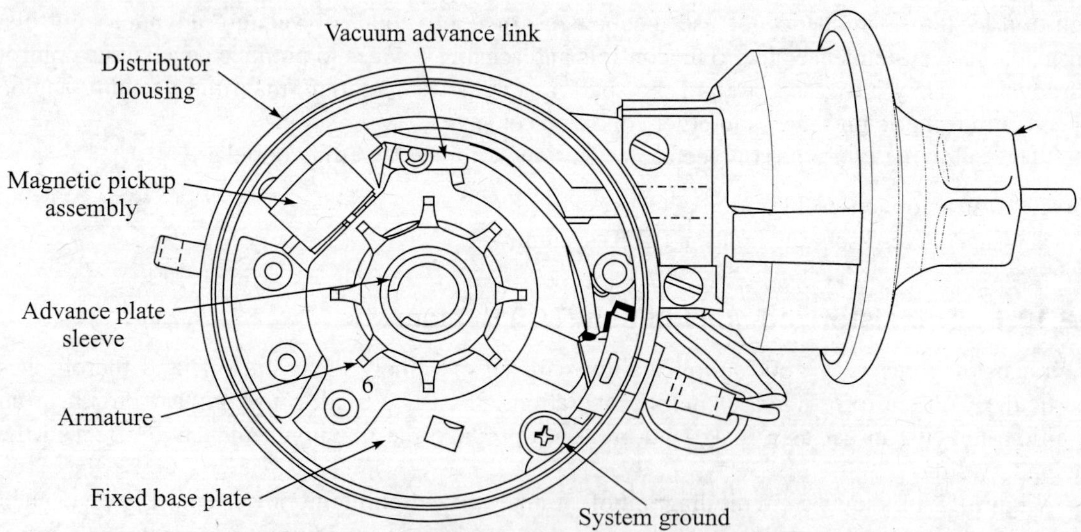

Figure 12.12 Top view of an electronic distributor with cap and rotor removed.

because it causes a change in the magnetic field around the coil. The coil therefore almost instantly sends a voltage pulse to the electronic control unit as a tooth passes by. The control unit then opens the circuit from the battery to the ignition coil. The magnetic field in the coil collapses and a high voltage surge is produced in the secondary winding of the coil. This surge is carried by the high voltage cables to the distributor cap, and then through the rotor and the spark-plug cable to the plug that is ready to fire. The number of teeth on the armature is the same as the number of cylinders.

The TCI system extends the life of the spark plug, improves ignition of lean mixtures, reduces maintenance and increases reliability and life. In order to ignite the mixture over a wider range of engine operation, the higher output voltage (about 35 kV) is required to jump a wider gap.

Electronic control unit contains the regulator, the pulse-shaping circuit, the dwell angle control system, the driver stage and the switching output stage.

1. **Regulator:** It is a voltage stabilizer. The supply voltage from the battery is closely regulated, so that the generated input pulse can be accurately reshaped and extended to provide the desired dwell angle.
2. **Pulse-shaping circuit:** The function of this circuit is to convert the control alternating voltage of the pulse generator into rectified rectangular current pulses.
3. **Dwell angle control system:** The dwell angle control system automatically advances the point at which the dwell period begins with increasing engine speed. Consequently, the duration for the ignition coil's primary current flow is extended.
4. **Driver stage:** The purpose of the driver's stage is to enlarge the dwell angle control current pulse, so that it can act as an input signal to the switching output stage.
5. **Switching output stage:** The function of this stage is to regulate the process of energizing and interruption of the ignition coil's primary winding current supply.

12.10.2 Capacitive-discharge Ignition (CDI) System

Figure 12.13 shows a schematic diagram of a capacitive-discharge ignition system. It is used to create a high-tension voltage spark across the spark-plug electrodes. When ignition is timed to occur, a thyristor power switch conducts and completes the capacitor-primary winding circuit of the ignition coil. Instantly, the capacitor (C) will discharge through the primary winding (P). The sudden flow of current in the primary winding induces a very high voltage in the secondary winding (S) and since the spark-plug forms a part of the secondary winding circuit, this voltage pulse will be dissipated at the plug air gap in the form of a spark, which is about 0.1 ms. In this system, the spark is strong but short due to fast capacitive discharge. This may lead to ignition failure when the mixture is very lean.

The capacitive-discharge ignition trigger box contains the capacitor, the thyristor power switch, the charging device, the pulse-shaping unit and the control unit.

(i) **Pulse-shaping circuit:** It is already described under the transistorized-coil ignition (TCI) system.

(ii) **Charging device:** It increases the amount of energy stored in the capacitor. It acts as a charging transformer. It steps up the 12 V battery potential to around 400 V and a diode (D) is included in the circuit to prevent the electric charge from flowing back into the charging device while the capacitor (C) remains charged.

(iii) **Control unit:** The control unit provides a small current pulse timed to open and close periodically the ignition coil primary winding circuit by means of the thyristor power switch.

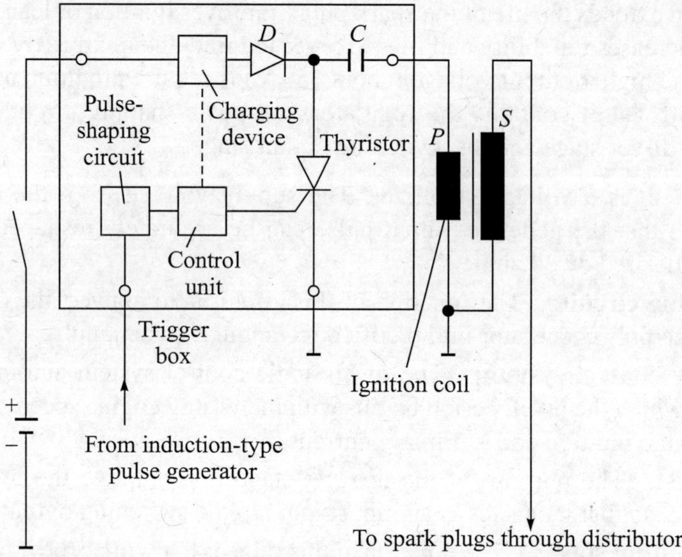

Figure 12.13 Circuit diagram of a capacitive-discharge ignition system.

12.11 FACTORS AFFECTING SPARK-ADVANCE

Some of the important factors affecting spark-advance are:

1. ***Mixture strength:*** For weaker mixtures the rate of flame propagation is lower, so the complete combustion period will be greater. In order to obtain the maximum power from weak mixtures the spark should therefore be advanced.
2. ***Compression ratio:*** With the increase in compression ratio the charge density will be increased, which will increase the rate of flame propagation. The spark-advance must therefore be reduced as the compression ratio is increased.
3. ***Engine speed:*** As the engine speed increases the combustion duration in terms of degrees crank-angle also increases. Therefore, the angle of spark-advance must increase as the speed increases.
4. ***Turbulence:*** As the degree of turbulence is increased, more quantity of mixture passes the ignition point in a certain time, so the effect is similar to that of increased flame speed. Increasing the turbulence should therefore require a retardation of the ignition timing.
5. ***Engine load:*** At light loads a partial vacuum is developed in the intake manifold, resulting in less quantity of mixture during the compression stroke. There is a large gain in efficiency at light loads by advancing the ignition timing.
6. ***Type of fuel:*** A fuel having lower flame speed will need a greater spark-advance for maximum power and economy.

The spark must therefore be automatically regulated to ensure maximum power and economy at different speeds and loads. In order to achieve these the engines are fitted with the spark-advance mechanisms.

12.12 SPARK-ADVANCE MECHANISMS

There are two general types of spark-advance mechanisms employed in spark-ignition engines to advance and retard the ignition timing automatically in relation to engine speed and operating conditions after the initial timing is set. They are the centrifugal advance system and the vacuum advance system.

12.12.1 Centrifugal-advance Mechanism

When the engine is idling, the spark is timed to occur just before the piston reaches the top dead centre on the compression stroke. At higher speeds, it is necessary to deliver the spark to the combustion chamber somewhat earlier. This gives ample time for mixture to burn and deliver its power to the piston. To provide this advance, a centrifugal-advance mechanism is used (Figure. 12.14). It consists of an advance cam integral with the ignition cam, a pair of governors or advance weights, two springs, and a plate attached to the distributor shaft. All are located beneath the ignition cam and breaker plate.

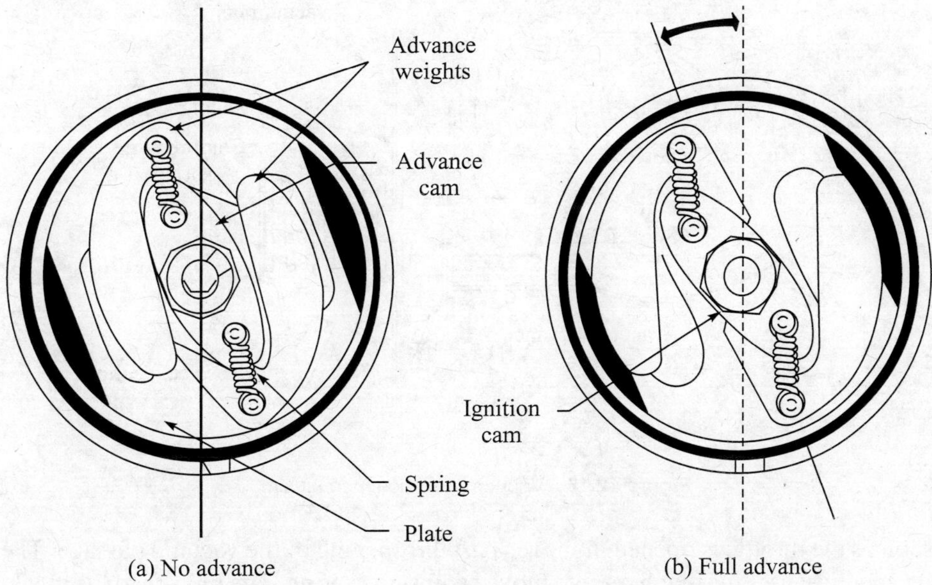

Figure 12.14 Centrifugal-advance mechanism.

At idling speed the parts are arranged as shown in Figure 12.14(a). As the engine speed increases the advance weights are thrown out against spring tension. They are pivoted on their pins. As they swing out they push the advance cam so that it rotates the ignition (breaker) cam ahead of the distributor shaft. This advance (Figure 12.14(b)) causes the cam to open and close the contact points earlier in the compression stroke at high speeds, since the rotor is also advanced, it comes into position earlier in the cycle. The timing of the spark to the cylinder thus varies from no advance at idling speed to full advance at high speed. At this point the weights reach the outer limits of their travel. In some designs the ignition timing is advanced as much as 28 degrees of crank angle by this method alone.

12.12.2 Vacuum-advance Mechanism

Under part load, a partial vacuum is developed in the intake manifold. It causes less air-fuel mixture to be admitted into the engine cylinder, resulting in low volumetric efficiency. Thus, during the compression stroke less mixture will be present. This mixture will burn more slowly when ignited. In order to obtain full power from it, the spark should be advanced. To obtain this spark advance, a vacuum-advance mechanism is used.

A vacuum-advance mechanism is shown in Figure 12.15. It contains a spring-loaded airtight diaphragm. The diaphragm is connected by a linkage to the breaker plate. The breaker plate is supported on a bearing so that it can turn a few degrees with respect to the distributor housing. The spring-loaded side of the diaphragm is connected though a vacuum line to a point just above the throttle valve in the carburettor, so that this opening is on the atmospheric side of the throttle valve when the throttle is in idling position. There is no vacuum advance in this position.

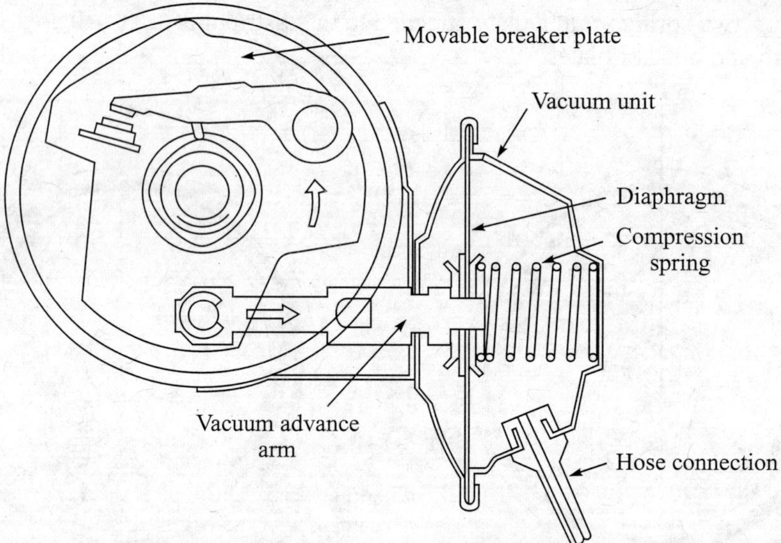

Figure 12.15 Vacuum-advance mechanism.

As soon as the throttle is opened, it moves past the opening of the vacuum passage. The intake manifold vacuum causes the diaphragm to move against the spring. The linkage to the breaker plate then rotates the breaker plate. The entire breaker-plate assembly rotates in a direction opposite to the direction of rotation of the ignition cam. This action causes the contact points to move around and enables the cam to open the point contacts sooner, causing the spark to be advanced. The spark then appears at the sparking plug earlier in the compression stroke. As the throttle is opened wider, there is less vacuum in the intake manifold and this results in less vacuum advance. At wide open throttle, there is no vacuum advance at all.

The combination of centrifugal advance and vacuum advance mechanisms gives the engine a practically perfect spark timing for all driving conditions. Some electronic ignition systems do not use mechanical devices to produce spark advance. Instead, the spark advance is produced electronically.

REVIEW QUESTIONS

1. What is the function of an ignition system? What are the different types of ignition systems?
2. What are the requirements of an ignition system of a spark-ignition engine?
3. What are the components of a battery-ignition system? Show the circuit diagram of this ignition system indicating the primary and secondary circuits.
4. Give a brief description of the battery and the ignition switch of a battery-ignition system.
5. What is the function of a ballast resistor? Where is it placed in the ignition system?
6. What is the function of an ignition coil? Describe it with the help of a diagram.
7. What are the differences between metal clad and core type ignition coils?
8. Describe the function and construction of contact breaker points. Explain the operation of this device with the help of a diagram.
9. Describe the function and construction of a condenser in an ignition system.
10. Describe the construction and function of a distributor in an ignition system with the help of a simple diagram.
11. Briefly explain the working of a battery-ignition system with the help of a circuit diagram.
12. What is the meaning of a firing order of a multi-cylinder engine? Why is a proper firing order important? What should be the recommended firing order for three, four and six cylinder engines?
13. Explain the meaning of dwell angle with the help of a diagram. What is the importance of a correct dwell angle? What are the factors that affect the dwell angle?
14. Describe the construction of a spark plug with the help of a diagram.
15. What are the requirements of a good spark plug?
16. Enumerate the factors that affect the establishment of a spark.
17. What are the disadvantages of running a too cold or a too hot spark plug? What are the factors on which the operating temperature of the plug depends? Show the relative paths of heat travel in hot and cold plugs. Where will you prefer to locate a hot or a cold spark plug?
18. Describe a magneto-ignition system with the help of a diagram.
19. What are the different types of magneto-ignition systems? How do they differ from each other?
20. Show and describe the relationship of break current with the number of sparks per minute for battery- and H.T. magneto-ignition systems.
21. Give a comparison of battery- and magneto-ignition systems.
22. What are the major problems associated with the conventional ignition system?
23. Describe the working of a transistorized-coil ignition system with the help of a circuit diagram. Show also the top view of an electronic distributor with cap and rotor removed. What are the advantages of this system?
24. Describe the capacitive-discharge ignition system with the help of a circuit diagram.
25. Describe the factors that affect spark-advance.
26. Describe the construction and working of a centrifugal-advance mechanism of an ignition system.
27. Describe the construction and working of a vacuum-advance mechanism of an ignition system.

13.1 INTRODUCTION

Friction power (fp) can be defined as the difference between the indicated power (ip) and the brake power (bp), i.e.

$$\text{fp} = \text{ip} - \text{bp} \tag{13.1}$$

The indicated power is the power delivered to the piston by the cylinder gases, and the brake power is the power measured as output at the crankshaft. All the indicated power transferred to the piston from the gases contained inside the cylinder is not available as brake power at the drive shaft. That portion of the power which is not available is usually termed friction power. The friction power is a sufficiently large fraction of the indicated power. It is about 10% at full-load and 100% at idle or no-load. Friction is an important factor taken into account while determining engine performance and efficiency. It directly affects the maximum brake torque and the minimum brake specific fuel consumption. A large part of the friction losses appears as heat in the coolant and lubricating oil which must be removed in the radiator and in the oil cooler system. Thus, the friction losses influence the size of the coolant system.

13.2 COMPONENTS OF ENGINE FRICTION

Engine friction losses in a standard engine can be divided into three main types: (a) rubbing losses, (b) pumping losses, and (c) auxiliary component losses.

13.2.1 Rubbing Losses

Rubbing losses are defined as those which result from relative motion between solid surfaces in the engine. They include friction between the piston rings, piston skirt, and the cylinder wall; friction in the wrist pin, big end, crankshaft, and camshaft bearings; friction in the valve mechanism; friction in the gears, pulley or belts, which drive the camshaft and engine accessories.

13.2.2 Pumping Losses

Pumping losses are defined as those which are associated with transporting fluids through the cylinder and they are made up of intake and exhaust pumping. Intake pumping means that fresh mixture is drawn through the intake system and into the cylinder, and the exhaust pumping means that the burned gases are expelled from the cylinder and out of the exhaust system. The pumping work is divided into two parts: one part is the throttling work, it includes the effect of restrictions outside the cylinder in the inlet and exhaust systems, i.e. air filters, carburettor, throttle valve, intake manifold, exhaust manifold and tail pipe, catalytic converter and muffler. The other part is the valve flow work. It corresponds mainly to pressure losses in the inlet and exhaust valves. As the load is reduced in an SI engine, the throttle restriction is increased, which increases the throttling work and decreases the valve flow work. The increase in throttling work is much more rapid than the decrease in valve flow work. Both throttling work and valve flow work increase as speed increases at constant load.

13.2.3 Auxiliary Component Losses

Auxiliary component losses include both rubbing and pumping losses due to driving of the engine accessories. These may include the fan, the water pump, the oil pump, the fuel pump, the generator, a power steering pump, and air-conditioner, etc.

As all the above losses are eventually dissipated as heat, the term friction work or friction power is therefore appropriate.

13.3 TOTAL FRICTION WORK

The total friction work, W_{tf}, is the sum of the rubbing friction work, W_{rf}, the pumping work, W_p, and the accessory work W_a.

$$W_{tf} = W_{rf} + W_p + W_a \tag{13.2}$$

Rubbing friction work is the work per cycle dissipated in overcoming the friction due to relative motion of adjacent components within the engine.

Pumping work is defined as the net work per cycle done by the piston on the gases during the inlet and exhaust strokes. It is represented by the area B in Figure 13.1. The area A in the diagram represents the gross indicated work. The pumping work is the negative work, i.e. the work done by the piston on the gas, and the gross indicated work is a positive work, i.e. the work done by the gases on the piston. Therefore, the net work per cycle is the difference between gross work and pumping work, i.e. area A – area B.

Accessory work is the work per cycle required to drive the engine accessories, e.g. pumps, fans, generator, etc.

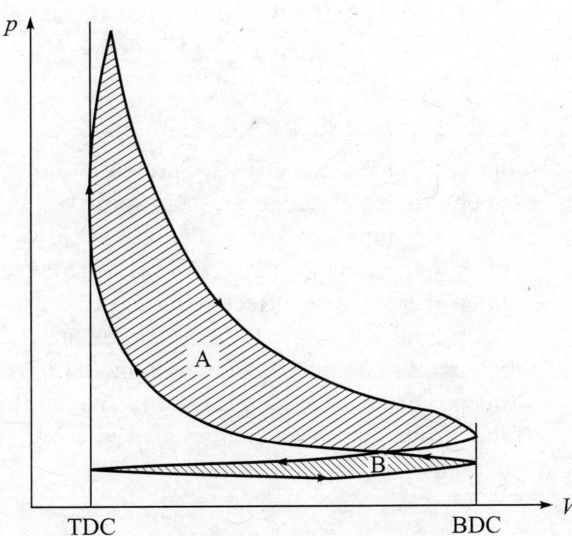

Figure 13.1 Pressure–volume (p–V) diagram of a four-stroke engine indicating pumping work (area B) and gross work (area A).

13.4 SOME MORE COMPONENTS OF ENGINE FRICTION

In two-stroke engines, a scavenging pump is used for the scavenging process. In such cases the power taken from the crankshaft to drive a scavenging pump is included in the friction power. In supercharged engines the power taken from the crankshaft to drive a supercharger is also included in the friction power. In some engines, an exhaust turbine is geared to the crankshaft. In such cases the power developed by the turbine will add to the brake power of the engine and could be classed as negative friction loss.

13.5 FRICTION MEAN EFFECTIVE PRESSURE

Engines are made up of different sizes and they run at different speeds, therefore, the most meaningful method of classifying and comparing frictional losses is in terms of mean effective pressure. The mean effective pressure (mep) can be related to work per cycle or power term as follows:

$$\text{Work per cycle, } W = (\text{mep})\, V_s \tag{13.3}$$

$$\text{Power, } \dot{W} = (\text{mep})\, V_s \left(\frac{N}{n}\right) \tag{13.4}$$

where

V_s = swept volume

N = engine speed

n = number of revolutions per cycle.

The friction mean effective pressure can be related to other mean effective pressures as follows:

$$\text{fmep} = \text{imep} - \text{bmep} - \text{amep} - \text{cmep} + \text{tmep} \tag{13.5}$$

and

$$\text{imep} = \text{gmep} - \text{pmep} \tag{13.6}$$

where

amep = auxiliaries mean effective pressure

bmep = brake mean effective pressure

cmep = mean effective pressure to drive the compressor as a supercharger or scavenging pump

fmep = friction mean effective pressure corresponding to rubbing losses only

gmep = gross mean effective pressure

imep = indicated mean effective pressure

pmep = pumping mean effective pressure

tmep = mean effective pressure recovered from the exhaust gas in a turbocharger.

If the turbine of a turbocharger is used to run the compressor only and the auxiliary work is ignored, then

$$\text{cmep} = \text{tmep} \qquad \text{and} \qquad \text{amep} = 0$$

$$\therefore \quad \text{fmep} = \text{imep} - \text{bmep} \tag{13.7}$$

The friction mean effective pressure can be related to engine speed by the following empirical relation.

$$\text{fmep} = A + BN + CN^2 \tag{13.8}$$

where

N = engine speed

A, B, C = empirical constants related to a specific engine.

The first term on the right-hand side of Eq. (13.8), constant A, accounts for boundary friction, i.e. metal-to-metal contact which occurs between the piston rings and the cylinder walls particularly at TDC and BDC, and in heavy-loaded bearings of the crankshaft.

The second term of Eq. (13.8), BN, is proportional to engine speed and accounts for the hydraulic shear that occurs between many lubricated components. The shear stress τ_s is given by

$$\tau_s = \mu \frac{du}{dy} \tag{13.9}$$

where

μ = dynamic viscosity of lubricating oil

$\frac{du}{dy}$ = velocity gradient between surfaces.

For a given viscosity and geometry, the velocity term du is proportional to the engine speed N.

The third term of Eq. (13.8), CN^2, is proportional to the square of the engine speed. This term accounts for the losses from turbulent dissipation in the intake and exhaust flows. Constants A, B and C depend upon the operating conditions of a given engine.

13.6 MECHANICAL FRICTION

All rubbing losses are the result of mechanical friction. It is the sum of resistance to motion of all the engine parts. The friction associated with the engine rubbing surfaces may be divided into the following classes:

1. Hydrodynamic or fluid-film friction
2. Partial-film friction
3. Rolling friction
4. Dry friction

13.6.1 Hydrodynamic or Fluid-film Friction

The hydrodynamic or fluid-film friction is associated with surfaces entirely separated by a film of lubricant. In this case the friction force entirely depends on the lubricant viscosity, which is a measure of the resistance to shear possessed by the oil film. This type of friction is the main component of the mechanical friction losses in an engine.

13.6.2 Partial-film Friction

Partial-film friction is associated with surfaces partially separated by a film of lubricant. In this case, some parts of the rubbing surfaces are lubricated and some parts of the surfaces are in contact.

During starting the engine bearing surfaces operate under this condition. However, in normal engine operation there is a very little metallic contact except between the piston rings and cylinder walls for a brief moment at the end of each stroke when the piston velocity is nearly zero. Thus partial-film friction is of little importance and contributes very little to engine friction.

13.6.3 Rolling Friction

The rolling friction is due to the rolling motion between the two surfaces. It is associated with ball-and-roller bearings and with cam-followers and tappet rollers. These bearings have a coefficient of friction which is nearly independent of load and speed. Rolling friction is negligible in comparison to total friction.

13.6.4 Dry Friction

It is not important in engines because some lubricant nearly always remains between the rubbing surfaces, even when an engine is not used for a long period of time. Dry friction can therefore be safely neglected while considering engine friction.

13.7 MECHANICAL FRICTION IN MAJOR ENGINE COMPONENTS

Mechanical friction in some important components of the engine is described in this section.

13.7.1 Piston Assembly Friction

A typical piston-and-ring assembly is shown in Figure 13.2.

The piston assembly is the dominant source of engine rubbing friction. The components that contribute to friction are: compression rings designed to seal against gas pressure, oil control rings

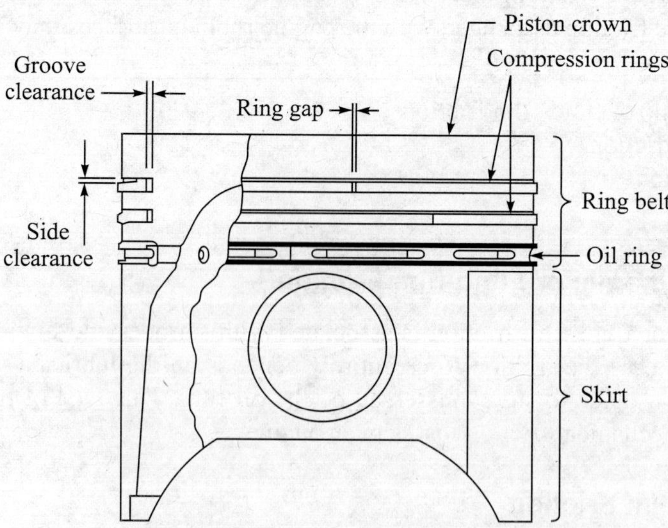

Figure 13.2 Piston-and-ring assembly.

designed to limit the passage of oil from the crankcase to the combustion space, piston skirt and piston pin. The forces acting on the piston assembly include the static ring tension which depends on ring design and materials, the gas pressure forces leaking into the grooves between the rings and piston which depend on engine load, and the inertia forces which are related to component mass and engine speed. The major design factors influencing piston assembly friction are ring width and its face profile, ring tension, ring gap which governs inter-ring gas pressure, liner temperature, ring clearance, skirt geometry and skirt-bore clearance.

The piston skirt is a load-bearing surface. It carries the side thrust when the connecting rod is at an angle to the cylinder axis, so it contributes to piston assembly friction. The side thrust is transmitted to the cylinder liner via the rings and piston skirt. It changes direction as the piston passes through TDC to BDC positions. Since the friction force changes sign at these locations and the gas pressure during expansion is greater than that during compression, the side thrust during expansion is greater. The gas force on the piston head is balanced by the load on the connecting rod. Since the connecting rod is usually at an angle to the cylinder axis, a resultant side thrust is produced as shown in Figure 13.3. As the cylinder pressures are increased, the side thrust will become greater and increase the friction between the piston and cylinder wall. An increase in side thrust may also increase the viscous friction of the piston because of the reduction in the thickness of the oil film on the thrust side of the piston.

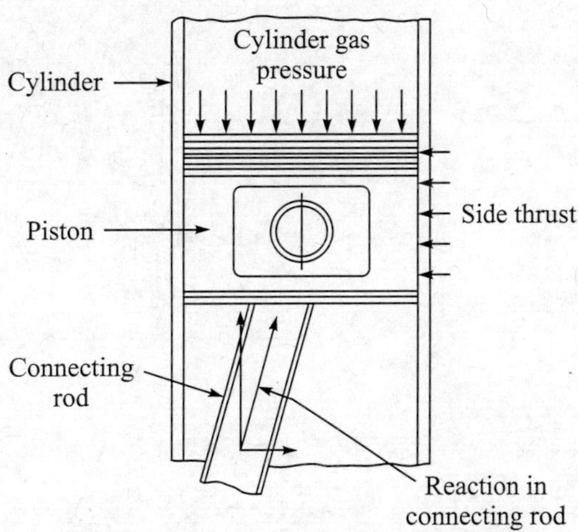

Figure 13.3 Side thrust due to cylinder gas pressure.

The friction forces are highest just before and after the top dead centre at the end of the compression stroke. The friction force is caused by the piston impulse, high side thrust and combustion gas pressure loading on the rings. Figure 13.4 shows the gas pressure behind the top piston rings.

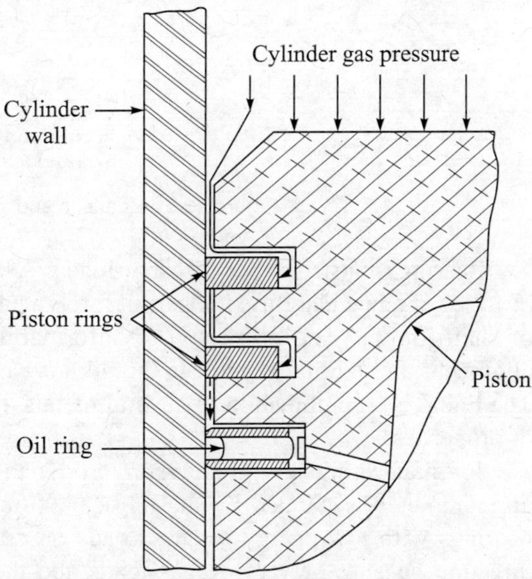

Figure 13.4 Cylinder gas pressure behind the top piston rings.

13.7.2 Bearing Friction

Bearings are used wherever there is a rotary motion between the engine parts. These bearings fit around the rotating shaft and are called sleeve bearings. The part of the shaft that rotates in the bearings is called a *journal*. The connecting rod and the crankshaft bearings are the main journal bearings and they are of the split type as shown in Figure. 13.5. The typical bearing-half is made of steel or bronze with linings of bearing material. The bearing material is soft, so that the bearing wears and not the expensive engine parts.

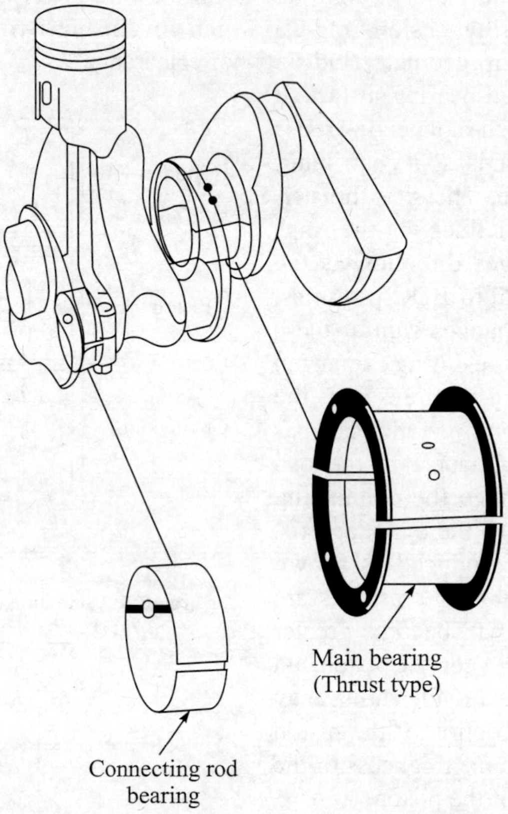

Figure 13.5 Bearings and their positions on the crankshaft.

Bearings must be able to carry loads. A bearing should resist fatigue. It should have the property of embeddability (i.e. permit foreign particles to embed in it in order to prevent scratching of shaft journal) and property of conformability (i.e. ability to conform to variation in shaft alignment and journal shape). It should not wear too fast. The bearing is usually made of steel. The lining is a combination of several metals, mixed or alloyed to provide the right combination of properties.

Journal bearings operate under hydrodynamic lubrication. The two surfaces in relative motion are completely separated by the lubricant film, therefore large loads can be carried by journal bearings with low energy losses. Loads on crankshaft journal bearings vary in magnitude and direction because of varying gas loads and the piston connecting rod mechanism. It results in eccentricity of the journal in the bearing. The minimum oil film thickness is determined from the journal eccentricity. Journal bearings are usually designed to provide minimum film thickness of about 2 μm.

13.7.3 Valve Train Friction

The critical contact regions in valve train, where friction has to be considered, are camshaft journal bearings, rocker arm fulcrum, and cam tappet interface. The valve train carries high loads over the entire speed range of the engine.

The loads acting on the valve train at lower speeds are mainly because of the spring forces and at higher speeds they are mainly because of the inertia forces of the component masses.

The valve train friction can be reduced by reducing the spring load and valve mass, by using the tappet roller cam followers, and by using the rocker arm fulcrum needle bearings. A low friction valve train design is shown in Figure 13.6.

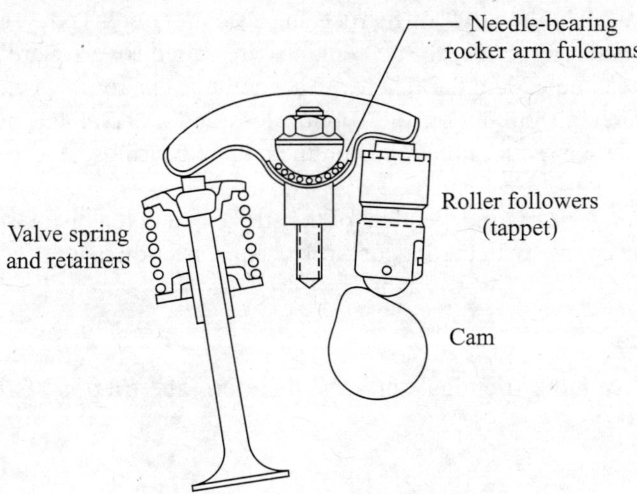

Figure 13.6 Low friction valve train.

13.8 BLOWBY LOSSES

It is the phenomenon that describes the escape of unburned air-fuel mixture and burned gases from the combustion chamber, past the piston and piston rings into the crankcase. It is not practical to fit the piston and piston rings to the cylinder closely enough to prevent blowby. Blowby losses increase with the increase in compression ratio and inlet pressure. As the engine speed is increased, the blowby losses are reduced. In case of worn-out piston rings, the blowby loss increases. The blowby loss should also be accounted in the overall frictional losses. A greater blowby loss reduces the engine power because some of the working mixture is lost, the temperature of the rings increases and consequently the rings lose their elasticity. The other detrimental effects are slogging of the rings, their sticking, and accelerated ageing of the oil.

13.9 EFFECT OF ENGINE VARIABLES ON FRICTION

The following are the important engine parameters that affect engine friction:

1. Stroke/bore ratio
2. Engine size

3. Piston size
4. Compression ratio
5. Gas pressure
6. Piston rings
7. Engine speed
8. Engine load
9. Oil viscosity
10. Cooling water temperature

Effect of stroke/bore ratio

Earlier, most engines were built with a long stroke and a smaller bore. Today, engines are designed with a shorter stroke and a larger bore. Such engines are called 'oversquare'. A 'square' engine has a bore and a stroke of equal lengths. There are several reasons for the swing to the oversquare engine. With the shorter piston stroke, the friction loss and wear of the piston rings are both reduced. The shorter stroke reduces the loads on the engine bearings. It also permits a reduction in engine height.

Mechanical friction is nearly independent of the stroke at a given piston speed but at the same rpm, it is larger for the engine with the longer stroke. Thus, friction is less for the small stroke-to-bore ratio.

Effect of engine size

The larger engines have more frictional surfaces, therefore, the friction losses are more in such engines.

Effect of piston size

A short piston with the non-thrust surfaces cut-away has reduced size and weight. It minimizes inertia loads and side thrust on the cylinder walls. Therefore, with a short piston the frictional losses are reduced.

Effect of compression ratio

At higher compression ratios the friction is higher because of the higher peak cylinder pressures. However, the mechanical efficiency may improve slightly because of the increased imep.

Effect of gas pressure

The gas pressure in the piston ring grooves considerably increases the force with which the ring is pressed against the liner. This squeezes out the oil and increases the work of rubbing friction. When the pressure behind the ring is high, the boundary friction appears between the ring and the liner. This is accompanied by excessive wear, primarily of the upper rings and the upper belt of the liner.

Effect of piston rings

The friction force of compression rings depends on the mean pressure of the gases in the grooves and on the width of the rings. The rubbing friction losses can be decreased by using fewer compression rings or by reducing their width.

Effect of engine speed

As the engine speed increases, the friction increases rapidly. It is therefore important to avoid high engine speeds in order to get good mechanical efficiency. A reduction in the total friction mean effective pressure reduces the fuel consumption. The mechanical efficiency at high speed can be improved by increasing the number of cylinders.

Effect of engine load

As the engine load increases the peak cylinder pressure also increases. Because of gas pressure behind the compression rings, the rubbing friction increases. The peak temperature also increases with the increase in engine load. It decreases the oil viscosity which reduces friction. Further, in the case of SI engines the throttling losses reduce as the throttle is opened more at higher loads. The total effect is to reduce frictional losses in SI engines. However, the frictional losses in CI engines are more or less independent of engine load because of absence of throttling losses.

Effect of oil viscosity

As the viscosity of the oil increases the friction loss also increases. With the increase in oil temperature, the viscosity decreases and therefore the friction loss reduces. Beyond a certain temperature, the local oil film is destroyed resulting in metal-to-metal contact that causes increased frictional loss.

Effect of cooling water temperature

As the cooling water temperature increases the viscosity of oil decreases. It results in reduced frictional loss. However, the frictional losses are higher during starting because of lower temperature of cooling water and lubricating oil.

13.10 SIDE THRUST ON THE PISTON

The forces acting on a piston are shown in Figure 13.7. The *X*-axis is taken along the centre-line of the cylinder and the *Y*-axis is taken in the radial direction outwards, with the zero at the centre-line. Any force in the *X*-direction is taken as positive in the direction of piston motion during the power stroke and is taken as negative in the reverse direction. A force balance in the *X*-direction results in the relation,

$$\sum F_X = m\frac{du_\mathrm{p}}{dt} = -F_\mathrm{r}\cos\phi + p\frac{\pi}{4}d^2 \pm F_\mathrm{f} \tag{13.10}$$

where

F_X = force in the *X*-direction
m = mass of the piston
u_p = instantaneous piston speed
F_r = force of the connecting rod
ϕ = angle which the connecting rod makes with the centre-line of the cylinder.
p = gas pressure in the combustion chamber
d = cylinder bore
F_f = friction force between the piston and cylinder walls.

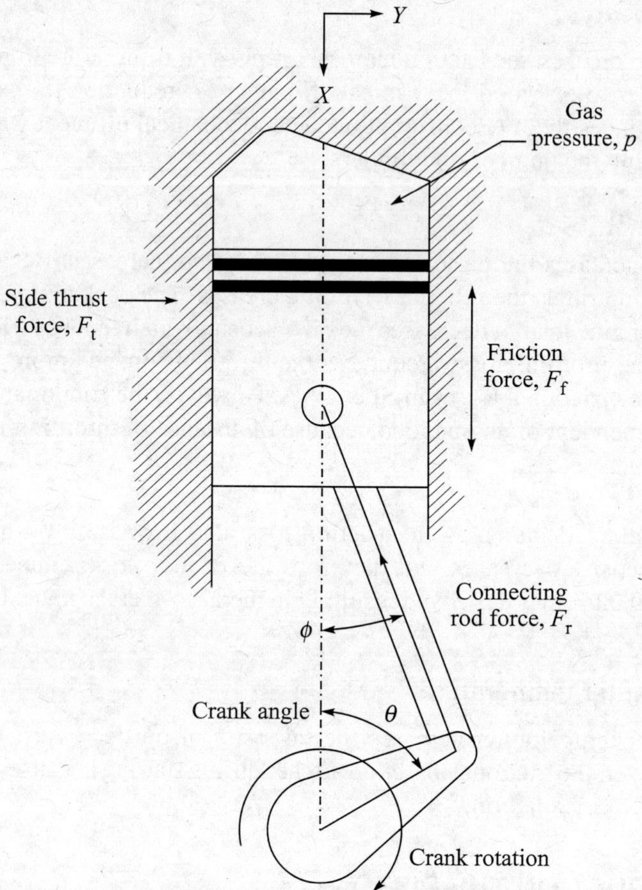

Figure 13.7 Forces acting on a piston.

The sign of F_f in Eq. (13.10) depends on the crank angle θ. When the piston is moving downwards, i.e. $0 < \theta < 180$, the friction force will be acting upwards and for this the negative sign is taken with F_f, and when the piston is moving upwards, i.e. $180 < \theta < 360$, the friction force will be acting downwards and for this the positive sign is taken with F_f.

There is no motion in the Y-direction, so a force balance in the Y-direction gives

$$\Sigma F_Y = 0 = F_r \sin \phi - F_t \tag{13.11}$$

where F_t is the side thrust on the piston. It is the Y-direction reaction to the force in the connecting rod and lies in the plane of the connecting rod.

Combining Eqs. (13.10) and (13.11), the side thrust on the piston can be written as

$$F_t = \left[-m \frac{du_p}{dt} + p \frac{\pi}{4} d^2 \pm F_f \right] \tan \phi \tag{13.12}$$

The side thrust F_t changes with the piston position ϕ, the acceleration of the piston $\frac{du_p}{dt}$, the pressure p, and the friction force F_f. They all vary during the engine cycle. During the power stroke the cylinder pressure is high, the side thrust will be on the left side in the plane of the connecting rod for an engine rotating as shown in Figure 13.7. This is called the major thrust side of the cylinder. During the compression stroke the cylinder pressure is not so high, the side thrust in the plane of the connecting rod will be on the other side of the cylinder (right side in Figure 13.7). This is called the minor thrust side. The side thrusts on the piston are less in planes turned circumferentially away from the plane of the connecting rod, reaching a minimum in the plane at right angles to the plane of the connecting rod.

The side thrust thus varies both along the length of the cylinder from TDC to BDC and also in the circumferential direction. Variation in side thrust causes variation in wear that occurs in the cylinder walls. The greatest wear occurs in the plane of the connecting rod on the major thrust side of the cylinder. Significant, but less, wear occurs in the plane of the connecting rod on the minor thrust side of the cylinder. Wear also varies along the length of the cylinder.

Side thrust can be reduced by reducing the piston mass and the piston skirt length. Less mass reduces the piston inertia and reduces the acceleration term in Eq. (13.12). Shorter piston skirts reduce rubbing friction because of the reduced surface contact area. A further reduction in side thrust can be achieved with the use of an offset wrist-pin. The wrist-pin is offset from the centre by 1 or 2 mm towards the minor thrust side of the piston without changing its vertical location. Reduction in side thrust reduces wear on the major thrust side.

Reduction in friction may be brought about by reducing the stroke length, but for a given swept volume the cylinder bore has to be increased. This increases heat loss due to a larger cylinder surface area. It also promotes knocking because of the greater flame travel distance. This is why most medium-sized automobile engines are close to a square ($L = d$) shape.

13.11 LUBRICATION

The reciprocating internal combustion engine has a large number of moving parts. Without an adequate film of lubricating oil between the surfaces of the reciprocating, oscillating and rotating metal parts, the force required to overcome the frictional resistance and the wear on the parts would be both very high. When relative motion takes place between the mechanical surfaces in contact, work is done against the frictional force and heat is produced at the surfaces, which may heat the parts to the point where melting or seizure takes place. In IC engines, during the combustion process, high temperature is experienced and during the cycle the temperature varies widely, moreover the bearing loads also fluctuate. All these make the lubrication problem more difficult. Inadequate or improper lubrication of the engine may cause serious engine troubles, such as scored cylinders, stuck piston rings, damaged bearings, engine deposits and sludge, and dirty spark plugs.

13.12 FUNCTIONS OF A LUBRICANT

Following are the main functions of a lubricant:

Reducing frictional resistance

A lubricant decreases friction by preventing direct contact of two rubbing surfaces by supplying some fluid or semi-fluid substances between them. It reduces friction power by reducing wear between the rubbing and bearing surfaces, thereby increasing the power output and the engine service life.

Cooling

A lubricant also acts as a coolant, carrying heat away from the bearings, cylinders and pistons.

Sealing

The lubricating film on the cylinder wall acts as a seal to prevent the gases of combustion from being blown by the piston rings and entering the crankcase.

Cleaning

A lubricant cleans carbon and other deposits from the surfaces of the piston, piston rings and bearings. It also eliminates dust and other contaminants.

Reduction in noise

A lubricant cushions the parts against vibration and impact, thus reducing the engine noise.

Protection against corrosion and wear

A lubricant protects the metallic surfaces against the corrosive action of combustion products such as water, SO_2, etc.

In order to fulfil the above functional requirements, the lubricant should be thin for cooling but thick for sealing. It should also be clean for proper lubrication, and for cleaning, it should be able to wash out deposits. So, a lubricant has to perform jobs which have contradicting requirements. Therefore, the selection of a proper lubricant is very important.

13.13 LUBRICATION PRINCIPLES

If one surface is moving and inclined to the other, the viscous drag of the oil tends to draw the lubricant into the space between the surfaces and builds up a wedge. This develops an oil-film pressure that can support a load. If the two surfaces were parallel or if they did not have relative motion, the oil-film pressure would not get developed and a load would not be supported by the lubricant.

There are three lubrication regions that are important for engine components. They are: hydrodynamic lubrication, mixed-film lubrication and boundary lubrication.

13.13.1 Hydrodynamic Lubrication (Full Film or Thick Film)

If the sliding surfaces are completely separated by a film of oil, there is no metal-to-metal contact and the wear on the surfaces is a minimum. This is termed full-film or hydrodynamic lubrication.

This type of lubrication is developed when there is a relative motion between the two inclined surfaces separated by an oil film as shown in Figure 13.8. A wedge-shaped film between the

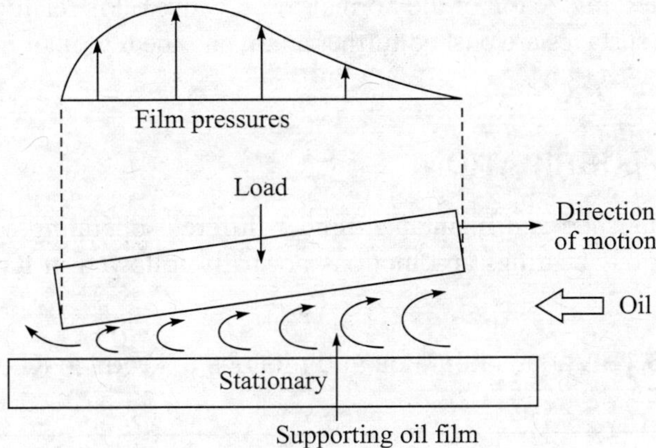

Figure 13.8 Oil pressure in wedge section (slider bearing).

surfaces is developed which is thicker at the leading edge than at the trailing edge. The oil contained into the separating space is thus squeezed into a smaller and smaller section, and therefore tries to escape. Because of the viscosity of the oil, escape is hindered, and a pressure is created. From the point of maximum pressure, oil tends to flow in all directions. This action is opposed by the viscous drag between the particles of the lubricating oil and permits additional oil to enter the separation. A hydrostatic pressure in the main body of the lubricant is thus developed which supports heavy load. This pressure can be controlled by the relative velocity of the surfaces, the viscosity of the lubricant, the normal force or load and by geometric configuration.

13.13.2 Boundary Lubrication (Thin Film)

Boundary lubrication is defined as the state in which two solid surfaces are virtually in contact and separated only by a film of lubricant insufficient to prevent occasional contact of the surfaces. In this case the depressions in the metal are filled with the lubricant and all the ridges are covered with at least a monolayer of oil molecules.

Boundary friction is associated with conditions of inadequate lubrication or excessive loading, whereby the lubricant is squeezed out from between the surfaces and the film thickness gets decreased to the extent where surfaces come in contact with each other, leading to wear and possible seizure.

Usually, the friction coefficients under the boundary lubrication conditions are from ten to twenty-five times greater than those for hydrodynamic lubrication. The coefficient of friction in the boundary lubrication region is independent of speed. Boundary lubrication can occur when the piston and piston rings are at the beginning and end of the stroke, between the piston pin and bushing, between gear teeth, and in many other locations.

13.13.3 Mixed-film Lubrication (Partial Film)

This region is in between boundary lubrication and hydrodynamic lubrication. Suppose the boundary lubrication region exists in the beginning between the two surfaces. When the relative motion (or the geometry) of the surfaces traps oil between them, the viscosity of the oil retards its escape from the squeezing action of the normal force. Since oil is relatively incompressible, a pressure is created, the surfaces are pushed further apart, and the region of mixed-film lubrication is entered.

13.14 BEARINGS LUBRICATION

Oil films must be established and maintained under different operating conditions in bearings of IC engines. The various bearings and motions normally followed in IC engines are given in Table 13.1.

Table 13.1 Various bearings and motions followed in IC engines

Motion	*Surface*
(a) Sliding contact	(a)
1. Rotating	1. Journal bearings—crankpins, crankshafts, camshafts, valve mechanisms, etc.
2. Oscillating	2. Journal bearings—piston pins, knuckle pins, rocker arm, etc.
3. Reciprocating	3. Slipper bearings or reciprocating bearings—piston, piston rings, valve stem, etc.
(b) Meshing contact	(b) Gears—crankshaft primary gear, camshaft gear, gear teeth on the flywheel, gears on distributor shaft, etc.

13.14.1 Rotating Journal Bearings

Journal is a part of shaft or axle that rests on supports called bearings. A journal bearing with exaggerated clearance is shown in Figure 13.9. When the journal is at rest, its weight allows line contact of shaft and bearing at A (Figure 13.9(a)). Since there is no relative motion between the bearing and the journal, there is no film-pressure and the oil cannot support the load. This results in a metal-to-metal contact. When the shaft begins to turn in a clockwise direction, the slight

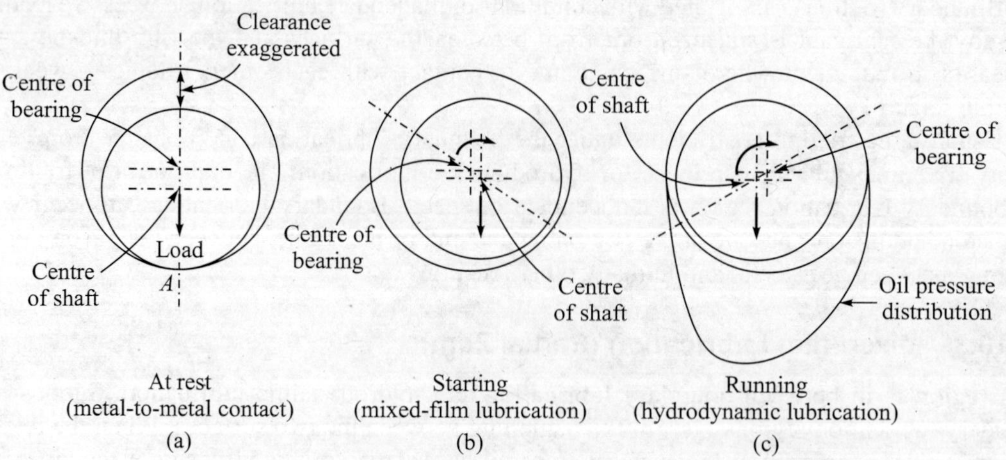

Figure 13.9 Action of lubricating oil in separating shaft and bearing.

movement enables the shaft to crawl upon an oil film to the right of the bearing (Figure 13.9 (b)). During crawling up, the lubricating oil adheres to the surfaces, the shaft operates in the region of thin film (boundary lubrication). With further rotation, oil is dragged into the wedged-shaped opening between the shaft and bearing and the pressure increases. Since the pressure on the right-hand side is greater than that of the left-hand side, the resultant force moves the journal away from the bearing wall, and boundary lubrication gives way to mixed-film lubrication (Figure 13.9 (b)). As the speed is increased to normal operating range, some of the lubricant adhering to the shaft is carried around with it. As the clearance under the shaft is smaller than that on the entering side, the oil being carried in cannot all be accommodated at this point. Viscous resistance to the return flow of this oil builds up a hydrodynamic pressure in the oil film which is great enough to float the shaft out of contact with the bearing and the journal centre moves to the left as shown in Figure 13.9(c). The journal bearing now operates in the region of full-film or hydrodynamic lubrication, thus minimizing the friction and wear.

Nearly all the wear on a journal bearing occurs during the starting of the engine as the bearing then operates in a thin-film region. It is therefore important to start the engine under no-load and at slow speed to keep the wear down. A fast start with a cold engine will increase the wear on the bearings. Wear during starting is also increased by abrasive particles present in the oil. Wear is decreased by eliminating these abrasive particles and by minimizing the duration of the thin-film operation at starting.

In the hydrodynamic lubrication region the friction force is produced by a shearing or sliding of the layers of oil film over each other. According to Newton's law of viscosity, shear stress is proportional to the velocity of one surface relative to the other and inversely proportional to the thickness of the oil film. Mathematically,

$$\tau \propto \frac{u}{h}$$

or

$$\frac{F}{A} \propto \frac{u}{h}$$

$\therefore$

$$F = \mu \frac{Au}{h} \tag{13.13}$$

where

τ = shear stress
u = velocity of one surface relative to the other
h = thickness of the oil film
F = friction force
A = area of the oil film being sheared
μ = viscosity of oil. It is the measure of the resistance to shear possessed by oil film.

Consider a cylindrical bearing as shown in Figure 13.10, having length L, shaft diameter D and oil film thickness h between the shaft and the bearing. Therefore,

$A = \pi DL$

$u = \pi DN$, where N is the shaft revolutions per unit time.

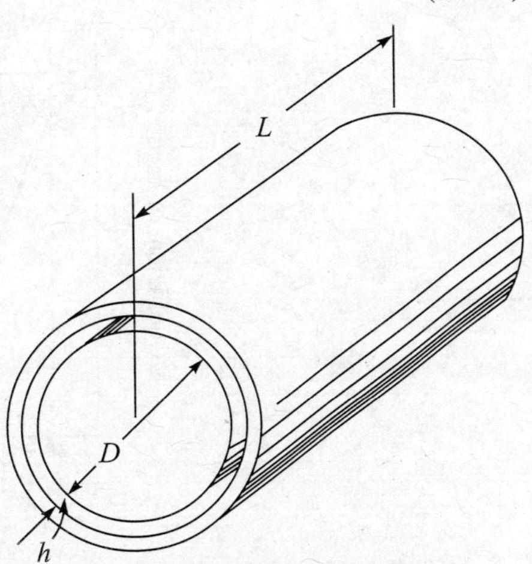

Figure 13.10 Cylindrical bearing.

$h = \frac{c}{2}$, where c is the diametral clearance between the bearing and the shaft.

The sum of the tangential friction forces acting on the shaft can be obtained from the relation (13.13), i.e.

$$F = \frac{\mu(\pi DL)(\pi DN)}{c/2} = \frac{2\pi^2 \mu D^2 LN}{c} \tag{13.14}$$

$$\text{Friction power loss} = Fu = \frac{2\pi^3 \mu D^3 L N^2}{c} \tag{13.15}$$

Let P be the loading per unit projected area of the bearing, then

$$P = \frac{W}{LD}$$

where W is the load or force acting normal to the surface.

$$\therefore \qquad W = PLD$$

$$\therefore \qquad \text{Coefficient of friction, } f = \frac{F}{W} = \frac{2\pi^2 \mu D^2 LN}{(c)(PLD)} = 2\pi^2 \left(\frac{\mu N}{P}\right)\frac{D}{c} \tag{13.16}$$

For a given bearing,

$$f \propto \frac{\mu N}{P} \tag{13.17}$$

$\frac{\mu N}{P}$ is a non-dimensional number on which the coefficient of friction of the bearing theoretically depends in the hydrodynamic lubrication region. Figure 13.11 shows the relation between the coefficient of friction f and the non-dimensional number $\frac{\mu N}{P}$ in hydrodynamic, mixed-film and boundary lubrication regions.

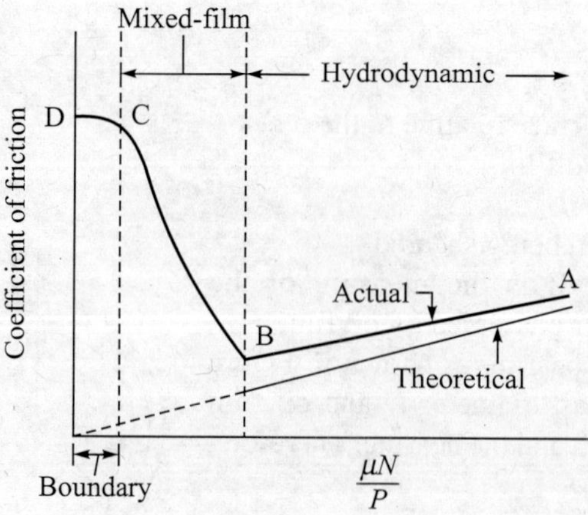

Figure 13.11 Relation between the coefficient of friction and $\frac{\mu N}{P}$ in different lubrication regions.

It is observed that as $\mu N/P$ decreases in the hydrodynamic lubrication region, the coefficient of friction also decreases; this decrease being in accordance with the theoretical results obtained from Eq. (13.17). It is represented by a linear relation AB. At a low value of $\mu N/P$ the hydrodynamic pressure in an actual bearing will not be sufficient to support the shaft load; the oil film becomes incomplete and enters into a mixed film region. In this region the coefficient of friction rises sharply as shown by curve BC. At point C the value of $\mu N/P$ is so low that the boundary lubrication region is reached. With boundary lubrication the coefficient of friction is high and relatively constant as shown by CD.

The point B represents the minimum film thickness required for normal operation without metal shearing. The conditions to the right of point B tend to be stable. Suppose, there is a temporary increase in oil temperature. With the increase in oil temperature, viscosity decreases and hence the value of $\mu N/P$ also reduces. It lowers the coefficient of friction which in turn reduces friction work and permits the bearing oil temperature to return to its former value. To the left of point B an increase in oil temperature increases the value of coefficient of friction, which results in a further increase in oil temperature and finally bearing failure may occur. It is therefore necessary to keep the value of $\mu N/P$ greater than that at point B for safety, but it should be kept in mind that this value should not be too high, which may result in excessive frictional loss.

The oil viscosity gets reduced with the increase in temperature. It is therefore necessary to replace the heated oil with fresh cool oil by an oil pump.

13.14.2 Oscillating Journal Bearings

An oscillating journal bearing as used with piston pins does not have tendency to form a wedge as in a rotating journal. In this case the load on the bearing reverses. The viscous oil between the approaching surfaces develops an oil-film pressure to support the load. Due to the excess pressure of the oil between the two surfaces, it squeezes out and flows to the unloaded side and gets ready to support the load when it is reversed.

13.14.3 Reciprocating Bearings

Piston, piston rings and valve stems operate with reciprocating motion. These parts are subjected to a very high flame temperature and to varying pressures throughout the cycle. It requires proper lubrication at extreme conditions. The lubricant on the cylinder walls must also act as a seal to prevent blowby losses. The piston rings and cylinders operate in a thin-film lubrication region for a great deal of time. A full film in these parts would result in excessive oil consumption, and high viscosity of lubricating oil required for sealing produces excessive friction.

In the thin-film region the surface finish and oilness of the lubricant play important roles in reducing wear. Tests indicate that a ground and lightly honed finish reduces the wear. With this type of finish the surfaces have flat spots with many small indentations scattered among them. These minute cavities serve as pockets for oil storage and help the oil to maintain a film on the surface. These cavities also provide better cooling of the lubricant as the surface area is largely increased.

13.14.4 Gear Teeth

The gears used in IC engines operate in the thin-film region. Gears are lubricated either by dipping them in the oil bath or by directing a jet of oil to the teeth as they disengage. The amount of oil must be regulated in the gear teeth. If too much oil is provided, it may be trapped between the teeth, creating an oil-film pressure which forces out the trapped oil and tends to push the gears apart. This results in vibration, high bearing loads and power loss.

13.15 PROPERTIES OF LUBRICANTS

Lubricants must possess certain basic properties to meet the lubrication requirement in IC engines. The important properties of lubricating oil are enumerated in the following subsections.

13.15.1 Viscosity

This is one of the most important properties of an oil and is used to grade lubricants. The viscosity of an oil is a measure of its fluid resistance to flow and is regarded as its internal fluid friction. It should be able to maintain an oil film between the bearing surfaces for desired load and speed to provide hydrodynamic lubrication conditions. In general, the more viscous the oil is, it can take up more bending load pressures and provide adequate sealing of the piston. On the other hand, if the oil is too viscous at low temperatures such as when the engine is cold, engine friction will be increased and cold starting will be difficult. Distribution of the high viscous oil will also be difficult. The viscosity of the oil at both low and normal oil temperatures—a spread of about 200°C—is therefore important.

The viscosity of oil is usually measured by a Saybolt viscosimeter. The viscosity is determined by measuring the time in seconds required for 60 ml of oil to flow by gravity through a capillary tube immersed in a constant temperature water bath. The temperature is usually 100°F (37.8°C), 130°F (54.4°C), or 210°F (98.9°C), depending upon whether the oil to be tested is light, medium, or heavy. The viscosity is expressed as Saybolt universal seconds (SUS).

13.15.2 Viscosity Index (VI)

It is advantageous to have an oil whose viscosity does not change much with the change in temperature, so that it may be used satisfactorily in cold starting, during warm-up and at normal operating temperatures. All oils do not become less viscous at the same rate as the temperature rises. The viscosity index is an empirical number which indicates the effect of temperature changes on viscosity.

To determine the viscosity index, the oil is cooled from 210°F (98.9°C) to 100°F (37.8°C), and the change in viscosity is measured. This change is compared with paraffin-base oil and naphthenic-base oil. The viscosity change of paraffin oil with the change in temperature is very small and arbitrarily the value of VI assigned to it is 100. The viscosity change of naphthenic oil with the change in temperature is large and the value of VI assigned to it is zero.

To increase the viscosity index, lubricating oils incorporate additives called viscosity-index improvers. These are high molecular weight compounds whose primary function is to reduce the viscosity variation with temperature. Certain oils containing additives and also some synthetic oils may have viscosity index greater than 100.

13.15.3 Pour Point

The pour point of an oil is found by cooling a sample in a test tube until no movement of the oil occurs for 5 seconds after the tube is tilted from the vertical to the horizontal position. The pour point is 5°F (2.8°C) above this temperature. The pour point indicates the temperature below which the oil loses its fluidity and will not flow or circulate in a lubricating system. The oils from paraffinic-bases tend to have a higher pour point than that of the naphthenic-base oils. In order to ensure flow of oil to the oil pump at low temperatures, an oil of low pour point should be used. Special processing and chemical additives are used to obtain lower pour points.

13.15.4 Flash and Fire Points

The flash point is defined as the lowest temperature at which an oil will vaporize sufficiently to form a combustible mixture of oil vapour and air above the surface of the oil. It is found by heating a quantity of the oil in a special container while passing a flame above the liquid to ignite the vapours. A distinct flash of flame occurs when the flash-point temperature is reached.

The fire point is the lowest temperature which must be reached before enough vapours can rise to produce a continuous flame above the liquid oil. This point is obtained if the oil is heated further, after the flash point has been reached. Both the flash and the fire points give a relative measure of the safety properties of lubricating oils, since a high flash point denotes that a high temperaure must be reached before the dangerous handling conditions are reached. These two temperatures must be high enough in a lubricating oil, so that the oil does not flash or burn in service.

13.15.5 Stability

The ability of an oil to resist oxidation that would yield acids, lacquers, and sludge is called stability. At high temperatures some oils have the tendency to break down chemically and form gummy deposits that stick to piston rings and other surfaces. Some oils form sludge in the presence of water and other products of combustion. The viscosity of the oil is changed by the presence of sludge that tends to clog the oil passages. Oil stability demands low temperature (under 100°C) operation and the removal of all hot areas from coming in contact with the oil.

13.15.6 Oilness

Some oils have the property of clinging to a metal surface by molecular attraction and forming a protective layer between the shaft and the bearing. Thus a thin layer of lubricant will be present even under extreme conditions. Oilness of the lubricant may be defined as a measure of the protective layer of oil film in the boundary lubrication region. The two oils having the same viscosity may differ in their coefficient of friction in the boundary lubrication region because one oil has more oilness than the other. More oilness will help to protect surfaces during starting before the normal flow of oil is established.

13.15.7 Corrosiveness

At high temperatures, especially if there is excessive blowby, acids may form in the oil which can corrode engine bearings and other parts. The oil should not be corrosive but should afford protection against corrosion. Corrosion inhibitors are added to the oil to prevent this corrosion. Rust inhibitors are also added which may displace water from metal surfaces. Some inhibitors also neutralize acids.

13.15.8 Detergency

The combustion process leaves deposits of carbon and other deposits on piston rings, valves and other parts. The deposits reduce the performance of the engine and speed up wear of parts. To prevent or slow down the formation of deposits, engine oil has the property of detergency to clean the deposits. It has also the ability of dispersing the particles, preventing them from clotting, and to keep them in a finally divided state. Without this property, the particles would tend to collect and form larger particles, which may block the oil passages.

13.15.9 Foaming

Any violent agitation in the crankcase causes engine oil to foam. It is because of the presence of air bubbles in the oil. This action accelerates oxidation and reduces the mass flow of oil to the bearings and other moving parts causing insufficient lubrication. In addition, foaming may cause some loss of oil through the crankcase breather. To prevent foaming, antifoaming additives are mixed with the oil.

13.16 ADDITIVES FOR LUBRICANTS

The oils obtained from refining by conventional methods are not completely satisfactory for use as lubricants in IC engines. Therefore, lubricants are highly refined and properties are improved by the additions of chemicals. In an endeavour to improve these properties, certain oil-soluble organic compounds containing inorganic elements such as phosphorus, sulphur, amine-derivatives and metals are added to the mineral-based lubricating oils. The main additives include the following:

(a) Detergent-dispersant
(b) Antioxidants and anti-corrosive
(c) Extreme-pressure additives
(d) Pour-point depressors
(e) Viscosity-index improvers
(f) Antifoam agents
(g) Oilness and film-strength agents
(h) Rust inhibitor.

Detergent-dispersant

The detergent-dispersant additives improve the detergent action of the lubricating oil by keeping the deposits in suspension in the oil. Since the additives are oil soluble, the deposits would be carried throughout the oil as suspension. The insoluble compounds are dispersed and this reduces the tendency for the compounds to stick to the metal surfaces to form sludge and lacquer.

Metallic salts or organic acids may be used as detergent-dispersant additives. The mechanism by which the additives work may be either from direct chemical reaction or from polar attraction as both the additives and the deposits in the engine are polar compounds.

Antioxidants and anticorrosives

The oxidation of lubricating oil increases at high temperatures. The oxidation also increases by the presence of certain metals, especially copper which acts as a catalyst for oxidizing hydrocarbons of the lubricants. Oxidation is undesirable because sludge and varnish are created. Acids are also formed which are corrosive. Additives are used to decrease the oxidation and corrosive characteristics of lubricating oils. Additives have a greater affinity for oxygen than the hydrocarbons have. Some additives combine chemically with the metal, and thus act as metal deactivators and corrosion shields.

The additives are generally in the form of sulphur and phosphorus compounds or amine and phenol derivatives. Zinc dithiophosphate is frequently used as an antioxidant and anticorrosive additive.

Extreme-pressure additives

Extreme-pressure (EP) additives are required at high load and speeds with high surface temperatures. Such additives combine with the metal surfaces to form a complex inorganic film which has a lower shear strength than that of the base metal, thereby reducing friction and preventing close contact and seizure of contacting surfaces at extreme pressures when the oil film is ruptured.

Zinc dithiophosphate, organic phosphates, acid phosphates, etc. serve as extreme-pressure additives to prevent scuffing or pitting.

Pour-point depressors

Lubricants contain paraffin compounds and when they are cooled, they form wax precipitates. The formation of wax at low temperatures reduces the fluidity of oil. Pour depressants are added to lower the pour point of the lubricating oil below that of the base oil which can sometimes have a higher pour point than what is acceptable in engine practice, and thus maintaining the fluidity of oil by preventing the formation of wax even during cold starting when the temperature is low. The additive prevents the growth of wax crystals in oil by providing a coating at reduced temperatures.

The pour-point depressors are invariably high molecular weight compounds. The additives are usually polymerized phenols, esters, alkylated naphthalene or methacrylate polymers.

Viscosity-index improvers

The viscosity-index of a lubricating oil can be improved by adding high molecule polymers. The additives increase the viscosity-index since they are less affected by temperature compared to oil, and thus help in increasing the resistance of an oil to change its viscosity with a change in temperature. The lubricating oil must posses high viscosity-index, so as to be satisfactory during cold starting as well as during normal operation and at high speeds and heavy load conditions.

Methacrylate polymers, butylene polymers, polymerized olefins or iso-olefins, etc. can be used as viscosity-index improvers.

Anti-foam agents

The anti-foam agents prevent the formation of foam by reducing surface tension, which allows air bubbles to separate from the oil more rapidly. The most effective anti-foam additives are silicone polymers.

Oilness and film-strength agents

Oilness and high film strength are important in partial-film or thin-film lubrication. Many organic compounds of sulphur, phosphorus and chlorine are used as additives to improve the film-strength of a lubricant.

Rust inhibitors

Rust inhibitors prevent rusting of ferrous engine parts during storage, and from acidic moisture accumulated during cold engine operation. Metal sulphonates, fatty acids and amines act as rust inhibitors.

13.17 SAE VISCOSITY NUMBER

The SAE viscosity number for lubricating oils is recommended by the Society of Automotive Engineers. It depends solely on the viscosity of the oil. The different SAE numbers are 5W, 10W, 20W, 20, 30, 30, 40 and 50. The SAE numbers followed by W indicate the suitability of oil for use in winter and the viscosity is determined at 0°F (–17.8°C). The SAE numbers without W are applied to oils commonly used under warmer conditions and the viscosity is determined at 210°F (98.9°C). Each number represents a viscosity range expressed in minimum and maximum Saybolt universal seconds (SUS) at a particular test temperature. The increasing numbers indicate increase in viscosity.

The viscosity ranges in SUS of lubricating oils are shown in Table 13.2, which also lists the approximate corresponding kinematic viscosities in centistokes.

Multigrade oils, for example 10W–40, satisfy service requirements at low as well as at high temperatures. The first number indicates the viscosity at 0°F (–17.8°C) and the second number at 210°F (98.9°C).

Table 13.2 SAE classification of lubricating oils

SAE viscosity No.	*Viscosity range* (SUS)				*Viscosity range* (centistokes)			
	0°F (–17.8°C)		210°F (98.9°C)		0°F (–17.8°C)		210°F (98.9°C)	
	Min.	Max.	Min.	Max.	Min.	Max.	Min.	Max.
5W	–	6,000	–	–	–	1,300	–	–
10W	6,000	12,000	–	–	1,300	2,600	–	–
20W	12,000	48,000	–	–	2,600	10,500	–	–
20	–	–	45	58	–	–	5.73	9.62
30	–	–	58	70	–	–	9.62	12.93
40	–	–	70	85	–	–	12.93	16.77
50	–	–	85	110	–	–	16.77	22.68

13.18 LUBRICATING SYSTEMS

The basic types of lubricating systems to meet the requirements for proper lubrication of the various types of internal combustion engines are as follows:

1. Petrol lubrication system
2. Wet-sump lubrication system
3. Dry-sump lubrication system

13.18.1 Petrol Lubrication System

In two-stroke SI engines, the air-fuel mixture passes through the crankcase on its way from the carburettor to the engine cylinders. The most heavily loaded components are in close contact with the mixture in the crankcase and then there is an unobstructed passage from the crankcase to the combustion chamber. It is thus not possible to maintain a reservoir of oil in the crankcase. The oil would be picked up by the passing air-fuel mixture, carried to the engine cylinder and burned.

The fact that the petrol and oil supplies seemed to be almost inseparably intermingled, deliberate pre-mixing of the oil and fuel resulted in the adoption of the petrol lubrication system. The lubricant is mixed with the fuel in the supply tank and after emerging from the carburettor jet in droplets it is carried into the inlet ports along with the fuel-air mixture. Initially some oil is deposited on the piston skirt, adjacent to the inlet port, and some will impinge on the cylinder wall opposite to the port. Further separation will take place in the crankcase as the mixture is compressed forming a film over the whole of the interior. As the mixture passes up the transfer ports, a further amount of oil in suspension is taken with it, much of it being burnt along with the explosive charge and thus tends to increase the carbon deposits on the ports, the piston head and the combustion chamber walls. However, enough oil is left behind to lubricate the main bearings, the big and small end bearings of the connecting rod and the cylinder walls.

The advantages of the petrol system are low cost, simplicity, coupled with the fact that oil supply is metered quite closely to the requirements of the engine. The oil supply is regulated by the throttle opening and is therefore proportional to the load or output.

The disadvantages of this system are that a large portion of oil is wasted, it causes deposits on exhaust port and also causes spark-plug fouling.

13.18.2 Wet-sump Lubrication System

There are three types of wet-sump lubrication systems used in engines—splash, pressurized, or a combination of these two.

Splash system

Many small four-stroke engines used in lawn mowers, golf carts, etc. use splash distribution of oil. A schematic diagram of this system is shown in Figure 13.12.

The oil is pumped by a low pressure oil pump from the large capacity oil sump to small troughs placed under each connecting rod. Oil levels in these troughs are maintained constant by supplying excess of oil and providing an overflow to each trough. The caps on the big-end bearings of the connecting rods are usually provided with scoops or dippers. When the piston reaches BDC,

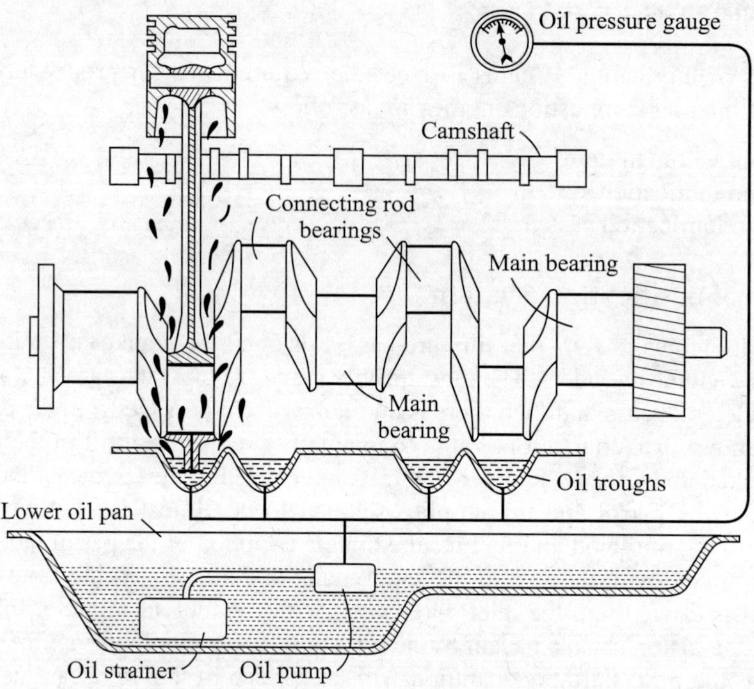

Figure 13.12 Splash lubrication system.

the corresponding connecting rod scoop dips into the trough and directs the oil through a hole in the cap to the big-end bearing. From the scoop the oil is splashed to all engine parts required to be lubricated including main bearings, camshaft bearings, lower portion of the cylinder walls, piston, wristpins, valves, timing gears and cams. The surplus oil eventually returns into the oil sump by gravity for recirculation.

Pressure-feed system

High-speed stationary engines, marine engines and most automobile type of engines have a pressure-feed system. A schematic diagram of this system is shown in Figure. 13.13. The oil is kept in an oil sump at the bottom of the crankcase. In this system, the engine parts are lubricated by oil fed from the oil pump at 2 to 4 bars. The oil pump draws oil from the sump through an oil filter and delivers to an oil gallery or header through an oil pressure control valve to maintain the desired pressure in the line. From the main oil gallery it is branched to the main and the camshaft bearings. The header supplies oil to the main bearings of crankshaft through the connecting passages. From the main bearings it is forced through holes drilled in the crankshaft to the connecting rod bearings. From the connecting rod bearings the lubricant is forced through holes drilled in the connecting rods to the wristpins. Through a passage in the camshaft bearing the oil flows to the tappet bridges. As the oil passages of tappets and tappet bridges line up during tappet motion, rocker arms and valve stems are pulse lubricated through the tappets and push rods. A typical automobile engine has oil passages built into the connecting rod, valve stems, push rods, rocker arms, valve seats, engine block and many other moving components. In addition, oil-spray holes on the upper part of

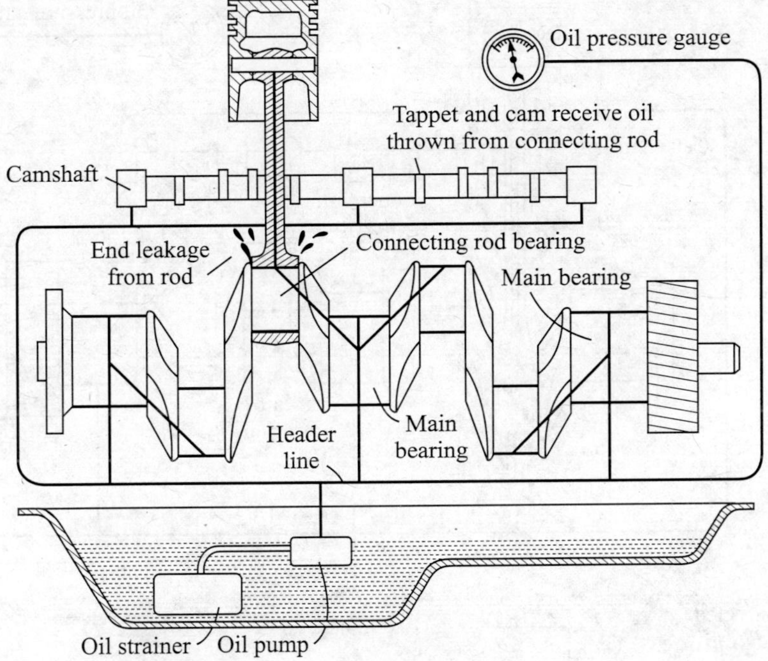

Figure 13.13 Pressure-feed lubrication system.

the connecting rod bearings are used to lubricate the thrust side of the cylinder wall. The oil from the ends of the connecting rod bearings is thrown up to lubricate the cylinder walls and cams. The gears of the main timing train are splash lubricated. Excess oil escaping from bearing ends and falling from pistons, cylinders and other parts is returned to the oil sump for circulation.

The splash and pressure lubrication system

It is a combination of the splash and pressure lubrication systems. It is more simple and less expensive to install than the complete pressure system. It enables higher bearing loads and engine speeds to be employed than those for the plain splash system. This system is applicable to medium-speed stationary engines.

The splash and pressure lubrication system is shown in Figure 13.14. In this system the oil is drawn from the sump by means of a gear pump and then delivered to the main and camshaft bearings. The big-end bearings of the connecting rods are lubricated by the splash system using scoops and troughs through slots cut in the lower ends of the connecting rods. The other parts of the engine are lubricated by splash or spray of oil thrown up by dippers.

13.18.3 Dry-sump Lubrication System

Aircraft engines and motor cycles involve appreciable changes in inclination. The surging of the relatively large quantity of oil in the sump would result in improper lubrication of the cylinders and increased frictional resistance due to churning of the oil. These disadvantages can be overcome by using a dry-sump lubrication system as shown in Figure 13.15.

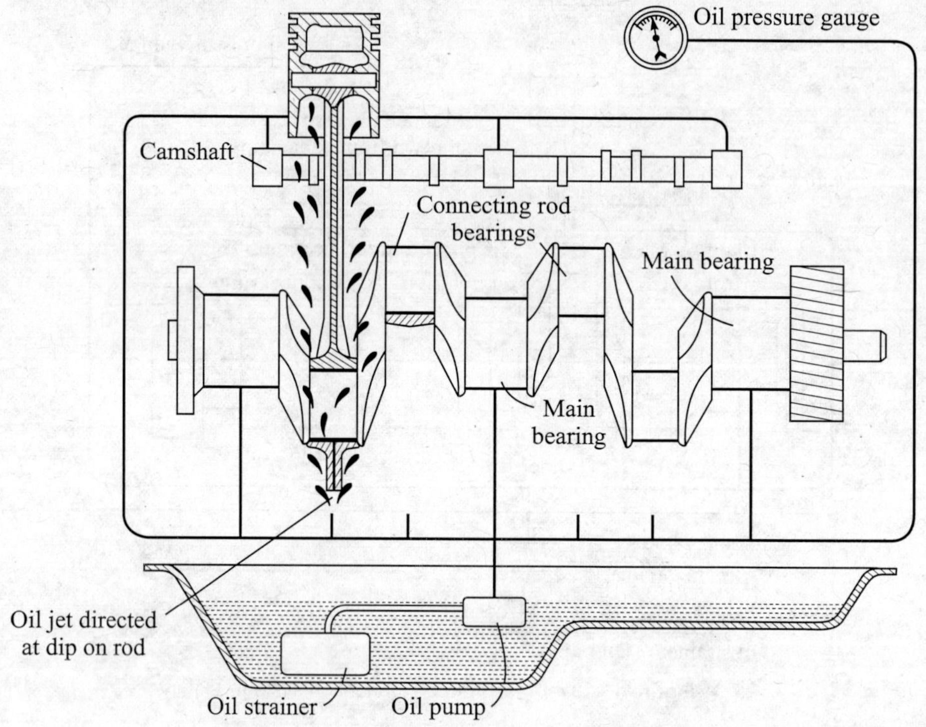

Figure 13.14 The splash and pressure lubrication system.

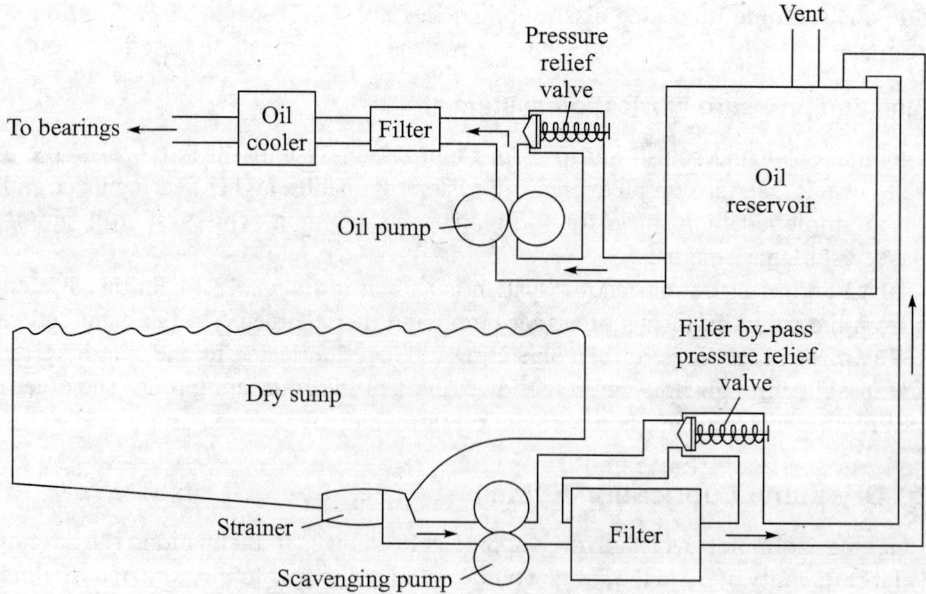

Figure 13.15 Schematic diagram of a dry-sump lubrication system.

In this system, the oil is stored in a separate oil reservoir or tank. The used oil drippings from the cylinders and bearings into the sump are constantly removed by an oil scavenging pump, then passed through a filter and returned to the oil reservoir. This reservoir is sometimes provided with cooler fins. If the filter becomes clogged, the pressure relief valve opens and permits the oil to by-pass the filter. Another and smaller capacity oil pressure pump draws its supply from the reservoir, and after passing the oil through a filter and oil cooler delivers it to the main, big-end and other bearings. A pressure relief valve in this line maintains the supply pressure of oil constant. As the pressure rises above the set value, the relief valve opens and through this valve excess oil returns and releases the pressure. In this system the capacity of the scavenging pump is greater than that of the oil pump, so the oil is prevented from accumulating in the sump. The majority of engines using dry sump have a full pressure feed system.

13.19 CRANKCASE VENTILATION

The pressure inside the combustion chamber is high during later part of compression stroke and early part of expansion stroke. Although the piston and its rings are designed to form a gas-tight seal between the sliding piston skirt and cylinder walls, a small amount of gases escapes past the piston from the combustion chamber and enters into the crankcase. These gases are called blowby gases, which are unburnt hydrocarbons and products of combustion. These gases can dilute and contaminate the engine lubricating oil, cause corrosion to critical parts and contribute to sludge build-up. At higher engine speeds, blowing gases increase crankcase pressure that can cause oil leakage from sealed engine surfaces. The greater the wear on the piston rings and cylinder walls, the greater will be the blowby into the crankcase.

The crankcase ventilation system removes these blowby gases from the crankcase before damage occurs and reduces the pressure inside the crankcase to obtain more power during expansion stroke.

Nowadays, a closed type positive crankcase ventilation (PCV) is used for this purpose. Before PCV was invented, blowby vapours were simply vented to the atmosphere through a road draft tube.

13.19.1 Open PCV System

Open PCV systems were used in 1963 in most cars. An open PCV system draws air through a mesh filter inside the oil filler cap or a breather on a valve cover. The flow of fresh air through the crankcase helps to pull moisture out of the oil. The oil life is increased and the sludge is reduced. The only drawback of this early open PCV system was that blowby vapours could still backup at high engine speeds and loads and escape into the atmosphere through the oil filler cap or valve cover breather.

13.19.2 Closed PCV System

In 1968, closed PCV systems were added to most cars. The breather inlet was relocated inside the air cleaner housing, so if pressure backed up, it would overflow into the air cleaner and get sucked down the carburettor. No vapour would escape into the atmosphere.

It uses manifold vacuum to draw crankcase vapours back into the intake manifold. Typically, blowby production is greatest during high load operations and very low during idle and light load operations. Since the characteristics of manifold vacuum do not match the flow requirements needed for proper crankcase ventilation, a PCV valve is used to regulate blowby flow back into the intake manifold.

During idle condition and deceleration, blowby production is very low, but intake manifold vacuum is very high. This causes the pintle inside the PCV valve to fully retract against spring tension. The positioning of the pintle provides a small vacuum passage and allows low blowby flow to the combustion chamber.

During acceleration and high load operations, blowby production is very high. The pintle extends out further from the restrictor, allowing the maximum flow of blowby into the combustion chamber. During extremely high engine loads, if blowby volume exceeds the ability of the PCV valve to draw in the vapours, the excess blowby flows through the breather hose to the air cleaner housing where it can enter the combustion chamber.

When the engine is shut off or it backfires, spring tension closes the valve completely, preventing the release of blowby into the intake manifold. The valve closes during a backfire to prevent the flame from travelling into the crankcase where it could ignite the enclosed fuel vapours.

13.20 ENGINE PERFORMANCE AND LUBRICATION

The viscosity of the lubricating oil has a considerable effect on the engine performance. If the viscosity is too high, there will be loss of power in shearing and in pumping the oil. It will result in reduced torque and power output of the engine, and increased fuel consumption. On the other hand, if the viscosity is too low, there will be poor sealing between the piston rings and the cylinder walls. It will result in increase in blowby with a consequent increase in oxidation of the crankcase oil, and increased oil consumption.

Bearings should operate in full-film lubrication region. Bearings should not be operated at low speeds and high loads with very light oils which will lower the value of the non-dimensional number $\mu N/P$ to such an extent that lubrication in mixed-film region will be reached. This region is avoided by avoiding high loads at low engine speeds through shifting to a lower gear in the transmission system of the vehicle.

With the increase in engine speed and load, the temperature in the cylinder increases which results in a decrease in oil viscosity, and consequently a greater quantity of oil passes through the piston rings resulting in increased consumption of the lubricating oil. This can be improved by increasing the viscosity index of oil by suitable additives and by using multigrade oils.

Oil consumption is also increased by uneven wear from top to bottom of the cylinder which encourages flexing of the piston rings.

EXAMPLE 13.1 An engine has 8 cm bore, 7.5 cm stroke and 15.2 cm connecting rod length. The skirt length of the piston is 6.2 cm. The engine performs a power stroke and the connecting rod makes a right-angle with the crank. The instantaneous piston speed at this moment is 8.2 m/s and the pressure in the cylinder is 3000 kPa and the compressive force in the connecting rod is 8.0 kN. The clearance between the piston and cylinder wall is 0.004 mm. The dynamic viscosity of the lubricating oil is 0.006 Pa·s. Calculate the friction force on the piston, and the thrust force on the cylinder wall at this moment.

Solution: *Given:*

$$\text{Bore, } d = 8 \text{ cm} = 0.08 \text{ m}$$
$$\text{Stroke, } L = 7.5 \text{ cm} = 0.075 \text{ m}$$
$$\text{Connecting rod length, } l = 15.2 \text{ cm} = 0.152 \text{ m}$$
$$\text{Skirt length of the piston, } h = 6.2 \text{ cm} = 0.062 \text{ m}$$
$$\text{Compressive force in connecting rod, } F_r = 8 \text{ kN} = 8000 \text{ N}$$
$$\text{Pressure, } p = 3000 \text{ kPa}$$
$$\text{Clearance, } \Delta y = 0.004 \text{ mm} = 0.004 \times 10^{-3} \text{ m}$$
$$\text{Dynamic viscosity, } \mu = 0.006 \text{ Pa·s}$$
$$\text{Piston speed, } \Delta u = 8.2 \text{ m/s}$$
$$\text{Shear stress, } \tau_s = \mu \frac{\Delta u}{\Delta y} = \frac{0.006 \times 8.2}{0.004 \times 10^{-3}} = 12{,}300 \text{ N/m}^2$$

Contact area between the piston and the cylinder wall,

$$A = \pi dh$$
$$= \pi \times 0.08 \times 0.062$$
$$= 0.01558 \text{ m}^2$$

Friction force on the piston, $F_f = \tau_s A$

$$= 12{,}300 \times 0.01558$$
$$= \boxed{192 \text{ N}} \quad \textbf{Ans.}$$

$$\text{Crank length, } r = \frac{L}{2} = \frac{0.075}{2} = 0.0375 \text{ m}$$

$$\text{From Figure 13.16, } \tan\phi = \frac{r}{l} = \frac{0.0375}{0.152} = 0.2467$$

$$\therefore \quad \phi = 13.86°$$

$$\text{Side thrust, } F_t = F_r \sin\phi$$
$$= 8000 \sin 13.86°$$
$$= \boxed{1916 \text{ N}} \quad \textbf{Ans.}$$

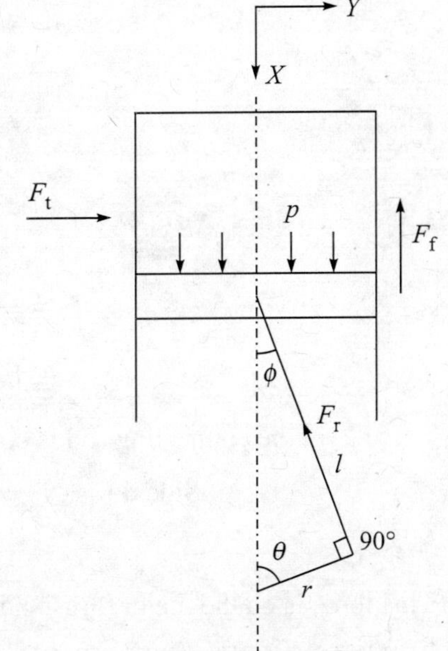

Figure 13.16 Example 13.1.

EXAMPLE 13.2 A four-cylinder IC engine has a 6.5 cm bore, 6 cm stroke and a connecting rod length of 12 cm. In the power stroke of the cycle for one cylinder at a crank position of 90° aTDC, the pressure in the cylinder is 50 bar and the sliding force on the piston is 900 N. The piston acceleration at this point is zero.

Calculate: (a) The force in the connecting rod and the side thrust on the piston at this point.
(b) The side thrust on the piston at this point, if the wrist pin is offset 2 mm to reduce the side thrust. Assume that the friction force is the same as above.

Solution: *Given:*

Cylinder bore, $d = 6.5 \text{ cm} = 0.065 \text{ m}$

Stroke, $L = 6 \text{ cm}$

Connecting rod length, $l = 12 \text{ cm}$

$\theta = 90°, p = 50 \text{ bar}$

Friction force, $F_f = 900 \text{ N}$

(a)
$$F_t = F_r \sin\phi \qquad \text{(i)}$$

$$-F_r \cos\phi + p\frac{\pi}{4}d^2 - F_f = 0 \qquad \text{(ii)}$$

$$r = \frac{L}{2} = \frac{6}{2} = 3.0 \text{ cm}$$

$$\sin\phi = \frac{r}{l} = \frac{3.0}{12.0} = 0.25$$

$$\cos\phi = \sqrt{1-\sin^2\phi} = \sqrt{1-0.0625} = 0.9682$$

$$F_r \cos\phi = p\frac{\pi}{4}d^2 - F_f$$

$$= 50\times 10^5 \times \frac{\pi}{4}(0.065)^2 - 900$$

$$= 16592 - 900 = 15{,}692 \text{ N}$$

$$\therefore \quad F_r = \frac{15692}{0.9682} = 16{,}207 \text{ N}$$

$$= \boxed{16.207 \text{ kN}} \quad \textbf{Ans.}$$

(Force F_r in the connecting rod is compressive.)

Side thrust, $F_t = F_r \sin\phi = 16.207 \times 0.25$

$$= \boxed{4.052 \text{ kN}} \quad \textbf{Ans.}$$

(This force is on the major thrust side.)

(b) Wrist pin is offset by 2 mm as shown in Figure 13.17.

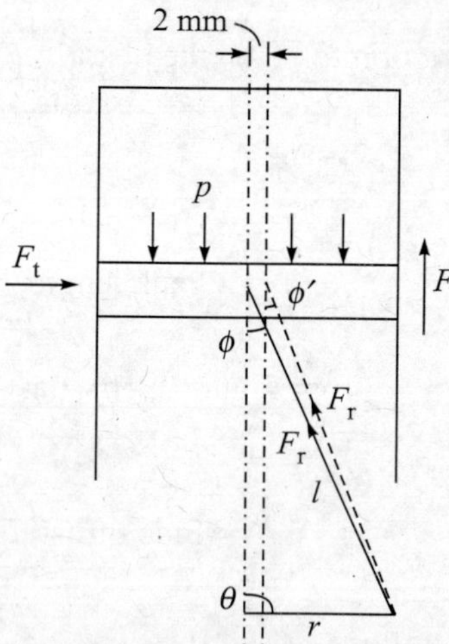

Figure 13.17 Example 13.2.

$$\sin\phi' = \frac{3.0-0.2}{12} = \frac{2.8}{12} = 0.2333$$

$$\cos\phi' = \sqrt{1-\sin^2\phi'} = \sqrt{1-(0.2333)^2} = 0.9724$$

$F_r \cos\phi' = 15{,}692 \text{ N}$ $\quad$ ($\because$ p, d and F_f remain the same as above)

$$\therefore \quad F_r = \frac{15{,}692}{0.9724} = 16{,}137 \text{ N} = 16.137 \text{ kN}$$

Side thrust, $F_t = F_r \sin\phi' = 16.137 \times 0.2333$

$$= \boxed{3.765 \text{ kN}} \quad \textbf{Ans.}$$

REVIEW QUESTIONS

1. Define friction power, the indicated power and the brake power.
2. What are the different types of engine friction losses in an engine? Briefly describe them.
3. Define rubbing friction work, the pumping work, the accessory work and the total friction work.
4. Define the friction mean effective pressure. How is it related with other mean effective pressures? Give the relation of friction mean effective pressure with the engine speed mentioning the significance of each term.
5. Briefly explain the following terms associated with mechanical friction: (a) Hydrodynamic friction, (b) Partial-film friction, (c) Rolling friction, and (d) Dry friction.
6. Show with the help of a schematic diagram a typical piston and ring assembly. What components of this assembly contribute to friction? What forces act on the piston assembly, and what are the major design factors that affect the piston assembly friction?
7. How is side thrust conveyed to the piston skirt in an engine? Explain it with the help of a diagram. What will be the effect of increasing the side thrust on viscous friction of the piston?
8. What is a journal bearing? What are the functions and properties of a journal bearing?
9. What are the critical contact regions in a valve train of an IC engine? How can the valve train friction be reduced?
10. Describe the term blowby losses. What are the factors that increase the blowby losses? What are the effects of increased blowby losses on engine performance?
11. What are the important engine parameters that affect the engine friction? Briefly describe them.
12. Derive an expression to evaluate side thrust on the piston of an IC engine during the cycle.
13. What do you understand by major thrust side and minor thrust side? How does the side thrust vary along the length of the cylinder and in the circumferential direction? What will be the effect of these variations on wear?
14. How can the side thrust on the piston during a cycle be reduced?
15. Why is it that most medium-sized automobile engines are close to square?
16. What are the functions of a good lubricant?
17. What is the principle of lubrication? What are the three lubrication regions that are important for engine components?
18. Describe the hydrodynamic lubrication region, the boundary lubrication region, and the mixed-film lubrication region.
19. What are the various bearings and their motions followed in IC engines?
20. Describe the types of lubrication regions that exist in rotating journal bearings during rest, starting and normal running conditions.
21. Why is it important to start an engine under a no-load condition?
22. Derive an expression to evaluate the coefficient of friction in a cylindrical bearing in the hydrodynamic lubrication region.

23. Describe with the help of a diagram the relation between the coefficient of friction and the non-dimensional number $\mu N/P$ for a bearing in different lubrication regions.
24. Briefly describe the type of oil films produced in oscillating journal bearings, reciprocating bearings and gear teeth.
25. What are the basic properties of lubricants used in IC engines?
26. Define viscosity of oil. What will be the effect of more viscous oil on the operation of an engine? How is the viscosity of oil measured?
27. Define viscosity index. How is it determined?
28. Define pour point, and flash and fire points.
29. Briefly describe the following properties of lubricating oil: stability, oilness, corrosiveness, detergency, and foaming.
30. What are the main groups of additives used in IC engines to improve the properties of lubricating oil?
31. Describe the following groups of additives and mention the chemicals used: detergent-dispersant, antioxidant and anticorrosives, extreme pressure additives, pour-point depressors, viscosity-index improvers, anti-foam agents, oilness and film-strength agents, and rust inhibitors.
32. Give a brief description of the SAE viscosity number.
33. Describe the petrol lubrication system. What are the advantages and disadvantages of this system?
34. Describe with the help of diagrams, the following wet-sump lubrication systems: (a) Splash system; (b) Pressure-feed system; and (c) The splash and pressure lubrication system.
35. Describe the dry-sump lubrication system with the help of a diagram.
36. What is the purpose of crankcase ventilation? Describe positive crankcase ventilation systems.
37. What are the effects of a viscosity of a lubricating oil on the engine performance?
38. What will be the effect of increasing the engine speed and the load on oil viscosity?

14.1 INTRODUCTION

The heat transfer process in internal combustion engines has always played an important role in engine design. One aspect of heat transfer concerns the loss of energy from the combustion chamber gases, which reduces the amount of piston work, and another aspect relates to the durability of engine components exposed to high temperature gases. In modern engine design, the amount of cooling applied should be optimized with respect to the balance between these two considerations, i.e. every effort should be made to keep engine cooling at the minimum level compatible with the temperature limits of the material employed.

About 35% of the total chemical energy that enters an engine as fuel is converted to useful crankshaft work, and about 30% of the fuel energy is carried away from the engine to the exhaust. The heat to coolant generally amounts to about a quarter to one-third of the chemical energy supplied in the fuel. About half of that is the result of heat transfer with the cylinder, and most of the remainder passes through the exhaust port walls from the outflowing charge, if these walls are not insulated. When the port walls are insulated, as they now often are to promote reactions that remove hydrocarbons and carbon monoxide from the exhaust, this contribution of heat transfer is greatly reduced, and the coolant heat load correspondingly drops appreciably.

Heat transfer between the working fluid and the internal surface of the engine is unsteady. During the working cycle of the engine, the heat flux between the working fluid and the surface varies both in time and space. The estimation of overall heat transfer rate to coolant requires evaluation of the time and space average heat transfer to the entire internal surface. The estimation of thermal stress in such components presents difficulties due to the piston or cylinder head requiring a detailed evaluation of the spatial distribution of time-average heat flux.

14.2 NECESSITY OF ENGINE COOLING

The cooling of an engine is necessary for the following reasons:

1. The peak gas temperature during the combustion process of an IC engine is of the order of 2500 K. The temperature of the inside surface of the cylinder walls is usually kept below 200°C to prevent deterioration of the oil film. High temperature of the lubricating oil may

result in physical and chemical changes in the oil and cause wear and sticking of the piston rings, scoring of the cylinder walls, or seizure of the piston. It is therefore necessary to provide cooling for the walls of combustion space.

2. During the process of converting thermal energy into mechanical energy, high temperatures are produced in the cylinders of the engine as a result of the combustion process. A large portion of the heat generated in the combustion chamber is transferred from combustion gases to the cylinder head and walls, and piston and valves. Heats absorbed by these components increase their temperatures. The temperature distributions are uneven causing uneven expansion of various engine parts, hence causing thermal stresses in the components of the engine. High thermal stresses cause fatigue and cracking of the components. Therefore, the temperature must be kept less than about 400°C for cast iron and about 300°C for aluminium alloys.
3. The temperature of the cylinder head must also be kept below 220°C. If the cylinder head temperature is high, this may lead to overheated spark-plug electrodes causing preignition in SI engines.
4. Spark plug and valves must be kept cool to avoid knock and preignition problems which result from overheated spark-plug electrodes or exhaust valves. Preignition results in a loss of efficiency and increases the cylinder head temperature to such an extent that engine failure or complete loss of power may result.

14.3 DISADVANTAGES OF OVERCOOLING

Excessive cooling is undesirable for the following reasons:

1. Starting of the engine will be difficult at low temperatures. The engine must be kept sufficiently hot to ensure smooth and efficient operation.
2. Vaporization of the fuel will be reduced at low temperatures, preventing formation of a homogeneous mixture with air. It may cause poor combustion and also increase fuel consumption.
3. Excessive cooling provided to the combustion chamber walls will lower the average combustion gas temperature and pressure and reduce the work per cycle transferred to the piston. Thus, the specific power and efficiency are reduced by excessive cooling.
4. Friction will be increased because of higher viscosity of lubricating oil at lower temperatures.
5. The sulphurous and sulphuric acids are formed from the oxidation of sulphur present in the fuel during the combustion process. These acids may condense at low temperatures and corrode the cylinder surfaces. To prevent condensation of acids, the coolant temperature should be greater than 70°C.

Thus, removing heat is highly critical in preventing an engine and engine lubricant from thermal failure. On the other hand, it is desirable to operate an engine as hot as possible to maximize thermal efficiency.

14.4 ENGINE TEMPERATURE DISTRIBUTION

Figure 14.1 shows a typical temperature distribution in a spark-ignition engine operating at steady state. The hottest regions are around the spark plug, the exhaust valve and port, and the piston

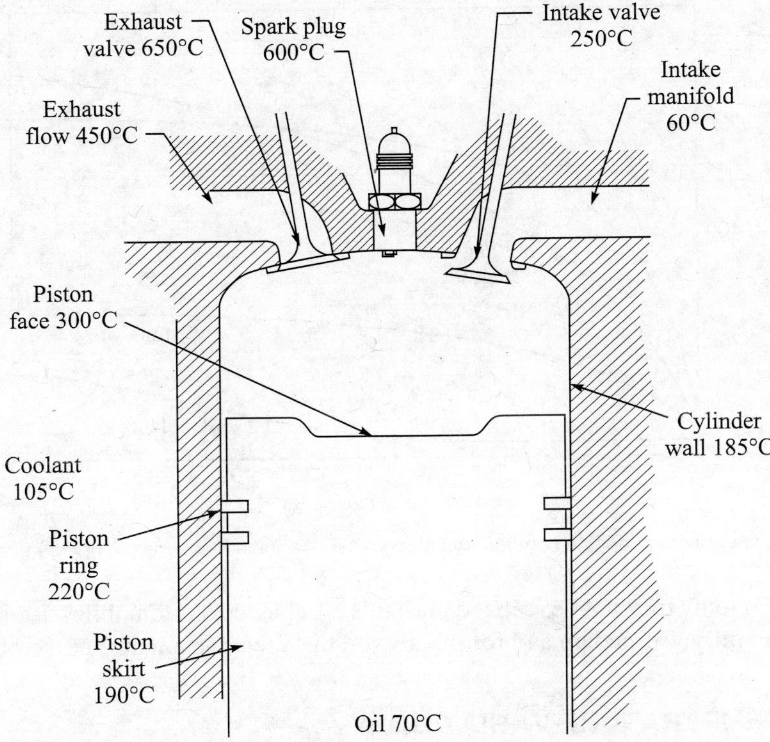

Figure 14.1 Typical temperature distribution in a spark-ignition engine.

face. These regions are exposed to the high temperature combustion gases and are difficult places to cool.

The spark plug fastened through to combustion chamber wall creates a disruption in the surrounding water jacket in water cooled engines and disrupts the cooling fin pattern in air-cooled engines, causing a local cooling problem.

The exhaust valve and port operate hot because they are exposed to the pseudo-steady flow of hot exhaust gases and create a difficulty in cooling similar to the one that the spark plug creates. The valve mechanism and connecting exhaust manifold make it very difficult to route coolant or allow a finned surface to give effective cooling.

The piston face is difficult to cool because it is separated from the water jacket in water-cooled engines or separated from the outer finned cooling surfaces in air-cooled engines. Most of the heat which enters the piston head flows out through the piston rings into the cylinder walls.

14.5 ENGINE WARM-UP

Figure 14.2 shows the increase in the temperature of engine components of a typical SI engine with time, after the cold engine is started. The start-up time to reach steady-state conditions depends upon the weather and may take a few minutes. Some parts of the engine reach steady state sooner than the other. Engines are built to operate best at steady-state conditions, and full power and

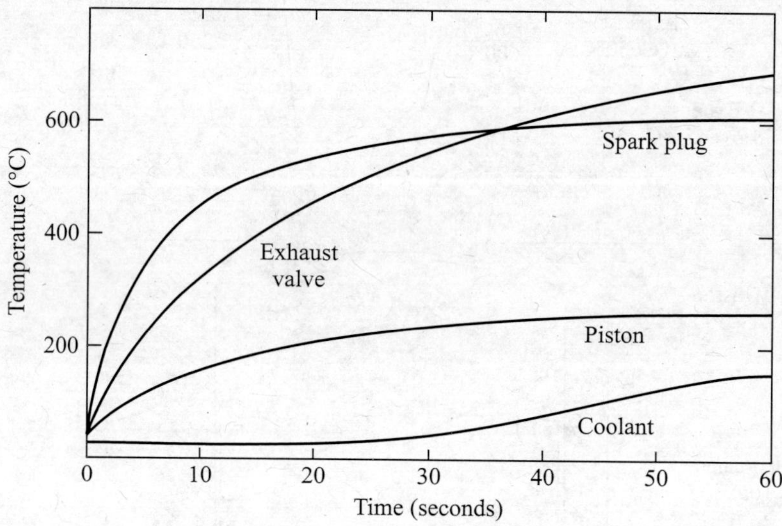

Figure 14.2 Temperatures of engine components of a typical SI engine as a function of time after cold start-up.

optimum fuel economy may not be realized until this is achieved. Automobiles used for short trips are generally not fully warmed up and for this reason they become a major cause of air pollution.

14.6 GAS TEMPERATURE VARIATION

The temperature of the gases in the cylinder varies appreciably during the different processes in a cycle. Figure 14.3 shows the typical variation of the gas temperature. The temperature at the beginning of the induction process is that of the residual gases present in the clearance space.

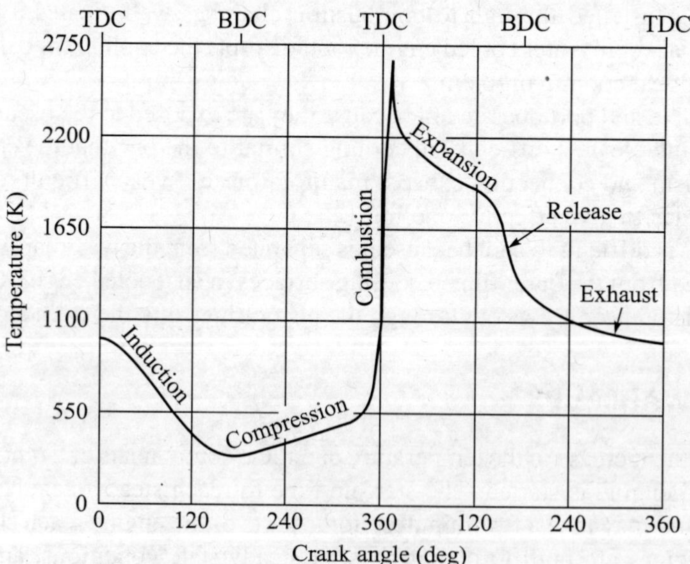

Figure 14.3 Gas temperature variation.

The temperature falls rapidly as the cool charge is inducted during the induction stroke. The temperature rises during the compression process and is very rapidly increased to the maximum by the combustion process. The expansion process decreases the temperature. Heat transfer to cylinder walls continues during the expansion stroke, but the rate quickly decreases. Expansion cooling and heat losses reduce the gas temperature within the cylinder during this stroke from a maximum temperature of the order of 2700 K to an exhaust temperature of about 800 K. During release when the exhaust valve opens, the temperature of the gases drops rapidly. During the exhaust stroke, heat transfer to the cylinder walls continues but at a greatly reduced rate. At this time the cylinder gas temperature is much lower, as is the convection heat transfer coefficient. There is no swirl or squish motion at this time, and turbulence is greatly reduced, resulting in a much lower convection heat transfer coefficient.

14.7 HEAT TRANSFER CONSIDERATIONS

Heat transfer occurs when a temperature difference exists. The combustion of the charge within the cylinder of the engine results in high temperature differences and causes heat transfer. Heat is transferred by conduction, convection and radiation.

14.7.1 Conduction

Heat is transferred by molecular motion through solids and through fluids at rest due to a temperature difference. Heat transfer by conduction per unit area per unit time, q, called heat flux, in a steady state is given by Fourier's law,

$$\dot{q} = -k\nabla T \tag{14.1}$$

where k is the thermal conductivity.

For a steady one-dimensional temperature variation,

$$\dot{q}_x = \frac{\dot{Q}}{A} = -k\frac{dT}{dx} \tag{14.2}$$

Heat is transferred by conduction through the cylinder head, the cylinder walls, the piston, the piston rings, the engine block and manifolds. Heat originates from the hot gases in the vicinity of the metal parts.

14.7.2 Convection

Heat is transferred through fluids in motion and between a fluid and a solid surface in relative motion. When the motion is produced by forces other than the gravity, the term forced convection is used. Heat is transferred by forced convection during the whole cycle between the in-cylinder gases and the cylinder head, the valves, the cylinder walls and the piston. Heat is also transferred by forced convection from the cylinder head and cylinder walls to the coolant, and from the piston to the lubricant. Heat lost from the engine to the environment is also by convection. The temperature of the incoming charge in the intake manifold is raised by convective heat transfer. During the exhaust process, substantial convective heat transfer occurs to the exhaust valve, the exhaust port and the exhaust manifold.

The heat transfer per unit area per unit time, $\dot{q}$ (heat flux) by forced convection in steady flow is given by

$$\dot{q} = h(T - T_w) \tag{14.3}$$

where h is the convective heat transfer coefficient, T is the temperature of the flowing fluid stream and T_w is the wall temperature.

For many flow geometries such as flow through pipes or over a plate, McAdams and others have shown by dimensional analysis and experiments that h is given by the relation of the form:

$$\left(\frac{hL}{k_g}\right) = \text{constant} \times \left(\frac{\rho v l}{\mu}\right)^m \left(\frac{c_p \mu}{k_g}\right)^n \tag{14.4}$$

where

h = coefficient of heat transfer
L = characteristic length
k_g = thermal conductivity of the gases
ρ = density
v = velocity of gases
μ = viscosity of gases
m, n = exponents

The term hL/k_g is recognized as Nusselt number, and $\rho vl/\mu$ is a Reynolds number for the cylinder gas motion. The term $c_p\mu/k_g$ is called the Prandtl number and for gases it varies little and is about 0.7. It may therefore be placed into the constant in expression (14.4), so that

$$\frac{hL}{k_g} = \text{constant} \times \left(\frac{\rho v l}{\mu}\right)^m \tag{14.5}$$

Since the Prandtl number is constant, we have $k_g \propto \mu c_p$, and substituting it in the expression (14.5),

$$\frac{hL}{\mu c_p} = \text{constant} \times \left(\frac{\rho v L}{\mu}\right)^m$$

$$\therefore \qquad h = \text{constant} \times c_p \times (\rho v)^m \left(\frac{L}{\mu}\right)^{m-1} \tag{14.6}$$

Substituting the value of h from Eq. (14.6) in Eq. (14.3), we get heat flux as

$$\dot{q} = \text{constant} \times c_p (\rho v)^m \left(\frac{L}{\mu}\right)^{m-1} (T - T_w) \tag{14.7}$$

This expression cannot be used when boiling occurs at the surface as may be the case on the coolant side in water-cooled engines.

The value of the constant in expressions (14.6) and (14.7) depends upon the geometry of the combustion chamber, the extent of carbon deposits, and other details of engine design. The value of the exponent m is about 0.8 for gases moving in pipes and it is closer to 0.6 or 0.7 in respect of engine cylinders.

14.7.3 Radiation

Heat transfer by radiation occurs through the emission and absorption of electromagnetic waves having wavelengths in the vissible range (0.4 to 0.7 μm) and the infrared range (0.7 to 40 μm). The radiative heat transfer within the cylinder is due to high temperature gases and the soot particles.

In SI engines the flame front is slightly luminous and the gaseous products are formed in the reaction at an intermediate step in the combustion process. Heat transfer in SI engines due to radiation amounts to about 10% of the total heat transfer. This is due to poor emitting properties of gases, which emit only at specific wavelengths. Nitrogen and oxygen which make up the major part of the gases before combustion, radiate very little, while carbon monoxide and water vapour of the products do contribute more to radiation heat transfer.

In ÇI engines the flame is highly luminous, and soot particles which are mostly carbon are also formed with gaseous products at an intermediate step in the combustion process. The solid carbon particles are good radiators at all wavelengths, and radiate about 20–35% of the total heat transfer. A large percent of radiation heat transfer to the walls occurs early in the power stroke. At this point the combustion temperature is maximum, and with thermal radiation potential equal to T^4, a very large heat flux is generated. At this point, heat flux of the order of 10 MW/m^2 can be experienced in a CI engine.

The radiation from soot particles in the diesel engine flame is about five times the radiation from the gaseous combustion products. Radiative heat transfer in SI engines is small compared with the convective heat transfer. However, the radiative heat transfer in CI engines is quite significant. Heat transfer to the surrounding from the external hot surfaces of the engine is also by radiation.

A basic concept in the theory of radiant heat transfer is that of a blackbody, namely the one which will absorb all the radiation that falls upon it and reflects none of the radiation falling on it. Such a body also radiates the maximum possible flux at every wavelength. As the temperature of the blackbody increases, the radiation increases at every wavelength, but the peak moves to shorter wavelengths. The radiant heat flux from one plane blackbody at absolute temperature T_1 to another parallel to it at absolute temperature T_2, across a space containing no absorbing material, is given by

$$\dot{q} = \sigma (T_1^4 - T_2^4) \tag{14.8}$$

where σ is the Stefan–Boltzmann constant 5.67×10^{-8} W/(m^{2}K^4)

Real bodies are not, in general, blackbody radiators, but reflect radiation to an extent which depends on the wavelength. Gases depart far from the blackbody behaviour. This aspect is dealt with by introducing a multiplying factor, called *emissivity* ε, to Eq. (14.8). Similarly, a shape factor is introduced to account for the variation in the angle of incidence of the radiation over any actual surface.

14.8 HEAT TRANSFER IN INTAKE SYSTEM

The intake manifold of an IC engine is normally hotter than the charge flowing through it. Heat transfer from the wall of the intake manifold to the flowing charge is by convection and the heat flux $\dot{q}$ is given by

$$\dot{q} = h(T_w - T_g) \tag{14.9}$$

where T is the temperature and h is the convective heat transfer coefficient.

Engines with carburettor or throttle body injection of the fuel have heated intake manifolds to assist in the evaporation of the fuel. Various methods are used to heat these manifolds. In some engines the intake manifold is placed above the exhaust manifold. Some engines may use hot water in a water jacket surrounding the manifold. Electrical heating can also be used to heat the intake manifold. Some systems may have hot spots in specific locations to heat the charge.

If the intake manifold is hot, it vaporizes the fuel earlier and allows more time for the mixing of fuel with air, resulting in a more homogeneous mixture. However, increasing the temperature reduces the volumetric efficiency of the engine as the density of air is reduced due to reduction in the mass of air displaced by the added fuel vapour. A compromise is made by vaporizing some of the fuel in the intake system and the rest in the cylinder during compression or even during combustion. Another reason to limit the heating of the inlet charge in the intake manifold is to keep the temperature at the start of compression stroke to a minimum, which reduces all temperatures throughout the cycle and thus the possibility of engine knock is reduced. Heating of the intake manifold is not required in engines with multipoint port injectors. With these injectors, finer fuel droplets are obtained which are easier to vaporize and the fuel is sprayed directly on to the back of the intake valve face. This accelerates evaporation and cools the intake valve.

14.9 HEAT TRANSFER IN COMBUSTION CHAMBERS

All the three modes of heat transfer play an important role in the heat transfer through the combustion chamber cylinder wall of an IC engine. A schematic temperature profile during heat transfer through a cylinder wall is shown in Figure 14.4.

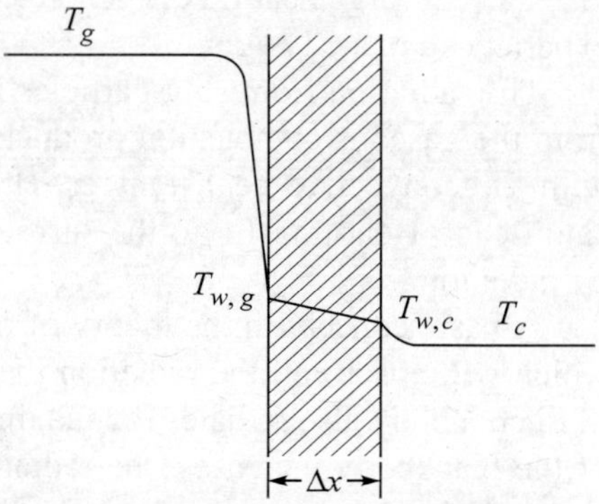

Figure 14.4 A schematic temperature profile across a cylinder wall.

For one-dimensional steady flow, the following equations relate the heat flux $\dot{q}$ to the temperatures:

Heat transfer on the gas side surface of the cylinder is given by

$$\dot{q} = \dot{q}_{\text{convective}} + \dot{q}_{\text{radiative}} = h_g(T_g - T_{w,g}) + \sigma\varepsilon(T_g^4 - T_{w,g}^4) \tag{14.10}$$

where

h_g = convective heat transfer coefficient on the gas side
T_g = average gas temperature in the cycle
$T_{w,g}$ = cylinder wall temperature on the gas side
σ = Stefan–Boltzmann constant
ε = emissivity.

The radiation term is generally negligible for SI engines, for which

$$\dot{q} = \dot{q}_{\text{convective}} = h_g(T_g - T_{w,g}) \tag{14.11}$$

Heat transfer through the cylinder wall is due to the conduction only. Heat flux through the cylinder wall is

$$\dot{q} = \dot{q}_{\text{conductive}} = \frac{k(T_{w,g} - T_{w,c})}{\Delta x} \tag{14.12}$$

where

k = thermal conductivity of the material of cylinder wall
$T_{w,g}$ = cylinder wall temperature on the gas side
$T_{w,c}$ = cylinder wall temperature on the coolant side
Δx = cylinder wall thickness.

Heat transfer on the coolant side is due to the convection only. Heat flux on the coolant side is given by

$$\dot{q} = \dot{q}_{\text{convective}} = h_c(T_{w,c} - T_c) \tag{14.13}$$

where

h_c = convective heat transfer coefficient on the coolant side
$T_{w,c}$ = cylinder wall temperature on the coolant side
T_c = average temperature of the coolant in the cycle.

For steady-state heat flux, $\dot{q}$ remains the same in relations (14.11), (14.12) and (14.13),

$$\therefore \qquad \dot{q} = h_g(T_g - T_{w,g}) = \frac{k(T_{w,g} - T_{w,c})}{\Delta x} = h_c(T_{w,c} - T_c) \tag{14.14}$$

From the above relations, the following equations are obtained:

$$T_g - T_{w,g} = \frac{\dot{q}}{h_g} \tag{14.15a}$$

$$T_{w,g} - T_{w,c} = \dot{q}\frac{\Delta x}{k} \tag{14.15b}$$

$$T_{w,c} - T_c = \frac{\dot{q}}{h_c} \tag{14.15c}$$

Adding Eqs. (14.15(a), (b) and (c)), we get

$$T_g - T_c = \dot{q}\left(\frac{1}{h_g} + \frac{\Delta x}{k} + \frac{1}{h_c}\right)$$

$$\therefore \qquad \dot{q} = \frac{T_g - T_c}{\dfrac{1}{h_g} + \dfrac{\Delta x}{k} + \dfrac{1}{h_c}} \tag{14.16}$$

The gas temperature T_g in the cylinder varies widely over an engine cycle. It is maximum during the combustion process and minimum during the intake process. Early in the intake stroke it can even be less than the wall temperature, thus momentarily reversing the heat transfer direction. Thus the heat flux changes continuously from a small negative value during the intake process to a positive value of order several megawatts per square metre early in the expansion process while the combustion process is still in progress. The temperature T_c and the convective heat transfer coefficient h_c on the coolant side of the wall will be fairly constant over a cycle. The convective heat transfer coefficients h_g on the gas side have large temporal and spatial variations during an engine cycle due to changes in gas motion, velocity, swirl, turbulence, etc. The thermal conductivity k of the cylinder wall is a function of wall temperature and will be fairly constant.

14.10 HEAT TRANSFER IN EXHAUST SYSTEM

The exhaust flow convective heat transfer rates are the largest in the entire cycle due to very high gas velocities developed during the exhaust blowdown process and due to the high gas temperature. The exhaust temperatures of SI engines are generally in the range from 400°C to 600°C. The exhaust temperatures of CI engines are lower due to their greater expansion ratio and are generally in the range from 200°C to 500°C. Exhaust gas heat transfer affects emissions burn-up in the exhaust system. It influences turbocharger performance, and it contributes significantly to the engine requirements.

14.11 PISTON COOLING

It is difficult to cool the piston. The piston face is exposed to hot combustion gases and it cannot be cooled by the coolant of the engine water jacket or by the air flowing over the external finned surface. One method used to cool the piston is by splashing or spraying lubricating oil on the back surface of the piston crown. This lubricating oil absorbs heat from the piston and then flows back into the oil reservoir in the crankcase, where it is again cooled. Piston is also cooled by the conduction of heat from the piston face. There are two conduction paths, one through the piston rings to the cylinder walls and into the coolant in the surrounding water-jacket and another down the connecting rod to the oil reservoir. Thermal conductivity through the piston body is high but the conductivity is poor through the lubricant film between the surfaces.

The piston rings have an important influence upon the piston temperature. They act as thermal bridges between the material of the piston and the cylinder walls. The temperature of the piston is influenced by the number of rings, their wall pressures, and the dimensions and materials of the rings.

In an air-cooled cylinder, the hottest point of the piston is not often at the centre, but at a point nearer the hottest side of the wall. Even in a water-cooled cylinder the spark plug may have a very marked heating effect on the wall in its vicinity and, consequently, on the piston temperature distribution. In an extreme case the temperature at the edge of the piston nearest the spark plug may even be greater than that at the centre of the piston. The use of a material having a high conductivity and a thicker piston head will reduce the temperature difference between the centre and edge of the piston. However, this raises the temperature at the edge and the temperature at the upper ring. Piston temperatures are higher with the larger bore engines and increase with an increase in load. Figure 14.5 shows a typical variation of isotherms in a piston. Isotherms are axi-symmetric.

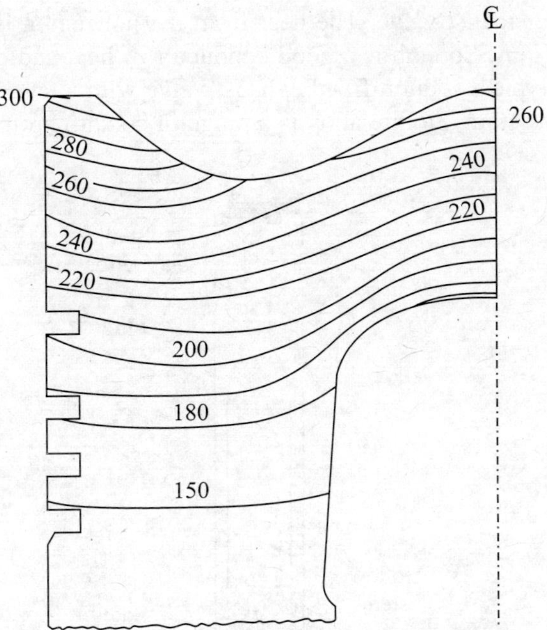

Figure 14.5 Typical variation of isotherms in a piston.

Aluminium pistons generally operate 30°–80°C cooler than the cast-iron pistons due to their higher thermal conductivity, thus reducing the knocking problems. Many modern engines employ pistons made of ceramic which is a poor conductor but can tolerate much higher temperatures. Some very large engines use water-cooled pistons.

14.12 VALVE COOLING

The elementary poppet valve consists of a head and a stem. The head is subjected to a uniform gas pressure loading which depends upon the maximum combustion pressure and is subjected to a concentrated load at the centre due to the valve-spring force when the valve is seated.

Heat flows into both the inlet and exhaust valves from the hot gases in the combustion chamber. The inlet valve runs much cooler than the exhaust valve, because it comes in contact with a comparatively cooler mixture during the induction process. The exhaust valve is not only subjected to the combustion temperatures but it also comes in contact with the hot exhaust products during the release and exhaust processes. Consequently, the exhaust valve runs hot, and the cooling of this valve and its seat is of great importance.

The following features help to keep down the temperature of the exhaust valve: A heavy head to allow rapid travel of heat from the centre to the seat, a wide seat to reject the heat, a large stem, and a heavy valve guide extending fairly close to the head.

To improve the conductivity of exhaust valves it is now customary to make the valve stems, and often the heads as well, hollow and to partly fill this space with sodium which melts and transfers an appreciable amount of heat from the head to the cooler stem of the valve during operation. The sodium melts at high temperature (near about 750°C) and is thrown up and down due to the motion of the exhaust valve. The heat from the hot region is thus transferred to the comparatively colder regions. Sodium is a good conductor of heat and has a high boiling point. Figure 14.6 illustrates a typical sodium-filled exhaust valve with a tapered plug seal placed into position to seal the hollow stem. Stellite alloy is a range of cobalt-chromium alloys designed for wear resistance.

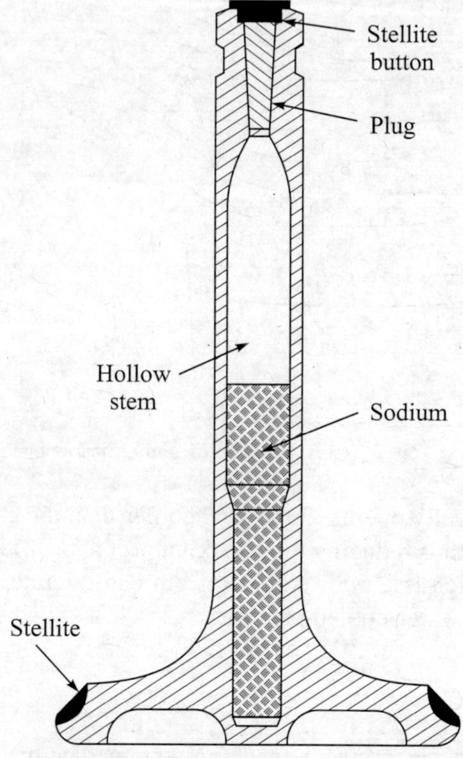

Figure 14.6 Sodium-cooled exhaust valve.

14.13 EFFECT OF OPERATING VARIABLES ON HEAT TRANSFER

Heat transfer within the different surfaces of the engine combustion chamber and the temperature distribution depend on certain variables as discussed in the following subsections.

14.13.1 Mixture Strength

Mixture strengths have marked influence upon the heat transmitted to the cylinder walls. The hottest cylinder is obtained in an SI engine, and the maximum heat transfer occurs at the equivalence ratio for maximum power of about $\phi = 1.1$. Heat transfer decreases as ϕ is leaned or enriched from this value. However, maximum heat transfer as a per cent of the fuel's chemical energy will occur at the stoichiometric condition, $\phi = 1.0$, and heat transfer decreases from this maximum value for richer and leaner mixtures as shown in Figure 14.7.

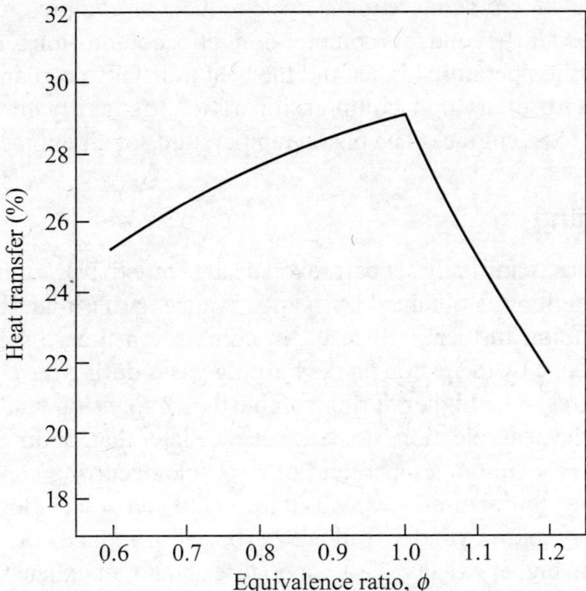

Figure 14.7 Percentage heat transfer to combustion chamber walls of an SI engine as a function of the equivalence ratio.

Weaker mixtures give less heat to cylinder walls and more to the exhaust gases, thus resulting in increased exhaust valve temperature. Also, an increase in the equivalence ratio beyond that for full power reduces the temperature of the cylinder surfaces because of incomplete combustion. This temperature is further reduced by the latent heat of evaporation of the greater quantity of fuel entering the cylinder.

In CI engines, the equivalence ratio variation is incorporated directly in the load variation effects.

14.13.2 Compression Ratio

In SI engines, as the compression ratio is increased up to about 10, heat transfer to the coolant is slightly reduced. Beyond this value of the compression ratio, knock may occur, and heat transfer is increased slightly. The magnitude of the change in heat transfer to the coolant with the compression ratio is very little. These changes in heat transfer occur mainly because of the change in gas properties. An increase in compression ratio increases the cylinder gas pressure, peak burned gas temperatures, gas motion and the surface-to-volume ratio close to top dead centre. It makes the combustion faster. Combustion duration will be short. Engine parts are exposed to high combustion temperature for a short time. However, the piston and the spark-plug electrode temperatures increase due to the higher peak combustion temperature at higher compression ratios. The gas temperature, late in the expansion stroke and during the exhaust stroke, is reduced because of the greater expansion of the gases and greater exposure of the larger cylinder wall area at higher compression ratios. This results in decreased head and exhaust valve temperatures. The temperature of the exhaust gas will also be much lower, and thus the heat rejected during blowdown will be less. The average gas temperature in the cycle reduces as compression ratio is increased up to 10, but when knock occurs beyond this compression ratio, combustion temperature is very high. It makes the average gas temperature higher and the heat transfer to coolant increases.

The CI engines, with their high compression ratios, generally have lower exhaust gas temperatures compared to SI engines. The piston temperatures of CI engines are generally higher.

14.13.3 Spark Timing

Spark timing in SI engines is normally set before top dead centre (bTDC), and for which maximum brake torque (MBT) condition is obtained. A spark advance earlier than this optimum condition (advancing the spark timing further) will result in combustion time losses on the compression stroke. The extra work done by the piston on the burning gases during the compression stroke will cause the gas temperatures to be higher during combustion, expansion, and blowdown processes, and thereby increasing the heat rejection. A spark advance later than optimum (retarding the spark timing) will result in the maximum temperature of the cycle occurring considerably after the top dead centre, since the late ignition timing extends the combustion process longer into the expansion stroke, with consequently more cylinder wall area exposed. It reduces the maximum burned gas temperature but results in higher exhaust gas temperature and hotter exhaust valves and ports. Heat loss to exhaust increases. However, the average gas temperature will be lower, resulting in less heat transfer at retarded spark timing. These variations of heat transfer as a percentage of fuel's chemical energy with spark timings are shown in Figure 14.8.

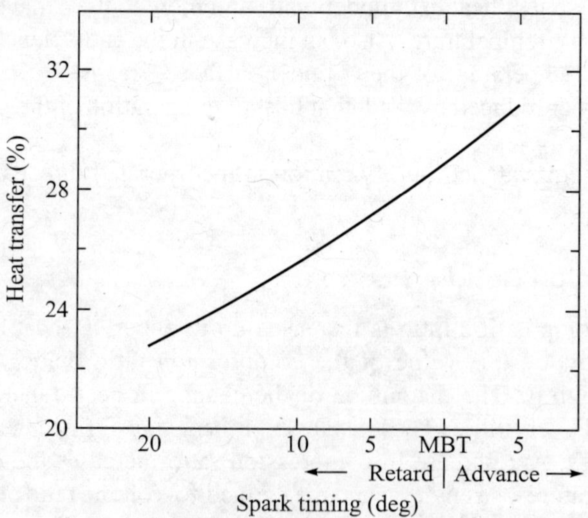

Figure 14.8 Percentage heat transfer to combustion chamber walls of an SI engine as a function of spark timing.

14.13.4 Engine Size

If two geometrically similar engines of different sizes are run under the same operating conditions, the heat loss fluxes to the coolant per unit area will be about the same, but the absolute heat loss of the bigger engine will be greater due to its larger surface areas.

A bigger engine generates more power with increased thermal efficiency. The energy generated depends upon the cylinder volume which goes up with length cubed, while the heat losses depend upon the surface area which goes up with length squared. This makes the bigger engines more efficient under similar conditions. It is therefore desirable to have a combustion chamber with a high volume-to-surface area ratio for good thermal efficiency. This is one reason why modern overhead valve engines are more efficient than the older valve in block L-head engines that had large combustion chamber surface areas. Also, a single open combustion chamber will have less percentage heat loss than the one with a split dual chamber that has a large surface area.

14.13.5 Engine Speed

Heat transfer during the intake and exhaust strokes and even during the early part of the compression stroke increases with increase in speed. This is because of the increased gas velocity in the intake and exhaust system with a resulting rise in turbulence and convective heat transfer coefficient.

Gas velocities within the cylinder during combustion and expansion are fairly independent of engine speed. Therefore, heat transfer by convection does not depend on the engine speed during this part of the cycle. Radiation, which is only important during this portion of the cycle, is also independent of speed. The rate of heat transfer (kJ/s) during this part of the cycle is therefore constant. However, at higher speeds the time taken to complete a cycle is less, therefore heat transfer per cycle reduces and the thermal efficiency increases.

Figure 14.9 shows the engine temperatures for various components as a function of engine speed for a typical SI engine. It is observed that as the engine speed increases the steady-state temperatures of all the components within the engine will increase.

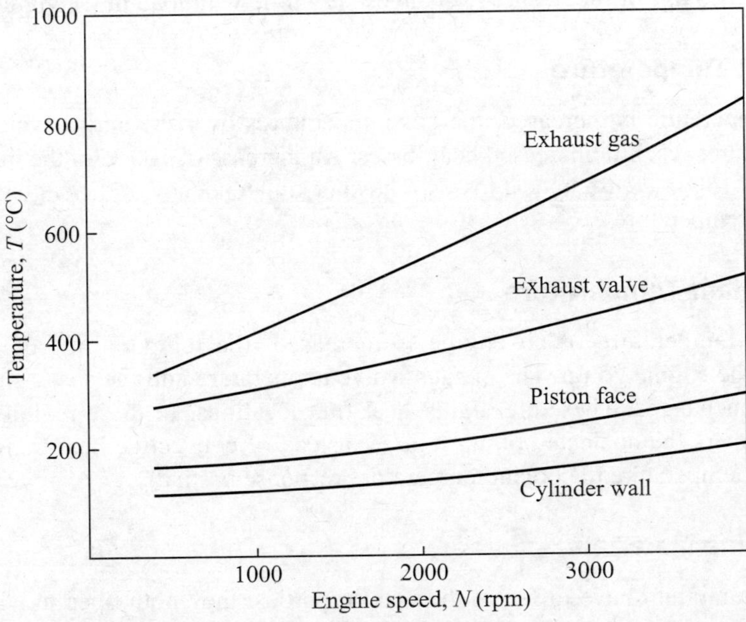

Figure 14.9 Engine temperatures as a function of engine speed for a typical SI engine.

At higher engine speeds there is less time available for heat transfer per cycle (kJ/cycle), which makes the engine to run hotter. Therefore, the rate of heat transfer (kJ/s) from the engine parts to the coolant increases with increase in engine speed.

14.13.6 Load

As the load on the engine increases the engine speed goes down for a given throttle position in SI engines. In order to maintain the engine speed constant, the throttle is opened more. It causes less pressure drop across the throttle valve, resulting in increased pressure and density in the intake system, consequently the mass flow rate of air and fuel mixture entering the engine increases. Reynolds number is proportional to mass flow rate, and the heat transfer coefficient increases with the increase in Reynolds number. Therefore, an increase in load increases the rate of heat transfer (kJ/s). However, heat loss as a percent of the fuel's chemical energy per cycle goes down slightly as engine load increases. Engine temperatures increase with load.

When the load on a CI engine is increased, more fuel is injected in order to run the engine at constant speed. The mass of air remains the same, so the total mass of the mixture increases very little, that too during the later part of the cycle. This slight increase in mass does not affect the convective heat transfer coefficient and therefore the heat transfer coefficient within the engine is fairly independent of engine load. However, steady-state temperatures depend upon the engine load. At light loads, less fuel is injected and burned, therefore, temperatures are lower, causing less heat transfer. At heavy loads, more fuel is injected and burned, therefore steady-state temperatures are higher, causing more heat transfer. Combustion of the richer mixture at heavy loads also generates more solid carbon soot. This further increases heat transfer by radiation. The amount of fuel and, consequently, the amount of energy in fuel per cycle increases with load. Therefore, the heat loss as a percentage of fuel's energy changes very little with load in CI engines.

14.13.7 Inlet Temperature

As the inlet temperature is increased, the gas temperatures over the entire cycle also increase. Higher temperatures result in increased heat losses. An increase of 100°C in the inlet temperature gives around 10–15% increase in heat losses. The knocking tendency of the engine increases with increased cycle temperatures.

14.13.8 Coolant Temperature

As the coolant temperature of an engine is increased, the temperatures of all the cooled components of the engine go up. The exhaust valve temperature and the spark plug temperature remain almost unchanged. These are higher heat flux locations. In these regions, heat transfer to the coolant enters the nucleate-boiling regime instead of convective heat transfer, where the change in metal temperature to coolant temperature response is small.

14.13.9 Engine Materials

Cast iron and aluminium have different thermal properties, they both operate with combustion chamber surface temperature in the range from 200°C to 400°C, this range is low relative to burned gas temperatures. Heat losses from the working fluid are reduced in the case of a substance which can operate at much higher temperatures. The thermal conductivity of ceramic materials, such as silicon nitride and zirconia, is lower than that of cast iron or aluminium, resulting in very high temperatures. The ceramic materials are able to tolerate the higher temperatures. With these thermally insulating materials it is possible to reduce the heat transfer through the wall by a substantial amount.

Ceramic coatings on cylinder walls, pistons and exhaust valves are most suitable for CI engines for improving the thermal efficiency. However ceramic coatings are not suitable for SI engines because such coatings facilitate the transfer of heat from the hot walls to the incoming charge, thereby increasing the unburned mixture temperature and leading to knock early in the cycle.

14.13.10 Knock

Gas pressure and temperature increase substantially above the normal combustion levels when the engine is knocking. Knock increases the local heat fluxes from the gas to regions of the piston, the cylinder head, and the liner in contact with the end gas. Extensive high local pressures and higher material temperatures due to knocking can cause surface damage to piston crowns and valves.

14.13.11 Swirl and Squish

Increased gas velocities due to swirl or squish motion result in a higher convective heat transfer coefficient. This results in a better heat transfer to the walls. Use of a shrouded valve to increase gas velocities within the cylinder increases heat transfer.

14.14 COOLING SYSTEMS

There are two basic types of cooling systems used in reciprocating IC engines to absorb and dissipate the heat from the hot cylinder walls. The two types of cooling systems are direct or air-cooling and indirect or liquid cooling.

14.15 AIR-COOLED SYSTEM

The air-cooled system is mainly used in small engines and in some medium-sized engines. Most small-engine tools and toys like lawnmowers, chain saws, model airplanes, etc. are air cooled. This system occupies less space, and allows the weight and cost of these engines to be kept low. Some motorcycles, automobiles, stationary engines and aircraft also have air-cooled engines because of their lower weights.

In air-cooled engines the current of air flows across the external surfaces of the cylinder walls and the cylinder head to remove the necessary heat and hence to prevent them from overheating. The film heat transfer coefficient of air is low compared to that of liquid. The rate of heat transfer is, therefore, improved by increasing the surface area in the air-cooled system. This is accomplished by providing extended surfaces, called cooling fins, on the outer surfaces of the engine, such as cylinder walls and the cylinder head as shown in Figure 14.10. The outer surfaces of the engine and fins are made of good heat-conducting metals to promote maximum heat transfer. On mobile vehicles, like motorcycles and aircraft, the forward motion of the vehicle supplies the air flow

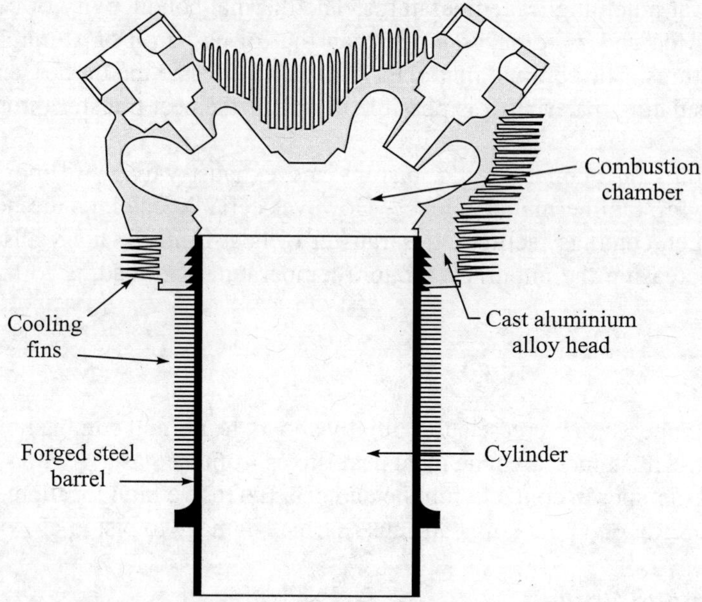

Figure 14.10 Cylinder and cylinder head of an air-cooled aircraft engine.

across the finned surfaces. Deflectors and duct work are often added to direct the flow to critical locations. Automobile engines usually have fans to increase the rate of air flow.

14.15.1 Cooling Fins

The necessary cooling surface area for an air-cooled engine is provided by radial or longitudinal fins on the cylinder wall and they are mounted normally on the cylinder head.

The dimensions and the number of fins are selected in such a way that enough heat is dissipated without causing the overheating of the cylinder. At the same time the heat dissipated should not be too much to cause overcooling, resulting in a loss of thermal efficiency.

The heat dissipating capacity of fins depends upon their size and shape, and is directly proportional to the temperature difference. As heat is gradually dissipated from the fin surface, the temperatures of the fins and, hence, their heat transfer rates decrease from roots of the fins to their tips. The thickness of the fin may be reduced towards the tip because of the decreased heat transfer rate. The material of a fin is used most efficiently if the drop in temperature from the root to the tip is constant per unit of length. A comparison of fins of different cross-sections is shown in Figure 14.11. The fin, marked A, is the one with

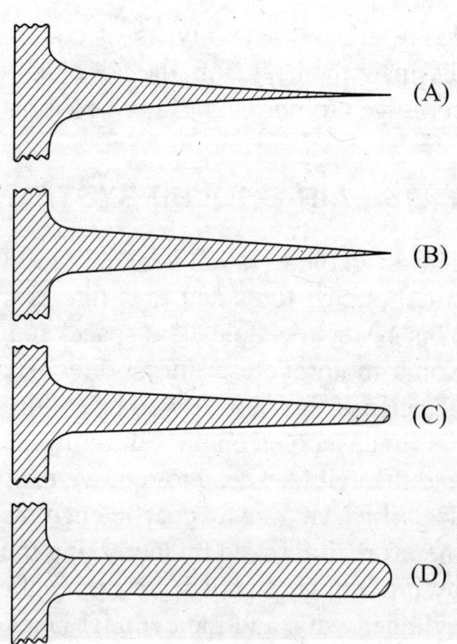

Figure 14.11 Fins of different cross-sections.

concave, or hollow, sides of parabolic shape terminating in a sharp edge. It is theoretically an ideal shape which gives maximum heat dissipation for a given weight of fin metal. The nearest practical approach is the triangular fin, marked B, but this is open to the objections of castings or machining difficulties and the risk of damage to the fin edges at the tips of the fins. From all points of view, therefore, the truncated conical fin with rounded edges, marked C, is the best practical compromise. The rectangular section, marked D, has the smallest temperature drop, dissipates less heat, and being heavier, represents rather an inefficient use of material. Also, the increased thickness of the outer portions of the fins tends to interfere with the circulation of the cooling air between the fins. Therefore, such a shape is not used. The drop in temperature from root to tip in each type is shown by curves in Figure 14.12.

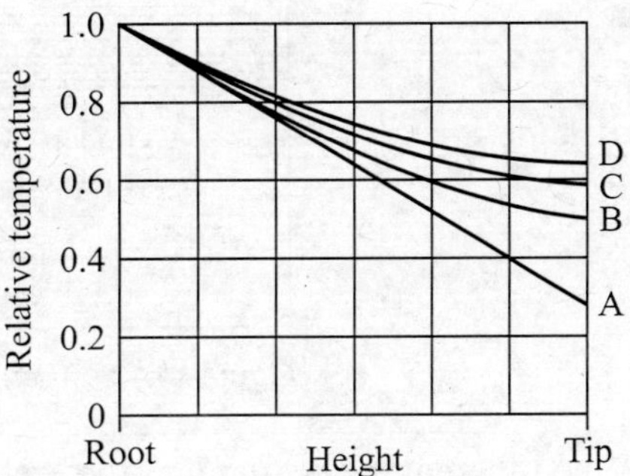

Figure 14.12 Relative temperatures of fins from root to tip.

14.16 LIQUID-COOLED SYSTEMS

In liquid-cooled systems, water is generally used as a cooling medium. However, other liquids or a mixture of water and other liquids may also be used in the system to prevent freezing of the coolant at lower temperatures. In this system the cylinder walls and heads are surrounded with jackets through which the cooling liquid circulates and absorbs the heat from the hot metal walls of the engine. The liquid is then cooled by means of an air-cooled radiator system, cooling tower or cooler and recirculated through the engine jackets. Thus, the liquid coolant absorbs heat from the cylinder and rejects it to the air stream.

Liquid cooling can be carried out by the following methods:

(a) Direct or non-return system
(b) Thermosyphon or natural circulation system
(c) Forced or pump circulation system
(d) Evaporative cooling system

14.16.1 Direct or Non-return System

This system is used where ample quantity of water is available. The cooling water enters the jackets surrounding the cylinder and cylinder heads. Water absorbs heat from the cylinder, the cylinder head, the valve ports and seats and other hot spots, and the resulting hot water is rejected into the sink. The water so rejected is not recirculated in the jacket for cooling the engine parts. Therefore, plenty of fresh cold water is required for cooling.

14.16.2 Thermosyphon or Natural Circulation System

In this system, there is a natural circulation of water around the cooling system. This system is illustrated in Figure 14.13. The principle employed depends upon the fact that when water is heated, its density decreases and it tends to rise to the top of the system, while the colder water tends to sink. When the engine is at rest, all of the water in the system is at the same temperature.

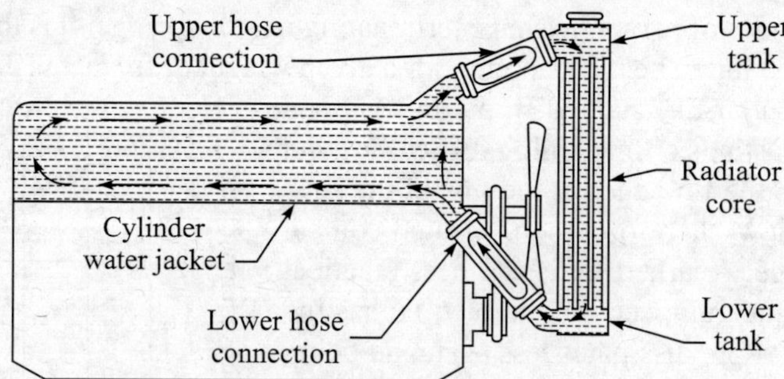

Figure 14.13 Thermosyphon or natural circulation cooling system.

When the engine is running, the water around the cylinder walls and the cylinder heads in the jacket becomes heated and flows in an upward direction to the upper tank of the radiator, its place being then taken by colder water from the lower part of the radiator. The hot water in the top of the radiator flows downwards by gravity through narrow tubes, where it is rapidly cooled by the rush of cold air though the radiator core. Thus, heat is rejected from water to the air and water reaches the bottom of the radiator at an appreciably lower temperature.

The rate of circulation of water in this system is low. It is therefore suitable only for small engines where the loads are light and only a relatively simple cooling system is necessary.

14.16.3 Forced or Pump Circulation System

A forced or pump circulating system is shown in Figure 14.14. It is almost similar in construction to the thermosyphon system except that a pump is used for the circulation of the coolant to increase the rate of flow of the coolant. This system is suitable for large automobiles.

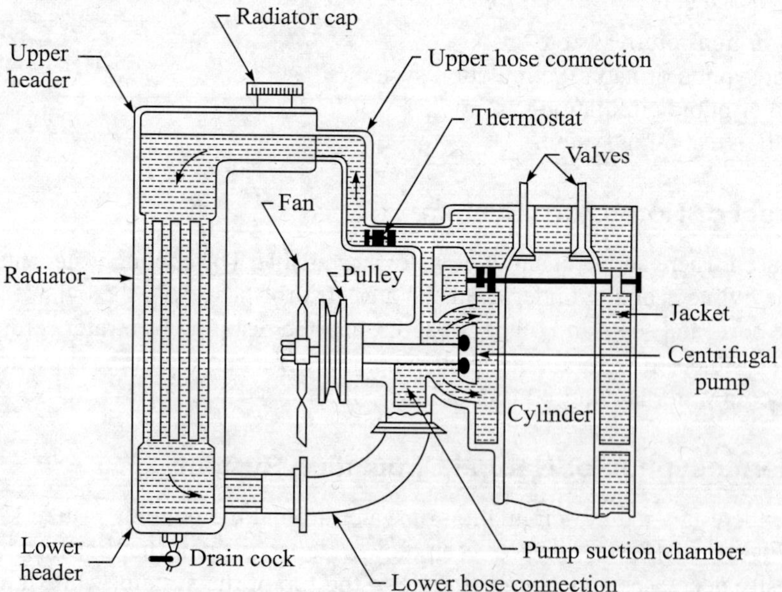

Figure 14.14 Layout of a pump circulating cooling system.

The coolant is circulated through jackets around the combustion chambers, cylinders, valve seats, and valve stems by a centrifugal pump driven by the engine. The coolant is passed through the radiator where it is cooled by air drawn through the radiator by a fan and by the air flow developed by the forward motion of the vehicle. Often the fan and the pump are mounted and driven on a common shaft. After passing through the radiator, the coolant is recirculated to the engine jackets through the radiator outlet hose, through the pump and the cylinder inlet passage. A thermostat is used to control the temperature of the coolant, as required for proper cooling.

This system has a number of component parts: pump, water jacket, radiator, fan, and thermostat.

Pump

A centrifugal type of pump is normally used for the cooling system. The hub and pulley assembly of the pump is driven by a fan belt, or the pump is mounted on a common shaft with the fan. It has a casing, an impeller, the inlet and outlet pipes and a sealing arrangement. The casing is generally of volute type that helps in building up the pressure. The impeller may be open or shrouded, on which several blades are mounted. Since the impeller is always submerged in the water, a seal must be placed on the shaft to prevent leakage. The sealing arrangement incorporates gland and packing, and prevents any possible leakage.

Water or coolant enters into the eye of the pump through the inlet pipe from the bottom of the radiator and it is caught by the rotating vanes of the impeller and is thrown by centrifugal force through the outlet to the water jackets. Because of the clearance between the impeller blades and the pump casing, centrifugal pumps are non-positive in action and will not build up dangerous pressures if an obstruction occurs in the radiator or other parts of the system.

Water jacket

The water passages between the double walls of the cylinders and the cylinder heads are formed as water jackets. These jackets are usually cast as an integral part of the cylinder block and the cylinder head. The jacket should cover the entire length of the stroke in order to reduce unequal expansion of the cylinder, thus reducing the thermal stresses. It also helps to prevent breakdown of the lubricating oil film due to excessive temperatures. In multi-cylinder engines, headers are usually provided to furnish equal distribution of the coolant to all cylinders. The header is supplemented by tubes and ducts which give the high rate of flow around critical sections of the engine.

Radiator

The purpose of the radiator is to cool the water that has absorbed heat from the engine. It is a heat exchanger in which the water passing downwards through it in thin streams is efficiently cooled by the forced flow of atmospheric air over a large surface area around the pipes or tubes carrying the water.

The radiator consists of an upper tank, containing the filler tank, and a lower tank. Between the tanks is the core which divides the water into thin streams. In passing through the core, the heat from the water is transferred to the metal walls and to the air streams which are forced through by a fan. The upper tank is connected to the water outlet from the engine jacket by a rubber hose, and the lower tank is connected by a hose to the jacket inlet through the pump as shown in Figure 14.14.

Fan

A fan is used to draw air through the radiator to cool the water. Such a fan usually has four blades, spaced unequally around their spiders to promote quieter operation. It is mounted on ball bearings to reduce friction as much as possible. The fan takes power from the engine and is driven by a V-belt. The fan bracket is so constructed that the tension on the belt is adjustable. At all times the belt should be under sufficient tension to prevent slippage. In the belt-driven fan, power is absorbed from the engine even when it is not required. The fan is required only at low engine speeds when the coolant is hot. At speeds above about 50 km/h the relative air flow into the radiator is usually sufficient.

The amount of power absorbed in driving the cooling fan increases rapidly with the fan speed, and may amount to as much as 8–10% of the brake power, in extreme cases. If a thermostat in the hottest part of the cooling system is used to actuate a device to switch off the fan when the temperature drops below a given limit and to switch it on when the temperature exceeds the normal value, an appreciable saving in fan power can be achieved. Various controlled-speed fan drives are now available. These include the electric motor driven fan with its switch controlled by a thermostat, the magnetic clutch fan drive, the viscous fluid drive, the variable fan blade pitch method, etc.

Thermostat

For efficient engine operation the coolant temperature should not be allowed to drop below a certain minimum value. It is achieved by means of a thermostat installed in the coolant loop. A thermostat is a thermally activated valve. When the thermostat is cold, it is closed and does not allow any fluid to flow through the radiator. The fluid recirculates in the jacket through a bypass passage. It happens when the engine is being started or when it is cold. As the engine warms up, the thermostat also warms up, and thermal expansion opens the flow passage and allows circulation of coolant through the radiator. The higher the temperature, the greater the flow passage opening, with the greater resulting coolant flow. The coolant temperature is, therefore, controlled fairly accurately by the opening and closing of the thermostat.

14.16.4 Pressure Cooling System

Older automobiles were operated at atmospheric pressure using water only. The temperature of water was restricted to a higher limit of about 82°C in order to avoid boiling for safety reasons. For better efficiency of the engine it was necessary to increase the coolant temperature. This was done by pressurizing the coolant loop and adding ethylene glycol to the water. The ethylene glycol raised the boiling temperature of the fluid. Pressurizing the system further raises the boiling temperature of the fluid. The normal coolant system pressures are about 2 bar. The amount of heat dissipated under these conditions would be greatly increased, so that a proportionately smaller radiator could be fitted.

In the cooling systems of this type, the pressure type radiator cap fits over the radiator filler tube and seals tightly around the edges. The cap contains two valves—the pressure valve and the vacuum valve. Figure 14.15 shows the radiator pressure–vacuum cap, showing the operation of the vacuum valve and the pressure valve. The pressure valve consists of a valve held against a valve seat by a calibrated spring. The spring holds the valve closed so that pressure is produced in

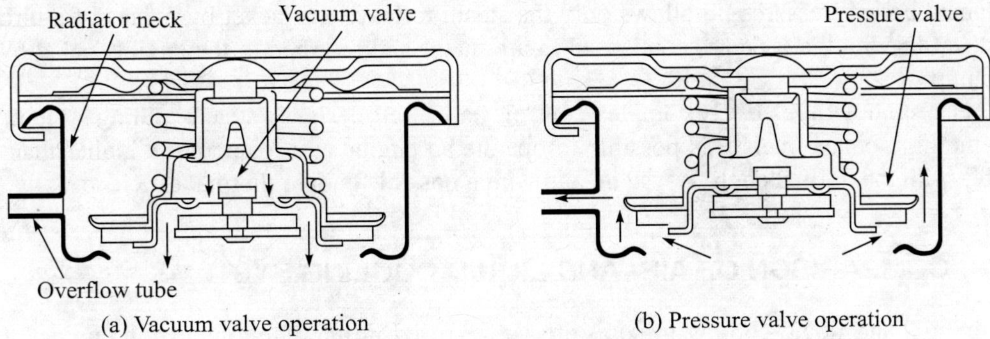

(a) Vacuum valve operation (b) Pressure valve operation

Figure 14.15 Radiator pressure-vacuum cap.

the cooling system. If the pressure rises above that for which the system is designed, the pressure valve is raised off its seat. This relieves the excessive pressure. The vacuum valve prevents the formation of a vacuum in the cooling system when the engine is stopped and begins to cool. If a vacuum forms, the atmospheric pressure from the outside causes the small vacuum valve to open and admit air into the radiator. Without a vacuum valve, the pressure within the radiator might drop so low that atmospheric pressure would collapse the radiator.

In the cooling system with an expansion tank, the radiator pressure cap is more or less installed permanently. The radiator overflow tube in the pressure cap is connected to the expansion tank. If coolant is required to be added to the cooling system, it is put into the expansion tank.

14.16.5 Evaporative Cooling System

It is based on the principle that when the coolant vaporizes, it absorbs the latent heat of vaporization from the metal surfaces to be cooled. The quantity of heat thus removed is much more and it is also possible to use a much smaller quantity of cooling water and a smaller radiator. The schematic of an evaporative cooling system is shown in Figure 14.16. In this system a positive displacement type gear pump is used, in preference to the usual centrifugal one, to return the water from the condensed steam from the base of the radiator to the cylinder jackets. The water jacket is

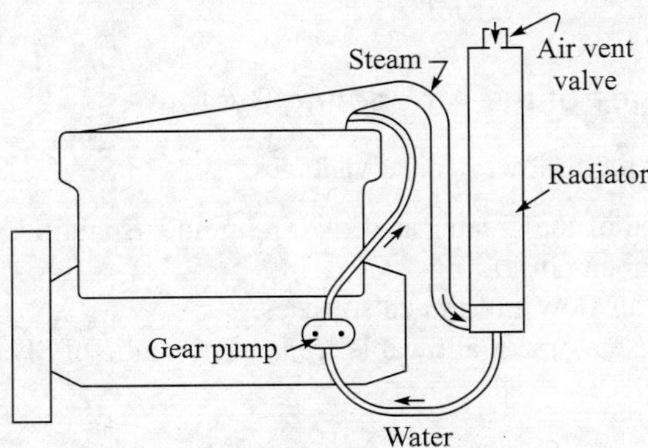

Figure 14.16 Evaporative cooling system.

maintained under pressure and allows only the steam to leave the jacket by means of a throttle. The steam leaving the jacket is condensed in a radiator and returned to the engine jacket by the gear pump.

This system with relatively complex controls has a well-designed steam-cooling system, and using the high octane fuels it is possible to operate an engine at a temperature higher than that possible with the normal cooling system, and with a possible gain in thermal efficiency.

14.17 COMPARISON OF AIR- AND LIQUID-COOLING SYSTEMS

Both the air- and liquid-cooling systems are widely used in internal combustion engines. It is, therefore, important to know the advantages and disadvantages of these two systems over one another.

14.17.1 Advantages of the Air-cooling System

When compared with the liquid-cooled engine, the air-cooled engine has the following advantages:

1. A simpler design of the engine is available because of the absence of a more complicated cylinder block with jackets of the liquid type of cooling system.
2. The cooling system is much simpler due to the absence of cooling liquid, pipes, pump, radiator, etc.
3. There is no danger of coolant system failures and leakage, owing to the absence of pump, radiator joints, tubing, etc. The air-cooling system is therefore less liable to breakdown, and is also easier to maintain.
4. The engine is a self-contained unit with an integral cooling system; it requires no external cooling components such as radiator, header tank, etc.
5. The air-cooled engine does not suffer from the problem of freezing under very cold weather storage conditions. This problem is encountered in the water-cooled engines.
6. The air-cooled engine warms up more rapidly and can be operated under load in far less time after starting from the cold condition compared with a liquid-cooled one.
7. The air-cooled engine runs hotter, resulting in reduced carbon deposits on combustion chamber walls.

14.17.2 Disadvantages of the Air-cooling System

The air-cooled engine has the following disadvantages:

1. It is noisier because of the greater air flow requirements and because of the absence of a water jacket to dampen noise.
2. It needs a directed air flow and finned surfaces.
3. It runs hotter, so the maximum allowable compression ratio of SI engines has to be lowered to avoid knocking.
4. It is less efficient.
5. Cooling in the air-cooled engine is less uniform.

14.17.3 Advantages of the Liquid-cooling System

The advantages of the liquid-cooling system are summarized as follows:

1. Cooling of the cylinder barrels and cylinder heads is more uniform due to the flow of coolant through jackets. It reduces the temperatures of the cylinder head and valve seats, permitting the use of higher compression ratios in SI engines for a given cylinder size. It results in higher power output per cylinder volume.
2. The liquid-cooled engine can be made more compact with appreciably smaller frontal area to reduce air resistance.
3. The fuel consumption of high compression liquid-cooled engines is lower than that for air-cooled engines.
4. The liquid-cooled engine with its cooling system is not necessarily required to be located in front of the vehicle. Its location does not affect its cooling.
5. The cooling of very large engines with the liquid-cooling system does not impose any serious design problem. In case of high powered air-cooled engines, considerable difficulty arises to get enough circulation of air for cooling purposes.

14.17.4 Disadvantages of the Liquid-cooling System

The water-cooled engine has the following disadvantages:

1. The engine requires complicated cylinder blocks with jackets.
2. The cooling system requires a pump, a radiator, and a fan, etc. These components increase the size, weight and cost of the cooling system.
3. The cooling system requires more maintenance. In the event of failure of the cooling system or leakage of the coolant, serious damage is likely to be caused to the engine.
4. The engine takes considerable time for warming-up.
5. The power absorbed by the pump for water circulation and fan is considerable and this affects the brake power of the engine.

14.18 MODERN COOLING CONCEPTS

A number of alternative engine cooling methods have been tested and developed. A few of them are described here.

Engines with dual-circuit cooling have dual water jackets. Such engines have separate cooling circuits for cylinder block and for cylinder head. The coolant temperature around the engine block is kept higher, which reduces oil viscosity and also reduces frictional losses associated with the piston and cylinder wall. The coolant temperature around the cylinder head is kept low to reduce knocking. This helps in increasing the compression ratio of the engine. It has been observed that with dual-circuit cooling, there is a reduction in the fuel consumption and hydrocarbon emissions but the NO_x emission on the other hand increases.

In order to reduce the cylinder size and weight of a small engine the cooling fins or water jackets can be eliminated by using lubricating oil as a coolant. The oil cools the cylinders by being circulated to circumferential passages built into the cylinder walls. It provides adequate cooling with more uniform temperature distribution. An oil-cooler system would be required with this type of engine.

In case of a leakage in the coolant system, some automobiles offer a safety feature on the engines by firing only half the number of cylinders at a time. The remaining cylinders do not get fuel and continue to pump air. Each cylinder fires for a period of time and pumps cooling air at other times. This technique helps to cool the engine enough and prevents overheating, and permits driving a long distance at moderate speeds with no coolant in the cooling system.

14.19 ADIABATIC ENGINES

The adiabatic process is defined as a no heat loss process, and hence an adiabatic engine implies a no heat loss engine. It should be noted that the engine is not adiabatic or without loss in the true thermodynamic sense, however the engine is without the conventional forced cooling system and strives to minimize the heat loss. This type of engine usually does not have a cooling jacket or finned surfaces. A small increase in the power output can be gained by decreasing the heat losses from the engine cylinders. Most of the reduced heat loss energy ends up in the enthalpy of the exhaust. Additional power can be obtained by converting the exhaust gas thermal energy to useful power through the use of turbomachinery.

In the adiabatic engine concept the engine parts are insulated with materials which can withstand higher temperatures. A preliminary cross-section of an adiabatic engine is shown in Figure 14.17. Recent advancements in key engineering areas such as high temperature materials, high temperature tribology and advanced ceramics have made it possible to build a minimum cooled adiabatic engine without mechanical or thermal failure. The concept centres around insulating as much of the combustion chambers and cylinder walls as possible, thereby reducing the heat rejection. The insulated chamber is expected to have other beneficial effects of diesel operation, including wide fuel tolerance, lower compression ratio, smoother combustion, improved cold start and emission characteristics. Adiabatic engines can be made smaller and

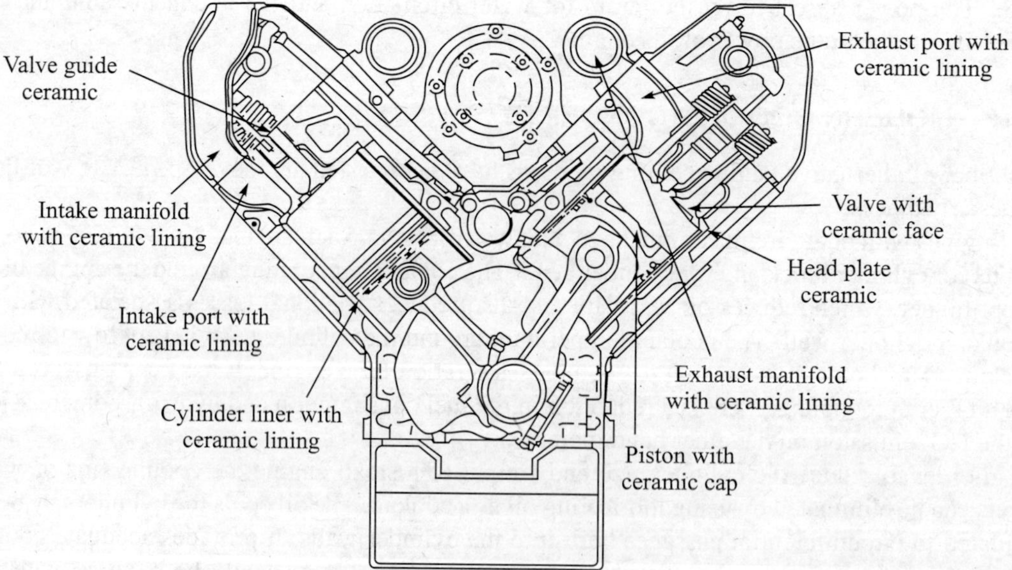

Figure 14.17 Cross-section of an adiabatic engine

lighter than the conventional engines. Vehicles can be made more aerodynamic with a lower drag coefficient because there is no radiator. This also gives greater flexibility in engine location and positioning.

Adiabatic engines are all of the compression ignition type. They cannot be used as a spark-ignition engine. In an SI engine, the thermal coating will increase the wall temperatures which can lead to unwanted knocking or pre-ignition.

REVIEW QUESTIONS

1. Why is it necessary to cool internal combustion engines?

2. What are the disadvantages of overcooling an internal combustion engine?

3. Show the temperature distribution in the different engine components of a typical SI engine under steady-state condition.

4. Name some of the hottest regions inside the engine which are difficult places to cool. Explain, why these regions cannot be efficiently cooled.

5. Show the temperature variations of spark plug, exhaust valve, piston and coolant in a typical SI engine during the warm-up period.

6. Show and discuss the temperature variation of the gases inside the cylinder during the different processes in a cycle of a four-stroke engine.

7. Define the conduction mode of heat transfer. Give the expression to evaluate heat flux under steady-state condition. Name the parts of the IC engine in which the heat transfer mode is by conduction.

8. Define the convection mode of heat transfer. Discuss the role of convective heat transfer in IC engines. Give the expression to evaluate heat flux by forced convection in the steady flow.

9. Give the non-dimensional relation given by McAdams to evaluate the convective heat transfer coefficient for flow through pipes. Recognize the non-dimensional numbers used in the expression. Evaluate the heat flux.

10. Define the radiation mode of heat transfer. Give an expression to evaluate heat flux by radiation. Explain, why heat transfer by radiation in a CI engine is more significant than that in an SI engine.

11. Discuss the heat transfer in the intake system of an IC engine. Why is it necessary to heat the intake manifold? What are the various methods used to heat the intake manifold? Give reasons for limiting the heating of inlet charge in the manifold.

12. Derive an expression to evaluate heat flux in the combustion chamber of an IC engine. Discuss the variations of gas temperature, coolant temperature, convective heat transfer coefficient on the gas side and those on the coolant side during an engine cycle.

13. Briefly describe the heat transfer phenomenon in the exhaust system of an IC engine.

14. Why is it difficult to cool the piston of an IC engine? What are the methods of cooling the piston? How does the piston ring influence the piston temperatures? Show a typical variation of isotherms in a piston. Explain, why the hottest point is not often at the centre of the piston.

15. How is the exhaust valve of an IC engine cooled?
16. Show and discuss the variation of percentage heat transfer to combustion chamber walls of an SI engine as a function of the equivalence ratio.
17. Discuss the effect of compression ratio, spark timing and engine size on heat transfer.
18. Discuss the effect of engine speed on heat transfer. Show the variation of engine temperatures for various components as a function of engine speed for a typical SI engine.
19. Discuss the effect of load on heat transfer for SI and CI engines.
20. Discuss the effect of inlet temperature, coolant temperature, engine materials, knock, and swirl and squish on heat transfer.
21. Describe with the help of a diagram the air-cooled system of an IC engine. Where does this type of system find its applications?
22. Compare with the help of diagrams the cooling fins of different cross-sections. Show the relative temperature drops from root to tip in each type of fin. Which is the most suitable type of fin in practice? Support your answer with reasons.
23. Describe with the help of a diagram the construction and working of a thermosyphon liquid-cooling system. Mention all its limitations.
24. Describe with the help of a diagram the construction and working of a forced circulation liquid-cooling system.
25. Briefly describe the pump, the water jacket, the radiator, the fan, and the thermostat used in a forced circulation liquid-cooling system.
26. How is the forced circulation liquid cooling system pressurized? What are the advantages of the pressure cooling system? Describe with the help of diagrams the operation of the valves of the radiator cap.
27. Describe the working principle of an evaporative cooling system with the help of a diagram.
28. What are the advantages and disadvantages of an air-cooling system?
29. What are the advantages and disadvantages of a liquid-cooling system?
30. Briefly describe some of the modern alternative engine cooling methods.
31. Give the concept of an adiabatic engine. Explain, why this concept is suitable only for CI engines and not for SI engines.

CHAPTER 15 Air Capacity and Supercharging

15.1 INTRODUCTION

The power output of an engine depends primarily on the chemical energy in the mixture that is supplied to the engine per unit time and also depends on the efficiency with which the engine converts this energy into work. Fuel cannot burn without the presence of air. Therefore, the rate of fuel burned depends upon the air capacity of the engine, that is the amount of air which the engine is capable of drawing per unit time. Air capacity is practically unaffected by the presence of fuel. The reason for this is that the increase in mixture volume by fuel vapour is compensated by the decrease in volume due to reduction in temperature caused by the evaporation of fuel in the inlet manifold. Hence, an engine running at constant speed will draw approximately the same amount of air per unit time, whether the fuel is introduced into the air stream or not.

Increasing the energy input requires the induction of more charge per cycle or per unit time. An increase in engine speed increases the mixture input unless the increase in speed results in a greater decrease in volumetric efficiency, a parameter used to measure the effectiveness of an engine's induction process. Engine speed is also limited by thermal stresses. Supercharging is used to increase the power output by forcing the charge into the engine at pressures above atmospheric. Any change in design or operating conditions that increases the volumetric efficiency may increase the engine performance, provided that knocking, overheating or failure of some part does not occur.

15.2 EFFECT OF AIR CAPACITY ON INDICATED POWER

The indicated thermal efficiency, η_i, is defined as the ratio of the indicated power to the heat energy supplied by the fuel per unit time, i.e.

$$\eta_i = \frac{\text{ip}}{\dot{m}_f \times \text{CV}} \tag{15.1}$$

where

ip = indicated power, in kW

$\dot{m}_f$ = mass of fuel burned per second, in kg/s

CV = calorific value of fuel, in kJ/kg.

$$\therefore \qquad \text{ip} = \dot{m}_f \times \eta_i \times \text{CV}$$

If F is the fuel/air ratio,

$$F = \frac{\dot{m}_f}{\dot{m}_a}$$

$$\therefore \qquad \text{ip} = \dot{m}_a \times F \times \eta_i \times \text{CV} \qquad (15.2)$$

The principal factors influencing the thermal efficiency include the fuel/air ratio, the engine speed, the load or throttle opening, the nature of fuel, the dimensions of the engine, the design of the combustion chamber and the valve timing.

At low engine speeds, owing to the relatively longer time period for contact of the heated gases with the cylinder walls the heat losses would be greater than those at higher speeds and therefore the thermal efficiency will be lower. With progressive increase in the engine speed the relative heat losses would tend to diminish, but as the turbulence effect increases with engine speed, thus bringing a greater proportion of the gases into contact with the cylinder walls, these two effects will tend to balance each other at a certain engine speed, and thereafter the efficiency will be practically independent of any further speed increase.

If for each throttle position the ignition timing is adjusted so as to allow the engine to develop its best output for the given throttle opening, then the indicated thermal efficiency will remain practically constant over the range of throttle openings.

With optimum ignition timing the indicated thermal efficiency of a given engine with a given fuel depends therefore exclusively on the fuel/air ratio, so that at constant fuel/air ratio the value of $F \cdot \eta_i \cdot \text{CV}$ will be constant. At values of F between 0.075 and 0.085, the indicated thermal efficiency, η_i, decreases with increased values of F in such a way that $F \cdot \eta_i \cdot \text{CV}$ remains approximately constant over the range. The indicated power is therefore proportional to air capacity $\dot{m}_a$ for any constant value of F or for values of F varying between 0.075 and 0.085, i.e.

$$\text{ip} \propto \dot{m}_a \qquad (15.3)$$

15.3 IDEAL AIR CAPACITY

The volume of air which will be drawn ideally by the engine per minute will be equal to the number of suction strokes per minute times the displaced volume. In a four-stroke engine the ideal volume of air in m^3 admitted per cylinder per minute in the engine is given by

$$(\dot{V}_a)_{\text{ideal}} = \frac{NV_d}{2} \qquad (15.4)$$

where

N = engine speed, in rpm

V_d = volume displaced by piston, in m^3.

The ideal air capacity in kg/min is given by

$$(\dot{m}_a)_{\text{ideal}} = (\dot{V}_a)_{\text{ideal}} \times \rho_{a,i}$$

$$= \frac{N}{2} V_d \, \rho_{a,i} \tag{15.5}$$

where $\rho_{a,i}$ is the density of air at the inlet condition.

15.4 VOLUMETRIC EFFICIENCY

Volumetric efficiency is a measure of the effectiveness of an engine's induction process. The induction system includes the air-filter, the carburettor and the throttle plate in a spark-ignition engine, the intake manifold, the intake port, and the intake valve. They all restrict the amount of air which an engine of given displacement can induct. Volumetric efficiency is only used with four-stroke cycle engines which have a distinct induction process. Volumetric efficiency of a normally aspirated engine is the ratio of the actual volume flow rate of air into the engine's cylinder ($\dot{V}_a$) at the atmospheric pressure and temperature conditions surrounding the engine to the rate at which the volume is displaced by the piston ($\dot{V}_d$). The volumetric efficiency of a supercharged engine is based on the intake manifold pressure and temperature conditions. The volumetric efficiency is given by

$$\eta_V = \frac{\dot{V}_a}{\dot{V}_d} = \frac{\dot{m}_a / \rho_{a,i}}{V_d \, N/2}$$

$$= \frac{2\dot{m}_a}{\rho_{a,i} \, V_d \, N} \tag{15.6}$$

where

$\dot{m}_a$ = actual mass of air admitted per unit time, in kg/min

$\rho_{a,i}$ = inlet air density, in kg/m^3

V_d = volume displaced by the piston, in m^3

N = engine speed, in rpm

The rate of air mass admitted per minute '$\dot{m}_a$' and the mass of air inducted into the cylinder per cycle 'm_a' can be related by

$$\dot{m}_a = m_a \frac{N}{2} \tag{15.7}$$

Therefore, an alternate equivalent definition for volumetric efficiency is

$$\eta_V = \frac{m_a}{\rho_{a,i} \, V_d} \tag{15.8}$$

Using expressions (15.5) and (15.6), the volumetric efficiency can also be obtained as

$$\eta_V = \frac{\dot{m}_a}{(\dot{m}_a)_{\text{ideal}}} \tag{15.9}$$

Therefore, the volumetric efficiency may also be defined as the actual mass of air drawn into the engine during a given period of time to the theoretical mass which should have been drawn in during the same period of time, based upon the total piston displacement of the engine and the temperature and pressure of the surrounding atmosphere.

Typical maximum values of volumetric efficiency for naturally aspirated engines at wide open throttle lie in the range from 80 to 90%. The volumetric efficiency of the CI engine is somewhat higher than that of the SI engine, since the restriction in the intake system of a CI engine is generally less because of the absence of carburettor and throttle plate.

15.5 EFFECT OF VARIABLES ON VOLUMETRIC EFFICIENCY

The variation of volumetric efficiency with engine speed is shown in Figure 15.1. It is desirable to have maximum volumetric efficiency in order to obtain maximum power from the given size of the engine. The volumetric efficiency is maximum at a certain engine speed and decreases both at higher and lower speeds. There are many physical and operating variables which are responsible for this variation of volumetric efficiency with engine speed.

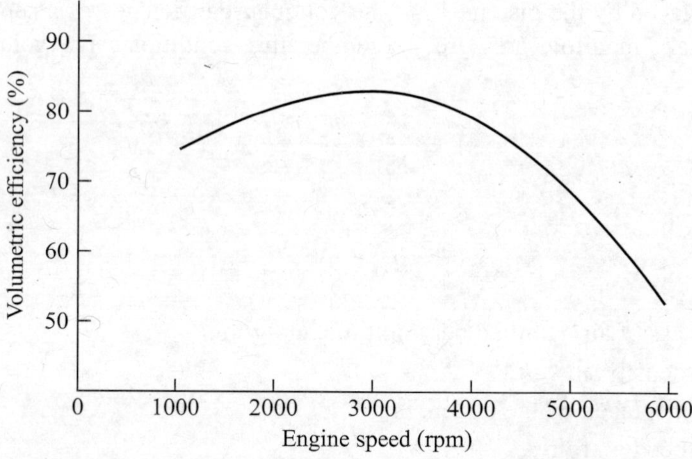

Figure 15.1 Volumetric efficiency as a function of engine speed.

Volumetric efficiency is affected by the fuel, the engine design and the operating variables. The effects of some of the important variables on volumetric efficiency are described below.

15.5.1 Fuel

Volumetric efficiency depends upon the type of fuel and how and when it is added, the fuel/air ratio, the fraction of fuel vaporized in the intake system, and the fuel heat of vaporization.

In engines equipped with carburettors or throttle body injection system, fuel is supplied early in the intake system. This results in a lower value of volumetric efficiency, because the fuel starts to evaporate early and more fuel vapour displaces the incoming air. This problem is not faced in the case of multipoint injectors. In the multipoint port injection system the fuel is injected at the intake valve ports. It will have higher volumetric efficiency because fuel evaporation does not occur until it enters the cylinder and hence no air is displaced in the intake manifold. In engines that inject fuel directly into the cylinders after the intake valves are closed, there is no loss of volumetric efficiency due to evaporation of fuel.

Gaseous fuels, like hydrogen and methane, displace more incoming air than liquid fuels and will therefore have less volumetric efficiency. With these fuels the intake manifold should be operated cool which will gain back some lost volumetric efficiency.

Fuels, like methyl alcohol and ethyl alcohol, require smaller air/fuel ratios. They experience a greater loss in volumetric efficiency. Alcohols have high heat of vaporization, causing greater evaporation cooling. This cooling creates a denser air flow for a given intake pressure, allowing more air to enter the cylinder. So, the engine will regain some of the lost volumetric efficiency.

From the above discussion it is clear that volumetric efficiency increases when the fuel vaporizes later in the intake system. However, this affects the mixing process of air and fuel promoting cylinder-to-cylinder maldistribution, and some of the unvaporized fuel settles on the cylinder walls, where it gets by the piston rings and dilutes the lubricating oil in the crankcase. It is, therefore, recommended that in a carburetted engine, around 60 % of the fuel should evaporate in the intake manifold and the rest of the evaporation should take place in the cylinder during the compression stroke and the combustion process.

15.5.2 Heat Transfer in the Intake System

The temperature of the intake system is generally higher than the surrounding air temperature and will therefore heat the incoming air. This lowers the density of air, which reduces the volumetric efficiency. At lower engine speeds, the rate of air flow in the intake system is low, therefore the air remains in contact with the hot intake system for a longer time. This reduces the air density, which lowers the volumetric efficiency at the low engine speed, as shown in Figure 15.1.

The temperature of the charge in the intake system can be reduced by injecting a small amount of water into the intake manifold. This will improve the volumetric efficiency.

15.5.3 Valve Overlap

Near the TDC, at the end of the exhaust stroke and in the beginning of the intake stroke, both intake and exhaust valves are open simultaneously for a very short time. During this period, some of the exhaust gas may enter into the intake system. During the suction stroke this exhaust gas gets carried back into the cylinder with the intake air-fuel charge, displacing some of the incoming air. This lowers the volumetric efficiency. This problem is more at the lower engine speeds when the real time of valve overlap in milliseconds, is greater. Therefore, a lower engine speed and a greater valve overlap period will both reduce the volumetric efficiency.

15.5.4 Viscous Drag and Restrictions

As the air passes through the air filter, the carburettor, the throttle plate, the intake manifold, and the intake valve, pressures drop because of viscous drag in the passage and because of flow restrictions. Therefore, the pressure of the air entering the cylinder is less than the surrounding atmospheric air pressure, which reduces the volumetric efficiency. Viscous drag increases with the square of flow velocity, resulting in reduced volumetric efficiency at higher speeds as indicated in Figure 15.1. Sharp corners and bends in the intake system should be avoided to reduce pressure losses. One of the greatest flow restrictions is the flow through the intake valve. This restriction can be reduced by increasing the intake valve flow area, which can be increased by providing a larger diameter of the intake valve or by providing two or even three intake valves per cylinder.

Smooth walls of the intake system reduce the viscous drag, but in some low performance, high fuel-efficient engines, the walls of the intake manifold are made rough to enhance turbulence in order to get better air-fuel mixing. High volumetric efficiency is not important in these engines.

15.5.5 Timing of Intake Valve Closing

The inlet valve is timed to close after BDC. The ideal time for closing the inlet valve is when the pressure inside the engine cylinder becomes equal to the intake manifold pressure. If the valve is closed earlier to this point, the air that would be still entering the cylinder is stopped and a loss of volumetric efficiency is experienced. If the valve is closed after this point, the air is forced out of the cylinder, resulting again in loss of volumetric efficiency.

The point at which the pressure equalization occurs depends highly upon the engine speed. At high engine speeds the flow rate of air increases which results in a greater pressure drop across the intake valve. Real cycle time also reduces at high speeds. Both these factors require the intake valve to close at a later cycle position. At low engine speeds the valve should close at an earlier position in the cycle. The closing position of the intake valve is designed for one particular speed, depending on the use for which the engine is designed. For a constant-speed industrial engine it imposes no problem but for an automobile engine that operates over a wide speed range, it is a compromise. The result of this single position valve timing is to reduce the volumetric efficiency of the engine at both high and low speeds. The solution to this problem is to use a variable valve timing (VVT) design.

15.5.6 Intake Tuning

In the intake manifold, pressure waves are created. When these primary waves reach the end of the manifold they are reflected back along the manifold. The pressure pulses of the primary waves and the reflected waves can reinforce or cancel each other, depending on whether they are in or out of phase. If the length of the intake manifold and the flow rate are such that the pressure waves reinforce at the intake valve, slightly more air will enter into the cylinder. The system is said to be tuned and the volumetric efficiency is increased. On the other hand if the flow rate of air is such that primary waves and reflected waves are out of phase at the intake valve, slightly less air will enter into the cylinder, thus reducing the volumetric efficiency. The intake manifold is normally of constant length, so it is tuned only for one particular engine speed. At other speeds the system is not perfectly tuned and the volumetric efficiency will be less at both higher and lower engine speeds.

In some modern engines, manifolds are fitted which can be varied in length. The air is directed to that length of the manifold for which the system is almost tuned for a particular engine speed.

15.5.7 Exhaust Residual

During the exhaust stroke, all the burned gases do not go out of the cylinder. Some of these gases, called exhaust residuals, are trapped in the clearance space. The amount of the residual gases depends upon the compression ratio, and somewhat on the location of the valves and valve overlap. During the suction stroke, these residual gases mix with the fresh charge and displace some incoming air which reduces the volumetric efficiency. Moreover, the hot residual gases heat the incoming air. The air density is thus reduced which lowers the volumetric efficiency. In turn, the residual gases are cooled and contracted which permit more air to enter into the cylinder, thus improving the volumetric efficiency slightly.

15.5.8 Exhaust Gas Recirculation (EGR)

Exhaust gas recirculation is used in the modern automobile engines to reduce the oxides of nitrogen in the exhaust. It displaces some incoming air and also heats the air, lowering its density. The volumetric efficiency is thus reduced by EGR.

15.6 SUPERCHARGING

Supercharging is a method of increasing the power output of the engine without increasing its weight and size. This can be made possible by increasing the density of the charge in the engine cylinder, which ensures greater amount of the charge aspirated into the same stroke volume. The density of the charge can be appreciably increased by raising the pressure of the air or mixture above that of the surrounding atmosphere by a compressor. So, supercharging can be defined as the admittance of the more charge into the engine cylinder than what the engine can take during the normal suction stroke.

The main object of supercharging is to obtain more power by burning the larger amount of fuel or to reduce the weight and size of the engine for a given power output.

15.6.1 Uses of Supercharged Engines

Supercharged engines are widely used at high altitudes. At a higher altitude the suction pressure of the charge in the cylinder reduces, which in turn reduces the power output. To overcome this reduction in power, the engines are generally supercharged. In the case of a motor vehicle engine climbing a hill, the power output reduces gradually. Therefore, with increasing height the degree of supercharged should be increased. Supercharging is necessary in the case of racing-car engines and aircraft engines where the specific output is of prime importance. Supercharging also reduces the bulk of the engine, so it is used where a limited space is only available for the engine.

Superchargers are a natural addition to aircraft piston engines that are intended for operation at high altitudes. As an aircraft climbs to higher altitude, air pressure and air density decrease. The output of a piston engine drops because of the reduction in the mass of air that can be drawn into the engine. For example, the air density at 9100 m is about one-thirds of that at sea level, and thus only one-thirds of the amount of air can be drawn into the cylinder, with enough oxygen to provide efficient combustion for only a third as much fuel. So, at 9100 m, only one-thirds of the fuel burnt at sea level can be burnt. A supercharger can be thought of increasing the density of the air by compressing it and forcing more air than normal into the cylinder so that the power output can be raised.

15.6.2 Factors which Increase the Power Output by Supercharging

The following factors are responsible for increased power output by supercharging:

Increase in power due to increase in inlet pressure of the charge

The output of the supercharged engines is increased mainly due to the increased density of the charge. This is obtained by increasing the inlet pressure of the charge. The pressure is increased by a compressor. After compression the temperature of the charge increases. As at high temperatures the density of the charge decreases, it is therefore essential to cool the charge in the aftercooler before it enters into the cylinder. This enables the cylinder to hold a comparatively denser charge, and hence the power output increases due to the effect of the aftercooler.

Increase in power due to the additional filling of the engine cylinder

As the inlet pressure of the charge is increased, the pressure inside the engine cylinder increases. Due to this high pressure of the charge in the engine cylinder, the residual gases present in the clearance space are compressed. This facilitates the additional filling of the cylinder. Thus the density as well as the volume of the charge increases. Hence the power output is increased.

Increase in power due to the positive gas exchange work

In the case of the naturally aspirated engine, the suction pressure is less than the exhaust pressure. So, a negative work is done during the gas exchange process. In the supercharged engine the inlet pressure is higher than the exhaust pressure, therefore, a positive work is done during the gas exchange process. This is because, during the exhaust stroke the piston has to move only against the back pressure in both the cases whether the engine is naturally aspirated or supercharged, but during the suction stroke the pressure is higher in the supercharged engine, hence there is additional energy delivered to the piston and therefore more power is obtained.

15.7 METHODS OF SUPERCHARGING

The following are the three basic methods used for supercharging:

15.7.1 Mechanical Supercharging

Figure 15.2(a) shows a method of mechanical supercharging. In this method, a blower or a compressor, driven by the engine, is used to provide the compressed air to the engine. It is the most common method of supercharging.

15.7.2 Turbocharging

In this method, a turbocharger—a compressor and a turbine mounted on a single shaft—is used to increase the inlet air density. The exhaust gases from the engine having sufficient energy drive the turbine. The turbine and the compressor are on a single shaft, so the turbine drives the compressor, which raises the inlet air density before the air enters into the cylinders.

Typical arrangements of different turbocharging systems are shown in Figures 15.2(b) to (f). Figure 15.2(b) is the most common arrangement. Figure 15.2(c) shows the combination of an engine driven compressor and a turbocharger. This arrangement is used in large marine engines. Figure 15.2(d) shows the arrangement for the two-stage turbocharging. This system provides very high boost pressures, 4 to 7 bar, to obtain higher power output. Figure 15.2(e) shows the arrangement of turbocompounding. In this system, two turbines are used. The second turbine is directly geared to the engine drive shaft, thus increasing the engine power and efficiency. Figure 15.2(f) shows the arrangement of charge cooling with a heat exchanger after compression and prior to entry to the cylinder. Cooling of the charge further increases the air or mixture density.

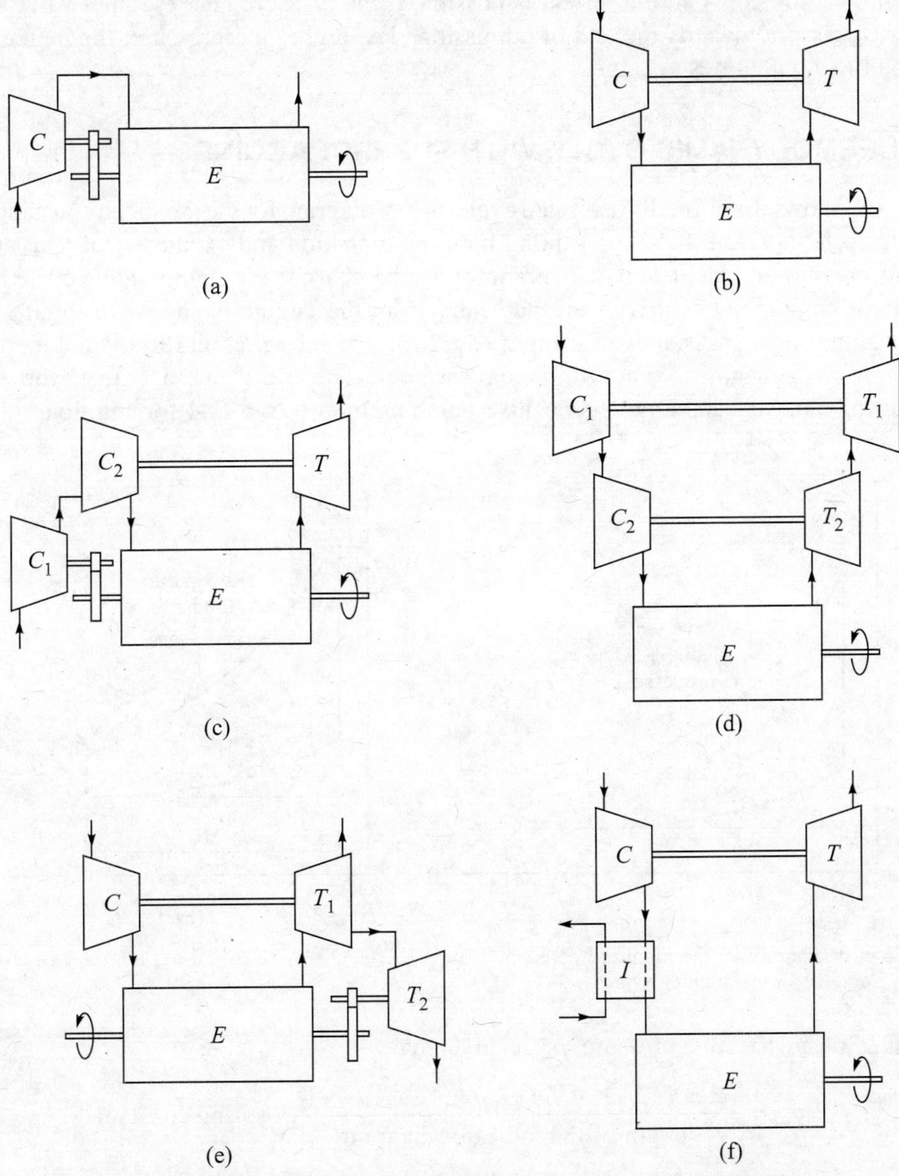

Figure 15.2 Supercharging methods: (a) mechanical supercharging, (b) turbocharging, (c) engine-driven compressor and turbocharger, (d) two-stage turbocharging, (e) turbocharging with turbocompounding, and (f) turbocharger with intercooler. E—Engine, I—Intercooler, T—Turbine, C—Compressor.

15.7.3 Pressure Wave Supercharging

The oscillatory motion of the gases in the pipes of high speed engines during intake and exhaust produces a pressure wave which can be used to increase the mass of the charge admitted into the cylinder. If, for example, the exhaust system is so adjusted that a rarefaction is built up near the exhaust valve towards the end of exhaust when the valves overlap, more burnt gases will flow out of the cylinder. As a result of better scavenging of the cylinder, a greater amount of fresh charge will be admitted. A similar effect is possible if the pressure near the inlet valve is above atmospheric pressure towards the end of admission. The higher the pressure, the better will the cylinder be filled with a fresh charge.

15.8 THERMODYNAMIC CYCLE WITH SUPERCHARGING

Figure 15.3(a) shows the normally aspirated cycle on *p-V* diagram for the idealized constant volume four-stroke cycle with line 1–5 representing both the induction and exhaust strokes at about the ambient air pressure p_a. An ideal *p-V* diagram for a supercharged constant volume cycle in which the blower or compressor is driven mechanically from the engine is shown in Figure 15.3(b). The physical arrangement is shown in Figure 15.2(a). The inlet pressure to the engine is p_i and the exhaust from the engine is at atmospheric pressure p_a. The effects are to increase the pressures reached during the cycle and to give a positive pumping loop 1–5–6–7–1 for addition to the main working loop 1–2–3–4–1.

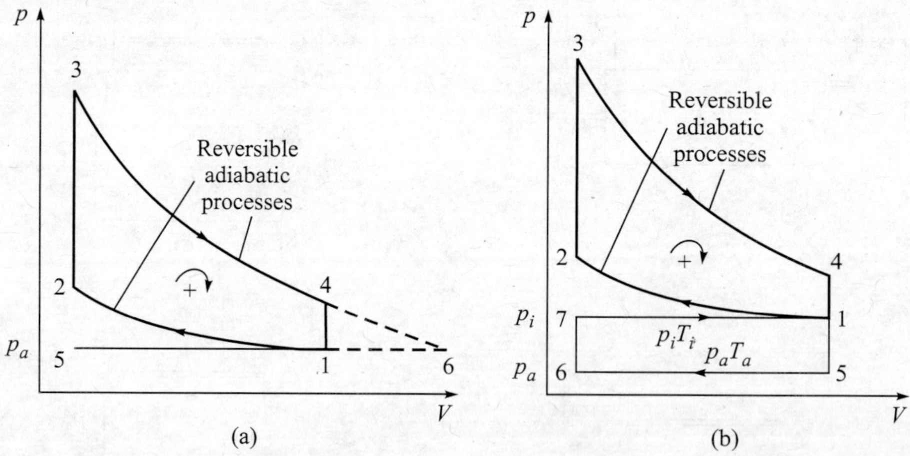

Figure 15.3 *p–V* diagram for the idealized constant volume four-stroke cycle (a) for the normally aspirated cycle and (b) for the supercharged cycle.

Indicated mean effective pressure, p_{mi}, is given by

$$p_{mi} = \frac{(\text{area } 1\text{–}2\text{–}3\text{–}4\text{–}1) + (\text{area } 1\text{–}5\text{–}6\text{–}7\text{–}1)}{\text{length of the indicator diagram}} \times \text{spring constant} \tag{15.10}$$

$$\text{Indicated power, ip} = \frac{p_{m_i} LAN}{120} [\text{W}] \tag{15.11}$$

where

L = stroke length, in m

$A = \frac{\pi}{4}d^2$, d = bore, in m

N = engine speed, in rpm

p_{mi} = indicated mean effective pressure, in N/m^2.

The power required to drive a blower or a compressor connected to the engine must be subtracted from the engine output to obtain the net brake power of the supercharged engine.

For the mechanically driven blower or compressor,

$$\text{brake power, bp} = (\eta_{\text{mech}} \times \text{ip}) - (\text{power to drive the blower}) \tag{15.12}$$

In Figure 15.3(a) at point 4 the exhaust valve opens and the exhaust process starts at a pressure substantially greater than the ambient pressure, p_a. The cylinder charge is suddenly exhausted by a free expansion and a considerable amount of energy released during combustion is lost as it does not contribute to power output of the engine. The attraction of turbocharging is evident, as the energy lost in this way is used to drive a turbine wheel integral with a compressor wheel which delivers compressed air or charge to the cylinder. The additional work available from the gas is indicated, after continuing the reversible adiabatic expansion line 3–4 down to the pressure p_a at 6, by the area 4–6–1–4. The physical arrangement is shown in Figure 15.2(b) and there is no mechanical connection to the engine. The turbocharger combination is a free running unit with approximately equal mass flow rates over the turbine and compressor wheels reaching an equilibrium speed in the range from 20,000 to 80,000 rpm.

The pressure in the intake manifold is p_i which is increased from atmospheric pressure p_a to p_i in the compressor. The exhaust pressure in the exhaust manifold of the engine before entering the turbine is p_e, such that $p_i > p_e$. The expansion process in the turbine reduces the pressure from p_e to p_a. The T-s diagram for the turbocharger is shown in Figure 15.4.

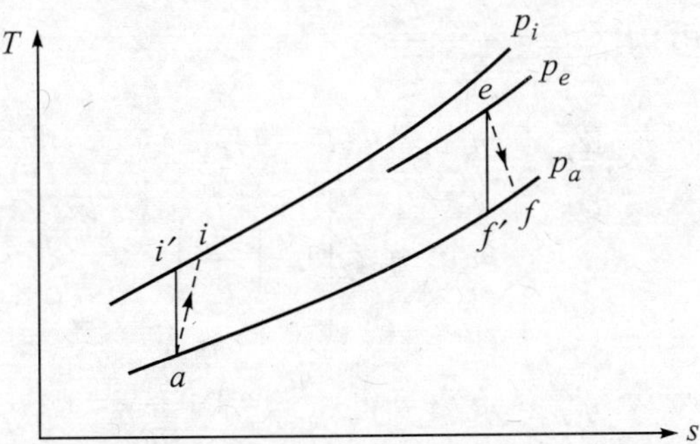

Figure 15.4 *T-s* diagram for the turbocharger.

In order to obtain the compressor work input, the flow energy equation is used between the points a and i. Neglecting the kinetic energy and potential energy changes and assuming irreversible adiabatic compression,

Work input to compressor, $W_C = H_i - H_a$

Power input to compressor, $\dot{W}_C = \dot{m}_a c_{p_a} (T_i - T_a)$ (15.13)

Isentropic efficiency of compressor,

$$\eta_C = \frac{T_i' - T_a}{T_i - T_a}$$

$$\therefore \quad T_i - T_a = \frac{T_i' - T_a}{\eta_C} \tag{15.14}$$

$$\therefore \quad \dot{W}_C = \dot{m}_a c_{pa} \frac{T_i' - T_a}{\eta_C}$$

$$= \dot{m}_a c_{pa} T_a \frac{\frac{T_i'}{T_a} - 1}{\eta_C} \tag{15.15}$$

The process *ai'* is isentropic process,

$$\therefore \quad \frac{T_i'}{T_a} = \left(\frac{p_i}{p_a}\right)^{(\gamma_a - 1)/\gamma_a} \tag{15.16}$$

$$\therefore \quad \dot{W}_C = \dot{m}_a c_{pa} T_a \frac{\left(\frac{p_i}{p_a}\right)^{(\gamma_a - 1)/\gamma_a} - 1}{\eta_C} \tag{15.17}$$

Similarly, the turbine work output,

$$W_T = H_e - H_f$$

Power output,
$$\dot{W}_T = \dot{m}_e c_{p_e} (T_e - T_f) \tag{15.18}$$

Isentropic efficiency of turbine,
$$\eta_T = \frac{T_e - T_f}{T_e - T_f'} \tag{15.19}$$

$$\therefore \quad \dot{W}_T = \dot{m}_e c_{p_e} \eta_T (T_e - T_f')$$

$$= \dot{m}_e c_{p_e} \eta_T T_e \left[1 - \left(\frac{p_a}{p_e}\right)^{(\gamma_g - 1)/\gamma_g}\right] \tag{15.20}$$

$$\dot{m}_e = \dot{m}_a + \dot{m}_f \qquad \frac{\dot{m}_e}{\dot{m}_a} = 1 + \frac{\dot{m}_f}{\dot{m}_a} = 1 + \frac{\text{F}}{\text{A}}$$

where $\frac{\text{F}}{\text{A}}$ is the fuel air ratio $= \frac{\dot{m}_f}{\dot{m}_a}$.

Also, $\dot{W}_C = \dot{W}_T \times \eta_M$, where η_M = mechanical efficiency of the drive.

$$\therefore \quad \frac{\dot{m}_a c_{p_a} T_a}{\eta_C}\left[\left(\frac{p_i}{p_a}\right)^{(\gamma_a - 1)/\gamma_a} - 1\right] = \eta_M \, \dot{m}_e \, c_{p_e} \, \eta_T T_e \left[1 - \left(\frac{p_a}{p_e}\right)^{(\gamma_g - 1)/\gamma_g}\right]$$

or
$$\left[\left(\frac{p_i}{p_a}\right)^{(\gamma_a-1)/\gamma_a}-1\right]=\left[1-\left(\frac{p_a}{p_e}\right)^{(\gamma_g-1)/\gamma_g}\right]\frac{c_{p_e}}{c_{p_a}}\left(\frac{T_e}{T_a}\right)\left(1+\frac{\text{F}}{\text{A}}\right)\eta_O \tag{15.21}$$

where, $\eta_O = \eta_M \eta_T \eta_C$ = overall efficiency of the turbocharger.

15.9 SUPERCHARGING OF SPARK-IGNITION ENGINE

The purpose of supercharging is to increase the power output without much sacrifice of the engine efficiency and fuel economy. As the supercharging is done by increasing the inlet pressure of the charge, the power output is theoretically increased, but from the practical point of view there are several limitations of this process. As the inlet pressure is increased, the charge density inside the cylinder increases. Due to the burning of these charges, more heat is developed inside the engine cylinder and the cooling system may not carry away the extra amount of heat. This causes overheating of the cylinder walls which may get damaged. The high inlet pressure also increases the peak pressure. So to withstand this high pressure the engine cylinder must be of high strength. It is observed that before cooling or structural problems occur, knocking is heard from the engine cylinder. Knocking is highly undesirable, because this causes rapid loss of power and complete failure of the piston.

In order to avoid knocking in the supercharged engine, it is essential to reduce the maximum pressure of the cycle. At high peak pressure the ignition delay is reduced which increases the tendency of the charge to knock. The peak pressure of the cycle can be reduced by lowering the compression ratio. The efficiency of the cycle depends on the compression ratio.

By lowering the compression ratio, the thermal efficiency of the engine is reduced and the specific fuel consumption gets increased.

The peak pressure of the cycle can also be reduced by retarding the spark advance. This causes poorer combustion, so the power is again reduced.

The knocking tendency of the engine also depends on the fuel/air ratio. Near the stoichiometric mixture ratio with slightly less air, the strongest knocking appears. Therefore, the engine can be safely supercharged by using a very lean or a very rich mixture. By using a very lean mixture there is a loss of power and by using a very rich mixture there is a loss of fuel economy.

Considering the above factors the spark-ignition engines are rarely supercharged, except in the cases where the engine output is of prime importance and the efficiency and fuel economy are not of much importance. Aircraft and racing car SI engines are normally supercharged.

To obtain greater power output by supercharging without the loss of efficiency and fuel economy, the knocking tendency of the fuel is suppressed by lengthening the ignition delay.

By lowering the inlet temperature of the charge, the temperature of the last part of the charge to burn reduces, hence the ignition delay is lengthened. The ignition delay of fuel can also be lengthened by the addition of a small quantity of tetraethyl lead.

The injection of water in the intake manifold of the engine cylinder also reduces the knocking tendency. It is probable that water increases the length of the ignition delay by chemical means as well as by reducing the temperatures within the cylinder.

Thus engines can be safely supercharged by using a lower compression ratio, retarded ignition timing, rich mixtures, higher octane fuels and lengthening the ignition delay.

Figure 15.5 shows the comparison of performance characteristics of supercharged and unsupercharged SI engines. The beneficial effects with regard to brake mean effective pressure (bmep) and brake power (bp) in the case of the supercharged engine are well illustrated. The brake specific fuel consumption (bsfc) is slightly higher for the supercharged engine.

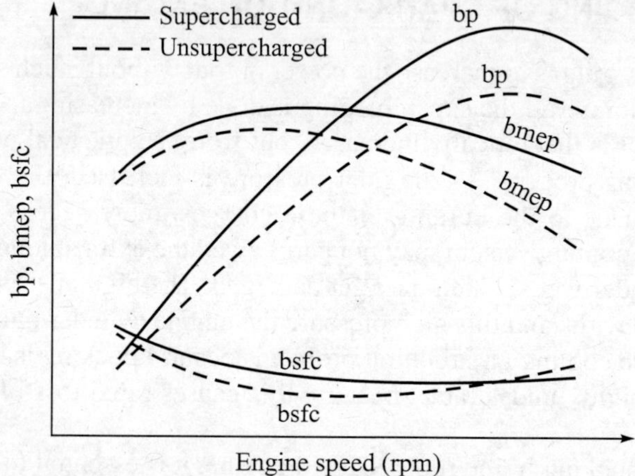

Figure 15.5 Comparison of performance of supercharged and unsupercharged SI engines.

15.10 SUPERCHARGING OF COMPRESSION-IGNITION ENGINE

The compression-ignition engine can be safely supercharged without any combustion difficulties. The increase in inlet pressure reduces the delay period. The delay period in the compression-ignition engine is the time interval between the injection and combustion of first part of the fuel. As the delay period is reduced by increasing the inlet pressure, the rate of pressure rise, i.e. dp/dt reduces, and the maximum pressure in the cylinder increases. Due to the high rate of pressure rise, a violent pounding noise known as Diesel knock is produced. As the rate of pressure rise is reduced at high inlet pressure, the knocking and preignition are absent and the engine runs smoother. It is essential to keep the temperature of the supercharge as low as possible in order to get high volumetric and thermal efficiency. However, an increase in temperature does not give rise to knocking or preignition. If the supercharge pressure is higher, the engine becomes less sensitive to either the cetane number or the volatility of the fuel. Hence, a wider range of fuels can be used as long as high supercharge pressure can be maintained at all speeds and loads.

Supercharging of compression ignition engine is limited by thermal loading. By increasing the strength of the materials of piston and valves, a higher degree of supercharging of the compression ignition engine is possible.

15.11 ADVANTAGES OF SUPERCHARGING OVER HIGH COMPRESSION

Both supercharging and high compression ratio can increase the power output of an engine. They can be compared with the help of p–V diagrams as shown in Figure 15.6. Diagram A relates

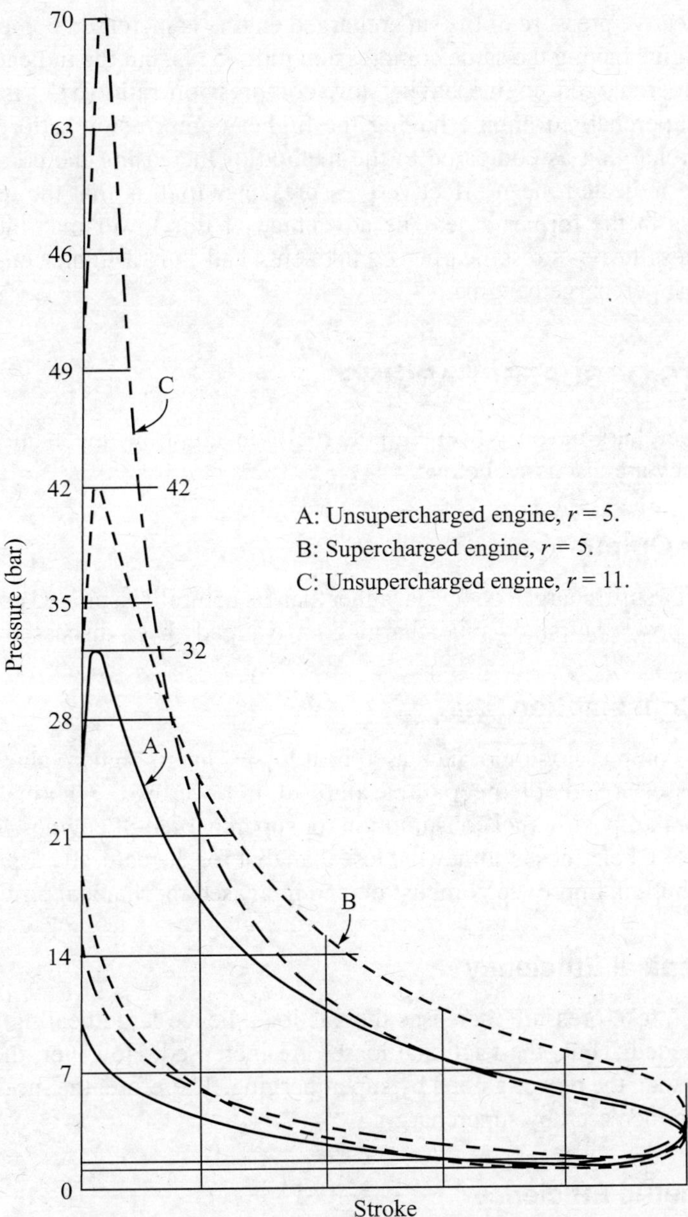

Figure 15.6 Supercharging compared with high compression.

to an unsupercharged engine of compression ratio 5:1, diagram B refers to the same engine, but supercharged 50%, and diagram C to an unsupercharged engine in which the compression ratio has been increased to 11:1 so as to give practically the same net indicated mean effective pressure as that of the supercharged engine. It is observed that the area of diagram B is greater than that of diagram A, and the areas of the indicator diagrams of B and C are almost equal. Therefore, the indicated mean effective pressure of the supercharged engine is increased compared to that of the unsupercharged engine having the same compression ratio (5 : 1), but the indicated mean effective pressure of the supercharged engine having low compression ratio (5 : 1) is almost the same as that of the unsupercharged engine having the higher compression ratio (11 : 1). The great advantage of supercharging as compared to the method of increasing the compression ratio for obtaining the same indicated mean effective pressure, or output, is that the maximum pressure is considerably less in the former case. The advantage of this lower maximum value of peak pressure leads to less heavy stress bearing components and thus a lighter engine for the same output favours the supercharged engine.

15.12 EFFECTS OF SUPERCHARGING

The effects of supercharging on power output, fuel consumption, mechanical efficiency and volumetric efficiency are discussed below:

15.12.1 Power Output

The power output of a supercharged engine is higher than its naturally aspirated counterpart. Factors which increase the power output by supercharging have already been discussed in Section 15.6.2.

15.12.2 Fuel Consumption

The use of a lower compression ratio, increased heat losses due to higher values of specific heats and dissociation losses at higher temperatures, all result in lowering the thermal efficiency and in producing higher brake specific fuel consumption for supercharged SI engines. The brake specific fuel consumption for CI engines is somewhat less than that for the naturally aspirated engines due to better fuel distribution, improved combustion and increased mechanical efficiency.

15.12.3 Mechanical Efficiency

An increase in the intake pressure increases the gas load, hence large bearing areas and heavier components are needed. Thus, the frictional losses are increased. However, the increase in frictional losses is less than the power gained by supercharging. Therefore, the mechanical efficiency of the engine is also increased by supercharging.

15.12.4 Volumetric Efficiency

Supercharging increases slightly the volumetric efficiency of an engine, since the residual gases in the clearance are compressed by the inducted charge which is at a pressure higher than the exhaust pressure. The rate of increase of volumetric efficiency becomes progressively less as the supercharging is increased, since as the intake pressure is increased the contraction of the residuals becomes proportionately less. Further, the rate of increase in volumetric efficiency with the increase in the intake pressure is higher at lower compression ratios, since under these conditions the volume occupied by the residuals is more and the possibility of contraction increases.

15.13 SUPERCHARGING LIMITS

The permissible amount of supercharging depends on the knocking characteristics of the air-fuel mixture, and the ability of the engine to withstand increased load and thermal stresses. Usually one of these limits is reached earlier than the other depending on the type of engine used. For SI engines the knock limit is usually reached first, while for CI engines the load and thermal stresses limits are reached first.

15.13.1 Supercharging Limits of SI Engines

The degree of supercharging in SI engines is mainly limited by the knock. Supercharging reduces ignition delay which increases the knocking tendency. The knock limit is mainly dependent on the type of fuel, the air/fuel ratio, the ignition timing, the valve timing, and cooling.

For volatile petroleum fuels of high octane number the knocking tendency is reduced at very rich and at very lean mixtures. Fuels having the same octane number may have different response to supercharging. For alcoholic fuels, the knocking tendency is reached at rich mixtures. It is because of the fact that these fuels have high latent heat of vaporization, thus producing cooling effect.

In engines using gasoline as a fuel, the strongest knocking occurs near the stoichiometric mixture with slightly less air. Therefore, the engine can be supercharged without knock by using a very lean or a very rich mixture. Usually, the supercharged SI engines are run on rich mixtures to control knock because the lean limits of non-knocking are narrow and require very accurate control of the mixture ratio. By using a very lean mixture, there is a loss of power because of irregular and intermittent engine operation, and by using a very rich mixture there is a loss of fuel economy.

Spark-ignition timing also affects the knock limit of the SI engine. At higher intake pressure and temperature the ignition must be retarded to reduce the peak pressure and temperature of the cycle, thus reducing mechanical and thermal load on the engine.

Although the knock limit in SI engine is reached earlier than the thermal load limits, the latter is also very important to control. The thermal load can be reduced by better cooling or by increasing the valve overlap period, so that the clearance volume is effectively scavenged.

15.13.2 Supercharging Limits of CI Engines

The factors imposing limits on the degree of supercharging the CI engine of a given capacity are: (a) maximum permissible cylinder pressure, i.e. mechanical loading, and (b) maximum permissible piston temperature, i.e. thermal loading.

The specific weight of the engine depends on the maximum cylinder pressure. As the pressure increases the specific weight of the engine should increase to withstand the high mechanical loading. Therefore, any increase in maximum pressure decreases the reliability of the engine. It also increases the rate of the heat release during combustion and hence the thermal load on the engine is increased too. The increased piston temperature may cause distortion of the piston and cylinder head, scuffing of the piston rings, and may even create lubricating troubles that cause heavy liner wear, etc.

15.14 ENGINE MODIFICATIONS FOR SUPERCHARGING

The following engine modifications are recommended for trouble free supercharging:

1. The valve overlap period should be increased to allow complete scavenging of the exhaust gases from the clearance space.
2. The compression ratio should be reduced in order to increase the clearance volume. The effect of this is to reduce mechanical and thermal loading on the engine.
3. In the SI engine the spark timing should be retarded in order to reduce the maximum pressure of the cycle.
4. In the CI engine the injection system should be modified to supply increased amount of fuel. This will require a larger nozzle area than that in the naturally aspirated engine.
5. In turbocharged engines the exhaust valve should open a little earlier to supply more energy to the turbocharger, and the exhaust manifold should be insulated to reduce heat losses.

EXAMPLE 15.1 A four-stroke CI engine has a capacity of 2.8 litres running at 3000 rpm. The average indicated power developed is 12.5 kW per m^3 of free air inducted per minute. It has a volumetric efficiency of 85% referred to free air conditions of 1.013 bar and 15°C. A blower driven mechanically from the engine is used as a supercharger. It has an isentropic efficiency of 74% and works through a pressure ratio of 1.6. Assume that at the end of induction the cylinders contain a volume of charge equal to the swept volume at the pressure and temperature of the delivery from the blower. Calculate the volumetric efficiency of supercharged engine, the increase in brake power by supercharging and express this increase in percentage. Assume all mechanical efficiencies to be 78%.

Solution: Refer to Figure 15.7:

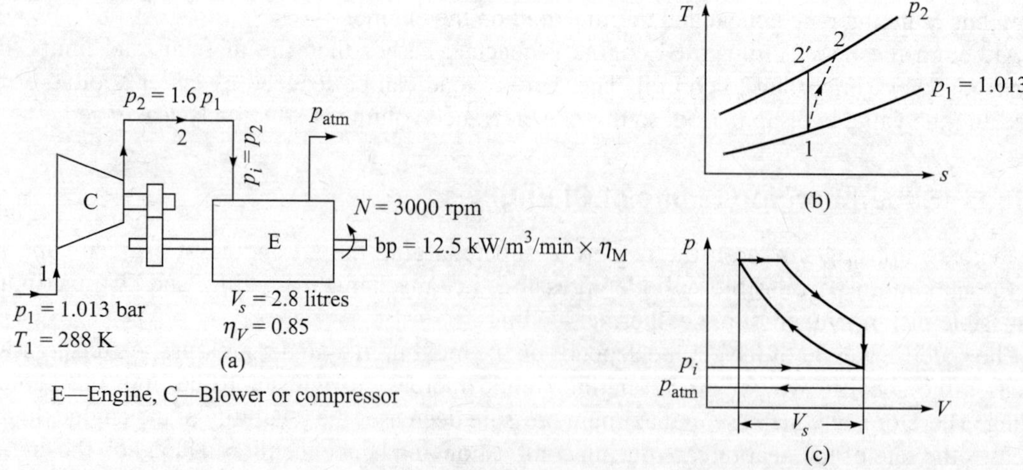

Figure 15.7 Example 15.1.

$$\text{Swept volume, } V_s = 2.8 \text{ litre} = 0.0028 \text{ m}^3$$

$$\text{Volume swept by the piston per min.} = V_s \frac{N}{2}$$

$$\therefore \qquad \dot{V}_s = 0.0028 \times \frac{3000}{2} = 4.2 \text{ m}^3/\text{min}$$

$$\text{Unsupercharged induced volume} = \eta_v \dot{V}_s$$

$$= 0.85 \times 4.2 = 3.57 \text{ m}^3/\text{min}$$

Blower delivery pressure, $p_2 = 1.6 p_1 = 1.6 \times 1.013 = 1.621$ bar

Temperature after isentropic compression, $T_2' = T_1 \left(\dfrac{p_2}{p_1}\right)^{(\gamma-1)/\gamma}$

$$= 288 \times (1.6)^{0.286} = 329.4 \text{ K}$$

Isentropic efficiency of blower,

$$\eta_i = \frac{T_2' - T_1}{T_2 - T_1}$$

$$\therefore \qquad \text{Blower delivery temperature, } T_2 = T_1 + \frac{T_2' - T_1}{\eta_i}$$

$$= 288 + \frac{329.4 - 288}{0.74} = 343.9 \text{ K}$$

The blower delivers 4.2 m^3/min at 1.621 bar and 343.9 K.

$$\text{Equivalent volume at 1.013 bar and 15°C} = \frac{4.2 \times 1.621 \times 288}{343.9 \times 1.013} = 5.628 \text{ m}^3/\text{min}$$

$$\text{Volumetric efficiency of supercharged engine} = \frac{5.628}{4.2} = 1.34 = \boxed{134\%}$$ **Ans.**

$$\text{Increase in induced volume} = 5.628 - 3.57 = 2.058 \text{ m}^3/\text{min}$$

$$\therefore \qquad \text{Increase in ip from air induced} = 12.5 \times 2.058 = 25.73 \text{ kW}$$

Increase in ip due to increased induction pressure

$$= \frac{(p_i - p_{\text{atm}}) \times \dot{V}_s}{60 \times 1000} = (1.621 - 1.013)10^5 \times \frac{4.2}{60} \times \frac{1}{1000}$$

$$= 4.256 \text{ kW}$$

$$\text{Total increase in ip} = 25.73 + 4.256 = 29.99 \text{ kW}$$

$$\therefore \qquad \text{Increase in engine bp} = \text{ip} \times \eta_M = 29.99 \times 0.78$$

$$= 23.39 \text{ kW}$$

The power required to drive the blower must be deducted from this output.

$$\text{Mass of air delivered per second by blower} = \frac{p_2 \times (\dot{V}_s/60)}{R T_2}$$

$$\therefore \qquad \dot{m} = \frac{1.621 \times 10^5 \times 4.2}{60 \times 287 \times 343.9} = 0.115 \text{ kg/s}$$

$$\text{Power input to blower} = H_2 - H_1 = \dot{m}c_p(T_2 - T_1)$$
$$= 0.115 \times 1.005\,(343.9 - 288)$$
$$= 6.46 \text{ kW}$$

$$\therefore \quad \text{Power required to drive the blower} = \frac{6.46}{\eta_M} = \frac{6.46}{0.78} = 8.28 \text{ kW}$$

$$\therefore \quad \text{Net increase in bp} = 23.39 - 8.28 = \boxed{15.11 \text{ kW}} \quad \textbf{Ans.}$$

$$\text{The bp of unsupercharged engine} = 12.5 \times 3.57 \times \eta_M = 44.63 \times 0.78$$
$$= 34.81 \text{ kW}$$

$$\therefore \quad \text{Percentage increase in bp} = \frac{15.11}{44.63} \times \frac{100}{0.78} = \boxed{43.4\%} \quad \textbf{Ans.}$$

EXAMPLE 15.2 The following data relates to a four-stroke CI engine operating at the sea level condition of 1.013 bar and 10°C:

$$\text{bp} = 275 \text{ kW}$$
$$\text{speed} = 1800 \text{ rpm}$$
$$\text{air/fuel ratio} = 20 : 1$$
$$\text{bsfc} = 0.24 \text{ kg/kWh}$$
$$\eta_V = 80\%.$$

The engine is fitted with a mechanical supercharger so that it may be operated at an altitude where the atmospheric pressure is 0.75 bar. The power consumed by the supercharger is 9% of the total power produced by the engine, and the temperature of the air leaving the supercharger is 30°C.

Assume that the air/fuel ratio and the volumetric efficiency remain the same. Calculate the following:

(a) Engine capacity and bmep of the unsupercharged engine.
(b) Increase in air pressure required in the supercharger to maintain the same net output of 275 kW.

Solution: (a) For the unsupercharged engine:

$$\text{bsfc} = \frac{\dot{m}_f}{\text{bp}} = 0.24 \text{ kg/kWh}$$

$$\text{Mass of fuel consumed, } \dot{m}_f = \text{bsfc} \times \text{bp}$$
$$= 0.24 \times 275 = 66.0 \text{ kg/h} = 1.1 \text{ kg/min}$$

$$\therefore \quad \text{Mass of air used, } \dot{m}_a = \dot{m}_f \times \frac{\text{A}}{\text{F}} = 1.1 \times 20 = 22 \text{ kg/min}$$

$$\text{Volumetric efficiency, } \eta_V = \frac{\text{actual mass of air taken in, per cycle}}{\text{mass of air corresponding to swept volume}}$$

$$= \frac{m_a}{\rho_{a,i} V_d}$$

$$\therefore \quad \dot{m}_a = m_a \times \frac{N}{2}$$

or $$m_a = \frac{2}{N} \times \dot{m}_a = \frac{2 \times 22}{1800} = 0.02444 \text{ kg/cycle}$$

$$\rho_{a,i} = \frac{p}{RT} = \frac{1.013 \times 10^5}{287 \times 283} = 1.247 \text{ kg/m}^3$$

$$\therefore \quad V_d = \frac{m_a}{\rho_{a,i} \eta_V} = \frac{0.02444}{1.247 \times 0.8} = 0.0245 \text{ m}^3$$

Hence, Engine capacity, $V_d = \boxed{0.0245 \text{ m}^3}$ **Ans.**

$$\text{bp} = p_{m_b} (LA) \frac{N}{2 \times 60} \times \frac{1}{1000} \text{ kW}$$

$$= p_{m_b} V_d \frac{N}{2 \times 60} \times \frac{1}{1000}$$

$$\therefore \quad \text{bmep, i.e. } p_{mb} = \frac{\text{bp} \times 120 \times 1000}{V_d \times N} \text{N/m}^2$$

$$= \frac{275 \times 120 \times 1000}{0.0245 \times 1800 \times 10^5} = \boxed{7.483 \text{ bar}}$$ **Ans.**

(b) For the supercharged engine:

Gross power produced by engine = 275 + (0.09 × gross power)

$\therefore$ (1 – 0.09) × gross power = 275

$$\therefore \quad \text{Gross power} = \frac{275}{0.91} = 302.2 \text{ kW}$$

$\because$ For power output of 275 kW, the mass of air required is 22 kg/min.

$\therefore$ For power output of 302.2 kW, the mass of air required is

$$\dot{m}_a = \frac{22}{275} \times 302.2 = 24.18 \text{ kg/min}$$

Now, $$m_a = \dot{m}_a \times \frac{2}{N} = \frac{24 \cdot 18 \times 2}{1800} = 0.02687 \text{ kg/cycle}$$

and $$\eta_V = \frac{m_a}{\rho_{a,i} V_d}$$

Here, $\rho_{a,i}$ is the density of air at the inlet condition, that is, the outlet condition of supercharger.

$$\therefore \qquad \rho_{a,i} = \frac{p_2}{RT_2} = \frac{p_2}{287 \times 303} \text{ kg/m}^3; \text{ kg/m}^3; \quad \text{where } p_2 \text{ is in N/m}^2.$$

$$\therefore \qquad \eta_V = \frac{0.02687 \times 287 \times 303}{p_2 \times 0.0245}$$

$$\therefore \qquad p_2 = \frac{0.02687 \times 287 \times 303}{0.0245 \times 0.8} = 1.192 \times 10^5 \text{ N/m}^2$$

$$= 1.192 \text{ bar}$$

∴ Increase in air pressure required in the supercharger is given by

$$= 1.192 - 0.75 = \boxed{0.442 \text{ bar}} \quad \textbf{Ans.}$$

EXAMPLE 15.3 Two identical four-stroke SI engines have swept volume of 3 litre. One engine is normally aspirated and develops a bmep of 9.0 bar at 4000 rpm. Its indicated thermal efficiency is 30% and its mechanical efficiency is 90%.

The other engine is fitted with a supercharger and develops a bmep of 12.0 bar at 4000 rpm. Its compression ratio is lowered to avoid knocking. Its indicated thermal efficiency is 25% and its mechanical efficiency is 91%.

The mass of the naturally aspirated engine is 200 kg and that of the supercharged engine is 220 kg. If both engines are supplied with sufficient fuel for a test of x hours, what is the maximum value of x, if the specific mass, i.e. the ratio of mass of engine plus fuel to bp, of the supercharged engine is always to be less than that of the unsupercharged engine. The calorific value of the fuel is 44,000 kJ/kg.

Solution: For the naturally aspirated engine:

$$\text{bp} = \frac{p_{m_b} LAN}{120 \times 1000}$$

$$= \frac{9 \times 10^5 \times 3 \times 10^{-3} \times 4000}{120 \times 1000} = 90 \text{ kW}$$

$$\text{ip} = \frac{\text{bp}}{\eta_M} = \frac{90}{0.9} = 100 \text{ kW}$$

$$\text{Indicated thermal efficiency, } \eta_i = \frac{\text{ip}}{\dot{m}_f \times \text{CV}}$$

$$\therefore \qquad \dot{m}_f = \frac{\text{ip}}{\eta_i \times \text{CV}} = \frac{100}{0.3 \times 44{,}000} = 7.576 \times 10^{-3} \text{ kg/s}$$

$$\text{Mass of fuel in } x \text{ hours} = 7.576 \times 10^{-3} \times 3600x$$

$$= 27.27x \text{ kg}$$

$$\therefore \qquad \text{Specific mass} = \frac{200 + 27.27x}{90}$$

For the supercharged engine:

$$\text{bp} = \frac{p_{m_b} LAN}{120 \times 1000}$$

$$= \frac{12 \times 10^5 \times 3 \times 10^{-3} \times 4000}{120 \times 1000} = 120 \text{ kW}$$

$$\text{ip} = \frac{\text{bp}}{\eta_M} = \frac{120}{0.91} = 131.9 \text{ kW}$$

$$\dot{m}_f = \frac{\text{ip}}{\eta_i \times \text{CV}} = \frac{131.9}{0.25 \times 44000} = 0.012 \text{ kg/s}$$

$$\text{Mass of fuel in } x \text{ hours} = 0.012 \times 3600x = 43.2x \text{ kg}$$

$$\therefore \qquad \text{Specific mass} = \frac{220 + 43.2x}{120}$$

According to the given condition,

$$\frac{220 + 43.2x}{120} < \frac{200 + 27.27x}{90}$$

or $$1.833 + 0.36x < 2.222 + 0.303x$$

or $$0.057x < 0.389$$

or $$x < 6.825$$

$$\therefore \qquad x_{\max} = 6.825 \text{ h} = \boxed{6\text{ h } 49 \text{ min } 30 \text{ s}}$$ **Ans.**

EXAMPLE 15.4 A centrifugal compressor is driven by a supercharged four-stroke cycle CI engine. The engine, consisting of four cylinders having 100 mm bore and 120 mm stroke, runs at 3000 rpm and develops an output torque of 120 N m. The mechanical efficiency of the engine is 85 %. The entire output of a supercharged engine is used to drive the compressor. The air enters the compressor at 15°C and 1 bar. The compressed air is then cooled in an after-cooler, where heat is rejected at the rate of 1200 kJ/min. After cooling, a part of the compressed air is used for supercharging the engine and the rest is used by the consumer. The air leaves the after-cooler at 55°C and 1.7 bar. The supercharged engine has a volumetric efficiency of 90% based on the induction manifold condition of 55°C and 1.7 bar.

Determine:

(a) The imep of the supercharged engine.
(b) The rate of air consumed by the engine in kg/h.
(c) The rate of air flow available to the consumer in kg/h.

Solution: Refer to Figure 15.8:

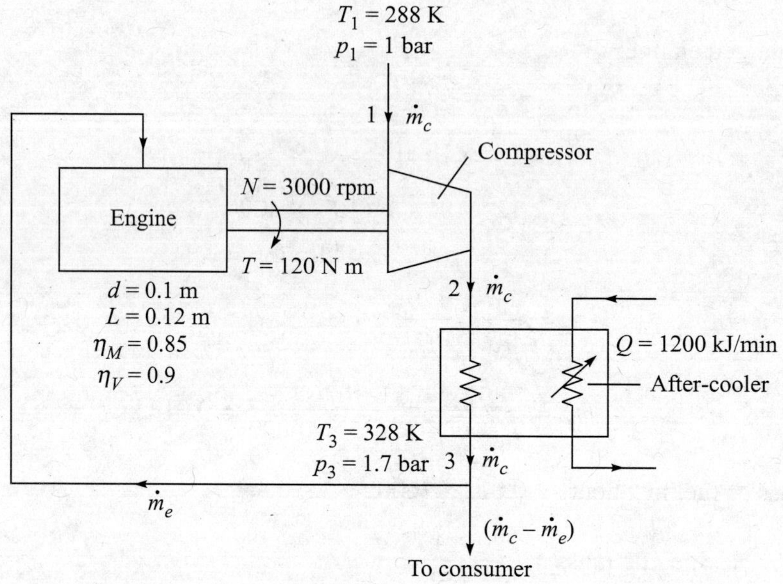

Figure 15.8 Example 15.4.

$$\text{bp} = \frac{2\pi NT}{60} = \frac{2\pi \times 3000 \times 120}{60 \times 1000} = 37.7 \text{ kW}$$

$$\text{ip} = \frac{\text{bp}}{\eta_M} = \frac{37.7}{0.85} = 44.35 \text{ kW}$$

$$\text{ip} = p_{m_i} \frac{LAN}{2 \times 60} \times \text{no. of cylinders}$$

or
$$44.35 = p_{m_i} \times 0.12 \times \frac{\pi}{4}(0.1)^2 \times \frac{3000}{120} \times 4 \times \frac{1}{10^8}$$

∴
$$p_{mi} = \frac{44.35 \times 10^3 \times 4 \times 120}{0.12 \times \pi (0.1)^2 \times 3000 \times 4} = 4.706 \times 10^5 \text{ N/m}^2$$

$$= \boxed{4.706 \text{ bar}}$$ **Ans.**

Engine swept volume, $V_s = \frac{\pi}{4} d^2 L$

Volume swept by the piston per min,

$$\dot{V}_s = \frac{\pi}{4} d^2 L \frac{N}{2} \times \text{no. of cylinders}$$

$$= \frac{\pi}{4} \times (0.1)^2 \times 0.12 \times \frac{3000}{2} \times 4 = 5.655 \text{ m}^3\text{/min}$$

Rate of volume flow of air into the engine, $\dot{V} = \eta_V \times \dot{V}_s$

$$= 0.9 \times 5.655$$

$$= 5.09 \text{ m}^3/\text{min}$$

Rate of mass flow of air into the engine, $\dot{m}_e = \dfrac{p\dot{V}}{RT} = \dfrac{1.7 \times 10^5 \times 5.09}{287 \times 328} = 9.192 \text{ kg/min}$

$$= \boxed{551.5 \text{ kg/h}} \quad \textbf{Ans.}$$

Power output of engine = power absorbed by the compressor

= gain in enthalpy per unit time in the compressor

$$= \dot{m}_c(h_2 - h_1)$$

$$= \dot{m}_c c_{p_a}(T_2 - T_1)$$

$$\therefore \quad \dot{m}_c \times 1.005(T_2 - 288) = 37.7 \text{ kJ/s} \quad \text{(i)}$$

where $\dot{m}_c$ is in kg/s.

Energy balance in the after-cooler,

$$\dot{m}_c c_{p_a}(T_2 - T_3) = \frac{1200}{60} \text{ kJ/s}$$

$$\text{or} \quad \dot{m}_c \times 1.005(T_2 - 328) = 20 \quad \text{(ii)}$$

Dividing equation (i) by (ii),

$$\frac{T_2 - 288}{T_2 - 328} = \frac{37.7}{20}$$

$$\therefore \quad T_2 = 373.2 \text{ K}$$

From Eq. (i),

$$\dot{m}_c = \frac{37.7}{1.005 \times (373.2 - 288)} = 0.44 \text{ kg/s}$$

$$= 1584 \text{ kg/h}$$

Rate of air flow available to the consumer = 1584 – 551.5

$$= \boxed{1032.5 \text{ kg/h}} \quad \textbf{Ans.}$$

EXAMPLE 15.5 A six-cylinder 4.5-litre 4-stroke supercharged engine operating at 4000 rpm has an overall volumetric efficiency of 150%. The compressor has an isentropic efficiency of 90% and a mechanical efficiency of 85% in its link with the engine. The compressed air is delivered to the engine cylinders at 57°C and 1.8 bar. The ambient conditions are 17°C and 1 bar.

Calculate:

(a) The rate of heat rejection from the after-cooler.
(b) The power absorbed by the supercharger from the engine.

Solution: Refer to Figure 15.9:

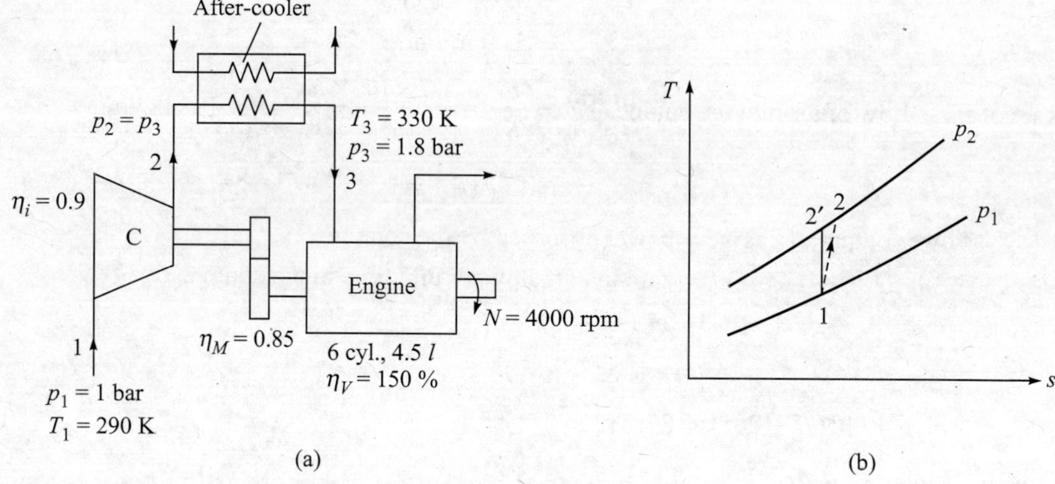

Figure 15.9 Example 15.5.

$$\frac{T_2'}{T_1} = \left(\frac{p_2}{p_1}\right)^{(\gamma-1)/\gamma}$$

or
$$T_2' = 290\left(\frac{1.8}{1.0}\right)^{0.286} = 343 \text{ K}$$

Now,
$$\eta_i = \frac{T_2' - T_1}{T_2 - T_1}$$

or
$$T_2 = T_1 + \frac{T_2' - T_1}{\eta_i} = 290 + \frac{343 - 290}{0.9} = 349 \text{ K}$$

For engine,

Swept volume, $V_s = 0.0045 \text{ m}^3$

$\therefore$
$$\dot{V}_s = V_s \times \frac{N}{2 \times 60} = \frac{0.0045 \times 4000}{2 \times 60} = 0.15 \text{ m}^3\text{/s}$$

Volume of air induced, $\dot{V}_a = \eta_V \times \dot{V}_s = 1.5 \times 0.15 = 0.225 \text{ m}^3\text{/s}$

The overall volumetric efficiency is referred at ambient conditions.
Density of air at ambient condition,

$$\rho = \frac{p}{RT} = \frac{1.0 \times 10^5}{287 \times 290} = 1.2 \text{ kg/m}^3$$

Mass of air induced, $\dot{m}_a = \rho \times \dot{V}_a$

$$= 1.2 \times 0.225 = 0.27 \text{ kg/s}$$

$$\text{Heat rejected from after-cooler} = \dot{m}_a c_p (T_2 - T_3)$$
$$= 0.27 \times 1.005(349 - 330)$$
$$= \boxed{5.156 \text{ kJ/s}} \quad \textbf{Ans.}$$

$$\text{Power needed to run the compressor} = \dot{m}_a c_p (T_2 - T_1)$$
$$= 0.27 \times 1.005\,(349 - 290) = 16 \text{ kW}$$

$$\text{Power absorbed from the engine} = \frac{16}{\eta_M} = \frac{16}{0.85} = \boxed{18.82 \text{ kW}} \quad \textbf{Ans.}$$

EXAMPLE 15.6 A turbocharger fitted to a four-stroke CI engine draws in air at a pressure of 0.98 bar and a temperature of 17°C. The air is delivered to the engine at 1.8 bar. The air/fuel ratio is 20 : 1. The temperature of the exhaust gases leaving the engine is 577°C and the pressure is 1.6 bar. The turbine exhausts at 1.03 bar. The isentropic efficiencies of compressor and turbine are 80% and 85% respectively. The c_p for air and the c_p for gas are 1.005 and 1.15 kJ/(kg K) respectively, and $\gamma_g = 1.33$.

Determine:

(a) The temperature of the air leaving the compressor
(b) The temperature of the gases leaving the turbine
(c) The mechanical power used to run the turbocharger when expressed as a percentage of power generated in the turbine.

Solution: Refer to Figure 15.10:

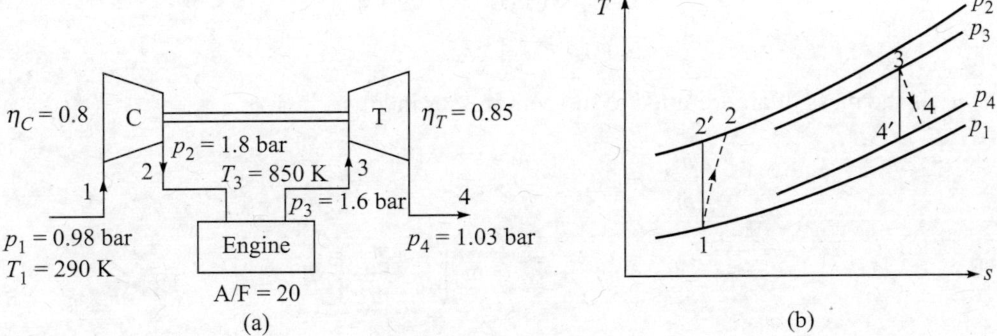

Figure 15.10 Example 15.6.

$$T_1 = 17 + 273 = 290 \text{ K}$$
$$T_3 = 577 + 273 = 850 \text{ K}$$

For compressor,

$$\frac{T_2'}{T_1} = \left(\frac{p_2}{p_1}\right)^{(\gamma-1)/\gamma}$$

or $$T_2' = 290 \times \left(\frac{1.8}{0.98}\right)^{0.286} = 345\text{ K}$$

Now, $$\eta_C = \frac{T_2' - T_1}{T_2 - T_1}$$

or $$T_2 = T_1 + \frac{T_2' - T_1}{\eta_C} = 290 + \frac{345 - 290}{0.8} = 359\text{ K} = \boxed{86°\text{C}} \quad \textbf{Ans.}$$

For turbine,

$$\frac{T_4'}{T_3} = \left(\frac{p_4}{p_3}\right)^{(\gamma-1)/\gamma}$$

or $$T_4' = 850\left(\frac{1.03}{1.6}\right)^{0.33/1.33} = 762\text{ K}$$

Now, $$\eta_T = \frac{T_3 - T_4}{T_3 - T_4'}$$

or $$T_4 = T_3 - \eta_T(T_3 - T_4')$$

$$= 850 - 0.85(850 - 762) = 775\text{ K} = \boxed{502°\text{C}} \quad \textbf{Ans.}$$

Power required by the compressor,

$$\dot{W}_C = \dot{m}_a c_{p_a}(T_2 - T_1)$$

$$= \dot{m}_a \times 1.005(359 - 290)$$

$$= 69.35\,\dot{m}_a\text{ kW}$$

where $\dot{m}_a$ is the mass of air admitted to the compressor in kg/s.

Now, $$\frac{\dot{m}_a}{\dot{m}_f} = 20$$

Mass flow of gas, $$\dot{m}_g = \dot{m}_a + \dot{m}_f = \dot{m}_a\left(1 + \frac{\dot{m}_f}{\dot{m}_a}\right)$$

$$= \dot{m}_a\left(1 + \frac{1}{20}\right) = \frac{21}{20}\dot{m}_a\text{ kg/s}$$

Power developed by the turbine,

$$\dot{W}_T = \dot{m}_g c_{p_g}(T_3 - T_4)$$

$$= \frac{21}{20}\dot{m}_a \times 1.15(850 - 775) = 90.56\,\dot{m}_a\text{ kW}$$

∴ Percentage of turbine power used to run the compressor

$$= \frac{69.35}{90.56} \times 100 = \boxed{76.6\%} \quad \textbf{Ans.}$$

EXAMPLE 15.7 Determine the cooling effect due to the evaporation of fuel on charge temperature in a turbocharged SI engine for the following two cases:

(a) Carburettor placed before the supercharger.
(b) Carburettor placed after the supercharger.

The air/fuel ratio is 14 : 1, the evaporation of fuel causes a 23°C drop in mixture temperature. The isentropic efficiency of the compressor is 75% and its pressure ratio is 1.3. The ambient air is at 15°C.

Take for mixture, c_{pm} = 1.05 kJ/(kg K) and γ = 1.33. Compare the supercharger work in both the cases.

Solution: (a) *Carburettor placed before the compressor:* Refer to Figure 15.11.

$$T_1 = 15 + 273 = 288 \text{ K}$$

$$T_2 = T_1 - 23 = 288 - 23 = 265 \text{ K}$$

$$\frac{T_3'}{T_2} = \left(\frac{p_3}{p_2}\right)^{(\gamma-1)/\gamma}$$

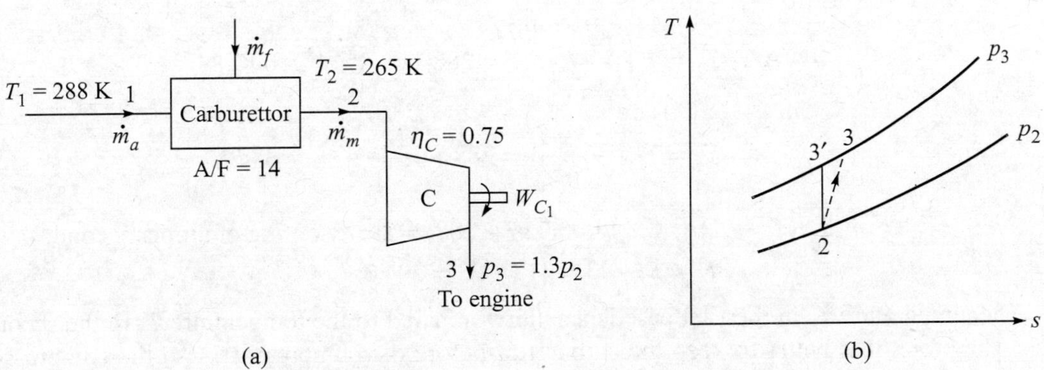

Figure 15.11 Example 15.7: carburettor placed before the compressor.

∴ $$T_3' = 265 \times (1.3)^{0.33/1.33} = 283 \text{ K}$$

Now, $$\eta_C = \frac{T_3' - T_2}{T_3 - T_2}$$

or $$T_3 = T_2 + \frac{T_3' - T_2}{\eta_C} = 265 + \frac{283 - 265}{0.75} = 289 \text{ K}$$

Power required by the compressor,

$$\dot{W}_{C_1} = \dot{m}_m c_{p_m} (T_3 - T_2)$$

$$= \left(1 + \frac{1}{14}\right) \times 1.05(289 - 265)$$

$$= 27.0 \text{ kW/kg of air per second}$$

(b) *Carburettor placed after the compressor:* Refer to Figure 15.12.

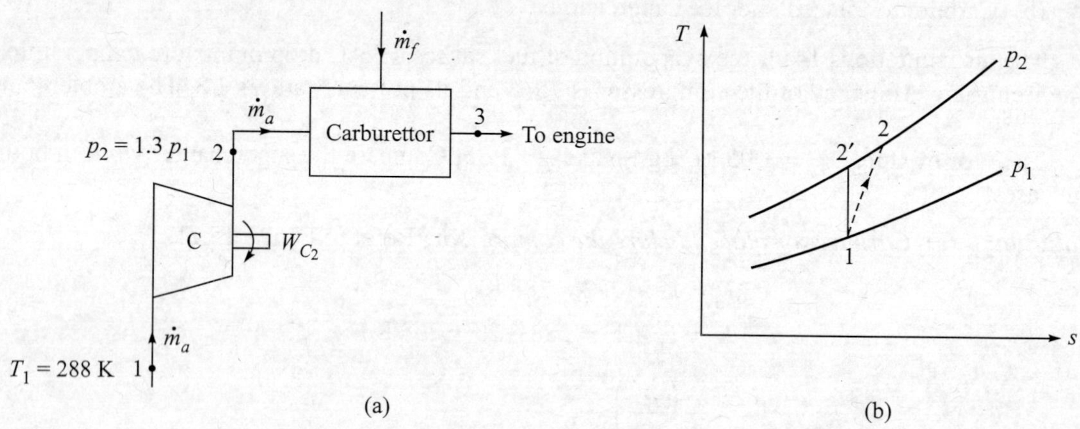

Figure 15.12 Example 15.7: carburettor placed after the compressor.

$$T_2' = T_1\left(\frac{p_2}{p_1}\right)^{(\gamma-1)/\gamma} = 288(1.3)^{0.286} = 310 \text{ K}$$

or

$$T_2 = T_1 + \frac{T_2' - T_1}{\eta_C} = 288 + \frac{310 - 288}{0.75} = 317 \text{ K}$$

$$\dot{W}_{C_2} = \dot{m}_a c_{p_a} (T_2 - T_1) = 1 \times 1.005(317 - 288) = 29.15 \text{ kW/kg of air per second.}$$

$$T_3 = T_2 - 23 = 317 - 23 = 294 \text{ K}$$

The temperature T_3 in the first case is smaller compared to the temperature T_3 in the second case. Therefore, it is better to keep the carburettor before the compressor. Also the compressor work in the first case is smaller.

$$\text{Saving of power in the first case} = \frac{(29.15 - 27.0) \times 100}{29.15} = \boxed{7.38\%}$$ **Ans.**

REVIEW QUESTIONS

1. What do you understand by air capacity of the engine? Why does the air capacity practically not change by the presence of fuel?
2. What are the principal factors that affect the thermal efficiency of the engine? What is the effect of engine speed on thermal efficiency?

3. Show that the indicated power is proportional to the air capacity under certain conditions. Mention the conditions clearly.
4. Define ideal air capacity of a four-stroke engine.
5. Define volumetric efficiency. Express it in different forms.
6. Discuss the effect of engine speed on volumetric efficiency.
7. Discuss the effect of the following variables on volumetric efficiency:
   (a) Fuel
   (b) Heat transfer in the intake system
   (c) Valve overlap
   (d) Viscous drag and restrictions
   (e) Timing of inlet valve closing
   (f) Intake tuning
   (g) Exhaust residual
   (h) Exhaust gas recirculation
8. Define supercharging of an IC engine. What are its objectives? What is the effect of altitude on power output of the engine?
9. What are the factors that can increase the power output by supercharging?
10. What are the uses of supercharged engines?
11. Briefly describe the typical arrangements of mechanical supercharging and also the different turbocharging systems.
12. Briefly describe the pressure wave supercharging system.
13. Show and discuss the ideal thermodynamic cycle of the supercharged IC engine on *p–V* diagram. Give expressions to evaluate imep, ip, and bp.
14. What are the advantages of a turbocharged engine? Draw the *T–s* diagram for a turbocharger. Evaluate the power input to compressor and power output from turbine of a turbocharger.
15. Describe supercharging of SI engines. Compare the performance of supercharged and unsupercharged SI engines.
16. Describe supercharging of CI engines.
17. Which engine is more suitable for supercharging—SI engine or CI engine? Give a clear explanation.
18. Compare supercharging versus high compression with the help of *p–V* diagrams, and indicate the advantages of supercharging over a high compression ratio.
19. Discuss the effects of supercharging on the following:
   (a) Power output
   (b) Fuel consumption
   (c) Mechanical efficiency
   (d) Volumetric efficiency
20. What are the supercharging limits of SI and CI engines?
21. What are the modifications recommended for supercharging an IC engine?

PROBLEMS

15.1 A four-stroke CI engine having 15 cm bore and 18 cm stroke develops an average indicated power of 15 kW/m^3 of free air inducted per minute when running at 4000 rpm. It has a volumetric efficiency of 80 % referred to the free air conditions of 1.013 bar and 15°C. The engine is supercharged with the help of an engine driven blower. The blower raises the pressure to 1.7 bar with an isentropic efficiency of 75%. Assume that at the end of induction the cylinders contain a volume of charge equal to the swept volume corresponding to discharge pressure and temperature of the blower. Take all mechanical efficiencies as 80%. Calculate:

(a) The blower delivery temperature
(b) The increase in ip from air inducted
(c) The increase in ip due to increased induction pressure
(d) The power required to drive the blower
(e) The net increase in bp by supercharging
(f) The percentage increase in bp by supercharging.

[**Ans.** (a) 349 K, (b) 55.82 kW, (c) 7.284 kW, (d) 13.79 kW, (e) 36.69 kW, (f) 60.1%]

15.2 The following data relates to a four-stroke CI engine of capacity 24 litre operating at sea level conditions of 1.013 bar and 12°C.

$$\text{bp} = 260 \text{ kW}$$
$$\text{speed} = 1600 \text{ rpm}$$
$$\text{air/fuel} = 18:1$$
$$\text{bsfc} = 0.245 \text{ kg/kWh}$$

Determine the volumetric efficiency of the engine under the above conditions. The engine is now fitted with a supercharger, so that it may be operated at an altitude where the atmospheric pressure is 0.7 bar. The delivery pressure and temperature from the supercharger are 1.19 bar and 32°C respectively to maintain the same net output of 260 kW. Assume that the air/fuel ratio and volumetric efficiency remain the same. Estimate the power consumed by the supercharger. What will be the percentage of this power in relation to the total power output from the engine? [**Ans.** 80.4%, 25.2 kW, 8.84%]

15.3 Two identical four-stroke SI engines having 180 mm bore and 200 mm stroke run at 4500 rpm. One engine is normally aspirated and develops an imep of 11.0 bar. Its brake thermal efficiency is 32 % and its mechanical efficiency is 88%. The other engine is fitted with a supercharger and develops an imep of 13.0 bar. Its brake thermal efficiency is 30% and its mechanical efficiency is 90%. The mass of naturally aspirated engine is 750 kg and that of the supercharged engine is 800 kg. If both engines are supplied with sufficient fuel for a test of h hours, what is the maximum value of h, if the specific mass (ratio of mass of engine plus fuel to bp) of the supercharged engine is always to be less than that of unsupercharged engine. The calorific value of the fuel is 44,000 kJ/kg. [**Ans.** 28.12 h]

15.4 A 4.0 litre four-stroke supercharged engine is operating at 3500 rpm. The compressor has an isentropic efficiency of 85 % and a mechanical efficiency of 88% in its link with the engine. The compressed air is delivered to the engine cylinders at 52°C, the pressure ratio in the compressor is 1.9 bar. Heat rejected from the after-cooling is 6 kJ/s. The ambient conditions are 15°C and 1.013 bar.

Calculate:

(a) The overall volumetric efficiency of the engine.

(b) The power absorbed by the supercharger from the engine.

[Ans. (a) 133.7% (b) 14.9 kW]

16.1 INTRODUCTION

For providing or verifying the new design concepts of an internal combustion engine, it is essential to measure a number of quantities with accuracy. It is also necessary to ensure that the measuring equipment is reliable, is easily checked, and can be put into use quickly. In order that different types of engines or different engines of the same type may be compared, engine tests are carried out under various conditions. The performance and emission characteristics of a given engine can be deducted theoretically, but it is necessary to check the theoretical deductions by tests under actual working conditions. Comparison of actual results with the theoretical ones and thereafter the analysis of the causes of deviation leads to improvement in design.

16.2 MEASUREMENT OF BRAKE POWER

The brake power is the power output of the engine. Measurement of the brake power involves the determination of the torque and the angular speed of the engine output shaft. The torque measuring device is called *dynamometer*. The engine is connected to a dynamometer which can be loaded in such a way that the torque exerted by the engine can be measured. It is capable of providing an adjustable and measurable torque opposing that of the engine.

The dynamometer may be of the absorption or the transmission type. The absorption dynamometer converts the work done by the engine on the dynamometer into heat. In the transmission type of dynamometer the torque transmitted by the driving shaft is measured directly and consequently it does not itself absorb any of the engine's work output. The majority of engine test-work utilizes an absorption type dynamometer. The important types of absorption dynamometers are described below.

16.2.1 Prony Brake

Figure 16.1 shows a prony brake. It consists of a frame with brake shoes, often made of two blocks of wood, each of which embraces slightly less than one-half of the rotating drum rim. The drum is attached to the output shaft of the engine. The two blocks can be drawn together by means of nuts and bolts, cushioned by springs, so as to increase the pressure on the drum. A load bar extends from

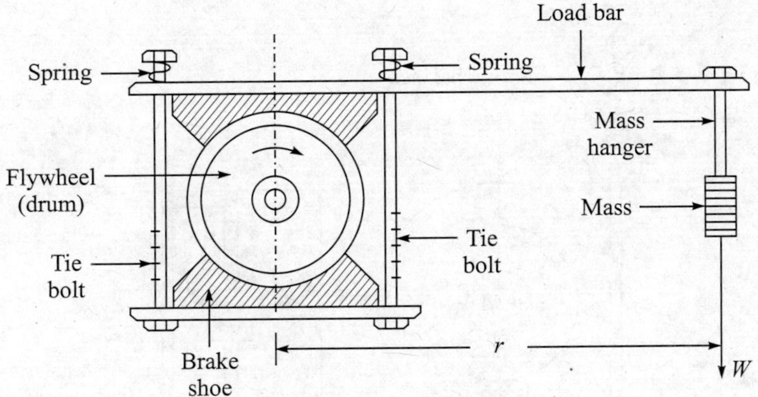

Figure 16.1 Prony brake.

the top of the brake and a weight is hanged to the end of the load bar. The weight W is adjusted so as to maintain the load arm horizontal. The torque is given by Wr, where r is the shortest distance between the centre line of the weight and the line passing through the centre of the pulley and is known as the *arm length*. Knowing the speed of the wheel, the power absorbed can be calculated as

$$\text{bp} = \frac{2\pi NT}{60} = \frac{2\pi NWr}{60} \tag{16.1}$$

This type of dynamometer must be cooled, since the power absorbed is converted into heat. It may be cooled by providing a water trough to the drum rim.

This brake is not suitable for absorption of a large amount of power as wear of the blocks and reduction in the coefficient of friction between the drum and the friction material with the rise in temperature require continuous tightening of the bolts.

16.2.2 Rope Brake

Figure 16.2 shows a rope brake dynamometer. It is directly coupled to the engine output shaft. In this brake, two or more ropes rest on the rim of a pulley. The ropes are spaced evenly across the width of the rim by means of wooden blocks positioned at different points around the rim. The total pull on the slack ends of the ropes is registered on a spring balance, while the pull on the tight end is provided by dead weights. The power absorbed is due to friction between the rope and the drum. Friction torque on the pulley may be increased by increasing the dead load by the addition of weights.

The drum requires cooling. This dynamometer is easy to fabricate but is not very accurate because of changes in the friction coefficient of the rope with temperature.

The brake power is given by

$$\text{bp} = \frac{2\pi RN(W - S)}{60 \times 1000} \text{ [kW]} \tag{16.2}$$

where

W = dead weight, in newton

S = spring balance reading, in newton

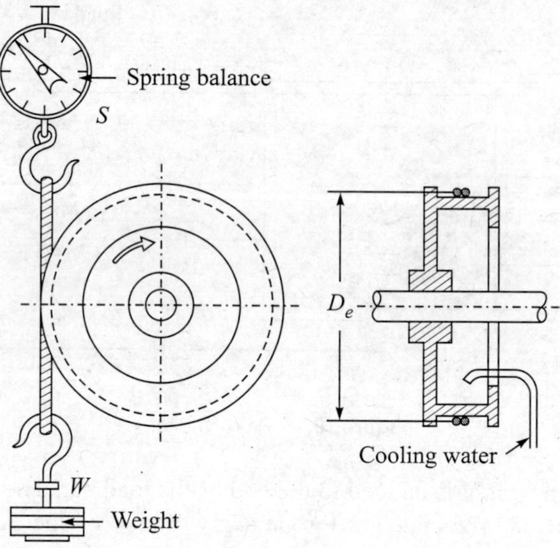

Figure 16.2 Rope brake dynamometer.

N = engine speed, in rpm

R = effective radius of the pulley, in metre = $\dfrac{D+d}{2} = \dfrac{D_e}{2}$

D = diameter of the brake pulley, in metre

d = diameter of the rope, in metre

D_e = effective diameter of the pulley, in metre

16.2.3 Hydraulic Dynamometer

A section of the Heenan–Froude hydraulic dynamometer is shown in Figure 16.3. It consists of a shaft carrying a rotor which revolves inside a watertight casing. The rotor has a number of radial vanes set obliquely to its axis, and the spaces between these rotor vanes form cups of semi-elliptical section. The casing is also provided with stator vanes forming a similar set of cups. The rotor vanes face forward in the direction of rotation, whereas the stator vanes face in the opposite direction. The rotor vanes direct the water outwards by centrifugal force towards the stator vanes which redirect it back into the rotor. This highly turbulent process repeats itself again and again. The change of momentum experienced by the water as it changes direction exerts a reaction force on the casing, tending to drag it round with the rotor. The engine power is absorbed by a continuous series of whirling eddies in the water. This power absorption results in a rise in the temperature of the circulating water. The heat thus generated is disposed off by arranging a continuous flow of water through the dynamometer.

The load on the engine may be varied by inserting or withdrawing thin sluice plates between the rotor and the casing by a handwheel. These plates are responsible for the formation of the power absorbing eddies.

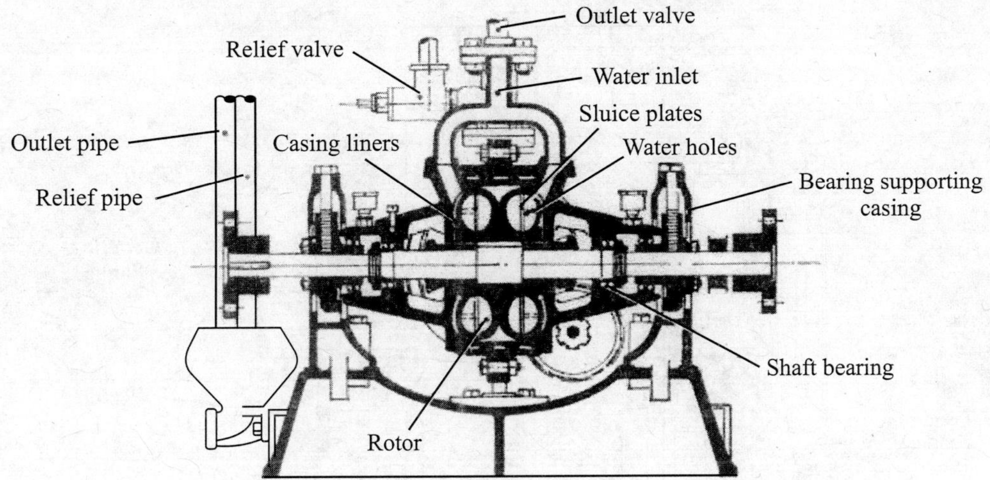

Figure 16.3 Section of the Heenan–Froude hydraulic dynamometer.

The rotor shaft rotates in bearings carried by the casing. Watertight glands are provided where the rotor shaft enters and leaves the casing. The casing itself is supported on a large ball-bearing trunnions, and an arm is attached to it at which the torque is measured.

The power of the engine is absorbed by shaft bearing friction, gland friction and the continuous change in direction and momentum of the water. They all turn the casing in the direction in which the engine is running, therefore, the total turning effort is measured by a single reading of a spring balance attached to the arm.

When the engine is at rest and with the water flowing through the dynamometer casing, the spring balance reading is set to zero. It ensures that the dynamometer is accurately balanced on its trunnions. With engine running and loads applied by the dynamometer, the readings of the spring balance should be taken only when the balance arm is horizontal, and for this purpose a pointer is attached to a fixed part of the framework.

16.2.4 Eddy Current Dynamometer

An eddy current dynamometer is illustrated in Figure 16.4. The dynamometer unit consists of a toothed rotor. It is mounted on a shaft running in bearings which rotates within a casing with a fine clearance between water cooled steel loss plates of high permeability steel. The stator or casing is supported on ball bearing trunnions, so that any tendency of the stator to rotate is read on a fixed scale. Two field coils connected in series are secured in the casing. When these coils are supplied with a direct current, a magnetic field is created in the casing across the air gap at either side of the rotor. When the rotor turns in the magnetic field, eddy currents are induced in the stator by the principle of electromagnetic induction and tend to rotate the stator, creating a braking effect between the rotor and the casing. Load changes are possible by varying the direct current supplied to coils.

The rotational torque exerted on the casing can be measured by a spring balance attached to the arm. It may also be measured by a strain gauge load cell incorporated in the restraining linkage between the casing and the dynamometer bed plate as shown in Figure 16.4.

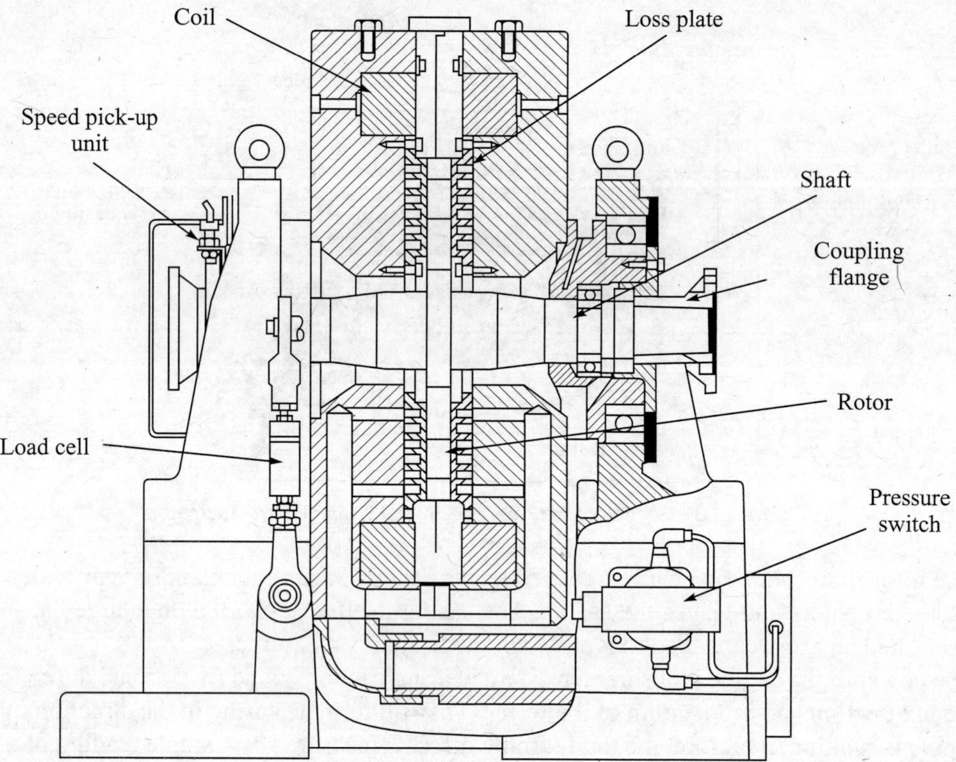

Figure 16.4 Eddy current dynamometer.

To prevent overheating of the dynamometer, a pressurized water-cooling system for the casing is used. Pressure switch regulates the flow of cooling water. Water passes from the inlet to the casing via a flexible connection, permitting movement of the casing. Water passes through the loss (grooved) plates in the casing, positioned on either side of the rotor, and absorbs the heat generated. Heated water discharges from the casing through a flexible connection to an outlet flange. While all hydraulic dynamometers must operate with a free discharge of cooling water, eddy current dynamometers can be used as a pressurized cooling system. Pressure switch ensures a constant flow to the dynamometer despite variations in thermal load.

16.2.5 Swinging Field DC Dynamometer

The electric or swinging field dynamometer measures the torque most accurately over a wide range of power and speed. It can also be used as a motor for starting the engine and for measuring friction and pumping losses.

Basically, this type of dynamometer is a dc shunt motor, which by means of suitable switching arrangements, may run either as a generator or as a motor. The field is separately excited either from the mains or from a battery which maintains a constant voltage. The whole machine is supported by its framework in trunnion bearings. The frame carries a torque arm for balancing the engine torque. The engine power is absorbed by generating electrical energy in the armature circuit, which is dissipated in an external resistor or it may be fed into the mains.

16.3 INDICATED POWER

This is the power developed in the engine cylinder by the working fluid by exerting pressure on the piston.

The operation of the entire cycle of a reciprocating internal combustion engine may be represented by a diagram where the instantaneous gas pressure within the cylinder is plotted versus the piston position in terms of degrees crank angle or volume of charge entrapped above the piston. This diagram is called the *indicator diagram* and can be obtained from an engine by means of a mechanical or electronic indicator.

16.3.1 Mechanical Indicator

A mechanical indicator is shown in Figure 16.5. By means of this indicator the relationship between the pressure and volume of the gases in the cylinder is recorded graphically throughout the cycle by combining two motions, one drawing exactly, but to a reduced scale, the motion of the piston, and the other, at right angles to the first representing pressure variations proportional

Figure 16.5 A mechanical indicator.

to the gas pressure in the cylinder. A sheet of paper is wrapped around a drum and the drum oscillation is linked to the piston displacement by a cord, thus reproducing the piston stroke. The pressure inside the cylinder is recorded on the graph paper by the pressure recording mechanism. It consists of a small spring-loaded piston working in a cylinder which is in direct connection with the engine cylinder through a valve. The motion of the spring-loaded piston is suitably multiplied at the recording pencils by a system of links and levers.
When the paper is unwrapped from the drum, the diagram represents the *p–V* curve. The area of the diagram can be measured by counting the number of squares on the graph paper or with the help of a planimeter, and the indicated work per cycle is calculated.

The indicator diagram has a positive loop (the area between the compression and expansion curves) and a negative loop (the area between the suction and exhaust curves). The positive loop gives the gross work done by the piston during the cycle and the negative loop represents the pumping loss due to admission of fresh charge and removal of exhaust gases. A typical *p–V* diagram taken by a mechanical indicator is shown in Figure 16.6

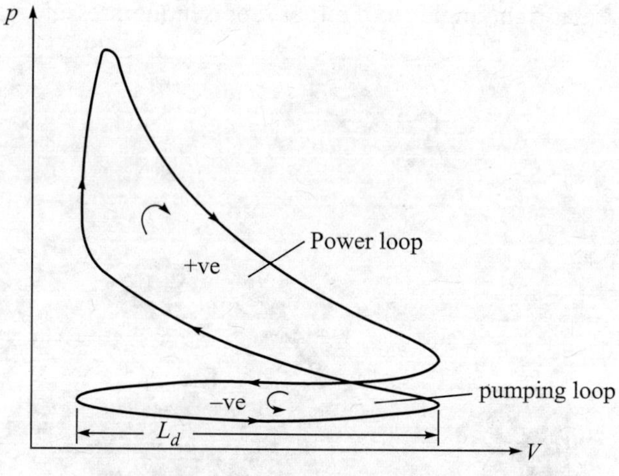

Figure 16.6 *p–V* diagram.

Suppose

A_d = the net area of the diagram, in mm^2
= area of the positive loop – area of the negative loop
S = stiffness of the spring, in bar/mm
L_d = length of the indicator diagram, in mm

The indicated mean effective pressure, p_{mi}, is given by

$$p_{mi} = \frac{A_d}{L_d} \times S \text{ [bar]} \tag{16.3}$$

The indicated power developed by the engine is given by

$$\text{ip} = \frac{p_{m_i} \times 10^5 (LAN_f)}{60 \times 1000} \text{[kW]} \tag{16.4}$$

where

p_{mi} = indicated mean effective pressure, in bar

L = stroke, in m

$A = \frac{\pi}{4}d^2$, d is the bore, in m

N_f = number of working strokes per minute

$N_f = N$ for two-stroke, $N_f = \frac{N}{2}$ for four-stroke

N = engine speed, in rpm.

The mechanical indicators are not very accurate because of the inertia effects of the moving parts, the friction, the backlash due to slackness of the joints of the mechanism, the stretch in the cord producing the stroke motion, the periodicity of the spring, and the temperature effects.

16.3.2 Electronic Indicator

The indicated power of an engine can be accurately obtained by the actual indicator diagram taken inside the engine cylinder by means of an electronic indicator. The measuring equipment and the data recording system are shown in Figure 16.7.

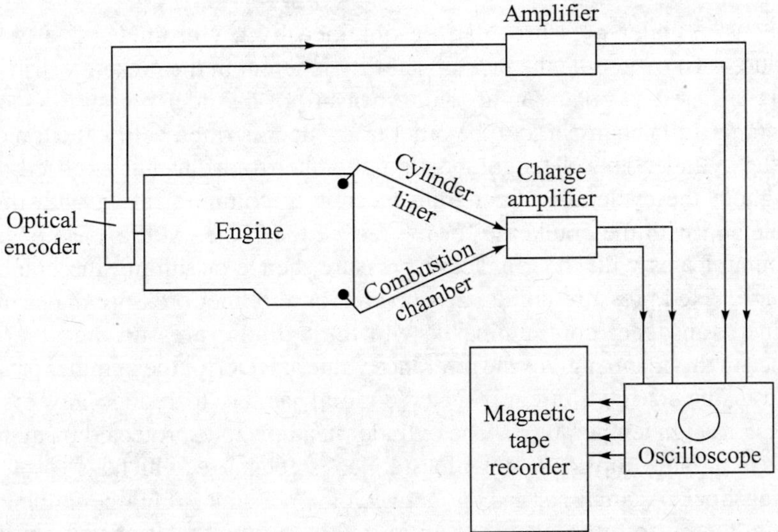

Figure 16.7 Pressure and crank-angle measuring equipment and the recording system.

Rotary encoder

The crank angle of the engine can be obtained by using a rotary encoder. The encoder has 720 equidistant transparent slots around the periphery of a black glass disc at 0.5° intervals. There is also a separate slot for recording the TDC. The photo-cell of the encoder picks up the light from a quartz-iodine bulb placed on the opposite side of the glass disc and produces square wave electrical pulses of suitable voltage by amplification and signal conditioning.

For mounting the encoder, the piston of the engine is brought to the TDC position by rotating the flywheel. The rotary encoder is then mounted on a bracket in such a position that it is possible to couple the encoder shaft with the engine crank shaft. The encoder shaft thus gets rotated with the crank shaft of the engine.

Pressure transducers

Depending upon the pressure range, different types of pressure transducers are used. For cylinder pressures, the piezo-electric transducers are used. For intake and exhaust systems where the pressures are low, the strain gauge transducers are used.

Piezo-electric transducers

Piezo-electric transducers utilize a crystal or quartz, rochelle salt or similar material having piezo-electric properties. The crystal, correctly oriented, is arranged so that the deflection of the diaphragm compresses the crystal, producing an electrical charge. This charge, being proportional to the deflection of the diaphragm, is also proportional to the cylinder pressure at any instant. The charge must, however, be amplified by means of a charge amplifier, before applying to the cathode ray oscilloscope.

Measurement of cylinder pressure

The instantaneous cylinder pressures can be obtained by a vibrometer quartz piezo-electric pressure transducer. To overcome the lack of stability inherent in the piezo-electric charge output, two transducers are used to obtain an absolute measure of cylinder pressure. One transducer is located in the combustion chamber and the other transducer is fitted below the top of the cylinder block face in the cylinder liner. The combustion chamber transducer is exposed to the cylinder pressure throughout the cycle. The liner transducer is in communication with the cylinder for part of the cycle and with the crankcase for the rest of the cycle. At the TDC position, the liner transducer communicates with the crankcase pressure, hence, assuming the crankcase pressure to be atmospheric, fixed pressure points are obtained on the liner pressure diagram. At the BDC position the liner transducer communicates with the cylinder pressure and the liner pressure diagram obtained in this manner gives the reference value at BDC for the cylinder pressure diagram.

Both the transducers have integral water cooling jackets to reduce any distortion of the diaphragm owing to high temperatures. The cylinder head piezo is protected from the combustion flame by means of a flame trap consisting of a stainless steel gauze, which also protects it form the effect of thermal shock. A small gap is kept between the transducer and the flame trap.

Signals from the piezo transducers are amplified by a vibrometer charge amplifier and then recorded on a tape.

The piezo transducers are calibrated with the help of a dead-weight tester, charge amplifiers and a digital voltmeter (DVM). One amplifier is assigned to each transducer. The amplifiers are adjusted to read zero volts on the DVM when there is no pressure applied to the transducers. Maximum combustion chamber pressure and liner pressures are estimated and these pressures are applied to the respective transducers by means of a dead-weight tester. Gain settings of the amplifiers are adjusted to obtain approximately 1 V on the DVM. Different pressures are then applied to the transducers and the corresponding DVM readings are recorded. A straight line relationship is obtained between the DVM readings and the applied pressures.

Data recorder

A magnetic tape recorder can be used for recording the analog signals. Different signals are recorded on a tape at different channels. Two instantaneous pressure signals from the transducers and two signals from the rotary encoder are recorded—one for the half-degree interval marking and another for the TDC.

Oscilloscope

A dual beam oscilloscope is used to monitor every signal to be recorded on the tape recorder. This is achieved by using coaxial T-junctions, so that the signals could be simultaneously fed to the oscilloscope and data recorder.

Data acquisition

The signals from the data recorder are fed into the analog digital converter (ADC) through the amplifier units where all signals are magnified.

The rotary encoder which generates an electrical pulse every 0.5° CA, also generates an additional electrical pulse for every TDC. This pulse, when transmitted, initiates the computer recording of the data. As there are two TDC pulses every cycle, a modification to the circuit is made to ensure that only the TDC pulse after the compression stroke is allowed to be transmitted to the computer.

Typical variations of liner pressure and cylinder pressure versus crank angle are shown in Figures 16.8 (a) and (b) respectively.

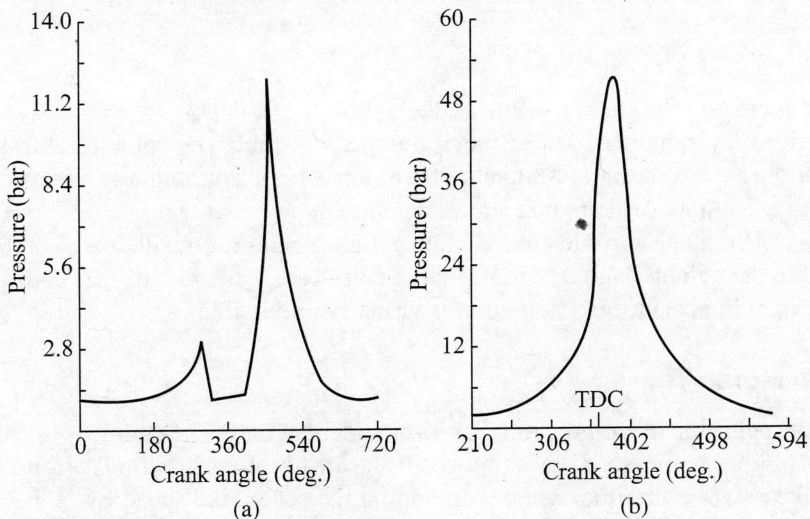

Figure 16.8 Typical (a) liner pressure and (b) cylinder pressure curves.

The high speed indicator diagram of pressure versus crank angle can be converted to a plot of pressure versus piston displacement by using the geometry of the piston, the connecting rod and the crank as shown in Article 1.15.

Thus, the p–V diagram can be constructed. The area of the p–V diagram represents the work done per cycle, from which the indicated power can be calculated.

16.3.3 Willan's Line Method

This is a method of determining the friction power and hence the indicated power (ip = bp + fp) of an unthrottled compression ignition engine. This method is not suitable for use with petrol engines. It is based on the fact that at light loads a relatively small amount of fuel is pumped into the air charge. Hence, there is plenty of air available for complete combustion within the engine cylinder. Therefore, at a given engine speed in the light load region, a straight line law exists between the rate at which fuel is consumed and the engine load or brake power. This straight line is Willan's line and is shown in Figure 16.9. By extrapolation, the fuel flow rate to give zero brake power can be determined. This is the fuel flow rate necessary to overcome friction, and consequently, the amount of negative brake power at zero rate of fuel consumption represents the friction power. From this, the indicated power and mechanical efficiency can be evaluated.

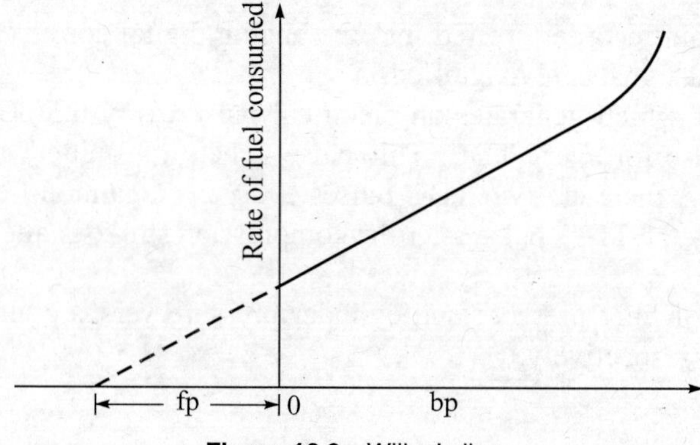

Figure 16.9 Willan's line.

The rapid increase in slope of the line at the high load end denotes a progressive reduction in combustion efficiency as more and more fuel is pumped into the given volume of air. It is therefore important that the extrapolation of Willan's line is carried out as accurately as possible, and that sufficient readings at light load are taken to define the line.

Since a petrol engine is throttled to maintain a high fuel/air ratio with load, combustion is not complete within the cylinder and a plot of brake power versus the rate of fuel consumption does not yield a straight line. Hence extrapolation is virtually impossible.

16.3.4 The Morse Test

This method is applicable to reciprocating multi-cylinder engines. The engine is run at a particular speed and the torque is measured by cutting out the firing of each cylinder in turn and noting the fall in brake power each time, while maintaining the set engine speed by reducing load. The observed difference in brake power between all cylinders firing and with one cylinder cut out is the indicated power of the cut-out cylinder.

If there are k cylinders and all are firing,

$$\text{ip} = \text{bp} + \text{fp}$$

or

$$\sum_{i=1}^{k} \text{ip}_i = \sum_{i=1}^{k} \text{bp}_i + \sum_{i=1}^{k} \text{fp}_i \tag{16.5}$$

With the first cylinder cut out, it will not produce ip and theoretically there will be no contribution to bp from the first cylinder. However, there will be almost the same fp.

$$\sum_{i=2}^{k} \text{ip}_i = \sum_{i=2}^{k} \text{bp}_i + \sum_{i=1}^{k} \text{fp}_i \tag{16.6}$$

Subtracting Eq. (16.6) from Eq. (16.5) yields the ip of the first cylinder, i.e.

$$\text{ip}_1 = \sum_{i=1}^{k} \text{ip}_i - \sum_{i=2}^{k} \text{ip}_i = \sum_{i=1}^{k} \text{bp}_i - \sum_{i=2}^{k} \text{bp}_i \tag{16.7}$$

Thus, the ip of each cylinder in turn can be obtained, and hence the sum of these values will give the ip of the engine with all *k* cylinders firing.

$$\text{ip} = \text{ip}_1 + \text{ip}_2 + \cdots + \text{ip}_k \tag{16.8}$$

It is assumed that the friction power remains constant and has the same value in both Eqs. (16.5) and (16.6). Strictly, this cannot be true. The temperature and pressure of the cut-out cylinder will be low. A reduced temperature will cause an increase in viscous drag on the piston, however a reduced pressure will reduce the frictional force on bearings and piston rings. These two effects tend to cancel out, but cannot do so exactly.

A petrol engine cylinder can be cut out by shorting the spark plug with a screw driver placed between its terminal and the engine frame, or a special high tension switch can be used. With CI engines, it may be possible to hold the fuel pump plunger off its cam with a suitable tool, so as to prevent fuel delivery to a particular cylinder.

16.3.5 Motoring Test

In the motoring test the engine is first run at a given speed and load conditions for sufficient time so that the temperature of the engine components, lubricating oil and cooling water reaches a steady state. A swinging field type electric dynamometer is used to absorb the power during this period. The ignition is then switched off and by suitable electric switching devices the dynamometer is converted to run as a motor. The motoring is done to crank the engine at the same speed at which it was operating previously. The test is conducted as rapidly as possible. The torque is measured under firing and under motoring conditions from which the bp and fp are evaluated. Then the ip and mechanical efficiency are determined.

The friction power determined by this method is reasonably good, but not very accurate. Although the coolant temperature will change little during changeover, the piston and cylinder wall temperatures will change markedly, since the engine is prevented from firing during the test. The temperature of the working parts within the engine is low, and it is the temperature of the working parts which affects the viscous drag and hence the friction power. Also in absence of the exhaust blow-down, the pumping losses are not representative.

The motoring method is suitable for assessing the relative contribution to the friction power of the many moving parts within an engine. Components such as piston rings, valve gear, the camshaft and all accessories can be removed in turn and the motor torque measured.

16.4 FUEL CONSUMPTION

The rate at which an engine consumes a liquid fuel may be determined in several different ways. There are two basic types of fuel measuring devices: (i) gravimetric and (ii) volumetric.

16.4.1 Gravimetric Fuel-flow Measurement

Fuel consumption is best determined by weighing. This method is shown in Figure 16.10. It is used for the measurement of fuel by weight, using a null method when fuel is actually delivered to the engine under running condition. The null method is the most reliable method, which employs a weigh-bridge. On one side of the weigh-bridge a container filled with fuel is placed and on the other side weights are placed. The balance is adjusted until the fuel container is slightly heavier than the balancing weights. As the engine is running, the fuel is consumed by the engine. When the fuel in the container balances the weights, indicated by a pointer passing a line, a stop watch is started. A known weight is removed from the weigh-bridge and the stop watch is stopped when the bridge is balanced again. Thus, the time for the consumption of a known mass of fuel is known. The container on the weigh-bridge is refilled by opening a valve on the fuel tank.

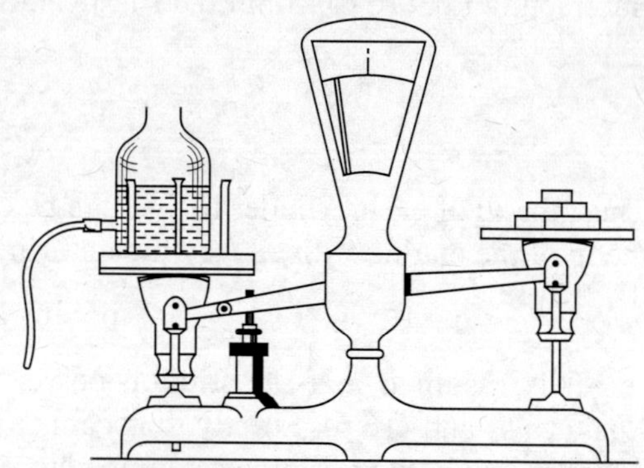

Figure 16.10 Fuel-weighing scales.

16.4.2 Volumetric Type Flowmeters

Figure 16.11 shows a device for volumetric measurement of liquid fuel which consists of a glass tube containing four knife-edged spacers. The spacers are so positioned as to contain an accurately calibrated volume of fuel.

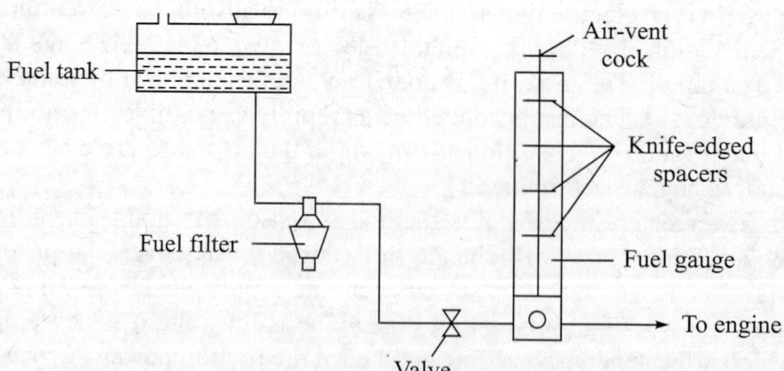

Figure 16.11 Volumetric measurement of liquid fuel.

The fuel gauge is first filled to a level above the top spacer by opening the air-vent cock on the top of the gauge. The valve connecting the gauge to the fuel supply tank is then closed. The engine thus draws fuel from the gauge. As the fuel level passes the top spacer, a stop watch is started. The measurement may be terminated when the fuel passes any one of the lower spacers. As the fuel passes the lower spacer the stop watch is stopped.

The time for Vcc of fuel is recorded and fuel consumption is calculated as follows.

$$\text{Density of fuel} = \rho \text{ [g/cc]}$$

$$\text{Time to consume } V\text{cc of fuel} = t \text{ [s]}$$

$$\therefore \quad \text{Fuel consumption} = \frac{V}{t} \text{[cc/s]}$$

$$= \frac{V}{t}\rho\text{[g/s]}$$

$$= \frac{V}{t}\rho \times \frac{3600}{1000}\text{[kg/h]} \quad (16.9)$$

The above method of measurement of fuel consumption intoduces a source of human error in recording the timing of the falling fuel column, particularly when the flow rates are very high or very low. An electrical timing system is much favoured and is described below.

Figure 16.12 shows an electronic timing system of an automatic volumetric type of fuel flow measuring device. It consists of a tube of measured volume A, which has photocell B and light source C fitted in tubular housings. An equalization chamber D is connected to the measuring tube via the air tube E. An equalization pipe F supplies fuel to the equalization chamber. There is a magnetic valve G, when it opens, the fuel flows directly to the engine from the fuel tank and when it closes the engine takes fuel from the measuring device.

The level of the fuel as it falls in the measuring tube during measurement is sensed by two small transducers arranged in positions corresponding to the calibration marks. Each transducer consists of a caliper type housing surrounding the measuring tube and carrying a small light source and a photocell, the two being diametrically opposite across the measuring tube. When the tube contains fuel the light source is concentrated upon the sensitive area of the photocell, so that the latter passes current

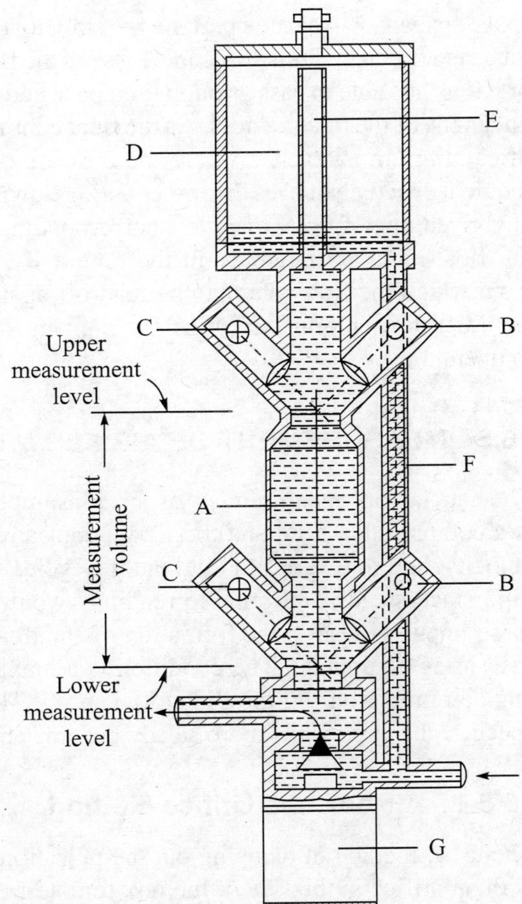

Figure 16.12 Electronic timing system of fuel flow measuring device.

in an electric circuit. Conversely, when the tube is empty, light from the source is diffused and thus the photocell does not sense the light. The two transducers and the clock relay are incorporated into a switching circuit. This is so arranged that, as the fuel level falls past the upper transducer, the clock relay receives an impulse which starts the clock. As the fuel level passes the lower transducer, the clock relay receives a second impulse which stops the clock. Thus the time for a known volume of fuel consumed is obtained.

16.4.3 Rotameter

Figure 16.13 shows a rotameter. It consists of a glass tube having an internal bore which increases uniformly from bottom to top. A light conical float is located inside the tube and has an external diameter equal to the internal bore of the bottom of the tube. When a liquid or gas is fed under pressure to the bottom of the tube the float rises because the pressure on its lower surface overcomes its own weight. Because of the tapered bore of the tube, an annular space appears between the periphery of the float and the bore of the tube, and the area of this space increases as the float rises, the liquid or gas is thus able to pass through this space and the pressure on the underside of the float reduces. As the float continues to rise, a point of equilibrium is reached where the pressure on the underside of the float exactly balances the forces acting downwards upon it. For any given rate of flow, there is a corresponding position at which the float remains stationary in the tube and a scale is provided from which the flow rate may be read off against the position of the float. Calibrations are done for a given fluid at some given temperature.

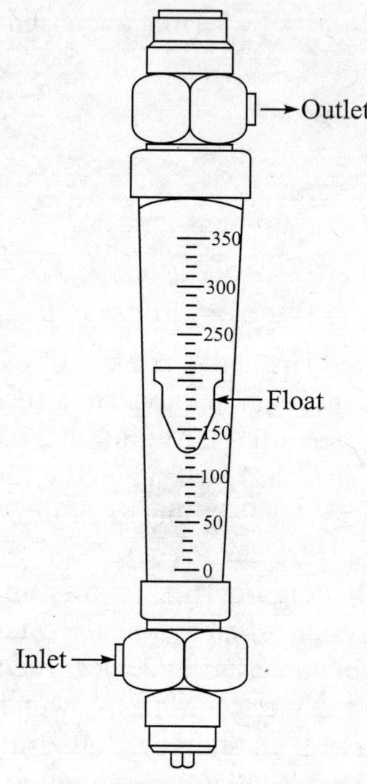

Figure 16.13 Rotameter

16.5 MEASUREMENT OF AIR FLOW RATE

The satisfactory measurement of air consumption is a more difficult task. It is because the air is a compressible fluid and the flow is pulsating due to the cyclic nature of the engine. The air velocity therefore varies throughout the cycle. This prevents the use of an orifice in the induction pipe since steady and reliable readings would not be obtained. Under certain circumstances, these impulses lead to the formation of standing pressure waves in the induction system and the measuring equipment. The condition is a maximum in the case of a single cylinder four-stroke engine running on full throttle at low speed. The greater the number of cylinders and higher the speed of the engine, the more steady becomes the air flow.

16.5.1 Air-box and Orifice Method

The usual method of damping out the pulsations is to fit an air box of suitable size to the engine with an orifice in the side of the box remote from the engine. The pressure difference across the orifice plate is steady if the system is correctly designed, and is measured by means of a suitable manometer. The pressure drop across the orifice plate is a measure of the air flow rate. A schematic arrangement of the method is shown in Figure 16.14.

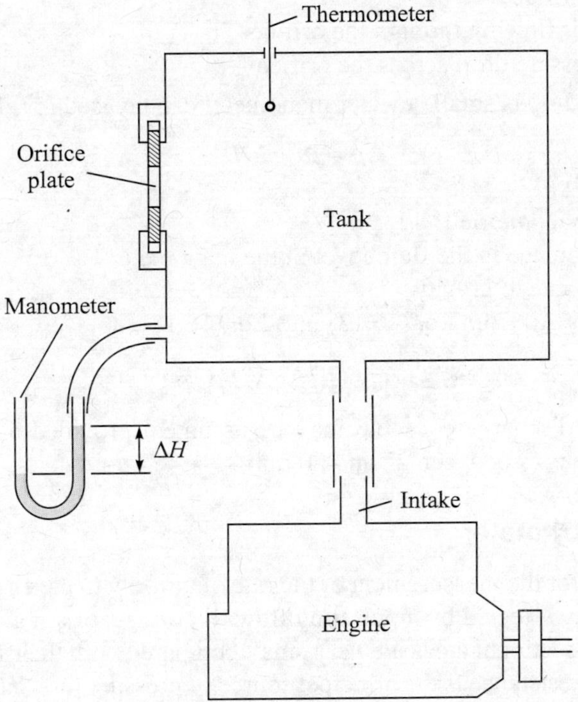

Figure 16.14 Tank and orifice.

As the orifice need not throttle the air flow, the pressure drop across the orifice is generally small and the incompressible form of energy equation may be used. Therefore, from Bernoulli's equation,

$$z_1 + \frac{p_1}{\rho_1 g} + \frac{v_1^2}{2g} = z_2 + \frac{p_2}{\rho_2 g} + \frac{v_2^2}{2g} \tag{16.10}$$

Planes 1 and 2 indicate the upstream and the downstream of the orifice plate. Here, $z_1 = z_2$ and the velocity of approach v_1 can be neglected. v_2 is the velocity of air and can be represented as v_{air}. Flow is assumed to be incompressible, therefore,

$$\rho_1 = \rho_2 = \rho_{air}$$

$$\therefore \quad v_{air} = \sqrt{\frac{2(p_1 - p_2)}{\rho_{air}}} = \sqrt{\frac{2\Delta p}{\rho_{air}}} \tag{16.11}$$

Substituting this into the continuity equation,

$$\dot{m}_{air} = \rho_{air}\, v_{air}\, A_o\, C_d \tag{16.12}$$

$$= C_d A_o \sqrt{2\Delta p\, \rho_{air}} \tag{16.12a}$$

where

C_d = coefficient of discharge of the orifice
A_o = area of the orifice
ρ_{air} = density of air flowing through the orifice
$\Delta p = p_1 - p_2$ = pressure drop across the orifice

Since the pressure drop is small, a water manometer can be used for effective measurement.

$$\Delta p = \rho_w \, g \Delta H \tag{16.13}$$

where

ρ_w = density of manometric fluid
ΔH = vertical difference in the fluid level in the manometer
g = acceleration due to gravity

Substituting the value of Δp from Eq. (16.13) into Eq. (16.12),

$$\dot{m}_{air} = C_d A_o \sqrt{2g\Delta H \rho_w \rho_{air}} \tag{16.14}$$

The coefficient of discharge $C_d = 0.6$ may be assumed, provided a sharp-edged orifice is employed and its diameter is between 25 and 50 mm.

16.5.2 Viscous Flowmeter

The viscous flowmeter for the measurement of the rate of air flow to the engine is shown in Figure 16.15. It is not adversely affected by a pulsating flow. It consists of a matrix of honeycomb type whose individual cross-sectional areas are very small compared with their length, so that the flow is viscous. The viscous resistance is the principal source of pressure loss and the kinetic effects are small. This gives a linear relation between the pressure difference and flow. In order to maintain its accuracy the element is cleaned frequently by using carbon tetrachloride, since the chief source of error arises from surface contamination. An inclined manometer filled with alcohol may be used to measure the pressure drop across the element. The viscous flowmeter is calibrated by using a steady flow suction rig having an orifice plate.

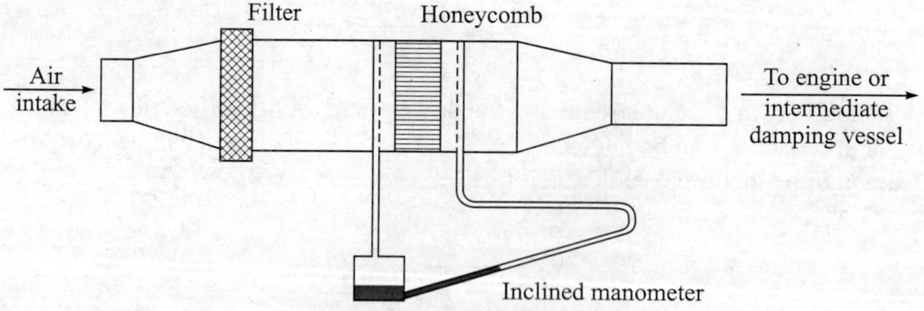

Figure 16.15 Viscous flowmeter

16.6 SPEED MEASUREMENT

Revolution counters, tachometers and electronic stroboscopes are commonly used for speed measurement.

The revolution counter is a simple mechanical device which may be integral with the dynamometer. Hand-held revolution counters are also used which may be held in frictional contact with a rotating shaft to indicate the number of revolutions made. The time is separately recorded. The method is suitable for very slow rotational speeds.

A tachometer gives a direct reading of rotational speed and may be operated mechanically or electrically. For very high rotational speeds an electronic transducer monitors the rotating shaft and produces electrical impulses at a frequency proportional to shaft speed. These impulses are fed to a gated electronic counter which sums the number of impulses over a fixed period of time. The rpm is digitally displayed by the counter.

A light sensitive device such as a photo-transistor may also be used for speed measurement. One-half of the periphery of the shaft is painted black to destroy its reflecting properties. Thus for every shaft revolution, the transistor will receive one burst of reflected light. The transistor will function as a high frequency switch, allowing current to flow only during its reception of reflected light. The electrical pulses so produced are counted and the shaft speed is displayed.

An electronic stroboscope may also be used to measure the rotational speed of the shaft. It is most satisfactory in the high range of speed measurement. The stroboscope produces brilliant flashes of light of very short duration. A single line is marked on the rotating shaft. The stroboscope is adjusted to match the frequency of the shaft. Under this condition the shaft appears to be stationary. The flash frequency is the rotational speed of the shaft.

16.7 SPARK-TIMING MEASUREMENT

The stroboscope may be used to examine the spark timing. For spark-timing measurement, flash may be initiated by an external electrical impulse by winding the stroboscope triggering lead, once or twice round the spark plug high-tension lead. Each time the plug fires, a small current is simultaneously induced in the coil of the trigger lead, so that the plug spark and the light flash from the stroboscope occur at the same instant.

Stroboscope light is directed on to a semi-circular scale, graduated in degrees. The mid-point of the scale is marked zero and it indicates the TDC position. The apparently stationary pointer marked on the crankshaft pulley moving over the scale will indicate the ignition point in relation to TDC.

16.8 COMBUSTION PHOTOGRAPHY AND FLAME SPEED DETECTION

Figure 16.16 shows the Ricardo's high speed colour photographic technique for the qualitative combustion studies. The cylinder of the engine has a transparent window of either quartz or perspex. A high-speed rotating prism camera is mounted on a rigid support. It can take up to 16,000 frames per second. Aluminized mirrors are arranged in such a manner so that the camera may simultaneously record the combustion process and the position of crankshaft in degrees. The fuel is normally doped with copper oleate to ensure that low-luminosity flames become visible. From these films, the combustion process inside the engine cylinder can be studied.

The location of the flame with time can be directly obtained experimentally by using an ionization probe. It consists essentially of a wire in an annulus. The wire is insulated from the body of the probe. When the flame reaches the annulus it becomes conducting as an electrical circuit is

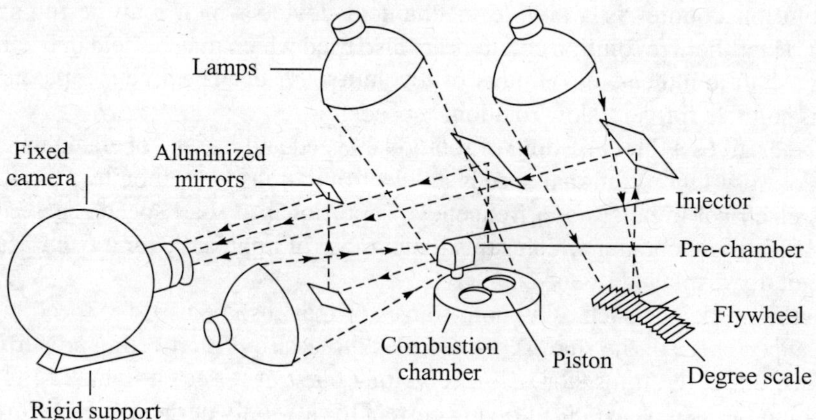

Figure 16.16 Layout of optical system and camera.

completed between the wire and the probe body. A dc voltage is normally applied to the wire and to the electronic timing circuit. When the flame arrives at the probe the timing circuit is triggered and the time is measured.

By locating a sufficient number of probes in the combustion chamber a quantitative picture of the flame propagation may be made. The minimum voltage required to trigger the ionization probe depends upon the air/fuel ratio.

16.9 PERFORMANCE CHARACTERISTICS

Variable speed characteristics and constant speed characteristics for SI engines and CI engines are described in this section.

16.9.1 Variable Speed Characteristics

SI engines

Figure 16.17 shows the performance characteristics of a typical spark-ignition engine at different engine speeds operating at full throttle setting. The variations of mean effective pressure and power with engine speed are shown in Figure 16.17. The difference between the ip and bp at any speed represents the fp, which increases with speed. The maximum values of ip and bp occur at different speeds. The ip after attaining the maximum value at a certain speed falls because of a sharp reduction in volumetric efficiency at higher speeds. The volumetric efficiency is influenced by gas temperature, valve timing, valve mechanism dynamics, and the pressure pulsation patterns in the induction and exhaust manifolds. The bp curve is also affected by the sharp reduction in volumetric efficiency at higher speeds, but the bp further decreases due to an increase in the fp, therefore, the bp reaches its maximum at a speed lower than that at which the ip reaches its maximum.

The variation of volumetric efficiency with speed is shown in Figure 15.1. The indicated mean effective pressure shows a maximum in the engine's mid-speed range. Since the full-load indicated specific fuel consumption varies little over the full speed range, the variation of full-load imep

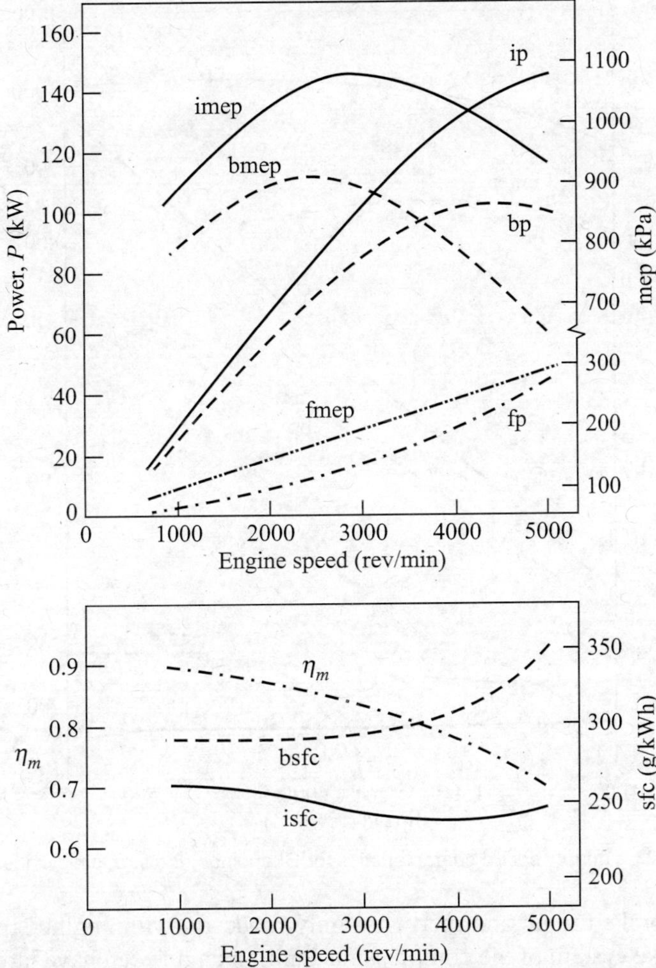

Figure 16.17 Variable speed characteristics of SI engines at wide open throttle (3.8 litre, six-cylinder engine).

with speed is also due to the variation in volumetric efficiency. Since the friction mean effective pressure increases almost linearly with increasing speed, the friction power fp will increase more rapidly. Hence the mechanical efficiency decreases with increasing speed. Thus, the bmep peaks at a speed lower than that at which imep peaks.

At part load at fixed throttle position, these parameters behave similarly. However, at higher speeds the torque and mep decrease more rapidly with increasing speed than at full load. At lower speeds the throttle open area is reduced which chokes the flow of the charge. The pumping component of total friction also increases as the engine is throttled, which in turn decreases the mechanical efficiency.

CI engines

Figure 16.18 shows the full-load indicated and brake power, the mean effective pressure and the specific fuel consumption for naturally aspirated direct-injection compression-ignition engines.

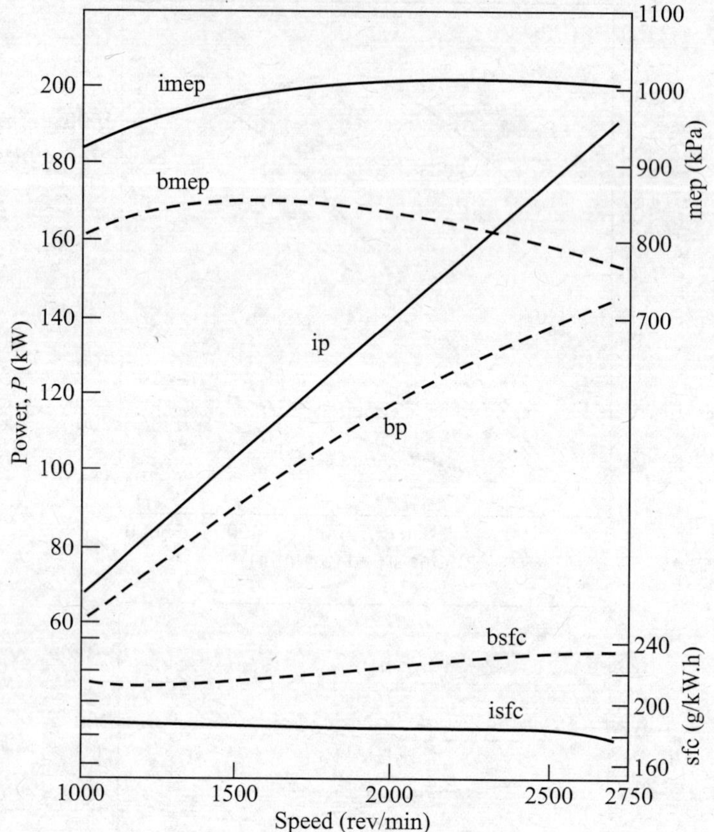

Figure 16.18 Variable speed characteristics for CI engines (8.4 litre, six-cylinder engine).

For CI engines the brake torque and mep vary only modestly with engine speed except at high speeds since the intake system of the compression-ignition engine can have larger flow areas than the intake system of the spark-ignition engines. The decrease in torque and bmep with increasing speed is mainly due to the increase in friction mean effective pressure with speed. The part load torque and the brake mean effective pressure characteristics have a shape similar to the full-load characteristics.

16.9.2 Constant Speed Characteristics

SI engines

SI engines are quantity governed by the opening or closing of a throttle valve which regulates the mass flow of charge to cylinders. The test is carried out on SI engines with constant speed, constant throttle opening and constant ignition setting. The only variable is the air/fuel ratio. The brake specific fuel consumption (bsfc) is plotted versus the brake mean effective pressure (bmep) and a hook curve or consumption loop is obtained. For a single-cylinder engine at full throttle the curve is defined as in Figure 16.19. The air/fuel ratio is a minimum at A (i.e. the richest mixture). As the air/fuel ratio is increased the bmep increases until a maximum is reached at B (usually

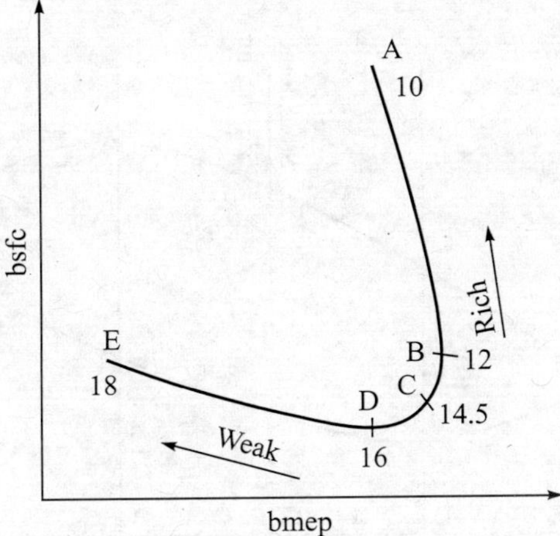

Figure 16.19 Consumption loop for SI engine.

for an air/fuel ratio between 11:1 and 13:1). Any further increase in the air/fuel ratio produces a decrease in bmep with increasing economy until the position of maximum economy is reached at D. Beyond D, for increasing air/fuel ratios, both the bmep and bsfc are adversely affected. Near the point A, with very rich mixtures, the engine could be running unsteadily and there may be combustion of the mixture in the exhaust system. Near the point E, with very lean mixtures, the engine could be running unsteadily and the combustion may be so slow that the gases continue burning in the clearance volume until the next induction stroke begins. It causes popping back through the carburettor. The point C is the point of chemically correct or stoichiometric air/fuel ratio, and this is about 14.5 : 1. The mixture strengths range between those at B and D, which are for maximum power and maximum economy respectively.

The consumption loop for part-throttle openings at constant speed is shown in Figure 16.20. The brake mean effective pressure and the brake specific fuel consumption may be plotted versus the air/fuel ratio as shown in Figure 16.21.

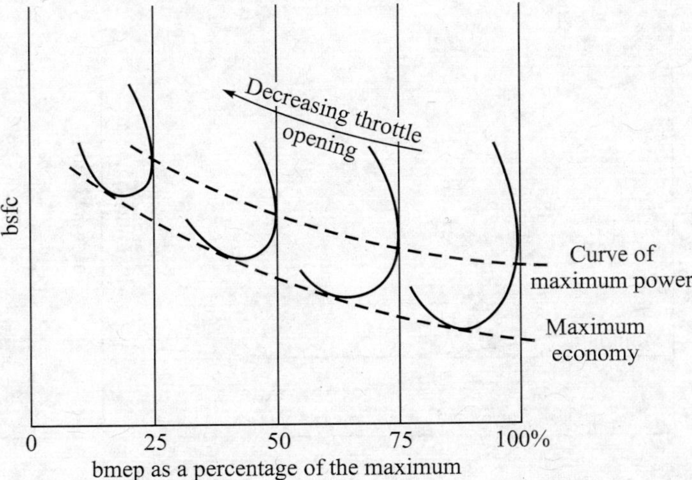

Figure 16.20 Consumption loop for part throttle.

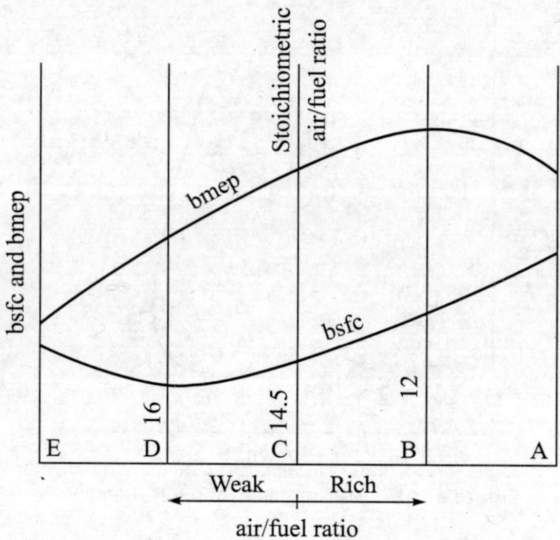

Figure 16.21 bmep and bsfc versus the air/fuel ratio.

CI engines

CI engines are quality governed. When adjusting the fuel supplied to a CI engine the limiting condition is given by the smoke limit, which is the appearance of black smoke in the exhaust. Engines should not be operated with mixtures rich enough to produce smoke. These mixtures may produce higher power output but the efficiency will be lower and the engine may soon become dirty. The smoke limit is reached with the air-fuel ratio of about 16:1. A leaner mixture is therefore recommended for the satisfactory running of the CI engines.

Figure 16.22 shows a consumption loop for a CI engine. The bsfc curve is reasonably flat over a wide range of values of bmep. The minimum bsfc and hence the maximum brake thermal

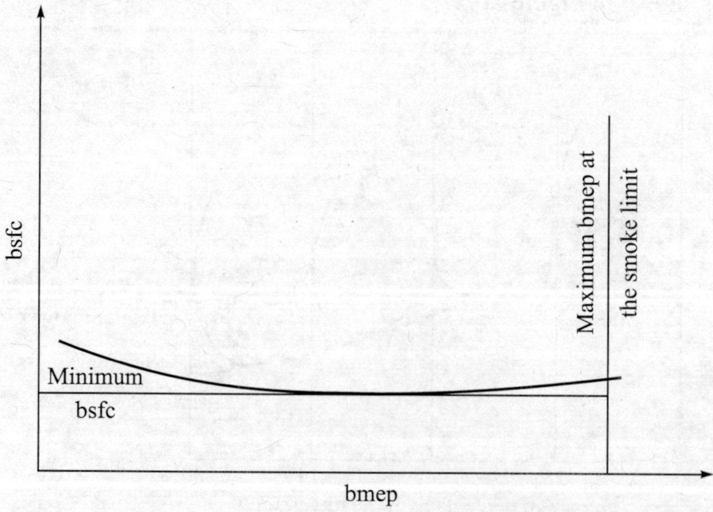

Figure 16.22 Consumption loop for CI engine.

efficiency is obtained at part load, i.e. less than the maximum bmep. It shows that the part load operation of a CI engine is better than that of an SI engine. This condition often prevails in motor vehicle engines.

16.9.3 Performance Maps

It is a single set of curves to represent the performance of an engine. This map is independent of the size of the engine. It is, therefore, possible to use the same set of curves to predict the performance of any one of a series of geometrically similar engines. The performance map may be constructed by using bmep as ordinate and piston speed as abscissa. The map also shows the curve of full-throttle bmep versus the piston speed. Any bmep under this curve can be obtained by varying the inlet pressure by means of the throttle in SI engines or by varying the quantity of fuel injected per cycle in the case of CI engines. Lines of constant bsfc are drawn on the plot, so that, for the given piston speed the bsfc can be instantly determined at any bmep.

The lines of constant brake power per unit piston area are independent of the particular engine and they pass through all points where the product of the piston speed and bmep is constant.

$$\text{Mathematically, bp} = \frac{\text{bmep} \times L \times A \times N_f}{60} \tag{16.15}$$

where L is the stroke length, A is the area of cross-section of the piston and N_f is the number of firings per minute.

$N_f = N$ for the two-stroke engine and $N_f = \frac{N}{2}$ for the four-stroke engine, where N is the rpm.

Piston speed,
$$s = 2\,L\,N \tag{16.16}$$

$$\therefore \quad \frac{\text{bp}}{A} \propto \text{bmep} \times s \tag{16.17}$$

Figure 16.23 shows the performance map for SI engines and Figure 16.24 shows that for the CI engines. The minimum bsfc point is marked by A in both the curves. By moving upwards from this point the bsfc increases in both the engines. In SI engines it is because of mixture enrichment from the action of the economizer and because of the poorer distribution at full throttle. In CI engines it is because of the increased fuel waste (smoke) associated with the high fuel/air ratios at high loads. Moving to a lower bmep from point A, the bsfc increases because of reduced mechanical efficiency.

Moving to the right of point A to higher piston speeds, the bsfc increases owing to increased friction.

Moving to the left from point A to a lower piston speed, the bsfc increases in SI engines because of the increased heat loss per cycle, poor distribution at low manifold velocities, and lowered efficiency due to the automatically retarded spark used for detonation control at low engine speeds. At very low speeds (not shown in Figure 16.24) the CI engines may also have increased bsfc because the injection equipment cannot be set to give completely satisfactory characteristics over the entire speed range.

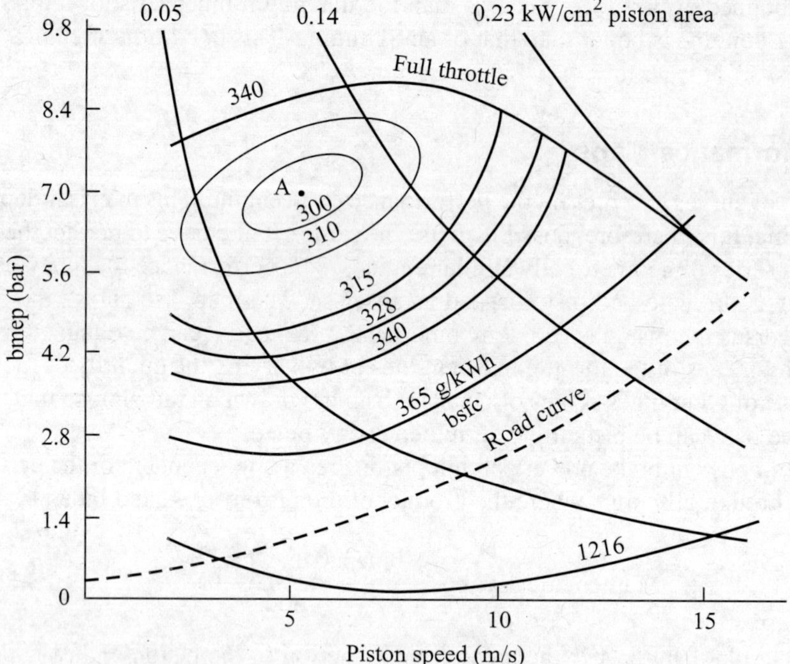

Figure 16.23 Form of performance map for a petrol engine.

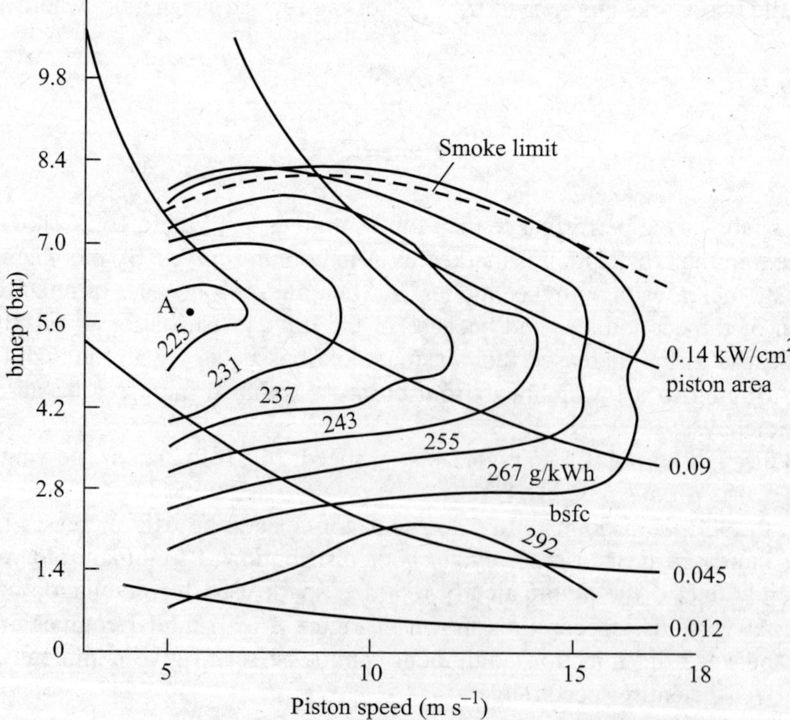

Figure 16.24 Form of performance map for a diesel engine.

16.10 HEAT BALANCE SHEET

The engine under test is considered to be enclosed within an imaginary boundary. When the engine runs for a certain period, the conditions become stabilized for a particular load. When this state is reached the rate at which the energy enters the boundary becomes equal to the rate at which the energy goes out of it.

The energy enters the boundary as:

(a) Heat energy of the fuel
(b) Sensible heat of air
(c) Sensible heat of circulating water entering the jacket
(d) Sensible heat of surrounding air entering the boundary by radiation.

The energy leaves the boundary as:

(a) Brake power
(b) Sensible heat of exhaust gases leaving the engine cylinder
(c) Sensible heat of circulating water leaving the jacket
(d) Sensible heat of surrounding air leaving the boundary.

By equating the rate at which the energy enters the imaginary boundary to the rate at which the energy goes out of the boundary, we get

Heat supplied by fuel to the engine = brake power + heat carried away by circulating water + heat carried away by exhaust gases + unaccounted losses obtained by difference (16.18)

Heat supplied by fuel to the engine = mass of fuel supplied per second × calorific value of fuel

i.e.

$$\dot{Q}_s = \dot{m}_f \times CV \text{ [J/s]} \tag{16.19}$$

$$\text{Brake power} = \frac{2\pi NT}{60} \text{ [W]} \tag{16.20}$$

Heat carried away by circulating water

The heat carried away by cooling water may be determined by measuring the rate of flow of water through the cooling water jacket and the rise in temperature of the water during the flow through the engine. The inlet and outlet temperatures of the water are measured by thermometers inserted in the pockets. The outlet temperature of the cooling water is usually limited to about 80°C to prevent the formation of steam pockets. The rate of flow of water is measured by collecting the water in a container for a specified period of time or it may be measured directly with the help of a calibrated rotameter.

The rate of heat carried away by cooling water is given by

$$\dot{Q}_w = \dot{m}_w c_{p_w} (T_{w_o} - T_{w_i}) \tag{16.21}$$

where

$\dot{m}_w$ = rate of mass flow of water
c_{pw} = specific heat of water
T_{wo} = outlet temperature of cooling water
T_{wi} = inlet temperature of cooling water.

Heat carried away by exhaust gases

The energy to exhaust can be obtained by means of the exhaust gas calorimeter. It is shown in Figure 16.25. It is simply a heat exchanger in which the exhaust gas is cooled by circulating water.

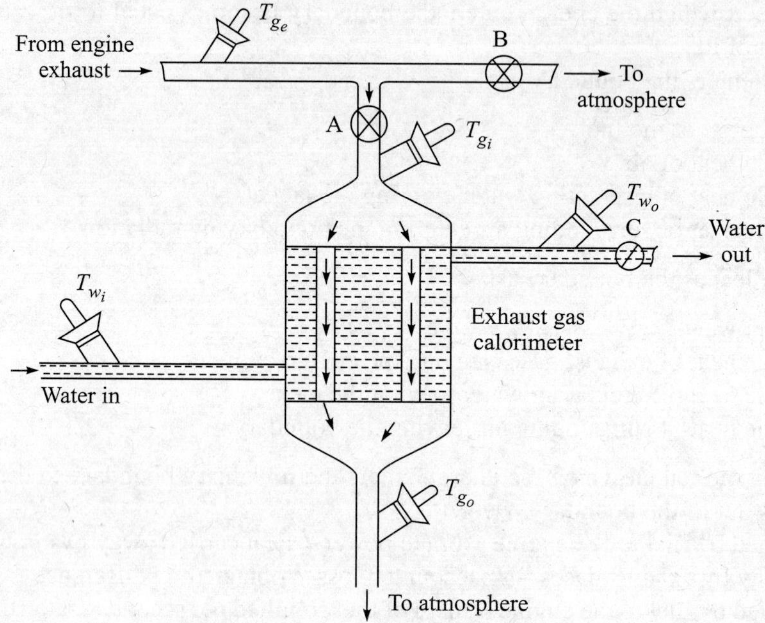

Figure 16.25 Exhaust gas calorimeter.

The exhaust gases from the engine exhaust are passed through the exhaust gas calorimeter by closing the valve B and opening the valve A. The hot exhaust gases are cooled by the water circulated in the calorimeter. The water flow rate is adjusted with the help of valve C to give a measurable temperature rise to circulating water.

It is assumed that the calorimeter is well insulated, then;

heat lost by exhaust gases = heat gained by circulating water

i.e.
$$\dot{m}_g c_{p_g} (T_{g_i} - T_{g_o}) = \dot{m}_w c_{p_w} (T_{w_o} - T_{w_i}) \tag{16.22}$$

where

T_{gi} = temperature of the exhaust gases entering the calorimeter
T_{go} = temperature of the exhaust gases leaving the calorimeter
T_{wi} = temperature of the cooling water entering the calorimeter
T_{wo} = temperature of the cooling water leaving the calorimeter
$\dot{m}_g$ = mass flow rate of exhaust gases (unknown)
$\dot{m}_w$ = mass flow rate of water circulated through the calorimeter
c_{pg} = specific heat of exhaust gases
c_{pw} = specific heat of water

$\therefore$
$$\dot{m}_g = \frac{c_{p_w}}{c_{p_g}} \left(\frac{T_{w_o} - T_{w_i}}{T_{g_i} - T_{g_o}} \right) \dot{m}_w \tag{16.23}$$

As all the quantities on the RHS are known, the gas flow rate can be determined.

Heat carried away by the exhaust gases is given by

$$\dot{Q}_g = \dot{m}_g c_{p_g} (T_{g_e} - T_a) \tag{16.24}$$

where

T_{ge} = temperature of exhaust gases leaving the exhaust valve

T_a = ambient temperature.

The exhaust gases are exhausted to the atmosphere during normal operation by closing the valve A and opening the valve B. When it is required to use the exhaust gas calorimeter, the valve A is opened and the valve B is closed. The heat carried away by the exhaust gases is also given by

$\dot{Q}_g$ = heat carried away by water passing through the calorimeter + heat in exhaust gases above the atmospheric temperature after leaving the calorimeter.

$$\dot{Q}_g = \dot{m}_w c_{p_w} (T_{w_o} - T_{w_i}) + \dot{m}_g c_{p_g} (T_{g_o} - T_a) \tag{16.25}$$

Unaccounted losses

A part of heat is lost by convection and radiation. Part of the power developed inside the engine is also used to run the accessories such as lubricating pump, cam shaft, and water circulating pump. These factors cannot be measured precisely and so these are known as unaccounted losses. The friction power is also not included in heat balance because the friction heat appears partly in the heat to the jacket cooling water, and partly in exhaust. It also raises the temperature of the lubricating oil.

The unaccounted heat energy is calculated by taking the difference between the heat supplied (Eq. (16.19)) and the sum of Eqs. (16.20), (16.21) and (16.24). Heat balance sheet may be prepared in kJ per second, per minute or on per hour basis.

The results of the above calculations are presented in a table and this is known as 'heat balance sheet'. It is a general practice to include the heat distribution as percentage of heat supplied in the heat balance sheet.

The form of heat balance sheet is shown in Table 16.1.

Table 16.1 Heat balance sheet

Heat input per minute	kJ/min	%	*Heat expenditure per minute*	kJ/min	%
Heat supplied by the combustion of fuel	Q_s	100	(a) Heat in bp	a	–
			(b) Heat carried by cooling water	b	–
			(c) Heat to exhaust gases	c	–
			(d) Unaccounted heat losses	$Q_s - (a + b + c)$	–
Total	Q_s	100	Total	Q_s	100

Heat carried away by exhaust gases may be split into heat carried away by dry exhaust gases and heat taken away by water vapour. The heat carried away by the water vapour can be calculated when the pressure of steam formation and the specific heat of superheated steam are known. The fuel analysis should also be known. This method of calculation is illustrated in Example 16.10.

EXAMPLE 16.1 A single cylinder four-stroke gas engine with a bore of 175 mm and a stroke of 320 mm, is governed on the hit and miss principle. The indicator cards give a working loop mean effective pressure of 6.5 bar, and a pumping loop mean effective pressure of 0.4 bar when the engine runs at 510 rpm at full-load. Diagrams from the dead cycle give a mep of 0.65 bar. The engine was run with no-load at the same speed, and a mechanical counter recorded 55 firing strokes per minute. Calculate the full-load brake power and the mechanical efficiency of the engine.

Solution: The friction power is assumed to be constant at a given speed, and is independent of load. It is calculated from the no-load test.

At full-load, the net imep, $p_{mi} = 6.5 - 0.4 = 6.1$ bar

At no-load, the number of working cycles per minute = 55

$$\text{Therefore at no-load, dead cycles per minute} = \left(\frac{510}{2} - 55\right) = 200$$

$$\text{fp} = \text{ip} - \text{bp}$$

$$= (\text{ip} - \text{bp})_{\text{working cycles}} + (\text{ip} - \text{bp})_{\text{dead cycles}}$$

bp at no-load is zero.

$$\therefore \quad \text{fp} = \text{ip}_{\text{working cycles}} + \text{ip}_{\text{dead cycles}}$$

$$\text{ip} = \text{ip of positive loop} - \text{ip of negative loop}$$

For dead cycles, the ip of positive loop is zero and the ip of negative loop is the pumping power of dead cycles.

$$\therefore \quad \text{fp} = \text{ip}_{\text{working cycles}} - \text{pumping power of dead cycles}$$

Now,

$$\text{ip}_{\text{working cycles}} = p_{mi}\, LAN_f$$

$$= 6.1 \times 10^5 \times 0.32 \times \frac{\pi}{4}(0.175)^2 \times \frac{55}{60} \times \frac{1}{1000}$$

$$= 4.3 \text{ kW}$$

$$\text{Pumping power of dead cycles} = 0.65 \times 10^5 \times 0.32 \times \frac{\pi}{4}(0.175)^2 \times \frac{200}{60} \times \frac{1}{1000}$$

$$= 1.67 \text{ kW}$$

$$\therefore \quad \text{fp} = 4.3 - 1.67 = 2.63 \text{ kW}$$

At full-load the engine fires regularly in every two revolutions, therefore, there are $\frac{510}{2} = 255$ firing strokes per minute.

$$\text{Full-load ip} = 6.1\times10^5\times0.32\times\frac{\pi}{4}(0.175)^2\times\frac{255}{60}\times\frac{1}{1000}$$

$$= 19.95 \text{ kW}$$

$$\text{Full-load bp} = \text{Full-load ip} - \text{fp}$$

$$= 19.95 - 2.63 = \boxed{17.32 \text{ kW}} \quad \textbf{Ans.}$$

$$\text{Mechanical efficiency, } \eta_M = \frac{\text{bp}}{\text{ip}} = \frac{17.32}{19.95} = 0.868 = \boxed{86.8\%} \quad \textbf{Ans.}$$

EXAMPLE 16.2 A four-cylinder SI engine has a bore of 60 mm and a stroke of 85 mm. It runs at 3000 rpm and is tested at this speed against a brake which has a torque arm of 0.35 m. The net brake load is 160 N and the fuel consumption is 6.6 l/h. The specific gravity of the fuel used is 0.78 and it has a lower calorific value of 44000 kJ/kg. A Morse-test is carried out and the cylinders are cut out in the order 1, 2, 3, 4 with the corresponding brake loads of 114, 110, 112 and 116 N respectively. Calculate for this speed the bp, the bmep, the brake thermal efficiency, the bsfc, the ip, the mechanical efficiency, and the imep.

Solution: Bore, $d = 0.06$ m; stroke, $L = 0.085$ m; $N = 3000$ rpm

Torque arm, $r = 0.35$ m; $W = 160$ N

$$\dot{V}_f = 6.6 \text{ l/h} = \frac{6.6\times10^{-3}}{3600} = 1.833\times10^{-6} \text{ m}^3\text{/s}$$

$$\rho_f = 780 \text{ kg/m}^3; \quad \text{CV} = 44{,}000 \text{ kJ/kg}$$

$$\dot{m}_f = \rho_f\times\dot{V}_f = 780\times1.833\times10^{-6} = 1.43\times10^{-3} \text{ kg/s}$$

$$\text{Torque, } T = Wr = 160\times0.35 = 56.0 \text{ N m}$$

$$\therefore \quad \text{bp} = \frac{2\pi NT}{60} = \frac{2\pi\times3000\times56}{60\times1000} = \boxed{17.59 \text{ kW}} \quad \textbf{Ans.}$$

$$\text{Also,} \quad \text{bp} = \frac{p_{m_b} LANn}{2\times60}$$

$$\therefore \quad p_{mb} = \frac{120(\text{bp})}{LANn} = \frac{120\times17.59\times1000}{0.085\times\frac{\pi}{4}(0.06)^2\times3000\times4}\times\frac{1}{10^5}$$

$$= \boxed{7.32 \text{ bar}} \quad \textbf{Ans.}$$

$$\text{Now,} \quad \eta_b = \frac{\text{bp}}{\dot{m}_f\times\text{CV}} = \frac{17.59}{1.43\times10^{-3}\times44{,}000} = 0.28 = \boxed{28\%} \quad \textbf{Ans.}$$

$$\text{and} \quad \text{bsfc} = \frac{\dot{m}_f}{\text{bp}} = \frac{1.43\times10^{-3}\times3600}{17.59} = \boxed{0.293 \text{ kg/kWh}} \quad \textbf{Ans.}$$

From Morse test,

$$\sum_{i=2}^{4} \text{bp}_i = \frac{2\pi \times 3000 \times (114 \times 0.35)}{60 \times 1000} = 12.53 \text{ kW}$$

$$\text{ip}_1 = 17.59 - 12.53 = 5.06 \text{ kW}$$

Similarly,

$$\text{ip}_2 = 17.59 - \frac{2\pi \times 3000 \times (110 \times 0.35)}{60 \times 1000} = 17.59 - 12.1 = 5.49 \text{ kW}$$

$$\text{ip}_3 = 17.59 - \frac{2\pi \times 3000 \times (112 \times 0.35)}{60 \times 1000} = 17.59 - 12.32 = 5.27 \text{ kW}$$

$$\text{ip}_4 = 17.59 - \frac{2\pi \times 3000 \times (116 \times 0.35)}{60 \times 1000} = 17.59 - 12.75 = 4.84 \text{ kW}$$

$$\therefore \quad \text{ip} = \text{ip}_1 + \text{ip}_2 + \text{ip}_3 + \text{ip}_4$$

$$= 5.06 + 5.49 + 5.27 + 4.84 = \boxed{20.66 \text{ kW}} \quad \textbf{Ans.}$$

$$\eta_M = \frac{\text{bp}}{\text{ip}} = \frac{17.59}{20.66} = 0.851 = \boxed{85.1\%} \quad \textbf{Ans.}$$

and

$$p_{mi} = \frac{p_{m_b}}{\eta_M} = \frac{7.32}{0.851} = \boxed{8.6 \text{ bar}} \quad \textbf{Ans.}$$

EXAMPLE 16.3 The following details were noted in a test on a single-cylinder four-stroke oil engine: bore = 150 mm; stroke = 160 mm; speed of the engine = 500 rpm; fuel consumption = 0.0475 kg/min; calorific value of fuel = 42,000 kJ/kg; the difference in tension on either side of the brake pulley = 400 N; brake circumference = 2.2 m; length of the indicator diagram = 50 mm; area of positive loop of indicator diagram = 475 mm^2; area of negative loop = 25 mm^2; spring constant = 0.8333 bar per mm. Calculate: (a) the bp, (b) the ip, (c) the mechanical efficiency, (d) the brake thermal efficiency, (e) the indicated thermal efficiency, and (f) the brake specific fuel consumption.

Solution: (a) Difference in tension on either side of brake pulley, $W = 400$ N.

$$\text{Arm length, } r = \frac{\text{circumference}}{2\pi} = \frac{2.2}{2\pi} = 0.35 \text{ m}$$

$$\text{Torque, } T = Wr = 400 \times 0.35 = 140 \text{ N m}$$

$$\therefore \quad \text{bp} = \frac{2\pi NT}{60} = \frac{2\pi \times 500 \times 140}{60} = 7330 \text{ W} = \boxed{7.33 \text{ kW}} \quad \textbf{Ans.}$$

(b) Mean height of indicator diagram $= \dfrac{\text{net area of indicator diagram}}{\text{length of indicator diagram}}$

$$= \frac{475 - 25}{50} = 9 \text{ mm}$$

$$\text{imep} = \text{mean height of indicator diagram} \times \text{spring constant}$$

$$= 9 \times 0.8333 = 7.5 \text{ bar}$$

$$\therefore \quad \text{ip} = \frac{\text{imep} \times L \times A \times N}{2 \times 60}$$

$$= 7.5 \times 10^5 \times 0.16 \times \frac{\pi}{4}(0.15)^2 \times \frac{500}{2 \times 60} \times \frac{1}{1000} = \boxed{8.835 \text{ kW}} \quad \textbf{Ans.}$$

(c) $$\eta_M = \frac{\text{bp}}{\text{ip}} = \frac{7.33}{8.835} = 0.83 = \boxed{83\%} \quad \textbf{Ans.}$$

(d) $$\eta_b = \frac{\text{bp}}{\dot{m}_f \times \text{CV}} = \frac{7.33 \times 60}{0.0475 \times 42{,}000} = 0.2204 = \boxed{22.04\%} \quad \textbf{Ans.}$$

(e) $$\eta_i = \frac{\text{ip}}{\dot{m}_f \times \text{CV}} = \frac{\eta_b}{\eta_M}$$

$$= \frac{0.2204}{0.83} = 0.266 = \boxed{26.6\%} \quad \textbf{Ans.}$$

(f) $$\text{bsfc} = \frac{\dot{m}_f}{\text{bp}} = \frac{0.0475 \times 60}{7.33} = \boxed{0.389 \text{ kg/kWh}} \quad \textbf{Ans.}$$

EXAMPLE 16.4 An eight-cylinder four-stroke SI engine of 80 mm bore and 100 mm stroke is tested at 4500 rpm on a dynamometer which has 55 cm arm. The dynamometer scale reading was 40 kg. The time for 100 cc of fuel consumption is recorded as 9.5 seconds. The calorific value of fuel is 44,000 kJ/kg. Air at 1 bar and 27°C was supplied to the carburettor at the rate of 6 kg/minute. Assume specific gravity of fuel to be 0.7. Clearance volume of each cylinder is 65 cc. Determine the bp, the bmep, the bsfc, the bsac, the air/fuel ratio, the brake thermal efficiency, the volumetric efficiency, and the relative efficiency.

Solution: $$\text{bp} = \frac{2\pi NT}{60} = \frac{2\pi \times 4500 \times 0.55 \times 40 \times 9.81}{60 \times 1000}$$

$$= \boxed{101.7 \text{ kW}} \quad \textbf{Ans.}$$

$$\text{bmep} = \frac{\text{bp} \times 60}{LA\frac{N}{2}n}$$

$$= \frac{101.7 \times 1000 \times 60}{0.1 \times \frac{\pi}{4}(0.08)^2 \times \frac{4500}{2} \times 8} = 6.744 \times 10^5 \text{ N/m}^2$$

$$= \boxed{6.744 \text{ bar}} \quad \textbf{Ans.}$$

$$\dot{m}_f = \frac{100 \times 0.7 \times 3600}{9.5 \times 1000} = 26.53 \text{ kg/h}$$

$$\text{bsfc} = \frac{\dot{m}_f}{\text{bp}}$$

$$= \frac{26.53}{101.7} = \boxed{0.261 \text{ kg/kWh}} \quad \textbf{Ans.}$$

$$\text{bsac} = \frac{6 \times 60}{101.7} = \boxed{3.54 \text{ kg/kWh}} \quad \textbf{Ans.}$$

$$\text{Air/fuel ratio} = \frac{3.54}{0.261} = \boxed{13.56} \quad \textbf{Ans.}$$

$$\eta_b = \frac{\text{bp}}{\dot{m}_f \times \text{CV}} = \frac{101.7 \times 3600}{26.53 \times 44000} = 0.314 = \boxed{31.4\%} \quad \textbf{Ans.}$$

Volume flow rate of air at intake condition,

$$\dot{V}_a = \frac{\dot{m}_a RT}{p} = \frac{6 \times 287 \times 300}{1 \times 10^5} = 5.166 \text{ m}^3\text{/min}$$

Swept volume per minute, $\dot{V}_s = \frac{\pi}{4} d^2 L \frac{N}{2} n$

$$= \frac{\pi}{4} \times (0.08)^2 \times 0.1 \times \frac{4500}{2} \times 8 = 9.048 \text{ m}^3\text{/min}$$

Volumetric efficiency, $\eta_V = \frac{5.166}{9.048} = 0.571 = \boxed{57.1\%}$ **Ans.**

Swept volume per cylinder, $V_s = \frac{\pi}{4} d^2 L$

$$= \frac{\pi}{4} \times (0.08)^2 \times 0.1$$

$$= 5.027 \times 10^{-4} \text{ m}^3 = 502.7 \text{ cc}$$

Clearance volume, $V_c = 65$ cc

Compression ratio, $r = \frac{V_s + V_c}{V_c} = \frac{502.7 + 65}{65} = 8.734$

Air-standard efficiency based on Otto cycle $= 1 - \left(\frac{1}{r}\right)^{\gamma - 1}$

$$= 1 - \left(\frac{1}{8.734}\right)^{0.4} = 0.58 = 58\%$$

$$\text{Relative efficiency} = \frac{\text{brake thermal efficiency}}{\text{air-standard efficiency}}$$

$$= \frac{0.314}{0.58} = 0.541 = \boxed{54.1\%} \quad \textbf{Ans.}$$

EXAMPLE 16.5 The power output of a six-cylinder four-stroke engine is absorbed by a hydraulic dynamometer for which the law is $\frac{WN}{20,000}$ kW, , where the brake load W is in newton and the speed N is in rpm. The air consumption is measured by an air-box with a sharp-edged orifice system. The following observations are made:

Orifice diameter = 30 mm
Coefficient of discharge = 0.6
Pressure drop across the orifice = 14 cm of Hg
Bore = 100 mm
Stroke = 110 mm
Brake load = 540 N
Engine speed = 2500 rpm
C/H ratio by mass = 83/17
Ambient pressure = 1 bar
Time taken for 100 cc of fuel consumption = 18 s
Ambient temperature = 27°C
Fuel density = 780 kg/m^3

Calculate the volumetric efficiency, the bmep, the bp, the torque, the bsfc and the percentage of excess air.

Solution: Density of air, $\rho_a = \frac{p}{RT} = \frac{1 \times 10^5}{287 \times 300} = 1.161 \text{ kg/m}^3$

$$\text{Mass flow rate of air, } \dot{m}_a = C_d A_o \sqrt{2g\Delta H \rho_{Hg} \rho_a}$$

$$\text{Volume flow rate of air, } \dot{V}_a = \frac{\dot{m}_a}{\rho_a} = C_d A_o \sqrt{2g\Delta H \rho_{Hg} / \rho_a}$$

$$\therefore \quad \dot{V}_a = 0.6 \times \frac{\pi}{4}(0.03)^2 \sqrt{2 \times 9.81 \times 0.14 \times 13600/1.161}$$

$$= 0.0761 \text{ m}^3\text{/s}$$

$$\text{Swept volume per second, } \dot{V}_s = \frac{\pi}{4} d^2 L \frac{N}{2 \times 60} \times n$$

$$= \frac{\pi}{4}(0.1)^2 \times 0.11 \times \frac{2500}{2 \times 60} \times 6$$

$$= 0.108 \text{ m}^3\text{/s}$$

$$\text{Volumetric efficiency, } \eta_V = \frac{\dot{V}_a}{\dot{V}_s} = \frac{0.0761}{0.108} = 0.705 = \boxed{70.5\%}$$ **Ans.**

$$\text{bp} = \frac{WN}{20000} = \frac{540 \times 2500}{20000} = \boxed{67.5 \text{ kW}}$$ **Ans.**

$$\text{bmep} = \frac{\text{bp}}{LA\dfrac{N}{2\times 60}n} = \frac{67.5\times 1000\times 120}{0.11\times \dfrac{\pi}{4}(0.1)^2\times 2500\times 6} = 6.25\times 10^5 \text{ N/m}^2$$

$$= \boxed{6.25 \text{ bar}} \quad \textbf{Ans.}$$

$$\text{bp} = \frac{2\pi NT}{60}$$

$$\therefore \quad T = \frac{60\times \text{bp}}{2\pi N} = \frac{60\times 67.5\times 1000}{2\pi\times 2500} = \boxed{257.8 \text{ N m}} \quad \textbf{Ans.}$$

$$\text{Mass rate of fuel, } \dot{m}_f = \frac{100}{18}\times 0.78\times \frac{1}{1000}\times 3600 = 15.6 \text{ kg/h}$$

$$\therefore \quad \text{bsfc} = \frac{\dot{m}_f}{\text{bp}} = \frac{15.6}{67.5} = \boxed{0.231 \text{ kg/kWh}} \quad \textbf{Ans.}$$

Stoichiometric oxygen required per kg of fuel

$$= 0.83\times \frac{32}{12} + 0.17\times \frac{8}{1} = 3.573 \text{ kg/kg fuel}$$

$$\therefore \quad \text{Stoichiometric air required} = \frac{3.573}{0.23} = 15.54 \text{ kg/kg fuel}$$

$$\text{Actual mass flow rate of air} = \dot{V}_a \times \rho_a$$

$$= 0.0761\times 1.161 = 0.0884 \text{ kg/s}$$

$$\text{Actual air/fuel ratio} = \frac{0.0884\times 3600}{15.6} = 20.4$$

$$\therefore \quad \text{Percentage of excess air} = \frac{20.4 - 15.54}{15.54}\times 100 = \boxed{31.3\%} \quad \textbf{Ans.}$$

EXAMPLE 16.6 A four-stroke cycle gas engine has a bore of 200 mm and a stroke of 300 mm. The compression ratio of the engine is 5.5. The engine runs at 400 rpm. The imep is 4.5 bar, the air/gas ratio by volume is 6 : 1 and the calorific value of the gas is 12,000 kJ/m³ at NTP. At the beginning of the compression stroke the temperature is 67°C and the pressure is 0.97 bar. Calculate the ip, the thermal efficiency and the relative efficiency.

Solution: Swept volume, $V_s = \dfrac{\pi}{4}d^2 L$

$$= \frac{\pi}{4}(0.2)^2 \times 0.3 = 9.425\times 10^{-3} \text{ m}^3$$

$$\text{Compression ratio, } r = \frac{V_s + V_c}{V_c} = \frac{V_s}{V_c} + 1$$

$$\therefore \quad V_c = \frac{V_s}{r-1} = \frac{9.425\times 10^{-3}}{5.5-1} = 2.094\times 10^{-3} \text{ m}^3$$

Total cylinder volume, $V = V_s + V_c = (9.425 + 2.094) \times 10^{-3}$

$= 11.519 \times 10^{-3}\ \text{m}^3$

Volume of the gas in the total cylinder volume,

$$V_g = \frac{1}{7} \times 11.519 \times 10^{-3} = 1.646 \times 10^{-3}\ \text{m}^3$$

$$\left(\frac{pV}{T}\right)_{\text{NTP}} = \left(\frac{pV}{T}\right)_{\text{working}}$$

$$\therefore \quad \text{Volume of gas at NTP} = \left(\frac{pV}{T}\right)_{\text{working}} \times \left(\frac{T}{p}\right)_{\text{NTP}}$$

$$= \frac{0.97 \times 1.646 \times 10^{-3}}{340} \times \frac{273}{1.013}$$

$$= 1.266 \times 10^{-3}\ \text{m}^3$$

Heat supplied by the fuel = volume of gas at NTP × calorific value

$$= 1.266 \times 10^{-3} \times 12{,}000$$

$$= 15.19\ \text{kJ/cycle}$$

$$= 15.19 \times \frac{N}{2 \times 60} = 15.19 \times \frac{400}{2 \times 60} = 50.63\ \text{kJ/s}$$

Now,

$$\text{ip} = \frac{p_{m_i} LAN}{2 \times 60}$$

$$= 4.5 \times 10^5 \times 0.3 \times \frac{\pi}{4}(0.2)^2 \times \frac{400}{2 \times 60} \times \frac{1}{1000}$$

$$= \boxed{14.14\ \text{kW}}$$ **Ans.**

$$\text{Indicated thermal efficiency, } \eta_i = \frac{\text{ip}}{\text{heat supplied by fuel}}$$

$$= \frac{14.14}{50.63} = 0.279 = \boxed{27.9\%}$$ **Ans.**

$$\text{Air-standard efficiency} = 1 - \left(\frac{1}{r}\right)^{\gamma - 1}$$

$$= 1 - \left(\frac{1}{5.5}\right)^{0.4} = 0.494 = 49.4\%$$

Relative efficiency based on indicated thermal efficiency

$$\frac{\text{indicated thermal efficiency}}{\text{air-standard efficiency}} = \frac{0.279}{0.494} = 0.565 = \boxed{56.5\%}$$ **Ans.**

EXAMPLE 16.7 A six-cylinder four-stroke CI engine develops 130 kW power output at 1800 rpm. The calorific value of the fuel is 42,000 kJ/kg and its percentage composition by mass is 86% carbon, 13% hydrogen and 1% non-combustibles. The absolute volumetric efficiency is 85%, the indicated thermal efficiency is 38% and the mechanical efficiency is 80%. The air consumption is 110% in excess of that required for stoichiometric combustion. Estimate the volumetric composition of dry exhaust gas and determine the bore and stroke of the engine, taking a stroke/bore ratio as 1.2. Assume the density of air at the given conditions as 1.3 kg/m³.

Solution: Stoichiometric air/fuel ratio $= \left(0.86 \times \dfrac{32}{12} + 0.13 \times \dfrac{8}{1}\right) \times \dfrac{1}{0.23}$

$$= 14.49$$

Actual air/fuel ratio $= (1 + 1.1) \times 14.49 = 30.43$

Molecular weight of air $= 0.23 \times 32 + 0.77 \times 28 = 28.92$ kg/kmol

Let x kmol of air be supplied per kg of fuel. The combustion equation per kg of fuel can be written as

$$\frac{0.86}{12}C + \frac{0.13}{2}H_2 + 0.21xO_2 + 0.79xN_2 = aCO_2 + bH_2O + cO_2 + dN_2$$

From carbon balance, $\dfrac{0.86}{12} = a = 0.07167$

From hydrogen balance, $\dfrac{0.13}{2} = b = 0.065$

From oxygen balance, $0.21x = a + \dfrac{b}{2} + c$

$$= 0.07167 + 0.0325 + c$$

$$= 0.10417 + c$$

Number of kmol of air per kg of fuel $= \dfrac{30.43}{28.92} = 1.052$

∴ $x = 1.052$

∴ $0.21 \times 1.052 = 0.10417 + c$

or $c = 0.1168$

From nitrogen balance, $0.79 \times 1.052 = d$

or $d = 0.8311$

The volumetric composition of dry exhaust gas is given by

Constituent	kmol	*volume* (%)
CO_2	0.07167	7.03
O_2	0.11680	11.46
N_2	0.83110	81.51

$$\text{ip} = \frac{\text{bp}}{\eta_M} = \frac{130}{0.8} = 162.5 \text{ kW}$$

$$\eta_i = \frac{\text{ip}}{\dot{m}_f \times \text{CV}}$$

$$\therefore \quad \dot{m}_f = \frac{\text{ip}}{\eta_i \times \text{CV}} = \frac{162.5}{0.38 \times 42000} = 0.01018 \text{ kg/s}$$

$$\dot{m}_a = \dot{m}_f \times \text{actual}(\text{A/F})$$

$$= 0.01018 \times 30.43 = 0.3098 \text{ kg/s}$$

$$\rho_a = \frac{\dot{m}_a}{\dot{V}_a}$$

$$\therefore \quad \dot{V}_a = \frac{\dot{m}_a}{\rho_a} = \frac{0.3098}{1.3} = 0.2383 \text{ m}^3/\text{s}$$

Swept volume per second, $\dot{V}_s = \dfrac{\dot{V}_a}{\eta_{\text{vol}}} = \dfrac{0.2383}{0.85} = 0.2804 \text{ m}^3/\text{s}$

Now, $\dot{V}_s = \dfrac{\pi}{4} d^2 L \dfrac{N}{2 \times 60} \times n$

$$\therefore \quad 0.2804 = \frac{\pi}{4} d^2 \times 1.2d \times \frac{1800}{2 \times 60} \times 6$$

or $d^3 = 3.3057 \times 10^{-3} \text{ m}^3$

$\therefore \quad d = 0.149 \text{ m} = \boxed{149 \text{ mm}}$ **Ans.**

$\therefore \quad L = 1.2d = 1.2 \times 149$

$= \boxed{178.8 \text{ mm}}$ **Ans.**

EXAMPLE 16.8 The following observations were made during a trial of a single-cylinder, four-stroke gas engine having cylinder diameter of 180 mm and stroke of 240 mm:

Duration of trial = 30 min
Total number of revolutions = 9000
Total number of explosions = 4450
Gross imep = 5.35 bar
Pumping imep = 0.35 bar
Net load on brake wheel = 40 kg
Diameter of the brake wheel drum = 0.96 m
Diameter of the rope = 4 cm
Volume of gas used = 2.6 m^3
Pressure of gas = 136 mm water of gauge
Density of gas = 0.655 kg/m^3
Ambient temperature = 17°C

Calorific value of gas at NTP = 19 MJ/m^3
Total air used = 40 m^3
Pressure of air = 720 mm Hg
Temperature of exhaust gas = 340°C
Specific heat of exhaust gas = 1.1 kJ/(kg K)
Cooling water circulated = 80 kg
Rise in temperature of cooling water = 30°C

Draw up a heat balance sheet and estimate the mechanical and indicated thermal efficiencies of the engine.

Solution:

$$\text{ip} = p_{mi}\, LAN_{\text{explosion}}$$

$$= (5.35 - 0.35) \times 10^5 \times 0.24 \times \frac{\pi}{4}(0.18)^2 \times \frac{4450}{30 \times 60} \times \frac{1}{1000}$$

$$= 7.549 \text{ kW}$$

$$\text{bp} = \frac{2\pi NT}{60} = \frac{2\pi \times N \times W \times r}{60} = \frac{\pi N W d_e}{60}$$

where d_e is the effective diameter.

$$\text{bp} = \pi \times \frac{9000}{30 \times 60} \times 40 \times 9.81 \times (0.96 + 0.04) \times \frac{1}{1000}$$

$$= 6.164 \text{ kW}$$

$$\text{Pressure of the gas supplied} = 760 + \frac{136}{13.6} = 770 \text{ mm of Hg}$$

$$\left(\frac{pV}{T}\right)_{\text{NTP}} = \left(\frac{pV}{T}\right)_{\text{working}}$$

$$\text{Volume of the gas used at NTP} = \left(\frac{pV}{T}\right)_{\text{working}} \times \frac{273}{760}$$

or

$$V_g = \frac{770 \times 2.6}{290} \times \frac{273}{760} = 2.48 \text{ m}^3$$

$$\text{Heat supplied by gas per minute at NTP} = \text{Volume of gas per minute} \times \text{CV}$$

$$= \frac{2.48}{30} \times 19{,}000$$

$$= 1571 \text{ kJ/min}$$

$$\text{Heat equivalent of bp} = 6.164 \times 60 = 369.8 \text{ kJ/min}$$

$$\text{Heat lost to cooling medium} = \dot{m}_w c_{p_w} \Delta T_w$$

$$= \frac{80}{30} \times 4.18 \times 30 = 334.4 \text{ kJ/min}$$

Total air used = 40 m^3 at 720 mm of Hg

$$\text{Volume of air used at NTP} = \left(\frac{pV}{T}\right)_{\text{working}} \times \left(\frac{T}{p}\right)_{\text{NTP}}$$

$$= \frac{720 \times 40}{290} \times \frac{273}{760} = 35.67 \text{ m}^3$$

$$\dot{V}_a = \frac{35.67}{30} = 1.189 \text{ m}^3/\text{min}$$

$$\rho_a \text{ at NTP} = \frac{p}{RT} = \frac{1.013 \times 10^5}{287 \times 273} = 1.293 \text{ kg/m}^3$$

$$\text{Mass of air used, } \dot{m}_a = \dot{V}_a \times \rho_a$$

$$= 1.189 \times 1.293 = 1.537 \text{ kg/min}$$

$$\text{Mass of gas at NTP} = V_g \times \rho_g = \frac{2.48}{30} \times 0.655 = 0.0541 \text{ kg/min}$$

$$\text{Total mass of exhaust gas, } \dot{m}_{\text{ex}} = 1.537 + 0.0541 = 1.591 \text{ kg/min}$$

$$\text{Heat lost to exhaust gas} = \dot{m}_{\text{ex}} c_{p_{\text{ex}}} \Delta T_{\text{ex}}$$

$$= 1.591 \times 1.1\,(340 - 17) = 565.3 \text{ kJ/min}$$

$$\text{Unaccounted heat loss} = 1571 - (369.8 + 334.4 + 565.3)$$

$$= 301.5 \text{ kJ/min}$$

Heat balance sheet

Heat input	kJ/min	%	*Heat expenditure*	kJ/min	%
Heat supplied by fuel	1571	100	(a) Heat in bp	369.8	23.5
			(b) Heat lost to cooling water	334.4	21.3
			(c) Heat to exhaust gases	565.3	36.0
			(d) Unaccounted losses	301.5	19.2
Total	1571	100	Total	1571.0	100.0

$$\eta_M = \frac{\text{bp}}{\text{ip}} = \frac{6.164}{7.549} = 0.8165 = \boxed{81.65\%} \quad \textbf{Ans.}$$

$$\eta_i = \frac{\text{ip}}{\text{heat supplied by fuel}}$$

$$= \frac{7.549 \times 60}{1571} = 0.288 = \boxed{28.8\%} \quad \textbf{Ans.}$$

EXAMPLE 16.9 The following observations were made during a test on an oil engine:

Brake power = 30 kW

Fuel used = 10 kg/h

Calorific value of fuel = 42,000 kJ/kg

Jacket circulating water = 9 kg/min
Rise in temperature of cooling water = 60°C
The exhaust gases are passed through the exhaust gas calorimeter for determining the heat carried away by exhaust gases.
Water circulated through exhaust gas calorimeter = 9.5 kg/min
Rise in temperature of water passing through the calorimeter = 40°C
Temperature of exhaust gases leaving the calorimeter = 80°C
Air/fuel ratio on mass basis = 20
Ambient temperature = 17°C
Mean specific heat of exhaust gases = 1.0 kJ/(kg K)
Specific heat of water = 4.18 kJ/(kg K)

Draw up a heat balance sheet on kJ/min and percentage basis.

Solution: Heat supplied by the fuel $= \dot{m}_f \times \text{CV}$

$$= \frac{10}{60} \times 42{,}000 = 7000 \text{ kJ/min } \times 42{,}000$$

$$= 7000 \text{ kJ/min}$$

Heat equivalent of bp $= 30 \times 60 = 1800$ kJ/min

Heat carried away by the jacket cooling water $= \dot{m}_w c_{p_w} (T_{\text{out}} - T_{\text{in}})$

$$= 9 \times 4.18 \times 60 = 2257 \text{ kJ/min}$$

Mass of exhaust gases formed = mass of fuel + mass of air

$$\therefore \quad \dot{m}_g = \dot{m}_f + \dot{m}_a$$

$$= \frac{10}{60} + \frac{10}{60} \times 20 = 3.5 \text{ kg/min}$$

Heat carried away by exhaust gases

$$= \dot{m}_w c_{p_w} (T_{w_o} - T_{w_i}) + \dot{m}_g c_{p_g} (T_{g_o} - T_a)$$

$$= 9.5 \times 4.18 \times 40 + 3.5 \times 1.0(80 - 17)$$

$$= 1588.4 + 220.5 = 1809 \text{ kJ/min}$$

Heat unaccounted for $= 7000 - (1800 + 2257 + 1809)$

$$= 7000 - 5866 = 1134 \text{ kJ/min}$$

Heat balance sheet

Heat input	kJ/min	%	*Heat expenditure*	kJ/min	%
Heat supplied by fuel	7000	100	(a) Heat in bp	1800	25.70
			(b) Heat lost to cooling water	2257	32.25
			(c) Heat to exhaust gases	1809	25.85
			(d) Unaccounted losses	1134	16.20
Total	7000	100	Total	7000	100.00

EXAMPLE 16.10 A four-stroke cycle gasoline engine has four cylinders of 8.5 cm bore and 9.5 cm stroke. The engine is coupled to a brake having a torque radius of 35 cm. At 3000 rpm with all cylinders firing, the net brake load is 430 N. When each cylinder in turn is cut-off, the average net brake load produced at the same speed by the remaining three cylinders is 300 N. Estimate the indicated mep of the engine. With all cylinders firing the fuel consumption is 0.24 kg/min, and the calorific value of fuel is 44,000 kJ/kg. The cooling water flow rate is 65 kg/min and the temperature rise is 12°C. The air/fuel ratio is 15, the temperature of the exhaust gas is 450°C, the temperature of the test room is 17°C, the barometric pressure is 76 cm of Hg, the proportion of hydrogen by mass in fuel is 15.5%, the mean specific heat of dry exhaust gases is 1 kJ/(kg K), the specific heat of superheated steam is 2 kJ/(kg K). Determine the imep, the indicated thermal efficiency, the bsfc and the volumetric efficiency based on atmospheric conditions. Draw up a heat balance sheet on kJ/min and percentage basis.

At 76 cm of Hg, $T_s = 100°\text{C}$ and $h_{fg} = 2257$ kJ/kg

Solution:

$$\text{bp} = \frac{2\pi NT}{60}$$

$$= \frac{2\pi \times 3000 \times 430 \times 0.35}{60 \times 1000} = 47.29 \text{ kW}$$

The bp of the engine when each cylinder is cut-off in turn

$$= \frac{2\pi \times 3000 \times 300 \times 0.35}{60 \times 1000} = 32.99 \text{ kW}$$

$$\text{ip per cylinder} = 47.29 - 32.99 = 14.3 \text{ kW}$$

$$\text{ip of the engine} = 4 \times 14.3 = 57.2 \text{ kW}$$

Now,

$$\text{imep} = \frac{\text{ip} \times 60}{LA\dfrac{N}{2} \times n}$$

$$= \frac{57.2 \times 1000 \times 60}{0.095 \times \dfrac{\pi}{4}(0.085)^2 \times \dfrac{3000}{2} \times 4} = 10.61 \times 10^5 \text{ N/m}^2$$

$$= \boxed{10.61 \text{ bar}} \quad \textbf{Ans.}$$

Indicated thermal efficiency,

$$\eta_i = \frac{\text{ip}}{\dot{m}_f \times \text{CV}} = \frac{57.2 \times 60}{0.24 \times 44000} = 0.325 = \boxed{32.5\%} \quad \textbf{Ans.}$$

$$\text{bsfc} = \frac{\dot{m}_f}{\text{bp}} = \frac{0.24 \times 60}{47.29} = \boxed{0.3045 \text{ kg/kWh}} \quad \textbf{Ans.}$$

Swept volume per minute, $\dot{V}_s = \dfrac{\pi}{4} d^2 L \times \dfrac{N}{2} \times n$

$$= \frac{\pi}{4}(0.085)^2 \times 0.095 \times \frac{3000}{2} \times 4$$

$$= 3.234 \text{ m}^3/\text{min}$$

$$\dot{m}_a = \frac{\text{A}}{\text{F}} \times \dot{m}_f = 15 \times 0.24 = 3.6 \text{ kg/min}$$

$$\rho_a = \frac{p}{RT} = \frac{1 \times 10^5}{287 \times 290} = 1.2 \text{ kg/m}^3 = 1.2 \text{ kg/m}^3$$

$$\dot{V}_a = \frac{\dot{m}_a}{\rho_a} = \frac{3.6}{1.2} = 3.2 \text{ m}^3/\text{min} = 3.0 \text{ m}^3/\text{min}$$

$$\text{Volumetric efficiency, } \eta_V = \frac{\dot{V}_a}{\dot{V}_s} = \frac{3.0}{3.234} = 0.928 = \boxed{92.8\%} \quad \textbf{Ans.}$$

$$\text{Heat supplied by fuel} = \dot{m}_f \times \text{CV}$$

$$= 0.24 \times 44{,}000 = 10{,}560 \text{ kJ/min}$$

$$\text{Heat equivalent of bp} = 47.29 \times 60 = 2837 \text{ kJ/min}$$

$$\text{Heat lost to cooling water} = \dot{m}_w c_{p_w} \Delta T_w$$

$$= 65 \times 4.18 \times 12 = 3260 \text{ kJ/min}$$

$$\text{Mass of water vapour} = 9 \times 0.155 \times 0.24 = 0.335 \text{ kg/min}$$

Mass of dry exhaust gas = mass of air + mass of fuel – mass of water vapour

$$\therefore \quad \dot{m}_{\text{ex}} = 3.6 + 0.24 - 0.335 = 3.505 \text{ kg/min}$$

Heat carried away by dry exhaust gas

$$= \dot{m}_{\text{ex}} c_{p_{\text{ex}}} (T_{\text{ex}} - T_a)$$

$$= 3.505 \times 1.0(450 - 17)$$

$$= 1518 \text{ kJ/kg}$$

Heat lost in steam = enthalpy of superheated steam above 17°C.

$$= 0.335\,[4.18(100 - 17) + 2257 + 2.0(450 - 100)]$$

$$= 1107 \text{ kJ/min}$$

$$\text{Unaccounted loss} = 10560 - (2837 + 3260 + 1518 + 1107)$$

$$= 1838 \text{ kJ/min}$$

Heat balance sheet

Heat input	kJ/min	%	*Heat expenditure*	kJ/min	%
Heat supplied by fuel	10,560	100	(a) Heat in bp	2837	26.86
			(b) Heat lost to cooling water	3260	30.88
			(c) Heat to dry exhaust	1518	14.37
			(d) Heat lost in steam	1107	10.48
			(e) Unaccounted loss	1838	17.41
Total	10,560	100	Total	10,560	100.00

REVIEW QUESTIONS

1. Explain the importance of engine testing.
2. Explain with neat sketches the construction and working of a prony brake and rope brake for the measurement of brake power. Compare their merits and demerits.
3. Describe the construction and working of a hydraulic dynamometer.
4. Describe the construction and working of an eddy current dynamometer.
5. Briefly describe a swinging field dc dynamometer.
6. How is the indicator diagram obtained with the help of a mechanical indicator? Explain the method to evaluate ip from the indicator diagram. What are the shortcomings of a mechanical indicator?
7. Briefly describe the construction and working of a rotary encoder for marking crank angle and TDC.
8. Briefly describe the construction and working of a piezo-electric transducer. Why is it necessary to cool the transistor? Explain the method of calibrating the transducer.
9. Describe with the help of a schematic diagram the method to obtain the indicator diagram from electronic indicators.
10. Deduce an expression to evaluate piston displacement from the TDC position in terms of crank angle, crank radius and connecting rod length.
11. Describe the Willan's line method to determine the ip of a CI engine. Explain its limitation towards the high load end. Why is this method not applicable to SI engines?
12. Describe the Morse test for determining the indicated power of a multi-cylinder IC engine. What is the main assumption made in this test?
13. Describe the motoring test for the measurement of friction power and comment on its accuracy.
14. Describe with the help of diagrams a gravimetric method and a volumetric method for measuring the fuel flow rate in IC engines. Compare their merits and demerits.
15. Describe the construction and working of an automatic volumetric fuel flow measuring device.
16. Describe with the help of a diagram the construction and working of a rotameter.
17. Describe the air-box and orifice method for the measurement of air flow in IC engines.
18. Describe the viscous flowmeter for the measurement of air flow in IC engines. What are the precautions necessary to be taken in the use of this flowmeter?
19. Briefly describe the different methods used for the measurement of engine speed.
20. Briefly explain a method used in the measurement of spark-timing.
21. Explain how the flame propagation during the combustion process can be measured using the high speed photographic technique.
22. Draw the typical variable speed characteristic curves for an SI engine at wide open throttle. Discuss the nature of the curves.
23. Draw the typical variable speed characteristic curves for a CI engine and discuss the nature of the curves.
24. Draw and discuss the nature of consumption loop for an SI engine. Show the variation of the loop at varying throttle openings.

25. Show the variation of bmep and bsfc versus the air/fuel ratio for SI engines.
26. Draw and discuss the nature of consumption loop for a CI engine. What do you understand by smoke limit?
27. Give the importance of performance maps. Draw and discuss the performance map for SI engines.
28. Draw and discuss the performance map for CI engines.
29. Describe with the help of a diagram the working of an exhaust gas calorimeter for the evaluation of heat lost to exhaust gases.
30. Explain, with mathematical expressions, how the heat balance sheet is prepared for IC engines.

PROBLEMS

16.1 A four-cylinder four-stroke racing engine of capacity 3 litres has a bore of 100 mm and a compression ratio of 12. When tested with a dynamometer of arm length 0.45 m, a maximum load of 650 N was obtained at 5000 rpm, and at the peak speed of 6500 rpm the load was 575 N. The minimum fuel consumption was 18 ml/s at a speed of 5000 rpm. The specific gravity of the fuel may be taken as 0.75 and the calorific value 44,000 kJ/kg. Calculate the maximum bmep, the maximum bp, the minimum bsfc and the maximum brake thermal efficiency at maximum torque.

[**Ans.** 12.25 bar, 176 kW, 0.318 kg/kWh, 25.8%]

16.2 A four-cylinder four-stroke SI engine having a bore of 75 mm and a stroke of 85 mm, was tested at full throttle at 3500 rpm over a range of mixture strengths. The following readings were taken during the test:

Brake load (N)	: 170	173	177	178	177	170	167
Fuel consumption (ml/s)	: 2.50	2.45	2.60	2.95	3.35	3.80	4.00

Take the brake arm length as 0.5 m and the relative density of the fuel as 0.725. Calculate the corresponding values of bmep and bsfc. Plot the consumption loop and obtain from it the corresponding values for maximum power and maximum economy.

[**Ans.** 7.45 bar, 0.236 kg/kWh, 7.24 bar, 0.202 kg/kWh]

16.3 A four-cylinder four-stroke SI engine has an output of 65 kW at 2500 rpm. A Morse test is carried out and the brake torque readings are 175, 170, 172 and 171 N m respectively. The specific fuel consumption for normal running of the engine at this speed is 0.365 kg/kWh. The calorific value of the fuel is 44,000 kJ/kg. Calculate the mechanical and brake thermal efficiencies of the engine. [**Ans.** 81.4%, 22.4%]

16.4 A single-cylinder four-stroke gas engine has a capacity of 7.5 litre and is governed by the hit and miss principle. The indicator cards give a working loop mep of 7.0 bar, and a pumping loop mep of 0.5 bar when the engine runs at 600 rpm at full load. Diagrams from the dead cycle give an mep of 0.75 bar. The engine was run with no-load at the same speed, and a mechanical counter recorded 60 firing strokes per minute. Calculate the full-load brake power and the mechanical efficiency of the engine. [**Ans.** 21.75 kW, 89.2%]

16.5 The following details were noted in a test on a single-cylinder four-stroke oil engine: displacement volume = 3 litre; speed of the engine = 600 rpm; fuel consumption = 2.9 kg/h; calorific value of fuel = 42,500 kJ/kg; difference in tension on either side of the brake pulley = 450 N; arm length = 0.35 m; length of the indicator diagram = 60 mm; area of positive loop of indicator diagram = 580 mm^2; area of negative loop = 30 mm^2; spring constant = 0.85 bar per mm. Calculate: (a) the bp, (b) the ip, (c) the mechanical efficiency, (d) the brake thermal efficiency, and (e) the brake specific fuel consumption.

[**Ans.** (a) 9.896 kW (b) 11.69 kW (c) 84.7% (d) 28.9% (e) 0.293 kg/kWh]

16.6 A four-cylinder four-stroke engine has a bore of 90 mm and stroke of 100 mm. The air consumption has to be measured by an air-box with a sharp-edged orifice of 40 mm diameter having coefficient of discharge equal to 0.6. If the engine runs at 3000 rpm, estimate the pressure drop across the orifice in terms of cm of water. Assume that volumetric efficiency at this speed is 0.8, ambient pressure is 1.013 bar and the temperature is 21°C.

[**Ans.** 27.9 cm of H_2O]

16.7 A four-cylinder, four-stroke diesel engine develops 85 kW at 2000 rpm with a specific fuel consumption of 0.235 kg/kWh, and the air/fuel ratio of 25:1. The analysis of the fuel is 87% carbon and 13% hydrogen, and the calorific value is 42,500 kJ/kg. The jacket cooling water flows at 0.25 kg/s and its temperature rise is 48 K. The exhaust temperature is 320°C. Draw up an energy balance for the engine. Take c_p = 1.05 kJ/(kg K) for the dry exhaust gas, and c_p = 1.9 kJ/(kg K) for superheated steam. The ambient temperature is 20°C, and the exhaust gas pressure is 1.013 bar.

[**Ans.** bp = 36%, cooling water = 21.3%, dry exhaust = 18.4%, steam = 8.3%, radiation and unaccounted = 16%]

CHAPTER 17 Exhaust Emissions

17.1 INTRODUCTION

Little attention was paid to pollution from motor vehicles until the peculiar climatic conditions of Los Angeles led to the formation of a photochemical smog which was found to be related to the rapidly rising concentration of motor vehicles, and thus the Californian Motor Vehicle Pollution Control Board (MVPCB) was set up in 1960. Its main abatement proposals resulted in the incorporation of crankcase and exhaust emission control devices and systems for minimizing evaporative emissions in gasoline engine vehicles.

The Californian problem naturally drew attention of the US Federal authorities and of other countries, particularly of those with an interest in exporting motor cars or accessories to the United States. These widespread and varied interests encouraged urgent and intensive research into the whole field of automotive engine emissions.

Starting in 1961 in California and in 1964 throughout the United States, emittants from the crankcase on all new vehicles were led back to the carburettor and burned in the combustion chamber. Fuel evaporation controls were put on all new cars in 1970 in California and in 1971 throughout the United States. Thus, the only emittant source from automotive vehicles remained the exhaust.

The most serious pollutants recognized from the gasoline engine exhaust are carbon monoxide (CO), hydrocarbons (HC) and oxides of nitrogen (NO_x). Diesel engine runs with very lean fuel-air mixture, so in the exhaust the concentrations of CO and HC are much lower than the typical SI engine levels. NO_x concentrations are comparable to those from SI engines. In addition to these pollutants, the exhaust from diesel engines contains particulate matter, which consists primarily of soot with some additionally absorbed hydrocarbon materials. Many countries have introduced strong legislation to prohibit the emission above certain levels.

Vehicle emission standards have been made very stringent during the last few years. Advancements in combustion systems, electronic management of fuel and engine, and exhaust after-treatment have all made the achievement of these targets possible. The gasoline direct injection engine became a reality during the mid 1990s. Future US emission standards have provided an impetus for the development of hybrid electric and fuel cell vehicles, which are close to commercial production. In addition to development of alternative energy vehicles, future challenges for research in IC engines are in the area of exhaust after-treatment under lean engine operation, and in the area of further improvements in the control of engine breathing and fuel introduction.

This chapter describes the techniques for the measurement of exhaust emissions, causes for pollutant formation and the developments in emission control devices.

17.2 MEASUREMENT OF EXHAUST EMISSIONS

The measurement of exhaust emissions is very important for the control of air pollution from IC engines. Carbon monoxide concentrations are measured by infrared absorption, NO_x concentrations are measured by chemiluminescence and unburned hydrocarbons are measured by flame ionization detector.

17.2.1 Non-dispersive Infra-red (NDIR) Analyzer

The NDIR analyzers are used for measuring the concentrations of carbon monoxide and carbon dioxide. This device is based on the principle that the infrared energy of a particular wavelength, peculiar to a certain gas, will be absorbed by that gas. The infrared energy of other wavelengths will be transmitted by that gas. Carbon dioxide absorbs infrared energy in the wavelength band of 4 to 4.5 microns (μm) and transmits the energy of the surrounding wavelengths. The carbon monoxide absorption band is between 4.5 and 5 microns (μm).

Nitric oxide (NO) has also a weak absorption band, allowing it to be analyzed by NDIR, but lack of sensitivity and interference by water vapour do not give high accuracy with low concentrations.

A schematic arrangement of the IR analyzer is shown in Figure 17.1. A wideband infrared radiation source consists of a heated wire, which is placed in a quartz tube mounted in the source block. Radiation from the source is reflected within the mounting block and passes out of a symmetrical pair of rectangular apertures as two parallel beams into the two separate cells—a sample cell and a reference cell. These cells are internally highly polished and gold plated to ensure high transmission of radiation. After passing through these cells the infrared radiation is received in two separate detector cells, which are full of the gas whose concentration is to be measured. The two detector cells contain equal amounts of this gas and are separated by a flexible diaphragm.

The sample cell is a flow-through tube that receives a continuous stream of the mixture of gases to be analyzed. When the particular gas to be measured is present in the sample, it absorbs the infrared radiation at its characteristic wavelengths. The percent of radiation absorbed is proportional to the molecular concentration of the component of interest in the sample. The sample cells may be divided by quartz windows into various lengths to give different ranges of sensitivity. The quartz windows do not absorb infrared energy in the region of interest. Low concentrations are best measured by longer cells so that more molecules of interest are present. The unused sample cells are generally flushed with a non-infrared absorbing gas such as oxygen or nitrogen, or with a gas free of the components being measured, e.g. fresh air for carbon monoxide analyzers.

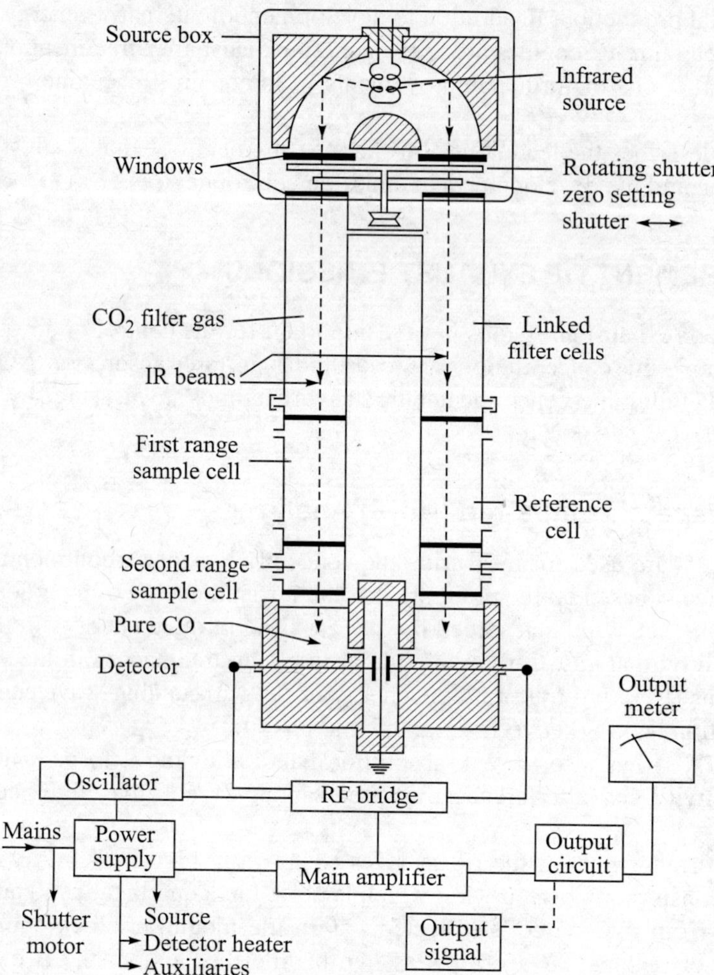

Figure 17.1 Schematic diagram of IR analyzer.

The reference cell is sealed and is physically identical to the sample cell. It is filled with an inert gas (usually nitrogen) which does not absorb the infrared energy of the characteristic wavelength of the species of interest.

The radiant energy, after passing through the cells, heats the gas in the corresponding chamber of the detector. Since no radiant energy is absorbed in the reference cell, the corresponding chamber in the detector is heated more and its pressure becomes higher than that in the other chamber. This pressure differential causes the diaphragm to move and vary the capacitance. Therefore, the variation in the capacitance is proportional to the concentrations of the species of interest in the exhaust sample.

The radiation from the source is interrupted by a rotating two-bladed shutter driven by a synchronous motor. The shutter is placed between the infrared source and the cells. When the shutter blocks the radiation, the pressure in the two compartments of the detector is equal because there is no energy entering either of the chambers of the detector. This allows the diaphragm to return to its neutral position. As the shutter alternatively blocks and unblocks the radiation, the diaphragm fluctuates causing the capacitance to charge cyclically. This sets up an ac signal, which is impressed on a carrier wave provided by a radio-frequency oscillator (amplifications of ac signals have better drift-free characteristics than the amplifications of dc signals). Additional electronic circuitry in the oscillator unit demodulates and filters the resultant signal. This signal is then amplified and rectified to a dc signal which is measured by a meter or recorder. The final dc signal is a function of the concentration of the species of interest in the exhaust sample.

To set the zero point, a non-infrared-absorbing gas, e.g. dry air, is passed through the instrument. For the other points on the scale, calibrating gases with known concentrations are passed through the analyzer.

An error in the NDIR readings may arise if the exhaust sample contains other species that can absorb radiation at the same frequencies that the gas in the detector will absorb. In order to minimize this interference, a large concentration of the interfering gas is placed in the filter cells. The analyzer zero is then set with this large concentration of the interfering gas.

17.2.2 Flame-ionization Detector (FID)

Some hydrocarbons have an infrared absorption at 3.4 microns, but some others, notably aromatics, have almost none. Only about 50 % of exhaust hydrocarbons is measured by NDIR, therefore, this method is not suitable for the measurement of HC concentrations.

The flame ionization detector is mainly used to measure the unburned hydrocarbon concentrations in the exhaust gases. It is based on the principle that pure hydrogen-air flames produce very little ionization, but if a few hydrocarbon molecules are introduced the flames produce a large amount of ionization. The ionization is proportional to the number of carbon atoms present in the hydrocarbon molecules.

A schematic arrangement of the instrument is shown in Figure 17.2. It consists of a burner assembly, an ignitor, an ion collector and electric circuitry. The burner consists of a central capillary tube. Hydrogen, or a mixture of hydrogen and nitrogen, enters one leg of the capillary tube and the sample enters through another leg. The length and bore of the capillary tubes are selected to control the flow rates. The mixture of $H_2 - N_2 - C_nH_m$ then flows up the burner tube. The air required for combustion is introduced from around the capillary tube. The combustible mixture formed in the mixing chamber is ignited by a hot wire at the top of the burner assembly and a diffusion flame stands at the exit to the burner tube. An electrostatic field is produced in the vicinity of the flame by an electric polarizing battery. This causes the electrons to go to the burner jet and the positive ions go to the collector. The collector and the capillary tube form part of an electric circuit. The flow of ions to the collector and the flow of electrons to the burner complete the electrical circuit. The dc signal produced is proportional to the number of ions formed and the number of ions is proportional to the number of carbon atoms in the flame. The dc signal generated is attenuated by a modulator and then fed to an ac amplifier and a demodulator. The signal is then recorded on a meter. The meter is calibrated directly in amount of hydrocarbon concentrations. To calibrate, the samples of known concentration of hydrocarbons are fed to the instrument and the meter readings are adjusted accordingly.

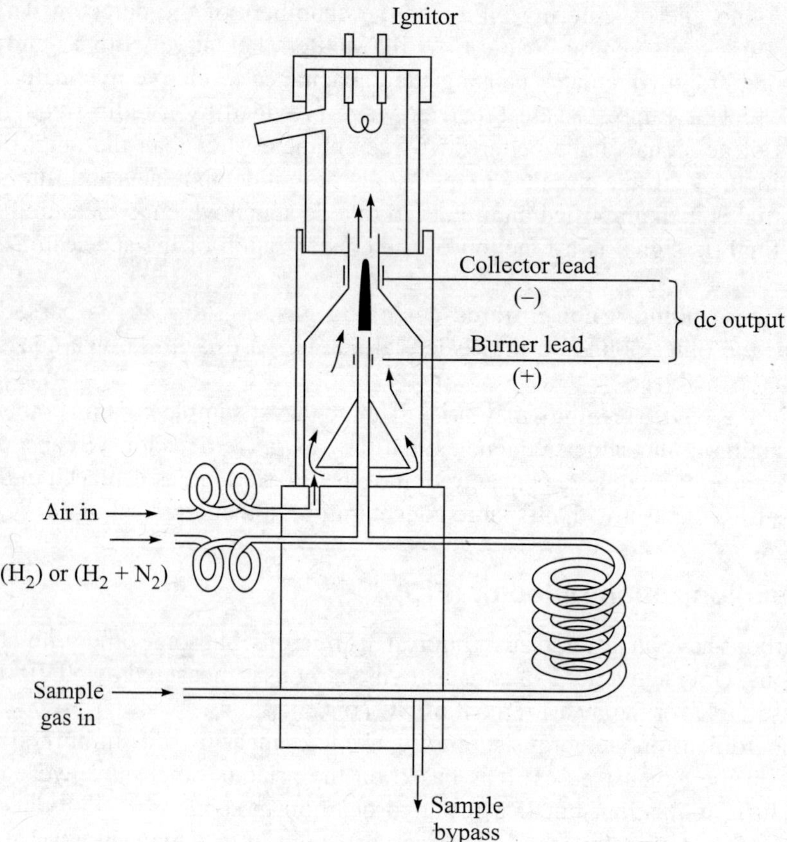

Figure 17.2 Flame ionization detector.

17.2.3 Chemiluminescence Analyzers (CLA)

The chemiluminescent analyzer measures the nitric oxide (NO) concentrations. This technique is based on the principle that NO reacts with ozone (O_3) to give some NO_2 in an electronically excited state. These excited molecules on decaying to the ground state emit red light (photons) in the wavelength region from 0.6 μm to 3 μm, i.e.

$$NO + O_3 \rightarrow NO_2^* + O_2$$
$$NO_2^* \rightarrow NO_2 + h\nu$$

where h is Planck's constant and ν represents a photon of light.

The oxides of nitrogen (NO_x) from the engine exhaust comprise mainly a combination of nitric oxide (NO) and nitrous oxide (NO_2). By converting any exhaust NO_2 to NO in a thermo-catalytic converter before supplying the exhaust gas to the analyzer, the value of total nitrogen oxides (NO_x) can be obtained.

A schematic arrangement of the chemiluminescent instrument is shown in Figure 17.3. The vacuum pump controls the pressure in the reaction chamber and draws ozone and the exhaust sample. The ozone is produced by an electric discharge in oxygen at low pressure. An NO_2-to-NO

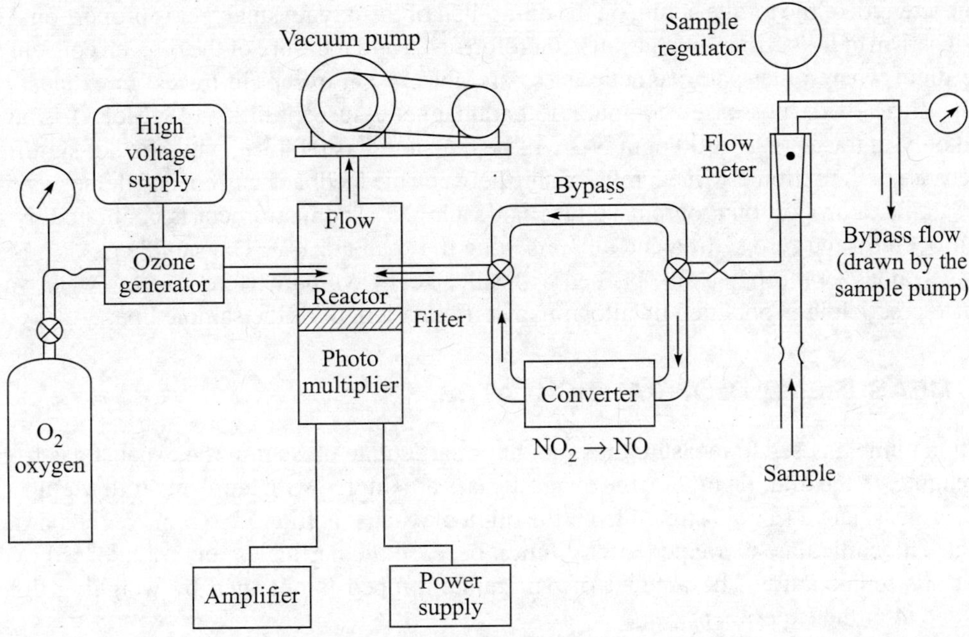

Figure 17.3 Chemiluminescence analyzer.

converter is also shown in the diagram. An arrangement is made by using a bypass line, so that it may be possible to measure only the NO concentrations or NO + NO_2, i.e. NO_x concentrations in the combustion engine exhaust.

A mixture of a gas sample and ozone enters a reaction chamber (reactor) which is maintained at a very low absolute pressure. The reaction of the ozone and nitric oxide when heated under vacuum at 600°C produces some electronically excited molecules of NO_2. The electronically excited molecules on decaying, emit light. The light can readily be detected accurately by a photo-multiplier. The signal is then amplified and fed to a recorder.

Many parameters affect light emission in the reactor, it is therefore essential to calibrate the analyzer regularly. Pure nitrogen may be used for zero setting. The zero control is adjusted until the digital voltmeter reads zero, the nitrogen gas is then disconnected and a standard NO/N_2 mixture is connected.

The NO/NO_x switch is set to 'NO' mode and the span control is used to adjust the NO reading to correspond with the standard. For the NO_x reading the NO/NO_x function switch is pressed to initiate the NO_x mode.

17.2.4 Oxygen Analyzer

Measurement of oxygen is useful in order to know the quality of mixture with which the engine runs. The oxygen analyzer is based on the principle that oxygen exhibits strong paramagnetism which distinguishes oxygen from most of the other common gases. It is observed that when a glass sphere filled with oxygen at the end of a rod is suspended by a silk fibre in a magnetic field, it is attracted by the magnet, whereas a hollow quartz sphere, being diamagnetic, is repelled.

The strength of the torque acting on the dumb-bell of the oxygen analyzer is proportional to the paramagnetism of the surrounding gas. It is therefore used as a measure of the oxygen concentration in a medium where other paramagnetic gases are not present except in traces. In exhaust gases, only the nitric oxide possesses comparable paramagnetic susceptibility (43% of O_2), but it is present only in traces (e.g. 1000 ppm NO will be registered as 0.043%, which is not significant).

Because of the extremely linear relationship between the feedback current and the susceptibility of the sample, a proportional output voltage is developed. The instrument is calibrated by using pure nitrogen for the zero setting and air for setting the span at 21%. The sample gas is passed to the oxygen analyzer through a tube packed with silica gel to avoid interference from water vapour. Adequate precooling is obtained by allowing a sufficient length of the sample line.

17.3 MEASUREMENT OF PARTICULATES

A dilution tunnel is used to measure the amount of particulate present in the exhaust gas from the diesel engine. In the dilution tunnel, the exhaust gases are diluted with ambient air to a temperature of 52°C or less, and a sample stream from the diluted exhaust is filtered to remove the particulate material. The particulate is trapped after dilution because the particulate gets condensed over the filter at this temperature. The amount of particulate trapped is obtained by weighing the filter before and after the experiment.

17.4 MEASUREMENT OF EXHAUST SMOKE

Smoke-meters are used to measure the intensity of exhaust smoke. Smoke-meters may measure either the relative quantity of light that passes through the exhaust gas (Hartridge smoke-meter), or the relative smudge left on a filter paper (Bosch smoke-meter).

17.4.1 Hartridge Smoke-meter

It is based on the principle that the intensity of a light beam is reduced by smoke which is a measure of smoke intensity. A schematic diagram to illustrate the principle of this smoke-meter is shown in Figure 17.4. Light from a source is passed through a standard length of a tube where the exhaust gas sample is continuously supplied from the engine and at the other end of the tube the transmitted light is measured by a photo-electric cell.

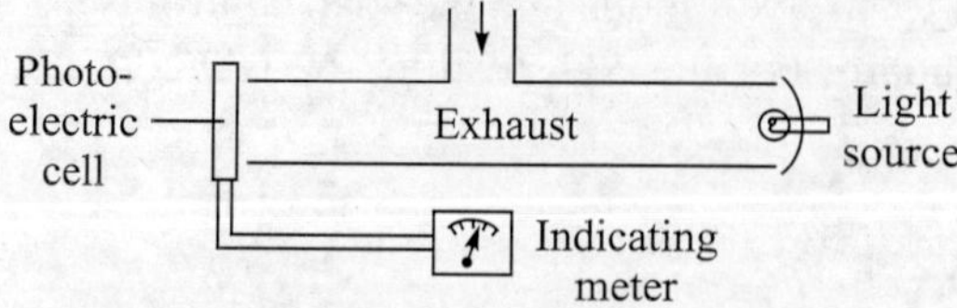

Figure 17.4 Principle of Hartridge smoke-meter.

The photoelectric cell converts the light intensity to an electric signal, which is amplified and recorded on a meter. The intensity of smoke is expressed in terms of smoke density. It is defined as the ratio of electric output from the photoelectric cell when an exhaust sample is passed through the tube to the electric output when clean air is supplied.

17.4.2 Bosch Smoke-meter

It is based on the principle that when a certain quantity of exhaust gas passes through a fixed filter paper, some smoke smudge is obtained on it, which is a measure of smoke intensity. A schematic diagram to illustrate the principle of this instrument is shown in Figure 17.5.

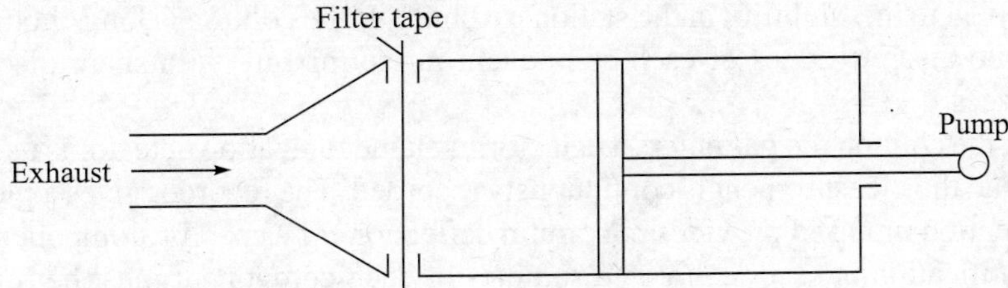

Figure 17.5 Principle of Bosch smoke-meter.

A fixed quantity of the exhaust gas from the engine is introduced into a tube, where it passes through a fixed filter paper. Depending upon the smoke density, some quantity of smudge is deposited on the filter paper, which can be evaluated optically. A pneumatically-operated sampling pump and a photo-electric unit are used for the measurement of the intensity of smoke smudge on the filter paper.

17.5 GAS CHROMATOGRAPHY

Gas chromatography is used for the detailed analysis of exhaust gas mixtures and for a definite identification of the different components present in the mixture. It is the most suitable method by which the amount of each exhaust hydrocarbon compound can be measured. A block diagram of a gas chromatograph is shown in Figure 17.6. The most important part of the instrument is a column. It is a packed tube containing a solid or a liquid phase which have certain absorbing properties. With a solid stationary phase it is called an *adsorption chromatography* and with a liquid stationary phase it is called a *partition chromatography*. The analysis of exhaust gas hydrocarbons involves partition chromatography. The gas to be sampled is injected into an inert gas stream called

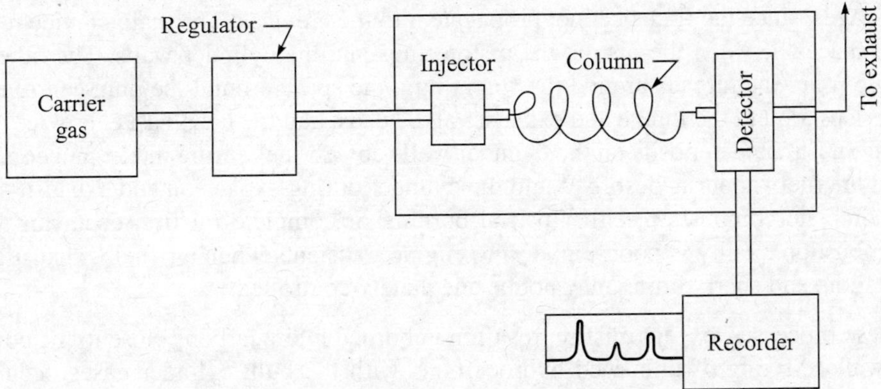

Figure 17.6 Main components of the gas chromatograph.

the carrier gas. It is a moving phase. The rate of flow of the carrier gas is regulated to a constant value in order to identify the components by their retention time. As the gas flows down the column, chemical components in the moving phase migrate into the stationary phase and then back again into the moving phase. The constituents tend to leave the column one at a time with relatively sharp separations, each constituent having a unique retention time within the column that is more or less proportional to its solubility in the stationary phase. If the column is long enough the sample will separate into discrete zones of each component in the mixture with sharp discontinuities at the zone interfaces.

On leaving the column the gas enters a detector. A flame ionization detector is used to measure hydrocarbons and the measurement is continuously recorded. The recorder gives a plot of arbitrary response vs retention time and provides maximum deflection for a specific component. Before this method of identification is used, calibrated samples of each component must be run through the apparatus to provide a standard retention time.

17.6 POLLUTANT FORMATION

17.6.1 Hydrocarbons (HC)

There are some unburned or partially burned hydrocarbons in the exhaust. The amount is insignificant from an energy standpoint, but it is objectionable from the viewpoint of its odour, its photochemical smog, and from the standpoint of its having a carcinogenic effect. The products of photochemical smog cause watering and burning of the eyes, and affect the respiratory system, especially when the respiratory system is marginal for other reasons.

Hydrocarbon emissions from SI engines

The most widely accepted causes for hydrocarbon emissions in exhaust gases of spark ignition engines are:

1. Flame quenching at the combustion chamber walls, leaving a layer of unburned fuel-air mixture adjacent to the walls.
2. Crevices in the combustion chamber, small volumes with narrow entrances, which are filled with the unburned mixture during compression, and remains unburned after flame passages, since the flame cannot propagate into the crevices. The main crevice regions are the spaces between the piston, the piston rings and the cylinder walls. The other crevice regions are the threads around the spark plug, the space around the plug centre electrode, crevices around the intake and exhaust valve heads, and the head gasket crevice.
3. The oil film and deposits on the cylinder walls absorb fuel during intake and compression, and the fuel vapour is desorbed into the cylinder during expansion and exhaust.
4. Incomplete combustion, either partial burning or complete misfire, occurring when the combustion quality is poor, e.g. during engine transients when air-fuel, exhaust gas recirculation, and spark timing may not be adequately controlled.

All these processes, except misfire, result in unburned hydrocarbons close to the combustion chamber walls. Mixing of unburned hydrocarbons with the bulk cylinder gases occurs during expansion and the exhaust blowdown processes. During the blowdown process a high concentration of hydrocarbons is released from the cylinder through the exhaust valve. During the exhaust stroke the piston pushes most of the remaining fraction of the cylinder mass with its high hydrocarbon concentration into the exhaust. The residual gases in the cylinder thus contain a high concentration of hydrocarbons. Unburned hydrocarbons are thus exhausted in two pulses, the first peak is obtained with the exhaust blowdown and the second occurs towards the end of the exhaust stroke.

Hydrocarbon emissions from CI engines

The CI engines operate with an overall fuel-lean equivalence ratio, therefore they emit only about one-fifth of the hydrocarbon emissions of an SI engine. The following are the major causes for hydrocarbon emissions in the exhaust of CI engines:

1. The diesel fuel contains components of higher molecular weights on average than those in a gasoline fuel, resulting in higher boiling and condensing temperatures. This causes some hydrocarbon particles to condense on the surface of the solid carbon soot generated during combustion. Most of this is burned as mixing continues and the combustion process proceeds but a small amount is exhausted out of the cylinder.
2. The air-fuel mixture in a CI engine is heterogeneous with fuel still being added during combustion. It causes local spots to range from very rich to very lean and many flame fronts exist at the same time unlike the homogeneous air-fuel mixture of an SI engine that essentially has one flame front. Incomplete combustion may be caused by undermixing or overmixing. With undermixing, in fuel-rich zones some fuel particles do not find enough oxygen to react with, and in the fuel-lean zones some local spots will be too lean for combustion to take place properly. With overmixing, some fuel particles may be mixed with burned gases and it will therefore lead to incomplete combustion.
3. A small amount of liquid fuel is often trapped on the tip of the injector nozzle even when injection stops. This small volume of fuel is called sac volume. This sac volume of liquid fuel is surrounded by a fuel-rich environment and therefore it evaporates very slowly causing hydrocarbon emissions in the exhaust.
4. CI engines also have hydrocarbon emissions for some of the same reasons as SI engines do, e.g. flame quenching, crevice volume, oil-film and deposits on the cylinder wall, misfiring, etc.

17.6.2 Carbon Monoxide (CO)

Carbon monoxide is toxic. The haemoglobin in the blood, which carries oxygen to the different parts of the body, has a higher affinity for carbon monoxide than for oxygen. The percent carboxy haemoglobin gradually increases with time to an equilibrium value which depends upon the carbon monoxide concentrations.

Carbon monoxide is generated in an engine when it is operated with a fuel-rich equivalence ratio as there is not enough oxygen to convert all carbon to carbon dioxide. For fuel-rich mixtures, CO concentrations in the exhaust increase steadily with the increasing equivalence ratio. The engine runs rich when it is started or when it is accelerated under load. For fuel-lean mixtures, CO concentrations in the exhaust are very low and are of the order 10^{-3} mole fraction. Poor mixing, local rich regions, and incomplete combustion create some CO. The SI engines often operate close to stoichiometric at part load and operate fuel rich at full load. Under these conditions, the CO emissions are significant. However, CI engines operate well on the lean side of stoichiometric and therefore produce very little CO emissions.

17.6.3 Oxides of Nitrogen (NO_x)

The oxides of nitrogen tend to settle on the haemoglobin in the blood. The most undesirable toxic effect of oxides of nitrogen is their tendency to join with the moisture in the lungs to form dilute nitric acid. NO_x is one of the primary causes of photochemical smog (smoke + fog). Smog is formed by the photochemical reaction as follows:

$$NO_2 + \text{energy from sunlight} \rightarrow NO + O + \text{Smog}$$

Monoatomic oxygen reacts with O_2 to form ozone (O_3) as follows:

$$O + O_2 \rightarrow O_3 \tag{17.1}$$

Ozone is harmful to lungs and other biological tissues. It is harmful to crops and trees. It reacts with rubber, plastics and other materials causing damage.

Most of the oxides of nitrogen comprise nitric oxide (NO), a small amount of nitrogen dioxide (NO_2) and traces of other nitrogen oxides. These are all grouped together and the group is called NO_x.

NO_x is mostly formed from atmospheric nitrogen. There are a number of possible reactions that form NO. NO forms in both the flame front and the post flame gases. In high-temperature combustion over a fairly wide range of equivalence ratio, the thermal or Zeldovich mechanism can be used. This mechanism consists of the following two reactions.

$$O + N_2 \rightarrow NO + N \tag{17.2}$$

$$N + O_2 \rightarrow NO + O \tag{17.3}$$

which can be extended by adding the reaction

$$N + OH \rightarrow NO + H \tag{17.4}$$

The three-reaction set is referred to as the extended Zeldovich mechanism.

N, O, OH are formed from the dissociation of N_2, O_2 and H_2O vapour at high temperatures that exist in the combustion chamber (2500–3000 K). The higher the combustion reaction temperature, the more diatomic nitrogen (N_2) will dissociate to monatomic nitrogen (N) and more NO_x will be formed. At low temperatures, a very small quantity of NO_x is created.

The flame temperature is maximum slightly at the rich equivalence ratio (ϕ = 1.1) but maximum NO_x is formed slightly at a lean equivalence ratio ($\phi = 0.95$). At this condition the flame temperature remains very high but excess oxygen helps in the formation of more NO_x.

The most important engine variables that affect NO_x emission are the fuel/air equivalence ratio, the burned gas fraction (EGR and residual gas fractions) and combustion duration within the cylinder. NO_x is reduced in modern engines with fast-burn combustion chambers. If ignition spark is advanced, the cylinder temperature will be increased and more NO_x will be produced. CI engines with divided combustion chambers and indirect injection (IDI) tend to generate higher levels of NO_x.

17.6.4 Particulates

The particulates from SI engines are lead, organic particulates including soot and sulphates. Gasoline may contain some sulphur, which is oxidized within the engine cylinder to form SO_2. This SO_2 is oxidized to SO_3 which combines with water to form a sulphuric acid aerosol. Leaded gasolines emit lead compounds. Soot emissions (black smoke) can result from combustion of overly rich mixtures. In properly adjusted spark-ignition engines, soot in the exhaust is not a significant problem.

Diesel particulates consist mainly of combustion generated carbonaceous material (soot) on which some organic compounds have been absorbed. Most particulates are generated in the fuel rich zones within the cylinder during combustion due to incomplete combustion of fuel hydrocarbons; some particulate matter is contributed by the lubricating oil. These are undesirable odorous pollutants. Maximum particulate emissions occur when the engine is under load. At this condition, maximum amount of fuel is injected to obtain maximum power from the engine. It results in a rich mixture and poor fuel economy. At temperatures above 500°C, soot particulates appear as clusters of a large number of solid carbon spheres with individual diameters of about 15 to 30 nm. As the temperature decreases below 500°C during expansion, the particles become coated with HC and with traces of other components.

The words particulates and soot are often used synonymously, but there is a difference in nature between these two emissions. Dry soot is usually the carbon that is collected on a filter paper in the exhaust of an engine. The unit of measurement of soot is usually the Bosch Smoke Number, which is assessed by the reflectance of a filter paper on which the soot has been collected. Particulates contain more than simply the dry soot; they are the soot particles on which the other compounds, often the polycyclic aromatic hydrocarbons (PAH), have condensed. The PAH compounds have a tendency to be carcinogenic. The level of particulates increases with the sulphur content in the fuel. Particulates are measured by trapping the particles on glass-fibre filter papers placed in a dilution tunnel, and then weighing the quantity.

17.7 EFFECT OF OPERATING VARIABLES ON SI ENGINE EXHAUST EMISSIONS

The important operating variables affecting the exhaust emission from SI engines are discussed in the following subsections.

Equivalence ratio

The equivalence ratio ϕ is one of the most important variables in determining the emissions from SI engines. Figure 17.7 shows qualitative variation of CO, HC and NO concentrations with equivalence ratio. Leaner mixtures give lower emissions until the combustion quality becomes poor and eventually misfire occurs. Under this condition, HC emissions rise sharply and engine operation becomes erratic. During warm-up, when the engine is cold, the rich mixture is supplied. With a fuel-rich mixture there is not enough oxygen to react with all the carbon, resulting in higher emissions of CO and HC in the exhaust. At part load conditions, lean mixtures could be used which would produce lower CO and HC emissions and moderate NO emissions. The maximum value of NO_x for a given engine load and speed invariably occurs at an equivalence ratio less than 1.0. The maximum cycle temperature with this lean mixture is lower than that with a richer mixture, but the available oxygen concentration is much higher. The relative amounts of CO, HC and NO_x depend on engine design and operating conditions. They are of the followings order: CO, 1–2% or 200 g/kg fuel; HC, 3000 ppm as C_1 (parts per million as CH_4) or 25 g/kg fuel; and NO_x, 500–1000 ppm or 20 g/kg fuel.

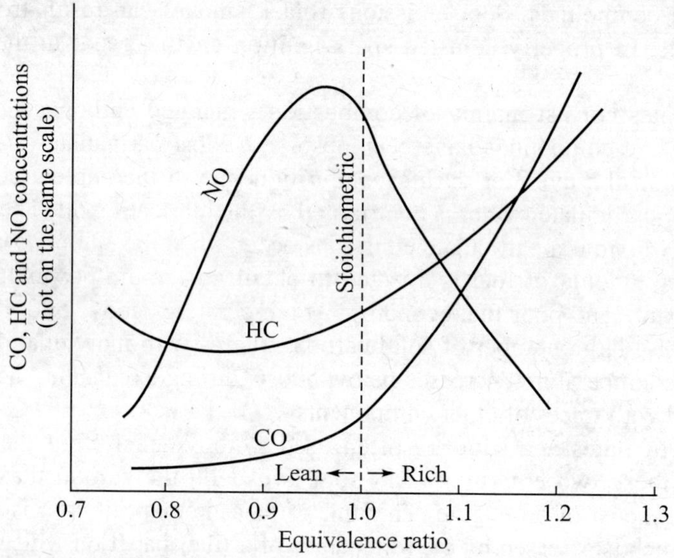

Figure 17.7 Qualitative variation of CO, HC and NO concentrations in the exhaust of an SI engine versus equivalence ratio.

Engine speed

An increase in the engine speed improves the combustion process within the cylinder by increasing turbulent mixing and eddy diffusion. This promotes after-oxidation of the quenched layer and reduces HC concentrations. At higher speeds, the exhaust port turbulence also increases which promotes exhaust system oxidation reactions of HC through better mixing, resulting in reduced HC. Speed has no effect on CO concentration because oxidation of CO in the exhaust is kinetically limited rather than mixing limited at the normal exhaust temperatures.

An increase in speed results in an increase in flame speed due to turbulence, and reduces the heat loss per cycle, which increases the temperature and therefore the NO_x concentration increases.

Spark timing

HC emission reduces with the spark retard because of increase in the exhaust temperature, which promotes HC oxidation. Spark retard has little effect on CO concentration except at very retarded timing where the lack of time to complete CO oxidation leads to increased CO emissions. This increase is offset to some extent by the higher exhaust temperatures, resulting in improved CO clean-up in the exhaust system.

By retarding, the spark combustion temperature decreases which then decreases the formation of NO_x.

Compression ratio

By decreasing the compression ratio the clearance volume largely increases with little increase in surface area, resulting in a large reduction in surface-to-volume (*S*/*V*) ratio. Because HC emissions arise primarily from quenching at the wall surfaces of the combustion chamber, a reduced *S*/*V* ratio reduces the concentration of HC emissions. CO concentration is not affected by changes in the *S*/*V* ratio. Also, with reduced compression ratios the exhaust gas temperature is increased because of less expansion. This improves the exhaust system after-reaction and lowers the HC emissions even more. However, reducing the compression ratio results in lower thermal efficiency and reduced engine power. A lower compression ratio also reduces the concentration of NO_x emission by decreasing the maximum cycle temperature.

17.8 DESIGN AND OPERATING VARIABLES THAT DECREASE HC CONCENTRATION FROM THE EXHAUST OF SI ENGINE

Hydrocarbon concentration from the exhaust of an SI engine can be decreased by the following methods:

Increasing the exhaust gas temperature

By increasing the exhaust gas temperature the oxidation reaction of HC increases, if sufficient oxygen is present in the exhaust, and this lowers the HC emission. The exhaust gas temperature can be increased by changing the following variables:

(a) Decreasing the compression ratio
(b) Retarding the spark
(c) Increasing the temperature of the coolant (heat lost to the cylinder walls reduces, the gas temperature increases)
(d) Increasing the speed
(e) Increasing the charge pressure
(f) Insulating the exhaust manifold.

More oxygen in the exhaust

Enough oxygen in the exhaust is necessary in order to carry out oxidation reaction of hydrocarbons. It can be achieved by:

(a) Using lean fuel-air mixture. However, the mixture should not be too lean to cause misfire.
(b) If rich fuel is used or enough oxygen is not available in the exhaust, some air may be added to the exhaust manifold from the atmosphere.

Lesser mass in the quench region

The mass of the hydrocarbons in quench regions can be reduced by:

(a) Decreasing the surface-to-volume ratio
(b) Increasing the turbulence during combustion
(c) Increasing the temperature of the coolant and the temperature of the charge
(d) Decreasing the compression ratio
(e) Decreasing the deposits at the walls (increased deposits may increase the surface area of the combustion chamber because of their irregular porous surface, thus increasing the mass of quenched HC).

More time for reaction

The reaction time can be increased by:

(a) Decreasing the speed
(b) Using a more homogeneous mixture
(c) Increasing the exhaust pressure
(d) Increasing the exhaust manifold volume and lengthening the flow passages by using baffles.

The variables discussed above are not all independent of each other. They may also oppose or aid those variables which influence HC and CO concentrations. The CO concentration in the exhaust is primarily a function of the air/fuel ratio and does not depend upon the engine load or speed.

17.9 DESIGN AND OPERATING VARIABLES THAT REDUCE NO_x CONCENTRATION FROM THE EXHAUST OF SI ENGINE

The concentration of NO_x from the exhaust of an SI engine can be reduced by the following methods:

Decreasing the combustion temperature

The temperature of the combustion products can be decreased by:

(a) Decreasing the compression ratio
(b) Retarding the spark
(c) Avoiding knock
(d) Decreasing the inlet temperature of the charge
(e) Decreasing the speed more heat loss, less turbulence, lower inlet temperature
(f) Decreasing the inlet pressure of the charge more heat loss, greater residual gas retention
(g) Exhaust gas recirculation EGR
(h) Increasing the humidity of air, or water injection
(i) Using a very rich or very lean air/fuel ratio.

Decreasing the availability of oxygen in the flame front

The oxygen availability in the flame front can be reduced by:

(i) Using rich mixtures
(ii) Decreasing the homogeneity of the mixture
(iii) Using a stratified charge
(iv) Sing the divided combustion chambers.

17.10 EFFECT OF OPERATING VARIABLES ON CI ENGINE EXHAUST EMISSION

The CI engines operate with lean mixtures and therefore produce very little CO and HC concentrations.

The important operating variables that affect the NO_x concentration from direct injection CI engine are given below:

Injection timing

Injection advance results in a longer delay period because the fuel is injected in air at lower pressures and temperatures. A longer delay period causes more fuel evaporation and mixing in the lean flame region (LFR), resulting in a high NO_x concentration. The NO_x formation in the other regions of the spray increases with injection advance due to the high temperatures reached. Late injection is one of the effective ways of reducing NO_x emissions. However, this results in a loss of brake mean effective pressure (bmep) and in an increase of brake specific fuel consumption (bsfc).

Fuel/air ratio

An increase in the fuel/air ratio (on leaner side of the stoichiometric) results in an increase in the maximum average gas temperature, and in an increase in NO_x concentration in the exhaust.

Type of fuel

The delay period is longer for the fuel with low cetane number and therefore more fuel is present in the lean flame region (LFR) when combustion starts. This larger quantity of fuel produces a higher gas temperature. Due to combustion early in the cycle, more NO_x is formed in the lean flame region.

Intake air charge dilution

The dilution of intake air charge with exhaust gases or water-injection reduces the combustion temperature and hence the NO_x emission.

17.11 CONTROL OF EXHAUST EMISSIONS

Attempts were made to control the emission of carbon monoxide, hydrocarbons and oxides of nitrogen by modifications of engine design. But it was found that using the engine modification approach only, the emission levels of these gases were not reduced to the desired level appreciably. Hence, the emission engineers were bound to switch over to other alternatives. Devices developed to achieve control of exhaust emissions include catalytic converters—oxidizing catalysts for HC and CO, reducing catalysts for NO_x and three-way catalysts for all three pollutants; thermal reactors for HC and CO, and traps or filters for particulates.

17.11.1 Catalytic Converters

For engines with sufficiently low level of emission of oxides of nitrogen, only a single converter to oxidize carbon monoxide and hydrocarbons may be used. In order to get an efficient oxidation of CO and HC, it is necessary that exhaust gases have sufficient oxygen with them, which is in general not found, particularly when an engine operates on a rich air-fuel mixture. Hence, secondary air is injected in the stream of exhaust gas before the catalytic converter. A representative diagram of an engine with the oxidation catalytic converter and secondary air is shown in Figure 17.8.

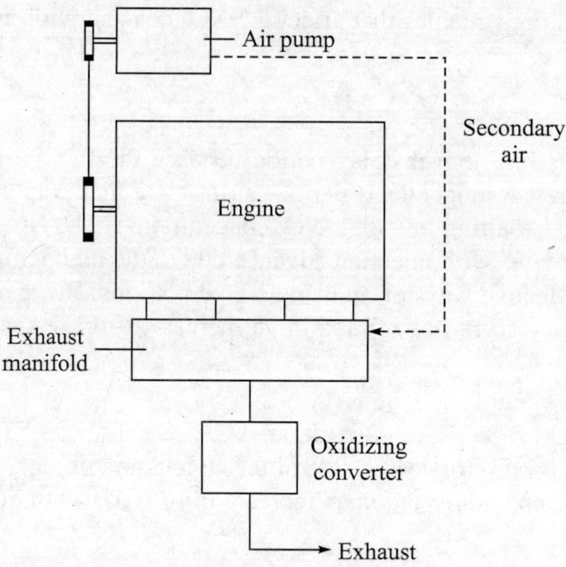

Figure 17.8 Catalytic converter system with only the oxidizing converter.

When emissions of oxides of nitrogen are high, it is necessary to use a reducing catalytic converter. Since a reducing catalytic converter requires an atmosphere having low oxygen concentration, it is necessary to operate the engine with a slightly rich air-fuel mixture. Excess oxygen needed in the oxidizing converter is obtained by injecting secondary air between the two converters. A schematic diagram of an engine with a dual catalytic converter and secondary air is shown in Figure 17.9.

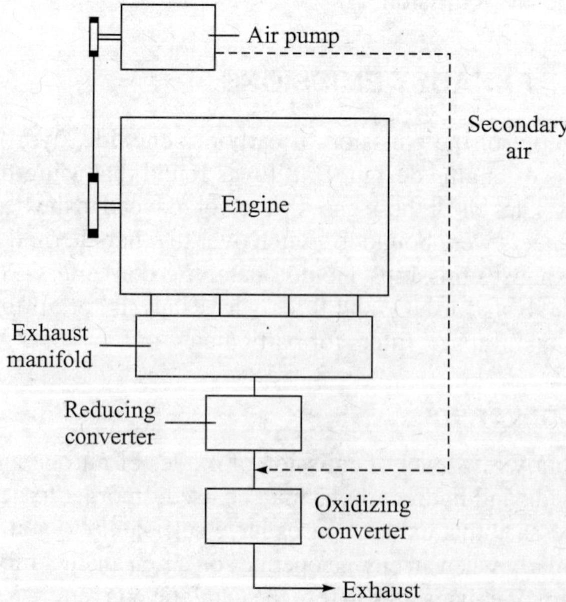

Figure 17.9 Catalytic converter system having both oxidizing and reducing converters.

It would be ideal if the oxidation of CO and HC and the complete reduction of NO could be achieved simultaneously in a single catalytic converter. In practice, if a proper catalyst is chosen, it is possible to decrease the concentration of hydrocarbons, carbon monoxide and oxides of nitrogen in a single converter. However, this requires an extremely close control of the input conditions to the combustion chamber, particularly of the air/fuel ratio. If there is too high a concentration of oxygen in the exhaust gases, reduction of nitric oxide will not occur. If there is too little oxygen, then oxidation of CO and hydrocarbons is not stoichiometrically possible.

The operation is feasible by a system in which the air/fuel ratio supplied to the engine is controlled by a feedback loop through an oxygen sensor in the exhaust. Catalysts known as three-way conversion (TWC) catalysts, have been developed for such systems. Because of an essentially stoichiometric engine operation, fuel economy and engine performance are near optimum.

Oxidation catalysts

Various materials have been found to be effective as oxidation catalysts. Among them, platinum and palladium have been found to be very effective. Other materials known to be effective for promoting the oxidation reaction are some transition metal-oxide catalysts such as copper, cobalt, nickel, and iron chromate as well as vanadium and manganese.

It is well known that platinum (Pt) and palladium (Pd) are very efficient catalyst for promoting the oxidation of exhaust hydrocarbon and carbon monoxide. No large difference in the emission control capabilities of platinum versus palladium has been found. Platinum catalysts are more effective at the higher metal loadings used, but the lower palladium loadings are more effective than the corresponding platinum catalyst. Palladium is a relatively poor poison-resistant catalyst but it has a high thermal sintering resistance and a low light-off temperature. The low light-off temperature is necessary to assure good conversion of the exhaust gases before the converter is fully warmed up. A mixture of platinum and palladium is most commonly used. The temperature needed to sustain the oxidation process is 250–300°C.

NO_x reduction catalyst

The NO reduction can be carried out under rich conditions where an excess of reducing species are available. Since both NO and CO are present in the exhaust, the reduction of NO by CO is possible in the presence of a suitable catalyst. Iron, nickel, copper and their alloys and oxides and many others are found to be suitable for NO_x reduction. Ruthenium (Ru) containing catalysts have a unique ability to reduce NO_x selectively to molecular nitrogen. Ruthenium is quite resistant to the common catalyst poisons which are present in the automobile exhaust, such as lead and sulphur. Ruthenium is combined with alkaline earth metal oxides such as CaO, BaO and SrO to form stable ruthenates. MgO has been found to stabilize Ru against volatilization and agglomeration under oxidizing conditions at temperatures up to 850°C.

The reduction catalyst system requires a follow-up oxidation catalyst to remove the remaining CO and HC. Such a system requires the addition of secondary air from an air pump before the oxidation catalyst. By this two-bed system, called the dual catalytic converter, all the three pollutants—NO, CO and HC can be removed from the exhaust. One of the difficulties with this dual reactor approach is that partial reduction of nitric oxide by hydrogen may occur in the first converter.

$$2NO + 5H_2 \rightarrow 2NH_3 + 2H_2O \tag{17.5}$$

The ammonia produced by this reaction may then be re-oxidized in the second converter, which is given by

$$2NH_3 + (5/2)O_2 \rightarrow 2NO + 3H_2O \tag{17.6}$$

This reoxidation reaction in the second converter reduces the net decrease in nitric oxide expected to be realized by the combination of the two converters.

Three-way catalysts

A three-way catalyst removes all the three pollutants (NO, CO and HC) simultaneously. There is a narrow range (about 0.1) of air/fuel ratios near stoichiometric in which high conversion efficiencies for all the three pollutants are achieved. An oxygen sensor in the exhaust is used to indicate whether the engine is running on rich or lean side of stoichiometric, and for providing a signal for adjusting the fuel system to achieve the desired air-fuel mixture. As the fuel flow varies, the feedback system adjusts the air/fuel ratio which oscillates around the stoichiometric mixture in an approximately periodic manner. Because of these cyclic variations, it is desirable that the catalyst should be able to reduce NO when a slight excess of oxygen is present and remove CO and HC when there is a slight deficiency of oxygen. Rhodium is the main ingredient used in commercial catalysts to remove NO under slightly lean of stoichiometric. Palladium and platinum promote the oxidation of CO and HC. The catalyst materials most commonly used for three-way conversion are platinum, palladium and rhodium.

It is desirable that catalytic converters have an effective lifetime comparable to that of the automobile. However converters loose their efficiency with age due to thermal degradation and poisoning of the active catalyst material. Thermal degradation occurs in the temperature range of 500–900°C. A number of impurities contained in fuel and lubricating oil find their way into the engine exhaust and poison the catalyst material. These include lead and sulphur from fuels, and zinc, phosphorus, antimony, calcium and magnesium from oil additives.

Substrates

The catalysts must be supported on some support material, called the substrate. For noble metal catalysts, such as platinum or palladium, support materials are required to make a physically durable catalyst. For transition metal-oxide catalysts, a support improves physical durability and thermal stability. Catalysts such as copper chromite are known to sinter at elevated temperatures, but if they are supported, this effect does not occur as readily. Support materials also affect catalyst density and the diffusion of reactants and products within the catalyst particle. Catalyst density is particularly important because of its effect on catalyst warm-up.

The most widely used substrates to support catalysts are pellets (beads) and monoliths fabricated from porous ceramics (honeycomb).

Spherical pellets or short cylindrical extrudes are supported within a basket as a catalyst element in these units. To attain reasonably low pressure drops through these elements, the flow rate through the bed must be low. The settling of pellets within the basket and attrition of pellets lead to problems with bypassing of part of the waste stream and poor flow distribution. The Pt/Al_2O_3 catalysts supported on ceramic honeycomb are more active than pelleted types. Essential features of a typical catalyst element on honeycomb substrate are shown in Figure 17.10.

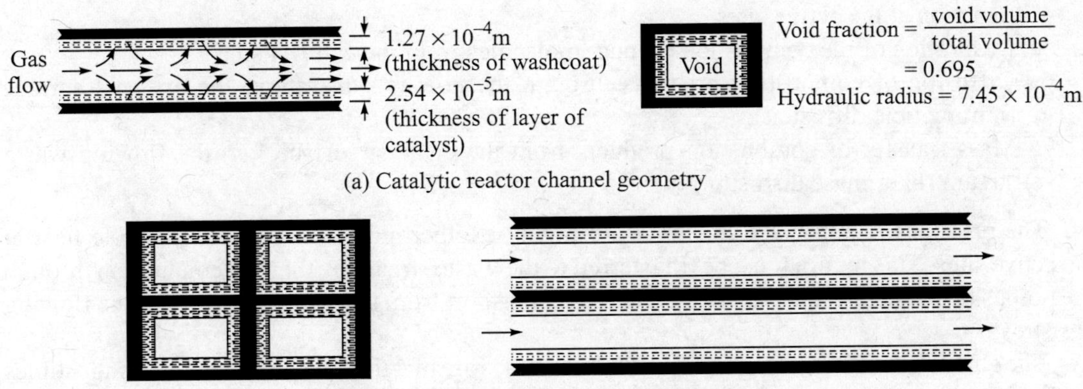

Figure 17.10 Essential features of a typical catalyst element on honeycomb substrate.

The monolithic or honeycomb type substrates are much better than the pelleted types. They have several attractive features such as less attrition, faster thermal response, and easier packaging compared to pelleted substrates. The monoliths have faster light-off characteristics because of faster thermal response and low pressure drop which permits utilizing a higher unit mass flow rate converter design. However, the monoliths may have several disadvantages, such as occasional internal failures due to faster thermal response and more rapid deactivation by potential trace poisons due to low internal surface area which makes the active sites more accessible to lead, zinc and copper.

The monolith substrates provide greater freedom than the bead substrates when designing converters for particular vehicle installations owing to their low pressure drops and adaptability to vehicle space requirements. Bead substrates provide high mass transfer rates but are less adaptable than monoliths to vehicle space requirement.

Mechanism

In a catalytic converter, the reactants flow through the bed or matrix and the reaction products are carried away by the flow of a waste stream. The reactions which convert reactants to product molecules, however, take place at discrete catalytic sites distributed throughout a thin, porous substrate layer coated on the surface of the matrix. The overall process of converting reactants to products in the flowing stream involves a number of sequential steps including transport processes between the gas stream and catalytic sites as well as the actual steps in the reaction at the catalyst sites. These steps are:

1. Mass transfer of reactants to the external surface of the catalyst (gas-particle interphase diffusion).
2. Diffusion of reactants into the pores of the substrate layer (intraparticle diffusion).
3. Adsorption of reactants on the active catalytic sites.

4. Reaction at the active sites.
5. Desorption of the combustion product molecules from the catalyst sites.
6. Diffusion of combustion products through the porous substrate to the external surface (intraparticle diffusion).
7. Mass transfer of combustion products from the catalyst surface into the flowing waste stream (interphase diffusion).

The reactions taking place in step 4 are highly exothermic, liberating considerable heat at the active sites. This in turn must be transferred to the waste stream by thermal conduction through the porous layer and by convective or conductive transport from the catalyst surface to the flowing gas stream..

The qualitative effects of these sequential processes are that the concentrations of combustibles are lower within the catalyst at the active sites than those in the gas stream flowing over its surface, and that the combustion product concentrations and the temperatures are higher within the catalyst. The concentration and temperature differences which develop, provide the driving forces for the various diffusion and heat and mass transfer steps in the overall sequence.

17.11.2 Thermal Reactor

A thermal reactor is a device that promotes the oxidation of unburned HC and CO by providing the exhaust gases at a high temperature for a sufficient period of time to allow the desired degree of reaction to occur. Thermal reactors for vehicles are normally larger than catalytic converters because they must have higher residence time to offset lower reaction rates. Reactors may work with or without air injection depending upon engine carburetion. When secondary air is injected, the reactor must also promote the mixing of this air with the effluent gases from the combustion chambers. For CO oxidation reaction to occur at a useful rate, the temperature must be held above 700°C. For the oxidation of hydrocarbons, a temperature above 600°C is required. It is therefore necessary for a thermal reactor to operate at a high temperature and to be large enough in order to provide adequate dwell time to promote the occurrence of these secondary reactions. The effectiveness of the reactor depends on its operating temperature, the availability of sufficient oxygen mixed through the reacting gases and the reactor volume. NO_x emission cannot be reduced with a thermal reactor; in fact it may increase it. However, these reactors employing rich carburetion emit relative low NO_x because the engine emission of NO_x is low with rich mixtures. With a lean air/fuel ratio, as lean as 17 or 18:1, the thermal reactors without air-injection may reduce NO_x.

A thermal reactor is essentially an enlarged exhaust manifold connected to the engine immediately outside the exhaust ports. Its function is to promote rapid mixing of the hot exhaust gases with any secondary air injected into the exhaust port. It is required with fuel-rich engine operation to produce a net oxidizing atmosphere. A thermal reactor retains the gases at a high temperature for a sufficient period of time to oxidize much of the HC and CO which exits the cylinder. A schematic diagram of a thermal reactor for a four-cylinder engine is shown in Figure 17.11. The reactor must be designed to reduce heat losses and increase residence time. A thin steel liner acts as the core of the reactor inside a cast-iron outer casing; with suitably arranged flow paths, this construction minimizes heat losses by thermally isolating the core.

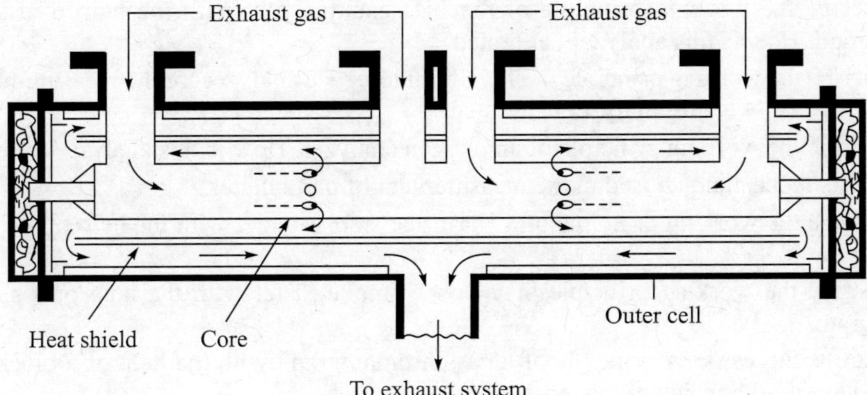

Figure 17.11 Exhaust thermal reactor for HC and CO oxidation.

17.11.3 Particulate Traps

Diesel engine particulate emissions are reduced by using the trap oxidizer. The objective is to trap or filter the particulates and then subsequently oxidize or burn them so that the filter is cleaned and reusable. Types of particulate filters include: ceramic monoliths, alumina-coated wire mesh, ceramic foam, silicon carbide, ceramic fibre mat, woven silica-fibre wound on a porous tube. Each of these has a different inherent pressure loss and filtering efficiency. Diesel particulate matter ignites at about 500–600°C. This is above the normal temperature of the exhaust. Therefore, either the temperature of the exhaust gas is raised by some means or a catalytic coating is used on the trap which reduces the ignition temperature to about 200°C.

The trapping mechanism consists of impaction, interception and diffusion. Impaction and diffusion are the primary trapping mechanisms of a wire-mesh of ceramic fibre trap. The larger sizes of particulates impact on filaments of the mesh and adhere to the filament surface and subsequently adhere to particulate materials already collected on the filaments. Some of the smaller sizes of the particulate migrate to these surfaces by diffusion and are retained. Interception is the primary trapping mechanism of a porous material trap.

The filters become loaded with particulates after sometime and further trappings of the particulates become difficult, as all the exposed surface area of the filter is filled with particulates and there is no place left in the filter for further trapping. Hence for the purpose of continuous use, the filters have to be cleaned over a period of time or cleaned almost at the same time as the particulates are trapped. This process is called *regeneration*. The regeneration techniques include electric heaters, fuel additives and in-line fuel burners.

REVIEW QUESTIONS

1. What are the major pollutants from the exhaust of gasoline and diesel engines? What are the detrimental effects of these pollutants?
2. Describe the working principle of the NDIR gas analyzer with the help of a schematic diagram. How is this analyzer calibrated?

3. Describe the working principle of the FID gas analyzer with the help of a schematic diagram. How is this analyzer calibrated?
4. Describe the working principle of chemiluminescence analyzer for the measurement of NO and NO_x. How is this analyzer calibrated?
5. Describe the working principle of an oxygen analyzer. How is this analyzer calibrated?
6. What is the technique used in the measurement of particulates?
7. Describe the working principle of a Hartridge smoke-meter with the help of a simple line diagram.
8. Describe the working principle of a Bosch smoke-meter with the help of a simple line diagram.
9. Describe the working principle of a gas chromatograph with the help of a block diagram and identify the main components.
10. What are the major causes for the formation of HC emission in the exhaust of SI engines?
11. What are the major causes for the formation of HC emission in the exhaust of CI engines?
12. What are the major causes for the formation of the CO in the exhaust of IC engines?
13. How is the NO_x formed in the exhaust of IC engines? What are the important engine variables that affect NO_x emissions?
14. What are the main constituents of particulates from SI and CI engines? Explain the difference between particulates and soot.
15. What are the effects of the following variables on SI engine exhaust emissions: equivalence ratio, engine speed, spark timing and compression ratio?
16. What are the design and operating variables which may decrease the formation of HC in the exhaust of an SI engine? Briefly explain them.
17. What are the design and operating variables which may decrease the formation of NO_x in the exhaust of an SI engine? Briefly explain them.
18. What are the effects of the following variables on CI engine exhaust emissions: injection timing, fuel/air ratio, type of fuel and intake air charge dilution?
19. What are the requirements of using an oxidation catalytic converter to minimize CO and HC emissions from the exhaust? Mention the important oxidation catalysts.
20. What are the requirements of using a dual catalytic converter? Mention the important NO_x reduction catalysts. What is the major problem faced by the dual catalytic converter?
21. What are the requirements of a three-way catalytic converter? Mention the important three-way catalysts.
22. What are the causes of poisoning the catalyst of a converter? How can it be reduced?
23. What is substrate of a catalytic converter? Compare the pelleted and honeycomb types of substrates. Show with the help of a diagram the essential features of a typical catalyst element on honeycomb substrate.
24. Describe the mechanism of gas flow in a catalytic converter.
25. What is the purpose of thermal reactor? What are the requirements for the proper working of this reactor? Describe its working principle with the help of a diagram.
26. What is the objective of a particulate trap? Mention the important types of filters used. Describe the trapping mechanism and regeneration of the trap.

CHAPTER 18 Alternative Potential Engines

Some of the important alternative potential engines with their relative merits and demerits are briefly described in this chapter.

18.1 STRATIFIED-CHARGE ENGINE

Stratified-charge engines are designed to have a different air/fuel ratio at different locations within the combustion chamber. The whole air-fuel mixture is distributed in layers or strata of different mixture strengths across the combustion chamber, while the overall mixture is lean, resulting in higher thermal efficiency and reduced pollution. It is a hybrid internal combustion engine that combines the best features of the SI and CI engines. The following are the important features of this engine:

1. Injecting the fuel directly into the combustion chamber during the compression stroke. It avoids the knock that limits the performance of SI engines with their premixed charge.
2. Igniting the fuel as it mixes with air with a spark plug to provide direct control of the ignition process. A rich mixture that ignites readily is desired around the spark plug, while the major volume of the combustion chamber is filled with very lean mixture. This gives good fuel economy. Special intake systems are necessary to supply this non-homogeneous mixture.
3. Combinations of multiple valves and multiple fuel injectors, along with variable valve and injection timing are used to accomplish the desired results.
4. Engine power is controlled by varying the amount of fuel injected per cycle. Air flow is unthrottled, it raises the volumetric efficiency, and the pumping work to introduce fresh charge in the cylinder is also minimized.

Stratified-charge engines do not require high antiknock quality of fuel as required by SI engines and also do not require high ignition quality of fuel as required by CI engines. They are usually fuel-tolerent and will operate with a wide range of liquid fuels.

Many different types of stratified-charge engines have been developed. A direct-injection stratified-charge engine and a jet-ignition or torch-ignition stratified-charge engine have all been commercially used. The Texaco Controlled Combustion System (TCCS) is illustrated in

Figure 18.1. The combustion chamber is of bowl-in-piston type. A high degree of air swirl is created during the suction stroke and enhanced in the piston bowl during the compression stroke to achieve a rapid mixing of fuel and air. A circular swirl is imparted to the air during the suction stroke by port design and by a shrouded inlet valve. Fuel is injected about 30° bTDC tangentially into the hot wall of the deep spherical piston bowl during the latter stages of compression. The fuel is carried around the wall of the bowl by the swirling flow, evaporated off the wall and mixed with the air. A long-duration spark discharge ignites the developing fuel-air jet as it passes the spark plug. The flame spreads and consumes the fuel-air mixture. Mixing continues and the final stages of combustion are completed during expansion.

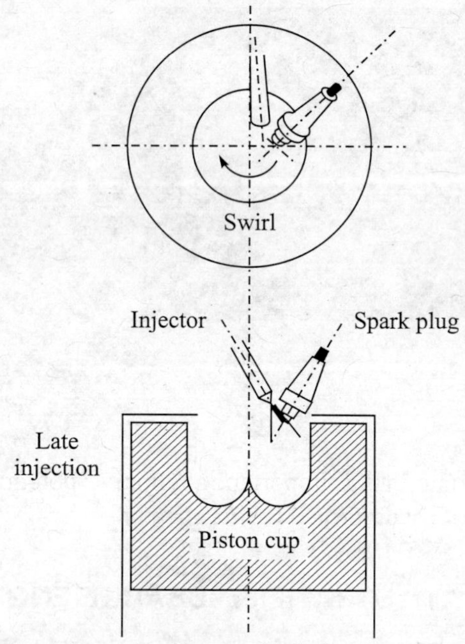

Figure 18.1 The Texaco Controlled Combustion System (TCCS).

The operating principles of a jet-ignition or torch-ignition stratified-charge engine with a three-valve carburetted system are illustrated in Figure 18.2. It uses a small prechamber fed during intake with an auxiliary fuel system to obtain an easily ignitable mixture around the spark plug. A separate carburettor and intake manifold feed a fuel-rich mixture through a separate small intake valve into the prechamber which contains the spark plug. The mixture strength is too rich to get ignited with the available air. At the same time, a very lean mixture is fed to the main combustion chamber through the main carburettor and intake manifold. After the intake valve is closed, the lean mixture from the main chamber is compressed and enters into the prechamber. The mixture strength in the prechamber becomes slightly rich but easily ignitable. After combustion starts in the prechamber, the rich burning mixture issues a jet through the orifice into the main chamber, entraining and igniting the lean main chamber charge. The overall mixture strength is leaner than the mixture that could normally be burned with conventional SI engines.

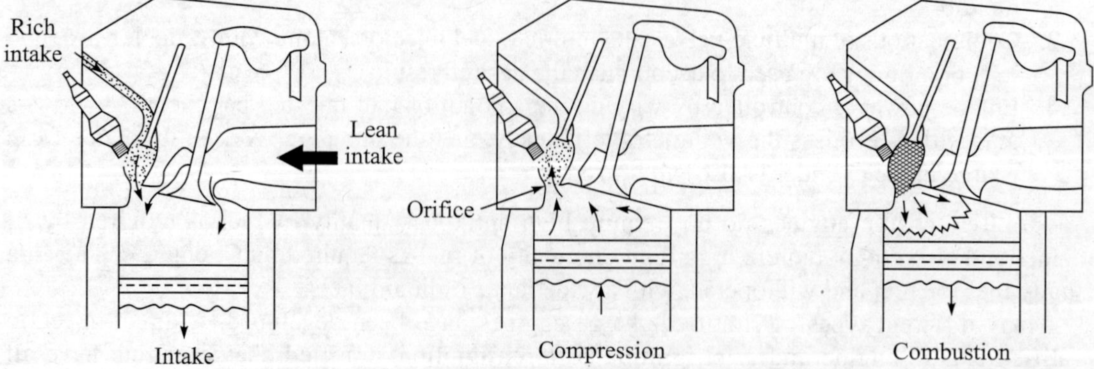

Figure 18.2 Torch-ignition, stratified-charge, spark-ignition engine.

Advantages

The advantages of the stratified-charge engine compared to the conventional SI engine can be summarized as follows:

1. A stratified charge obtained by injecting fuel late on the compression stroke decreases knock at the given compression ratio and air/fuel ratio.
2. A cheaper fuel having low octane rating can be used at higher compression ratios.
3. Fuel economy at part load is improved.
4. Load control can be achieved without air throttling.
5. It is possible to use multifuels and obtain good performance.
6. NO, CO and HC pollutions in the exhaust are low.

Disadvantages

The disadvantages of the stratified-charge engine compared to the conventional SI engine can be summerized as follows:

1. Maximum output is reduced due to the presence of air in the cylinder.
2. The speed range is inferior.
3. The cost of injection system is additional.
4. The design complications of injection system and spark ignition system are increased.

18.2 WANKEL ENGINE

Wankel engine is a rotary combustion engine originated by a German engineer, Felix Wankel, in 1924. It is used where compactness and higher engine speeds are required, such as in racing cars. The engine is well balanced and runs smooth. However, problems related to higher heat transfer, sealing and leakage exist. The major components of the Wankel rotary combustion engine are shown in Figure 18.3(a). It consists of three rotating parts: the triangular-shaped rotor having the internal ring gear, the output shaft with eccentric and the flywheel. The stationary parts are two-lobe centre housing, called the stator, with the intake and exhaust ports, two side housings with the fixed timing gear on one side of the housing, called the sun gear. The rotor revolves directly on the eccentric within its stator in such a manner that its apexes always make contact with the surface of the stator. The internal timing gear of the rotor meshes with the fixed timing gear to maintain the correct phase relationship between the rotor and eccentric shaft rotations. The rotor with its planetary motion about the sun gear drives the output shaft three times faster.

The two-lobe stator is an epitrochoid curve, derived from a family of curves based upon the path of a point on the radius of a circle which rolls upon the outside of a fixed circle.

Breathing is through ports in the centre housing. The combustion chamber lies between the stator and rotor surface and is sealed by proper sealing materials at the apex of the rotor and around the perimeters of the rotor surfaces. There are distinct induction, compression, ignition, power and exhaust processes, similar to a four-stroke cycle. The operation of the Wankel rotary engine for one of the chambers defined by rotor surface AB is shown in Figure 18.3(b).

As the apex A of the rotor passes over the intake port, the entrapped volume between the rotor surface AB and the chamber of the centre housing expands, thereby, the fuel-air mixture is inducted into the chamber.

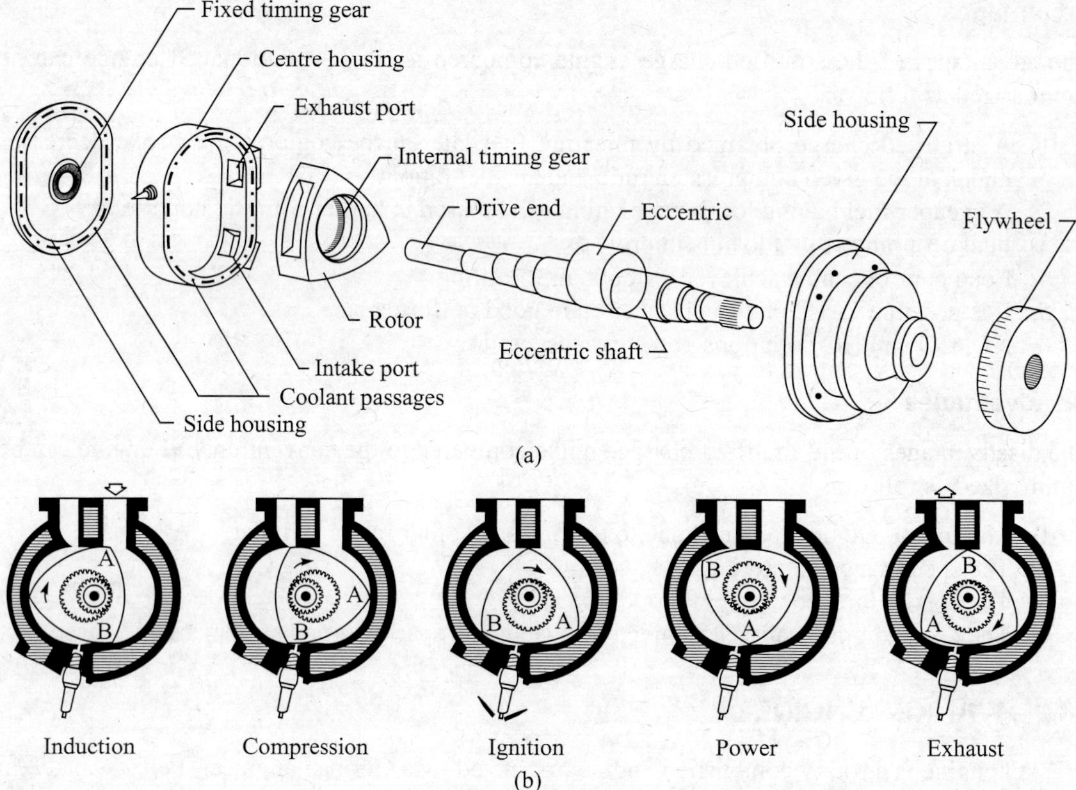

Figure 18.3 (a) Major components of the Wankel rotary engine; (b) induction, compression, power, and exhaust processes of the four-stroke cycle for the chamber defined by rotor surface AB.

As the rotor turns further, the enclosed volume is isolated from the intake port and the charge is compressed. The spark ignites the compressed charge and the expansion of the burned gases produces the power. Finally, when the apex B passes the exhaust port in the opposite end of the chamber, the exhaust process is executed and one thermodynamic cycle is completed. The remaining two chambers defined by the other rotor surfaces undergo exactly the same sequence. Since the rotor has three sealed sections, there are three power pulses for each rotor revolution.

Advantages

Some of the advantages of the Wankel engine over the reciprocating engine are:

1. The engine for a given specific power output (power/weight) is appreciably smaller in size and less in weight than the size and weight of an equivalent reciprocating engine.
2. The engine can be easily balanced. The engine contains no reciprocating parts, so that no unbalanced forces occur.
3. The engine has fewer parts, so it is simpler as it has no valves or valve gear.
4. The friction mean effective pressure is lower. It is thus possible to obtain higher speeds and power outputs.

5. Three combustion processes per rotor revolution allow faster warm-up of the engine.
6. Fuels having lower octane number can be used.
7. This engine is claimed to be cheaper when mass produced.

Disadvantages

Some of the disadvantages of the Wankel engine over the reciprocating engine are:

1. High compression ratios are difficult to achieve.
2. The surface-to-volume ratio is high which yields large quench areas, crevices, blowby and mixture leaks. It causes inefficient combustion resulting in a higher HC emission.
3. It is difficult to obtain good rotor sealings.
4. Engine mean torque is lower at lower road speeds of the vehicle.
5. Engine braking effect is reduced when installed in the vehicle.
6. Specific oil consumption is higher.
7. Specific fuel consumptions, particularly in the lower speed range, are higher.
8. There is a possibility of chamber distortion due to close proximity of cool inlet and exhaust ports.
9. Cooling and lubrication are difficult due to high temperatures produced because of three combustion processes per revolution of the rotor.
10. More frequent replacement of spark plug is required.
11. The higher speed range, e.g. up to 9000 to 10000 rpm, requires new transmission designs in vehicles.

18.3 FREE-PISTON ENGINE

The free-piston engine is an attempt to combine the high thermal efficiency of a reciprocating engine with the high specific power output (power/weight) of a turbine. It is a combination of an opposed piston diesel engine and turbine. The diesel engine acts as an efficient method of generating the hot gases under pressure. The engine does not produce shaft power, but the high pressure exhaust gases from the engine are used to operate the gas turbine. The useful power output is obtained from the turbine shaft. The engine is called free-piston engine because the connecting rods, the camshaft, the crankshaft and multiple valve assemblies are absent.

Figure 18.4 shows the schematic diagram of a free-piston engine. The main elements of the engine are the opposed diesel piston *A*, the compressor and bounce piston *B* and the fuel-injection nozzle *C*. It is a highly supercharged two-stroke opposed piston diesel engine with the hot gases of combustion expanding down to atmospheric pressure in the turbine. High pressure gas is generated in the diesel cylinder combustion chamber by injecting and burning the fuel in the centre of the diesel cylinder. The expansion of the gases of combustion forces the opposed pistons apart. Each of the diesel pistons is made integral with the compressor pistons, and this arrangement acts as a single-stage air compression on its inner face, and as an air bounce cushion on its outer face. The energy stored in the bounce or air cushion cylinder by the outward movement of the piston is utilized to drive the pistons inwards, compressing the air in both the air cylinder and in the diesel cylinder. Air at 5 to 7 bar passes through valves from the compressor cylinder into a central air space from which it enters the diesel cylinder through the intake ports. When the two diesel pistons

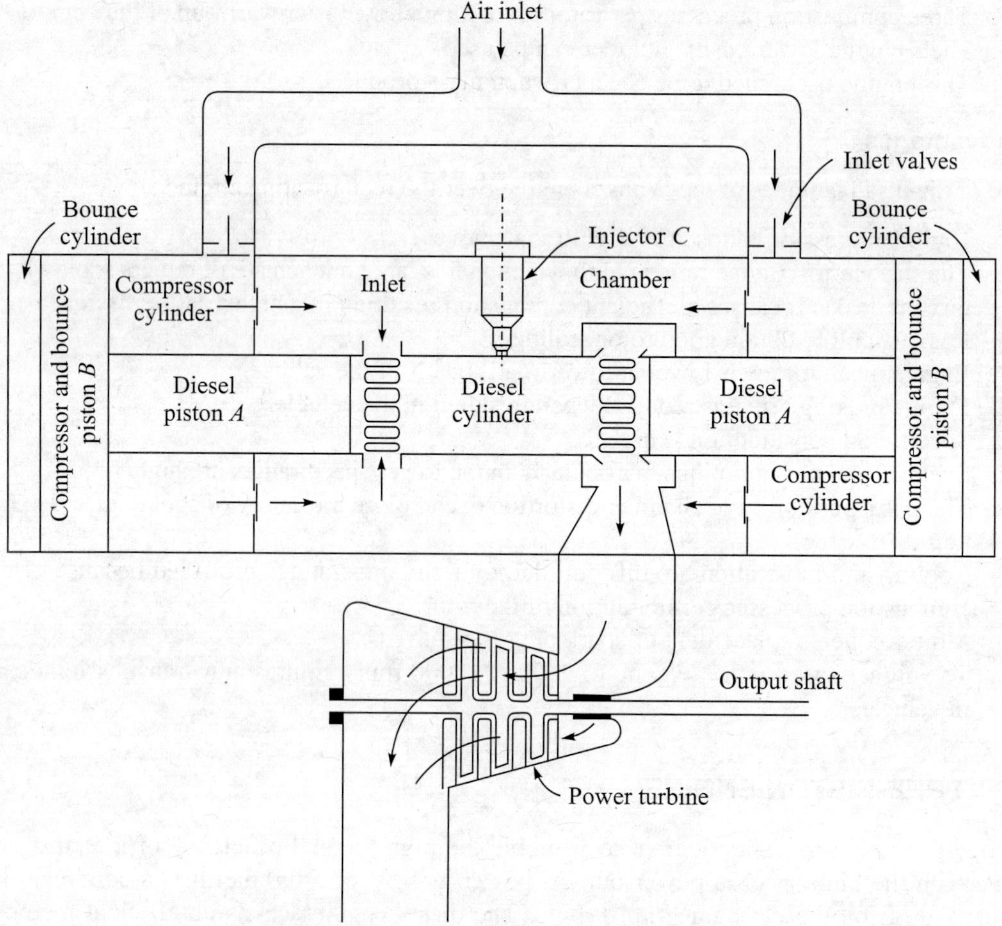

Figure 18.4 Free-piston engine.

are closest together, the air charge in the combustion chamber is fully compressed and at this moment fuel is injected in atomized form from a fuel injection nozzle of the usual diesel engine type into the space between the pistons acting as the combustion chamber. The combustion of this fuel raises both the temperature and pressure of the combustion chamber contents, the pistons being forced outwards until one of the pistons uncovers the exhaust port and a little later the other piston uncoveres the air inlet port. This causes the compressed air to flow into the cylinder and force the remaining exhaust gases out of the exhaust port. Thus, the air scavenges and supercharges the diesel cylinder. The mixture of the scavenging air and the gases of combustion at a temperature around 800 K pass through the exhaust ports and are expanded down to atmospheric pressure in the turbine. All the useful shaft power of the plant is produced by the turbine.

The stroke and compression ratio are not fixed but are determined by the amount of fuel injected, which depends upon the engine load. By injecting more fuel, a larger stroke results, which causes greater compression in the two bounce cylinders and therefore a greater compression on the return strokes of the two diesel pistons.

Advantages

The free-piston engine has the following advantages:

1. It has higher thermal efficiency and lower specific fuel consumption compared to those of a gas turbine in the smaller power class.
2. It has no crankshaft, connecting rods, valves and valve mechanism.
3. It has no piston side thrust, and the inertia forces are not applied to bearing surfaces.
4. The engine is perfectly balanced, with quieter and practically vibrationless performance.
5. It has higher volumetric efficiency than that of the conventional diesel engine.
6. It has higher specific power output than that of a comparable diesel engine.
7. A number of gas generating units can be connected in parallel with a single power turbine.
8. It exhibits greater tolerance to low-ignition quality fuel than that exhibited by a comparable conventional engine because the compression stroke will usually continue until ignition occurs. Thus, the compression ratio adjusts to the fuel's ignition requirements.
9. It requires relatively low temperature gas to the power turbine, since the energy to drive the compression process has been taken away.
10. Thermal distortion, corrosion and deposit problems are radically reduced, thus leading to a long life with low maintenance.

Disadvantages

The principal disadvantages of the free-piston engine are:

1. The specific fuel consumption of a free-piston engine is higher at rated output compared to that of a conventional diesel engine.
2. It has relatively poorer fuel economy at part load and particularly at very light loads.
3. The free-piston engine operates over a much narrower range of speeds than the conventional crankshaft engine and its idling speed cannot be reduced to that of the reciprocating type engine.
4. It requires more maintenance effort, particularly for pistons, piston rings, and cylinders and non-return valves, compared to that required by the gas turbine.
5. Starting of the free-piston engine may prove difficult at times, since there is no flywheel to provide the energy for successive cycles.
6. It has reduction gear in addition to a turbine and gas generating unit. The geared power turbine running on hot gases is an expensive equipment, and therefore the whole power plant is larger and heavier.

18.4 STIRLING ENGINE

The stirling engine is based on an idea first proposed by Robert Stirling, a Scottish, in 1816. This engine could not compete with petrol and diesel engines because of its lower specific power output. However, it has a unique feature in that if the ideal Stirling thermodynamic cycle with perfect regeneration could be achieved practically, the efficiency of a frictionless Stirling engine would be equal to that of Carnot cycle, which is the maximum for any heat engine. The Stirling engine powered vehicle has the potential for simultaneously delivering excellent fuel economy, extremely low emissions and low noise level. Stirling engine using rhombic drive has been developed to a stage which could compete with diesel and petrol engines.

The conventional petrol and diesel engines burn the fuel intermittently inside the engine cylinder, whereas the Stirling engine burns the fuel continuously outside the engine cylinder, i.e. it is an external continuous combustion engine. The distinctive features of the Stirling engine are:

1. Fuel, air and products of combustion remain outside the engine cylinder. The heat generated by the combustion of fuel is transferred to the working fluid present inside the engine cylinder.
2. The working fluid is completely sealed within the engine cylinder and repeatedly undergoes a thermodynamic cycle to operate the engine. The working fluid may be air, hydrogen, helium, or any other suitable gas.
3. As the working fluid is completely sealed, there is no gas exchange process and, therefore, there is no need of intermittently operating valves or ports.
4. After the expansion process of the working fluid, the intermittent flow heat exchanger stores a large amount of heat, which could otherwise be rejected to the atmosphere, and subsequently returns it to the working fluid after compression, thus accomplishing thermal regeneration.

Figure 18.5 shows the ideal thermodynamic cycle of a Stirling engine on p–V and T–s diagrams. It consists of two reversible constant temperature and two reversible constant volume processes.

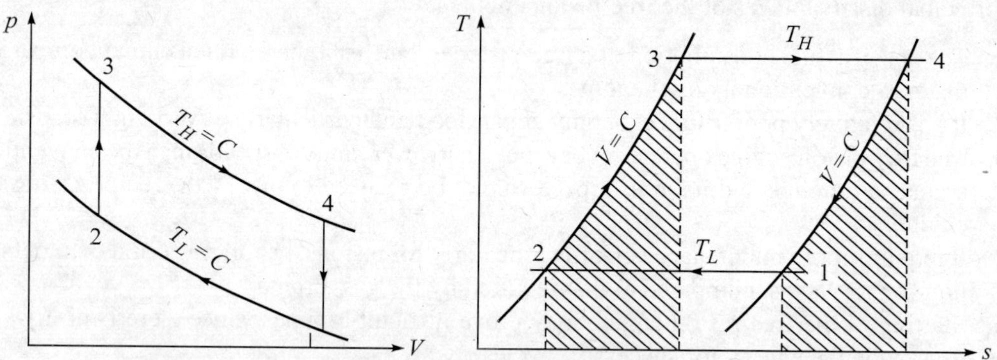

Figure 18.5 Stirling cycle.

(a) Isothermal compression work W_{1-2} with simultaneous heat rejection Q_{1-2},

$$W_{1-2} = Q_{1-2} = mRT_L \ln\frac{V_1}{V_2} \tag{18.1}$$

(b) Constant volume heat addition, $W_{2-3} = 0$,

$$Q_{2-3} = m\, c_V\,(T_H - T_L) \tag{18.2}$$

(c) Isothermal expansion work W_{3-4} with simultaneous heat addition Q_{3-4},

$$W_{3-4} = Q_{3-4} = mRT_H \ln\frac{V_4}{V_3} = mRT_H \ln\frac{V_1}{V_2} \tag{18.3}$$

(d) Constant volume heat rejection, $W_{4-1} = 0$,

$$Q_{4-1} = m\, c_V\, (T_H - T_L) \tag{18.4}$$

The efficiency of the Stirling cycle is less than that of the Carnot cycle because heat transfer takes place during constant volume processes as well. However, if a perfect regenerative system is used, such that:

$Q_{2-3} = Q_{4-1}$, i.e. the area under the curve 2–3 is equal to the area under the curve 4–1 in T–s diagram, then the cycle efficiency becomes

$$\eta = \frac{m\,R\,T_H \ln \dfrac{V_1}{V_2} - m\,R\,T_L \ln \dfrac{V_1}{V_2}}{m\,R\,T_H \ln \dfrac{V_1}{V_2}}$$

i.e.

$$\eta = \frac{T_H - T_L}{T_H} \tag{18.5}$$

So, the regenerative Stirling cycle has the same efficiency as the Carnot cycle.

Figure 18.6 illustrates a Stirling engine with the piston displacer system and rhombic drive. It consists of a cylinder and two pistons—one is called the displacer piston and the other is called the power piston. The cylinder is divided into two portions. The upper portion between the cylinder head and the displacer piston is a hot (expansion) space and the lower portion between the displacer piston and the power piston is a cold (compression) space. Both the pistons reciprocate within the cylinder. The reciprocating motion of the power piston causes volume change of the working fluid;

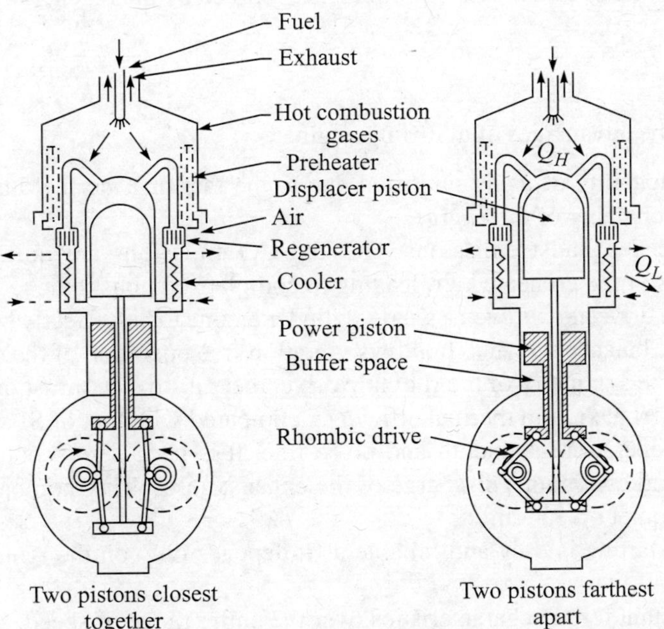

Figure 18.6 Stirling engine with the piston displacer system.

this is how the process of compression and expansion is accomplished. The displacer piston only displaces the working fluid from one portion to the other. Its motion cannot change the volume of the working fluid. When the displacer piston is at the top of the cylinder, the gas is in the cold space and when the displacer piston is moved down, the gas from the cold space flows to the hot space through the regenerator and gets heated. Similarly, when the displacer piston is moved up, the working gas is transferred from the hot space to the cold space through the regenerator and gets cooled.

The relative motion of the power piston and the displacer piston is such that the Stirling cycle, comprising four processes, is executed.

Consider Figure 18.5. At position 1, the power piston is at BDC and the displacer piston is at TDC. The whole of the working fluid is in cold space and maximum volume of the working gas is present in the cylinder. During the process 1–2, isothermal compression is obtained by moving the power piston from BDC to TDC without moving the displacer piston. During this process heat is rejected. At position 2, the power piston remains stationary at TDC and the displacer piston moves down, forcing the working fluid from the cold space to the hot space through the regenerator. Since the motion of the displacer piston does not change the volume of the working fluid, the heat supplied process 2–3 occurs at constant volume. During this process the working fluid gets heat from the regenerator, the two pistons are closest together and the cylinder volume is minimum. During the process 3–4, isothermal expansion is obtained as the power piston is moved from TDC to BDC without moving the displacer piston. During this process heat is supplied to the working fluid by burning the fuel outside the cylinder. During the process 4–1, the power piston remains at BDC, while the displacer piston moves from BDC to TDC and displaces the working fluid from the hot space to the cold space at constant volume through the regenerator. During this process, heat is rejected from the working fluid and absorbed by the regenerator. The cycle is thus completed.

Advantages

The following are the advantages of a Stirling engine:

1. It has the capability of using multifuel due to the fact that the fuel burns in the external combustion chamber of the engine.
2. It has reduced exhaust emissions. HC and CO emissions are quite low due to high temperatures of the preheated air, leading to complete combustion.
3. The rhombic drive allows even a single-cylinder engine to be perfectly balanced by suitable placement of balance weights. It allows vibration free operation of the engine.
4. Because of the symmetry of the rhombic drive, there is no side thrust on the bearings.
5. It gives higher maximum thermal efficiency compared with that of SI and CI engines.
6. The engine operation is smooth and noise free. It is because of continuous combustion of the fuel and continuous discharge of the exhaust gases. The rhombic drive also allows smooth and quiet operation.
7. The engine starting is easy and reliable as it depends only on the ignitions of fuel in the burner.
8. It gives constant torque characteristics over the entire range of speed. It can be used for a variety of applications, such as automobiles, ship propulsion, etc.

Disadvantages

The following are the disadvantages of a Stirling engine:

1. The design of a Stirling engine is complex because of the use of rhombic drive, regenerator, heaters, and coolers.
2. It requires a large amount of cooling water compared to that required by SI and CI engines. For automotive applications, the size of the radiator increases.
3. It requires a blower in order to force the air through the pre-heater and combustion chamber.

18.5 VARIABLE COMPRESSION RATIO (VCR) ENGINE

Today's modern engines have to satisfy customer's requirements for a high power output as well as for low fuel consumption in the part load range. Both these objectives can be accomplished by designing an engine with a variable compression ratio, which can be controlled as a function of load and speed. In fixed compression ratio engines, the high peak pressure problem, imposing thermal load, is encountered when the specific power output of the engine is increased by supercharging. Increasing the engine power by increasing the engine speed imposes dynamic loads and increased wear, thereby reducing reliability and life. In variable compression ratio engines the compression ratio is reduced at full-load to allow the supercharger to increase the intake pressure without increasing the peak cycle pressure. The compression ratio is sufficiently increased for good starting and for part load operation. This variable compression ratio concept can be used for both SI and CI engines.

All engine concepts with a variable compression ratio require major modifications of the basic engine or of engine components; some of the feasible engine concepts are:

1. Auxiliary chamber in cylinder head
2. Piston with variable compression height
3. Vertically sliding cylinder head
4. Adjustable connecting rod length
5. Eccentric main crankshaft bearings.

Auxiliary chamber in cylinder head

The concept of an auxiliary combustion chamber for application of variable compression ratio, requires an adjustment of the additional volume by the position of a small auxiliary piston carried by a worm drive. The position of the piston can be controlled as a function of load and speed by an electric motor coupled to the worm gearing. If the compression ratio is at maximum the auxiliary chamber piston is flush with the surface of the main combustion chamber. At all lower compression ratios the piston is pulled back opening the auxiliary chamber to a greater volume.

Piston with variable compression height

The British Internal Combustion Engines Research Institute (BICERI) variable compression ratio engine is based upon the piston design shown in Figure 18.7.

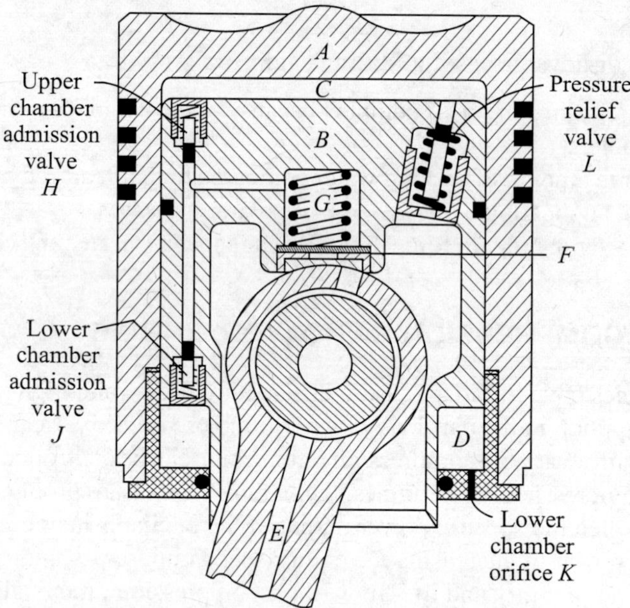

Figure 18.7 VCR piston with variable compression height.

The position of the piston *A* relative to the carrier *B* and the connecting rod is controlled hydraulically. Oil flows from the connecting rod passage *E* through the flapper valve *F* loaded with spring *G* and enters the chambers *C* and *D* through the admission valves *H* and *J*, respectively. Oil can escape from chamber *C* when the relief valve *L* opens. The opening of the valve *L* is set at the maximum desired combustion pressure. Oil from chamber *D* escapes continuously through the orifice *K*. During the latter part of the exhaust stroke and the beginning of the intake stroke, inertia moves piston *A* over the carrier *B* upwards, thus enlarging the chamber *C* while diminishing the chamber *D*. Oil enters the chamber *C* and escapes from the chamber *D*. The upward movement of piston *A* is counteracted when the combustion pressure exceeds the setting of the relief valve. The relief valve *L* opens and oil escapes. An equilibrium position is attained when the amounts of oil escaping and entering in chamber *C* are equal. When the load is less having low combustion pressure, the chamber *C* gradually fills to its maximum volume, the piston *A* reaches to the maximum height and maximum compression ratio is achieved. Conversely, with high load having high combustion pressure, the chamber *C* decreases in volume quickly, resulting in lowering of piston *A* over carrier *B* and thus reducing the compression ratio.

Vertically sliding cylinder head

The cylinder head assembly is raised or lowered by means of a fine screw threaded adjustment mechanism. The clearance volume is reduced as the cylinder head is lowered and the clearance volume is increased as the cylinder head is raised. There is no change in the stroke volume. Thus the compression ratio of the engine is increased by lowering the cylinder head and the compression ratio is reduced by raising the cylinder head. Here the compression ratio is changed manually, so the engine is ideal for laboratory studies.

Advantages

The following are the main advantages of the variable compression ratio engine:

1. It is very compact and has a high power-to-weight ratio without any penalty on specific fuel consumption.
2. It has a lower rate of pressure rise throughout the load range.
3. The fuel economy at part load increases because of using higher compression ratio.
4. It has a good cold starting and idling performance because of using higher compression ratio at these conditions.
5. Due to higher compression ratios at starting and at part load operation, the VCR engine has good multifuel capability.

Disadvantages

The following are some of the disadvantages of the variable compression ratio engine:

1. The design of the engine components is complicated.
2. Since the surface area-to-volume ratio of the combustion chamber rises progressively with higher compression ratios, the efficiency losses due to friction and pumping increase at part load.
3. HC emission also increases at part load.

18.6 DUAL-FUEL ENGINES

Many large stationary engines use two fuels, one is gaseous and the other a liquid fuel. The two fuels can be taken in widely varying proportions to run an engine. Such an engine is usually called dual-fuel engine.

The gaseous fuel requires a high compression ratio to burn efficiently because of its high self-ignition temperature. For this reason, mainly diesel engines have been used as duel-fuel engines. Dual-fuel engines are modified diesel engines wherein a primary fuel that is normally gaseous is inducted along with air and is compressed. A small quantity of diesel, called the pilot, is injected near the TDC. The pilot self ignites and becomes the ignition source for the induced charge. The flame propagates from the ignition centres formed by the pilot fuel droplets into the homogeneous primary fuel-air mixture to complete the combustion process.

The engine can be switched over from dual-fuel operation to diesel operation or back while the engine is running. Basically the change-over is affected simply by opening or closing a gas valve and letting the governor regulate the diesel admission. The fuel combination can be varied, using any proportion of diesel and gas, from 5 per cent diesel and 95 per cent gas to operation on diesel alone. The minimum diesel admission acts as a pilot flame to ensure ignition and smooth burning of the gas, particularly at small loads.

The important feature of the dual-fuel engine is that the combustion starts in a fashion similar to a CI engine but propagates by flame fronts in a manner similar to an SI engine.

Working principle

A schematic arrangement of the dual-fuel control system for a four-stroke engine is shown in Figure 18.8. During the suction stroke the gas admission valve and inlet valve are opened.

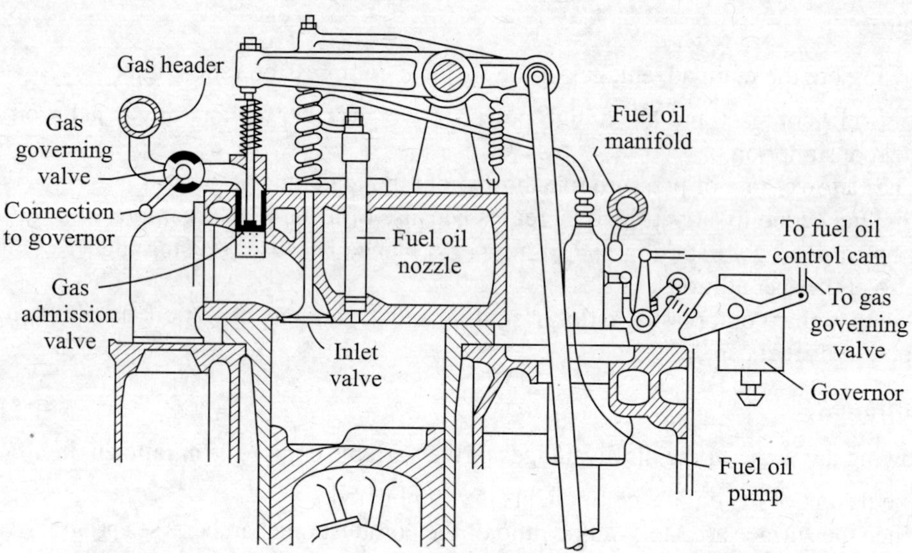

Figure 18.8 Schematic of dual-fuel control system for a four-stroke engine.

A regulated amount of gas flows through the gas governing valve, operated by a governor, and enters into the intake manifold. Here, the gas is mixed with a relatively high velocity of air and forms a homogeneous mixture of air and gaseous fuel. As the inlet valve is opened during the suction stroke, the mixture enters into the cylinder. On the compression stroke, the homogeneous mixture is compressed. Close to TDC, a small pilot charge of diesel oil is injected through a multihole nozzle. A number of ignition nuclei appear in the spray envelopes with consequent rapid start of combustion. The multiple flames spread throughout the combustion chamber, and ignite the gas-air mixture.

When the engine load is decreased, the dual-fuel operation reduces both the quantity of gaseous fuel and the air. The inlet air is throttled to maintain the optimum gas/air ratio. If this is not done, flame propagation would be slow at leaner gas/air ratios and flame propagation might also fail.

Advantages

The following are the advantages of a dual-fuel engine:

1. Cheap gases available from various sources can be used as a main fuel.
2. Gaseous fuels do not leave residue after combustion, therefore the exhaust is clean. The air pollution from the exhaust of the engine is very much reduced.
3. The clean combustion of gaseous fuels results in reduced wear of engine parts.
4. The maximum pressure in the cylinder is less than that for a diesel engine, thereby reducing the blowby past piston. This reduces the contamination and consumption of the lubricating oil.
5. The engine can run on varied proportions of gaseous and liquid fuels. With gaseous fuels the diesel oil requirement is hardly 5 %. The engine can also be easily run with diesel oil alone.

Disadvantages

The following are the disadvantages of a dual-fuel engine:

1. The efficiency of dual-fuel engines at part loads is poor because of increased delay periods at lean mixture strengths.
2. The effective compression ratio is less than that for the diesel engine.
3. A dual-fuel control system is required, additionally, which increases the cost of the engine.
4. The small amount of fuel in the pilot charge cannot be handled effectively by the relatively large injection plungers required for diesel operation. Hence it is necessary to use either two pumps or double-plunger pumps consisting of a pilot plunger and a main plunger.

18.7 MULTI-FUEL ENGINES

A multi-fuel engine is one which would operate satisfactorily on a wide variety of fuels ranging from diesel oil to lighter fuels like gasoline.

The main reasons for the development of a multi-fuel engine have been the military requirements. The need is of an engine which would satisfactorily run on any automotive fuel in case of emergency.

A multi-fuel engine must fulfil the following requirements:

1. It must have good combustion efficiency. This implies that the heat losses should be minimum.
2. The temperature of the combustion chamber must be comparatively higher in order to burn low ignition quality of fuel like gasoline.
3. The engine must be capable of starting at low temperature without any external aid.
4. The engine must have low exhaust emissions and low noise levels.

Based upon the above requirements, a multi-fuel engine must have the following design features:

Higher compression ratio

All gasolines have a low cetane number, which is an indication of their low ignition quality for use in a diesel engine. So, the compression ratio of multi-fuel engines is raised to about 23 : 1 in order to increase the temperature at the end of compression stroke for proper burning.

Large stroke-to-bore ratio

A large stroke-to-bore ratio gives a compact combustion chamber resulting in a hot combustion chamber under all conditions of speed and load.

Combustion chamber

(a) An open-combustion chamber is preferred compared to a pre-combustion chamber, because additional heat losses occur in the throat of the latter chamber.

(b) A spherical combustion chamber may be preferred because of the lowest surface-to-volume ratio. However, air utilization is not optimum.

(c) A combustion chamber of M-type is most suitable because of high degree of air utilization. It gives better fuel economy and quieter operation.

18.7.1 Suitability of Other Engines as Multi-fuel Units

The suitability of other engines as multi-fuel engines is discussed below:

Two-stroke opposed piston engine

The two-stroke opposed piston engine has a relatively large stroke-to-bore ratio, providing a favourable and compact combustion chamber for working as a multi-fuel engine.

Variable compression ratio (VCR) engine

A VCR engine has good ability to run on a variety of fuels at different engine speeds and loads. It also overcomes difficulty of cold starting.

Stratified-charge engine

Stratified-charge engines have good multi-fuel ability.

REVIEW QUESTIONS

1. What do you understand by a stratified charge? What are the important features of a stratified-charge engine?
2. Describe with the help of a simple diagram the operating principle of a stratified-charge engine of Texaco Controlled Combustion System.
3. Describe with the help of a simple diagram the operating principle of a jet-ignition stratified-charge engine.
4. What are the advantages and disadvantages of a stratified-charge engine?
5. Describe with the help of a simple diagram the construction and working principle of a Wankel engine.
6. What are the advantages and disadvantages of a Wankel engine?
7. Describe with the help of a diagram the construction and working principle of a free-piston engine.
8. What are the advantages and disadvantages of a free-piston engine?
9. What are the distinctive features of a Stirling engine?
10. Prove that the thermal efficiency of a Stirling cycle with perfect regeneration is the same as the Carnot efficiency operating between two fixed temperatures.
11. Describe with the help of a diagram the construction and working principle of a Stirling engine.
12. What are the advantages and disadvantages of a Stirling engine?
13. What is the concept of a variable compression ratio engine? What are the different methods adopted for the variation of compression ratio in an engine? Describe any one of them.

14. Describe the constructional details and operation of a VCR piston with variable compression height.
15. What are the advantages and disadvantages of a variable compression ratio engine?
16. What is a dual-fuel engine? What are the main features of a dual-fuel engine?
17. Describe the working principle of a dual-fuel engine with the help of a schematic diagram.
18. What are the advantages and disadvantages of a dual-fuel engine?
19. What is a multi-fuel engine? What are the requirements of any other engine to operate as a multi-fuel engine?
20. What are the design features of a multi-fuel engine?
21. Mention some of the engines which may be operated successfully as a multi-fuel unit.

19.1 INTRODUCTION

Internal combustion engines have been used in automobiles for a very long time. Combustion principles of both SI and CI engines are now well understood. In a conventional SI engine nearly stoichiometric homogeneous mixture of fuel and air is ignited by a spark. A flame rapidly propagates from the point of ignition to the end of the combustion chamber. The thermal efficiency of this engine is comparatively lower than that of CI engine. This is the main drawback of this engine. Exhaust pollution can be reduced to a very low level by using catalytic converters.

The combustion process in CI engine is very complex. Overall a lean air-fuel mixture with high compression ratio is used. Autoignition occurs at several locations in the combustion chamber, where there is a combustible mixture. The thermal efficiency of the CI engine is higher than that of SI engine because of the higher compression ratio. The exhaust emissions from CI engines with heterogeneous mixture are unburned hydrocarbons, oxides of nitrogen, aldehydes, carbon monoxide, smoke particulates and odour. CI engines produce less CO and HC emissions compared to SI engines, but NO_x and particulate matters or soot are the most noticeable emissions. These emissions need to be controlled as they have adverse effect on health and environment.

In the past, the prime objective of the engine development was focused on improving the engine performance. Several methods were adopted in order to increase the power and efficiency of the engine. Some of these methods are using supercharger and turbocharger, higher compression ratio, multipoint port fuel injection (MPFI) in SI engines and light weight higher strength materials. Later, engine development was focused on reducing air pollution from the vehicles. Exhaust after-treatment systems and exhaust gas recirculation were used to combat emissions to some extent.

In order to achieve higher engine output, higher efficiency and very low exhaust emissions advanced low temperature combustion technology with electronic management system is used.

19.2 ADVANCED LOW TEMPERATURE COMBUSTION (LTC)

The increase in thermal efficiency and decrease in exhaust emissions from the engine can be achieved by using a low temperature combustion strategy. Some of the important and successfully used low temperature combustion modes are Homogeneous Charge Compression Ignition (HCCI),

Premixed Charge Compression Ignition (PCCI) and Reactivity Controlled Compression Ignition (RCCI).

In all the low temperature combustion (LTC) modes air-fuel mixture is premixed and the combustion temperature is controlled to keep low. At low temperature NO_x formation is reduced. Autoignition and combustion phasing are controlled precisely by using Electronic Management System. LTC modes use a high rate of Exhaust Gas Recirculation (EGR), high intake pressure, high compression ratio, Variable Valve Timing (VVT), lean mixture and fast burn rate to reduce NO_x and particulate matters to a very low level.

HC and CO emissions are high in LTC modes. Due to low combustion temperature there is a possibility of fuel condensation and flame quenching on the surface of the combustion chamber. Fuel droplets may get trapped in the crevice regions during compression and comes out without burning during exhaust stroke. Impingements of the fuel spray on combustion chamber walls make the soot control challenging.

However, HC and CO emissions can be controlled by using oxidation catalytic converter but higher brake specific fuel consumption (bsfc) and CO_2 emissions are still of great concern. CO_2 is one of the greenhouse gases, which is trapped in the atmosphere and is responsible for global warming.

19.3 HOMOGENEOUS CHARGE COMPRESSION IGNITION (HCCI) ENGINE

In HCCI mode of combustion homogeneous air-fuel mixture is prepared either in the intake manifold or it can also be obtained by direct injection of the fuel in the cylinder. The mixture is then compressed until the temperature is high enough for autoignition. There is no well defined combustion initiator such as spark in spark-ignition engines or fuel injection timing in compression-ignition engines.

HCCI combustion occurs homogeneously throughout the mixture almost at the same time. Thus the autoignition of the whole charge takes place without a flame front. Therefore, there is no fuel rich diffusion as obtained in conventional diesel engine combustion. This eliminates the high flame temperature region in the combustion chamber. The NO_x and particulate emissions are therefore reduced significantly.

The start of combustion and the rate of heat release should be controlled in HCCI combustion engine over a wide range of engine speed and load. In order to control the rate of heat release to a safer and acceptable limit, the lean fuel mixture is significantly diluted with EGR. The engine is operated unthrottled. The pumping work reduces and fuel economy is improved. The peak burned gas temperature is also reduced by charge dilution, which reduces the heat loss and improves the thermal efficiency.

Thus, HCCI engine significantly reduces NO_x and particulate emissions and at the same time it keeps the thermal efficiency high. Various types of fuel can be used successfully. However, CO and HC emissions are high and it is difficult to start the engine when it is cold.

19.3.1 HCCI Challenges

HCCI combustion seems to be of great advantage but it imposes some difficulties. Its operating range is narrow and combustion phasing is difficult to control particularly at light loads. At high loads the main difficulty is to control the high rate of heat release, which may damage the engine parts.

During compression stroke a significant amount of unburned charge is trapped in the crevice regions of the combustion chamber. This trapped charge re-enters the combustion chamber during the expansion stroke but it does not burn because the combustion chamber temperature is not so high. This will result in increased CO and HC emissions in the exhaust. In this case oxidation catalytic converter is also not much effective as the exhaust gas temperature is too low to take part in oxidation reaction.

There are several methods to control the HCCI mode of combustion for the development of successful HCCI engine.

19.3.2 HCCI Control

Combustion process is difficult to control in HCCI engine. In SI engine it is controlled by spark timing and in CI engine it is controlled by fuel injection timing. In HCCI engine homogeneous mixture is compressed to an extent where high pressure and temperature initiates autoignition. There is no well defined combustion initiator as present in SI and CI engines. So, in order to initiate and control the combustion process in HCCI engine various methods are employed. These are discussed below.

Compression ratio

Raising the compression ratio would seem to be an obvious solution to triggering autoignition of gasoline at near idle loads where fuel/air ratio is lean. However, it may cause severe knock problems and combustion generated noise at full load. Variable compression ratio would seem to be a potential solution to this problem.

Fuel-injection system

It is necessary to produce a homogeneous mixture of fuel and air in HCCI engines. In order to get almost a homogeneous mixture early and long mixing timings are required. This can be obtained by using port fuel injection (PFI) system. However, PFI system has some limitations. It will not be possible to use higher compression ratio which is required for higher thermal efficiency and it has no control on combustion phasing. Engine with higher compression ratio can be used by early direct-injection of the fuel in the combustion chamber. This will extend the light load limit of the HCCI engine. Combustion phasing can also be controlled by direct-injection by changing injection timing. By altering the injection timing from early in the intake stroke to late in the compression stroke, it is possible to achieve optimum combustion phasing over a range of intake air temperature, engine loads and speeds.

Intake charge temperature and mixture dilution

Sufficiently high intake charge temperature and mixture dilution are necessary requirements for controlled combustion in HCCI engine. High charge temperature is required for initiating the ignition, sustaining the chemical reactions and leading fuel to autoignition. Proper mixture dilution is required to control the rate of heat release during combustion. Both of these requirements can be fulfilled by recirculating certain amount of exhaust gas and mixing it with fresh charge in the intake manifold of the engine.

Exhaust gas recirculation (EGR)

The main challenge for HCCI operation is to obtain sufficient thermal energy to initiate autoignition of the mixture late in the compression stroke at lower load. The most practical means to do this in a gasoline HCCI engine, where compression ratio is limited, is through the use of high levels of recirculated exhaust gases. At higher load EGR is the most effective way to control the rate of pressure rise to a safer limit.

Variable valve timing (VVT)

By using variable valve timing mechanism the effective compression ratio can be altered. With the change in compression ratio the amount of trapped residual gases is changed. Thus the temperature, pressure and mixture concentration entering into the cylinder vary. These parameters contribute for controlling combustion process in HCCI engine.

19.3.3 Advantages and Disadvantages of HCCI Engine

Advantages

The following are the advantages of homogeneous charge compression ignition engine:

1. HCCI engines may reduce fuel consumption up to 30%.
2. HCCI engines run on lean fuel, therefore higher compression ratio like conventional diesel engines can be used while using gasoline as a fuel. This will produce higher thermal efficiency of the engine.
3. Homogeneous charge produces lower emissions. Since peak temperatures are significantly lower than those in typical SI engines, NO_x levels are almost negligible. In addition the premixed lean mixture does not produce soot.
4. HCCI engines can operate on gasoline, diesel and most alternative fuels.
5. The part load efficiency of HCCI engine is higher than that of SI engine, because the engine is operated unthrottled.
6. In HCCI, the entire reactant mixture ignites almost simultaneously. Since very little or no pressure difference exists between different regions of the gas, there is no shock wave propagation and hence no knocking.

Disadvantages

The following are the disadvantages of a homogeneous charge compression ignition engine:

1. Ignition timing is difficult to control as there is no well defined combustion initiator as present in SI and CI engines.
2. It is difficult to control the high rate of heat release and pressure rise during combustion process. These may cause damage to the engine.
3. It is not easy to start the engine when it is cold. Cold starting requires additional means.
4. HCCI engine is difficult to operate at very low and at very high loads. It operates successfully within certain range of loads. At light loads it is restricted by lean flammability limits and at high loads the operation is restricted by high rate of pressure rise, causing knock.
5. CO and HC emissions are higher. It is because of incomplete combustion of the charge trapped in the crevice regions.

19.4 PREMIXED CHARGE COMPRESSION IGNITION (PCCI) ENGINE

In HCCI engine preparation of homogeneous mixture and control of combustion phasing are difficult. In order to overcome these difficulties PCCI mode of combustion is used. Here also an advanced low temperature combustion technology is used. In PCCI combustion a part of the fuel is injected into the cylinder well before ignition, which is premixed with the induced charge (air+EGR) to form a lean homogeneous mixture. Premixing is possible either by port fuel injection or by early direct injection. This part of fuel undergoes combustion process just like HCCI combustion. The mixture is compressed and it ignites at multiple locations in the chamber. The remainder of the fuel is injected near TDC, this part of injection is called late direct injection and fuel undergoes combustion process just as in the case of conventional CI engines. In this way combustion phasing is controlled. Thus, in PCCI mode of combustion a two stage split injection strategy is adopted. The first injection is called the pilot injection, and the second injection is called the main injection. The fuel-injection significantly affects the PCCI combustion. Increasing the fuel injection pressure up to 700 bar improves the combustion process. Increasing fuel injection pressure beyond certain limit may result in poor engine performance because of engine knocking.

The mixture preparation in PCCI engine is not very difficult as in HCCI engine, because only a part of fuel undergoes HCCI type of combustion and remaining fuel undergoes combustion process as in conventional CI engine which can easily be controlled by injection timing.

19.4.1 Factors Affecting Engine Performance and Emission

Following are the factors affecting engine performance and emissions:

First injection timing

When the first injection timing is advanced, the premixing timing of air-fuel is prolonged. This improves the mixing process and the mixture becomes more homogeneous. As there is no over-rich mixture in any region, the combustion temperature reduces which decreases NO_x emissions. With the advancement of the first injection, the cylinder pressure reduces which causes wall-wetting and the whole combustion process is advanced causing increased pumping work, which deteriorates combustion efficiency and increases incomplete combustion products. With the advancement of the first injection timing the fuel is trapped in the crevice regions and it is not fully burned during combustion process. Therefore CO and HC emissions increase in the exhaust.

Second injection timing

When the second injection timing is retarded the combustion becomes late, causing low charge temperature which reduces NO_x emissions. It also prolongs diffusion combustion which results in increased soot formation. As the charge temperature is not high enough to burn the second injection fuel completely, CO and HC emissions generally increase. Retarding the second injection timing also deteriorates the engine performance because a part of diffusive combustion shifts after TDC.

Injection quantity

Increasing the first injection quantity of fuel increases more fuel-air mixing which is advantageous for autoignition of first part of fuel injected. This releases more heat during this period of combustion. Further, as the first injection quantity of fuel is increased the ignition delay is shortened and the burned gas temperature during the second stage of combustion is increased, the NO_x emissions increase. In order to reduce NO_x emissions EGR technique is used. CO and HC emissions also increase because increased first injection fuel quantity results in wall-wetting and more fuel is trapped in the crevice regions causing incomplete combustion. Wall-wetting during the first injection timing increases the soot formation, but the higher combustion temperature and smaller proportion of diffusive combustion during the second injection of fuel decreases the soot formation. In general increasing the first injection quantity increases the soot formation. Therefore, the first injection quantity should not increase more than 40% of the total injected quantity of fuel.

Fuel spray angle

As the fuel spray angle is decreased, the spray will be more in axial direction and less in the radial direction of the cylinder. So, less fuel reaches to the cylinder wall. This causes more fuel to atomize before reaching the cylinder wall and hence wall impingement and wall-wetting are reduced. These phenomena reduce CO and HC emissions. If the fuel spray angle is reduced beyond a certain limit, the mixing process of air and fuel is affected and the performance of the engine goes down.

Blending ratio

The physical and chemical properties of different fuels are different. When the blending ratio of two different fuels is changed, the fuel spray, atomization, combustion and emission characteristics change. For oxygenated fuel, which has high oxygen content, the heat of evaporation is high and the heating value is low, so the heat release during combustion is reduced and the combustion temperature is lowered. This reduces NO_x emissions but CO, HC and soot emissions increase. In the blend of gasoline and diesel fuel, as the proportion of gasoline is increased, the ignition delay is prolonged resulting in better fuel-air mixing. In addition, the fuel vaporization increases which decreases the cylinder charge temperature resulting in reduced NO_x and soot emissions, but CO and HC emissions increase.

Injection pressure

Increase in injection pressure of the fuel up to a certain limit increases the thermal efficiency of the engine and provides better oxidation of soot, CO and HC emissions. This is because a higher injection pressure will result in a better air-fuel mixing, less fuel-rich regions, higher heat release rate and higher temperature. The combustion duration also becomes shorter. A further increase in injection pressure causes longer spray penetration and more serious cylinder wall wetting resulting in more soot, CO and HC emissions. NO_x emissions increase with increasing injection pressure due to a higher combustion temperature.

19.4.2 Comparison of Conventional Diesel, HCCI and PCCI Mode of Combustion

A comparison of the key characteristics of conventional diesel, HCCI and PCCI modes of combustion are presented in Table 19.1.

Table 19.1 Comparison of conventional CI, HCCI and PCCI modes of combustion

Combustion mode	*CI*	*HCCI*	*PCCI*
Injection timing	Just before TDC	Early injection to form homogeneous charge	Part of the fuel is injected early and remaining part just before TDC
Ignition	Autoignited by controlling injection timing	Autoignited by chemical kinetics	First part of fuel is autoignited by chemical kinetics and remaining part by injection timing
Combustion strategy	Diffusion from rich to lean charge	Combustion of the whole charge almost at the same time because of premixed homogeneous charge	First part undergoes combustion like a premixed charge and remaining by diffusion
Combustion temperature	High	Low	Low
NO_x	High because of high combustion temperature	Low because of low combustion temperature and dilution by EGR	Low because of low combustion temperature and dilution by EGR
Soot	High due to diffusion combustion	Low due to lean homogeneous charge	Generally low

19.5 REACTIVITY CONTROLLED COMPRESSION IGNITION (RCCI) ENGINE

RCCI engine makes use of two fuels of different reactivity. It needs both a low-reactivity fuel and a high reactivity fuel. Examples of fuel pairings are gasoline and diesel, ethanol and diesel, gasoline and gasoline with small addition of a cetane number booster (di-tert-butyl peroxide), but gasoline and diesel are the most common fuels. Gasoline is the low reactivity fuel having low cetane number and diesel is a high reactivity fuel having high cetane number. The low reactivity fuel (gasoline) is firstly premixed with air and then charged into the cylinder through the intake port using a port injector during the intake stroke as the piston moves down from TDC.

During the compression stroke, a direct fuel injector injects high reactivity fuel (diesel) into the combustion chamber, creating a mixture of air, gasoline and diesel. As piston gets closer to the top, a bit more diesel is injected into the mixture and this sets off ignition. This is called a cool flame because it is not very hot. This injection causes the diesel and gas mixture to ignite and that causes the remaining gas to ignite. Thus, the reactivity stratification is formed and the reactivity controlled combustion is accomplished.

Combustion phasing, rate of pressure rise and heat release can be controlled by changing the proportion of the two fuels of different reactivity and by adjusting the injection timings. The combustion process of RCCI engine also gets affected by changing the compression ratio and the geometry of the piston bowl. The evaporation, mixing and combustion process depend upon the type of fuels used in RCCI engines.

RCCI engine gets the advantages of best features of HCCI and PCCI modes of combustion. It reduces engine emissions without using after-treatment methods. By selecting the proper reactivity of fuels, their relative amounts and injection timing, it is possible to achieve optimized power and efficiency, controlled temperatures to reduce NO_x emissions and controlled equivalence ratio to reduce soot formations.

The main benefits of the RCCI mode of combustion include:

1. Reduced emissions of NO_x and particulate matters
2. Increased power output
3. Increased thermal efficiency
4. Costly after-treatment equipment are not required.

The main drawback of the RCCI mode of combustion is its requirement for multiple and separate fuel supply systems, which include separate fuel tanks, fuel lines and fuel filters.

19.5.1 Engine Performance and Emission Characteristics

Engine performance

Thermal Efficiency

With the increase in load the thermal efficiency increases. The thermal efficiency of RCCI engines is higher than that of conventional CI engines. In RCCI engines two fuels of different reactivity in suitable proportions are used, which ensure complete combustion of the charge and the combustion temperature is low which reduces the heat loss, resulting in higher thermal efficiency.

Emission Characteristics

Carbon mono-oxide (CO)

CO emission increases with the increase in load. The CO emission from RCCI engines is higher than that from conventional CI engines. The formation of CO depends upon injection timing, injection pressure, quality and quantity of the charge, and combustion temperature. The in-cylinder temperature is low in RCCI engines. This decreases the reaction rate causing more CO formation. Another reason for increased CO is that some fuel is trapped in the crevice regions due to early injection of the fuel causing incomplete combustion.

Hydrocarbons (HC)

The HC emission from RCCI engines is higher than that from conventional CI engines. Trapped fuel in the crevice regions due to early injection of some part of the fuel causing incomplete combustion is the main cause of increased HC.

Oxides of Nitrogen (NO_x)

NO_x emissions from RCCI engines are less than that from conventional CI engines. This is mainly because of low combustion temperature developed during RCCI mode of combustion. The other main factors on which NO_x formation depends upon are oxygen concentration and reaction time during combustion process.

Smoke

Smoke emission is generally less in RCCI mode of combustion. Smoke formation depends on fuel properties, such as density, viscosity, droplet size and equivalence ratio.

19.5.2 Comparison of Different Modes of Combustion

A comparison of the key characteristics that differentiate SI, CI, HCCI, PCCI and RCCI is shown in Table 19.2.

Table 19.2 Comparison of different modes of combustion

Combustion mode	*SI*	*CI*	*HCCI*	*PCCI*	*RCCI*
Fuel	Gasoline or high octane fuel	Diesel or high cetane fuel	Gasoline, diesel and alternative fuels	Gasoline, diesel and alternative fuels	Fuels of different reactivity
Ignition	Spark-ignition	Compression ignition, autoignited	Compression ignition, autoignited	Compression ignition, autoignited	Compression ignition, autoignited
Charge quality	Nearly stoichiometric	Lean fuel	Lean fuel and charge dilution	Lean fuel and high charge dilution	Fuels reactivity and air-fuel ratio stratification
Combustion strategy	Air-fuel premixed before ignition	Combustion proceeds with diffusion	Air-fuel premixed before ignition	Premixed and diffusion	Premixed and stratified
Control of heat release rate	Flame speed	Vaporization and mixing	Chemical kinetics	Chemical kinetics and injection timing	Chemical kinetics and fuel reactivity
Combustion temperature	High	Substantially high	Low	low	Low
Pollutants	High CO and HC, low NO_x and PM	High NO_x and PM. NO_x can be reduced by EGR	Low NO_x and PM, high CO and HC	Low NO_x and PM, high CO and HC	Very low NO_x and PM, High CO and HC

19.6 FLEXIBLE FUEL VEHICLE (FFV)

Flexible fuel vehicles commonly known as Flex-fuel vehicles have Internal Combustion Engines which can run on single fuel or more than one fuel. Normally, gasoline blended with either ethanol or methanol is used. Flex-fuel and dual fuel vehicles are different. In dual-fuel vehicles two fuels are stored in separate tanks and the engine runs on one fuel at a time, whereas in flex-fuel vehicles they are stored in the same tank.

Any proportion of ethanol can be blended with gasoline which can burn in the combustion chamber of flex-fuel engine. This is possible by adjusting fuel-injection timing and spark timing automatically by using fuel composition sensor and electronic control unit (ECU). Inside the fuel system, there is a small processor which detects the blend proportions of the two fuels and adjusts accordingly the ignition time and prepares the required air-fuel mixtures automatically. The flex-fuel vehicles can also run only on gasoline if required.

The fuel system and engine parts of conventional ICE are not suitable for flex-fuel engine. In order to use the blended fuels effectively, engine requires modifications. The main difficulty is that the ethanol is more corrosive. The high percentage of ethanol reduces the life span of fuel tank, fuel system, exhaust system and other parts of the engine. In order to protect these parts manufacturing cost increases with the increase in ethanol blend.

Technology exists to run the vehicle on any mixture of gasoline and ethanol. The flex-fuel engines may run on pure gasoline as well as on 100% ethanol. However, upper limit is set for a blend of 85% anhydrous ethanol with 15% gasoline. It is to avoid cold starting problem of the engine caused by low temperature and also to avoid unburned ethanol in the exhaust.

Higher ethanol blends in flex-fuel vehicles show improved acceleration. Performance characteristics such as cylinder pressure, heat release rate, brake thermal efficiency are improved due to clean burning of ethanol. Over a wide range of ethanol blend CO and HC emissions decrease without much affecting the formation of NO_x.

19.6.1 Technical Aspect

Ethanol is highly corrosive, which requires all parts of the vehicle where the ethanol is either stored or flows must be built in such a way so that they could withstand corrosiveness. The overall operation of the engine is monitored by Electronic Control Unit (ECU). For variable gasoline to ethanol ratio ECU sets proper ignition timing, controls combustion process and sticks to emission standards.

Ethanol contains acids, gums and moisture, so it is highly corrosive. Metals, plastics and rubber seals which come in contact with ethanol must be protected against corrosion. Corrosion resistant materials are used in fuel tank, fuel lines, filters, pressure regulator, fuel pump, injectors, gaskets and seals. The materials used for engine parts, such as valves, valve seats, piston and piston rings must withstand the corrosiveness of ethanol. Internal surfaces of intake manifold and exhaust manifold must also be protected against corrosion. Another challenge with ethanol is that it easily breaks-down and has tendency to wash away the engine lubricant, causing wear and tear to the engine.

Calorific value of ethanol is lower than that of gasoline. So, the presence of ethanol in the blend lowers the overall calorific value. In order to get the improved performance from the flex-fuel engine increased quantity of fuel injection is required. This needs fuel injectors with larger holes to increase the fuel flow rate, otherwise there will be a sharp reduction in power output. It also requires a fuel pump with a higher flow rate.

Laminar flame speed of ethanol is slightly higher than that of gasoline, so the flame propagation speed in the combustion chamber increases. The combustion duration is decreased and the combustion phasing is improved. This increases indicated mean effective pressure and thermal efficiency of the flex-fuel engine.

Octane number of ethanol is higher than that of gasoline, so the octane rating of the fuel increases with increased percentage of ethanol in the blend. Higher compression ratio can be used with higher octane number without knocking, which increases thermal efficiency and reduces fuel consumption. However, flex-fuel engine imposes knocking problem when it runs on gasoline. This problem is overcome by finding ways to reduce the temperature at the end of compression stroke. This can be accomplished by using Exhaust Gas Recirculation (EGR), water injection in the intake manifold and Variable Valve Timing (VVT). By using VVT effective compression ratio can be changed that prevents knock in the engine.

Latent heat of evaporation of ethanol is higher than that of gasoline, so the cold starting of the engine will be difficult especially when higher percentage of ethanol is present in the blend. It requires a secondary gasoline assisted cold starting system.

19.6.2 Engine Components

Apart from the regular components like internal combustion engine, battery, intake and exhaust system, the following additional components are used in the flex-fuel engine:

1. **Electronic Control Unit (ECU):** All engine parameters are controlled by this electronic unit. It monitors overall operation of the engine. It controls the gasoline assisted pilot-injection during cold starting, helps to prepare proper fuel-air mixture, optimizes spark timing and fuel injection during all operating conditions.
2. **Oxygen Sensor:** The oxygen sensor is fitted in the exhaust system. The ECU gets the signal from the oxygen sensor which senses the amount of oxygen that is flowing in the exhaust pipe. On the basis of this, ECU sends signal to the electronic fuel injectors, which stay open to prepare a stoichiometric fuel-air mixture. The quality of the mixture is checked by a fuel sensor.
3. **Intake manifold pressure and temperature sensors:** These sensors give information of intake pressure and temperature to ECU which calculates the amount of air and fuel flow in the intake system and adjusts the opening of injectors accordingly. These sensors with fuel and oxygen sensors optimize the opening of the electronic fuel injector.

19.6.3 Advantages and Disadvantages of Flexible-fuel Engine

Advantages

1. Flexible-fuel engines are designed in such a way, so that they can use any proportion of ethanol blend in gasoline.
2. Ethanol has cleansing property, so the carbon deposits on valve seats, in fuel injectors and in combustion chamber reduce.
3. Ethanol burns cleaner than gasoline, so the flex-fuel engine emits less pollutant. It also emits less CO_2 reducing greenhouse gases, so it is environmental friendly.
4. Flex-fuel engine uses advanced technology to control and optimize the whole operation of the engine by using ECU and various sensors.
5. Ethanol can be sustainably produced from sugarcane, corn, molasses, broken rice and crop residues.
6. A good amount of corn and sugarcane are produced in India every year. A large amount of ethanol can be produced from these ingredients. Use of ethanol in automobiles will definitely cut the fuel import bill drastically.

Disadvantages

1. Ethanol is highly corrosive as it contains acids, gums and moisture. It absorbs dust easily. Use of higher proportion of ethanol may corrode and damage the engine.
2. Flex-fuel engine requires compatible materials to avoid corrosion, which increases the manufacturing cost.
3. Ethanol is a good solvent. It could easily wipe out the protective lubricating oil film layer, which could cause wear and tear to the engine.
4. Ethanol has high latent heat of vaporization in comparison to gasoline, which causes cooling of the charge and makes cold starting of the engine difficult.
5. Calorific value of ethanol is less in comparison to gasoline, the vehicle gives less mileage per gallon.

19.7 GASOLINE DIRECT INJECTION COMPRESSION IGNITION (GDCI) ENGINE

GDCI engine uses gasoline as a fuel, however it does not use spark plug for ignition as in conventional gasoline spark ignition engine. It is a pure autoignited compression ignition engine. In order to get higher thermal efficiency the compression ratio is kept high. So, using gasoline as a fuel and compressing it like a diesel fuel by utilizing high compression ratio, the engine knocks. In order to avoid knock the low temperature combustion strategy is adopted, which provides diesel like thermal efficiency and low exhaust emissions for all operating conditions of the engine.

19.7.1 GDCI Engine Concept

A part of the gasoline fuel is injected at low pressure which vaporizes and mixes quickly with air and it is called partially premixed charge. The remaining fuel is injected intermittently at high pressure. This part of injection is called multiple-late-injection.

The first part of the fuel is injected at low pressure during intake stroke. This premixed charge is lean, which prolongs the ignition delay. The ignition delay period can further be increased by cooled exhaust gas recirculation (EGR). At high load, late inlet valve closing (IVC) can be used to reduce cylinder pressure and temperature, which also increases ignition delay. Therefore, the charge does not ignite early. It ignites near TDC after all fuel injection is completed.

The multiple late injection part of fuel is injected into a piston bowl at high pressure. The charge is stratified and injection is completed late on the compression stroke, which controls the heat release rate. Also there is no end gas exists due to late injection. This improves the combustion efficiency and eliminates the possibility of combustion knock. Moreover, at this point cylinder pressure and temperature are high enough to vaporize the late injected fuel quickly which reduces liquid penetration and avoids wall-wetting.

In GDCI engine heat release is controlled during combustion process. Combustion temperature is not that high as in conventional diesel engine. Thus, GDCI engine utilizes low temperature combustion strategy and reduces both NO_x and PM emissions.

19.7.2 GDCI Injection Strategy

In GDCI engine, mixing the fuel with air and cooled EGR, and controlling the rate of combustion mainly depend upon the injection system. The intake of the fresh air in the intake manifold is unthrottled and it is diluted with cooled EGR depending on the engine load. The boost system delivers the correct air and EGR mixture in the chamber. The contour of the inlet passage does not induce a swirl.

The injector is located at the centre of the chamber. The chamber shape is symmetrical shallow bowl in the crown of the piston. This provides a quiescent chamber with low squish and low or no swirl. Spray angle is very important in the injection system. Wider spray angle is normally preferred. Spray pattern of the fuel covers most of the combustion chamber without impinging fuel on the walls. All these are required in the injection strategy to avoid over-mixing of fuel and air and create stratification of the charge.

Spray pattern and multiple injections produce stoichiometric mixture locally in certain region but overall mixture is lean. The thermal efficiency of the engine is increased.

19.7.3 Combustion Strategy

GDCI engine is also known as partially premixed compression ignition (PPCI) or partially premixed combustion (PPC) engine. Low temperature combustion strategy is adopted in the combustion process of GDCI engine. Ignition delay of premixed charge is prolonged, so that the combustion takes place near TDC. Stratified charge establishes stable ignition and controlled heat release. Autoignition takes place near TDC or preferably just after TDC. Autoignition is not that violent as in the case of homogeneous mixture. The ignition takes place when the piston passes through TDC, which results no negative work on the crankshaft. The ignition process in GDCI engine is gradual, so it does not impose high load on the engine structure as it is in the case of HCCI engine where the ignition is violent and combustion is instantaneous.

The combustion strategy in GDCI engine enables gasoline to be used in CI engine, which develops high specific power output. The thermal efficiency of the engine is higher in comparison to SI engine. Exhaust pollution is also low.

REVIEW QUESTIONS

1. Why the advanced low temperature combustion strategy is used?
2. Mention some of the important low temperature combustion modes and its common features.
3. How a homogeneous charge compression ignition (HCCI) can be achieved in an IC engine? Mention the characteristics of HCCI engine.
4. How a HCCI engine can be controlled to improve fuel economy and reduce emissions?
5. Mention the main reasons by which HCCI combustion is limited in the IC engine.
6. Mention the possible challenges which HCCI engines may face. How these challenges can be overcome?
7. Describe briefly different operating factors which are responsible for controlling the dynamic operation of the HCCI engine.

8. What are the advantages and disadvantages of HCCI engine?
9. What do you understand by premixed charge compression ignition (PCCI) mode of combustion? How the PCCI combustion mode can be achieved?
10. Describe the effect of advancing the first injection timing in PCCI on the engine performance and emissions.
11. Describe the effect of retarding the second injection timing in PCCI on the engine performance and emissions.
12. Describe the effect of injection quantity and fuel spray angle in PCCI on the engine emissions.
13. Describe the effect of blending ratio in PCCI on the engine emissions.
14. Describe the effect of injection pressure in PCCI on the engine performance and emissions.
15. Compare the key characteristics of conventional diesel, HCCI and PCCI combustion engine.
16. What is a reactivity-controlled combustion ignition (RCCI) engine? How does it work?
17. What are the key benefits and drawbacks of RCCI combustion mode?
18. Describe briefly the engine performance and emission characteristics of RCCI mode of combustion.
19. Compare the key characteristics that differentiate SI, CI, HCCI, PCCI and RCCI combustion modes.
20. What is a flexible-fuel engine? Describe its main features. How can it be realized in actual practice?
21. Describe the technical aspect of flexible-fuel engine.
22. Describe briefly the different key components of a flex-fuel engine.
23. What are the pros and cons of flex-fuel engines?
24. What is a gasoline direct injection compression ignition (GDCI) engine? How can a GDCI engine be realized in practice?
25. Describe the injection and combustion strategy used in a GDCI engine.

CHAPTER 20

Introduction to Electric Vehicles

20.1 AN OVERVIEW

In spite of using advanced combustion technology with electronic management system, the internal combustion engine (ICE) can never be developed to the extent where they emit zero emissions, and it can never be 100 per cent energy efficient. Internal combustion engines (ICEs) using fossil fuels are emitting harmful gases which are dangerous to our planet. The promising technology that drastically reduces emissions is the electric driven vehicles. These vehicles produce low noise and operate with high efficiency.

Although electric vehicles (EVs) are not the subject of internal combustion engine (ICE), but it is included here in order to understand the future trend in motor vehicles. Moreover, ICE is being used with electric drive as a hybrid electric vehicle. This chapter includes a short description of electric vehicle technologies. In general, EVs are classified as follows:

1. Battery Electric Vehicle (BEV)
2. Hybrid Electric Vehicle (HEV)
3. Fuel Cell Electric Vehicle (FCEV)

20.2 BATTERY ELECTRIC VEHICLE (BEV)

A Battery Electric Vehicle (BEV) is operated by using a large battery pack, electric motor and drive train. ICE is not used in this type of electric vehicle. Battery pack consists of a large number of rechargeable batteries, which are charged by plugging them into the electric supply. The recharging time depends upon the type of charger, the size and type of battery. The battery pack provides power to one or more electric motors, which rotate the wheels through drive trains.

BEVs are zero-emission vehicles, because they do not pollute air directly while running. In order to charge batteries electricity is required, while producing the electricity some pollutants may generate if the production of electricity is not by renewable energy sources. Even when these pollutants are taken into consideration BEVs are much cleaner than ICE vehicles.

As BEV does not use fossil fuels, there is no fuel supply system, engine and exhaust system as required in ICEs. All these make BEVs to be less complicated, light weight, easier to use and repair. A schematic diagram of battery electric vehicle is shown in Figure 20.1.

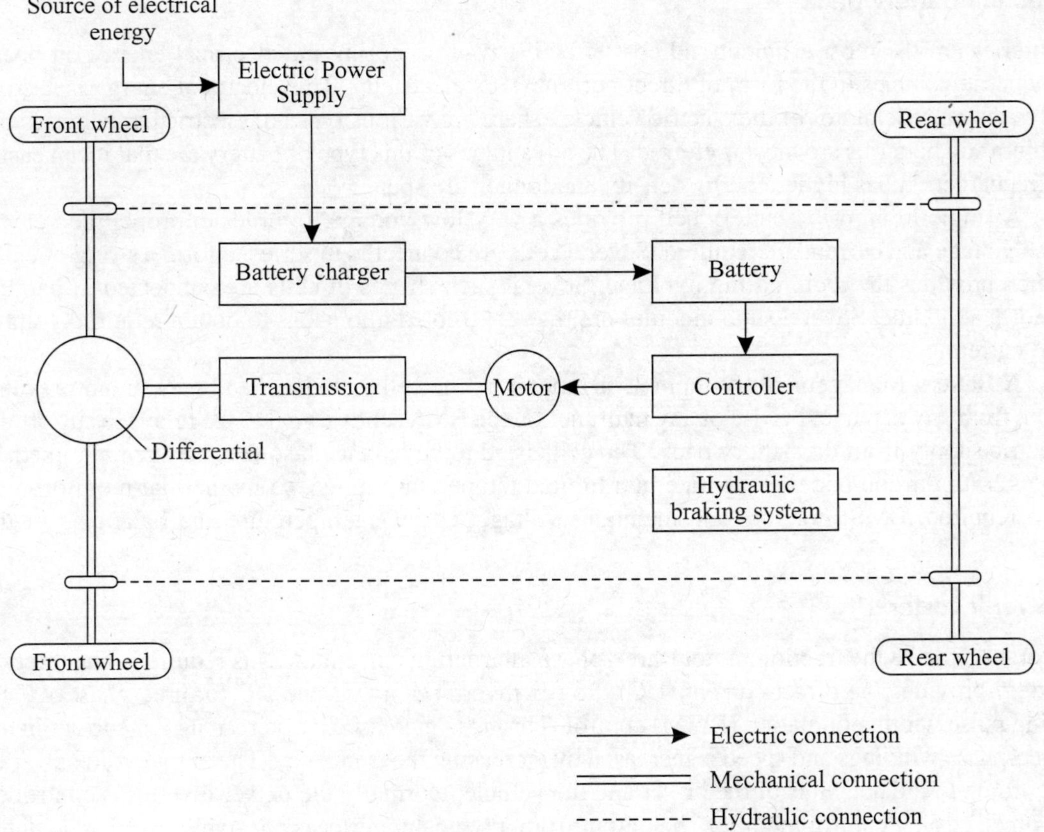

Figure 20.1 Schematic diagram of Battery Electric Vehicle (BEV).

20.2.1 Battery Electric Vehicle Components

The key components of a battery electric vehicle are:

1. Traction Battery Pack
2. Power Inverter
3. Controller
4. Electric Traction Motor
5. Battery Charger
6. DC/DC Converter
7. Auxiliary Battery
8. Energy Management System (EMS)
9. Thermal System
10. Charge Port
11. Transmission

Traction battery pack

Batteries are the most efficient and cost-effective way of storing the electrical energy on-board in electric vehicles in the form of direct-current (DC) electricity. This electrical energy is used to drive the electric motor of the electric vehicle. There are various types of electric car batteries but lithium-ion batteries are most preferred. The advantages of this type of battery are that it can easily be recharged, it has higher energy density and longer life span.

A single lithium-ion battery cell provides a very low voltage. In order to propel the vehicle high voltage and current are required. Several cells are connected in series to form a string of cells, which provides the required high voltage. Several such strings of cells are connected in parallel to form a module. Several such modules are then assembled into packs to obtain required voltage and current.

A Battery Management System (BMS) monitors the cells, modules and packs, and prevents them from any damage. In case of any malfunction the BMS sends signal to the relays to cut-off the electric supply from the battery pack. The cells need to be balanced, so that they remain in equal states. Lithium-ion batteries operate in a limited temperature range, so its thermal monitoring is also required. BMS monitors and maintains voltage, current, temperature and balancing of the cells.

Power inverter

In order to drive the traction motor, three-phase alternating current (AC) is required. The traction battery provides the direct-current (DC). Power inverter converts the DC to three-phase AC by using pulse width modulation (PWM) control. Torque is increased by increasing the current in the three-phase windings and speed is increased by increasing the frequency. The output voltage varies according to the demands of the driver and the vehicle. Normally the power inverter is controlled by an electronic control module. The output from a typical inverter is constantly being calculated using input signals from the accelerator pedal, the motor's shaft speed sensor, the motor's direction sensor, and the brake pedal. The inverter is essential to the operation of the motor and, if it fails, the motor cannot run. However, electric vehicles that use DC motor do not need an inverter.

When regenerative braking is used, the energy flows in the reverse direction. The electric motor charges the battery, so the conversion of AC to DC is required. For this purpose a rectifier is required. Bi-directional inverter can meet both the above requirements.

For automotive application the space in the vehicle is limited, therefore size of the inverter should be kept small. It can be achieved by using high power density technology in main circuit component parts, such as in power modules and capacitors.

Controller

Controller regulates the flow of electrical energy from battery and inverter to the electric motors. The controller gets the input signal from the car pedal set by the driver and sends frequency and voltage variation signals to the traction motor. The speed and torque produced by the traction motor is thus controlled. The controller responds to the movement of the brake pedal during regenerative braking. It also reverses the current flow to the motor when reverse drive is selected.

Modern controllers adjust motor speed through pulse width modulation (PWM). Pulse width is the length of time, in milliseconds, that a component is energized. Controllers rely on transistors to rapidly interrupt the flow of electricity to the motor. High electrical power (during high speed, acceleration, and/or heavy loads) is available when the intervals during which the current is stopped are short. During slow-speed driving, little power is needed, and the intervals of no current flow are larger.

Electric traction motor

The electric traction motor provides the power to rotate the drive wheels of the vehicle. It should be capable of meeting all the requirements, which a vehicle has to deal in actual practice. It requires frequent start and stop operation. It needs to accelerate or decelerate the vehicle. It has to run at variable load and speed. It requires high torque and low speed while climbing the hill and it requires low torque and high speed while cruising. Traction motor should be of small size, light weight, high power output and high efficiency. Figure 20.2 shows typical power-torque-speed characteristics of electric traction motors.

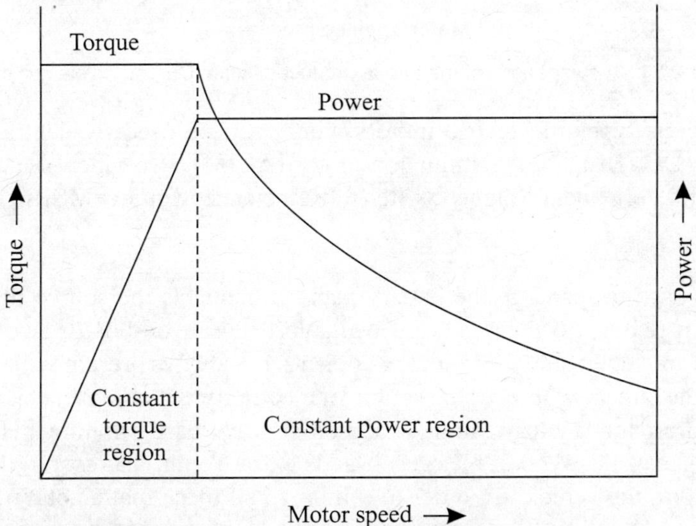

Figure 20.2 Typical power-torque-speed characteristics of electric traction motor.

Internal combustion engine (ICE) does not develop peak torque until it has reached a particular rpm. Once the peak torque is reached, it starts to decrease quickly. This is why IC engines are coupled to multispeed transmissions. The various available gear ratios allow the engine to run at speeds where it is most effective. With an electric motor, instant torque is available at any speed. The entire rotational force of the motor is available at the instant the accelerator pedal is pressed. Peak torque stay constant to nearly 4500 rpm, and then it begins to decrease slowly. A comparison of the torque produced by an IC engine and electric motor is shown in Figure 20.3.

The wide torque band, especially the available torque at low speeds, eliminates the need for multispeed transmissions. There is no need for a reverse gear, since switching the polarity of the stator will cause the rotor to turn in reverse. The absence of a typical transmission saves weight and makes the power-train less complex.

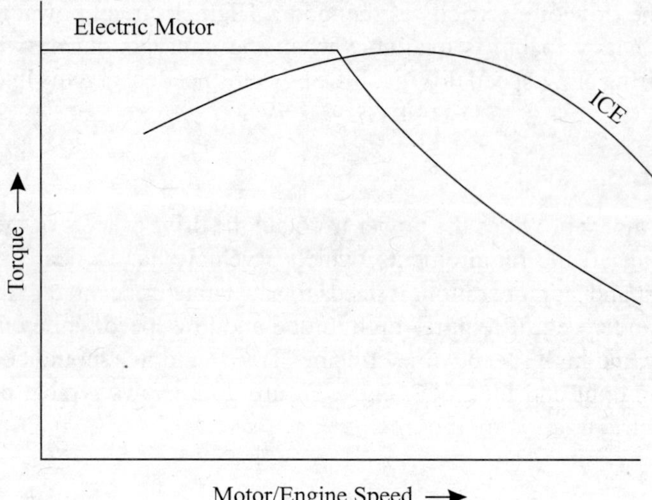

Figure 20.3 A comparison of the torque produced by an ICE and an electric motor.

There are various types of electric motors which can be used in electric vehicles. These include: Brushless DC (BLDC), AC induction motor (ACIM), Permanent Magnet Synchronous Motor (PMSM), and Permanent Magnet Switched Reluctance Motor (PMSRM).

Battery charger

Battery charger is used to recharge the battery pack. It connects the source of electricity to the battery pack. Mainly two types of chargers are available. They are on-board and off-board chargers. On board chargers are kept inside the vehicle. As charger is always present with the vehicle, it can be used to charge the battery wherever the source of electricity is available. This type of charger is more common at present as battery can be charged overnight easily at home. However, there are certain disadvantages of this type of charger. It adds extra weight and cost to the vehicle. It also occupies space within the vehicle. In order to minimize all these manufacturers normally install low power chargers which take long charging time. It takes several hours to recharge the battery. The charging time varies with the size and type of the battery and also depends upon the charger. New designs of chargers utilize sophisticated electronics to monitor the cells and regulate the charging voltage and current. The recharging time is reduced.

Off-board chargers are placed at fixed locations. As it is not inside the vehicle, it does not add extra weight and cost to the vehicle. These chargers are large and capable of charging the battery in short time by supplying more power. In order to charge the battery driver has to take the vehicle to fixed locations. The availability of these chargers depends upon the infrastructure developed in the country for this purpose.

DC/DC converter

The purpose of the DC/DC converter is to change high voltage DC power from the traction battery pack to the lower voltage DC power required to operate certain accessories, such as wipers, lights, horns, mirror movements, infotainment system, etc. The auxiliary battery also needs lower DC voltage for its recharging.

Auxiliary battery

It is advisable to have an auxiliary battery in addition to main traction battery pack. If the main battery is drained below certain level, then this battery helps to run vehicle accessories and bring back the vehicle home.

Energy management system (EMS)

Energy management system (EMS) acts as a brain of electric vehicles (EVs). All the required functions of EVs are monitored and controlled by EMS. It is a computer based system. It sets a suitable algorithm for charging the battery to its optimum condition. The output energy flow from the battery is also controlled. By doing so, the battery life and operating range of the vehicle increase. EMS is capable to detect the status of battery charge, which helps the driver to predict the available range of the vehicle.

EMS collects information from various sensors fitted to the system. Based upon these information EMS monitors and controls all the operations of the vehicle. Sensors are used to check temperatures, battery voltage, charging current, external lighting conditions, speed and acceleration of the vehicle. EMS automatically adjusts the head light brightness by sensing the external lighting conditions. Navigation system can be coupled with the EMS, which can help in locating the charging station. A block diagram of energy management system is shown in Figure 20.4.

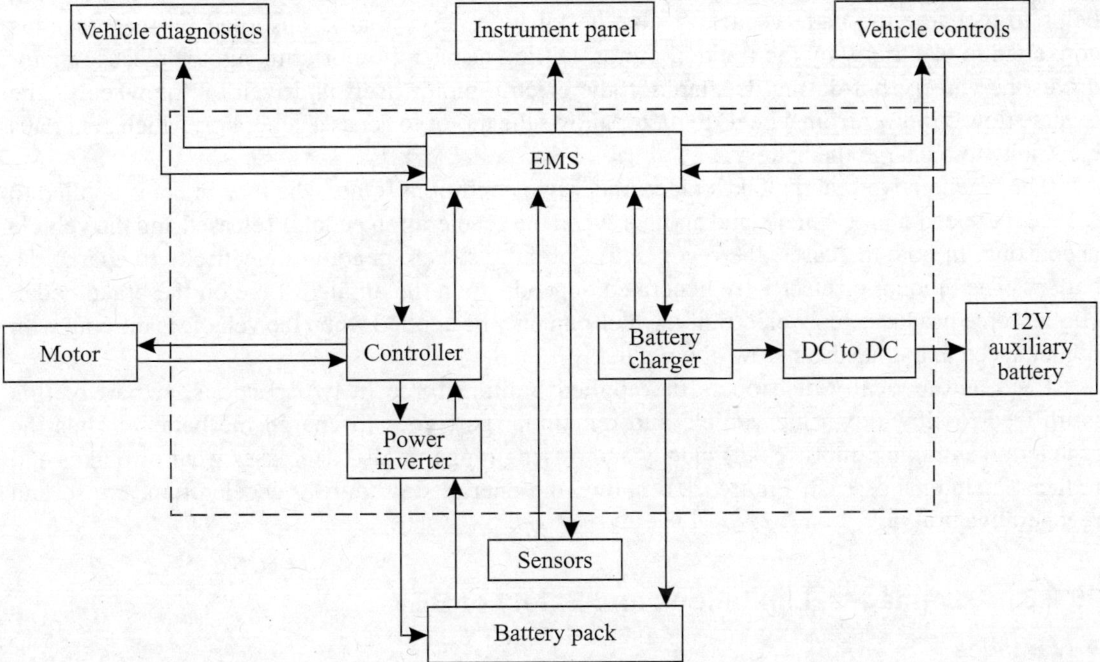

Figure 20.4 Block diagram of energy management system (EMS).

Thermal system

All electrical and electronic components such as electric motor, power electronics and other electrical components require a suitable temperature range for their optimum operations. The required temperature range of these components is maintained by combining forced air cooling, liquid cooling and thermoelectric heating. Thermal system together with EMS increases battery life and optimizes operating conditions.

Charge port

The charge port connects the electric vehicle to an external power supply in order to charge the traction battery pack. The charge port can be located either in front or rear part of the vehicle.

Transmission

The power of the electric motor is transmitted to the wheels through transmission system. The transmission system of electric vehicles is different than that of ICEs. In ICEs multispeed gear box is used, whereas EVs do not require multispeed transmissions. Thus, power loss during transmission reduces, which increases the transmission efficiency.

20.2.2 Regenerative Braking

In normal braking, when brake is applied the discs and the brake pads create friction which absorbs kinetic energy of the vehicle and generates heat which dissipates into the surroundings. This heat energy which is lost during normal braking can be utilized by using regenerative braking, which helps to recharge the battery pack. Thus, the kinetic energy that is absorbed by the brakes is converted to electricity by the traction motor. During acceleration or cruising, the electric motor drives the wheels, but during deceleration the electric motor itself is driven by the wheels. The reverse flow of power from wheels to motor allows the motor to act as a generator, which generates electricity to recharge the battery.

The regenerative braking takes place under two conditions. One, when the brake is applied to reduce the speed of the vehicle, and another when the acceleration pedal is released and the vehicle is coasting. In both the cases, the regenerative braking system generates electricity to charge the battery. The amount of electricity generated depends upon the applied force on the brake pedal. Higher force produces more electricity, which can only be applied when the vehicle is moving with higher speed and it needs to slow down quickly.

EMS automatically distributes the applied braking force in two portions. A part of this is utilized to slow-down the vehicle and remaining part goes to charge the battery. Thus the regenerative braking improves efficiency and driving range. It also decreases wear of brakes and reduces maintenance cost. Figure 20.5 shows the energy flow during acceleration, cruise and regenerative braking.

20.2.3 Advantages, Limitations and Safety of BEV

Advantages

Following are the advantages of Battery Electric Vehicles (BEVs):

1. BEVs do not use any fuel. It runs purely on electricity. So, there is no pollutant emitting into the atmosphere while running. It can be said as zero exhaust emission vehicles.

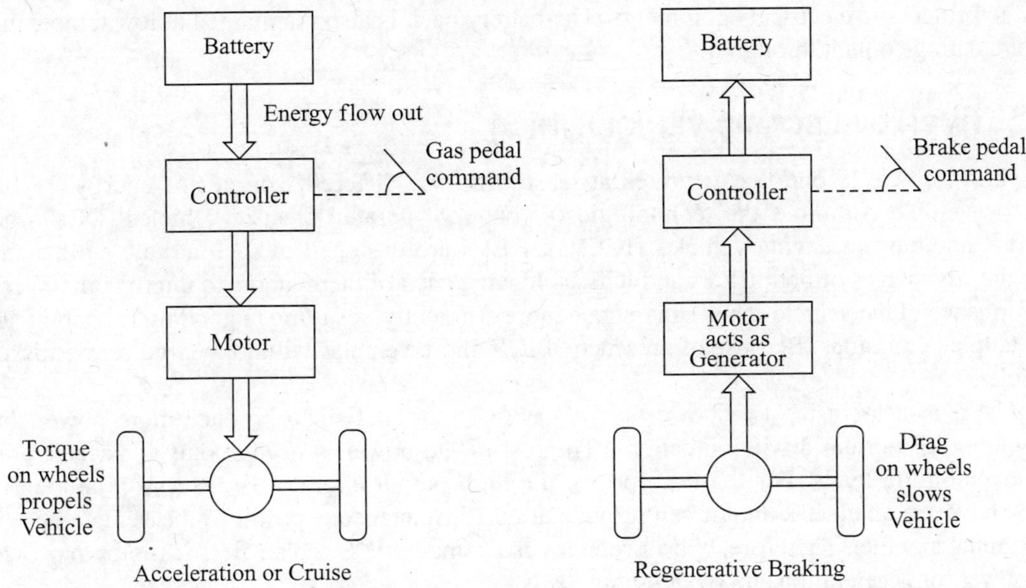

Figure 20.5 Energy flow during acceleration, cruise and regenerative braking.

2. Renewable sources such as solar, wind, hydro, geothermal, etc. can be used to produce electricity for charging the battery. The use of fossil fuel can be reduced.
3. Greenhouse gases (GHGs) can also be reduced by using renewable sources for electricity production. GHGs cause climate change, which threatens the global warming.
4. BEVs produce very low level of noise. The vehicle runs quietly at low speeds. So, there is no noise pollution.
5. The running cost of the vehicle is much less than that of ICEs.
6. BEVs have fewer moving parts, therefore maintenance cost is low.

Limitations

Following are the limitations of battery electric vehicles (BEVs):

1. The driving range of BEVs is limited. It depends upon the charge status, size and type of batteries.
2. The charging process of the batteries takes long time. It depends upon the types of charger and batteries.
3. Off-board charging stations will always be limited in number.
4. Battery pack needs to be changed after few years of using, as it loses the capacity of holding the charge. It is expensive to replace it with a new one.

Safety

BEVs are quite safe to operate. The safety standard of electric vehicles is measured and found to be of the same standard as ICE vehicles. The possibility of catching fire to lithium ion battery is rare. Extra safety measures are taken in order to avoid any such incidents. Fuses and circuit-breakers are used to protect the electrical equipments. The battery pack is also maintained at low temperature by circulating coolant through it.

20.3 HYBRID ELECTRIC VEHICLE (HEV)

A hybrid vehicle is one that operates at least with two different sources of energy. Hybrid electric vehicle combines the technologies of battery operated electric vehicles (BEVs) and internal combustion engine vehicles (ICEVs). HEVs are designed in such a way, so that they take the advantages of both BEVs and ICEVs. Ideally, each of them works to improve the overall performance of the vehicle. They can operate more efficiently, resulting in good fuel economy and low tailpipe emissions. Because of presence of ICE, these vehicles will not be true zero-emission vehicles.

The reason for using two power sources is that most ICEs can produce more power than it requires for normal driving situations. The rest of the power is needed only for acceleration and overcoming loads. For these purposes, the high output engines use excessive amount of fuel. However, an electric motor can provide almost instantaneous power to the vehicle without consuming any fuel. Therefore, hybrid vehicles use a smaller ICE and an electric motor to provide the power for acceleration and overcoming loads.

In a hybrid electric vehicle, motor is powered by batteries. These batteries are continuously recharged by a generator driven by ICE. The battery is also recharged by regenerative braking. This type of system is without plug-in charging as the batteries are charged while driving.

20.3.1 Plug-in Hybrid Electric Vehicle (PHEV)

Charging efficiency of batteries in HEVs without plug-in charging depends mainly on fuel used in ICE and regenerative braking. The total dependency of battery charging on these two methods can be avoided by using plug-in charging to the vehicle, called plug-in hybrid electrical vehicle (PHEV). This provides additional facility for charging the batteries from external power outlets.

The cost of charging the battery pack is less, since the electrical energy obtained from the main outlet is cheaper than that produced on-board. PHEVs extend the vehicle range, reduce pollution and utilize electricity from grid connection.

20.3.2 Photovoltaic Hybrid Electric Vehicle (PVHEV)

The driving range of HEV can be extended if the battery pack is allowed to charge continuously while running. The solar energy can be utilized to charge the battery continuously. Solar-driven electric vehicle with ICE is known as Photovoltaic Hybrid Electric Vehicle (PVHEV). The use of fuel is minimized and hence air pollution is reduced.

20.3.3 Fuel Cell Hybrids

The hybrid vehicle is the vehicle powered by two separate energy sources, but all hybrids do not have internal combustion engine (ICE) and electric motor. In fuel cell hybrids ICE is not used. It is a combination of fuel cell and traction battery pack. The electricity produced by the fuel cell can directly drive the electric motor or it can be stored in the traction battery pack. Hydrogen fuel cells generate electricity with zero emission.

20.3.4 Configurations of Hybrid Electric Vehicles (HEVs)

In HEVs, there are two power devices, ICE and electric motor. These two devices can be connected in series or parallel, or in combination of both series and parallel.

Series hybrid

A series hybrid drive train is shown in Figure 20.6. Here ICE and electric motor are placed in series. Only electric motor is providing power to drive wheels. ICE is not connected to the final drive.

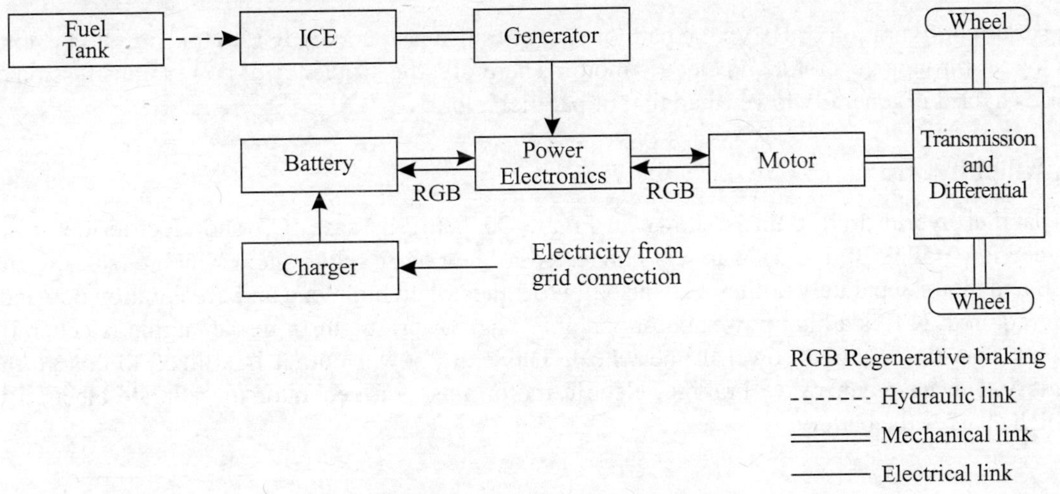

Figure 20.6 A series hybrid drive train.

ICE gives power to the generator, which generates electricity to power electric motor directly, which drives the wheel, and it also charges the battery. The battery can also be charged by external charger connected to the grid. Thus, the electric motor may receive electricity directly from ICE, or from the battery, or from both making use of power electronics. In this case, the engine is decoupled from the vehicle's transmission system; therefore engine speed does not depend on vehicle speed. The engine can be operated and controlled at its optimum speed to get the best fuel economy. Different combinations can be selected for different operating conditions of the vehicle. These combinations are as follows:

1. **Only battery:** The vehicle can run only on battery when the power demand is low and battery is sufficiently charged. Under this condition ICE-generator set can be turned off.
2. **Only engine-generator set:** When the vehicle is cruising on highways, the power demand is moderately high. Under this situation electric motor takes the power directly from engine-generator set. It may not draw electricity from battery.
3. **Combination of battery and engine-generator set:** During acceleration and high loads vehicle demands high power. The electric motor gets electricity from both battery and engine-generator set.
4. **Split of power from engine-generator set:** When the power demand for driving the vehicle is less than the optimum power produced by ICE and the battery is not sufficiently charged, the electric power produced by generator splits into two portions. A part of it is used to drive the electric motor and remaining part is utilized to charge the battery.

Advantages

In series configuration of HEVs, there is no effect of road load on ICE, as it is decoupled from the final drive. The engine runs smoothly and there is no abrupt change in operating conditions even when the road conditions are changing. This reduces exhaust emissions. As ICE is not connected to the final drive, ICE-generator set can be placed conveniently at any location. The design of series configuration is simple.

Disadvantages

In series configuration of HEVs the path of energy transmission from ICE to transmission is long as it goes through generator and electric motor. Therefore, the efficiency of power transmission of series hybrid is generally lower than that of parallel hybrids.

Parallel hybrid

A parallel hybrid drive train is shown in Figure 20.7. In this case ICE and electric motor are placed in parallel. Both are separately connected to the power split system which enables to run ICE and motor separately or they may also run together when high load and acceleration demands are required. ICE is a high power output engine. The power output from the engine is generally higher than the required power at the wheel. This extra power output is utilized to charge the battery. During regenerative braking, electric motor acts as a generator to generate electricity, which charges the battery.

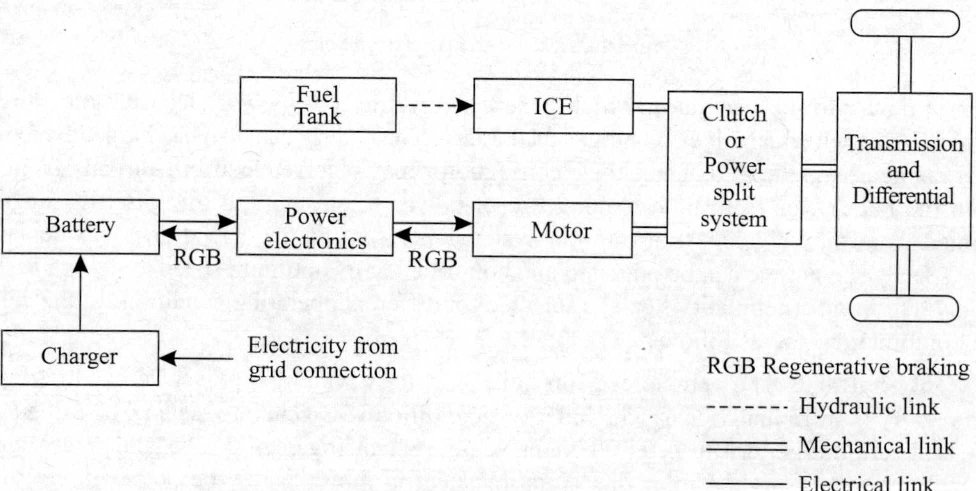

Figure 20.7 A parallel hybrid drive train.

The advantages of parallel hybrid in comparison to series hybrid are that it requires smaller ICE, smaller electric motor and smaller rating of battery. It has fewer electrical components. A separate electric generator is not required. All these reduce the cost of parallel hybrid. It is possible to obtain best efficiency as the operation of ICE can be controlled freely. However, there is a disadvantage that it requires a complex power split control system.

Series-parallel hybrid

A series-parallel hybrid drive train is shown in Figure 20.8. With the help of a mechanical coupler, it is possible to change this drive train as series or parallel hybrid. In series-parallel hybrid configuration, ICE is connected to the transmission system through a mechanical coupler, and electric motor is directly connected to the same transmission. Here mechanical coupler plays an important role. When ICE is coupled with the transmission shaft, the parallel hybrid configuration is obtained, and when the ICE is decoupled from the transmission shaft, series hybrid configuration is established. In series hybrid configuration ICE connects to the generator, which further connects to the power converter. The power convertor sends electricity to run the electric motor and also used to recharge the battery.

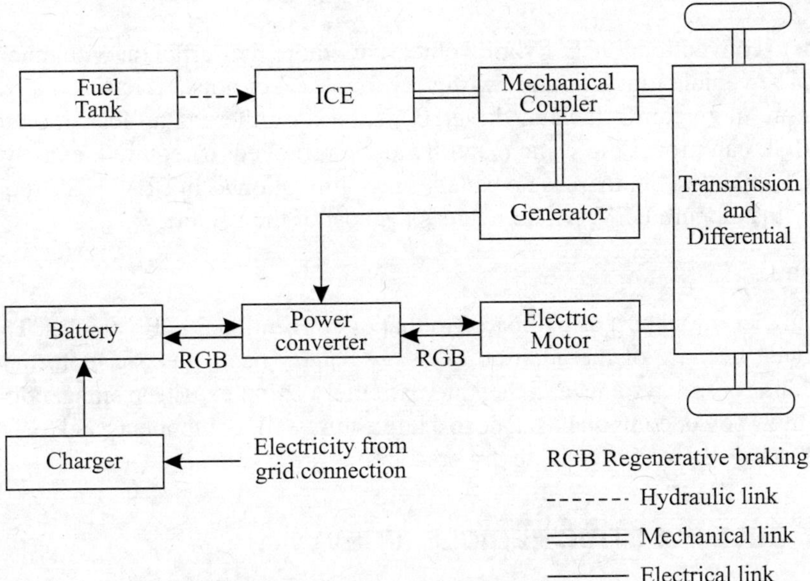

Figure 20.8 A series-parallel hybrid drive train.

Series-parallel hybrid configuration derives advantages from both series and parallel configurations depending upon the requirement, but the configuration is more complex because of the use of mechanical coupler and it increases the overall cost.

20.3.5 Hybrid Electric Vehicle's Technology

The feasibility of HEVs mainly depends upon the technology used in battery, electric motor, transmission and control system.

The energy density of battery pack is lower than that of gasoline or diesel. Therefore, battery pack is normally larger and heavier than a typical fuel tank.

Electric motor is added to a conventional ICE to make the vehicle efficient. The electric motor provides full torque at low speeds and it assists the ICE during acceleration, overtaking heavy loads or hill climbing. The motor acts as a generator in the case of regenerative braking.

Continuous variable transmission (CVT) system is normally used in this type of vehicle. CVTs do not have fixed gear ratios, the drive ratio changes continuously as the speed and load change.

Control systems are complex. Computers having fast processing speeds control the switching between the electric motor and ICE. There are many control systems, such as engine controller, battery management system, electric motor controller, transmission controller and brake system controller. Control systems have individual computers. They are linked together and communicate with each other by high speed communication buses. Control area network (CAN) communications take place between many control systems.

20.3.6 Advantages and Disadvantages of HEV

Advantages

In comparison to conventional ICE, hybrid vehicles are more fuel-efficient with much lower level of emissions. The vehicle runs smoothly with very low level of noise. It has also advantages over BEV. Drivers are more comfortable to drive HEVs, because there is no fear of battery drain. In that situation ICE can propel the vehicle and it can be refuelled, if required, exactly in the same way as pure ICE vehicles. So, there is no battery range limitation as in BEV. HEV requires a much smaller battery than a pure BEV, which reduces the cost of the vehicle.

Disadvantages

HEVs have a higher initial cost as compared to that of conventional ICE vehicles. These vehicles are much heavier, because of the addition of heavy battery pack, electric motor and generator. There are various control systems and they are extremely complex. Electromagnetic interference may cause by high-power components. Due to a large number of components and their complexity, reliability and safety reduce for which extra precautions are taken.

20.4 FUEL CELL ELECTRIC VEHICLE (FCEV)

FCEVs work on DC voltage generated by an on-board fuel cell assembly using hydrogen as their fuel or energy source. Hydrogen is stored in the fuel tank of the vehicle or it can be extracted from another fuel by a reformer. The electrical energy generated from the fuel cells can directly power the DC motor. Some of the energy can also be stored in battery pack. This stored energy can be utilized when extra power demand is required. It also supplies power to vehicle's accessories. FCEVs may also have regenerative braking system for charging the battery. An external source of energy is not required to charge the storage device.

A single fuel cell does not produce much power. Large number of fuel cells are combined together to get the amount of power needed to propel the vehicle. This assembly is called fuel cell stack.

Oxygen from air and hydrogen from the fuel tank are supplied continuously to fuel cells to generate electricity. The energy conversion in fuel cell is more efficient than that in internal combustion engine (ICE). In ICE combustion takes place, where the process is irreversible and electron exchange is uncontrolled. In fuel cells electron exchange is controlled electrochemically by using electrochemical device. The chemical energy is directly converted to electrical energy isothermally without undergoing the combustion process.

20.4.1 Fuel Cell

The construction and working principle of a hydrogen fuel cell (PEM type) is illustrated in Figure 20.9. A fuel cell produces electricity through an electrochemical reaction that combines hydrogen and oxygen to produce water and heat. A fuel cell consists of two electrodes: a negatively charged electrode (anode) and a positively charged electrode (cathode). They are coated with a catalyst and separated by an electrolyte. The catalyst used is normally platinum which promotes the chemical reaction in the fuel cell. It does not take part in chemical reaction, so it is not consumed. The electrolyte is most often a polymer membrane, called proton or ion exchange membrane. The polymer membrane serves as an electric insulator.

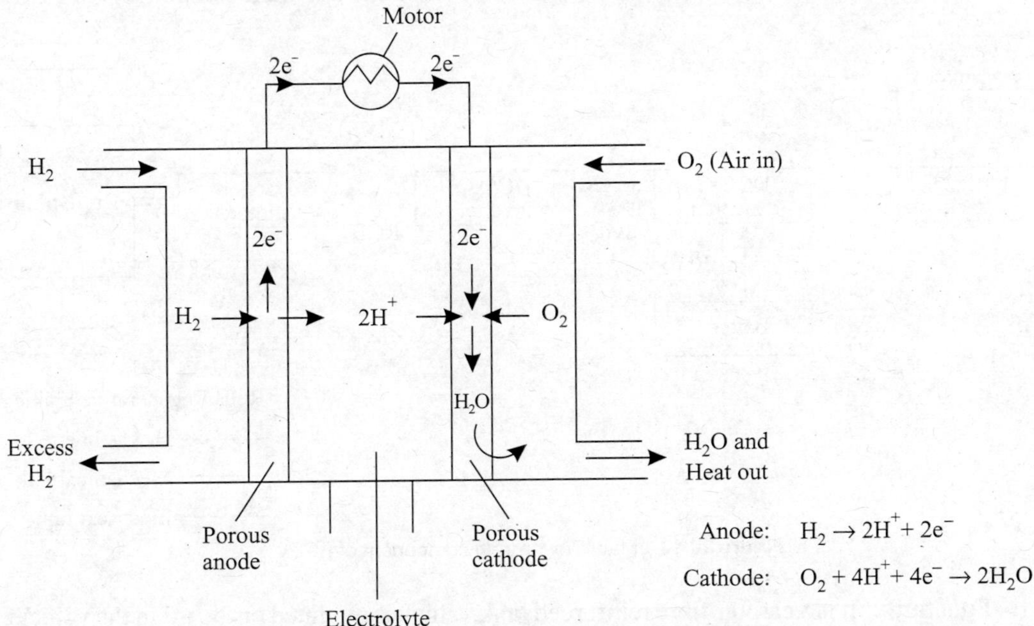

Figure 20.9 Hydrogen-oxygen fuel-cell (PEM type).

The working principle of fuel cell is the reverse of electrolysis. In electrolysis water molecules are separated into hydrogen and oxygen atoms by passing electric current through an electrolyte placed between two electrodes. In a fuel cell hydrogen and oxygen are fed into it from outside. Hydrogen fuel is stored in the fuel tank, it is fed into the cell from a pressurized tank and oxygen is taken from air. Air compressor may be used to deliver the oxygen. Catalyst is used to combine the hydrogen fuel with oxygen. At the surface of anode hydrogen is ionized and each hydrogen atom emits a negatively charged electron and a positively charged hydrogen ion (H^+). Electron cannot pass through the membrane. It moves towards anode. Anode which is rich in electrons, compared to cathode, is represented as negatively charged electrode. Free electrons then flow from anode to cathode through an external circuit and electricity is produced to drive the motor or generator. Hydrogen ions, H^+ (protons) flow through the electrolyte and reaches to the surface of cathode, where they combine with oxygen and free electrons to form water vapour. Some of this water vapour is used to humidify incoming hydrogen and oxygen that keeps the ion exchange membrane moist. The remaining water goes to the atmosphere. Some heat is also generated by the fuel cell. It can be captured and utilized to heat the fuel cell and passenger compartment.

In hydrogen fuel cell vehicles where hydrogen is directly used, the tailpipe emission is only water vapour. It is pollutant free. However, if hydrogen is extracted from a fuel by a reformer, some pollutants may generate.

20.4.2 Components of FCEV

The main power-train components of a fuel cell electric vehicle (FCEV) are shown in Figure 20.10. Many components of FCEV are similar to BEVs. Following are the main components of FCEVs:

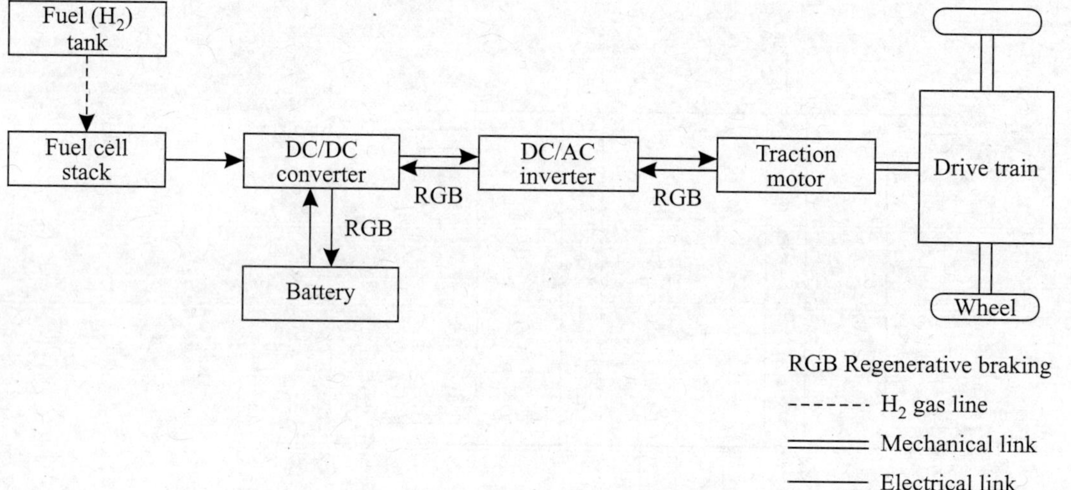

Figure 20.10 Main Power-train component of FCEV.

1. **Fuel tank:** It is a carbon-fibre reinforced tank, which is mounted on-board in the vehicle. It stores hydrogen gas at high pressure. Hydrogen is fed into the fuel cell from this pressurized tank.
2. **Fuel cell stack:** It is an assembly of large number of fuel cells that combines hydrogen and oxygen to produce electricity. For efficient operation of fuel cells following additional systems are used:
   (i) *Air-supply system*: An air compressor is used to draw air from atmosphere and supplies oxygen to the fuel cells.
   (ii) *Humidification system*: For efficient operation of the fuel cell, the electrolyte membrane must be kept moist. Therefore some of the water produced during chemical reaction in the fuel cell is re-circulated to humidify the incoming hydrogen and oxygen.
   (iii) *Thermal system*: Fuel cells operate best within a certain temperature range. Thermal system helps to maintain the temperature within that range. It also controls the temperature of other components.

3. **Storage battery:** Lithium-ion battery pack can be used on-board to store the energy either from fuel cell or from regenerative braking. Battery is not charged from external electrical source or grid.
4. **DC/DC converter and inverter:** The DC/DC converter converts higher voltage DC power from fuel cell or battery pack to low voltage DC power required to operate different accessories. Inverter converts DC power to AC power required to run traction motor.
5. **Traction motor and transmission:** Traction motor is driven by the electrical energy received from fuel cells and the motor drives the wheel through transmission system.
6. **Exhaust:** Water vapour is exhausted through tailpipe of the exhaust system of the vehicle.

20.4.3 Advantages and Disadvantages of FCEV

Following are the advantages and disadvantages of FCEV:

Advantages

1. Chemical reactions for energy conversion in fuel cells take place more efficiently than the combustion process in ICEs.
2. The performance of FCEV is excellent. It generates high torque at low speed and accelerates the vehicle quickly.
3. Hydrogen fuel can be filled in the fuel tank generally within five minutes, whereas BEVs require much longer time to recharge the battery, which may take several minutes.
4. Driving range of FCEVs is much longer than that obtained from BEVs.
5. The only exhaust is water vapour from hydrogen FCEVs. It does not pollute the environment.
6. Hydrogen FCEVs do not generate greenhouse gases, such as carbon dioxide. So, it does not contribute to harmful global warming.
7. FCEV does not produce noise pollution. It sounds just like BEV. It is much quieter in operation as compared to ICE.
8. Hydrogen can be produced from sources available in the country. Our dependency on oil import can be reduced.

Disadvantages

1. Initial cost of fuel cell vehicles is higher in comparison to other type of vehicles. It is because of the precious metal, platinum is required as a catalyst in the fuel cells. Electrolytes used are also expensive.
2. Immediate adoption of FCEV is not possible for public transport as the infrastructure for supply of hydrogen fuel is not developed. It can be adopted in future.
3. Production, transportation and storage of hydrogen fuel are complex and expensive in comparison to other conventional fuels.
4. The running cost of FCEVs is higher than that of BEVs.
5. Although there is no air pollution directly from the FCEV, but hydrogen production requires electricity and if this electricity is produced by using fossil fuels, air is likely to be polluted. A renewable source of producing electricity can be adopted to eliminate pollution.
6. In some temperature and humidity range FCEVs may not be as durable and reliable as ICEs.
7. Fuel cells become very hot while they operate. PEM cells require large radiator, that means more space is required in the vehicle.

20.5 TYPES OF FUEL CELLS

There are several types of fuel cells currently under development, each with its own advantages, limitations and potential applications. Different types of fuel cells are:

1. Polymer Electrolyte Membrane (PEM) Fuel Cell
2. Direct Methanol Fuel Cell (DMFC)
3. Alkaline Fuel Cell (AFC)
4. Solid Oxide Fuel Cell (SOFC)
5. Molten Carbonate Fuel Cell (MCFC)
6. Phosphoric Acid Fuel Cell (PAFC)

20.5.1 Polymer Electrolyte Membrane (PEM) Fuel Cell

Polymer electrolyte membrane (PEM) fuel cell is also called as Proton exchange membrane fuel cell. Figure 20.9 is the representation of this type of cell. Protons (H^+) flow from anode to cathode through a solid polymer electrolyte. Electrons flow through external circuit to produce electricity. At the cathode these protons combine with oxygen and electrons to produce water and heat. In order to promote chemical reactions platinum catalyst is used in porous carbon electrodes. The following electrochemical reactions occur in the PEM fuel cells:

$$\text{Anode reaction:} \quad 2H_2 \rightarrow 4H^+ + 4e^-$$

$$\text{Cathode reaction:} \quad O_2 + 4H^+ + 4e^- \rightarrow 2H_2O$$

$$\text{Net cell reaction:} \quad 2H_2 + O_2 \rightarrow 2H_2O$$

Solid polymer membrane warms-up quickly to activate fuel cells. For efficient operation of fuel cells they should operate relatively at low temperatures between 30 to 100°C. The best operating temperature is around 80°C. Because of low temperature operation, they are more durable and reliable. Fuel cells are quite compact and capable of producing high power output. The energy conversion of this cell is 40–60 %.

The cost of this type of cell is relatively high as it uses precious metal, platinum as a catalyst. In the presence of even a small quantity of CO, platinum gets poisoned and loses its activity. CO is likely to enter the cell with atmospheric air. Electrolyte membrane needs to be moist for which proper management of re-circulation of exhaust water is required. Thermal management is also required for maintaining the proper temperature of the cell.

This type of fuel cell is used most commonly in FCEVs. The vehicle speed can be controlled by controlling the power output from the fuel cells by varying the supply of hydrogen to the cell. PEM fuel cells can also be used for stationary and portable electric power generation.

20.5.2 Direct Methanol Fuel Cell (DMFC)

In this fuel cell methanol is used as a fuel to generate electricity. Polymer membrane, similar to the PEM cell, is used as an electrolyte. This cell uses a thin membrane lightly covered with a layer of platinum catalyst. A mixture of methanol and water is introduced into the cell at the anode. Methanol takes very little energy to release hydrogen. It is considered to be an ideal hydrogen carrier. Methanol is oxidized at the anode to form carbon dioxide (CO_2), protons (H^+) and electrons

(e^-). Protons flow through the electrolyte to the cathode. Electrons travel through the external circuit to generate electricity. Oxygen is introduced at the cathode from outside, where oxygen is reduced to water by gaining protons and electrons. The following electrochemical reactions take place in DMFC.

$$\text{Anode reaction: } 2CH_3OH + 2H_2O \rightarrow 2CO_2 + 12H^+ + 12e^-$$

$$\text{Cathode reaction: } 3O_2 + 12H^+ + 12e^- \rightarrow 6H_2O$$

$$\text{Net cell reaction is: } 2CH_3OH + 3O_2 \rightarrow 4H_2O + 2CO_2$$

Liquid ethanol is easier to store than hydrogen. It does not require pressurized fuel tank and it has much higher energy density than compressed hydrogen. The cells are light weight and compact. The cost of cells is lower because less platinum is needed compared to PEM fuel cell.

The operating temperature of DMFCs is in the range of 50–120°C. The energy conversion efficiency of this cell is 30–40% and response time for its activation is more in comparison to PEM fuel cells. This is because some of the methanol is likely to pass through the membrane. DMFCs also produce carbon dioxide as emission, which is not produced by other type of fuel cells.

This type of cells can be used for off-grid and portable operations like laptops and cell-phones. In order to power any of this equipment small size fuel cartridge filled with methanol can be installed within that and it can easily be connected to the fuel cells. The fuel cartridge is powerful and long lasting.

20.5.3 Alkaline Fuel Cell (AFC)

All fuel cells have two electrodes and they are separated by an electrolyte. Normally electrolytes are different in different fuel cells. The electrolyte used in AFC is a porous matrix saturated with aqueous alkaline solution of potassium hydroxide (KOH). The two electrodes are coated with a thin layer of catalyst. The operating temperature of this cell is from ambient temperature to 90°C. At this low temperature highly active catalyst is required, such as platinum.

Hydrogen as a fuel is introduced to the anode side of fuel cell. At anode, hydrogen combines with hydroxyl-ions (OH^-) to form water and electrons. Hydroxyl-ions are produced at the cathode and flows from cathode to anode through electrolyte. Electrons flow from anode to cathode through an external circuit and produce electricity. At cathode side of the cell, pure oxygen is introduced which combines with water and electrons and produces hydroxyl-ions (OH^-). It receives electrons from the external circuit and water by recirculation of the exhausted water from the anode. Overall AFC consumes hydrogen and pure oxygen and produces water, heat and electricity. Following are the electrochemical reactions taking place in AFC:

$$\text{Anode reaction: } 2H_2 + 4OH^- \rightarrow 4H_2O + 4e^-$$

$$\text{Cathode reaction: } O_2 + 2H_2O + 4e^- \rightarrow 4OH^-$$

$$\text{Net cell reaction: } 2H_2 + O_2 \rightarrow 2H_2O$$

AFCs are most developed fuel cells. They are highly powerful and efficient. They can be operated instantaneously without pre-heating even at very low temperatures. They are also resistant to humidity. The efficiency of this fuel cell is more than 60%.

The most significant disadvantage of this type of cell is that it requires pure oxygen. Normally, oxygen is taken from atmospheric air and the air always contains some carbon dioxide (CO_2).

The presence of CO_2 even in traces can combine with KOH to form potassium carbonate which degrades the electrolyte and reduces overall operation of the fuel cell. This fuel cell is not suitable for automobiles because of its high cost.

The AFC is the oldest design used by NASA in space shuttles to power electrical systems. The exhaust water from this cell is used as a drinking water in the spacecraft.

20.5.4 Solid Oxide Fuel Cell (SOFC)

The electrodes of SOFCs are made up of ceramic materials. Electrolyte used is a solid, non-porous ceramic material, called Yttria-Stabilized Zirconia (YSZ). The cell does not become active until it reaches a temperature around 800 to 1000°C. Hydrogen is used as a fuel. It is fed through anode, where it combines with negatively charged oxygen ion (O^-) and produces water and electrons. Water is exhausted and electrons move from anode to cathode through external circuit to produce electricity. Oxygen is fed through the cathode, where it combines with electrons to produce oxygen ion (O^-). The oxygen ion flows from cathode to anode through electrolyte. Following electrochemical reactions take place in the cell:

$$\text{Anode reaction:} \quad 2H_2 + 2O^{2-} \rightarrow 2H_2O + 4e^-$$
$$\text{Cathode reaction:} \quad O_2 + 4e^- \rightarrow 2O^{2-}$$
$$\text{Net cell reaction:} \quad 2H_2 + O_2 \rightarrow 2H_2O$$

SOFCs operate at very high temperature and they do not require precious metal catalyst for activation. This reduces the cost of the fuel cell. Because of the high temperature many fuel can be oxidized, which includes methanol and bio-fuel. These cells have high efficiency of the order of 60% and they are long lasting.

A high operating temperature also imposes some problems. It requires thermal protection to other components of the cell. Its response time for varying load is slow. Because of the high temperature, the cell is not suitable to drive the traction motor of the vehicle. However belt driven alternator in ICE can be replaced by this cell to recharge the battery and to power other electrically driven accessories.

20.5.5 Molten Carbonate Fuel Cell (MCFC)

The anode is made from an alloy of highly sintered nickel and chromium powder and cathode is an alloy of nickel oxide and lithium. The electrolyte used is a molten carbonate salt, typically a mixture of potassium carbonate and lithium carbonate. This electrolyte enables to transfer carbonate ions (CO_3^{2-}) from cathode to anode. MCFCs operate at temperatures over 800°C. Hydrogen as a fuel is supplied to the anode, where it combines with carbonate ions to form carbon dioxide, water and electrons. Electrons flow through an external circuit from anode to cathode and generate electricity. Carbon dioxide is re-circulated from anode to cathode and water is exhausted. Oxygen is supplied to the cathode, where it combines with carbon dioxide and electrons to form carbonate ions. These ions travel from cathode to anode through the electrolyte. Following electrochemical reactions take place in MCFC:

$$\text{Anode reaction:} \quad 2H_2 + 2CO_3^{2-} \rightarrow 2CO_2 + 2H_2O + 4e^-$$
$$\text{Cathode reaction:} \quad O_2 + 2CO_2 + 4e^- \rightarrow 2CO_3^{2-}$$
$$\text{Net cell reaction:} \quad 2H_2 + O_2 \rightarrow 2H_2O$$

MCFC operates at high temperature, so it has got ability to extract hydrogen from coal based fuel, natural gas and other hydrocarbons without the use of external reformer. It does not use precious metals as catalyst, so it is not expensive to manufacture. The energy efficiency of this cell is high, it is almost 60%.

Because of high operating temperature of the cell, it imposes some problems. It takes more time to reach this temperature and its response time with changing demand of electricity is slow. The electrolyte is of corrosive nature, it reduces the cell life.

This type of fuel cell is not suitable for automobiles. It is best suited for the generation of electricity for industrial purpose where coal, natural gas or bio-gases can be used as a fuel.

20.5.6 Phosphoric Acid Fuel Cell (PAFC)

In this fuel cell electrodes are made up of porous graphite substrate, where platinum particles are dispersed as a catalyst. Electrolyte used is the concentrated phosphoric acid (H_3PO_4), which is contained in a Teflon bounded silicon carbide matrix. The operating temperature of this cell is between 150 to 200°C. Hydrogen is fed through the anode where it breaks up in protons (Hydrogen ions, H^+) and electrons (e^-). Electrons flow from anode to cathode through an external circuit to generate electricity. Protons (H^+) pass through the electrolyte from anode to cathode. Oxygen is fed through the cathode where it combines with hydrogen ions and electrons to form water, which is exhausted into the atmosphere. Following electrochemical reactions take place in PAFC:

Anode reaction: $2H_2 \rightarrow 4H^+ + 4e^-$

Cathode reaction: $O_2 + 4H^+ + 4e^- \rightarrow 2H_2O$

Net cell reaction: $2H_2 + O_2 \rightarrow 2H_2O$

Atmospheric air can be used for the supply of oxygen. The presence of CO_2 in air does not impose any problem with acidic electrolyte. The catalyst used in PAFCs is platinum, which is a precious metal, so they are costly. The efficiency of the fuel cell is low. It can operate at 37 to 42% efficiency. This cell is likely to corrode the components of the cell with time because of acidic nature of the electrolyte. This type of fuel cell is not suitable for automobiles. It works fine where demand for power output is constant.

REVIEW QUESTIONS

1. How does a battery electric vehicle work (BEV)? Draw a schematic diagram of a BEV. What are the main components of BEV?
2. Describe the functions and construction of a traction battery pack in EVs. Why a battery management system is required?
3. What are the functions of a power inverter and a controller in EVs?
4. What are the functions of an electric traction motor? Explain typical power-torque-speed characteristics of an electric traction motor.
5. Give a comparison of the torque produced by ICE and electric motor.
6. What are the different types of electric motors used in EVs?
7. What are the different methods of charging the battery?

8. Describe the functions of the following components of EVs:
   (a) DC/DC converter
   (b) Auxiliary battery
   (c) Thermal system
   (d) Charger port
   (e) Transmission
9. How does an energy management system monitor and control the functions of electric vehicle? Explain with the help of a block diagram.
10. Explain the regenerative braking system in electric vehicles.
11. What are the advantages, disadvantages and safety issues of BEVs?
12. What are hybrid electric vehicles? Why do we need hybrid vehicles?
13. Why the plug-in hybrid vehicle is more beneficial than simple HEV?
14. What are the features and benefits of a photovoltaic HEV?
15. What are the features and benefits of a fuel cell hybrid?
16. Describe the series configuration of HEV. Show the drive train of this configuration. What are its advantages and disadvantages?
17. Describe the parallel configuration of HEV. Show the drive train of this configuration. What are its advantages and disadvantages?
18. Describe the series-parallel configuration of HEV. Show the drive train of this configuration. What are its advantages and disadvantages?
19. Describe briefly the hybrid electric vehicle's technology.
20. What are the advantages and disadvantages of HEVs?
21. How does a fuel cell electric vehicle (FCEV) work?
22. What is a fuel cell? Describe its construction and working.
23. What are the main components of a FCEV? Draw a block diagram of the main power-train components of FCEV.
24. What are the advantages and disadvantages of FCEVs?
25. What are the different types of fuel cells commonly used?
26. Describe the construction and operation of a PEM fuel cell. What are its advantages, disadvantages and uses?
27. Describe the construction and operation of a DMFC fuel cell. What are its advantages, disadvantages and uses?
28. Describe the construction and operation of an AFC fuel cell. What are its advantages, disadvantages and uses?
29. Describe the construction and operation of a SOFC fuel cell. What are its advantages, disadvantages and uses?
30. Describe the construction and operation of a MCFC fuel cell. What are its advantages, disadvantages and uses?
31. Describe the construction and operation of a PAFC fuel cell. What are its advantages, disadvantages and uses?

Bibliography

Adams, W.H., H.G Hinrichs, F.F. Pischinger, P. Adamis, V. Schumacher, and P. Walzer, "Analysis of the Combustion Process of a Spark Ignition Engine with a Variable Compression Ratio", SAE Paper 870610, 1987.

Agrawal, S.K., *Internal Combustion Engines*, New Age International, New Delhi, 2000.

Annand, W.J.D., "Heat Transfer in the Cylinders of Reciprocating IC Engines", Proc. I. Meek E., Vol. 177, No. **36**, 1963.

Baker, R.E., E.E. Daby, and J.W. Pratt, "Selecting Compression Ratio for Optimum Fuel Economy with Emission Constraints", SAE Paper 770191, 1977.

Balagurunathan, K., S. Sriram, and G Sakthinathan, "Ceramic Filters for Particulate Emission Control in DI Diesel Engine Exhaust", *Proc. of the XVI National Confernece on IC Engines and Combustion*, Narosa Publishing House, New Delhi, 2000.

Balagurunathan, K., P. Kumar, and M. Rajmohan, "Measurement of Particulate by Dilution Tunnel", *Proc. of the XVI National Conference on IC Engines and Combustion*, Narosa Publishing House, New Delhi, 2000.

Baruah, PC, R.S. Benson, and H.N. Gupta, "Performance and Emission Predictions for a Multi-cylinder Spark-Ignition Engine with a Catalytic Converter", SAE Paper 780672, 1978.

Benson, R.S. and PC. Baruah, "A Generalized Calculation for an Ideal Otto Cycle with Hydrocarbon-Air Mixture Allowing for Dissociation and Variable Specific Heats", *I. Mech. E and UMIST, 1976, IJMEE*, Vol. 4, No. **1.**

Benson, R.S. and N.D. Whitehouse, *Internal Combustion Engines*, Vol. 1 and 2, Pergamon Press, Oxford, 1979.

Benson, R.S., *Advanced Engineering Thermodynamics*, Pergamon Press, Oxford, 1977.

Benson, R.S., *The Thermodynamics and Gas Dynamics of Internal Combustion Engines*, Vol. 1 and 2 in: J.H. Horlock and D.E. Winterbone (Eds.) Vol. 2, Clarendon Press, Oxford, 1982.

Bose, P.K., K. Raju, and R. Chakraborty, "Effect of Ethanol and Methanol Blends on Performance and Emission Characteristics of a Four-Cylinder Four-Stroke SI Engine", *Proc. of the XVI National Conf. on IC Engines and Combustion*, Narosa Publishing House, New Delhi, 2000.

Bracco, F.V. and W.A. Sirignano, "Theoretical Analysis of Wankel Engine Combustion", *Combustion Science and Technology*, 1973.

Burghardt, M.D., *Engineering Thermodynamics with Applications*, Harper and Row, New York, 1978.

Burman, P.G and F. Deluca, *Fuel Injection and Controls for Internal Combustion Engines*, U.S.A., 1962.

Campbell, A.S., *Thermodynamic Analysis of Combustion Engines*, John Wiley, New York, 1979.

Das, L.M., "Hydrogen Engines: A View of the Past and a Look into the Future", *Int. J. Hydrogen Energy*, Vol. 15, No. **6**, 1990.

Das, L.M., "Fuel Induction Technique for a Hydrogen Operated Engine", *Int. J. Hydrogen Energy*, Vol. 15, No. **11**, 1990.

Deluchi, M.A., R.A. Johnston, and D. Sperling, "Methanol vs. Natural Gas Vehicles: A Comparison of Resources Supply, Performance, Emission, Fuel Storage, Safety, Costs, and Transitions", SAE Paper 881656, 1988.

Dhandapani, S. and Y. Robinson, "Experimental Investigation on Semi-direct Electronic Injection in Four-Stroke SI Engine by Virtual Instrumentation", *Proc. of the XVIII National Conf. on IC Engines and Combustion*, Combustion Institute (Indian Section), 2003.

Eastop, T.D. and A. McConkey, *Applied Thermodynamics for Engineering Technologists*, Longman Scientific & Technical, 1987.

Edwards, J.B., *Combustion Formation and Emission of Trace Species*, Ann Arbor, MI, 1974.

Erjavec, J., " Hybrid Electric and Fuel-Cell Vehicles", Delmar Cengage Learning, second edition, 2013.

Ferguson, C.R. and A.T. Kirkpatrick, *Internal Combustion Engines Applied Thermosciences*, 2nd ed., John Wiley & Sons, Inc., 2004.

Fleming, R.D and D.B. Eccleston, "The Effect of Fuel Composition, Equivalence Ratio, and Mixture Temperature on Exhaust Emissions", SAE Paper 710012, 1971.

Furuhama, S. and Y. Enomoto, "Heat Transfer into Ceramic Combustion Wall of Internal Combustion Engines", SAE Paper 870153, 1987.

Gajendra Babu, M.K., "Gasoline Fuel Injection Systems for Spark Ignition Engines", *Proc. of the XV National Conf. on IC Engines and Combustion*, Allied Publishers, New Delhi, 1997.

Ganesan, V., *Internal Combustion Engines*, Tata McGraw-Hill, New Delhi, 2000.

Gill, P.W., J.H. Smith, and E.J. Ziurys, *Fundamentals of Internal Combustion Engines*, Oxford & IBH Publishing Co., New Delhi, 1952.

Giri, N.K., *Automobile Mechanics,* Khanna Publishers, New Delhi, 1996.

Gorille, I., N. Rittmannsberger, and P. Werner, "Bosch Electronic Fuel Injection with Closed Loop Control", SAE Paper 750368, 1975.

Green, R.K. and N.D. Glasson, "High-Pressure Hydrogen Injection for Internal Combustion Engines", *Int. J. Hydrogen Energy*, Vol. 17, No. **11**, 1992.

Greene, A.B. and GG Lucas, "*The Testing of Internal Combustion Engines*", The English Universities Press Ltd., London, 1969.

Gupta, H.N., "Supercharging of IC Engines", *Indian and Eastern Engineer*, May 1973.

Gupta, H.N., "Control of Air-Pollution from Motor Vehicles", *Indian and Eastern Engineer*, April 1981.

Gupta, H.N., "Experimental Investigation of Fuel Maldistribution in a Multi-Cylinder Spark-Ignition Engine", *The J. of Thermal Engineering*, Indian Society of Mech. Engineers, Vol. 2, No. **2**, 1981.

Gupta, H.N., "Techniques for the Measurement of Instantaneous Pressure Inside the Cylinder, Intake and Exhaust Systems of an Internal Combustion Engine", *Journal of the Institution of Engineers* (India), Vol. 72, Jan. 1992.

Gupta, H.N. and L.R. Govil, "Study of Detonation with a Non-Conventional Fuel in SI Engine", *Proc. of the XIV National Conf. on IC Engines and Combustion*, Tata McGraw-Hill, New Delhi, 1995.

Gupta, H.N. and A.K. Rai, "The Effect of Exhaust Gas Recirculation on Performance and Emission of Four-Stroke Single Cylinder SI Engine", *Proc. of the XV National Conf. on IC Engines and Combustion*, Allied Publishers, New Delhi, 1997.

Gupta, H.N. and G Prasad, "Performance and Emission Prediction for a Natural Gas Fuelled Spark-Ignition Engine", *Proc. of the XVI National Conf. on IC Engines and Combustion*, Narosa Publishing House, New Delhi, Jan. 2000.

Gupta, H.N., "Effect of Flame Speed and Compression Ratio on the Performance and Emission Characteristics of Lean Burn SI Engine", *Journal of the Institution of Engineers* (India), Vol. 82, April 2001.

Gupta, H.N. and Anoop Gupta, "Simulation of Effect of Water Injection on the Performance of Hydrogen Fuelled SI Engine", *Proc. of the XVII National Conf. on IC Engines and Combustion*, Tata McGraw-Hill, New Delhi, Dec. 2001.

Gupta, H.N. and R. Rai, "Performance and Emission Prediction for a Natural Gas and Hydrogen Fueled Spark-Ignition Engine", *Proc. of the XVIII National Conf. on IC Engines and Combustion*, Combustion Institute (Indian Section), Dec. 2003.

Gupta, H.N. and P. Kalita, "Mathematical Modelling of SI Engine with Catalytic Converter", *Proc. of the XVIII National Conf. on IC Engines and Combustion*, Combustion Institute (Indian Section), Dec, 2003.

Gupta, K.M., *Automobile Engineering*, Vols. 1 and 2, Umesh Publications, 2001.

Haddad, S.D. and N. Watson, *Principles and Performance in Diesel Engineering*, Ellis Horwood, England, 1984.

Haynes, J.H., *Owner's Workshop Manual*, J.H. Haynes and Company, England, 1971.

Heinz, H., *Advanced Engine Technology*, Edward Arnold, 1994.

Heitner, J., *Automotive Mechanics—Principles and Practices*, D. Van Nostrand, New York, 1953.

Heywood, J.B., *Internal Combustion Engine Fundamentals*, McGraw-Hill, New York, 2000.

Homan, H.S., P.C.T. De Boer, and W.J. McLeant, "The Effect of Fuel Injection on NO_x Emissions and Undesirable Combustion for Hydrogen-Fuelled Piston Engines", *Int. J. Hydrogen Energy*, Vol. 8, No. **2**, 1983.

Jaaskelainen, H.E. and J.S. Wallace, "Performance and Emissions of a Natural Gas-Fuelled 16 Valve DOHC Four-Cylinder Engine", SAE Paper 930380, 1993.

Jaaskelainen, H.E. and J.S. Wallace, "Examination of Charge Dilution with EGR to Reduce NO_x Emissions from a Natural, Gas-Fuelled 16 Valve DOHC Four-Cylinder Engine," SAE Paper 942006, 1994.

Jones, J.B. and R.E. Dugan, *Engineering Thermodynamics*, Prentice-Hall of India, New Delhi, 1998.

Judge, A.W., *Carburettors and Fuel Injection Systems*, The English Language Book Society and Chapman and Hall, England, 1965.

Judge, A.W, *Modem Petrol Engines*, Chapman and Hall, England, 1965.

Judge, A.W., *Small Gas Turbines and Free Piston Engines*, Chapman & Hall, England, 1960.

Judge, A.W., *High Speed Diesel Engines*, Chapman & Hall, England, 1948.

Kalghatgi, GT, "Spark Ignition, Early Flame Development and Cyclic Variation in IC Engines", SAE Paper 870163, 1987.

Keck, J.C. and J.B. Heywood, "Early Flame Development and Burning Rates in Spark Ignition Engines and their Cyclic Variability", SAE Paper 870164, 1987.

Kempton, W., S.E. Letendre, "Electric Vehicles as a new power source for electric utilities", Transportation Research, 1997, Elsevier.

Kilmistra, J., "Performance of Lean-Burn Natural-Gas-Fuelled Engines—On Specific Fuel Consumption Power Capacity and Emissions", SAE Paper 901495, 1990.

Ladommatos, N. and R. Stone, "Conversion of a Diesel Engine for Gaseous Fuel Operation at High Compression Ratio", SAE Paper 910849, 1991.

Lancaster, D.R., R.B. Krieger, S.C. Sorenson, and W.L. Hull, "Effects of Turbulence on Spark Ignition Engine Combustion", SAE Paper 760160, 1976.

Lichty, L.C., *Internal Combustion Engines*, McGraw-Hill, New York, 1951.

Lichty, L.C., *Combustion Engine Processes*, McGraw-Hill, New York, 1967.

Li, J., W. Yang, D. Zhou, " Review on the management of RCCI engines", Renewable and Sustainable Energy Reviews, 2017, Elsevier.

Lucia, U., "Overview on fuel cells", Renewable and Sustainable Energy Reviews, 2014 , Elsevier.

Maleev, V.L., *Internal Combustion Engines*, McGraw-Hill, New York, 1945.

Marsee, F.J., R.M. Olree, and WE. Adams, "Compression Ratio Effects with Lean Mixtures", SAE Paper 770640, 1977.

Mathur, H.B., "Alternate Fuels for Automobiles: Properties, Problems and Possibilities", *Proc. of the XVI National Conf. on IC Engines and Combustion*, Narosa Publishing House, New Delhi, 2000.

Mathur, M.L. and R.P. Sharma, *A Course in Internal Combustion Engines*, Dhanpat Rai Publications, New Delhi, 1998.

Matsumoto, K., T. Inoue, K. Nakanishi, and T. Okumura, "The Effects of Combustion Chamber Design and Compression Ratio on Emissions, Fuel Economy and Octane Number Requirement", SAE Paper 770193, 1977.

Mehta, P.S., "Predictions of Combustion Chamber Geometry Effects in Direct Injection Diesel Engines", *Proc. of the XVIII National Conf. on IC Engines and Combustion*, Combustion Institute (Indian Section), Dec. 2003.

Meier, F, J. Kohler, W. Stolz, W.H. Bloss, and M. Al-Garni, "Cycle-Resolved Hydrogen Flame Speed Measurement with High Speed Schlieren Technique in a Hydrogen Direct Injection SI Engine", SAE Paper 942036, 1994.

Meyer, R.C., J.J. Cole, E. Klenzle, and A.D. Wells, "Development of a CNG Engine", SAE Paper 910881, 1991.

Moore, C.H. and T.L. Hoehne, "Combustion Chamber Insulation Effect on the Performance of a Low Heat Rejection Cummins V-903 Engine", SAE Paper 860317, 1986.

Morel, T, R. Keribar, P.N. Blumberg, and E.F. Fort, "Examination of Key Issues in Low Heat Rejection Engines", SAE Paper 860316, 1986.

Muranaka, S., Y. Takagi, and T. Ishida, "Factors Limiting the Improvement in Thermal Efficiency of SI Engine at Higher Compression Ratio", SAE Paper 870548, 1987.

Nag, P.K., *Engineering Thermodynamics*, Tata McGraw-Hill, New Delhi, 1981.

Noguchi, M., S. Sanda, and N. Nakamura, "Development of Toyota Lean Burn Engine", SAE Paper 760757, 1976.

Obert, E.F., *Internal Combustion Engines and Air Pollution*, Harper & Row, New York, 1973.

Parikh, PP., P.K. Banerjee, and S. Veerkar Shashikantha, "Design Development and Optimization of a Spark Ignition Producer-gas Engine", *Proc. of the XIV National Conf. on IC Engines and Combustion*, Tata McGraw-Hill, New Delhi, 1995.

Patterson, D.J. and N.A. Henein, "Emissions from Combustion Engines and Their Control", Ann Arbor, MI, 1974.

Patton, K.J., R.G Nitschke, and J.B. Heywood, "Development and Evaluation of a Friction Model for Spark Ignition Engines", SAE Paper 890836, 1989.

Prakash, G, A. Ramesh, and A.B. Shaik, "An Approach for Estimation of Ignition Delay in a Dual Fuel Engine", SAE Paper 1999-01-0232.

Pulkrabek, W.W., *Engineering Fundamentals of the Internal Combustion Engine*, Prentice-Hall of India, New Delhi, 2003.

Pye, D.R., *The Internal Combustion Engine*, Vol. 1, Oxford at the Clarendon Press, 1937.

Quader, A.A., "What Limits Lean Operation in Spark Ignition Engines—Flame Initiation or Propagation?", SAE Paper 760760, 1976.

Ricardo, H.R. and J.GG Hempson, "The High Speed Internal Combustion Engine", The English Language Book Society and Blackie & Son, England, 1969.

Rogers, G.F.C. and Y.R. Mayhew, *Engineering Thermodynamics Work and Heat Transfer*, Longman, London, 1976.

Rogowski, A.R., *Elements of Internal Combustion Engines*, McGraw-Hill, New York, 1953.

Samaga, B.S. and B.S. Murthy, "On the Problem of Predicting Burning Rates in a Spark Ignition Engine", SAE Paper 750688, 1975.

Samria, N.K., M. Vittalaiah, and A.A. Hadi, "Temperature Distribution and Thermal Stress Analysis in Diesel Engine Piston by Finite Element Method", *Proc. of the XIV National Conf. on IC Engines and Combustion*, Tata McGraw-Hill, New Delhi, 1995.

Sapre, A.R., "Properties, Performance and Emissions of Medium Concentration Methanol-Gasoline Blends in a Single-Cylinder Spark-Ignition Engine", SAE Paper 881679, 1988.

Sellnau, M., M. Foster, W. Moore, K. Hoyer, J. Sinnamon, and B. Clemm, "Advancement of Gasoline Direct Injection (GDCI) for US 2025 CAFÉ and Tier 3 Emissions",June 14, 2017 ERC Symposium.

Sharma, J.K., "Prediction of Rate of Injection Pattern for Diesel Fuel Injection System with Pintle Nozzle", *J. of the Institution of Engineers* (India), Vol. 71, 1990.

Taylor, C.F., The *Internal Combustion Engine in Theory and Practice*, Vol. 1, The Technology Press of the MIT, Cambridge, MA, and John Wiley & Sons, Inc., New York, 1959.

Taylor, C.F., *The Internal Combustion Engine in Theory and Practice*, Vol. 2, The MIT Press, Cambridge, MA, 1968.

Tirkey, J.V., H.N. Gupta, " Computer Simulation and Combustion diagnostics of multicylinder 4-stroke SI Engine using CNG as a fuel", *International Journal of Applied Engineering Research*, Vol. 3, No. 5, 2008.

Tiwari, S.B., H.N. Gupta, and N.K. Samria, "Study of Heat Transfer and Combustion Model in SI Engine by Simulation and Comparison with Experimental Data", *Proc. of the XVI National Conf. on IC Engines and Combustion*, Narosa Publishing House, New Delhi, Jan. 2000.

Turns, S.R., *An Introduction to Combustion—Concepts and Applications*, McGraw-Hill, New York 1996.

Weaver, C.S., "Natural Gas Vehicles—A Review of the State-of-the-Art", SAE Paper 892133, 1989.

Wentworth, J.T., "Effects of Combustion Chamber Shape and Spark Location on Exhaust Nitric Oxide and Hydrocarbon Emissions", SAE Paper 740529, 1974.

Whitney, K.E. and B.K. Bailey, "Determination of Combustion Products from Alternative Fuels—Part I: LPG and CNG Combustion Products", SAE Paper 941903, 1994.

Witze, P.O., "The Effect of Spark Location on Combustion in a Variable Swirl Engine", SAE Paper 820044, 1982.

Woschni, G, W. Spindler, and K. Kolesa, "Heat Insulation of Combustion Chamber Walls—A Measure to Decrease the Fuel Consumption of IC Engines", SAE Paper 870339, 1987.

Yamin, J.A.A., H.N. Gupta, B.B. Bansal, and O.N. Srivastava, "Effect of Combustion Duration on the Performance and Emission Characteristics of a Spark Ignition Engine Using Hydrogen as a Fuel", *International Journal of Hydrogen Energy*, 2000.

Yao, M., Z. Zheng, and H. Liu, " Progress and recent trends in homogeneous charge compression ignition (HCCI) engines", Progress in Energy and Combustion Science, 2009, Elsevier.

Index